Cybersecurity Engineering for Smart Grid

Developing a Security Program for Distributed Energy Resources and Systems

Dr. Mariana Hentea

CWP
Central West Publishing

This edition has been published by Central West Publishing, Australia

For more information about the books published by Central West Publishing, please visit https://centralwestpublishing.com

Disclaimer

Every effort has been made by the publisher and author while preparing this book; however, no warranties are made regarding the accuracy and completeness of the content. The publisher and author disclaim without any limitation all warranties as well as any implied warranties about sales, along with fitness of the content for a particular purpose. Citation of any website and other information sources does not mean any endorsement from the publisher and author. For ascertaining the suitability of the contents contained herein for a particular lab or commercial use, consultation with the subject expert is needed. In addition, while using the information and methods contained herein, the practitioners and researchers need to be mindful for their own safety, along with the safety of others, including the professional parties and premises for which they have professional responsibility. To the fullest extent of law, the publisher and author are not liable in all circumstances (special, incidental, and consequential) for any injury and/or damage to persons and property, along with any potential loss of profit and other commercial damages due to the use of any methods, products, guidelines, procedures contained in the material herein.

NATIONAL LIBRARY OF AUSTRALIA

A catalogue record for this book is available from the National Library of Australia

ISBN (print): 978-1-922617-47-7

Cybersecurity Engineering for Smart Grid

Developing a Security Program for Distributed Energy Resources and Systems

To my husband, Toma, and our children, Irina and Marius, with gratitude for ever.

"A man travels the world over in search of what he needs and returns home to find it."
(George A. Moore, Irish Writer)

Foreword

“The opposite of every great idea is another great idea.” (Niels Bohr, Danish Physicist)

Smart Grids and Distributed Energy Resources (DERs) are seen from the perspective of the more general topic of cyber-physical systems - hybridized systems that combine computer based communication, control, and command with physical equipment to yield improved performance, reliability and resiliency. Interdependencies among the power grid, the communications infrastructure, and the information infrastructure pose a particularly serious challenge to the design of secure and survivable applications.

Therefore, a power grid becoming smarter and more secure at scale raises the bar for the utilities and their partners. Organizations need to identify the right partnership strategy to surround themselves with an ecosystem of capabilities and support of Smart Grid goals and cybersecurity requirements.

Cybersecurity solutions for the power grid must be carefully engineered to not interfere with energy delivery functions. Of distinct concern is the integration of the new Smart Grid applications and DERs into the existing power system environment, and how any new cybersecurity vulnerabilities would be mitigated through technology or other measures.

In this book, the protection for Smart Grid and DERs systems is discussed from the perspective of the adequacy of technical approach for addressing interoperability and cybersecurity. The interoperability approach supports how to provide a description of the automation component interfaces (devices and systems) and how integration is supported to achieve interoperability. The cybersecurity approach supports how to provide a description of addressing cybersecurity concerns using risk management approaches.

When developing Security and Privacy programs for Smart Grid and DER systems, the functions of a security or privacy program should be based on Systems Engineering and Systems Security Engineering approaches as well as standards. These functions should include principles, models, methods, and best practices to design secure and resilient systems that support the goals of Smart Grid. This book is an attempt to engage Engineers and Security Professionals to design and support a unified security program. A unified security program for Information Technology and Industrial Control Systems should be based on continuous risk assessment and improvement to identify measures for preventing attacks. Engineers and Security Professionals are provided with information to understand the characteristics of industrial control systems and approaches to design adequate security for Smart Grid and DER systems. As argued in this report by W. Ware, Security Controls for Computer Systems, Report of the Defense Science Board Task Force on Computer Security, February 1970, providing adequate security in a system is inherently a system design problem.

I hope that readers of this book can bring new ideas for solutions and innovations to advance the design and implementation of Smart Grid and DER systems.

Acknowledgments

Although I am the only author of this book, the content is the product of learning from discussions with colleagues and friends about various topics and projects at work, interactions with researchers at conferences and workshops, meetings and presentations provided by professional societies, my published research works, presentations and talks at conferences, teaching courses in the university, leading research projects with students, meetings with IEEE, ISA, AITP, ISC2, ISSA members, etc.

I have been inspired by the works of other authors and abundance of research publications on technologies for Smart Grid and Distributed Energy Resources that require systematic engineering approaches to accelerate more advances on security and privacy protection.

My sincere thanks to Bernie Leung for reviewing and encouraging me to accomplish the writing during so many dynamic changes in cybersecurity.

My thanks also go to Central West Publishing editors and managers for their offer to publish my book.

I am grateful to all of those with whom I have had the pleasure to work on this and other related projects.

Mariana Hentea

Preface

"Great discoveries and improvements invariably involve the cooperation of many minds."
(Alexander Graham Bell, U.S. Inventor)

A retrospective at the recent media announcements and the Smart Grid manufactured products and services quickly underscores the fact that distributed energy resources (DERs) were a major theme of the recent past years' technology events, considerably increasingly more each year than just a year before. In fact, one could argue that past years marked a market success when it comes to the acceptance and adoption of DERs. While reliability is one most important factor for DER adoption, utilities also have other priorities such as quick response from DERs for load control, communications speed, accuracy of operations, security, flexibility.

Security requirements may or may not be within the scope of a single utility's responsibility, imposing the requirements upon cooperating interconnecting systems and the corresponding capabilities to support some aspects of security objectives. Security involves a system of systems approach in design and operations, and therefore security responsibility must extend to stakeholders and parties outside and in addition to the electric utility.

The need for an efficient and effective security program to protect the current power grid and future Smart Grid is recognized by the stakeholders including legislative policy makers, regulatory authorities, public at large and all variations of end consumers.

This book provides a systematic approach for the design of secure systems and implementation of a security and privacy program for the Smart Grid and DERs. The focus is on the aspects of building blocks to engineer the security of the Smart Grid, DERs, and systems. The book includes methodology guidelines and a knowledge base to apply analysis and risk management methods to identify the appropriate solutions to ensure and meet the security needs of Smart Grid, DERs and systems. Besides describing on how to apply methods and models to support security problem for the power grid as the most complex system, the book discusses Smart Grid implementation challenges and how to address ethics and security awareness to meet the Smart Grid stakeholders' needs. Also, the book describes how to blend Systems Engineering and Systems Security Engineering techniques with standards and best security practices.

This book helps to educate Engineers and Security Professionals in electrical subsector on how to:

- Design, implement, and maintain a security program adequate to power grid systems that includes both Information Technology (IT) systems and Industrial Control Systems (ICS)

- Recognize differences between security requirements of traditional IT systems and security of ICS, specifically Supervisory Control and Data Acquisition (SCADA) systems

- Provide a unified approach based on risk management for building a security and privacy program for Smart Grid and DER systems using Systems Engineering techniques, standards, and best practices

- Apply security principles and methods for the design of secure and resilient systems to ensure the high level of information assurance, availability and security necessary to maintain a reliable power system and consumer's confidence

- Describe how to validate the effectiveness of the cybersecurity controls
- Support resiliency of systems and how is addressed - through diligently engineered and coordinated efforts in cybersecurity activities
- Educate Security Professionals to gain skills on specific security requirements for ICS
- Educate Engineers to gain skills on building a security program for an organization and designing security features in the hardware and software products.

The key topics discussed in the book include:

- Principles of a unified security program that meets the Smart Grid objectives for security, interdependencies, and business mission
- Functions of a security and privacy program for Smart Grid based on Requirements Engineering and Security Engineering approaches
- Elements of an effective security program and Smart Grid cybersecurity strategy: proactive versus reactive, defining adequate enforcement mechanisms toward simplifying data and protection as opposed to counting compliance to standards
- Fundamentals of the enterprise security architecture, vulnerability management, costs, and risks
- Identify security problem, system attributes, security requirements, quality of security services
- Dependability and security relationships, taxonomy of dependability and security, dependable infrastructure, dependable critical infrastructure challenges
- Survivability approaches to power systems, reliability versus availability, reliability and availability challenges of IT technologies
- Applying Security Engineering principles to ensure resiliency of systems
- Understanding fault tolerance principles, fault tolerance and security relationships, fault tolerant systems, security and safety relationships
- Analysis of DER systems characteristics and contingency planning , simulation models
- Applying System Engineering methods and Systems Security Engineering approaches to Smart Grid and DERs, security analysis framework to Smart Grid and DER systems
- Core practices and Security Engineering design – Security by Design and Privacy by Design, Defense in Depth

- Risk management methods, risk management process, building innovative risk models for Smart Grid

- Developing security requirements for Smart Grid using frameworks, guidelines, and use cases; IT systems requirements versus ICS requirements, including critical issues and constraints of the Smart Grid and DER needs

- Developing Smart Grid systems from the perspective system of systems, system architecture principles, Smart Grid architecture reference model, different security architecture frameworks, secure coding and build security in, secure coding standards, testing criteria, economics versus security

- Maintenance methods, tasks and tools, best security management practices, Smart Grid maintenance – needs and challenges, security program performance, reporting, writing and updating technical documentation

- Ethics definition, various codes of ethics, technical and ethical issues in engineering Smart-Grids and DER systems, ethics and awareness, Smart Grid challenges and awareness, examples of certification and accreditation organizations.

The book describes the skills, training, and education needed to support the security of emerging Smart Grid, Distributed Energy Resources, and advanced technologies. It provides foundation for understanding the current developments and solutions, published research proposals, and future needs for cybersecurity research.

The book may be used to educate software manufacturers, current work force and future graduates (e.g., Engineers, Computer Science, Information Technology, Technology Management, Data Science) to gain skills and more knowledge on engineering approaches on designing and maintaining security and privacy programs for Smart Grid and DERs. Also, the material can be used to educate future graduates in Philosophy, Sciences, Law, Business about the regulations, engineering ethics, and general code of ethics as they apply to engineering the Smart Grid, DERs, and DER systems. Researchers may identify new problems that require future solutions.

Contents

Chapter 1

Security Program for Smart Grid

1.1 Smart Grid Overview

It is recognized that power industry relies increasingly on information to operate the power systems and both power system infrastructure and information infrastructure must now be managed. Figure 1.1 shows the relationship between these infrastructures. The management of the power system infrastructure is reliant on the information infrastructure as automation continues to replace manual operations, as market forces demand more accurate and timely information, and as the power system equipment ages. This dependency on information infrastructure requires a reliable information infrastructure that must be managed to the level of reliability needed to provide the required reliability of the power system infrastructure.

While recognizing that the different names used for the future grid have meaningful distinctions to some stakeholders, the term Smart Grid is used to describe a modernized grid that enables bidirectional flows of energy and uses two-way communication and control capabilities that will lead to an array of new functionalities. This European strategic deployment (SDD) of 2010 [SDD 2010] defines

A Smart Grid is an intelligent network that can intelligently integrate the actions of all users connected to it – generation, consumers and those that do both – in order to efficiently deliver sustainable, economic, and secure electricity supplies.

As the electricity grid continues to modernize, distributed energy resources (DERs) based on storage and advanced renewable technologies can help facilitate the transition to a smarter grid [EPRI 2015]. Also, the proposal for a Unified Smart Grid in U.S. introduces another perspective in the management of power and business systems for Smart Grids. A Unified Smart Grid is a wide area grid that is a National interconnected network relying on a high-capacity backbone of electric power transmission lines linking all the nation's local electrical networks that have been upgraded to Smart Grids [Wiki Unified]. Conceptually, it is a topology that has many access points with each node being a Smart Grid that could be a virtual power generation cluster, could be a local electric utility system, or could be a grid energy storage facility. In Europe, the analogous project is sometimes referred to as the Super Smart Grid. The Unified Smart Grid must be secure and rely on innovative security programs.

In addition, the migration to a Smart Grid encounters challenges that all need to be met as the Smart Grid evolves. Examples include:

- Satisfy a broad set of requirements for all stakeholders from consumers to utilities, service providers, and system operators.
- Respond to all aspects of Smart Grid complexity and response time sensitive to human response and interaction, while others need instantaneous, intelligent and automated responses.
- Plan for a gradual transition to Smart Grid and long coexistence of diverse technologies and support of design to avoid unnecessary expenses and unwarranted decreases in reliability, safety, or cybersecurity.
- Ensure cybersecurity of the systems in all aspects from technologies to humans.

- Build consensus on standards to deliver better results.
- Development and support of standards by using an open process based on benefits from the expertise and insights of a broad constituency.

Management's responsibility is to provide protection for all assets and resources. The resources consist of human, capital, hardware, and informational forms. More specifically, management needs to identify and evaluate business objectives, user needs, functionality requirements and objectives, and security risks. Management must ensure that a security program identifies vulnerabilities and recognizes the threats that can affect these resources and be assured that the necessary protective measures are implemented. Also, management must provide the necessary resources including funding and representatives to participate in the security program. Therefore, Information Security management principles, processes, and security architecture need to be applied to Smart Grid systems without exception.

This work [Wang 2013] argues that cybersecurity in the Smart Grid is still under development, especially because information security must be taken into account with electrical power systems. The authors believe that features of the Smart Grid communication network, such as heterogeneous devices and network architecture, delay constraints on different time scales, scalability, and diversified capabilities of embedded devices, make it indeed impractical to uniformly deploy strong security approaches all over the Smart Grid.

However, variants of Smart Grid implementations have already been rolled out in various jurisdictions across the United States as well as the rest of the world for several years. The window of opportunity to integrate security into the Smart Grid from the beginning is shrinking fast. It is necessary to understand the interdependency and mutual vulnerability of the wholesale electric grid and the wholesale electric market in maintaining the security and stability of the smart power grid. Market participants require to ensure protection of their critical cyber assets and to support an appropriate security program.

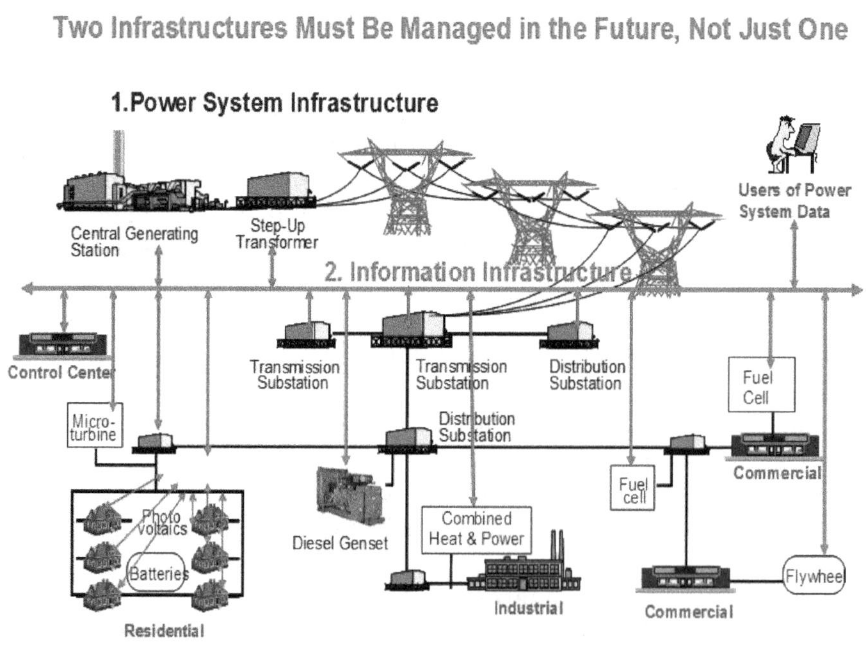

Figure 1.1 Two infrastructures must be managed, not just one. *Source:* [Cleveland 2012].

The increasing sophistication and number of cyber attacks to energy sector and critical infrastructure are national concerns that require better security and privacy protection, an educated work force of engineers in the area of security and privacy issues and security professionals in the area of industrial control systems, particularly developing and implementing security programs for the emerging Smart Grid and DER systems. More information about concepts and terms for cybersecurity, power grid, Smart Grid, and DERs can be found in Appendix A, Appendix B, and [Hentea 2021].

1.1.1 Defining a Security Program

The first action of a management program to implement Information Security is to have a security program in place. The need for an efficient and effective security program to protect the current power grid and future Smart Grid is recognized by the stakeholders including legislative policy makers, regulatory authorities, public at large and all variations of end consumers.

Security program is a combination of all aspects of managing security, ranging from the definition and communication of policies through implementation of best industry practices and ongoing operation and auditing [ISA99 Glossary]. A security program has to include both information technology (IT) and industrial control systems (ICS), but it has to distinguish between their similarities and differences. In addition, the security solutions should be tailored to the characteristics of secured entities.
Management must define a clear scope and to determine actual goals expected to be accomplished with a security program. Basically, a security program should include components such as objectives, management responsibilities, security controls (classified into three categories: administrative, technical or logical, physical), and information management systems as may be summarized by the following:

- Security program objectives
  - Protect the company and its assets
  - Manage risks by identifying assets, discovering threats and estimating the risk
  - Provide direction for security activities by framing of information security policies, procedures, standards, guidelines and baselines
  - Information classification
  - Security organization, and
  - Security education.
- Security management responsibilities
  - Determining objectives, scope, and policies, which are expected to be accomplished from a security program
  - Evaluate business objectives, security risks, user productivity, and functionality requirements
  - Define steps to ensure that all the above are accounted for and properly addressed.
- Administrative controls
  - Developing and publishing of policies, standards, procedures, and guidelines
  - Screening of personnel
  - Conducting security awareness training and
  - Implementing change control procedures
- Technical or logical controls
  - Implementing and maintaining access control mechanisms
  - Password and resource management
  - Identification and authentication methods
  - Security of the devices, and

  - Configuration of the infrastructure.
- Physical controls
  - Controlling individual access into the facility and different departments
  - Locking systems and removing unnecessary floppy or CD-ROM drives
  - Protecting the perimeter of the facility
  - Monitoring for intrusion and
  - Environmental controls.
- Information Security Management System (ISMS)
  A management system includes organizational structure, policies, planning activities, responsibilities, practices, procedures, processes and resources. A management system uses a framework of resources to achieve an organization's objectives. In terms of Information Security, a management system allows an organization to:
  - Satisfy the information security requirements of customers and other stakeholders
  - Improve an organization's plans and activities
  - Meet the organization's information security objectives
  - Comply with regulations, legislation and industry mandates and
  - Manage information assets in an organized way that facilitates continual improvement and adjustment to current organizational goals.

However, the methodologies for building an effective security program should be adapted to specific security needs of each organization. It is expected that each organization should develop its own cybersecurity strategy for the implementation of security program. For example, organizations would need to develop their own policy to protect their DER assets, employees, and general public who are at risk when human (intentional or unintentional) threats or natural disasters occur.

1.1.2 Unified Security Program

The need for improving the critical infrastructures is voiced in U.S. [EO 13636], Australia [TISN 2007]. This need requires more stringent functions and higher expectations from a security program. Figure 1.2 shows the industries that could significantly impact on the social or economic well-being of the nation or affect Australia's ability to conduct national defense and ensure National security. As the diagram shows, utilities and telecommunications are providing the underpinning support services for many critical services such as government, financial, transport, law enforcement, health and emergency services.

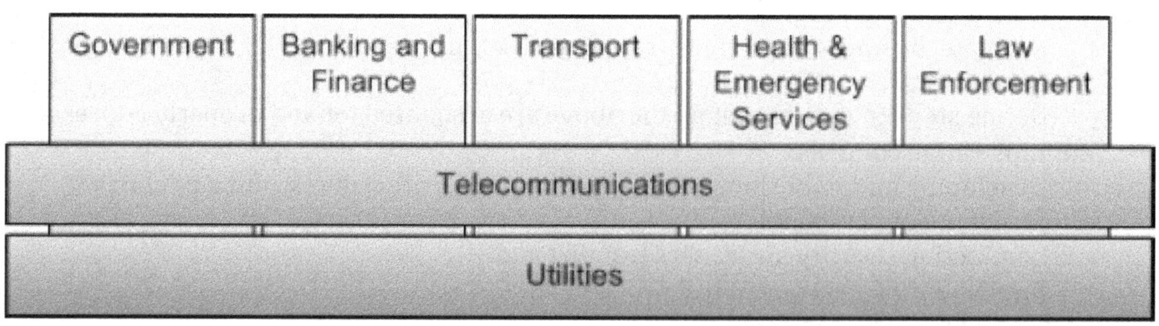

Figure 1.2 Critical infrastructure industries. *Source:* [TISN 2007]. Public Domain.

Security program is the plan that must cover security governance, planning, prevention, operations, incident response, and business continuity. In the electrical sector, one issue that needs to be addressed by management is the planning of a unified security program that includes both business and control systems. The main benefit is avoiding the separation and duplication of management functions and reducing the complexity of administration, operation, and budgeting. Further, there is a need to provide a unified approach for the functions of the security program without jeopardizing the reliability requirements for the power systems, ICS and DER systems. DER systems include both business and real time applications that share infrastructure and exchange information (see also the acronyms list in Appendix C).

In a large organization (more than one thousand devices), the security program functions can be grouped and performed by different specialized groups. Security programs in small organizations such as service providers for DER systems may have limited resources and staff may be required to support many functions. However, the functions of a security program have to be always performed with diligence and vigilance to ensure the security of information systems.

Although best practices are useful and can be used for some activities in the management of a security program, there is a need to lay down a foundation for security planning. Therefore, each organization needs to develop its own cybersecurity strategy for the implementation of a security program. Cybersecurity must address not only deliberate attacks launched by disgruntled employees, agents of industrial espionage, and terrorists, but also inadvertent compromises of the information infrastructure due to user errors, equipment failures, and natural disasters [NISTIR 7628r1]. As new capabilities are included in the Smart Grid, potential new privacy concerns emerge for which no legal mitigation currently always exists.

1.1.3 Privacy Program

A significant number of privacy breaches occur not because of an attack but through non-compliance with privacy policy or having no policy. For example, a laptop that has a copy of Personally Identifiable Information (PII) data becomes a privacy breach if the laptop is improperly disposed, lost, or stolen. The term PII refers to information which can be used to distinguish or trace an individual's identity, such as their name, social security number, biometric records, etc. alone, or when combined with other personal or identifying information which is linked or linkable to a specific individual, such as date and place of birth, mother's maiden name, etc. (see Federal Information Security Management Act of 2002 (FISMA) [FISMA 2002] and the [Privacy 1974].

Hence, measures for protection of privacy have to designed and implemented too. Thus, a privacy program has to be planned, designed, implemented, and maintained. The challenges of the Smart Grid include size, third party networks and customer privacy. All involved stakeholders (manufacturers, telecommunication companies, operators, and end-users) have to deal with security and privacy problems. Factors that should be considered in design of a privacy program include:

- Privacy rights continue to evolve by legislation, litigation, and regulation, and the data gathered will be subject to the relevant jurisdiction(s).
- Anonymization issues include: if private information is not properly anonymized, even data like electrical appliance usage or electric vehicle charging schedules may constitute a privacy violation. In electrical sector, the ownership and rights associated with PII are varying by jurisdiction. In some jurisdictions, the person owns their data, while in other jurisdictions, ownership is less clear. For example, a utility that gathers contact and other information for billing purposes may be restricted in use of the PII for any other purposes without consent of the customer possession of the data is not the same as ownership.

- Technologies and capabilities misuse such as the advancing of technologies (e.g., data mining and pattern recognition) can be used on identifying the identity of persons when customer data and energy data is analyzed. Recognizing electric signatures of smart appliances and developing detailed, time stamped activity reports, utilities or third party service providers can determine lifestyle details that could be legitimately characterized as PII in most jurisdictions.
- Dedicated privacy group with its own management is needed. Although in many organizations, security group is supporting the privacy requirements, the future commands for more responsibility and accountability for the implementation of data privacy specifically in smaller size enterprises [Shei 2013]. There are two aspects that organizations have to understand: security is only one aspect of privacy and privacy protection implies organization and business decisions.

Ensuring privacy requires a bundle of technologies, policies, culture, regulations, and harmony between many business units from security to legal to human resources to employees [Shei 2013]. Guidelines and recommendations for the protection of privacy data and harmonization of disparities in national privacy regulations are documented in [OECD 2013].

The government and agents of the government have a direct responsibility to protect civil liberties arising from cybersecurity activities. As referenced in the [NIST 2014a] framework, government or agents of the government that own or operate critical infrastructure should have a process in place to support compliance of cybersecurity activities with applicable privacy laws, regulations, and Constitutional requirements.

To address privacy implications, organizations may consider how, in circumstances where such measures are appropriate, their cybersecurity program might incorporate privacy principles include data minimization in the collection, disclosure, and retention of personal information material related to the cybersecurity incident; use limitations outside of cybersecurity activities on any information collected specifically for cybersecurity activities; transparency for certain cybersecurity activities; individual consent and redress for adverse impacts arising from use of personal information in cybersecurity activities; data quality, integrity, and security; and accountability and auditing.

As organizations assess the NIST framework [NIST 2014a], the following processes and activities are recommended to be considered as a means to address the above referenced privacy and civil liberties implications:

- Governance of cybersecurity risk
  - An organization's assessment of cybersecurity risk and potential risk responses considers the privacy implications of its cybersecurity program
  - Individuals with cybersecurity related privacy responsibilities report to appropriate management and are appropriately trained
  - Process is in place to support compliance of cybersecurity activities with applicable privacy laws, regulations, and Constitutional requirements
  - Process is in place to assess implementation of the foregoing organizational measures and controls.
- Approaches to identifying and authorizing individuals to access organizational assets and systems

- Steps are taken to identify and address the privacy implications of access control measures to the extent that they involve collection, disclosure, or use of personal information

- Awareness and training measures
  - Applicable information from organizational privacy policies is included in cybersecurity workforce training and awareness activities
  - Service providers that provide cybersecurity related services for the organization are informed about the organization's applicable privacy policies.
- Anomalous activity detection and system and assets monitoring
  - Process is in place to conduct a privacy review of an organization's anomalous activity detection and cybersecurity monitoring.
- Response activities, including information sharing or other mitigation efforts
  - Process is in place to assess and address whether, when, how, and the extent to which personal information is shared outside the organization as part of cybersecurity information sharing activities
  - Process is in place to conduct a privacy review of an organization's cybersecurity mitigation efforts.

Besides organizing the functions of the security program and privacy program, the management needs to organize and build security teams.

1.1.4 Security Team Diversity

In order to satisfy the numerous activities and different areas of security topics, management must establish a security team based on a diverse group of people that leverage their skills and responsibilities from many areas. Management needs to organize and build teams with people that leverage their skills from many areas as shown in Figure 1.3.

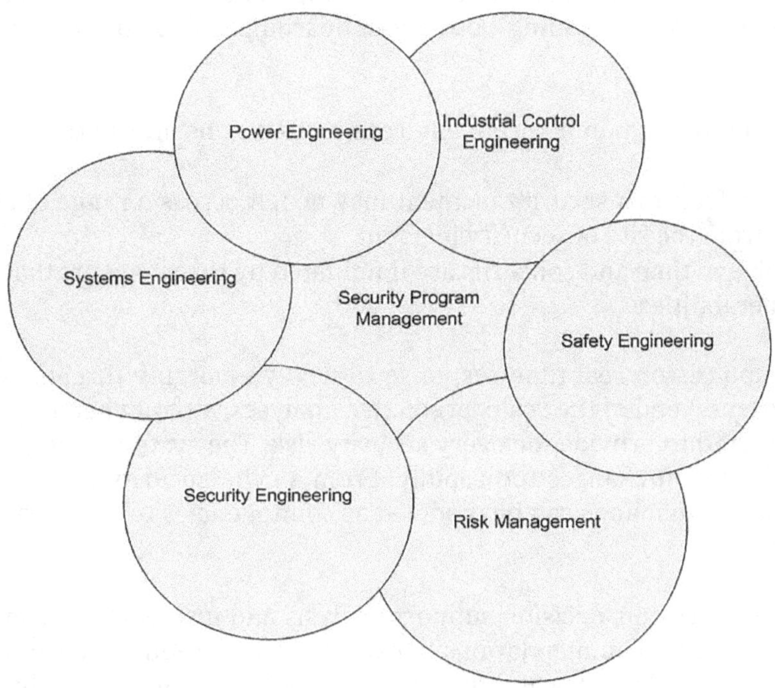

Figure 1.3 Management domains of security program for Smart Grid.

Besides Security Engineers and Security Professionals, the team should include Systems Engineers, Power Engineers, Industrial Control Engineers, Safety Engineers. Therefore, managers have to understand the goals and complexities of Smart Grid. Definitions of these domains are provided elsewhere. Each domain contributes on the development and implementation of Smart Grid and DER systems. Added to each domain goals are security goals making security management to intersect with these domains.

In addition to the specific functions of each domain, all these domains intersect with each other. Often the intersection of these domains may create interactions resulting in conflicts that have to be managed in a proper and optimal way. Therefore, security professionals need to understand all subtleties of common parts as well as differences in order to implement and manage a security program. For example, Security Engineering is a type of Risk Management, and this is the main characteristic shared with Safety Engineering. Security Engineering (SE) seeks to reduce the likelihood of future security incidents and the severity of their consequences should they occur. A range of security analytic techniques and risk reduction strategies are deployed to achieve this strategy. The current performance of a system, in terms of security incidents reported, is monitored and used as an input to future strategies. Measures of risk (future performance in terms of likelihood of occurrence and effects) and past achieved performance are the core measures of security.

There are similarities with Safety Engineering, but also differences of emphasis. Safety Engineering is mainly concerned with hazards arising from weaknesses in the system design, development, and operation. Issues external to the system are considered including environmental effects and exposure times, but these are viewed as relatively static. The main concerns tend to focus on failure scenarios that start with component failures within the system and lead rapidly and uncontrollably to accidents for given operational contexts. The traditional approach has placed emphasis on developing a safe product, having it certified as acceptably safe for operations, and then operating it within defined constraints. Current trends are moving towards a more through life approach in which a Safety Management System, used in the development phase of a system, is transferred to an operational support role, providing continuous learning and improvement of safety performance.

SE has to deal with a more dynamic threat environment and this affects the risk management approach in two ways:

- The harmful effects of a security incident may be felt across a range of different timescales and remote from the site of security incident.
- Threats evolve in time and concerns are dominated by threat agents that learn and adapt to system vulnerabilities.

There is greater emphasis on real time response to newly emerging threats. At the large system (and networked systems) end of the scale, predictive analyses, although an important part of planning, cannot be expected to provide for every security risk. The systems involved are too complex and the threat environment changes too rapidly. From a cybersecurity perspective, physical consequences of malicious commands can be modeled as contingencies to assess risk and develop mitigations well in advance.

Security management performs decision support analysis and justifies its recommendations to executive management. Such recommendations to executives may require coordination and planning to win security acceptance. The security management's success is measured by its ability to facilitate or support the delivery of reliable information technology (IT) services while ensuring adequate confidentiality, integrity and availability of the organization's information and information assets. In this context, adequate security means that a system achieves a level of confidentiality, integrity and availability that is commensurate with the information's value to the organization. However, the implementation of the security programs for the Smart Grid may face several challenges that have to be addressed.

1.1.5 Addressing Challenges

The World Economic Forum (WEF) makes recommendations to better combat cybercrime [WEF 2016]:

- Cooperation between public and private sectors by creating permanent and secured information sharing channels between law enforcement authorities and the private sector; real time sharing of information with both Computer Emergency Response Teams (CERTs) and law enforcement, related to hacking cases and to new modus operandi; sharing of experiences from investigations and prosecutions; sharing of technical prevention and protection measures; sharing of information on technological development trends and achievements; sharing of best practices related to IT education and training of end users.
- Creation of a common cybercrime taxonomy.
- Fostering of technology innovations and investment to meet global security challenges.
- Technology and infrastructure need to come together in a way that is meaningful and trustworthy to users that need networks that are fast, scalable and secure in order to interact with each other and with things through billions of simultaneous connections
- Legal and economic frameworks should be adapted to treat digital personal identity securely and to empower users to decide how their data is used and valued.
- High quality standards should be developed to support informed consumer decisions over their data, including its use by third parties.
- Consistency in terminology and classification of cyber related terms. While speed and constant innovation are often cited as reasons for the difficulty in combating cybercrime, a further impediment that should not be disregarded is the varying definitions and classification of cyber related terms. These can mean different things to different actors. Therefore, if common accord and action is to be achieved, consistency in terminology and classification must be reached first [WEF 2016].
- Without an integral security management approach that integrates the different security aspects (e.g., physical, logical, environmental, and safety) can result in some risks being overlooked [ENISA 2011a] .
- Comprehensive and robust set of security metrics for Industrial Control Systems (ICS) is essential for organizations to meet various business objectives [Knowles 2015].
- Education and awareness topics have to be included in a comprehensive security plan [ENISA 2011a]. Examples of topics include: proper understanding of the current information technology and cybersecurity issues; knowledge of differences between IT and ICS technologies, along with the process safety and associated management processes and methods; developing practices that link the skill sets of all the organizations to deal with cybersecurity collaboratively; active collaboration between IT security sector and ICS manufacturers is essential to improve ICS security. IT is the common term for the entire spectrum of technologies for information processing, including software, hardware, communications technologies and related services. In general, IT does not include embedded technologies that do not generate data for enterprise use [Gartner Glossary].

In addition to security and privacy functions, the objectives of the Smart Grid play an increasing role in the definition and realization of a security program for the Smart Grid and DER systems.

1.2 Objectives for Smart Grid

As defined in [ISO/IEC 27002], Information Security is the protection of information from a wide range of threats in order to ensure business continuity, minimize business risk, and maximize return on investments and business opportunities. There we need to clarify that information security is not equivalent with computer security or information assurance concepts that are frequently used interchangeably. Although there are common security goals for ensuring the security services, there are differences in the methodologies and techniques used. Information systems are decomposed in three main components: hardware, software, and communications.

While focused dominantly on information in digital form, the full range of information assurance encompasses not only digital but also analog or physical form. Information assurance as a field has grown from the practice of information security which in turn grew out of practices and procedures of computer security.

As described in [CENELEC 2014], Information Security for the Smart Grid refers to technical and organizational needs for sustainable state of the art Smart Grid Information Security, data protection and privacy, enabling the collection, utilization, processing, storage, transmission and erasure of all information to be protected for all participating actors. Smart Grid information security objective is to create a bridge between electrical grid operations and information security.

1.2.1 Security Objectives

The main principles in all security programs are availability, integrity, confidentiality, and non-repudiation of information. In general, for IT systems, the priority for the security objectives is confidentiality first, then integrity and availability. For industrial control systems, including power systems, the priorities of the security objectives are availability first, integrity second, and then confidentiality [P2030 2011].

One objective of the security program is the protection of assets and people. Assets may be tangible (e.g., computers, devices, facilities, supplies) or intangible (e.g., reputation, data, intellectual property). Besides security concerns, the computer systems including control systems raise the issue of safety causing harm and catastrophic damage when they fail to support applications as intended. In specific environments such as manufacturing, chemical plants, steel industry, and power grid, a security incident may have safety-related consequences. For example, the destruction of power grid systems and assets would have a debilitating impact on energy security, economic security, public health or safety.

The impact to safety resulted in a common framework for treating these incidents based on the standard [IEC 61508]. This standard is intended to be a basic functional safety standard applicable to all kinds of industry. However, the application of the standard to security may not produce good results because it cannot use the same statistical measures that have been designed for safety [Brandle 2008]. The authors argue that the distinction between security and safety is beginning to disappear as computer systems are more integrated.

Others argue that ICS security solutions and services should adapt to ICS security objectives defined by the SRA model (Safety, Reliability, Availability) in contrast to security objectives of IT systems defined by the CIA model (Confidentiality, Integrity, Availability) [ENISA 2011a]. In order to balance safety and security in ICS systems, security and safety should be included in the assurance framework [Knowles 2015]. Functional assurance moves away from the fail-secure approach of information security, emphasizing the ability of industrial control systems to fail safely and fail securely based on potential security risks. It is recommended that safety and security must be considered jointly in the functional assurance process.

Another objective is physical security which has to be integrated with cybersecurity. Physical security for the Smart Grid includes a different set of threats, vulnerabilities, risks, and types of security measures. A physical security program should comprise safety and security mechanisms. In addition, physical security mechanisms include site design and layout, building, facility, computer rooms, data center, windows, doors, fences, parking lots, environmental components, emergency response readiness, power control, fire protection, HVAC, guards, dogs, etc. Physical security mechanisms protect people, data, equipment, systems, facilities, and many other assets. However, several guidelines, documents, standards, and legislation can be used to design the necessary functions. Designing physical security program is complex and it could very difficult to achieve a quality security program without proper integration. Therefore, Security Professionals need to identify strategies, objectives, standards, and regulations that apply to the physical security program for the Smart Grid.

The security goal can be met through security objectives such as:

- Availability (of systems and data for intended use only); availability is frequently an organization's foremost security objective.
- Integrity (of system and data); integrity is commonly an organization's most important security objective after availability.
- Confidentiality (of data and system information); for many organizations, confidentiality is frequently behind availability and integrity in terms of importance; however, for some systems and for specific types of data in most systems (e.g., authenticators), confidentiality is extremely important.
- Non-repudiation means to ensure that a transferred message has been sent and received by the parties claiming to have sent and received the message.
- Accountability (to the individual level); it directly supports non-repudiation, deterrence, fault isolation, intrusion detection and prevention, and after action recovery and legal action.
- Safety, in particular, is an important consideration due to its potential to be negatively influenced by ICS security solutions [ENISA 2011a]; ICS devices must not only be designed for safety but also for security [Robles 2009].
- Assurance (that the other four objectives have been adequately met); assurance is the basis for confidence that the security and safety measures, both technical and operational, work as intended to protect people, the system, and the information it processes; however, assurance is a continuum; the amount of assurance needed varies between systems.

1.2.2 Interdependencies

Although security objectives seem to be independent of each other, there are dependencies that have to be understood. Achieving one objective without consideration of the others is seldom possible [NIST SP800-33]:

- Confidentiality is dependent on Integrity, in that if the integrity of the system is lost, then there is no longer a reasonable expectation that the confidentiality mechanisms are still valid
- Integrity is dependent on Confidentiality, in that if the confidentiality of certain information is lost (e.g., the super-user password), then the integrity mechanisms are likely to be bypassed

- Availability and Accountability are dependent on Confidentiality and Integrity, in that:
  - if confidentiality is lost for certain information (e.g., super user password), the mechanisms implementing these objectives are easily bypassable, and
  - if system integrity is lost, then confidence in the validity of the mechanisms implementing these objectives is also lost
- Non-repudiation is dependent on integrity and accountability is dependent on non-repudiation.
- All of these objectives_are interdependent with Assurance.

Figure 1.4 shows examples of security objectives dependencies.

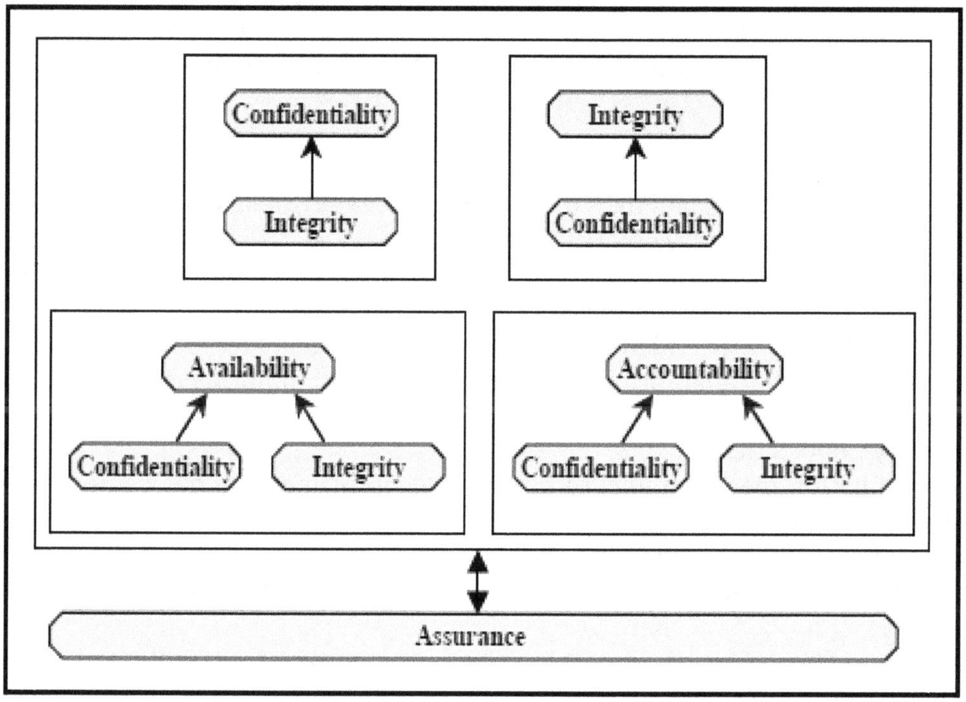

Figure 1.4 Example of security objectives dependencies. *Source*: [NIST SP800-33]. Public Domain.

1.2.3 Business Objectives

In addition to the objectives discussed above, the strategic planning of the security program should be outlined to consider the future Smart Grid needs as summarized in [Bossart 2013]:

- Possible developments of concepts for future modern grid
- Shifts in technologies and business models
- Technical, regulatory, and cultural challenges
- Work around where concepts are not fully developed
- Short and long term vision plans.

Specifically, management should focus to plan for cybersecurity to support the business objectives and modernization needs. Examples of needs include [Bossart 2013]:

- Changes in ancillary services requirements

  - Reserve margins could be reevaluated considering distributed and consumer assets
  - Solid state power electronics, energy storage, distributed generation (DG) demand change in technical approach for ancillary service
  - Growth of distributed and consumer assets imply change in ancillary service market
  - Reestimation of volt-amper reactive (VAR) value (reliability and power quality).
- Higher levels of power quality demands
  - Frequency and voltage control in narrow range
  - Established provider: electric service provider or consumer?
  - Tiered pricing and value
  - Address both critical loads (e.g., outages) and sensitive loads (e.g., power quality)
  - Move to conservation voltage reduction.
- Backup generation connected to grid
  - Additional source of distributed generation
  - Serves load and ancillary services
  - Serves critical loads of customer owned
  - Assists in stabilizing grid operation.
- Dynamic line rating increasing power flow
  - Real time rating of transmission capacity
  - Addresses transmission congestion and improves financial aspects of generation and transmission.
- Greater interaction between transmission and distribution
  - Distributed assets alleviate transmission congestion
  - Transmission and main grid distribution assets alleviate microgrid congestion
  - Better communication between transmission and distribution operations
  - Merging of wholesale and retail markets
  - Optimization of entire power system
  - Coordinate generation, transmission, and distribution to gain efficiency.
- Support of new and interesting dynamics of consumer behavior in Demand/Response (DR)
  - More participation in DR provides less benefit to each customer and consumer behavior study
  - Retention in DR programs
  - Level of participation in DR, limits of consumer behavior (e.g., hot days).
- Shift to more hybrid alternating current/direct current (AC/DC) systems in both transmission and distribution
  - DC microgrid connection to AC grid
  - Reduce number of power converters.
- Integration of systems and processes of electric service providers
  - Improved efficiency, communications, and collaboration
  - Improved operations, maintenance, outage maintenance system (OMS), billing, customer service, engineering, planning, trading, procurement, workforce scheduling
  - Weather forecast, adjust schedules, check inventories, assemble crews & tools
  - Integration of power, communications, and market software.
- Growth of smart microgrids
  - Greater flexibility in meeting power needs
  - Serve critical loads
  - Desire to become more self-sufficient.
- Growth of distributed energy resource aggregators
  - Demand and response
  - Distributed and consumer generation

- Energy storage (e.g., electric vehicles, community storage).

There has to be an understanding of what ancillary services are about. Ancillary services are necessary to support the transmission of electric power from seller to purchaser given the obligations of control areas and transmitting utilities within those control areas to maintain reliable operations of the interconnected transmission system. The United States Federal Energy Regulatory Commission (FERC) defines ancillary services as the specialty services and functions provided by the electric grid that facilitate and support the continuous flow of electricity so that supply will continually meet demand. This term is used to refer to a variety of operations beyond generation and transmission that are required to maintain grid stability and security. These services generally include, frequency control, spinning reserves and operating reserves [Hirst 1996]. Traditionally ancillary services have been provided by generators, however, the integration of intermittent generation and the development of Smart Grid technologies have prompted a shift in the equipment that can be used to provide ancillary services. The integration of power electronics, Distributed Energy Resources (DERs), and energy storage could ultimately be the answer for power quality [Yu 2006].

Basically, what needs to be understood is that the Smart Grid applications are dependent on the security of the information, and implicitly of the security program including information technology and communications networks that all require additional cybersecurity controls.

1.3 Developing the Security Program

A wide variety of organizations practice SE in the development of computer programs, whether as operating systems software, security managing and enforcing functions, software, middleware or applications programs that are used in power industry. Appropriate methods and practices are therefore required by product developers, service providers, system integrators, system administrators, and even security specialists. Some of these organizations deal with high level issues (e.g., ones dealing with operational use or system architecture), others focus on low level issues (e.g., mechanism selection or design), and some do both.

Therefore, any organization of energy sector, particularly each utility needs to intensify the efforts to develop security and privacy programs. A security program has to be built using the SE approach. This approach requires building systems to remain dependable in the face of malice, error, or mischance [Anderson 2008]. Also, the successful implementation of a security program requires certain basic functions that should be included in any budget allocation [Whitman 2014]. Each organization has to support a thorough analysis of existing security program components and identify gaps or define a security program that provides protection of current and future assets aligned with business objectives. The nature of dependencies and interdependencies among applications require more focus as follows.

1.3.1 Utilities Achieving Security

Like most corporations, electric utilities depend on computers to run their day to day operations such as business functions, engineering power flow, or monitoring and control of their systems. Utilities use computer technology, on the business side, especially for accounting and customer billing, and on the engineering side for analysis and design, monitoring and control of the generation-transmission-distribution of power. A new function, brokering buy-sell transactions, is spawning a new type of company called power brokers. DER systems are used by utilities either directly or via service providers. Therefore, many security concerns transcend beyond utilities and several aspects have to be considered such as:

- Business computers are used for accounting and billing, supply chain business, and storing relevant information about partners and customers in company wide databases. The integrity and availability of the information are essential to the day-to-day operations. In the past, limited external connectivity minimized security and survivability concerns. However, the trend towards increased connectivity is raising these concerns.
- Engineering computers store so much data (power and engineering designs) which is publicly available. Because the power systems are interconnected, the planning, design and operation of the grid must be coordinated. A free flow of this data between companies is at risk not only to internal sabotage by disgruntled employees but to external foes that have access to engineering expertise.
- Control center hosts SCADA systems and their more modern counterparts, energy management systems, which are the nerve centers and have large amounts of historical data as well as real time data on the power system. Until recently, these were stand-alone systems. However, nowadays there are security concerns due to several types of connections to the control center as well as utility connections to several types of DERs, DER systems, customers, and third-party providers.
- Embedded computers are found in a large variety of control equipment on the power system, much of it microprocessor based. These embedded computers are often located at remote unmanned substations. These computers could be accessed remotely for setting control and protection parameters. The savings in cost make it very tempting to do so. However, this opens up the possibility of security breaches where sensitive control equipment can be recalibrated by remote break-ins.

Guidance to end-to-end implementation of cybersecurity activities is provided in technical literature and several documents including guidelines in this white paper [NIST 2014b]. The guidelines describe the approach, which begins with the risk management process activities and identification of the business functions, then the associated mission and business processes and the associated Smart Grid systems and assets. Then, it is required to identify and select the appropriate Smart Grid high level security requirements needed to protect the Smart Grid systems, perform a gap assessment, create a plan to remediate identified gaps, and develop a plan to monitor and maintain Smart Grid high level security requirements as part of a repeatable risk management process.

The high-level Smart Grid organizational business functions from an executive viewpoint, establish the organization's risk tolerance, and identify the organizational cybersecurity risk management strategy. The executive cybersecurity risk management governance team identifies strategic goals and objectives (such as distribution management, meter to cash, etc.). Examples of such business functions when developing the security program may include:

- Power operations
- Metering to cash
- Customer services
- Corporate services.

As each organization may develop and implement its risk management programs at different levels, a framework guidance designed for any organization is described in [DOE 2015a]. This framework can be used whether the organization is the entire enterprise, or a business unit or process within the enterprise. The framework is designed to assist energy sector organizations to characterize their existing security program:

- Assess their current and target cybersecurity posture

- Identify gaps in their existing cybersecurity risk management programs, using the framework as a guide, and identify areas where current practices may exceed the framework
- Recognize existing sector tools (e.g., cybersecurity capability maturity model), standards, and guidelines that may support the framework implementation
- Effectively communicate the risk management approach and use of the framework to both internal and external stakeholders.

The concerns for security of power systems, financial transactions, contractual agreements, and personal information including corresponding proliferation of products, systems and services that maintain and protect information call organizations to practice security management in a more mature manner. For example, security products and systems typically come to market in one of two ways: through lengthy and expensive evaluation or without evaluation. Acquirers and users must rely solely on the security claims of the product or system developer or operator. Often, trusted products reach the market long after their features are needed and secure systems are being deployed that no longer address current threats.

One important function of a security program is managing the development of security requirements. The development of requirements can be accomplished by using the Requirements Engineering methods.

1.3.2 Requirements Engineering

There are more than one definition for the field of Requirements Engineering depending on the viewpoint of the definer. Requirements Engineering is the process of eliciting, documenting, analyzing, validating, and managing requirements [Laplante 2007] to the subfields of Systems Engineering and Software Engineering concerned with this process [Nuseibeh 2000].

A requirement can range from a high-level, abstract statement of a service or constraint to a detailed, formal specification. There is so much variability in requirements detail because of the many purposes that the requirements must serve.

From the perspective of Systems Engineering discipline, Requirements Engineering is concerned with determining the goals, functions, and constraints of hardware and software systems [Laplante 2007]. It is also concerned with the relationship of these factors to precise specifications of software behavior, and to their evolution over time and across software families [Zave 1997]. Another definition combines other definitions. Requirements Engineering is the branch of Engineering concerned with the real world goals for functions of, and constraints on systems. It is also concerned with the relationship of these factors to precise specifications of system behavior, and of their evolution over time and across families of related systems [Laplante 2014].

Security and safety considerations are constraints among other system constraints such as regulatory agencies and standards, hardware limitations, and interfaces to other applications. During the Systems Engineering process, architectures are generated to better describe and understand the system [DoD 2001].

To understand problem effectively, all the phases of Requirements Engineering are described thoroughly and independently (see Figure 1.5). In reality, all the phases are iterative, interleaved, and can take the whole duration of the entire software systems development life cycle [Shukla 2015].

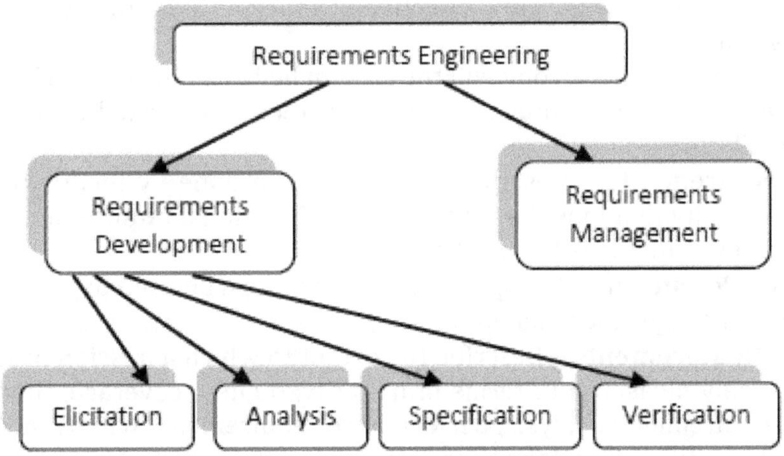

Figure 1.5 A hierarchical decomposition of Requirements Engineering. *Source*: [Shukla 2015]. © 2015, CAE.

Requirements development starts with the analysis and elicitation of the objectives and constraints of the organization. Requirements management includes supporting planning for requirements, integrating requirements as well as relationships with other information delivering against requirements, and changes for these [Shukla 2015]. The context of most activities is an organization in which development takes place or in which a system will operate.

One task of Requirements Engineering is stakeholder identification. From the security perspective, stakeholders, although not all, are interested in a secure system or product. Stakeholders are persons or organizations (legal entities such as companies, standards bodies) which have a valid interest in the system. They may be affected by it either directly or indirectly. Other stakeholders may include:

- Anyone who operates the system (normal and maintenance operators)
- Anyone who benefits from the system (functional, political, financial and social beneficiaries)
- Anyone involved in purchasing or procuring the system. In a mass market product organization, product management, marketing and sometimes sales act as surrogate consumers (mass-market customers) to guide development of the product
- Organizations which regulate aspects of the system (financial, safety, and other regulators)
- Organizations or entities responsible for systems that interface with the system under design.

The identification of stakeholders helps on determining the types of requirements that have to be analyzed and written. The general types of requirements include the following:

- **Customer requirements** - define the expectations of the system in terms of mission objectives, environment, constraints, and measures of effectiveness and suitability. The customers are those that perform the eight primary functions of Systems Engineering, with special emphasis on the operator as the key customer. Operational requirements define the basic need and, at a minimum, answer the questions posed such as:
  - Operational distribution or deployment: Where will the system be used?
  - Mission profile or scenario: How will the system accomplish its mission objective?
  - Performance and related parameters: What are the critical system parameters to accomplish the mission?
  - Utilization environments: How are the various system components to be used?

  - Effectiveness requirements: How effective or efficient must the system be in performing its mission?
  - Operational life cycle: How long will the system be in use by the user?
  - Environment: What environments will the system be expected to operate in an effective manner?
- **Functional requirements** - Describe what has to be done by identifying the necessary task, action or activity that must be accomplished. Functional requirements analysis will be used as the top-level functions for functional analysis.
- **Non-Functional Requirements** - Specify criteria that can be used to judge the operation of a system, rather than specific behaviors.
- **Performance Requirements** - Describe the extent to which a mission or function must be executed; generally measured in terms of quantity, quality, coverage, timeliness or readiness. During requirements analysis, performance requirements will be interactively developed across all identified functions based on system life cycle factors; and characterized in terms of the degree of certainty in their estimate, the degree of criticality to system success, and their relationship to other requirements.
- **Design Requirements** - Define the build, code, and buy requirements for products and how to execute requirements for processes expressed in technical data packages and technical manuals.
- **Derived Requirements** - Describe requirements that are implied or transformed from higher level requirements.
- **Allocated Requirements** - Describe a requirement that is defined by dividing or otherwise allocating a high-level requirement into multiple lower level requirements.
- **Security Requirements (IT and ICS)** - Do any of the following:
  - Address security goals (e.g., Confidentiality, Integrity, Availability, Non-repudiation, Accountability, Authenticity, etc.)
  - Prevent unauthorized disclosure of systems and information
  - Prevent unauthorized modification of systems and information
  - Prevent disruption of service and productivity.

Besides this classification, there are other classifications of requirements (see [DoD 2007], [DoD 2008], [Summerville 2005], [Laplante 2007], [Laplante 2014], [Shukla 2015]).

1.3.3 Issues in Requirements Engineering

Requirements Engineering is often performed in an ad hoc manner and the lack of appropriate requirements engineering practices can introduce major problems in the system life cycle [Zafar 2013]. Generally, only the functional requirements are identified whereas the nonfunctional or quality requirements are ignored. Without systematic requirements engineering, the requirements tend to be conflicting and inconsistent, which makes the system, in later phases of development and operation, vulnerable to serious defects [Nuseibeh 2000].

Many delivered systems do not meet their customers' requirements due, at least partly, to ineffective Requirements Engineering. This is often treated as a time consuming, bureaucratic and contractual process. Although this attitude is changing and Requirements Engineering is increasingly recognized as a critically important activity in any Systems Engineering process, there are still issues.

Quite often, there could be customer requirements that are conflicting, incomplete, unidentified, undecided, contradictory or ambiguous at the time of initial stage of development. However, any uncertainty in any type of requirements can affect the quality of the system and implicitly the quality and security of the system. More precisely, customer requirements are the result of interviews with stakeholders or can be obtained through modeling.

A revision of existing processes for Requirements Engineering identifies similar issues that have to be addressed by a preprocessing process. This process has to identify and detect the major problem during requirement analysis [Lal 2013]. This work suggests a suite of questions during the requirements analysis. However, the recommended subprocess is based on human expertise like many other tasks of Requirements Engineering process.

Another issue is missing the feasibility study [Lal 2013]. In some life cycle models, the Requirements Engineering process begins with a feasibility study activity, which leads to a feasibility report. If the feasibility study suggests that the product should be developed, then requirement analysis can begin. If requirement analysis precedes feasibility studies, which may foster outside the box thinking, then feasibility should be determined before requirements are finalized.
Requirements Engineering is not well practiced, and the fundamental challenge is recognizing that customers often confuse requirements and goals [Laplante 2014]. Goals are high level objectives of a business, organization, or system, but a requirement specifies how a goal should be accomplished by a proposed system. Goals evolve as stakeholders change their minds and refine and operationalize goals into behavioral requirements.

Another issue is lack of technical skills and multidisciplinary training. Whether viewed at the systems level or the software level, Requirements Engineering is a multidisciplinary, human centered process. The tools and techniques used in Requirements Engineering draw upon a variety of disciplines, so the Requirements Engineer may be expected to master skills from a number of different disciplines. In addition, many organizations even do not employ a Requirements Engineer [Nuseibeh 2000].

The application of standards to define processes in Requirements Engineering is not widely practice. Also, there are many standards and there could be confusion about choosing the most suitable standard from a set of competing standards. In addition, several standards go through continuous revisions or replacements. For example, the standard [ISO 29148] specifies the required information items that are to be produced through the implementation of the requirements processes and provides guidelines for the format of the required and related information items to ensure conformance to ISO/IEC/IEEE 15289 standard [ISO 15289]. It can be added to the existing set of requirements related life cycle processes defined by [ISO 12207] or ISO/IEC/IEEE 15288 (see [ISO 15288]) or can be used independently. However, this standard replaces older standards.

Issues such as Requirements Engineering, reliability, logistics, coordination of different teams, testing and evaluation, maintainability and many other disciplines necessary for successful system development, design, implementation, and ultimate decommission become more difficult when dealing with large or complex projects.

Based on generated requirements, the architecture of a system can be generated. Various organizations and researchers define systems architecture in different ways (see definitions in [ANSI 1471], [Jaakkola 2011]). An architecture can be defined in various contexts in the general field of Engineering [DoD 2001]. Architecture can be used as a general description of how the subsystems join together to form the system. It can also be a detailed description of an aspect of a system: for example, the Operational, System, and Technical Architectures used in Command, Control, Communications, Computers, Intelligence, Surveillance, and Reconnaissance (C4ISR), and software intensive developments.

The novelty of many software applications, the speed with which they need to be developed, and the degree to which they are expected to change, all play a role in determining how the systems development process should be conducted. The demand for better, faster, and more usable software systems will continue, and Requirements Engineering should therefore continue to evolve in order to deal with different development scenarios.

1.3.4 Issues in Security Requirements Engineering

Although the field of Requirements Engineering has made substantial progress including many published materials during the last two decades, there is not enough information about the practice and the application of Requirements Engineering approaches to Information Security and security of complex systems such as the Smart Grid.

Security Requirements Engineering is often the most critical task of managing secure system development. It is at the intersection of the most difficult technical issues, the most acute bureaucratic power struggles, and the most determined efforts at blame avoidance [Anderson 2008].
Security Requirements Engineering is a challenging and complex task [Zafar 2013]. Security requirements are often taken from a predefined set of standard security requirements, that are not properly integrated with other requirements of the system, and a close analysis and specification process is not carried out. This results in weak security requirements specification, which adversely affects the design and implementation of a system. During requirements gathering phase, it is recommended to accomplish the following [Harris 2013]:

- Security requirements
- Security risk assessment
- Privacy risk assessment
- Risk level acceptance.

Security requirements should be defined in the categories of Availability, Integrity, and Confidentiality [Harris 2013]. However, this approach is missing other objectives of Information Security including aspects of System Security. Some examples of system security are physical security, personnel security, and network security. Also, security requirements for control systems have to include real time constraints, safety, and reliability.

Systems Security Engineers have focused on security standards, so standards compliance can serve as evidence that security problems are addressed [Bayuk 2011]. Although standards provide guidelines and help on defining security issues, there is a need to build in security based on approach developing systemic security requirements and systems views of verification and validation [Bayuk 2011]. Compared to Systems Engineering methodologies, the available methodologies for Security Engineers have consistently lagged behind [Anderson 2008]. A new Systems Engineering security roadmap recommends that Systems Engineers and Security Engineers converge on empirical methods. Systems engineering is an interdisciplinary field of Engineering that focuses on how to design and manage complex engineering systems over their life cycle. At its core, Systems Engineering utilizes systems thinking principles to organize this body knowledge. Currently, the Systems Engineering field is undergoing a major transformation driven by the increasing complexity of systems and technical advances.

While in the past, security requirements were classified as non-functional requirements, they have the objective to address the security of the information and systems. Management needs to include development of security requirements from the beginning of a system's development. As long as Systems Engineers don't consider security a functional requirement, it won't likely rise to the top of the implementation checklist [Bayuk 2011].

Two fundamental concepts in computer and information security are the security model, which outlines how security is to be implemented - in other words, providing a blueprint - and the architecture of a computer system, which fulfills this blueprint.

1.4 Security Model – Definition and Components

A model is a way to describe a given set of components and how those components relate to each other in order to describe the main workings of an object, system, or concept [ISACA Glossary]. It is an abstract representation of reality in any form (including mathematical, physical, symbolic, graphical or descriptive form) to present a certain aspect of that reality for answering the questions studied [ISO 157004]. Security model is an important concept in the design and analysis of secure systems. In Information Security, the diverse models used include system architecture model, threat model, attack model, vulnerability model, network security model, network topology model for control system, security model, etc. A security model can refer to an information security model, security policy model, or organizational security model. Therefore, security professionals need to make this distinction and communicate to users and developers.

A security model may be founded upon a formal model of access rights, a model of computation, a model of distributed computing, or no particular theoretical grounding at all. Security models are used in security evaluation, sometimes for proofs of security. Thus, security models are an important concept in the design and analysis of secure computer systems. Information security models include methods that are used to authenticate security policies as they are intended to provide a precise set of rules that a computer can follow to implement the fundamental security concepts, processes, and procedures contained in a security policy. These models can be abstract or intuitive.

1.4.1 Security Policy Models

A security policy model is a formal description of a security policy. A security policy model is a symbolic representation of a policy that should be enforced in the system [Harris 2013]. It maps the abstract goals of the policy to information system terms by specifying explicit data structures and techniques necessary to enforce the security policy.

Several security policy models have been developed to enforce security policies. For example, a security model provides a deeper explanation of how a computer operating system should be developed to properly support a specific security policy. A computer security model is implemented through a computer security policy. A computer security policy defines the goals and elements of an organization's computer systems. The definition can be highly formal or informal. Security policies are enforced by organizational policies or security mechanisms. A technical implementation defines whether a computer system is secure or insecure. These formal policy models can be categorized into the core security principles of Confidentiality, Integrity and Availability [Krutz 2004]. A typical top-down approach used in SE can take the form of threat model – security policy – security mechanisms [Anderson 2008].

Therefore, it should be understood that a security policy is a set of rules and practices dictating how sensitive information is managed, protected, and distributed. A security policy expresses exactly what the security level should be by setting the goals of what the security mechanisms are to accomplish. This is an important element that has a major role in defining the design of the system. The security policy is a foundation for the specifications of a system and provides the baseline for evaluating systems. Security policies can be directed toward the organization itself or can address operating systems and applications. The different policies are similar but have different targets among organizations as opposed to individual computer systems. A security policy can also refer to organizational policy, specific issues, system, or information.

Also, the policy is the corner's stone of an organization's information security architecture [Hansche 2004]. A security policy could capture the security requirements of an enterprise or describe the steps that have to be taken to achieve security. A policy is important to establish both internally and externally what an organization's position on a particular topic might be. Policies are supported by standards, baselines, procedures, and guidelines.

A system provides trust by fulfilling and enforcing the security policy and typically deals with the relationships between subjects and objects. The policy must indicate what subjects can access individual objects, and what actions are acceptable and unacceptable. The definition of what trust means is derived from a framework and the security policy works as this framework for computing systems. For a system to provide an acceptable level of trust, it must be based on an architecture that provides the capabilities to protect itself from untrusted processes, intentional or accidental compromises, and attacks at different layers of the system. A majority of the trust ratings require a defined subset of subjects and objects, explicit domains, and the isolation of resources so their access can be controlled, and activities performed on them can be audited.

A security policy model is usually represented in mathematics and analytical ideas, which are then mapped to system specifications, and then developed by programmers through programming code. So we have a policy that encompasses security goals like each subject must be authorized to access each object. The security model takes this requirement and provides the necessary mathematical formulas, relationships, and structure to be followed to accomplish this goal. From here, specifications are developed per operating system type (e.g., Unix, Windows, or Macintosh), and individual vendors can decide how to implement mechanisms that meet these necessary specifications.

1.4.2 Security Policy Models: Examples, Strengths and Weaknesses

Individual systems and devices can have their own security policies. The systems' security policies and models they use should enforce the higher-level organizational security policy that is in place. A security policy outlines goals with no idea of how they would be accomplished. For example, a security model can provide a deeper explanation of how a computer operating system should be developed to properly support a specific security policy.

The first security models of 1970s and earlier were designed to meet the needs of multilevel security [Hansche 2004]. A multi-level security is defined as a class of system containing information with different sensitivities that simultaneously permits access by users with different security clearances and needs to know, but prevents users from obtaining access to information for which they lack authorization. This concept originated in the late 1960s when the U.S. Department of Defense wanted to develop a method of protecting classified information stored on computers.

Then and today, systems have been developed and continue to provide security controls for multilevel security uses. Multilevel security can be implemented in a number of ways. Several approaches evolved from the original idea of 1970s. Since about 2005, one approach uses virtual machines [Anderson 2008]. Virtualization products such as [VMware] and [Xen] provide multiple systems at different security levels on the same computer. For example, different databases requiring distinct security levels (High and Low) run on VMware on top of operating system (e.g., SELinux).

Formal security models, such as Bell-LaPadula and Biba models, are used to provide high assurance in security. Some security models enforce rules to protect Confidentiality, such as the Bell-LaPadula model. Other models enforce rules to protect Integrity, such as the Biba model. Informal models, such as Clark-Wilson, are used more as a framework to describe how security policies should be expressed and executed.

Many of these security issues must be thought through before and during the design and architectural phase for a product. Security is best if it is designed and built into the foundation of operating systems and applications and not added on as an afterthought. Once security is integrated as an important part of the design, it has to be engineered, implemented, tested, audited, evaluated, certified, and accredited. The security that a product provides has to be rated on the availability, integrity, and confidentiality it claims. Consumers then use these ratings to determine if specific products provide the level of security they require. The security models can be grouped into three types of models [Krutz 2004]:

- Access control models (e.g., access matrix, the state machine model, Take Grant model, and Bell-LaPadula model)
- Integrity models (e.g., Biba model, Clark-Wilson model)
- Information flow models (e.g., information flow model, noninterference model, and Brewer and Nash model).

However, some models of one type exhibit characteristics of other types.

1.5 Prominent Security Models

The following is a brief overview of the most prominent security models.

1.5.1 State Machine Model

The state machine model is used to describe the behavior of a system to different inputs. It is based on mathematical constructs that represent sets (subjects and objects) and sequences. The model captures the state of a system. States change only at discrete points in time, e.g., triggered by a clock or an input event. The state should capture the essential features of the system under investigation, e.g., the security of a computer system. A system that has employed a state machine model will be in a secure state in each and every instance of existence. The model employs Basic Security Theorem which says that if a system initializes in a secure state and all state transitions are secure, then every subsequent state will be secure no matter what inputs occur. The model was developed to keep the integrity of the system (computer, operating system, data, program, or process) intact at all times.

The model is used to check that all state transitions starting in a secure state yield a secure state. An inappropriate change to one variable could change or corrupt the system or a process. It will boot up into a secure state, execute commands and transactions securely, allow subjects to access resources only in secure states shut down and fail in a secure state in some contexts. Failing in a secure state is often extremely necessary. It is imperative that if anything unsafe takes place that the system can save itself and not make itself vulnerable. When a user sees error messages, or a system reboots or freezes, these are all safety measures.

1.5.2 Bell-LaPadula Model

The Bell-LaPadula model is an important historic milestone in computer security. The model is a state machine model capturing confidentiality aspects of access control. It is a subject-to-object model. Access permissions are defined through an access control matrix and through a partial ordering of security levels. Security policies prevent information flowing downwards from a high security level to a low security level. The model only considers the information flow that occurs when a subject observes or alters an object. The model's main goal is to prevent secret information from being accessed in an unauthorized manner. A system that employs the Bell-LaPadula model is called a multilevel security system because users with different clearances use the systems, and the systems process data with different classifications.

The level at which information is classified determines the handling procedures that should be used. The subject's clearance is compared to the object's classification; if the clearance is higher or equal to the object's classification, the subject can access the object without violating the security policy. The model uses subjects, objects, access operations (read, write, and read/write), and security levels. Subjects and objects can reside at different security levels and have relationships and rules dictating the acceptable activities between them. If properly implemented and enforced, this model has been mathematically proven to prevent data from a higher security level from flowing to a lower security level. It is an information flow security model also, which means that information does not flow in an insecure manner. The Bell-LaPadula model was developed to make sure secrets stay secret; thus, it provides and addresses confidentiality only.

This model does not address integrity of the data the system maintains - only who can and cannot access the data. Other issues that are not addressed include [TechTarget]:

- Management of access control, because there is no mechanism to modify access rights
- Covert channels
- File sharing used in more modern systems.

A covert channel is a mechanism that was not designed for communication, but which can be abused to allow information to be communicated down from high to low-level security.

There are other issues with the use of the Bell-LaPadula model [Anderson 2008]. One implication is that a process acquires the security labels of all the files it reads, and these become the default. To address this cumbersome issue, application software must be rewritten (or at least modified) to run on multilevel security platforms. Other issues include addressing mandatory access control and control of the interactions when separating users and processes; addressing the creation and destruction of subjects and objects.

1.5.3 Biba Model

Similar to Bell-LaPadula model, the Biba model is a state machine model that was developed after Bell-LaPadula model to address the integrity of data. It uses a lattice of integrity levels that prevents data from any integrity level from flowing to a higher integrity level. The model uses three main rules to provide this type of protection:

- Integrity axiom states that a subject cannot write data to an object at a higher integrity level (referred to as no write up)
- Simple integrity axiom states that a subject cannot read data from a lower integrity level (referred to as no read down)
- Invocation property states that a subject cannot request service (invoke) of higher integrity.

The Biba model is an information flow models because it is most concerned about data flowing from one level to another. However, Biba model has the same fundamental problems as Bell-LaPadula model [Anderson 2008]. It cannot accommodate real world operation very well without numerous exceptions. For example, Biba model does not allow trusted subjects to overwrite the security model, a needed capability in the real world systems. Also, the model cannot express many real integrity goals. The goals of Integrity include:

- Prevent unauthorized users from making modifications
- Prevent authorized users from making improper modifications (separation of duties)
- Maintain internal and external consistency of well formed transactions).

However, the Biba only addresses the first goal only.

1.5.4 Clark-Wilson Model

The Clark-Wilson model was developed after Biba and takes some different approaches to protect the integrity of information by preventing authorized users from making unauthorized modification of data, or commit fraud and errors within commercial applications. The model addresses all the goals of Integrity.

The Clark-Wilson model does not allow users to access and manipulate objects directly, but must access the object through a program (referred to as access triple: subject (user), program, and object). This provides another layer of protection between the subject and the object and further restricts the type of actions that can take place on that object, thus protecting the integrity of the object. The program has its own set of restrictions dictating what actions the users can and cannot perform on objects. This streamlines the way objects are protected and reduces the methods that could cause the data to be corrupted or changed in an undesirable manner.

Also, this model enforces separation of duties, which divides an operation into different parts and requires different users or rules to perform each part. This ensures that a critical task cannot be carried out by one entity. Auditing is also required in this model to track the information coming in from the outside of the system.

Finally, the Clark-Wilson model prevents authorized users from making modifications by requiring them to go through programs to modify objects. It also prevents authorized users from making improper modifications by enforcing separation of duties and maintains an audit log for external transactions.

1.5.5 Information Flow Model

Information flow model can deal with any kind of information flow, not only the direction of the flow.

The information flow model looks at insecure informational flow that can happen at the same level and between objects along with the flow between different levels. When the information flow model is used, a system is secure if there is no illegal information flow permitted. One way that the information flow model provides this type of protection is by ensuring that covert channels do not exist in the code.

Basically, information can flow from one security level to another or from one object to another in the same security level until a restricted operation is attempted. In most systems, once a restricted operation is attempted, the system looks at an access control matrix to see if this action has been specifically permitted for this specific subject or object.

Developers and architects have to ensure that the covert channels (storage and timing) do not exist. Information is restricted in its flow to only go to and from entities in a way that does not negate the security policy.

1.5.6 Noninterference Model

The concept of noninterference is implemented to ensure that any actions that take place at a higher security level does not affect, or interfere, with actions that take place at a lower level. Therefore, multilevel security properties can be expressed in another way.

The noninterference model does not concern itself with the flow of data, but with what a subject knows about the state of the system. So if an entity at a higher security level performs an action, it cannot change the state for the entity at the lower level.

If a lower level entity was aware of a certain activity that took place by an entity at a higher level and the state of system changed for this lower level entity, the entity might be able to deduce too much information about the activities of the higher state, which in turn is a way of leaking information. Users at a lower security level should not be aware of the commands executed by users at a higher level and should not be affected by those commands in any way. Commands and activities performed at one security level should not be seen or affect subjects or objects at a different security level.

Basically, the noninterference model allows ways to prevent subjects operating in one domain (e.g., financial) from affecting each other in violation of security policy. The real intent of this model is to address covert channels and inference attacks. An inference attack occurs when someone has access to some type of information and can infer something that he does not have the clearance level or authority to know.

1.5.7 Other Models

Other models include: Brewer and Nash model, Graham-Denning model, and Harrison-Ruzzo-Ullman model.

- Brewer and Nash model (also called Chinese Wall model) is a model that allows for dynamically changing access controls that protect against conflicts of interest. Brewer and Nash model allows controls to ensure that there is no conflict of interest; it provides a way of separating competitor's data within the same integrated database. If a user accesses one organization's data, the competitor's data can automatically be deemed off limits.
- Graham-Denning model is a model that creates rights for subjects, which correlate to the operations that can be executed on objects. It shows how subjects and objects should be created and deleted.
- Harrison-Ruzzo-Ullman model is a model that deals with access rights of subjects and the integrity of those rights. A subject can carry out only a finite set of operations on an object. The model provides a finite set of procedures to edit the access rights of a subject.
- Positive and negative security models [OWASP]. These models are more like simple solutions to enforce a broad security policy. A positive security model (also known as whitelist) is one that defines what is allowed and rejects everything else. This should be contrasted with a negative (or blacklist) security model, which defines what is disallowed, while implicitly allowing everything else.

However, there are issues that need to be understood about security models.

1.5.8 Summary of Issues

A summary of issues includes:

- **Costs** - The more expressive a security model is, both with respect to the security properties and the systems it can describe, is usually difficult and costly to verify security properties.
- **Limited Value** - Formal models and security proofs have limited value [Denning 1999]. Commercial systems would never be perfectly secure. They would always have weaknesses. Even in the domain of cryptography, the race will never end between the code makers and code breakers.
- **No Guarantee for Security** - Security models and formal methods do not establish security. They only establish security with respect to a model, which by its very nature is extremely simplistic compared to the system that is to be deployed, and is based on assumptions that represent current thinking. Even if a system is secure according to a model, the most common (and successful) attacks are the ones that violate the model's assumptions. Other issues are identified in [Denning 1999]:
    - Security models have theoretical limits; it cannot be always proven that a model satisfies certain security conditions
    - Security models based on strict mathematical properties can lead to systems that are totally unusable
    - Building systems from rigorous mathematical security models is extremely time consuming and costly; most commercial systems will not be based on formal models
    - Security models and formal methods do not establish security. Systems are hacked outside the models' assumptions
    - Provable security, even if it were achievable, is not a panacea.

To overcome the limitations of different models, the users look for innovative solutions.

1.6 Innovative Solutions

Therefore, new solutions proposed by researchers are identified. The need to satisfy both confidentiality and integrity prompted the designers in Healthcare industry [Tabor 2015] to look for a solution proposed by [Zhang 2008].

The concepts underlying these models, Biba model and the Bell LaPadula model, are opposite to each other, they cannot be used together, because the system will not be usable. The Biba model is a strict Integrity model that provides a high level of protection and simplicity in its design. But, a disadvantage is that the usability of the system is restricted, even more so when integrated into a system with confidentiality policies in place [Garnaut 2011].

Thus, the solution that could work for Healthcare industry is to adopt dynamic methods that allow to use these two models in the same time for the same system as described in [Zhang 2008]. The methods check the domain and privilege, and then assign the operations.

Also, within the realm of National security, there has been recent recognition of the role that integrity has in the overall security of a system besides confidentiality. A review of the associated literature including the prevalent data integrity models, evaluation mechanisms and integrity centric implementations. Specifically, the Biba and Clark-Wilson data integrity models are discussed along with their variations in [Garnaut 2011]. The authors identified key issues such as:

- There are several compromises that must be made during the selection or development of data integrity models for information systems.
- Biba model restricts the usability of the system, even more so when integrated into a system with confidentiality policies in place.
- Extensions to the original models (e.g., Biba model and Clark-Wilson model) were developed to improve their usability, but they are also more complex and could therefore be more difficult to understand and costly to formally verify their correctness.
- Difficulty to quantify is the associated performance cost with some of these new models.

Information Security Models bridge the gap between security policy statements (which explain which users should have access to data) and the operating system implementation (which allows an administrator to configure access control). The models help map abstract goals onto mathematical relationships that underpin whichever implementation is eventually chosen (Windows, Unix, MacOS, etc.). These models together are well worth understanding for they are the underpinning of modern access control implementations. Although there are opportunities for use of many and different security models, the selection of a model should be based on business needs, security objectives, and budgets.

1.6.1 Which Security Model to Choose?

Considering the Smart Grid needs, the models that could be used include: Information Flow Model, Noninterference model, Brewer and Nash model, and Harrison-Ruzzo-Ullman model. The idea of information security standards and models with measurable effects on the business becomes more present in practice and more respected by experts in security industry. Maturity model can be described as structured set of elements that describe certain aspects of improvement (maturity) in the organization. Therefore, it is the task of security team to evaluate the current processes used in an enterprise and recommend the use of the appropriate model based on paradigm goal and some criteria including benefits, weaknesses, standard compliance, metrics, improvement, license, and development costs.

1.6.2 Comparison of Models

Besides the Systems Security Engineering Capability Maturity Model (SSE-CMM) model described by [ISO/IEC 21827] standard, other maturity models and process models emerged that are used in the organizations. However, the life of many models as well as standards is unpredictable. There is a continuous process of improvement and innovation such that better models are adopted as the organizations structure their businesses and new trends in economy, management, and technology emerge. Some models are used in systems developed for the government, some models are adopted by commercial enterprises, some standards are adopted by a country. For example, versions of [ISO/IEC 21827] standard are adopted by other countries and identified as CAN/CSA-ISO/IEC 21827-09 in Canada or ISO/IEC 21827:2008 in United Kingdom.

A comparison of models based on a set of criteria such as paradigm goals, compliance to standards, inputs, outputs, metrics, approach, continuous improvement, license, and price is discussed in [Stevanović 2011]. Also, this work shows similarities and differences between ISO/IEC 21827 (SSE-CMM) and ISM3 models.

Essential differences are in the approach to standard implementation and the price of introduction. While the Information Security Management Maturity Model (ISM3) is oriented on business result achievement, which is on the top of the priority list, SSE-CMM is focused primarily on information security assurance, and the effect on business is measured during the implementation. The ISM3 model is, thus, focused on a top down approach, from the most important goal of the organization, to lower ones, including information security. SSE-CMM uses a bottom up approach, in which it goes from information security, as the basis, and other goals of the organization depend, in great extent, from it. The price of implementation depends on the introduction price (ISM3 is a commercial product) and SSE-CMM is free. The results of comparison are illustrated in Table 1.1

Table 1.1 Comparison of ISM3 and SSE-CMM.

Criterion	**ISM3**	**SSE-CMM**
Number of achievement levels	5	5
Paradigm	Process oriented	Process oriented
Goal	Admissible level of security maximal business result	Secure and quality ISMS with least expenses
Input	Yes	Yes
Output	Yes	Yes
Metrics	Yes	Yes
Compliance	Yes with ISO 27001	Partly with ISO 27001 Integrated in ISO 21827:2008
Approach	Top-down	Bottom-up
Extent	Strictly	Technological, behavioral
Improvement	Continuously through metrics	Cyclically through outputs
License	Commercial	Free
Last update	April 2009	June 2008

Source: [Stevanović 2011].

However, these drawbacks make these models not efficient due to:

- Static in nature
- Process oriented
- Not risk based
- Do not consider external factors
- Do not consider culture
- Do not plan for unpredictable situations.

Also, lifecycles are not always linear or cyclical. Certain events can shift an object from one state to another, in a nonlinear or non-cyclical fashion. While assessing and recommending the appropriate model, it is necessary to identify the components of information systems which are complex and have various tangible and intangible components. The components can be classified at the chosen level of abstraction according to structural and transactional features [ISM3 2009]:

- Structural features include various assets from which an information system may be built such as repositories (any temporary or permanent storage of information), interfaces (any input/output device), channels (physical or logical pathways for the flow of messages, including buses, LAN networks, etc.); a network is a dynamic set of channels; borders (define the limits of the system); structural objects exist in every logical and physical level.
- Transactional features include various assets from which an information system produces actual results such as services (any value provider in an information system, including services provided by BIOS, operating systems and applications), messages (any meaningful information exchanged between two services or a user and an interface).

Transactional assets are dynamic, such as running processes and moving messages. Static assets such as mail or program files stored in a repository are not considered. A service can collaborate with other services or lower level services to complete a task that provides value, like accessing information from a repository.

The existing security models while valuable do not answer each of these criteria for Confidentiality, Integrity, and Availability. For example, the traditional Information Security models have capabilities such as:

- Allow for access control
- Provide guidance for defense in depth solutions
- May assist with compliance
- Provide a tool to manage situational security.

1.6.3 Choosing a Model

Ultimately, the selection of a model should be based on the desired levels of improvement that gives to the business which depends on organizational processes and their specific goals and also on dedication of the resources to the general business goal. Processes must implement certain activities that support the general goal of an organization.

In the analysis and selection of any model, the information security culture and skills are also factors to be considered. The information security processes and controls influence the information security behavior and culture [Viega 2008].

1.7 A Unified Security Approach

The electric grid is fundamental to the economic and physical wellbeing of the nation, and emerging cyber threats targeting electricity systems highlight the need to employ techniques that integrate advanced security measures to protect critical assets.

Traditionally, cybersecurity for IT focuses on the protection of information and information systems from unauthorized access, use, disclosure, disruption, modification, or destruction in order to provide confidentiality, integrity, and availability. Cybersecurity for the Smart Grid requires an expansion of this focus to address the combined IT, ICS, and communication systems, and their integration with physical equipment and resources in order to maintain the reliability and the security of the smart grid and to protect the privacy of consumers.

The cybersecurity of the Smart Grid must include a balance of both electricity and cyber technologies as well as processes in IT and in ICS operations and governance. When practices from one sector, such as the IT or communications sector, are applied directly to the electricity sector, care must be taken because such practices may degrade reliability and increase risk. This is because the requirements for the electricity sector, for timing of communications, for example, may be different from the IT and communications sectors.

In the electricity sector, the historical focus has been on implementation of equipment that could improve electricity system reliability. Communications and IT equipment were formerly viewed as just supporting electricity system reliability. However, both the communications and IT sectors are becoming more critical to the reliability of the electricity system. As described in [NIST SP1108r3], cybersecurity for the Smart Grid is being expanded to address the following:

- Combined power systems
- IT and communication systems required to maintain the reliability of the smart grid
- Physical security of all components
- Reduced impact of coordinated cyber physical attacks, and
- Privacy of consumers.

In addition to the electrical infrastructure that produces and carries electric power, major elements of the Smart Grid include the IT, the ICS, and the communications infrastructure used to send command information across the grid. These elements are also used to exchange usage and billing information between utilities and their customers.
It is imperative that cybersecurity is viewed from a broader perspective that includes all these elements, many of which are critical assets. Therefore, cybersecurity based on a unified approach and a unified security program needs to be designed into the new systems that support the Smart Grid, and if possible, added into existing systems without impacting operations.

1.7.1 Unified Security Program - Principles

A unified security program assumes identification of many similarities and sensible differences between IT systems and ICS systems that require a focused management on both issues in the same time due to intertwining of traditional IT technologies in ICS systems. The Smart Grid is transforming utility operations and pushing IT across its traditional boundary into operations technology (OT) at a remarkable clip, rapidly blurring the distinction between the two categories [Meyers 2013]. In the context of the Smart Grid, a unified security program should be based on the integration of approaches needed for IT systems and ICS systems. This blending can ensure cybersecurity and privacy protection.

1.7.2 DER Integration: Unified Approach

Further, it is needed to provide a unified approach for the functions of the security program without jeopardizing the reliability requirements for the power systems and distributed energy resources (DER) systems. DER systems include both business and real time applications that share infrastructure and exchange information. Therefore, a unified security approach facilitates integration of various DER applications and systems.

Integrating DER operational information like equipment status, phasor measurements, distributed generation and storage status with business level information such as consumer preferences, energy usage and market pricing allows organizations to achieve higher end to end business and operational efficiency, reduce grid stress and improve overall enterprise productivity, cybersecurity, and privacy.

1.7.2.1 Converged OT and IT Systems

Common to most of the smart grid technologies is an increased use of communication and information technology, including an increased interaction and integration of formerly separated business (traditional IT) and industrial control systems. Competitive market forces make utilities rely heavily on a robust business information environment that requires interconnections among the control and business information system domains, the external internet, and other peer organizations.

Also, utility operations technology are transformed by an increase of IT systems so rapidly blurring the distinction between the two categories [Meyers 2013]. OT encompasses power equipment, control room applications, such as distributed supervisory control and data acquisition (DSCADA) systems that monitor the network, reaching out to devices as complex as substation gateways, or as simple as sensors. Figure 1.6 shows a unified view of converged OT and IT systems in CPS dominated Smart Grid. There are many changes such as:

- Increasing interdependence of various power generation, transmission, distribution, customer energy management and business functionalities with sensing, measuring, consuming, processing, controlling.
- Operating the grid with a diversity in interconnections (e.g., remote monitoring and testing, synchrophasor measurement system, field devices and equipment, asset management, corporate analysis and decision systems).
- An emerging trend for adaptivity in the ways control decisions dynamically interact and influence the entire sense-process-decide-control loop so transforming the power system from a very complicated machine to a complex system.
- Awareness for the improved security of Smart Grid as part of National critical infrastructure.

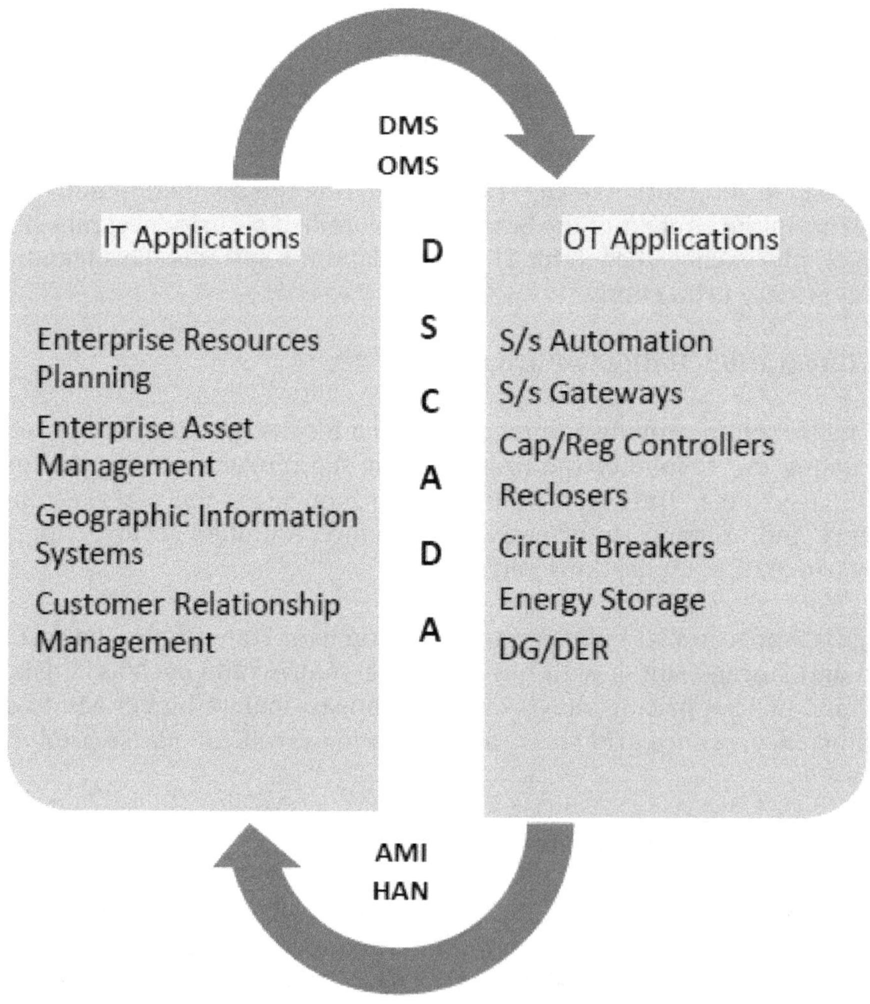

Figure 1.6 IT/OT convergence in CPS dominated Smart Grid.

These changes require secure automated information exchange, analysis and intelligent decision making distributed throughout the grid [Ray 2010]. Thus, an emerging convergence of OT and IT is pushing for new solutions. The authors propose a unified security risk management approach for Smart Grid. This work argues that adaptive orchestrations of situational awareness, decision intelligence and control activations ensure that the convergent system operates much more efficiently with much less reserve capacity (spinning reserves) and suffering much lower energy losses over the power network while enhancing reliability and safety.

1.7.2.2 Requirements

There we should understand that OT represents a broad category of components and applications that utilities depend on for safe and reliable generation and delivery of energy. With the emerging Smart Grid, IT systems increased in both quantity and quality. Examples of new IT systems include Improved Enterprise Resources Planning (ERP), Geographic Information Systems (GIS), and Customer Relationship Management (CRM) systems, along with office based productivity tools and mobile computing devices. In the past, Distribution Supervisory Control And Data Acquisition (DSCADA) and Distribution Management Systems (DMS) applications were developed and maintained in complete isolation from the rest of the IT infrastructure and were also exclusively focused on controlling a limited number of operating assets.

Now DSCADA and DMS applications are characterized by broader application and integration and quality requirements. In the same time, the outage management system (OMS) at many utilities has migrated away from the enterprise (IT system) and towards the operations domain. Modern OMS deployments now embed network intelligence to support restoration and switching and the trend is to be integrated into the other applications and workflow of the operations center. Thus, the convergence of IT and OT has implications for the communications infrastructure and operational applications that model, monitor, and manage the distribution network.

Also, priorities to fulfill security and quality requirements have to be coordinated to satisfy the needs of both IT and OT systems. While traditional enterprise business is focused on understanding cybersecurity in terms of goals and priorities for confidentiality, integrity, and availability of the electronic information and communication systems, the current security requirements of DER systems require to ensure these goals with a different priority (availability, integrity, confidentiality) and additional goals such as – safety, reliability, authenticity, accountability, trust, and privacy. For example, one factor in the determination of information reliability is the availability of information for a specific application. Some type of information has a high need for availability within a desired time frame (e.g., status of battery discharge), other information is not as critical and, if missed, can be retransmitted until it is received.

The security approach for cyber-physical systems demands additional security requirements such as ensuring continuous process for power delivery and accurate as well as dynamic pricing that are needed by the physical systems even when a cyber incident occurs. Such requirements are usually closely related to the models and states of the system, which are difficult to address by information security alone. Therefore, hybrid approaches such as the one based on both information security and system theory are necessary for securing cyber-physical systems [Mo 2012]. This work argues that combining cybersecurity and system theory approaches provides a better security incident detection and more robust security solutions for the Smart Grid.

Therefore, the Smart Grid requires a unified security approach when dealing with various systems. The following is a summary of topics that are the basis for a unified security approach and design of requirements.

1.7.3 Unified Approach based on Security Models and Frameworks

The unified approach requires the use of security models and frameworks to ensure quality designs good practices for security management. Power operation systems are increasingly being designed and implemented using IT standard hardware and software platforms and network protocols to take advantage of efficiencies such common platforms provide.

Interconnection of a wide array of traditional and emerging utility business and operation information systems such as energy management system (EMS), DMS, SCADA, MMS, advanced metering infrastructure (AMI), CIS, OMS enables utilities, customers and other service providers not only to improve cybersecurity and privacy management, but to monitor energy usage and grid status with higher precision and accuracy but also to analyze and activate distributed energy resources and storage options, construct pricing responses to appropriately reorient consumption patterns to balance available power generation capacity and demand continuously - all in real time. These activities require secure automated information exchange, analysis and intelligent decision making distributed throughout the Smart Grid. Common frameworks ensure interoperability and exchanges of information among various systems.

1.7.3.1 Regulatory Compliance

A unified approach could provide more visibility in managing the regulatory compliance. Several regulatory compliances require access controls and auditing that meet the policies requirements for power grid reliability, security, and safety. One example is an identity-based solution on controls for data access. Identity-based security gives visibility into user activity while enabling policy creation based on the user identity, meeting the requirements of regulatory compliances.

1.7.3.2 Communication Architecture

The transformation of power grid to a Smart Grid is pushing for unified solutions in many areas. A unified communication architecture to support both IT systems and OT is emerging to provide a unified environment and user experience [Meyers 2013].

As shown in Figure 1.7, Smart Grid communication structure comprises various infrastructures and various monitoring and control facilities, smart meters, software, and hardware. Secure communications and interoperability are essential in order to interconnect all these components and to exchange the information and control commands reliably and timely from one end to the other. For example, AMI infrastructure is a potential target of various threats as summarized in Figure 1.8. There is a high possibility of various security threats including data disclosure and data piracy that exist in the two-way communication using smart devices such as smart meter and AMI. Therefore, risks should be identified and mitigated accordingly. Appropriate access controls (e.g., authentication, authorization, firewall, intrusion detection and prevention, etc.) and cryptographic algorithms should be considered to protect data privacy and security of the information.

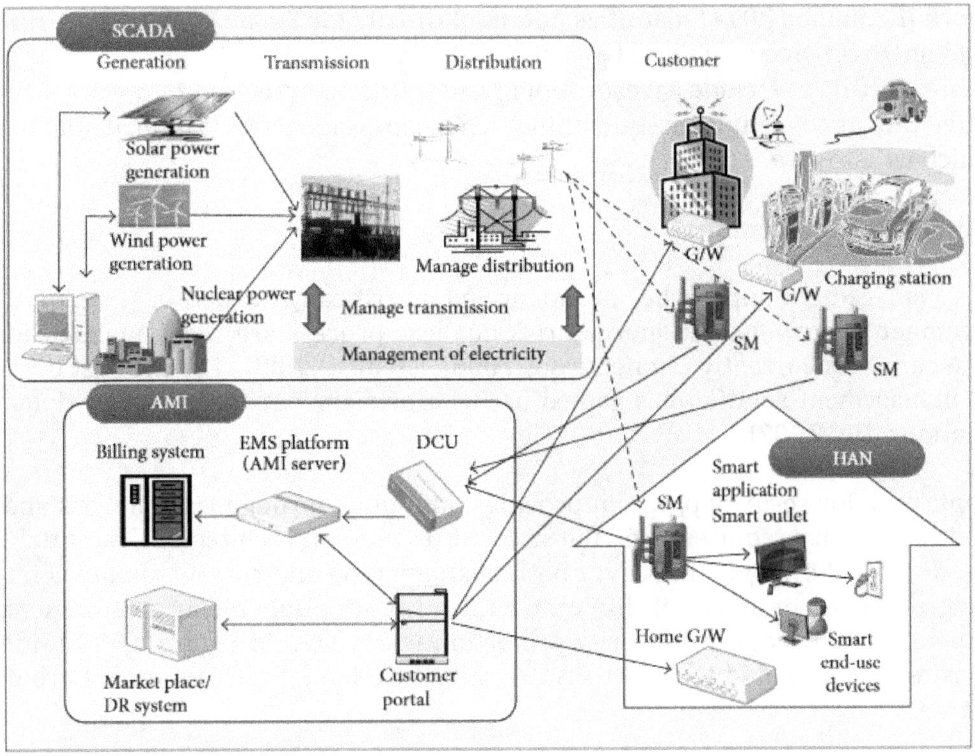

Figure 1.7 Communication structure of Smart Grid. *Source:* [Cho 2014]. Licensed under CC BY 3.0.

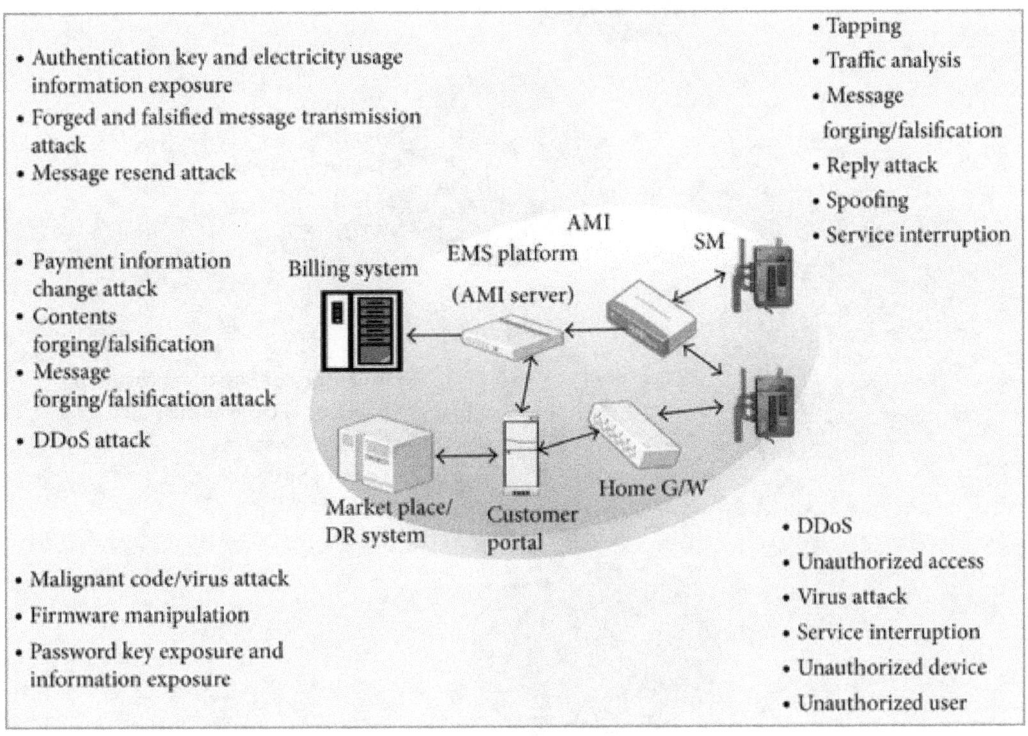

Figure 1.8 Security threats in AMI network. *Source:* [Cho 2014]. Licensed under CC BY 3.0.

Another work [Komninos 2014] identifies potential threats for Smart Grid / Smart Home. The authors select known threats from these two categories:

- Passive threats to include eavesdropping and traffic analysis
- Active threats to include masquerading, replay, message modification, denial of service and malicious software.

1.7.4 Unified Risk Management

Unified risk management approaches are identified as critically needed in [Ray 2010]. A unified risk management framework and merged risk analysis process are recommended for analyzing safety and security concurrently [Stoneburner 2006], [Hentea 2008]. For the past decade, the unified threat management solution has gained use as a primary network gateway defense solution for organizations [IDC 2007].

A unified approach for security program provides not only a through risk analysis and a rigorous development of security requirements, but it could provide an architecture essential for tackling security needs of DER systems that cover both business type and control applications. A diagram summarizing examples of assets of different nature (information, electric equipment, computer hardware, people, etc.) requiring security protection is depicted in Figure 1.9. Besides these assets, composite assets related to Smart Grid Architecture Model (SGAM) model are described in [ENISA 2014a].

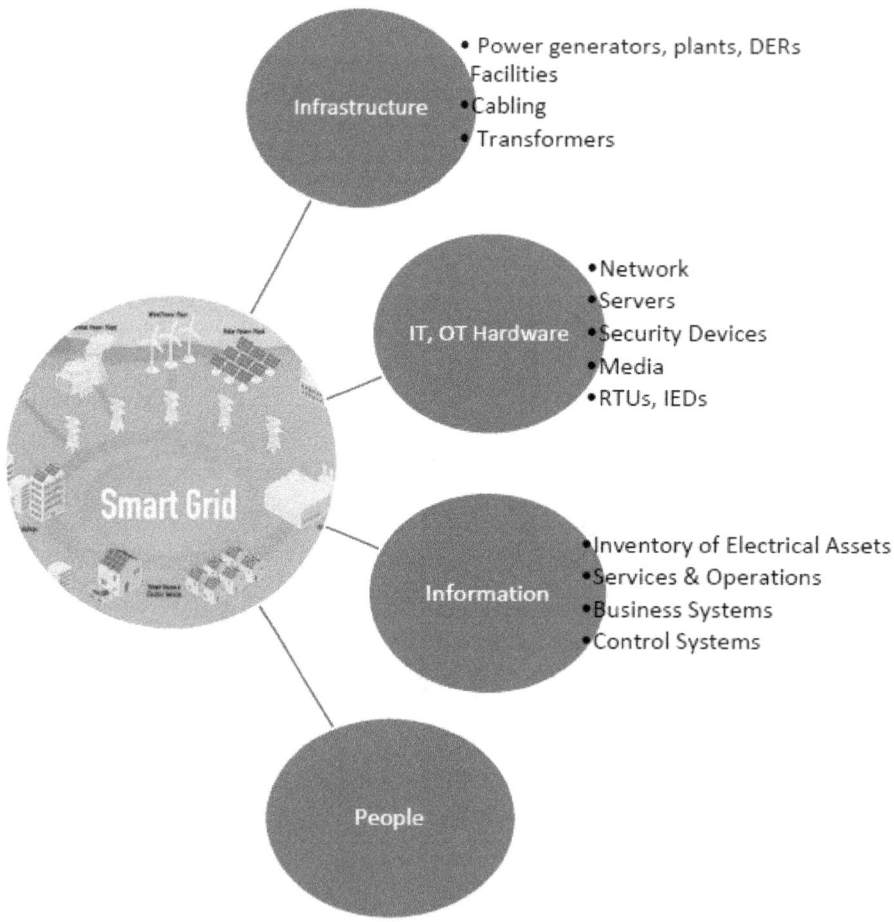

Figure 1.9 Overview of Smart Grid assets.

It is necessary to perform analysis for all the vulnerabilities, different performance requirements and security priorities of the various data and control flow through the entire information system or interconnected business and control systems without adversely impacting various business and control system performance requirements. Therefore, this approach has to be mitigated with other strategies business goals to ensure adequate security for DER systems and Smart Grid.
To determine whether an asset should be protected, and to what level, requires analysis of the potential threats against that asset and the likelihood that they will materialize. The level of protection is a function of the probability of an attack occurring and the effects of the attack should it succeed. If an attack is unlikely, protecting against it has a lower priority than protecting against a likely one.

This work [Komninos 2014] describes a detailed risk assessment of threats and use case scenarios for the Smart Grid/Smart Home environment using an adapted model based on the [NIST SP1108r1] conceptual model. The authors identify potential threats from known threats within the two categories passive (specifically, eavesdropping and traffic analysis) and active attacks (specifically, masquerading, replay, message modification, denial of service and malicious software). For describing the Smart Home, a conceptual separation is made between its internal and external environments. Focusing on prevention rather than detection, the authors use the assessment methods to identify and evaluate the risks based on different attack scenarios for these distinct elements:

- Smart Home - the attack scenarios target all entities and applications within Smart Home.
- Smart Grid - the attack scenarios target utility assets and applications.
- Smart Home/Smart Grid - the attack scenarios that initially are affecting or taking control of entities within the Smart Home, but then end up affecting entities within the Smart Grid.
- Smart Grid/ Smart Home - the attack scenarios that start by affecting entities within the Smart Grid and evolve in ways that affect the environment of the Smart Home.

For the assessment of the criticality and sensitivity of certain interactions and the evaluation of the impact level of threats against those interactions within the Smart Home/Smart Grid environment, the authors adopt the impact level assessment criteria of [FIPS 199] document. FIPS 199 characterizes potential impact of threats as Low, Moderate or High as said to be:

- Low - if the violation of one or more of the security goals described above is expected to have a limited adverse effect on Smart Home/Smart Grid operations, assets or individuals; limited adverse effect could mean degradation of an entity's capability to efficiently perform its primary functions, minor damage to assets, minor financial losses or minor harm to individuals.
- Moderate - if the violation of one or more of the security goals is expected to have a significant adverse effect on Smart Home/Smart Grid operations, assets or individuals; significant adverse effect could mean significant degradation of an entity's capability to efficiently perform its primary functions, significant damage to assets, significant financial losses or significant harm to individuals (not including loss of life or life threatening injuries).
- High - if the violation of one or more of the security goals is expected to have a severe or catastrophic adverse effect on Smart Home/Smart Grid operations, assets or individuals; severe or catastrophic adverse effect could mean severe degradation or loss of an entity's capability to perform its primary functions, major damage to assets, major financial losses or severe harm to individuals (that could even result in loss of life or life threatening injuries).

First, the authors [Komninos 2014] acquire a more comprehensive view regarding threats to both the Smart Home and Smart Grid as individual elements, then they identify some of the main threats that aim at their interaction. For each categories of elements, attack scenarios are analyzed and lists are built with information that include the scenario, threats, affected security attributes, and impacts. This approach allows to get a risk assessment of the individual elements and the dependencies between various applications and systems. This approach combined with a method to identify also the unknown threats could pave the way for an effective security management for the DER and Smart Grid systems. Also, the authors recommend security controls to protect security goals such as confidentiality, integrity, availability, authenticity, authorization, non repudiation and privacy.

1.7.4.1 Monitoring and Control

The unified approach would allow a utility to monitor and control services provided by third party service providers including hardware and software manufacturers. For example, compliance to security in supply chain should be monitored according to functions described in the unified security program following an approach for the entire utility organization. Also, the small business units that provide services to a utility or DER system should collaborate for establishing and maintaining compatible security approaches.

Many vendors, whether physical security or cybersecurity, continue to develop their own proprietary solutions that require expensive tie ins with other products they offer. The emergence of the Physical Security Information Management (PSIM) market provides great value to today's security organizations. By converging video surveillance and other physical security systems with IT and ICS security, taking in the enormous load of information and distilling it into the most pertinent insight, organizations can significantly improve cross organization communications and efficiencies while fortifying a comprehensive security posture.

PSIM is a category of software that provides a platform and applications to integrate multiple unconnected security applications and devices and control them through one comprehensive user interface. It collects and correlates events from existing disparate security devices and information systems (video, access control, sensors, analytics, networks, building systems, etc.) to empower personnel to identify and proactively resolve situations. PSIM integration enables numerous organizational benefits, including increased control, improved situation awareness and management reporting. Ultimately, these solutions allow organizations to reduce costs through improved efficiency and to improve security through increased intelligence.

The prime value that a PSIM solution can offer is when it is based on standards and can interconnect an organization's existing proprietary systems (e.g., analog or digital, cameras or alarms, firewalls or intrusion prevention systems) thus enabling the organization to leverage and build off its existing investments rather than a discard and replace approach.

1.7.4.2 Interoperability

In the context of integrating applications and information domains for business and control systems, the unified and interoperable security solutions are essential to assure communication and information security among varied systems of different media and topologies found across DER systems and Smart Grid.

1.7.4.3 Development

A unified view of security management functions drives development and deployment of security solutions which should satisfy domain specific and common requirements. Even though there are differences, similarities such as common platforms and infrastructure that support both business and control systems call for a unified approach, which can better identify differences between systems. For example, control systems need to fulfill very different risks, priorities, performance and reliability requirements.

In addition, the unified approach helps on providing more dynamic changes and improvements including adaptive solutions for situational awareness, decision intelligence and control actions such that DER systems operate much more efficiently while enhancing reliability and safety.

1.7.4.4 Secure Smart Grid Architecture

A modern Information Communication Technology (ICT) architecture of a system, like Smart Grid, involves both IT business and control network components, which share many common characteristics, regardless of how diversified their applications may be. In general, there are four zones domains that provide:

- External connectivity to the Internet, peer locations, and back up facilities
- External connectivity for corporate communications
- Control systems communications from external services
- Control systems operations, be they process based or SCADA.

Each of these zones requires a unique security focus. Thus, defensive strategies that secure each of the core zones can create a defensive strategy with depth, offering the administrators more opportunities for information and resources control, as well as introducing cascading countermeasures that will not necessarily impede business functionality.

ICT is often used as an extended synonym for IT, but is usually a more general term that stresses the role of unified communications and the integration of telecommunications (telephone lines and wireless signals), intelligent building management systems and audio visual systems in modern information technology [ITU T]. ICT term is used in Smart Grid documents by various organizations such as OECD, ENISA, ISO, ETSI, etc. (see acronym definitions in Appendix C).

1.7.4.5 Infrastructure

The Smart Grid goal is moving to a future architecture of the infrastructure that will be more robust, adaptive, and being able to selfheal in response to threats, failures, and other perturbations. Therefore, the implementation of the Smart Grid infrastructure can support intelligence to the grid, isolating the failures before they destabilize the whole system, intelligent adaptive islanding, real time, flexibility, etc.

The [NERC CIP] standards provide a set of requirements designed to secure the assets required for operating North America's bulk electric system. A bulk electric system (BES) is a large interconnected electrical system made up of generation and transmission facilities and their control systems [NERC Glossary]. A BES does not include facilities used in the local distribution of electric energy. If a bulk power system is disrupted, the effects are felt in more than one location. These standards and guidelines are mainly focused on cybersecurity including the physical security of cyber assets for the bulk electric systems.

However, it is also necessary to understand the interdependency and mutual vulnerability of the wholesale electric grid and the wholesale electric market in maintaining the stability of the Smart Grid. Market participants require to ensure protection of their critical cyber assets and to support an appropriate security program that must cover governance, planning, prevention, operations, incident response, and business continuity.

Market participants must define their security perimeters and identify their boundaries and defenses as well as protection of critical cyber assets in the security perimeter, protection of communication channels between security perimeters and other market participants (example: telemetry that is exchanged directly between market operation systems and power plants and substations). The electric power grid requires coordination within an infrastructure in order to operate properly because one bad link of chain may cause a disaster.

1.7.5 Addressing Issues

The increasing number of SCADA/EMS systems operating in the electric industry, the growing number of market participants, and the development of complex market models relying more on information technologies contribute to raising the interdependency between the operation of the power grid (including generation, transmission, and distribution) and the operation of the whole sale electric market. The electric market and the power system are every day more closely tight. The operation of one depends on the continuous and reliable operation of the other. In addition, the vulnerability of the power system is not mainly a matter of electric system or physical system, is more a matter of cybersecurity. Attacks (such as attacks upon the power system, attacks by the power system, and attacks through power system) to the Smart Grid infrastructures could bring huge damages on the economy and public safety.

1.7.5.1 Reliability and Availability

The reliability of a component or a system is the probability that the component or the system will be operating correctly without failure in a time interval. The correlation between reliability and security is a recurring theme; it has been shown that investment in software quality reduces the incidence of computer security problems, regardless of whether security was a target of the quality program or not.

The availability of a system is defined as the probability that the system is available at a specific time. Different from the reliability that focuses on a period when the system is free of failures, the availability is concerned with a point in time at which the system does not stay at the failed state. It can be viewed as the percentage of time the system is available to perform its expected task. Availability is indicative of service duration; reliability is suggestive of failure frequency. A system can have low reliability but have high availability.

1.7.5.2 Critical Infrastructure with Interdependencies

A joint program between Electric Power Research Institute (EPRI) and the Department of Defense (DoD) is addressing the cascading effect based on the strong interdependence of electric infrastructure with other infrastructures such as gas, telecommunications, transportation, financial, etc. by developing tools and techniques that allow a critical infrastructure such as power grid to self-heal in response to threats, failures, natural disasters or other perturbations.

Control systems such as SCADA are highly interconnected with IT systems within electric industry and with external infrastructures and economic sectors. Historically, control system security meant locating and identifying problems in a closed loop system; now unauthorized intrusion or attacks are evolving issues to be addressed.

The interdependencies are manifested at different levels. Security dependencies can occur and have all sorts of side effects. Risk assessment and management in large scale systems such as Smart Grid requires an understanding of how and to what degree the systems are interdependent.

1.7.5.3 Cyber-Physical Systems

In [NIST SP1108r3], the Smart Grids are viewed from the perspective of cyber-physical systems - hybridized to yield improved performance, reliability, resiliency, and user and producer awareness. Cyber-physical systems (CPS) combine power system operational equipment with cyber based control of that equipment. These systems provide the functions that the equipment is designed for and functions to protect that equipment against failures and often against certain types of human errors. Usually, these systems operate in a degraded mode if communications are lost or some other abnormal conditions occur.

One critical requirement is coping with threats and attacks, since power system equipment cannot just be shut off if an attack is occurring; it should continue to be functional as much as possible. There are many dependencies between the cyber-physical controls required to support the Smart Grid and the communication and computations that must be protected from cyber attack. A layered approach is introduced to evaluate risk based on the security of both the physical power applications and the supporting cyber infrastructure [Sridhar 2012]. The authors discuss the significance of cyber infrastructure security in conjunction with power application security to prevent, mitigate, and avoid the cyber attacks.

1.7.5.4 Human Health and Safety

Specifically, cyber physical systems and controls in power systems may affect human health and lives or interact with the environment. A cyber-physical system could be a threat to safety and could definitely not only bring significant damages to National economy but also cause great threat to the life of human beings. For example, a chemical battery spill from an electrical battery storage facility may affect both people and environment. As companies get more comfortable with the inevitability and value of a fully integrated security program and environment, security solutions can enable security personnel to reduce stress and manage more efficiently and effectively the flood of new threats emerging daily.

1.7.5.5 Smart Grid Systems: One Consolidated, Correlated View

One issue that needs to be addressed by management in the electrical sector is the planning of a unified security program that includes both business and control systems. The main benefit is avoiding the separation and duplication of management functions and reducing the complexity of administration, operation, and budgeting.

1.7.5.6 Continuous Security Process: Common Security Process

A common security process for both control systems and IT systems can help organizations to reduce the cost of information security by using proper planning and control of each activity to reduce risk, if not eliminate, the cost of an incident occurring for Smart Grid systems. The cost of an incident and the risk to the organization is unknown until an incident has occurred. Information security consists of measures to prevent the unauthorized use, misuse, modification, or denial of use of data, information, or capabilities. Figure 1.10 depicts the continuous process of information security comprised of key activities such as:

- **Risk Assessment** - determine the value of the information assets of an organization, the treats and vulnerabilities, and the importance of the overall risk to the organization; the assessment is accomplished by following the risk management approach.
- **Policy** - defines how security should be implemented; it defines the proper mechanisms to use to protect information and systems as well as physical security; it includes several as-

pects such technical aspects, best practices, preventative measures, employees, incident response, administration and management, etc.

- **Deployment -s**ecurity policies, standards and measures to be effective should be implemented by an organization practicing due care and due diligence.
- **Training** - awareness training is the mechanism to provide necessary information to employees.
- **Audit** - controls are configured and monitored correctly with regard to policy. Functions include policy adherence audits, periodic and new assessments, penetration tests.

However, this process includes a summarized set of activities. In reality, the assessment activity is part of risk management process which is a dynamic process compared to risk assessment that is a static process.

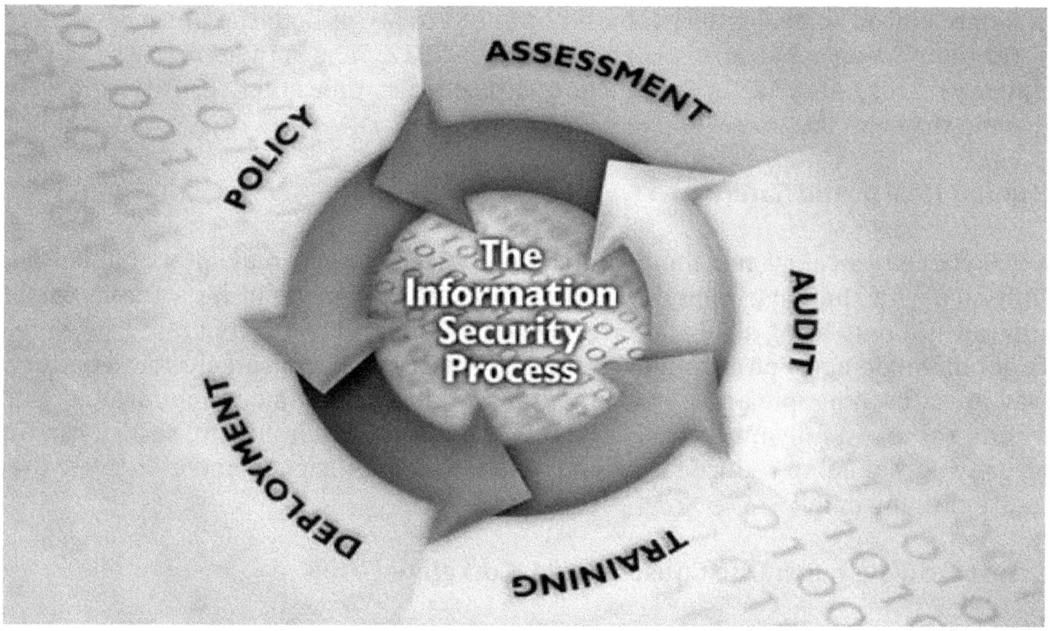

Figure 1.10 General security process – continuous cycle. *Source:* [Cleveland 2012]. © 2012, Xanthus International.

Taking the proper measures before an accident occurs is a proactive approach to information security. There is a need for a concise and unambiguous representation of the information security interdependent activities such that an improved process can be implemented.

Models can be used to specify the order of the activities with the structure of inputs, outputs, and controls of the functions, activities, and processes for information security management. A model provides support for better analysis and communication techniques for professionals involved in managing and improving security. Security is an ever-evolving process and is not static.

Analyzing the security of a control system or IT system requires understanding of the controls and mechanisms that enforce security policy. It also requires knowledge of the related assumptions and trust, which lead to the threats and the degree to which they may be realized. Such knowledge allows one to design better mechanisms and policies to neutralize these threats. A threat is a potential violation of security. Human beings are the weakest link in the security management of any system. Therefore, security policy is a statement of what is, and what is not, allowed. Any useful policy and controls must balance the benefits of the protection against the cost of designing, implementing, and using the controls and mechanisms. This balance can be determined by analyzing the risks of a security breach and the likelihood of it occurring.

1.8 Standards, Guidelines, and Recommendations

In the electricity sector, the historical focus has been on implementation of equipment that could improve electricity system reliability. Communications and IT equipment were formerly viewed as just supporting electricity system reliability. However, both the communications and IT sectors are becoming more critical to the reliability of the electricity system.

Each of these sectors has existing cybersecurity standards to address vulnerabilities and assessment programs to identify known vulnerabilities in their systems, but these vulnerabilities also need to be assessed in the context of the Smart Grid infrastructure. More, the Smart Grid may have additional vulnerabilities not only because of its complexity, but also because of its large number of stakeholders and highly time-sensitive operational requirements. Through the ongoing efforts of NIST and the public-private SGIP, DOE, and NERC standards are being reviewed for cybersecurity, and recommendations are made for how to include cybersecurity in future revisions and how to include cybersecurity in the implementations of the standards.

Further, a unified and generalized data and power system model that includes security attributes at the required levels could create more benefits to all stakeholders. The following are examples of initiatives and standards that support security program for the Smart Grid.

1.8.1 Cybersecurity Standards

Many disparate efforts continue to define new standards. There are many standards that are designed for legacy systems or overlap. Efforts to review and harmonize existing standards could contribute to faster advances in the implementation of the Smart Grid. Many standards are often developed over a time period of many months, with review cycles averaging every five years to determine if any updates are necessary. As a result, there are many standards that do not include cybersecurity, nor have up-to-date normative references to cybersecurity standards. Also, many standards are developed for short term [Cardenas 2008], [Knowles 2015].

1.8.2 SGIP Catalog of Standards

The Catalog of Standards (CoS) is a compendium of standards and practices considered to be relevant for the development and deployment of a robust, secure, and interoperable Smart Grid [SGIP CoS]. The CoS provides a framework of protocols and model standards for an interoperable Smart Grid. The information provided by each standard is a useful resource for utilities, manufacturers, regulators, consumers, and other smart grid stakeholders.

Included in the catalog is [RFC 6272], which identifies the key infrastructure protocols of the Internet Protocol suite for use in the Smart Grid. Security has to be integrated into every protocol extension, and consequently, every layer of the protocol stack to be useful.

Also, included in the CoS are these five standards, which will continue to evolve with the Smart Grid technology [NSF 2011]:

- [IEC 61970] and [IEC 61968] provide a Common Information Model (CIM) necessary for exchanges of data between devices and networks, primarily in the transmission (IEC 61970) and distribution (IEC 61968) domains. These standards facilitate the exchange of data between utilities, and allow the exchange of data between applications within a single company

- [IEC 61850] facilitates substation automation and communication as well as interoperability through a common data format. The standard specifies the requirements for how applications interact with and control devices, in addition to the standards for testing for conformity. It also creates common language for devices, results in fewer errors and greater capabilities, and implements modern networking technology in substations
- [IEC 60870-6] facilitates exchanges of information between control centers. It specifies the method of exchanging time critical data between utilities. [IEC 60870] standards define systems used for telecontrol (SCADA). Such systems are used for controlling electric power transmission grids and other geographically widespread control systems. By use of standardized protocols, equipment from many different suppliers can be made to interoperate. [IEC 60870] is a series of parts, defining general information related to the standard, operating conditions, electrical interfaces, performance requirements, and data transmission protocols.

- [IEC 62351] addresses the cybersecurity of the communication protocols defined by the other IEC standards.

In the context of [ISA/IEC 62443] standard (formerly ISA-99), the specific structure and content of policies, standards, and guidelines are left to the Industrial Automation Control System (IACS) asset owner. However, the [ISA/IEC 62443] family of documents organizes the subject areas associated with IACS processes in the following clauses categories/domains:

- Security Policy
- Organization of Security
- Asset Management
- Human Resources Security
- Physical and Environmental Security
- Communications and Operations Management
- Access Control
- Systems acquisition, development and maintenance
- Incident Management
- Business Continuity Management
- Compliance.

For a traditional information system, requirements and detail for the above clauses/domains can be found in [ISO/IEC 27002]. However, the [ISA/IEC 62443] standards describe a specific Security Management System for IACS; the unique requirements and enhancements can be found in [ISA TR 62443-2-2]. These requirements should be blended with security requirements for business systems as defined by [ISO/IEC 27000] series of standards. Also, there it should be used this technical report [ISO/IEC TR 27019]. This report extends the [ISO/IEC 27000] set of standards to the domain of process control systems and automation technology, thus allowing the energy utility industry to implement a standardized information security management system (ISMS) in accordance with [ISO/IEC 27001] that extends from the business to the process control level.

The ISA99 standards development committee (International Society of Automation) is focused on industrial automation and control systems security. This original and ongoing ISA99 work is being utilized by the International Electrotechnical Commission in producing the multi-standard IEC 62443 series [IEC 62443].

The ISA99/IEC 62443 standard is the worldwide standard for security of the ICS in the OT domain of organizations. The standard offers organizations handle to improve the digital security and safety of their process and SCADA environments. Implementation of the standard brings your organization to a higher level for security of the OT domain, the process or production environments (see more [ISA99]).

The ISA versions of the standards and reports in the series have names of the form "ISA-63443-x-y", while the IEC versions appear as "IEC 62443-x-y." The ISA and IEC versions of each document are released as closely together as possible (more information can be found in [ISA 62443SER]).

The 62443 series builds on established standards for the security of general-purpose information technology systems (e.g., the ISO/IEC 27000 series), identifying and addressing the important differences present in Industrial Automation and Control Systems (IACS). Many of these differences are based on the reality that cybersecurity risks with IACS may have Health, Safety or Environment (HSE) implications and the response should be integrated with other existing risk management practices addressing these risks.

Chapter 2

Developing Security Program

2.1 Smart Grid Framework

Various profit and non-profit organizations have developed approaches or frameworks for security management, security processes, security controls, and enterprise architecture. However, customizing a security framework is still common in commercial organizations. The specific needs of the Smart Grid and distributed energy resources (DER) systems can be better served by a framework designated as Smart Grid Framework.

This framework should include maturity models, standards, and best practices. Best practice is a feature of accredited management standards (e.g., [ISO 9000] and [ISO 14001]). Although best practices are the most used buzzword in many industries including security industry, it is necessary to perform an assessment of fitness in the context of Smart Grid applications because what may work in other industries, it may not be producing good results for the security of the Smart Grid systems. For example, best practices (e.g., security controls such as encryption algorithms and firewalls) for IT systems do not apply well for control systems.

Also, the Smart Grid framework should include aspects of process management, security management, information security governance, Security Engineering (SE) practices, and compliance to national policies and regulations. Because there are many similarities and differences between various best practices and approaches, there is a need to identify first the common approaches used by electric enterprises, then to customize a framework based on Smart Grid Reference model, standards, and the best practices.

An information security program model for Smart Grid should:

- Clearly articulate what is part of a security program and what is not
- Provide a means for understanding how components of a program function
- Predict the end result that will be achieved when change is introduced
- Enhance communications among individuals and groups who provide or benefit from information security program activities.

Also, an Information Security Management System (ISMS) shall provide the necessary organizational structures, processes, policies and procedures to be able to respond to the ever-evolving threat panorama, foster training and awareness rising among staff and deal with technological issues. An ISMS recommends risk management approaches.

Figure 2.1 shows the steps of an ISMS framework as recommended by the European Commission. However, a framework should include other management functions besides policies and risk management. A security program includes many other functions that should be planned and managed. For example, an incident response plan should include control systems, specifically Supervisory Control And Data Acquisition (SCADA) and Industrial Control Systems (ICS) networks. Although incident response planning is the responsibility of senior management for Information Technology (IT) systems [Whitman 2014], the planning for control systems is a collaborative process that involves also representatives from other departments such as [Clark 2016]:

- Human resources (e.g., define policy, communicate policy to employees, provide training and safety awareness about control environment)

- Legal (e.g., Federal, state, or industry regulations how to disclose a security incident)
- Engineering (e.g., define policy, authorize procedures for incident response and recovery)
- IT (e.g., define procedures for incident response and recovery)
- Security (e.g., define procedures for incident response and recovery)
- Product support (e.g., define policy, communicate incident to vendor, define procedures for incident response).

Although cybersecurity for Smart Grid is getting more attention, there should be a global wide perspective instead of focusing only on one perspective view (e.g., horizontal, vertical, or bidirectional). Also, cybersecurity should include all domains of Smart Grid (see [NISTIR 7628r1]) and the Information Security Management System (ISMS) framework should include technology, processes, people, and organization.

2.1.1 ENISA Framework

Recognizing the importance of security for the Smart Grid, a standard common framework for the Smart Grids in Europe is recommended by European Network and Information Security Agency (ENISA). This framework has to ensure a minimum level of harmonization on security and resiliency requirements across Europe and Member States, and to establish the basis for a minimum set of auditable controls across Europe. Also, this framework could allow National Regulatory Authorities (NRAs) to effectively measure the appropriate security controls and to make comparisons among different companies. This framework includes standards, guidelines, regulatory issues, Smart Grid Architecture Model (SGAM), and recommendations on how to secure Smart Grid systems [ENISA 2012]. These recommendations should be implemented by the public and private sector involved in the definition and implementation of the Smart Grids. These recommendations intend to provide useful and practical advice aimed at improving current initiatives, enhancing cooperation, raising awareness, developing new measures and good practices, and reducing barriers to information sharing.

Although ENISA recommendations are providing the road towards a conceptual framework, the organizations can expand the use of a known framework in the beginning phase. So a framework could be one that organizations have already used or they can implement quickly. Established frameworks such as Control Objectives for Information and Related Technologies (COBIT) version 5 or Open-Information Security Management Maturity Model (O-ISM3) can be adapted to create a uniform framework for Smart Grid.

2.1.2 COBIT 5 Framework

COBIT 5 is a framework for IT management and IT governance and an umbrella framework to connect to other information security frameworks, good practices and standards. It is based on a business-oriented model developed to address the complexity of security. It promotes a balance between protection and business. It includes elements and dynamic interconnections among these elements. The elements include Organization Design and Strategy, People, Process, and Technology. The dynamic interconnections include Culture, Architecture, Governance, Emergency, Enabling and Support, and Human Factors (see more information at [COBIT v5]).

COBIT 5 for Information Security describes the pervasiveness of information security throughout the enterprise and provides an overarching framework of enablers such as: Business Model for Information Security (BMIS) (ISACA), Standard of Good Practice for Information Security, and ISO/IEC 27000 Series of standards. COBIT 5 is one of the most commonly used frameworks to comply with Sarbanes-Oxley regulations. Companies that are publicly traded in the US are subject to the Sarbanes-Oxley Act of 2002.

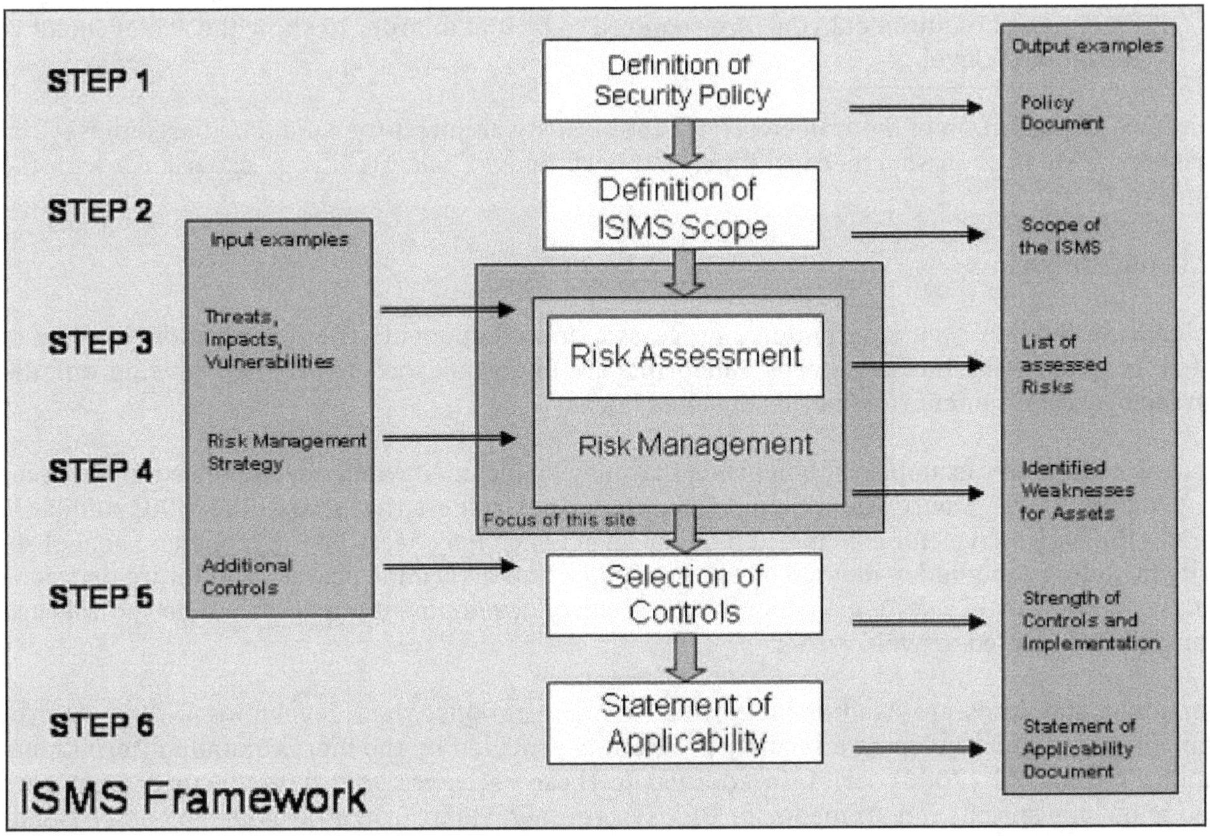

Figure 2.1 ENISA ISMS framework. *Source:* [ENISA ISMS]. © ENISA.

2.1.3 O-ISM3 Framework

O-ISM3 (Open Information Security Management Maturity Model) [O-ISM3] is a framework for managing information security and it is based on ISM3 model. It aims to ensure that security processes operate at a level consistent with business requirements and provides guidance on defining security metrics (Security Metrics, http://ism3.com/). O-ISM3 framework is compatible with [ISO 27001] standard series and Information Technology Infrastructure Library (ITIL) and COBIT frameworks.

However, additional standards are needed to complement these frameworks. The following are examples of standards that could benefit these frameworks:

- [ISO/IEC 15504] standard also known as SPICE (Software Process Improvement and Capability Determination) is a framework for the assessment of processes
- [ISO/IEC 38500] standard applies to the governance of management processes (and decisions) relating to the information and communication services used by an organization; it provides guiding principles for directors of organizations on the effective, efficient, and acceptable use of IT within their organizations
- [ISO/IEC 15026-2] standard specifies minimum requirements for the structure and contents of an assurance case to improve the consistency and comparability of assurance cases

and to facilitate stakeholder communications, engineering decisions, and other uses of assurance cases
- [ISO/IEC 15026-3] standard specifies the concept of integrity levels with corresponding integrity level requirements that are required to be met in order to show the achievement of the integrity level.

Therefore, the selection of the framework for the security architecture should be based on several attributes such as open source, capabilities, support, budget, and architects' experience with the framework.

2.2 Using Blueprints

A blueprint lays out security solutions, processes, and components the organization can use to match its security and business needs. Blueprints are important tools to identify, develop, and design security requirements for specific business needs.

Figure 2.2 illustrates examples of blueprints that may be included when developing a unified security program for the Smart Grid systems. This program considers that [ISO/IEC 27000] standards work at the higher level, for example at a policy level which may describe the scope and goal of security including a method of method of action to guide and determine present and future decisions concerning security in a system. In information security development (analysis and design), security models can be used to avoid errors.

Commonly, standards are used as blueprints by organizations when developing a security program. Guidance for developing a security program is provided by the most common international standards [ISO/IEC 27000] series. In addition to these reference standards, Security Engineers need to include specific requirements for DER systems and Smart Grid as provided by other standards developed by NIST, NERC, IEEE, etc.

Different standards can be used to develop blueprints. Generally, [ISO/IEC 27000] family includes sector specific guidance and can be used for initiating and developing blueprints. While some aspects of information security management changed in time since the ISMS concept was promoted, the emerging trends become more prevalent [e.g., [Hentea 2005]. Therefore, the policies and strategies need to be organized on providing an update of the security threats and impacts on users and organizations, global challenges and standardization impacts. In addition to technologies and Smart Grid trends, the strategies should be focused on evolving information security management infrastructure needs including emerging trends.

Other standards, models, and methods can be used to define more specific issues for Smart Grid systems or applications. A process control engineer needs to define blueprints using the profiles that are defined by standards such as [IEC 62351] and [ISA/IEC 62443] series of standards that provide guidance for implementing secure control systems. An auditor may check compliance with [NERC CIP] standards. Also, NIST standards such as [NIST SP800-12r1] and [NIST SP800-16r1] recommend a broad view and advice how to manage IT security; [NIST SP800-37r2] and [NIST SP800-53Ar5] standards can be used to address security and privacy issues specific to Federal Information Systems and Organizations.

The IETF [RFC 2196] provides guidance on network security, incident response, and security policies for information systems connected on the Internet. The [ISO 15408] standard can be used in designing the blueprint for the integration of applications. The international standard [ISO/IEC

21827] can be used to measure the maturity of ISO controls objectives so it can be used to design the blueprint for the assessment function.

Security foundation means core security program and architecture established. The security foundation presents a unified context for threat and vulnerability management, identity management, application integrity management, systems and network infrastructure, physical and environmental security, risk metrics formulation and measurement, information assets management, infrastructure blueprint, business continuity planning, logging, monitoring and reporting.

Strategic alignment means that all stakeholders are speaking in the same terms of process and defined security strategy. Security effectiveness is focused on a few critical objective indicators that enhance business and visibility, compliance monitoring (internal and external audit plans, regulatory, and self assessments). Business enablement is end-to-end transaction integrity, consistent application of solution models based on requirements. Process enhancement is evolutionary integration and consistent execution of security across the enterprise.

Strategic drivers like risk management and legal and regulatory requirements influence the security foundation on an ongoing basis along with process and business enablement functions. As newer methods and technologies emerge, the strategic drivers and security foundations dynamically realign the organization to combat newer and unknown (zero day) threats and vulnerabilities.

The blueprints are detailed descriptions of specific components for IT and control systems that can be part of a security enterprise architecture. An architecture is a tool used to conceptually understand the structure and behavior of complex entity through different views. In the past, system architectures were developed to meet the stakeholders' concerns, but security is a new concern and a system needs to include security requirements.

2.2.1 Mapping Security and Business Requirements

Cybersecurity for DER systems refers to protection of both power and Information and Communication Technology (ICT), processes in IT and control systems operations, and governance. The ICT supports DERs cyber infrastructure as well as IT systems, control systems, sensors, actuators, organization, and people. Because the DER systems include systems from the IT, telecommunications, and energy sectors, the security requirements should be analyzed for all three sectors as they interact in the DER system and with Smart Grid. The numerous and various interfaces between DERs and power grid are increasing the security risks. Also, the security requirements must accommodate the new business applications and Smart Grid trends. Therefore, security requirements need to be defined to protect new types of devices for the network that are integrated to support applications from monitoring and control of supply, to more efficient generation and consumption to more innovative energy storage. Examples of devices include:

- Sensors for medium and low voltage lines
- Sensors for low voltage circuit breakers
- New inverters to streamline the integration of renewable energy sources
- On load tap changers that enable transformers to become active elements and making the grid flexible
- Grid and consumer scale energy storage devices based on different technologies
- Smart meters and home area network devices are also blurring the lines between energy supply and energy distribution domains.

In addition, more sophisticated functionality in traditional equipment requires security improvements to existing devices. An increasing intelligence in solid state relays, controllers for reclosers and sectionalisers, regulators and capacitor controllers require analysis and management of security risks. For example, capacitor bank controllers can utilize remote, real-time parameters to influence switching, while monitoring and reporting on power quality through an integrated communications module.

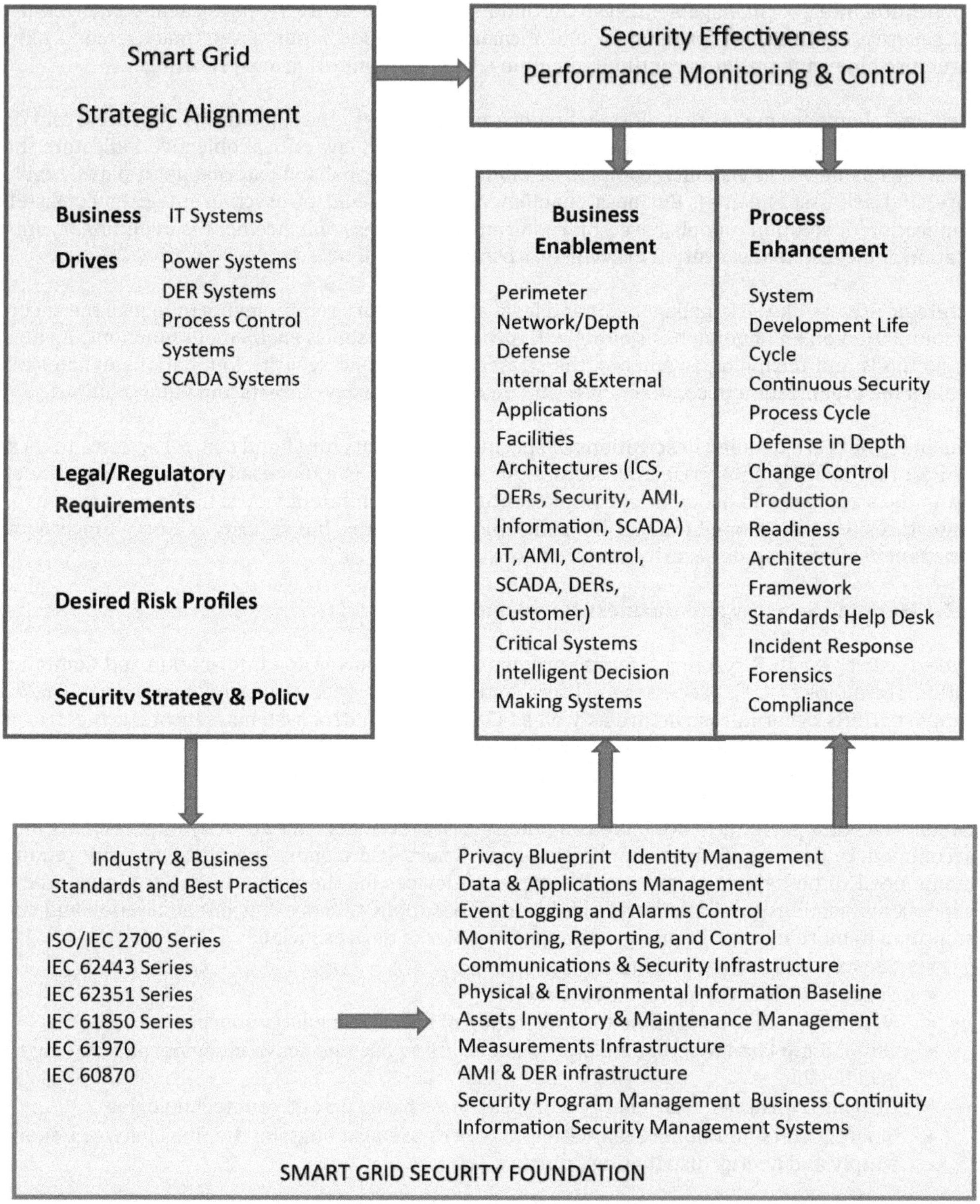

Figure 2.2 Blueprints for Smart Grid.

Thus, new business requirements require a refocus of security analysis. For example, a growing focus on Volt/VAR (see list of acronyms in Appendix C) control applications increases the number of capacitor and regulator controllers, as well as line voltage sensors in many cases. Feeder automation for self-healing and improved switching control can contribute to an increase in intelligent switching devices outside the substation.

Also, information technology in the operations domain is undergoing a steady transformation. In the past, Distribution Supervisory Control and Data Acquisition (DSCADA) and Distribution Management Systems (DMS) applications were developed and maintained in complete isolation from the rest of the IT infrastructure and exclusively focused on controlling a limited number of operating assets.

Currently, DSCADA and DMS applications are characterized by far broader application and integration requirements of both control and business systems. Therefore, the convergence of IT and operations technology (OT) has many implications for the operational applications that model, monitor, and manage the distribution network [Meyers 2013]. Implicitly, this convergence requires an analysis of security requirements from a different point of view. For example, more data and enhanced functionality mean more security requirements and faster decision making. On the other hand, added complexity increases the risk of operational errors. In many utilities, operators have to process higher volumes of information and need to select operating options from an increased set of alternatives.

For example, a security engineer can develop the enterprise security architecture using the Sherwood Applied Business Security Architecture (SABSA) model. A security architecture is the result of applying the System Engineering process. A complete system security architecture includes administrative security, communication security, computer security, emanations security, personnel security, and physical security. A complete security architecture needs to deal with both intentional, intelligent threats and accidental threats. Also, security goals have to be defined before the architecture of a system is created, and specific security views of the system need to be created to help guide the design and development phases. Sometimes security is integrated at the architecture phase.

When we talk about the security built in, that means security concerns are addressed during the development phase and not architecture phase. Security requirements can be categorized as [ISC2 2006]:

- Functional requirements define security behavior of the product, IT system, or control system. Examples of requirements include:
  - Determine logical and physical data architecture
  - Define user roles
  - Determine access needs
  - Determine integrity needs
  - Determine availability needs
  - Map users to data access
  - Define criteria to data access
  - Identify data owner
  - Identify security policies
  - Define security requirements for a particular business solution (e.g., portal, control, remote access, marketing, supply chain, manufacturing, etc.)
- Assurance requirements establish confidence that the security function will be performed as intended. Examples of requirements include:

- Determine level of concern (data protection level)
- Determine probability of data insecurity
- Determine business impact from data insecurity.

Many organizations depend on providers' services and vendors' products to support their business. Therefore, new categories of security requirements have to be included in security plans. A comprehensive list of security requirements are identified to guide the procurement of ICT services and products from vendors and outsourcing partners [ENISA 2014b]. For the purpose of this guide, the security requirements are classified into these categories:

- Governance and risk management
- Human resources security
- Security of systems and facilities
- Operations management
- Incident management
- Business continuity management
- Monitoring, auditing, and testing.

This guide is intended to help at the early phase of the procurement process. Also, it can be used by vendors in the development and security management of their products and services.

2.2.2 When and How to Use a Blueprint

A blueprint is used for guidance and when there is need for a quick development of a security function and assumptions are assumed (e.g., security policy is in place). Blueprints are used to identify, develop, and design security requirements for a particular business solution such as portal, enterprise resource planning, supply chain, manufacturing, etc. Security blueprints provide proven models for establishing cost effective, sustainable security business practices and solutions.

The blueprints are customized to fulfill the requirements of an organization or a business unit. Generally, a blueprint is a guide that directly addresses the needs of large enterprise customers, so in order to apply it directly to small and medium businesses, and even home offices, there is a need to perform analysis and business requirements design because they do so on a different scale.

Often, security requirements design phase is skipped. However, a security assessment and analysis should be current such that a security policy can be satisfied using the blueprint solutions. For example, if an organization does not use identity management, there is no need to build a blueprint for this. Therefore, not all aspects of a specific blueprint may apply, but all should be considered. Examples of security blueprint solutions for a privacy program are discussed elsewhere.

Another example of blueprint solution is the DHS' s document of 2011 for business continuity and contingencies [DHS 2011]. The document describes how to identify and harden critical information infrastructures as well as preparing for contingencies. Another example of blueprint is the security blueprint for the enterprise networks [Cisco SAFE]. Examples of other types of blueprints include architecture and infrastructure blueprints.

Security blueprint can include tailored security best practices that, in total, form a comprehensive security policy program and technical architecture [ISC2 2006]. Also, it can be composed of several security domains, that as a minimum, are mapped from ISO/IEC 27000 series of standards. Individual security blueprints reflect tailored requirements meeting the organization's specific requirements.

A security blueprint can include security solutions as shown in Figure 2.3. However, specific blueprints can be defined such as architecture blueprints and infrastructure blueprints [ISC2 2006]:

- Infrastructure security blueprints are based on individual security blueprints that reflect
  - Tailored requirements meeting the organization's specific requirements
  - Influenced by legal, regulatory, business, IT drivers.
- Architecture blueprints are based on tailored security best practices that, in total, form a comprehensive security policy program and technical architecture. They can be composed of several security domains, that at a minimum, are mapped from the [ISO/IEC 27000] standard series.

2.2.3 Utilities as Power Brokers

Like most corporations, electric utilities depend on computers to run their day to day operations such as business functions, engineering power flow, or monitoring and control of their systems. Utilities use computer technology, on the business side, especially for accounting and customer billing, and on the engineering side for analysis and design, monitoring and control of the generation-transmission-distribution of power. A new function, brokering buy-sell transactions, is spawning a new type of company called power brokers. DER systems are used by utilities either directly or via service providers. Therefore, many security concerns transcend beyond utilities and several aspects have to be considered such as:

- Business computers are used for accounting and billing, supply chain business, and storing relevant information about partners and customers in company wide databases.
- The integrity and availability of the information are essential to the day-to-day operations. In the past, limited external connectivity minimized security and survivability concerns. However, the trend towards increased connectivity is raising these concerns.

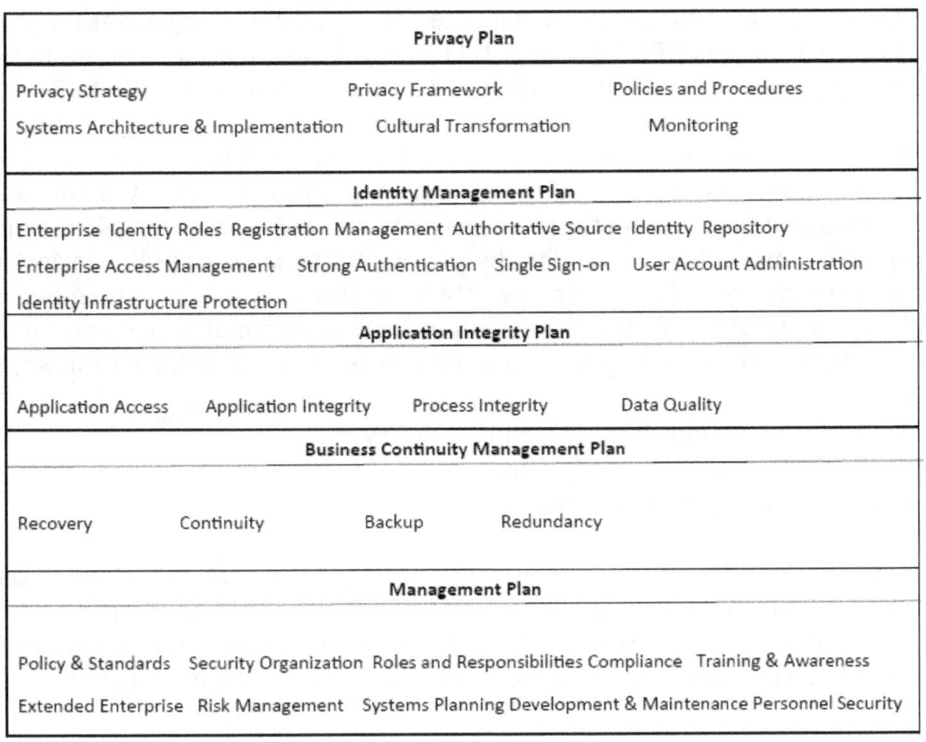

Figure 2.3 Examples of security architecture blueprints.

2.3 Elements of Effective Security Program

Achieving resilient energy systems requires more than a focus on compliance, instead a culture focused on achieving comprehensive and effective cybersecurity [NSF 2011]. A security program for Smart Grid is a combination of all aspects of processes, managing security, ranging from the definition and communication of policies through implementation of best industry practices and ongoing operation and auditing [ISA 62443-1-1]. The following is an overview of important components of a security program.

2.3.1 Policies

Security policy is a definition of what it means to be secure for a system, organization or other entity. For an organization, it addresses the constraints on behavior of its members as well as constraints imposed on adversaries by mechanisms such as doors, locks, keys and walls. For systems, the security policy addresses constraints on functions and flow among them, constraints on access by external systems and adversaries including programs and access to data by people.

A security program should include security policies, standards, guidelines, baselines, procedures, security awareness training, incident response plan, and a compliance program. These terms are described in [ISO/IEC 27000] series and specific aspects of implementation may be provided by Department of Energy (DOE) in USA. Security starts at a policy level, which are high level directives that provide the foundational goals for a system overall and the components that make it up from a security perspective. A quality information security program begins and ends with policy [Whitman 2014]. Policies must contain information on what is required and what is prohibited. For a policy to be effective, it must be properly disseminated to those responsible for implementing it.

Security policy is set of rules that specify or regulate how a system or organization provides security services to protect its assets [IEC TS 62443-1-1]. Security policy is a definition of what it means to be secure for a system, organization or other entity. It is a strategic tool that dictates how sensitive information and resources are to be managed and protected. For an organization, it addresses the constraints on behavior of its members as well as constraints imposed on adversaries by mechanisms such as doors, locks, keys and walls. For systems, the security policy addresses constraints on functions and flow among them, constraints on access by external systems and adversaries including programs and access to data by people. Often security policy is confused with security procedures which are definitions of exactly how practices are implemented and executed [IEC TS 62443-1-1]. Generally, policies include general organizational policy and functional implementing policies. Management is required to define three types of information security policies [Whitman 2014]:

- Organization information security program policy
- Issue specific information security policies
- System specific information security policies.

General organizational policy is the management's security statement. It is defined based on laws, regulations, requirements, organizational goals, and objectives. Once an overall organizational security policy has been approved by the governing body of the organization it is necessary to develop a supporting infrastructure of control objectives. This framework may include other functional policies such as:

- Email and Internet use policy
- Remote access policy

- Privacy policy.

Functional implementing policies are management's security directives. Policies can be implemented using standards, procedures, baselines, guidelines, and recommendations. These supporting elements will enforce the security policy principles on every business process and system. Policy development approaches and preparing a coverage matrix are described elsewhere (e.g., [Whitman 2014], [Barman 2002]).

More directed to systems, other types of security policy are defined to address operating systems, devices, and applications. Each component of a system has a responsibility to support and enforce the security policy of a system. Multilevel security policies are security policies that prevent information from flowing from a high security level to a lower security level. These policies permit a subject to access an object only if the subject's security level is higher than or equal to the object's classification.

More examples of various security policies are provided in [Greene 2014]. Any security program should distinguish between information security policies and organization of information security. The information security policies domain focuses on information security policy requirements and the need to align policy with organizational objectives. The organization of information security domain focuses on the governance structure necessary to implement and manage information security policy operations, across and outside of the organization. Risk management is a fundamental aspect of governance, decision making, and policy. Risk management is important (see standards, [ISO/IEC 27005] and [ISO/IEC 31000]). Further, user versions of information security policies (for example, acceptable use policy), vendor versions of information security policies, chief information security officer policy, information security risk management oversight policy, information security risk assessment policy, information security risk response policy, and others are discussed in [Greene 2014].

Security in information systems considers the protection of information and systems that manage it, against a wide range of threats in order to ensure business continuity, minimize risks and maximize the return on investment and business opportunities. Information systems are distinct from information technology in that an information system has an IT component that interacts with the process components. Information Security (INFOSEC) is defined as measures that implement and assure security services in information systems, including in computer systems and communication systems [RFC 4949].

Information security is the set of procedures implemented to prevent the unauthorized access, abuse, alteration, or denial of access to knowledge, data, or resources. This definition is broad because of the unique nature of information. Since information takes many forms and can be stored and used in many ways, the security of information presents distinctive problems.

The unique nature of information forces the information security professional to look at many different parameters when developing a sound security policy. This policy must take into account the many different forms of security to offer the most complete protection of knowledge, data, and resources. Physical, communications, emission, computer, and network security are all implemented when designing a complete information security policy. A high level description of what should be incorporated within a computer security policy is also described in several documents. For example, the DOE cyber policy is described in [DOE Policy]. The document provides guidelines for improving different aspects of cybersecurity policy such as increasing public awareness, increasing cybersecurity education, promoting cybersecurity as an enterprise leadership responsibility, sharing responsibility and partnership for cybersecurity between public and private sectors, effective partnership with the international community, and other issues.

In addition, there could be policies to satisfy regulatory requirements. In an effort to protect the citizens of the United States, legislators recognized the importance of written information security policies. Gramm-Leach-Bliley Act (GLBA), Health Insurance Portability and Accountability Act (HIPAA), Sarbanes-Oxley (SOX), Family Educational Rights and Privacy Act (FERPA), and the Federal Information Systems Management Act (FISMA) all require covered entities to have in place written policies and procedures that protect their information assets. They also require the policies to be reviewed on a regular basis. However, organizations that try to write their policies to match federal state regulations find the task difficult. Since regulations could have common elements, it is possible to write a set of information security policies based on a framework such as the [ISO/IEC 27002] standard that can be mapped to multiple regulatory requirements. Policy administrators would be required to include a cross-reference to specific regulatory requirements.

2.3.2 Security Elements

Analysis of a complex system often leads to the breakdown of its elements into three groups: people, processes, and technology. Applied to cybersecurity, these elements are called security triangle concept that is depicted in Figure 2.4. While this concept was an important driver in late 1990s and recommended in [ISA TS 62443-1-1], the evolving security management demands change such as security controls should employ more technology than relying on procedures and policing by humans [Schneier 2013].

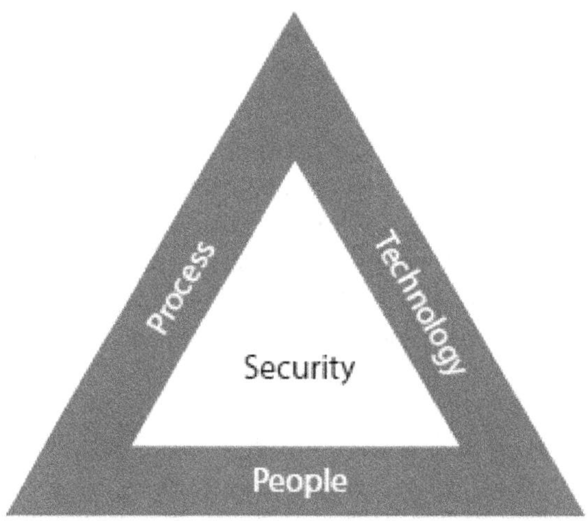

Figure 2.4 Security triangle.

The idea is that robust processes can often help to overcome potential vulnerabilities in a security product, while loose processes may lead to poor implementations that can render good technologies ineffective. Often, human weaknesses (e.g., errors, lack of education) can diminish the effectiveness of technology.

Also, security planning, policy and programs should address the protection of essential functions. These are functions or capabilities that are required to maintain health, safety, the environment and availability for the equipment under control) [ISA 62443-1-1]. Essential functions include, but are not limited to, the safety instrumented function, the control function and the ability of the operator to view and manipulate the equipment under control. The loss of essential functions is commonly termed loss of protection, loss of control and loss of view respectively. In Smart Grid and DER systems, additional functions such as history should be considered essential.

For IT security to be effective in protecting the interest of the organization, it must be aligned with physical and personnel security practices of the organization. This unified approach is the practice of information security at its core [TISN 2007].

Although security protection is determined by considering these issues (pointed by the standard ISA 62443-1-1), the organization factor must be included as the fourth node (see also BMIS and ISM3 models).

An effective security program contributes to improvements of situational awareness. A number of factors such as operator training, coordination and effectiveness in communications and decision making, adequate tools for system reliability assessment, visualization of security and reliability posture of the overall DER system following some disturbance events help on understanding of system modes of operations and interactions with power grid. Effectiveness is the extent to which planned activities are realized and planned results achieved [ISO 27001].

As the complexity and connectivity of DER systems to power grid continue to grow, for the system operators and coordinators, situational awareness and implicitly effectiveness of security program becomes more and more important. New methodologies needed for better awareness of system operating conditions can be established. Also, the capability of control centers can be enhanced with better situational awareness. This can be partially promoted by development of operator and control center tools which allow for more efficient proactive control actions as compared with the conventional preventative tools. Real time tools, which are able to perform robust real time system security assessment even with the presence of system wide structural variations, are very useful in allowing operators to have the better mental model of the system's security posture. Therefore, prompt control actions can be taken to prevent possible system wide outages such as the blackout incident of 2003 in USA.

Information technology benchmarking of cybersecurity requires measurements for comparing both, different IT systems and single IT systems in dedicated situations. The technical approach is a predefined catalog of security events (security incident and vulnerability (computing)) together with corresponding formula for the calculation of security indicators that are accepted and comprehensive.

Information Security Indicators have been standardized by the ETSI Industrial Specification Group (ISG) [ETSI ISG]. These indicators provide the basis to switch from a qualitative to a quantitative culture in IT security scope of measurements such as:

- External and internal threats (attempt and success)
- User's deviant behaviors
- Nonconformities and/or vulnerabilities (software, configuration, behavioral, general security framework).

Preliminary work on information security indicators have been done by the French Club R2GS [R2GS]. The first public set of the ISI standards (security indicators list and event model) have been released in April 2013 [ISI Portal].

Another key issue is implementing effective enterprise security governance. In [EAC 2014], an outline for energy sector executives and boards states that enterprise security program effectiveness for both physical and cyber is now a management concern and is called out as essential in the Department of Energy's Electric Sector Cybersecurity Capability Maturity Model (ES-C2M2). The [DOE ES-C2M2] framework comprises a maturity model (at its core is [DOE C2M2] model, an evaluation tool, and DOE facilitated self-evaluations.

Therefore, organization's business processes must be adjusted to take into account the risks introduced by today's dynamic and evolving security threats. Substantial changes to business are required to adequately safeguard critical energy infrastructure, satisfy regulatory requirements, and protect customer data and corporate reputations. Changes of this magnitude can only be accomplished with the full support and leadership of senior executives and a new approach to enterprise security governance. The characteristics of effective security program governance are identified as follows [EAC 2014]:

- Clearly defined responsibilities from the board of directors to senior leadership to employees
- Presence of an active Security Governance Board comprised of senior stakeholders from across the company
- An executive owner of Enterprise Security with purview over IT, OT and Physical security policy, designated Chief Security Officer (CSO) or similar
- Striving for 100% alignment with of security with business/mission
- Striving for 100% alignment with of security with business/mission
- Using measurement of key indicators to increase awareness and drive improvement (with maturity tools like the [DOE C2M2]).

In organizations large and small, effective enterprise security program governance should seek to drive a single view of the security and regulatory posture. Information, communication networks, and basic applications should be governed by security and privacy policies in the context of Smart Grid. This includes the information and communications between utility center, customer, and third party service provider.

2.3.3 System Level Programs

While a unified security program addresses the entire spectrum of security and privacy for an organization, system level programs ensure appropriate and cost effective security for each system of Smart Grid. This includes influencing decisions about what controls to implement, purchasing and installing technical controls, day-to-day security administration, evaluating system vulnerabilities, and responding to security problems. It encompasses all the areas included in the security program.

As implied by the name, an organization may typically have several system level security programs for a particular system of Smart Grid. In setting up these programs, the organization should carefully examine the scope of each system level program. System level computer security programs may address, for example, the information and computing resources within an operational element, a major application, or a group of similar systems (either technologically or functionally).

System level security program personnel are to be the local advocates for security and privacy. The system security manager should raise the issue of security with the assigned system manager and help develop solutions for security problems including specific security requirements for that system. For example, it may be checked the system's security requirements, vulnerabilities, threats, contingency plan been tested.

As part of a system level, site specific vulnerability and threats, both environmental and physical factors should be evaluated. Then risks should be assessed. The protection depends on the risk acceptance level and it is also derived from laws and regulations that are specific to Smart Grid, and particularly to each system, for example DER systems that could be owned by a utility organization, third party provider, or a customer. The physical controls and countermeasures are designed as a layered defense model by a team of individuals with various backgrounds from management, construction, environment, electronics, electrical, mechanical, safety, health, etc.
Many factors influence how successful a system level security program is. Many of these are similar to the organization's security program. However, additional considerations needed to be addressed by utilities or service providers include:

- Security and privacy plans for sensitive systems (such as real time control systems for microgrids or electrical storage systems) should have appropriate and cost effective security; system level security personnel should be in a position to develop and implement security plans.
- System specific security policy because many computer security policy issues need to be addressed on a system specific basis; the issues can vary for each system, although access control and the designation of personnel with security responsibility are likely to be needed for all systems. A cohesive and comprehensive set of security policies can be developed by using a process that derives security rules from security goals of the organization.
- Life cycle management that ensures that changes to the system are made concurrently with security and that accreditation is accomplished.
- Effective integration with system to ensure that system managers and application owners consider security in the planning and operation of the system; the system security manager/officer should be able to participate in the selection and implementation of appropriate technical controls and security procedures and should understand system vulnerabilities; also, the system level security program should be capable of responding to security problems in a timely manner.
- Separation from operations to ensure independence and authority; in many instances, operational components tend to be far larger and therefore more influential seek to include the security program in security operations; therefore, this conflict between the need to be a part of system management and the need for independence has several solutions.
- Independent audit function should complement an organization security program's compliance functions.

The general purpose of the security program, to improve security, causes it to overlap with other organizational operations as well as with other security controls. The organization or system level security program need to address most controls at the policy, procedural, or operational level. The organization security program normally produces policy and supporting procedures and guidelines concerning general and organizational security issues and often issues specific policies. However, the system level security program normally produces policy for that system.

2.4 Strategic, Tactical, and Operational Views

Traditionally, the design and operation of ICS and IT business systems are different and supported by different personnel. However, the need for reliable, safe, and flexible performance of Smart Grid and implicitly, DER systems, requires approaches to resolve conflicting technical, cultural, and operational differences in managing security principles based on strategic, tactical, and operations view.

An organization needs to develop a general security strategy and overall strategic plan for major systems. Then, strategic plans are used to create tactical plans, which are used to develop operational plans. Often, tactical plans are oriented around projects or systems. One approach for security is based on defense-in-depth strategic framework.

Cybersecurity for Smart Grid and DER systems requires a different approach than for typical IT systems. Traditionally, cybersecurity for IT systems focuses on the protection required to ensure the confidentiality, integrity, and availability of the electronic information communication systems. However, security of the Smart Grid and DERs must include a balance of both power and cyber system technologies and processes in IT and power system operations and governance. Poorly applied practices from one domain that are applied into another may degrade reliability [NISTIR 7628r1].

The overall cybersecurity strategy for the Smart Grid should be based on a risk mitigation strategy to use both domain specific and common requirements to ensure the interoperability of solutions across different parts of the infrastructure when developing security solutions [NIST 7628r1]. The primary goal of the cybersecurity strategy for DER systems should be on prevention. However, it also requires that a resilient response and recovery strategy to be developed in the event of a cyber attack on the power grid or DER system.

The security requirements and the supporting analysis that are described in [NISTIR 7628r1] document may be used by implementers of the DER systems and Smart Grid. This means that different entities such as utilities, equipment manufacturers, regulators have to be included as input to the risk assessment processes. The governing principle behind an ISMS is that an organization should design, implement and maintain a coherent set of policies, processes and systems to manage risks to its information assets, thus ensuring acceptable levels of information security risk.
Threfore, cybersecurity strategy for the Smart Grid and DER systems should address issues from a strategic, tactical, and operational view. A security program should contain all the elements necessary to provide overall protection to an organization and support a long-term security strategy.

2.4.1 Strategic View

The Smart Grid, regarded as the next generation power grid, uses two-way flows of electricity and information to create a widely distributed automated energy delivery network. Therefore, once a general cybersecurity strategy is defined for the Smart Grid and DER systems, strategic plans and security objectives have to be established. These objectives describe the long-term direction to be taken by an organization. For example, enabling technologies for three major systems, namely the smart infrastructure system, the smart management system, and the smart protection system require future directions and management objectives, such as improving energy efficiency, profiling demand, maximizing utility, reducing cost, and controlling emission.

Examples of various management methods to achieve these objectives including protection mechanisms which improve the reliability of the Smart Grid and explore the security and privacy issues in the smart grid are discussed in [Fang 2012]. In addition, there should be objectives for minimizing the costs, efficiency, scalability, ease of component reuse, interoperability, integration of security architecture with the enterprise (IT and ICS) architecture and its legacy systems.

Utilities have to develop strategies that can allow a utility to leverage its existing programs and make adjustments to harness new opportunities. Without a roadmap to the future, utilities may be easy overwhelmed by the variety and multi layered implications of new, often interdependent technologies [Deiuliis 2012].

Another issue is planning to implement an approach that supports a strategic alignment. When strategically aligned, the security functions become a business enabler that add value. Security is becoming an expected topic of discussion among decision makers and is given the same level of respect as other fundamental drivers and influencing elements of the business. Therefore, the integrated approach to information security should be considered versus a parallel approach. A parallel approach silos information security assigns responsibility for being secure to the IT department, views compliance as discretionary, and has little or no organizational accountability. An integrated approach recognizes that security and business success are intertwined.

Other objectives for the integration of diverse security related functions needed for DER systems should be considered too. The long-term view of strategic investing in the risk response needs for organizations can help to reduce the continuing focus on near term and adhoc solutions to support the information systems. Overall strategic planning is translated into lower level tactical and operational planning.

Most important, there should be no gaps in the level of information asset protection. Security should be implemented by the utilities and service providers to ensure protection of assets from utility centers and consumers. Overlaps in areas of security planning or management should be minimized. Another indicator is the level of integration for information assurance activities with security. Quite often published materials describe specific solutions or best practices under the term strategy, although they are solutions to a system or short-term solutions. Therefore, these materials are more likely in the categories of tactical planning or operational planning.

2.4.2 Tactical View

Tactical planning has a more short-term focus than strategic planning, usually one to three years. Each applicable strategic goal is broken into incremental objectives, each with a specific delivery date. Generally, tactical plans are oriented around projects. The tactical plans and activities guide the allocation of information protection needs which, in turn, influences and guides the allocation of the security controls to specific components and systems. The activities at this level influence the activities at the operational level. However, analysis of activities at the tactical level may also provide feedback to the strategic level.

Information security solutions often are designed, acquired and installed on a tactical basis. These plans should include activities such as: develop processes to support business functions; define the types of information needed to successfully execute the mission/business processes; define the criticality/sensitivity of the information; define the information flows both internal and external to organizations; incorporate security requirements into the mission/business processes; establish an enterprise architecture and information security architecture which promotes cost effective and efficient information technology solutions consistent with the strategic goals and objectives of the organization and measures of performance, etc. ([NIST SP800-39]).

The overall resilience objective of information systems is a key factor and performance measure in determining the potential survivability of missions/business functions under a cyber incident. Certain emerging information technologies may introduce inherent vulnerabilities into these information systems that have to be managed.

2.4.3 Operational View

Derived from the tactical plans, operational plans allow to organize the ongoing, day-to-day performance of activities. Operational planning may encompass activities such as selection, and configuration, and deployment of security technologies, implementation of security education, training, and awareness, monitoring and control, analysis of data and make decisions. Also, the operational plan may include identified coordination activities, progress reports, summaries, etc.

Information system owners, common control providers, system and security engineers, and information system security officers make risk-based decisions regarding the implementation, operation, and monitoring of organizational information systems. In addition, the activities at the operational level provide essential feedback to strategic and tactical levels.

Roles and responsibilities should be clearly defined for specific functions and conflict of interests should be avoided. This includes the relationships between various internal and external providers of information assurance. Often, conflicting interests between security and network personnel may affect the implementation of security mechanisms. For example, a network administrator goal is ensuring high availability and performance of the network and resources as well as providing functionality to user's requests. However, many times the network administrator is focused on performance at the cost of security.

Similarly, a process control engineer may ignore requirements for authentication of devices and processes. Security mechanisms commonly decrease performance in either processing or network transmission because of additional processing. Examples of these mechanisms include content filtering, virus scanning, intrusion detection prevention, anomaly detection, encryption systems, and access controls. Since security administration is not in the areas of network administration, security mechanisms may not be enforced by network administrator. Although a separate chain of command for these administrators is recommended [Harris 2013], a collaboration between the security and network personnel should be promoted to achieve a balance of objectives for both groups.

2.4.4 Smart Grid Strategy Overview

The cybersecurity strategy for the Smart Grid includes many aspects from development to operations and new approaches to risk management. Among the most important issues that have to be addressed include:

- Smart Grid risk assessment process is based on existing risk assessment approaches developed by both the private and public sectors and includes identifying impact, vulnerability, and threat information to produce an assessment of risk to the Smart Grid and to its domains and subdomains, such as homes and businesses. Each organization can use this document [NISTIR 7628r1] and other guidance information as it implements the security strategy and performs the security risk assessment.
- Since Smart Grid includes systems from the IT and ICS, telecommunications, and energy sectors, the risk assessment process has to be improved such that it is applied to all of these sectors as they interact in the Smart Grid.
- Cybersecurity must be included in all phases of the system development life cycle, from design phase through implementation, maintenance, and disposition/sunset.
- The strategy should be focused on prevention while addressing detection, response, and recovery. This overall strategy is potentially applicable to complex infrastructures.
- The organizations must design and implement a process that ensures continual movement from the current state to the future state. The future state will generally be a combination of one or more closing gaps between IT and SCADA systems, present and desired future security strategy, upgrades and replacements of hardware and software on both IT and SCADA systems.

- Management must plan recovery strategies if security incident occur because the power must remain on as much as feasible even when equipment is removed for repair. Also, collection of data should be planned to allow accurate time stamped event logging to capture as much information as possible about the attack sequences for both future protection and possible legal and forensic activities.
- Strategies should be developed for cyber-physical systems. Cyber-physical systems are physical processes integrated with networked embedded computing systems that monitor, control, and manage them. Examples include smart grids, industrial process control systems, unmanned autonomous vehicles, etc. CPSs, in general, have strong requirements for real-time operation, continuous availability, and fail-safe operation to function correctly and without causing harm to the physical systems whose processes they control and manage. Assessing the impact of coordinated cyber physical attacks to the Smart Grid requires expertise in cybersecurity, physical security, and the electric infrastructure.
- An acquisition strategy is needed for Systems of Systems (SoS). An acquisition done in an SoS context must be robust against change and it has to consider a wider spectrum of failures and mitigate the effects of changes in usage, technology, and in the individual components. Unanticipated interactions among systems can induce degraded service. As a consequence, security needs to consider also general system failures that usually have been associated with reliability. Therefore, a new approach to risk management has to be adopted across the life cycle to operations and sustainment [Creel 2013]. This work includes more recommendations for building an acquisition strategy (before acquisition begins) to help cope with the various challenges of an SoS development.
- As an entity of critical infrastructure, the operation center for Smart Grid should be continuously improved and proactive (see an example of the strategic architecture for a NASA operation center described in [Wang 2010], [Torres 2015]).
- One task in the development of cybersecurity strategy is planning for the development of security architecture. Volume 1 of [NISTIR 7628r1] describes Smart Grid cybersecurity strategy, high level requirements, and security architecture. The architecture has to incorporate a long-term strategy. Currently, most organization implement an adhoc or short term (tactical level) architecture.

In summary, a long-term security program has to include a new cybersecurity strategy that is proactive, agile, contextual, and risk based.

2.4.5 Developing Enterprise Security Architecture - Fundamentals

An enterprise security architecture includes the information security strategy consisting of policies, standards, and procedures and the way they are connected across an enterprise strategically, tactically, and operationally. Besides security, the benefits of enterprise security architecture include better interoperability and integration. Strategic alignment means the business drivers and the regulatory and legal requirements are met by the enterprise security architecture. Although enterprise architecture is based on a framework that is similar to a security program framework, it provides more granularity because it works its way from the policy level to the component level. When developing an enterprise architecture, it is required to support the following:

- Process enhancement meaning refinement and integration in system life cycles and day-to-day operations
- Business enablement meaning the core business processes are integrated into the security operating model
- Security effectiveness meaning how security solutions perform in meeting the defined requirements.

Implementing enterprise information security architecture generally starts with documenting the organization's strategy, then it flows down into the discrete information technology components such as:

- Organization charts, activities, and process flows of how the IT and ICS organization operates, including collaborative processes
- Organization cycles, periods and timing
- Suppliers of technology hardware, software, and services
- Applications and software inventories and diagrams
- Interfaces between applications that include events, messages and data flows
- Intranet, extranet, Internet, e-commerce, EDI links with parties within and outside of the organization
- Data classifications, databases and supporting data models
- Hardware, platforms, hosting (e.g., servers, network components and security devices and where they are kept)
- Local and wide area networks, Internet connectivity diagrams.

Enterprise information security architecture framework is only a subset of the enterprise architecture frameworks. The intermediate outcome of an architecture process is a comprehensive inventory of business security strategy, business security processes, organizational charts, technical security inventories, system and interface diagrams, and network topologies, and the explicit relationships between them. The inventories and diagrams are merely tools that support decision making. But this is not sufficient. It must be a living process.

The enterprise architecture is viewed as a service oriented architecture composed of business architecture, information architecture, technology architecture, and security architecture. The security architecture is often standards based, and business driven blueprints are used to develop and implement an enterprise security program, core security architecture and operating model architecture. The key components of the security architecture include:

- Management processes that leverage foundations for design, business processes, development life cycles and day-to-day operations, evolutionary integration of security measures across the enterprise.
- Compliance and monitoring to report security performance and effectiveness by measuring key performance metrics.
- Solution implementation which defines, develops, and implements security architecture consistent with the organization's defined risk profile to ensure end-to-end security services (e.g., availability, integrity, confidentiality, non-repudiation, etc.).

Concerns regarding the advances of security architecture are described as ad hoc, not integrated, not planned and costly [Ross 2012]. This study finds that:

- Information security solutions and architecture often designed, acquired and installed on a tactical basis
- No strategic dimension
- Organization builds up a mixture of technical solutions on an ad hoc basis
- No guarantee that they will be compatible and interoperable
- Solution is needed to base decisions on business requirements, including:
  - The need for cost reduction
  - Modularity
  - Scalability
  - Ease of component reuse
  - Operability

- Usability
- Interoperability both internally and externally
- Integration with the enterprise IT architecture and its legacy systems.

Therefore, a robust enterprise security architecture must adopt an architecture framework from a list of frameworks. However, the list of frameworks is quite impressive, so choosing the appropriate framework for security architecture requires careful and detailed feasibility studies because each framework has its own strengths and weaknesses. Various frameworks are discussed in [Ross 2012]. As discussed in this work, the security architecture should be systematically and strategically planned and executed. It should support a strategic view and should be designed based on security requirements and business needs. A framework is needed because it provides a generic problem space and a common vocabulary within which individuals can cooperate to solve a specific problem. Frameworks are not necessarily comprehensive, but they can be leveraged to provide at least a starter set of the issues and concerns that must be addressed in architecture development. According to [Marley 2003], a framework should support:

- Views provide the mechanisms for communicating information about the relationships that are important in an architecture
- Methods provide the discipline to gather and organize the data and construct the views in a way that helps ensure integrity, accuracy and completeness
- Training/experience that enables the application of method and use of tools.

A system or system concept has a system architecture that is communicated via by an architecture description that is governed by an architecture framework [MITRE Guide]. Architecture frameworks and architecture description languages (ADLs) are two mechanisms widely used in architecting. Instances of each can be specified by building on the concepts of architecture description presented in [ISO 42010] standard. The diagram of Figure 2.5 depicts the elements of the architecture framework.

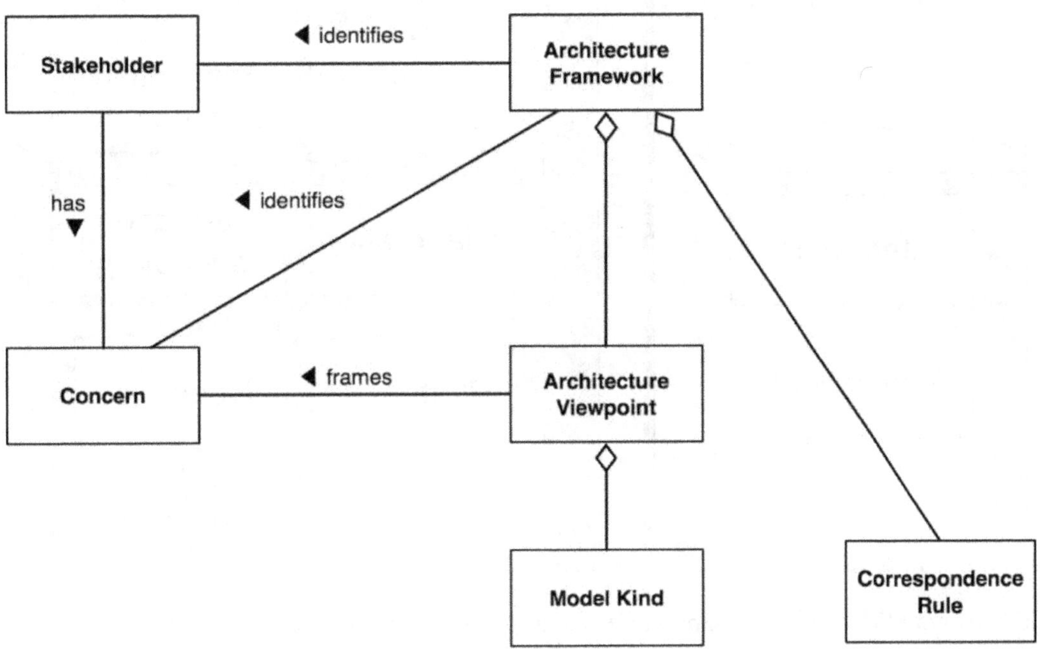

Figure 2.5 The contents of an architecture framework. Adapted from [Hilliard 2012]. © 2012, IEEE.

Many of the enterprise architecture frameworks differ in terms of their approach and level of detail [Urbaczewski 2006]. As concluded in this study, selecting an architecture framework and determining a best fit of a framework are dependent on specific stakeholder needs for a given project.

Although the best framework for the security architecture is still a need for many organizations to be defined [Ross 2012], the Security Professionals can adopt a framework and customize it for their needs. For example, the Huxham framework (see Figure 2.6) is an acceptable high level conceptual security architecture framework to satisfy the needs of an organization (e.g., utility), which implements Smart Grid systems. This security framework includes the relationships with other architectures (business, information and technology) and system management. The Zachman is the organizational model for an architecture. It is an evolution focused framework and supports legacy integration. It can be used for transitional architectures and finding commonality across businesses [Marley 2003].

Therefore, it is imperative to choose an architectural framework in the early stages of defining the security program because architecture frameworks address what it means to define and document an architecture [Marley 2003]. Also, the author argues that a framework provides a generic problem space and a common vocabulary within which individuals can cooperate to solve a specific problem.

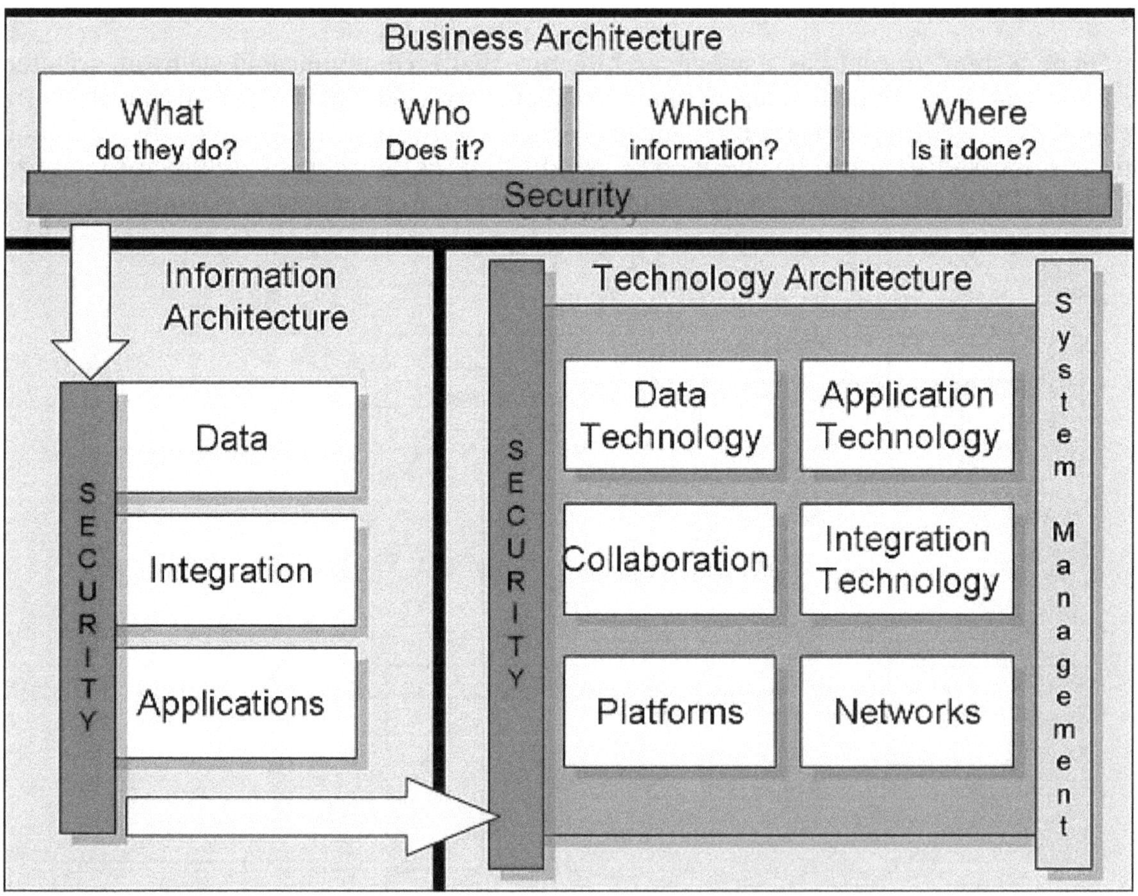

Figure 2.6 Huxham EISA framework. *Source:* [Huxham 2006]. Public Domain.

Frameworks are not necessarily comprehensive, but they can be leveraged to provide at least a starter set of the issues and concerns that must be addressed in architecture development.

A conceptual architecture that is planned in the early stages of any development and includes long term vision of cybersecurity allows an organization to advance the evolution of cybersecurity posture. This ensures a trusted and resilient information and communications infrastructure.

Therefore, organizations need to change the nature of investment from the implementation of existing solutions for a system to an added strategic focus on investing in leading edge information security technologies (essentially experimenting with innovative security solutions/technologies and being an early adopter) or investing in information security research and development efforts to address specific technology gaps.

2.5 Smart Grid Security Strategy: Proactive Versus Reactive

Organizations often have very different broad cybersecurity strategies. Strategies can generally be characterized along a spectrum ranging from proactive to reactive, where a proactive strategy implies that security compromises are anticipated, and safeguards are built into the IT system to prevent them; a reactive strategy implies that an organization is responding to known threats with typically established technologies so that security compromises can be addressed efficiently and effectively [Rowe 2006].

Security of DER systems must enable the business and support its objectives rather than becoming self serving. Utilities are becoming proactive and are looking at computing and data as an opportunity to improve load balance, better address outages, and shape consumer behavior to save energy.

Most security professionals use these two basic approaches to deal with security vulnerabilities: proactive and reactive approaches [Steinberger 2015]. It is reported that fewer security compromises result when an organization adopts a proactive strategy as opposed to a reactive strategy, but the frequency and extent of such compromises, realized or averted, were not disclosed [Rowe 2006]. Every modern organization realizes the value of dedicating some resources to the prevention of expensive damages that will likely occur if such preventive measures are not taken [Steinberger 2015]. Some organizations have started using measures to detect computer intrusions and then activate defensive measures when an attack is detected. Also, each organization may have plans in place to respond to cyber attacks and losses when the proactive measures either were not effective, or did not exist. Having an appropriate set of reactive responses prepared and ready to implement is just as important as having proactive measures in place.

2.5.1 Proactive Objectives

A proactive view must include objectives such as:

- Self healing infrastructure systems that require a vision towards more robust and adaptive infrastructure to be able to self heal in response to threats, failures, and other disturbances. There is a need to provide some intelligence to the grid, isolating the failures before they destabilize the whole system. The objective is to minimize the cascaded failures as a consequence of a failure within a domain. Also, it is needed to design a particular scheme that could answer to some possible contingencies.
- Effective and sustainable security for the entire system life cycle, including design, installation, operation, fault monitoring and diagnosis, maintenance, and control. Secure implementations of technology and procedures should be followed by using the developed standards and proposed guidelines for security policies.

- Dependable systems ensure that the systems are predictably dependable in the face of all sorts of malice, from bombers to botnets. In addition, as attacks shift from the hard technology to the people that operate it, computing systems must also be resilient to error, mischance and even coercion.
- Hazard analysis and safety requirements to assess risks and decide on a strategy to cope with them such as avoidance, constraint, redundancy. Another way of doing the hazard analysis is failure modes and effects analysis (FMEA), pioneered by NASA, which is a bottom up rather than top down approach. This involves tracing the consequences of a failure of each of the system's or domain's components all the way up to the effect on the mission. Also, safety must be built in as the system is developed, not retrofitted. Various factors and assumptions need be included in this analysis. Examples include:
  - Cyber attacks exploit vulnerabilities previously not modeled or unknown to a system.
  - SCADA networks were initially designed with little attention to security.
  - SCADA networks traditionally used dedicated telephone lines to send control messages to field devices from the control station and get the current status of the field equipment.
  - Modern SCADA networks, integrated with corporate networks and the Internet, have become far more vulnerable to unauthorized cyber attacks causing the national infrastructures at risk and easy targets of attacks by terrorists.
  - By sending a false control message from a computer connected to the Internet, an unauthorized intruder can manipulate traffic signals, electric power switching stations, chemical process control systems, or sewage-water valves, creating major concerns to public safety and health. It is necessary to address not only the individual vulnerabilities, but the breadth of risks that can interfere with critical operations.
- Develop a framework that allows security experts to program new types of detection and prevention functions or to extend the capabilities of existing functions, and fast. As suggested in [Chesla 2015], a paradigm shift in the design of security products need to take place. Such a shift would allow network administrators and network security experts to start designing their own security solutions, rather than just administrate existing solutions in their limited security protection space.
- Operational organizational structures have to recognize the importance of a cybersecurity culture to support a national cybersecurity strategy [Ghernouti 2010].
- A difficult set of decisions need to be made in determining how much resources (time, money, people) to dedicate to proactive approaches and how much to reactive approaches. These decisions can be further complicated by decisions about whether to use inhouse resources, or to outsource [Steinberger 2015]. Therefore, the cybersecurity strategies need to be factored in investment decisions and business goals. Every organization needs to find an appropriate balance between how many resources and money can be devoted to proactive measures and how much to devote to reacting to intrusions. In addition, expectations of future threats or compromises, vulnerabilities, and technical change influence the timing of investments and thus the implied costs and the benefits received [Rowe 2006].

2.5.2 Reactive Approaches

Reactive approaches include Disaster Recovery Plans, use of private investigation services and loss recovery specialists, reinstallation of operating systems and applications on compromised systems, or switching to alternate systems in other locations. While reacting to a security incident is mandatory, the operations personnel have an obligation to include the objective of preventing recurring problems, reducing hardware and software failures aspects.

There should be investigation of any unusual or unexplained occurrences, unscheduled initial program loads, deviations from standards, or other abnormal situations. Central monitoring systems and event management solutions can help on finding the root cause of problems.

The emerging technologies Security Operations, Analytics and Reporting (SOAR) support workflow management and automation, analytics and reporting. This enables security operations teams to automate and prioritize security operational activities and report data to inform better business decision making [Gartner 2015].

There are three primary SOAR technology types: security incident response, security operations automation, and threat and vulnerability management. It is argued that SOAR technologies fulfill many of the requirements to adopt an adaptive security architecture to support functions for information on more aspects – prevention, detection, response to be addressed by SOAR.

2.6 Smart Grid Threat and Vulnerability Management

Critical infrastructures, such as electricity generation plants, transportation systems, oil refineries, chemical factories and manufacturing facilities are large, distributed complexes that use Industrial Control Systems. These are command and control networks and systems designed to support industrial processes.

The largest subgroup of ICS is SCADA systems that are used in industrial control or in the control of the critical infrastructures. SCADA systems, in particular, perform vital functions in national critical infrastructures, such as electric power distribution, oil and natural gas distribution, water and wastewater treatment, and transportation systems. They are also at the core of transportation management. The disruption of these control systems could have a significant impact on public health, safety and lead to large economic losses. Several control applications can be labeled as safety critical because their failure can cause irreparable harm to the physical system being controlled and to the people who depend on it.

Most industrial control systems have a hierarchical structure. Figure 2.7 shows a common network architecture of layers described as [Cardenas 2008]:

- The physical infrastructure is instrumented with sensors and actuators that are connected via a field area network to programmable logic controllers or remote terminal units (RTUs), which in turn implement local control actions (regulatory control)
- The control network carries real time data between process controllers and operator workstations. The workstations are used in area supervisory control, planning the physical infrastructure set points
- The corporate network, at the higher level, is the site plant operations, which is in charge of production control, optimizing the process, and keeping a process history.

Also, the criticality of ICS in vital sectors, and high impact in case of disruption, makes them a major target for malicious activities. An unprecedented increase of reported cybersecurity incidents occurred in 2014 [ENISA 2015]. As shown in Figure 2.8, between 2009 and 2014, the number of reported cybersecurity incidents in the ICS-SCADA area increased more than 27 times. At the same time, more than half of the incidents (59% in 2013) were aimed at the energy and critical manufacturing sectors and around 55% involved advanced persistent threats. Still many ICS or SCADA cybersecurity incidents stay undetected or unreported.

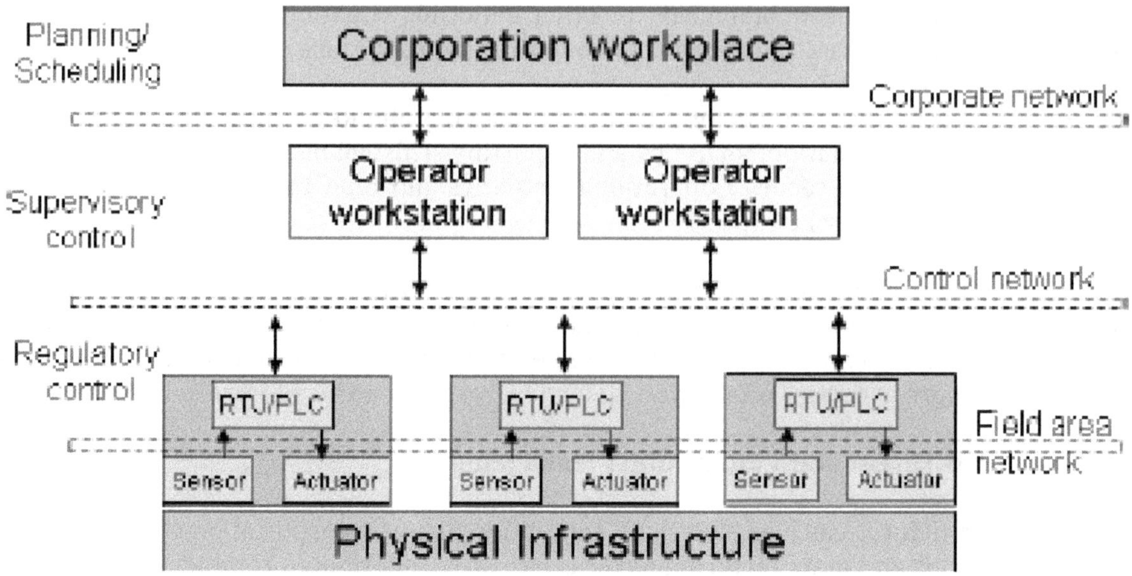

Figure 2.7 Architecture of control systems. *Source:* [Cardenas 2008]. Courtesy of the author.

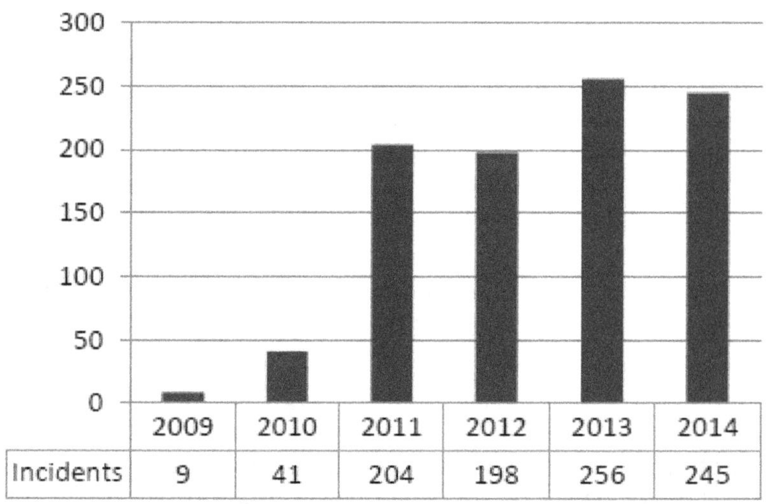

Figure 2.8 ICS-SCADA Cyber security incidents between 2009-2014. *Source:* [ENISA 2015]. © 2015, ENISA.

Large and highly skilled IT global workforce includes groups of people that generate attack vectors for computer based systems. Cybercrime is active and a number of attack tools and cybercrime networks can be used even by less skilled people; interest in cybercrime against control systems is rising.

Reports show that Industrial Control Systems Cyber Emergency Response Team (ICS-CERT) responded to more than 200 incidents between October 2012 and May 2013 [Prince 2013]. While these incidents occurred across all critical infrastructure sectors, the highest percentage of incidents (53 percent) reported to the organization occurred in the energy sector (53 percent). All of the incidents involved sophisticated threat actors who had successfully compromised and gained access to business networks (called corporate networks in Figure 2.10).

Another report [Symantec 2016] examines multiple facets, including targeted attacks, smartphone threats, social media scams, and Internet of Things (IoT) vulnerabilities, as well as attackers' tactics, motivations, and behaviors. This report describes key findings and trends from 2015:

- Increase of ransomware by 35 percent in 2015 and governments worldwide reported a huge increase in ransomware attacks in 2021 (total marked a 105% increase over in 2020 which is more than triple the number seen in 2019 [SonicWall 2022]). Cyber criminals are using encryption as a weapon to hold companies' and individuals' critical data hostage. Ransomware evolved to more sophisticated impacts and included more devices.
- The cyber attack against the power grid of Ukraine in December 2015 is the first attack with a high degree of sophistication that employed several types of hacking and malicious tools to cause a blackout to a large area of more than 230,000 customers and severe disruption causing many damages to infrastructure. Although the energy delivery service recovered and the lights were on after approximative 6 hours, the impact and loss was very high. The attack caused damages to 30 substations, control centers, hardware, firmware, SCADA systems, and software that made it most of these assets unusable and required to be replaced [Zetter 2016].
- The ransomware attack against the U.S. Colonial Pipeline in May 2021 impacted the computer systems causing a shut down of this major oil pipeline for almost one week until the ransomware was paid [Javers 2021].
- The attack against SolarWinds software in January 2021 caused disruptions to several services and damages to several financial and governmental agencies in U.S. [Vavra 2021].
- Increase of cyber scammers and evolving fake technical support scams ask users for money; Symantec blocked 100 million fake technical support scams in 2015.

The users and developers need to understand the threats, vulnerabilities, and risks associated with Smart Grid systems and infrastructure including the preventive and detective measures that are available to counter them. Since Smart Grid includes both, business systems and control systems, the threats and vulnerabilities have to be analyzed based on the characteristics of both systems. Therefore, threat and vulnerability management need to be more in the focus of managers and Security Engineers.

Also, it has to be understood that the characterization of threats and adversaries (including representative tactics, techniques, and procedures), or sources of vulnerability information may not be consistent with how some organizations conduct their missions/business functions. A source of threat/vulnerability information, that is useful for one mission/business function, may not, in fact, be useful for others.

2.6.1 Threat Management

Cyber threat management is emerging as best practice for managing cyber threats beyond basic risk assessment found in ISMS. Basically, it is a practice for enabling early identification of threats, motivations, data driven situational awareness, decision making and threat mitigating actions. A threat is potential for violation of security, which exists when there is a circumstance, capability, action, or event that could breach security and cause harm [ISA TR62443-1-2]. A threat is a possible danger that might exploit a vulnerability [RFC 4949]. A threat may be intentional (a possibility of an attack by an intelligent entity) or accidental threat (a possibility of human error or omission, unintended equipment malfunction, or natural disaster such as fire, flood, earthquake, windstorm, and other causes listed in [FIPS 31]. The Common Criteria [CC] standard characterizes a threat in terms of a threat agent, a presumed method of attack, any vulnerabilities that are the foundation for the attack, and the system resource that is attacked. The Common Criteria for Information

Technology Security Evaluation (abbreviated as Common Criteria or CC) is an international standard (see [ISO 15408]) for computer security certification.

Threat management is handled in the operations or system development. Threat management is handled through appropriate management processes and activities foreseen by laws, standards and methodologies. It includes several activities beginning with understanding the threat definition. Several definitions are available, see also [Wiki Threat]. Examples of activities include: threat analysis, threat classification, threat agents, motivations, origination, relationship to vulnerability and risk, threat modeling, etc.

Key to threat management are business continuity management plans in order to protect, maintain and recover business critical processes and systems. The threat management process can be verified by activities such as information security auditing and penetration testing.

2.6.1.1 Threats Identification

Security Engineer and Control Engineer need to identify threats including targeted threats to Smart Grid and DER systems. Also, they have to understand the common assumptions about threats to industrial control systems, and many other activities (e.g., detection, response, protection, and prevention) as practiced and documented in several publications [Anderson 2008], [Whitman 2014], [Harris 2013], [Kadrich 2007], [Krutz 2006], [Berger 2012], [Weiss 2010], [Pacyna 2009], [Hasselbring 2006], [Ray 2010], [Hentea 2008]. Threat analysis includes the identification of threats that may adversely impact the systems and applications. While many activities are discussed elsewhere, we mention a few activities that require more attention.

Threat intelligence feeds into a security monitoring system that supports data aggregation for improved incident handling [Torres 2015]. As described in this work, mature security operation centers (SOCs) continually develop the capability to consume and leverage threat intelligence from their past incidents and from information sharing sources, such as a specialized threat intelligence vendor, industry partners, the cybercrimes division of law enforcement, information sharing organizations (such as ISACs), or their security monitoring technology vendors. A security monitoring system's capability is to operationalize threat intelligence and use it to discover patterns in endpoint, log and network data, as well as associate anomalies. One objective an SOC is to make sure threats and incidents are [Stamp 2012]:

- Detected quickly
- Responded to swiftly and efficiently, and
- Remediated before they impact the business.

As shown in the Figure 2.9, during a cyber attack several events can occur causing a system to transition from an initial state to a final state. Events can occur very quickly and reach maximum impact with little warning or prior indication of an imminent impact.

Assuming that an initial state is a secure state, a final state is an unsecure state after a cyber attack realized by a series of events. A secure state is the state being or feeling secure, by having the ability to avoid being harmed at an irrecoverable level, by any risk, danger or threat [Fragkos 2005].

A class of events (intentional or unintentional, either rarely occur, or, in some cases, have never occurred), called High-Impact, Low-Frequency (HILF) events, has become the focus of risk managers and policy makers in the power industry [NERC 2010]. As described in this document, these events could have potential for catastrophic impacts on the electric power system. Examples of HILF risks include coordinated cyber, physical, and blended attacks, the high altitude detonation of a nuclear weapon, and major natural disasters like earthquakes, tsunamis, large hurricanes, pandemics, and geomagnetic disturbances caused by solar weather. HILF events may generate other events that increase their magnitude of impact. Deliberate attacks (including acts of war, terrorism, and coordinated criminal activity) pose especially unique scenarios due to their inherent unpredictability and significant national security implications.

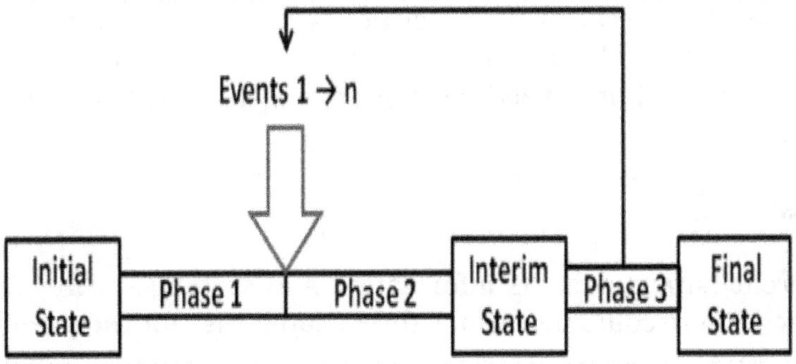

Figure 2.9 Critical threats cycle.

HILF events can occur very quickly and reach maximum impact with little warning or prior indication of an imminent risk. The ability to accurately flag suspicious activity will be contingent upon asset owners' ability to access up to date information on known threats and intrusion characteristics collected by the monitoring system.

Effective response and restoration from HILF events require fast initiation and mobilization exercised through prior planning. Priority should be given to collapsing the amount of free time an attacker has within the system (see Figure 2.10). Security controls (e.g., intrusion detection and prevention systems) should be integrated into all system paths to ensure attacks are quickly and accurately identified and notifications given to information technology personnel and system and plant control room operators.

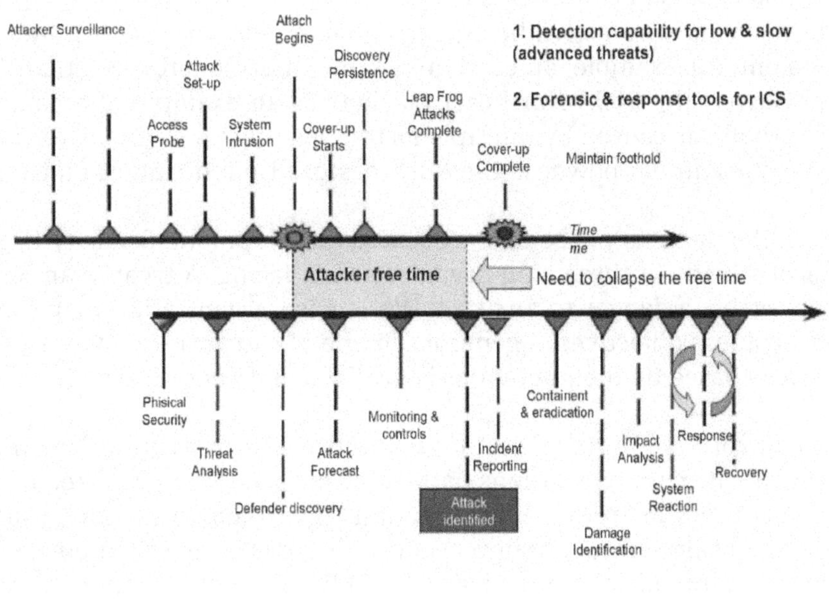

Figure 2.10 Typical attack time cycle. *Source:* [DOE 2010]. Public Domain.

In this work [Dundas 2012], the authors describe the sequence of events prior to the beginning of a targeted attack (RSA target), certain information that was extracted during the attack, motivation of the attack, and confirm that information taken from RSA was used as an element in an attempted broader attack against another organization. Therefore, one control action is reducing the attacker free time.

Lessons learned after a breach include focus on the critical assets and use of the context based security analytics fused with threat intelligence. Organizations need comprehensive visibility, agile analytics, actionable intelligence, and optimized incident management that may be realized by improving the SOC [Stamp 2012].

2.6.1.2 Threat Categories

Threats can refer to different targets such as information and systems, power sector, industrial control systems, and privacy. There could be various threat categories for the power sector.

The power sector is vulnerable to various disruptive events that require preparation for mitigating impacts and restoring service in a timely fashion. The following includes a list of threats [DOE 2015b]:

- Weather related are the most common type of disruptive events such as hurricanes, tropical storms, tornadoes, snow and ice storms, and flooding causing outages.
- Cyber terrorism from around the world can affect the U.S. grid causing power outages.
- Theft and physical attacks target electric assets; recently, a California substation was attacked, resulting in the shutdown of numerous giant transformers that supplied power to an extensive commercial and industrial customer base.
- Man made accidents such as vehicle crashes, software glitches, and other human errors can result in power outages; examples include, civilian vehicles crashing into utility poles or utility employees accidentally tripping wires while conducting routine maintenance.
- Supply/demand imbalance within a given area can result in power failures; for example, a sudden surge in demand due to extreme temperatures can cause power outages (e.g., Texas rolling blackouts in 2006 and 2011).
- Other natural events such as wildfires, earthquakes, and animals can interfere with electrical equipment; for example, an earthquake in California left more than 70,000 customers without power; the 2003 Northeast blackout is an example of a complex large scale outage event that was caused by multiple factors leading up to the historic outage that left 50 million people without power in the Northeastern United States and Canada.

To better understand the analysis phase in the security development, Security Engineer needs to know about the various kinds of threats an organization is facing. A treat is an object, person, or other entity that represents a danger to an asset. While each organization may categorize threats based on their own business objectives, a general scheme of categories may not be also unique (a categorization of threats based on their activities is available in [Hentea 2021]).

Cyber threats to a control system may include unauthorized access to a control system device and/or network using a data communications pathway. This access can be directed from within an organization by trusted users or from remote locations by unknown persons using the Internet. Threats to ICS can come from numerous sources, including hostile governments, terrorist groups, disgruntled employees, and malicious intruders [ICS-CERT Threat]. Another source, the known - GAO Threat Table, includes a description of various threat actors to ICS networks as shown in Table 2.1.

Table 2.1 GAO Threat table.

Threats	Description
Bot network operators	These are hackers; however, instead of breaking into systems for the challenge or bragging rights, they take over multiple systems in order to coordinate attacks and to distribute phishing schemes, spam, and malware attacks. The services of these networks are sometimes made available in underground markets (e.g., purchasing a denial of service attack, servers to relay spam, or phishing attacks, etc.).
Criminal groups	These seek to attack systems for monetary gain. Specifically, organized crime groups are using spam, phishing, and spyware/malware to commit identity theft and online fraud. International corporate spies and organized crime organizations also pose a threat to the United States through their ability to conduct industrial espionage and large scale monetary theft and to hire or develop hacker talent.
Foreign intelligence services	They use cyber tools as part of their information gathering and espionage activities. In addition, several nations are aggressively working to develop information warfare doctrine, programs, and capabilities. Such capabilities enable a single entity to have a significant and serious impact by disrupting the supply, communications, and economic infrastructures that support military power impacts that could affect the daily lives of U.S. citizens across the country.
Hackers	These break into networks for the thrill of the challenge or for bragging rights in the hacker community. While remote cracking once required a fair amount of skill or computer knowledge, hackers can now download attack scripts and protocols from the Internet and launch them against victim sites. Thus, while attack tools have become more sophisticated, they have also become easier to use. According to the Central Intelligence Agency, the large majority of hackers do not have the requisite expertise to threaten difficult targets such as critical U.S. networks. Nevertheless, the worldwide population of hackers poses a relatively high threat of an isolated or brief disruption causing serious damage.
Insiders	The disgruntled organization insider is a principal source of computer crime. Insiders may not need a great deal of knowledge about computer intrusions because their knowledge of a target system often allows them to gain unrestricted access to cause damage to the system or to steal system data. The insider threat also includes outsourcing vendors as well as employees who accidentally introduce malware into systems.
Phishers	Individuals, or small groups, who execute phishing schemes in an attempt to steal identities or information for monetary gain. Phishers may also use spam and spyware/malware to accomplish their objectives.
Spammers	Individuals or organizations who distribute unsolicited e-mail with hidden or false information in order to sell products, conduct phishing schemes, distribute spyware/malware, or attack organizations (e.g., denial of service).
Spyware/malware authors	Individuals or organizations with malicious intent carry out attacks against users by producing and distributing spyware and malware. Several destructive computer viruses and worms have harmed files and hard drives, including the Melissa Macro Virus, the Explore.Zip worm, the CIH (Chernobyl) Virus, Nimda, Code Red, Slammer, Blaster, Stuxnet, etc.
Terrorists	Terrorists seek to destroy, incapacitate, or exploit critical infrastructures in order to threaten national security, cause mass casualties, weaken economy, damage public morale and confidence. May use phishing, spyware/malware.

Source: [GAO 2005]. Public Domain.

Another source [NIST SP800-82r2] describes categories of threats to ICS based on type of threat source as shown in Table 2.2.

Table 2.2 Threats to ICS.

Type of Threat Source	Description	Characteristics
ADVERSARIAL Individual, Outsider, Insider, Trusted Insider, Privileged Insider, Group, Ad hoc, Established, Organization, Competitor Supplier, Partner, Customer, Nation- State	Individuals, groups, organizations, or states that seek to exploit the organization's dependence on cyber resources (e.g., information in electronic form, information and communications technologies, and the communications and information handling capabilities provided by those technologies	Capability, Intent, Targeting
ACCIDENTAL User, Privileged User/Administrator	Erroneous actions taken by individuals in the course of executing their everyday responsibilities.	Range of effects
ENVIRONMENTAL Natural or man made disaster, Fire, Flood/Tsunami, Windstorm/Tornado, Hurricane, Earthquake, Bombing, Overrun, Unusual Natural Event (e.g., sunspots) Infrastructure Failure/Outage, Telecommunications, Electrical Power	Natural disasters and failures of critical infrastructures on which the organization depends, but which are outside the control of the organization. Note: Natural and man made disasters can also be characterized in terms of their severity and/or duration. However, because the threat source and the threat event are strongly identified, severity and duration can be included in the description of the threat event (e.g., Category 5 hurricane causes extensive damage to the facilities housing mission-critical systems, making those systems unavailable for three weeks).	Range of effects

Source: [NIST SP800-82r2]. Public Domain.

A classification of possible threats against Smart Grids is shown in Table 2.3. Not only security threats have been included, but a wider scope has been considered for completeness in Annex II of [ENISA 2012] study. As recommended in this document, in order to succeed in managing risks it is important to firstly identify potential threats, at all levels, from natural disasters to very technical aspects. When identifying threats, it should be not only considering those targeting primarily infrastructure operations, but also those targeting the end consumer (e.g., privacy aspects) as well as national security related factors (e.g., government personnel); other national facilities, assets, and interests identified by national intelligence, etc.

The proliferation of advanced technologies used in the implementation of Smart Grid systems brings new forms of cyber attacks. Attacks that in the past would require exploits of multiple vulnerabilities to bypass multiple layers of control or the lack of interoperability among systems, may now be simplified to a single attack vector in Smart Grids [ENISA 2012]. In most cases, attackers will attempt networks in order to cause a denial of service, get sensitive information, or bypass perimeter security controls so as to gain access to internal networks. Social engineering attacks could easily exploit the human factors vulnerabilities. Physical attacks (e.g., theft, destruction of power equipment, cutting cables, etc.) could create serious power disruption and interruption of many services in various economy sectors.

Also, threat agents target products that do not have implemented adequate security features. Disparate regulations that emphasize compliance instead of cybersecurity culture contribute to manufacturing and distribution of unsecure products.

Table 2.3 ENISA Threat classification.

Threat Type	Examples of Threats
Technical	Malware, Non optimal processes, Weak innovation, Manipulation of device's internal electronics, Physical manipulation of devices' subcomponents, removable component replacement, Manipulation of home devices, Unauthorized firmware replacement, Compromised firmware update, Escalation of privileges, Sensible information interception, Alteration of information in transit, Traffic injection, Sensible information theft, Credentials discovery, Partial denial of service, Breakdown, Propaganda, Disclosure of information, Disinformation
Corporate Image and Information Management	Low quality information for decision making, Damage to Brand Image/reputation, Rumor, Bad patenting policies and procedures, Weak knowledge of regulations, Lack of comprehensive insurance coverage, Unfavorable contractual agreements
Legal, Social Aspects and Human Ethics	Non compliance with National and International regulations, Strike, Sabotage, Retention, Faked sickness, Incompetence, Bribery, Dishonest behavior, Employee unreliability, Error, Illicit action, Panic, Epidemics, Penuries
Organizational	Weak relations between management staff, Weak internal controls, Not respected management, Procedures are not followed, Illness, Badly controlled outsourcing, Low morals, Labor accidents
International Relations/Politics	War, Terrorism, Regional conflict, Organized crime, Kidnapping, Government corruption, Mass psychoses, Group anarchy, Riots
Marketing, Economical, Financial	Volatile market, Product/service boycott, Unsuccessful merger/acquisition, Bad product/service performance, Non adapted product, Unsatisfied client, Bad strategic decisions, Client dependence, High competition, Interrupted production, Negative Return on Investment (ROI), Debt, Low capital, Demands of stakeholders, Untrustworthy financial sources, Slowdown in economic growth, Fraud, Insufficient resources
Environment	Natural catastrophe, Pollution, Nuclear catastrophe, Biological disaster, Chemical disaster, Radio electric incident

Source: [ENISA 2012]. © 2012, ENISA.

2.6.2 Vulnerability Management

The top priority for private and corporate users in the power industry should be to reduce the risks of vulnerabilities' exposure by reducing the number of vulnerabilities and the window of opportunity to exploit vulnerabilities, reducing the complexity of patching programs, applying unified patching, increasing awareness, and embracing vulnerability management. In addition, the complexity of IT systems and control systems, lack of awareness, lack of a unified and automated way to deploy critical security updates – or patches - contributes to a lengthy process of removing the vulnerabilities. Unified technology is the new paradigm that is needed to allow users to automatically install security updates for a broad array of programs.

Vulnerability management is the cyclical practice of identifying, classifying, remediating, and mitigating vulnerabilities [Foreman 2010]. Vulnerability management is a segment of risk management process that expands from environment, political, physical layer to technology and supply chain. Figure 2.11 depicts key roles in risk management framework.

Vulnerability management includes many tasks that are described in standards and guidelines. Although vulnerability management involves complex processes, one process, called vulnerability coordination, is supported by ICS-CERT process [ICS-CERT 2016]. This process is intended for handling a vulnerability and involves five basic steps:

- Detection and collection

- Analysis
- Mitigation coordination
- Application of mitigation
- Disclosure.

This process and resolution are essential to a user, vendor, power engineer, security engineer, and management.

Vulnerabilities are characteristics of an installation, system, asset, application, or its dependencies that could cause it to suffer a degradation or loss (incapacity to perform its designated function) as a result of having been subjected to a certain level of threat or hazard [Robles 2008]. Vulnerabilities may exist in a business plan, corporate strategy, environment, software, firmware, technology, architecture, and systems. Vulnerabilities can be known or unknown.

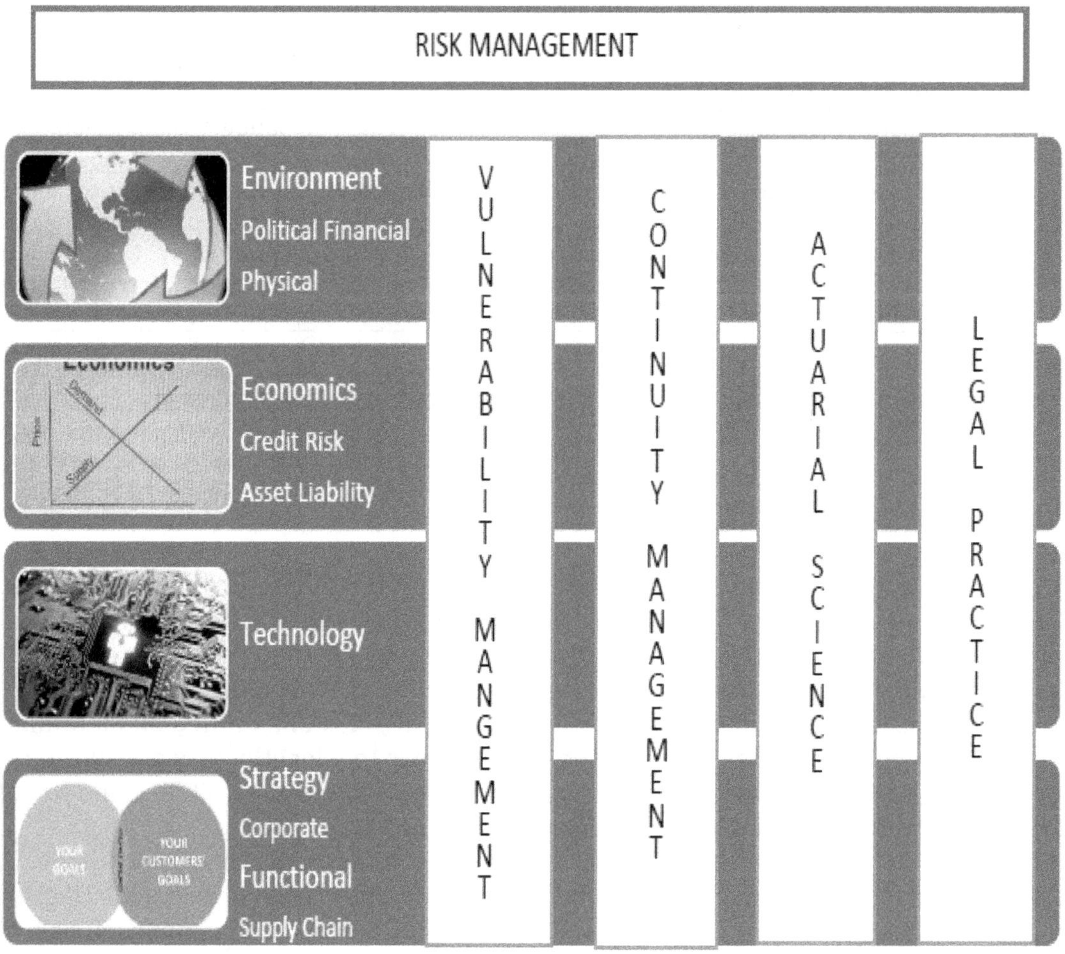

Figure 2.11 Roles in risk management framework.

2.6.2.1 Identifying Smart Grid Vulnerabilities

A taxonomy for security vulnerabilities should provide some benefits. The specific goals of such a taxonomy are identified [Bishop 1996]:

- Providing a historical record of the vulnerabilities in a form that system designers and implementers can use to anticipate flaws in their systems
- Describing the vulnerabilities in a form useful for detection
- Showing common characteristics in related flaws for prevention and elimination
- Enabling a security monitor to detect exploitation (or attempted exploitation) of the flaws.

A taxonomy, similar to the biological classification of plants and animals, allows one to classify each vulnerability as a unique ordered tuple [Bishop 1996]. As described in this work, the benefits include help on detecting new vulnerabilities, determining how many instances of a larger class of flaws are known, suggesting how to reduce or eliminate the flaws. Most importantly a taxonomy of vulnerabilities allows to characterize conditions under which a flaw arises, suggesting ways to detect new instances of the flaw. However, there is no generally accepted vulnerabilities taxonomy for computer systems.

Several categories of vulnerabilities are described in [NISTIR 7628r1]. Examples include:

- Security policies vulnerabilities are often introduced due to inadequate development of, implementation of, or the lack of security policies.
- Privacy policy vulnerabilities lead to unwanted exposure of employee or personal information, leading to both business risk and security risk.
- Risk management deficiencies can lead to vulnerabilities throughout the organization's security program; lack or misapplication of adequate risk assessment processes can lead to poor decisions based on inadequate understanding of actual risk.
- An inadequate continuity of operations or disaster recovery plan could result in longer than necessary recovery from a possible plant or operational outage.

Software continues to exhibit vulnerabilities, some are unknown and some publicly disclosed. The publicly disclosed vulnerabilities at any moment expose a system to potential attack. A method for identifying and analyzing these vulnerabilities uses public data from easily accessible sources.

In the analysis of the vulnerabilities, certain vulnerabilities may seem acceptable risks when considered in isolation. But an intruder can often infiltrate a network through a multi step intrusion, in which each step prepares for the next. Therefore, in order to determine potential exposure to attacks and their consequences, it is necessary to analyze relationships among the known vulnerabilities and the latest discovered vulnerabilities, threats, and risks [Feng 2008].

A list of vulnerabilities classes and specific vulnerabilities are described in [NISTIR 7628r1]. A vulnerability class is a grouping of common vulnerabilities which could adversely impact the operation of the electric grid. Vulnerability classes include classes of potential vulnerabilities for the Smart Grid. Individual vulnerabilities are classified by category. A vulnerability is defined as a weakness in an information system, system security procedures, internal controls, or implementation that could be exploited or triggered by a threat source. Although these classes of vulnerabilities are not an exhausted list, they can be used as a stimulus for vulnerability assessment and detailed risk analysis of real or proposed systems. Vulnerability assessment involves the identification of vulnerabilities present in information systems. Table 2.4 shows examples of vulnerability categories and classes of vulnerabilities.

Using the CWE and OWASP taxonomies as a guide, the [NISTIR 7628 r1] document describes classes and subclasses of vulnerabilities in platform software and firmware. The Common Weakness Enumeration (CWE) and the Vulnerability Categories defined by OWASP are two taxonomies which provide descriptions of common errors or oversights that can result in vulnerability instances. A taxonomy provides a way of describing the causes of vulnerabilities, which are largely independent of the operational environment, whereas the impact of these vulnerabilities may differ in a Smart Grid environment compared to a traditional IT enterprise.

However, discovering and reporting of vulnerability instances for computer systems occur continuously and the Common Vulnerability and Exposures (CVE) specification establishes a common identifier for known vulnerability instances, so the list of vulnerabilities for a specific system has to be continuously updated.

Smart Grid can be seen as a system of heterogeneous interconnected systems (e.g., virtual power plants, transmission grid, distribution grids, etc.), each one with its own set of communication equipment and technologies, intelligent devices, automated control elements and algorithms, processing applications, connection points, specific procedures, different organizational approaches, etc. This heterogeneity, diversity and complexity make highly difficult its protection and at the same time introduce new and complex (and likely vulnerable) interrelationships and dependencies which were not present in traditional power systems.

A brief list of Smart Grid's components that need to be considered as a source of vulnerabilities is suggested to include [ENISA 2012]:

- Operational systems such as generators, transformers, SCADA Systems and Energy/Distribution Management Systems (EMS/DMS), programmable logic controllers (PLCs), substations, smart meters, and other intelligent electrical devices (IEDs).
- Classic IT systems such as PCs, servers, mainframes, applications, databases, web sites, web services, etc., among which include the components of corporate infrastructure.
- Communications networks and protocols: Ethernet, Wi-Fi, PRIME, DLMS/COSEM, Zigbee, 4G, DNP3, 5G, etc.
- End points such as smart meters, EVs, smart phones and other mobile devices; taking into account both physical and logical aspects.

In addition, most of the ICT applications and technologies which are envisioned to be enablers for the success of the Smart Grids are also a source of new vulnerabilities. Given the ICS dominance in many functions, there is a strong reason to review their vulnerabilities.

2.6.2.2 Identifying ICT/OT Supply Chain Vulnerabilities

The term supply chain refers to the linked set of resources and processes between and among multiple levels of an enterprise, each of which is an acquirer that begins with the sourcing of products and services and extends through the product and service life cycle [NIST SP800-161].

The ICT/OT supply chain relies on a globally distributed, interconnected supply chain ecosystem that consists of public and private sector entities (e.g., acquirers, suppliers, developers, system integrators, external system service providers, and other ICT/OT related service providers). Examples of products and services that enterprises rely on the supply chain include, but are not limited to, provision of systems and system components, custom software, operational support services, hosting systems and services, and performing system support roles.

Procuring ICT/OT products directly from suppliers establishes a direct relationship between those suppliers and the acquirers. This relationship is also usually guided by an agreement between the acquirer and the supplier. However, commercial ICT/OT developed by suppliers are typically designed for general purposes for a global market and are not typically tailored to an individual customer's specific operational or threat environments. The supply chain represents a significant risk to industrial control systems (ICS). ICS manufacturers and software developers create their products in many different locations around the world. Purchasing commercial off-the-shelf (COTS) technologies increases the likelihood of receiving nongenuine equipment.

Table 2.4 Vulnerability classes.

Vulnerability Class	**Examples**
PEOPLE, POLICY AND PROCEDURE Policies and procedures are the documented mechanisms by which an organization operates, and people are trained to follow them. Policies and procedures lay the groundwork for how the organization will operate; adequate training ensures that people understand their role/responsibility in implementing the policy and procedures.	Training, policy and procedures, Risk management, Inadequate periodic security audits, Inadequate security oversight by management, Inadequate continuity of operations or disaster recovery plan, etc.
PLATFORM SOFTWARE/FIRMWARE VULNERABILITIES Software and firmware are the programmable components of a computing environment. Errors or oversights in software and firmware design, development, and deployment may result in unintended functionality that allows adversaries or other conditions to affect, via programmatic means, the confidentiality, integrity, and/or availability of information. These errors and oversights are discovered and reported as vulnerability instances in platform software and firmware.	Software development such as: Code quality vulnerability (CWE-398), Authentication vulnerability (CWE-287), Authorization vulnerability (CWE-284), Cryptographic vulnerability (CWE-310), Environmental vulnerability (CWE-2), Input and Output validation (CWE-20 and (CWE-116), etc.
PLATFORM Platforms are defined as the software and hardware units, or systems of software and hardware, that are used to deliver software based services. The platform comprises the software, the operating system used to support that software, and the physical hardware	Use of inadequate security architectures and designs, Inadequate malware protection, etc.
NETWORK Networks are defined by connections between multiple locations or organizational units and are composed of many differing devices using similar protocols and procedures to facilitate a secure exchange of information	Inadequate integrity checking, Inadequate network segregation, Inappropriate protocol selection description, etc.

Source: Adapted from [NISTIR 7628r1]. Public Domain.

In this process, risks may occur at different stages caused by intentional or unintentional threats that could exploit vulnerabilities of the entities and actors participating in the process including the ICT supply chain infrastructure.

Cybersecurity risk in supply chain refers to the potential for harm or compromise arising from a relationship with suppliers, their supply chains, and their supplied products or services. These adverse impacts materialize when a human or non-human threat successfully exploits a vulnerability tied to a system, product, service, or the supply chain ecosystem. Cybersecurity risks in the supply chain arise from threats that exploit vulnerabilities or exposures within products and services traversing the supply chain as well as threats exploiting vulnerabilities or exposures within the supply chain itself.

The ICT supply chain risks arise from the loss of confidentiality, integrity, or availability of information or information systems and reflect the potential adverse impacts to organizational operations (including mission, functions, image, or reputation), organizational assets, individuals, other organizations, and the Nation. An ICT supply chain compromise can occur anywhere within the system development life cycle (SDLC) of the product or service. For example, an organized crime enterprise introduces counterfeit products into the market resulting in a loss of customer trust and confidence; an adversary jeopardizes the confidentiality, integrity, or availability of a system or the information the system processes, stores, or transmits.

Supply chain cybersecurity vulnerabilities may lead to threats that could have a negative impact on the enterprise such as reputation damage, degraded services leading to customer dissatisfaction, theft of intellectual property or degradation of critical mission and business processes. Often, it may take years for the vulnerabilities to be exploited or discovered.

Managing cybersecurity risk in the supply chain is a complex undertaking that requires cultural transformation and coordinated, multidisciplinary approach across an enterprise. This document [NIST SP800-161] provides guidance on how to apply an effective cybersecurity supply chain risk management (C-SCRM) inside the enterprise as well as outside the enterprise (e.g., suppliers, developers, system integrators, external system service providers, and other ICT/OT related service providers) to actively collaborate, communicate, and take actions to secure favorable C-SCRM outcomes. This white paper [ISAGCA 2020] provides an overview of the security lifecycles (for product and automation solution) that are described in the ISA/IEC 62443 series of standards that specify the requirements for the security of industrial automation and control systems (IACS). Also, this document [DHS 2016] includes recommended practices for ICS security. It is recommended that C-SCRM process should be integrated into enterprise wide risk management processes described in [NIST SP 800-39].

Organizations in electricity industry have significant OT components throughout their enterprise architecture. They often use specialized service providers for secure implementation and maintenance of devices, systems, or equipment. Any enterprise or individual providing services which may include authorized access to an ICT or OT system should adhere to enterprise C-SCRM requirements. In addition, enterprises should apply special scrutiny to ICT/OT related service providers managing mission critical and/or safety relevant assets.

Smart Grid is connected with distributed energy resources (DERs), controllers, smart meters, thermostats, plug-in electrical vehicles, lights, refrigerators, and more appliances than can all be connected to the IoT. The Internet of Things (IoT) refers to the network of connected objects that are able to collect and exchange data in real time using embedded sensors. It is also important to consider various vulnerabilities of IoT devices and suppliers. Examples of vulnerabilities include: lack of security protection of the hardware or software in the IoT device that is offered by a manufacturer; unsecure protocols; lack of access management (e.g., authentication) and data integrity controls; lack of firmware validation; unsecure delivery mechanisms for the firmware or software updates; no notifications from the IoT vendor when a new version of the firmware or software is available; no periodic security check-ups for the IoT service; presence of incomplete security policies for the supply chain; weak security training for the stakeholders in an IoT ecosystem, etc.

Another issue is the lack of regulations for personal identifiable information (PII) data collection by sensors. There could be a violation of individuals' s right to privacy and be informed about their personal data being collected. Sensors may collect PII data that is used by different applications in an IoT ecosystem. There could be situations when a component designed and manufactured by a supplier is offered by multiple vendors in different countries so PII data collected by sensors could be exposed beyond the intended application.

Vulnerabilities in the supply chain are often interconnected and may also expose enterprises to a cascading cybersecurity risk in the supply chain. For example, a large scale service outage caused by an exploited vulnerability in IT management software at a major cloud services provider may cause service or production disruptions for multiple entities within an enterprise's supply chain and lead to negative effects within multiple mission and business processes.

Similarly, a vulnerability in the integration process for SolarWinds software was exposed that allowed an adversary to install malware thru a back door. The malware spread to its clients via updates and went undetected for months causing data theft and spying on organizations. The multi-stage attack lasted more than 18 months. It caused an impact that affected more than four-fifths of the Fortune 500 U.S. companies and hit virtually every major sector in the U.S. government and military [Vavra 2021].

Therefore, these recent cyber attacks – SolarWinds, Colonial Pipeline, and Microsoft Exchange- are warnings that public and private sector entities increasingly face sophisticated malicious cyber activity from both nation state actors and cyber criminals.

Enterprises can use a framework such as NIST cybersecurity framework [NIST 2018] implementation tiers to baseline their C- SCRM capabilities. This enables a useful context for an enterprise to track and increase their foundation for C-SCRM practices.

2.6.2.3 Vulnerability Disclosure

Security vulnerabilities in software are one of the primary reasons for security breaches, and an important challenge from knowledge management perspective is to determine how to manage the disclosure of knowledge about those vulnerabilities. However, there are questions raised in this situation that include:

- Should the vulnerability be kept secret (e.g., not announced to the public at large) until the vendor releases a patch?
- Should the public be informed about the vulnerability immediately after its discovery?
- Should there be some other mechanisms to disclose vulnerability information to the public?
- How patches should be released by vendors and updated by firms?
- What are the costs and risks?

Among all these questions that have to be answered, the vulnerability disclosure is one key issue that has to be understood by security team.

The security community has proposed several disclosure mechanisms, such as full vendor, immediate public, and hybrid, and has debated about the merits and demerits of these alternatives. The steps followed to handle dissemination of the knowledge of software vulnerabilities after a benign user identifies them are collectively known as the vulnerability disclosure process. The vulnerability information has relevance to the vendor, software users, and hackers. Therefore, the sequence and timing of announcement of the vulnerability, that is, how vulnerability knowledge is managed, affects the total cost of the vulnerability to the society. The perception about what constitutes responsible vulnerability disclosure has changed over time.

Responsible vulnerability disclosure addresses how a vulnerability identifier should disclose vulnerability information to appropriate people, at appropriate times, and through appropriate channels in order to minimize the social loss4 associated with vulnerabilities. Despite the consensus on the objective of responsible vulnerability disclosure, no such consensus exists on how the vulnerability knowledge should be disseminated to various stakeholders. The lack of consensus is partially because there is no common understanding of the impact of the vulnerability knowledge on different stakeholders. Disclosure of vulnerabilities is still a debatable issue [Wiki Disclosure].

Different practices that are also known include:

- Vendor disclosure – common practice, during the early days of software development, was to inform only the vendor about the discovered vulnerability and to keep it secret from the public until the vendor developed a patch. This practice gives the control of handling a vulnerability to the vendor and cannot force the vendor to develop a fix, some reported vulnerabilities have gone unfixed or have been fixed after a long delay. Unless the vendor is committed to develop a patch as soon as possible, full vendor disclosure puts the public at risk.
- Public disclosure - the tardiness of vendors has caused some users to adopt immediate public disclosure, which releases vulnerability knowledge immediately to the public after its identification. It gives vendors a strong incentive to fix the vulnerability as soon as possible to prevent a public embarrassment and it allows vulnerable customers to take some intermediate measures to reduce the risk of exploitations until the vendor releases a patch. It is argued that public disclosure and disseminating vulnerability knowledge to the public does not improve overall security because hackers have an opportunity to develop exploits and attack vulnerable systems before the vendor releases a patch to address the problem.
- Hybrid disclosure – is a middle ground practice in which both the vendor and benign user can compromise in the vulnerability disclosure process to make the process better for society. The benign user does not announce the vulnerability knowledge to the public immediately, but instead allows the vendor some time to develop a patch. If the vendor does not release its patch before the deadline, the public is informed about the vulnerability. It is argued that this approach is optimal. Over time, the Computer Emergency Response Team (CERT) has emerged as a third party coordinator to handle the hybrid vulnerability disclosure process.

A policy disclosure is promoted by CERT-IST of France. This policy [CERT-ICS Disclosure] is based on documents such as [IETF 2002], [NIAC 2004], and most current standard [ISO/IEC 29147].

In addition, a number of guidelines are currently available to govern this direct relationship between the vendor and the identifier, such as [OIS 2004]. Under this type of hybrid disclosure, the coordinator's role is limited to resolving any disputes between the vendor and the vulnerability identifier. Third party and security consulting firms are also actively involved in the vulnerability identification process. They usually have their own guidelines for responsible vulnerability disclosure as well as working directly with software vendors. These guidelines typically follow either full vendor disclosure or hybrid disclosure. The multitude of disclosure mechanisms characterized as responsible vulnerability disclosure creates chaos and confusion on the vendor side.

An analysis of the state of affairs in the world of software vulnerabilities, various techniques for disclosing these vulnerabilities, and the costs, benefits and risks associated with each approach is provided in [Cencini 2005]. This work reveals the secrets of security business and who benefits from security problems. Typically, it is believed that the first beneficiaries are attackers breaking into vulnerable computer systems and stealing money and valuable information from victims. However, there is another side to this security industry: the community of security professionals who build a reputation and earn a living finding and reporting security problems. The authors argue that working as a computer security professional can be quite lucrative, with the benefit of not having to break the law or compromise one's ethics – and quite often, the technical details and challenges of this legitimate work are not much different from those when the work is done for less legitimate purposes.

Although responsible disclosure policy tends to be one of the most widely agreed upon best practices, the biggest challenge facing any new vulnerability disclosure policy is universal adoption [Cencini 2005]. The authors argue that vulnerability disclosure policy is not one that a single nation can govern and a global approach towards adoption of the new policy is the best strategy. For example, a software vulnerability disclosure legislation passed in USA may not necessarily apply to another country. It has to be recognized that the Internet has brought the software industry to a global level, but certain regulations are not universal.

Another study on how vulnerabilities should be disclosed to minimize the social loss is reported in [Cavusoglu 2007]. The authors argue that the characteristics of the vulnerability (vulnerability risk before and after disclosure), cost structure of the software user population, and vendor's incentives to develop a patch determine the optimal (responsible) vulnerability disclosure. This study of various disclosure policies for the vulnerability disclosure process proposed by different stakeholders is focused on the identification of the optimal disclosure policy that minimizes the social loss. It concludes that none of the disclosure policies, immediate public, full vendor, or hybrid, is always optimal. Characteristics of the vulnerability—risk, user population—and vendor's incentives determine the optimal vulnerability disclosure policy. Another issue that has to be addressed is patch management. An optimal solution would be release of patches by vendors and updated by affected users in batches. It not clear that the social welfare is maximized when patch release and update cycles are synchronized. However, cost sharing and liability can help coordinate the patch management decisions of vendors and firms to achieve social optimality.

2.6.3 Costs and Risks

Organizations require to differentiate costs when addressing threats and vulnerabilities. The strategic investments required to address the security risk from more traditional adversaries (e.g., hackers conducting small group activities with limited capabilities) are considerably different than the investments required to address the risk associated with advanced persistent threats consistent with more advanced adversaries (e.g., nation states or terrorist groups with highly sophisticated levels of expertise and resources that seek to establish permanent footholds in organizations for purposes of impeding aspects of the organizational missions) [NIST SP800-39]. Based on this standard, one recommendation is to allocate budgets to address less sophisticated threats to ensure that needed safeguards and countermeasures (e.g., security controls, security services, and technologies) are obtained, implemented correctly, operating as intended, and producing the desired effect with regard to meeting information security policies and addressing known vulnerabilities. In addition to these basic investments, organizations need to invest in continuous monitoring processes to ensure that the acquired security controls, services, and technologies are operating effectively throughout the system development life cycle. When dealing with persistent threats, the nature of investment has to change from a tactical view to a strategic focus on a longer term.

Also, reducing the continuing focus on near term vulnerabilities discovered in information systems can be mitigated with a long term view of strategic investments in the risk response needs for organizations. This investment strategy is a change from vulnerability and patch management to a longer term strategy addressing information security gaps such as the lack of information technology products with the trustworthiness necessary to achieve information system resilience in the face of advanced persistent threats.

In some cases, disclosure of software vulnerabilities can help improve the overall security of computer systems, while in others, such disclosures can lead to costly widespread exploitation of security problems by black hat hackers. In any case, discovery of a software vulnerability is not free. The accumulated costs may include these aspects: costs incurred by software vendors to process, verify, fix, and distribute patches for a vulnerability; researchers spend considerable time and effort in locating software flaws; and users and administrators spend large amounts of time finding, testing and deploying patches to their systems.

Approaches to identify vulnerabilities are topics of current research and Smart Grid development projects that are funded by NIST, DOE, U.S. Department of Commerce, government, etc. One example is this research work [Conte Leon 2002]. A model based on graph theory is developed for EMS and SCADA systems. This model allows to calculate device vulnerability using the concept of a device visibility path. This information helps power substation operators and administrators identify and harden those areas of the control system that are most vulnerable to cyber attacks. However, addressing advanced persistent threats may require information security investments over the course of several years, as new security solutions and technologies transition from research to development to full deployment.

2.7 Technical and Management Documentation

An organization needs to allocate resources for generating and maintaining the technical and management documentation that includes information for all security aspects such as plans, processes, policies, requirements, designs, specifications, architecture, and operational procedures. For example, policies are living documents that constantly change and organizations need to properly disseminate them. Good management practices for technical and management documentation make for a more security resilient organization. Although the task of documentation is often viewed as an unpleasant task, it is an essential piece of business, and therefore an essential piece in disaster recovery and business continuity [Harris 2013].

For Federal Information Systems, the protection of a system must be documented in a system security plan. This is a requirement of the Office of Management and Budget (OMB) [OMB 130], and [FISMA 2002]. Although NIST document [NIST SP800-18r1] is addressing the Federal Information Systems, the guide can be used by commercial organizations to shape its documentation by including or adapting a system security plan as described in this document.

The system security plan is viewed as documentation of the structured process of planning adequate, cost effective security protection for a system. It should reflect input from various managers with responsibilities concerning the system. Additional information may be included in the basic plan and the structure and format organized according to organization's needs.
Currently, most of documentation is in an electronic form. Therefore, storage and security issues have to be addressed to avoid the loss or modification of the documentation.

2.8 Standards, Guidelines, and Recommendations

Besides cybersecurity standards that we presented in Chapter 1, there are more examples of initiatives and standards that support security program for the Smart Grid.

2.8.1 NIST Specifications

NIST published several specifications and guidelines for the development of the Smart Grid and

DER systems. For example, [NIST SP1108r4] and [NISTIR 7628r1] guidelines include sections on the overall security strategy for the Smart Grid, threats and vulnerability analysis, diagrams, interfaces, privacy, and requirements. In volume 1 of [NISTIR 7628r1], the electricity subsector has tailored [NIST SP800-39] to meet its unique attributes. This tailored approach is also included in the document, U.S. Department of Energy as Cybersecurity Risk Management Process (RMP), which provides the risk management framework and organizational structure needed before system specific controls identified in [NISTIR 7628r1] can be applied.

However, these documents are not prescriptive, nor mandatory. Rather they are advisory, intended to facilitate each organization's efforts to develop a cybersecurity strategy effectively focused on prevention, detection, response, and recovery. These are guides to assist organizations as they craft a cybersecurity strategy that includes also requirements to mitigate risks and privacy issues pertaining to Smart Grid customers and uses of their data.

2.8.2 NIST Cybersecurity Framework

In addition to the efforts focused on Smart Grid cybersecurity, there has also been additional attention to the cybersecurity of all the critical infrastructure sectors including the energy sector. The Cybersecurity Framework (CSF) is described in [NIST 2018] document. This document serves as a national level framework that is flexible enough to apply across multiple sectors. The CSF is intended to ensure that existing work within the sectors, including the energy sector, can be utilized within the framework. The existing voluntary cybersecurity standards for the Smart Grid including guidelines, and practices can be leveraged to address the CSF functions in the context of an organization's risk management program.

2.8.3 DOE Cybersecurity Framework

The U.S. Department of the Energy released this guidance [DOE 2015a] to help the energy sector establish or align existing cybersecurity risk management programs to meet the objectives of the CSF.

2.8.4 European Standards and Recommendations

In Europe, several reports provide surveys and studies regarding the use and development of security standards for the Smart Grid. Using as basis the standards and guidelines developed in USA, the European Union Agency for Network and Information Security (ENISA) extended these standards and included their specific standards for the development of Smart Grids in Europe. Also, ENISA provides surveys and studies for different aspects of cybersecurity. Examples of reports include:

- [ENISA 2015] is a study that reveals the current maturity level of ICS-SCADA cybersecurity in Europe and identifies good practices used by European Member States to improve this area.
- [ENISA 2011b] provides an overview of existing methods, procedures and guidelines in the area of control system (cyber) security. It takes into account activities of international organizations, important national activities in Europe and the US.
- [CENELEC 2014], [CENELEC 2017] characterize the security standards as focused into requirements (the What) and solutions (the How) standards. The stated list of standards covers requirements and solution standards that provide different level of detail. These standards are analyzed regarding their coverage following the classification approach as depicted in Figure 2.12.

While mapping a standard to the diagram is on an abstract level, it shows which scope and to what level of detail the standards address on each of the four quadrants. Moreover, also addressed is the relevance of the standards for organizations (Smart Grid operators) as well as products and services (product manufacturer and service providers). The mapping of the selected standards to the standards areas is done under the following terms:

- Details for operation - the standard addresses organizational and procedural means applicable for all or selected actors. It may have implicit requirements for systems and components without addressing implementation options.
- Relevance for products - the standard directly influences component and/or system functionality and needs to be considered during product design and/or development. It addresses technology to be used to integrate a security measure.
- Design details - the standard describes the implementation of security means in details sufficient to achieve interoperability between different vendor's products for standards on a technical level and/or procedures to be followed for standards addressing organizational means.
- Completeness -the standard addresses not only one specific security measure but addresses the complete security framework, including technical and organizational means.

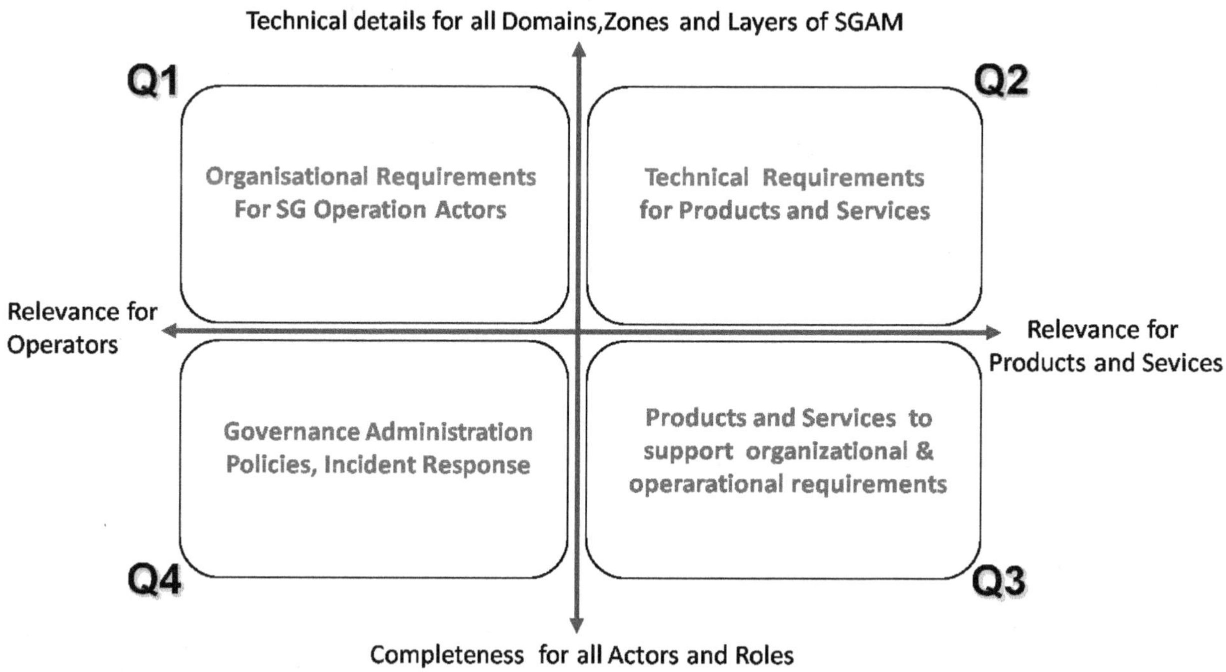

Figure 2.12 Standard areas. *Source:* [CENELEC 2017]. © 2017, CENELEC.

2.8.5 Using Additional Sources

This document [Cleveland 2013] provides additional standards that include more cybersecurity topics related to the Smart Grid. Also, this document includes standards that cover IT and business systems. It organizes the standards based on common topic and by the users (e.g., utilities, control systems, transmission systems, customer applications, DER systems). Figure 2.13 depicts a diagram of standards used with [IEC 61850] profiles for DER systems.

Several reports show the benefits and limitations of many standards including recommendations for new standards for the emerging Smart Grid applications (e.g., [Wang 2013], [Kezunovic 2011], [Hentea 2008], [Dong 2011]).

Others describe how to combine various approaches and models to define a secure Smart Grid [Uslar 2014]. Several published materials such as guidelines, white papers, survey reports, and research studies also contribute to addressing various aspects of cybersecurity for the Smart Grid. Therefore, identification and consultation of new publications is strongly recommended before any Smart Grid system is developed.

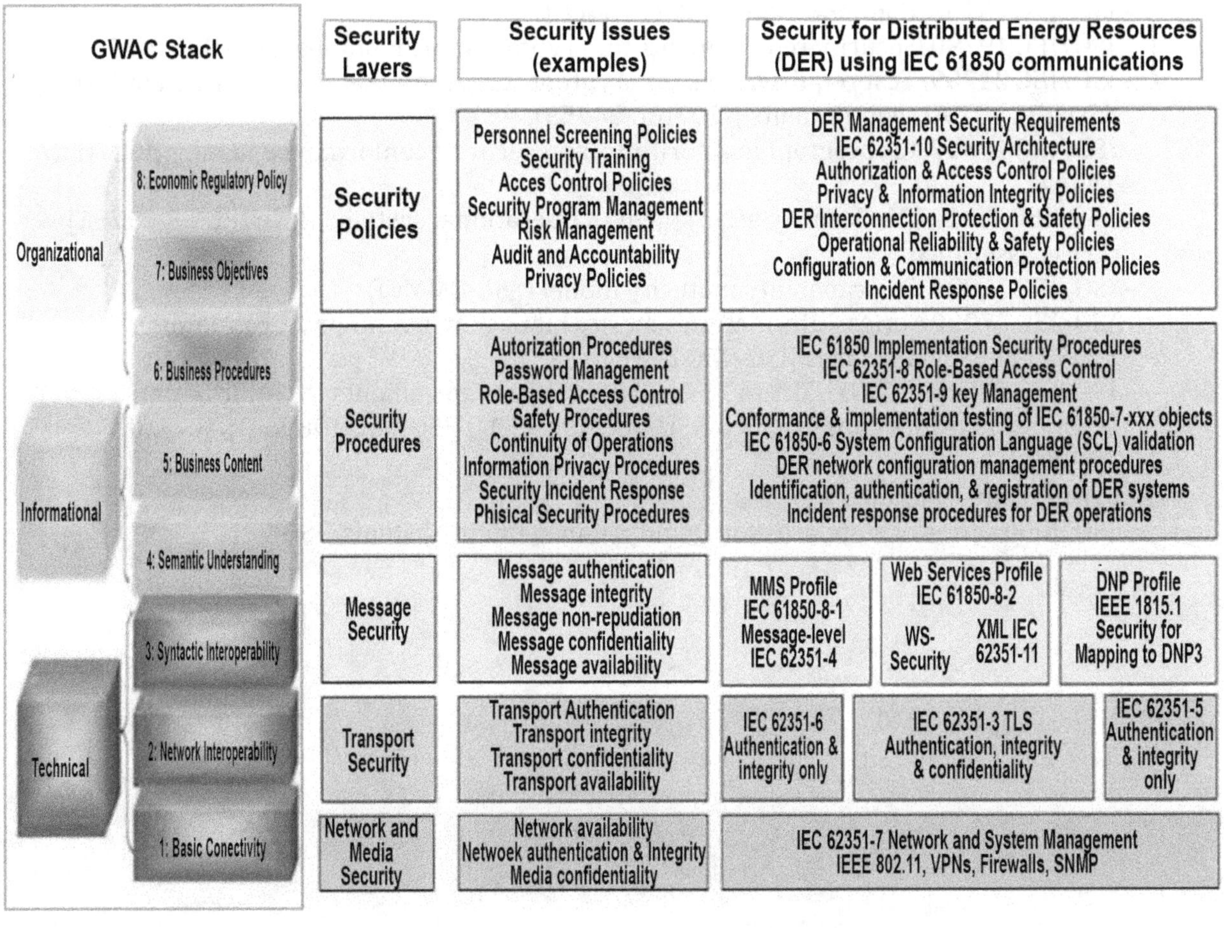

Figure 2.13 DERs Security with IEC 61850 and IEC 62351. *Source:* [Cleveland 2012]. © 2012, Xanthus International.

2.8.6 International Standards

Many countries have launched significant efforts to encourage the development of their own Smart Grid. As countries move forward with their individual initiatives, it is important that the development of standards is internationally coordinated. More information about ISO security evaluation, testing, specification, and certification of IT products that include cryptographic modules is presented in [Easter 2015].

The 62443 series of standards developed by ISA (International Society of Automation) and adopted by the International Electrotechnical Commission (IEC) provides a framework for mitigating vulnerabilities in Industrial Automation and Control Systems (IACS) associated with Industry 4.0 and Critical Infrastructure. The IEC 62443 series of standards can be utilized across industrial control systems for manufacturing and many industry sectors. It has been approved by many countries. IEC 62443 is evolving to become a key standard in the industrial control systems.

Other examples of standards with an international base include standards that support these activities:

- Product assurance
  - ISO/IEC 15408, Common Criteria [ISO 15408]
  - ISO/IEC 19790, Security requirements for cryptographic modules [ISO/IEC 19790]
  - ISO/IEC 24759, Test requirements for cryptographic modules, demonstrate conformity ISO/IEC 19790 requirements [ISO/IEC 24759]
  - ISO/IEC 18367, Cryptographic algorithms and security conformance testing [ISO/IEC 18367]
  - ISO/IEC TR 19791, Security assessment of operational systems [ISO/IEC TR 19791]
- Process assurance
  - ISO/IEC 21827, SSE capability maturity model (SSE-CMM®)
  - ISO/IEC 27000-series, Information security management and practices, known as ISMS Family of standards [ISO/IEC 27k]
  - ISO/IEC 19896-1, ISO/IEC 19896-2, Competency requirements for information security assurance conformance [ISO/IEC 19896-1], [ISO/IEC 19896-2], [ISO/IEC 19896]
- Environment assurance
  - ISO 9000 series of standards for Quality Management Systems [ISO 9000].

Chapter 3

Security Quality Problem

3.1 Security Engineering Principles

Security Engineering focuses on defining customer needs, security protection requirements, and required functionality early in the systems development lifecycle, documenting requirements, and then proceeding with design, synthesis, and system validation while considering the complete problem [CNNSI 4009]. Although many approaches and methods can be used to secure information systems, it is expected that certain intrinsic goals must be met whether the system is small or large or owned by a government agency or by a private corporation. The intrinsic expectations are described as generally accepted system security principles. A principle is a rule or standard, especially of good behavior [American]. Systems Security Engineering (SSE) applies scientific, engineering, and information assurance principles to deliver trustworthy systems that satisfy stakeholder requirements within their established risk tolerance [CNNI 4009]. SSE is a specialty engineering field strongly related to Systems Engineering.

SSE is a part of systems engineering that helps to ensure that the appropriate security principles, concepts, methods, and practices are applied during the system life cycle to achieve stakeholder objectives for the protection of assets caused by disruptions, hazards, and threats. It also helps to reduce system defects that can lead to security vulnerability and as a result, reduces the susceptibility of the system to adversity.

3.1.1 Evolving Security Principles

Using the OECD guidelines [OECD 1992] for the security of information systems as the base for the principles, NIST developed a set of principles encompassing broad areas (e.g., accountability, cost effectiveness, integration, etc. [NIST 2001]). The principles address computer security from a very high-level viewpoint. The principles are to be used when developing security programs and policy and when building new systems, practices or procedures.

The security principles evolved and defined in standards. This publication [NIST SP800-160v1] addresses the engineering-driven actions necessary to develop more defensible and survivable systems - including the components that compose and the services that depend on those systems. The document describes systems security engineering techniques, methods, and practices to be included in those systems and software engineering activities described in this standard, Systems and software engineering — System life cycle processes [ISO/IEC/IEEE 15288].

The appendix F of [NIST SP160v1] includes a taxonomy of security principles grouped in three categories: security architecture and design, security capability and intrinsic behaviors, and life cycle security.

For legacy control systems, it is recommended the organization applies security principles to system upgrades and modifications, to the extent feasible, given the current state of the hardware, software, and firmware components within the system [NIST SP800-53r5]. Large, integrated, complex systems require different layers of interoperability, from a plug or wireless connection to compatible processes and procedures for participating in distributed business transactions.

Similar security principles are provided by ISO (e.g., [ISO/IEC 27001]) and other associations such as The Open Group and a joint group of ISF, (ISC)2, and ISACA that have published their security principles grouped in categories of principles designed to help security practitioners worldwide [ISACA 2010], [ISC2 Blog], [OpenGroup 2011].

Principles of secure systems, a combination of experience and published literature, are often used and taught in education courses. For example, a summary of security principles include [Irvine 2001]:

- Modularity and layering - isolate functions and related data structures into compact logical units (modules); layering allows the relationships of these units to be better understood, so that dependencies are clear, and undesired logical looping can be avoided.
- Economy of mechanism - keep the system design as simple and small as possible.
- Complete - always invoked - a strong mechanism must be effective against all parties and under all circumstances.
- Non-bypassable - if a protection mechanism can be avoided, for example by using an alternate interface to access an object, then the mechanisms will be ineffective.
- Least privilege - each entity is given only enough privilege to complete the job at hand and not privileges that might go far beyond those associated with the task.
- Fail-safe defaults - the system should be in a secure state when it is initialized and when it fails; the system does not enter a state in which protection mechanisms are no longer in effect
- No security through obscurity - the foundations of a system's security should be based upon the quality of its design and the rigor of its engineering process.
- Usability - secure systems should be appealing to potential users.
- Attention to shared variables and resources - design of secure systems requires attention to the adversary's ability to monitor global variables and shared resources.
- Separation of privilege - the division of a privilege between multiple parties can decrease the likelihood of that privilege being abused.
- Privileged resources and tools - the commands and resources utilized by the control systems should not be available to general users; the means through which a process gains entry to the privileged state needs to be strictly controlled.
- Per user resource allocation - in multi-user systems, the resources allocated to one user must be protected from access by another user.
- Information hiding and encapsulation - access to a set of related data should be managed by a minimal number of functions.
- Analysis - the engineer may develop system security requirements by applying a threat model.

While most organizations who develop their own security architectures have their own inhouse security principles which guide their architecture designers [OpenGroup 2015], it is imperative to adjust the security principles based on new technical, social, and global trends.

Due to changes in technologies, rise of threats and sophistication, OECD continuously revised the security principles since the first publication in 1992 [OECD 1992]. The security principles defined in the most recent document [OECD 2015] replace the security principles of [OECD 2002]. Recognizing the perimeter of information systems and networks is increasingly blurred and that, as a consequence, the management of risks and the protection measures should extend to the more global ecosystem level, OECD defined the following security principles in two categories [OECD 2015]:

- General Principles
  - Awareness, skills and empowerment
  - Responsibility
  - Human rights and fundamental values
  - Co-operation
- Operational Principles
  - Risk assessment and treatment cycle
  - Security measures
  - Innovation
  - Preparedness and continuity

The eight principles are complementary and should be taken as a whole. All are indispensable and each of them will be ineffective if interpreted or implemented in isolation, or if one of them is neglected [OECD 2015].

As attacks shift from the hard technology to the people who operate it, systems must be resilient to errors and many kinds of threats. Therefore, this situation calls for organizations to apply security principles in a more mature and responsible manner.

3.1.2 Applying Security Engineering Principles to Smart Grid Systems

Organizations in energy sector may specialize in a particular type of technology use or a specialized context. Security Engineering deals with the development of detailed engineering plans and designs for security features, controls and systems of the Smart Grid. The main objective is to support functional and user requirements with an added dimension of constraints and restrictions to prevent misuse and malicious behavior. Each organization needs to perform a risk assessment to determine the applicability of the recommended requirements (e.g., [NISTIR 7628r1]) and other DOE documents including measurable goals, use cases, types of requirements as they apply to the specific business strategy and applications of each organization.

The following include recommendations of using the security principles to help the design of systems for the Smart Grid:

- Discover security needs
- Define system security concepts
- Design system's security
- Implement system
- Assess security effectiveness

Because many security products and systems come to market through lengthy and expensive evaluation or no evaluation, there is a need for:

- Continuity
- Repeatability
- Efficiency
- Assurance.

Also, there is a need to apply information security principles and to include the concepts of support, operations, constraints, and technology in the development phases that include [Kossiakoff 2011]: concept development, engineering development, and post development.

The basic information security principles recommended for critical infrastructure in [TISN 2007] are sound guidelines that may contribute to the improvement of the Security Engineering processes of any organization of the electric subsector.

While the primary focus of these principles remains on the implementation of technical countermeasures, these principles highlight the fact that, to be effective, a system security design should also consider non-technical issues, such as policy, operational procedures, and user education. Organizations have to understand that security risk is dynamic in nature. It includes aspects related to the cyber and physical environments, the people involved in the activity and the organizational processes supporting it.

Therefore, the security and privacy of Smart Grid and DER systems requires the application of Security Engineering principles. The OECD security principles [OECD 2015] and Security Engineering principles as described in [NIST SP800 160v1] should be adopted by organizations developing Smart Grid systems.

In addition, security practices described in several technical documents need to be considered. Examples of practices include but, are not limited to: develop layered protections; establish sound security policy, architecture, and controls as the foundation for design; incorporate security into the system development life cycle; delineate physical and logical security boundaries; ensure developers/integrators are trained on how to develop secure software for information systems (both IT and control systems); tailor security controls to meet organizational and operational needs; reduce risk to acceptable levels thus, enabling risk executives to make informed decisions.

In summary, the following are examples of the benefits of security principles applied to any Smart Grid system:

- Information Security to be integral to the organization's strategy
- Secure systems improve reliability and resilience of power grid
- Security engineering processes based on risk management principles
- Balanced set of security needs of internal and external stakeholders
- Confidence in the correctness and effectiveness of security mechanisms
- Operational impacts due to residual security vulnerabilities are tolerable
- Integration of all security aspects into a combined understanding of the trustworthiness of a system
- Commitment and continuous improvement.

Another requirement is that new systems should be designed so that present or legacy aspects do not unnecessarily limit future system evolution. The requirements for legacy integration should clearly specify the degree of conformance needed (e.g., minimum or full conformance) and the impact on the overall system must be considered. For example, a security issue in one system might have an undesired effect on another system even though the systems are only indirectly related. Confidentiality protection may be satisfied with encryption and authentication in one system, but these security technologies may impact the availability of real time data because of the delays in accessing the keys, processing the cryptographic algorithms, or authentication steps.

In addition, there is a need to understand that IT and ICS systems evolved to information and communication technology (ICT) systems that involve communications, computational, information and many other topics. Quite often, organizations develop isolated and closed information systems based on established old theories, which cannot produce solutions that are needed today and in years to come. An information system design should include both information design and communications design [Hanseth 2010], [Hanseth 2002].

3.2 Security Quality Attributes

Quality of data, information, information system, or service are common goals in the development of any system to support users and business objectives. Quality is the degree to which an object (entity) (e.g., process, product, or service) satisfies a specified set of attributes or requirements [EU Glossary]. However, this is not the single definition and other definitions are provided in standards such as [ISO 9000], [ISO 9001], [ISO 9004] and [ISACA 2019]. Quality management definitions [Praxiom] describe quality attributes are non-functional requirements used to evaluate the performance of a system.

Thus, the engineering perspective views quality as a complex quality factor that is composed of multiple quality subfactors. For example, security is a quality factor with the most prevalent subfactors such as confidentiality, integrity, and availability. Additionally, the integrity subfactor can be further divided into hardware, software, data, and communications integrity. Other security relevant quality subfactors include, but are not limited to, privacy and non-repudiation.

There are also quality subfactors that generally have been considered only by the system safety engineering, such as continuity, resiliency, and fault tolerance, that are now being assessed in terms of susceptibility to malicious intent and the resultant impact on the mission/business; all motivated by mission assurance concerns that span the entire spectrum of incidental and accidental misuse through to attack by an advanced persistent threat.

Therefore, Systems Security Engineering perspective ensures that all security relevant quality subfactors are satisfied by the engineered system and that the system achieves mission/business security objectives.

3.2.1 System Quality Attributes

The information system quality is defined as how well the system supports the business process. From a business point of view, it is crucial to understand the business process and its objectives in order to understand what role the information system has in the particular process. A list of system quality attributes is provided in [Wiki List]. Commonly, security is a subset of attributes: confidentiality, integrity, availability. However, in [ISO/IEC 25010] standard, security is a quality characteristic of a software product or system and includes the following attributes: confidentiality, integrity, accountability, authenticity, and non-repudiation. Security is often treated together with dependability, another subset which is an aggregate of availability, reliability, safety, integrity, and maintainability.

Quality attributes for a system are overall factors that affect run time behavior, system design, and user experience. The extent to which a system possesses a desired combination of quality attributes such as usability, performance, reliability, and security indicates the success of the design and the overall quality of the software. Quality attributes are the cross-cutting concerns. Attributes may conflict with each other and environment can influence the behavior of a system. Another major problem is that developers have to consider user needs too. Therefore, it should be a balance between design goals and user preferences. This requires analysis of many of the individual characteristics of quality that could be in conflict. For example, efficiency is often purchased at the price of portability, accuracy, understandability, and maintainability. Users generally find it difficult to quantify their preferences in such ambiguous situations. The ultimate goal is the ability to quantitatively evaluate and trade off multiple quality attributes to arrive at a better overall system. Characteristics are distinguishing features that can be inherent (permanent) or assigned. Also, characteristics can be classified in quantitative or qualitative categories.

Quality service attributes are to be considered as properties of the service delivered by a system to its users. The service delivered by a system is its behavior as it is perceived by its users; a user is another system (physical or human) which interacts with the former [Laprie 1992]. The service is initiated by some event, which is a stimulus to the system, signaling the need for the service. The stimulus can originate either within the system or external to the system.

A Six Sigma quality program serving utilities and all other energy service providers would allow to build a new electricity service culture based on customer service value, efficiency, and quality. Different surveys reveal that the electric sector is the only industry of its size that does not employ quality methods as a fundamental way of doing business, although the electric utilities require that all of their major suppliers utilize quality methods. A methodology based on Six Sigma quality principles and available technology to achieve a perfect power system that enhances the efficiency, reliability and security of the power system is described in [Kelly 2010].

The electric industry is a collaborative enterprise consisting of generators, distribution centers, transmission centers and control centers. Electric companies are broker power and share information amongst themselves. Control center computers communicate with computers at substations and transmission centers. Thus, utilities recognize dependence on the quality of information systems and information. The electric sector needs systems that meet quality features and have provisions to support essential services even in adverse situations such as a cyber incident. Therefore, the Smart Grid applications and DER systems should be designed to support data and system quality needs. Utilities need systems that ensure security and resiliency, privacy to satisfy the consumers and no exploitation of personal data.

3.2.2 Common Quality Attributes

From another perspective, a system is described by a list of quality attributes in this guide [Meier 2009]. Attributes are qualities of a system. These can be assessed to determine its overall dependability using qualitative (e.g., evaluate human, environment, or ethic issues) or quantitative (e.g., measurements) measures. This guide categorizes the attributes in four specific areas as follows:

- Design - Conceptual Integrity, Maintainability, Reusability
- Runtime – Availability, Interoperability, Manageability, Performance Testability, Reliability, Scalability
- System – Supportability, Testability
- User qualities – Usability.

As it is shown, security is not perceived a quality attribute in this guide [Meier 2009]. We can argue that software developers do not follow a common agreed standard in the development of products. Definitions of various attributes are described based on this guide. However, an application or system for the Smart Grid may need to identify other relevant quality characteristics and attributes. Also, different sources are describing quality attributes.

3.3 Increasing Quality of Software with Security Features

This work [Pasik-Duncan 2006] emphasizes the complexity of control systems and greater need to compliance for safety, quality of service, and security of systems and data. SCADA security design and information security management can be improved by applying a wide range of control principles and methods as well productivity control, involving decision making under uncertainty with increased levels of decision support [Hentea 2008]. Therefore, the improvements for SCADA security have to be broad - at the systems level, and detailed - at the component level.

Many companies around the globe consider that security is an important and growing aspect of software quality for control systems. New products are developed to normalize the events in SCADA application logs that are not understood or used in security monitoring. Software products that can detect the security events and provide a Security Event Monitoring (SEM) benefit SCADA systems [Hentea 2008].

The ultimate goal is the ability to quantitatively evaluate and trade off multiple quality attributes to arrive at a better overall system. How to develop a unifying approach for reasoning about multiple software quality attributes is discussed in [Barbacci 1995].

Quality attributes are important for the overall usability, performance, reliability, and security of a system. The quality of the system is measured by the extent to which it possesses a desired combination of desired quality attributes. When designing a system to meet any of the quality attributes requirements, it is necessary to consider the potential impact on other requirements. It is needed to analyze the tradeoffs between multiple quality attributes. The importance or priority of each quality attribute differs from system to system; for example, in a line of business system, performance, scalability, security, and usability are more important than interoperability, while in a packaged application or Smart Grid, interoperability is very important.

3.3.1 Security Quality Model

A number of quality models are being used by software engineers to evaluate the quality of the software product. In the software product development life cycle, the quality may refer to process quality, internal quality, external quality, and quality in use as described in [ISO/IEC 9126] (replaced by [ISO/IEC 25010]). Various quality models have been developed to evaluate internal and external quality of software product. All models are based on the defined hierarchic link of quality characteristics and sub characteristics. The quality in use model of a product is also based on certain characteristics for the evaluation of the software quality. The development of quality systems and certifying of organizations in compliance with the [ISO 9001] quality standards is one of the widely applied approaches in the world practice for assurance of process quality. Therefore, quality systems shall be developed according to the defined guidelines. However, no quality models are established for the evaluation of the process quality [Sproge 2012].

Quality models of varying degrees of completeness and usability are defined by different entities such as international standards, de facto industry standards, organization specific including software engineering books [Firesmith 2003]. Figure 3.1 shows a quality model as a hierarchical model consisting of quality factors (also known as quality attributes) containing quality subfactors. The model formalizes the concept of quality, and it decomposes quality into a taxonomy of relevant quality factors and subfactors. For each of these, it defines associated quality criteria and metrics that provide specific measurable descriptions of the quality that is being analyzed or specified.

As shown in Figure 3.1, the quality factors and subfactors are the primary ways in which the concept of quality is decomposed. However, quality is made specific with regard to systems when it is judged by application specific quality criteria and measured in terms of specific quality metrics.

In 2002, these groups (Utility Communication Architecture 2.0 and the International Electrotechnical Commission (IEC) 61850) has begun the work of establishing an infrastructure allowing power substations, program logic controllers, remote terminal units, intelligent electronic devices, and other devices to effectively and efficiently communicate over a utility intranet that is based on Internet standards using commercial of the shelf (COTS) components. Since the number of COTS based systems being built continues to increase, a model that ensures quality characteristics of such systems becomes a necessity for each organization.

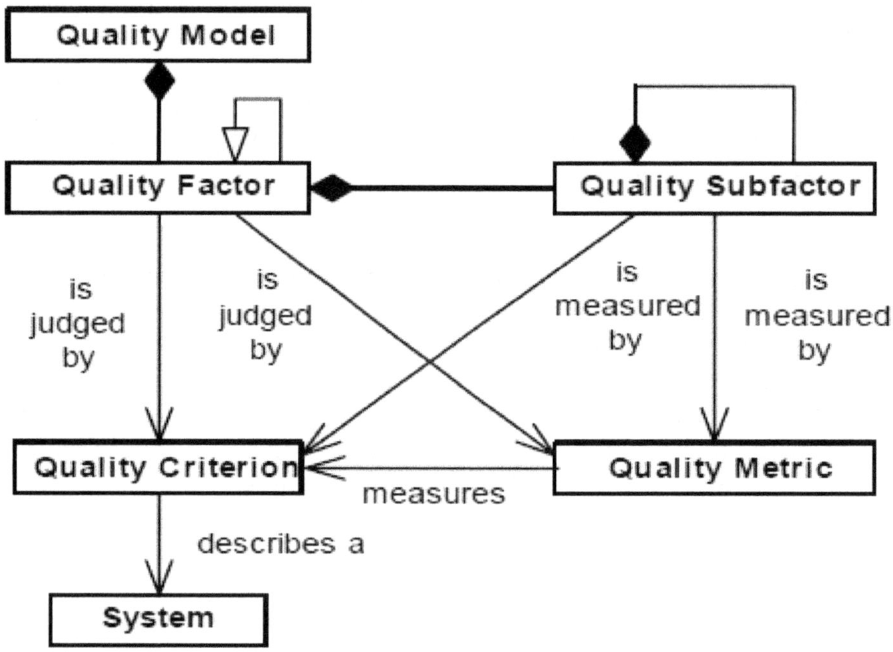

Figure 3.1. Information metamodel for quality models. *Source:* [Firesmith 2003]. © 2003, CMU.

However, selecting which one of the quality models to evaluate security attributes is a real challenge because each model is based on quality characteristics (or factors) that reflect different perspectives of quality and only a limited view of security attributes. Deficiencies and improper definitions are discussed in [Botella 2004], [Adnan 2006].

Software quality models (e,g., McCall's, Boehm, ISO 9126, FURPS, Dromey, ISO/IEC TR 15504-2 1998, Triangle and Quality Cube) are evaluated for the purpose of defining a new model specializing in evaluating and selecting COTS components [Adnan 2006]. Due to limitations found in these models, such as the tendency to ignore a certain quality feature like functionality or the failure to describe how the quality measurement is used in these models, a new model has been carried out in this work. The quality model for the COTS acquisition is based on [ISO 9126] quality model (see Figure 3.2).

Although issues are identified, this model supports a standard set of quality characteristics suitable for evaluating COTS components, along with newly defined sets of subcharacteristics associated with them. Also, stakeholders may have different interests such as:

Enduser → Functionality
Analysts → Reliability
Quality Assurance → Usability
Business Owner → Maintainability
Project Manager → Manageability.

Another analysis and comparison of different quality models concludes that [ISO 9126-1] quality model is the most useful for security quality because includes security function [Al-Qutaish 2010]. Also, the standard has been built based on an international consensus and agreement from all the country members of the ISO organization. However, the standard has been revised and replaced by [ISO/IEC 25010] in 2011 and its use is guided by this [ISO/IEC 25010] series of standards. This standard, compared to [ISO 9126], added security and compatibility as main characteristics.

A security quality model that provides a framework to identify known security defects, their fixes, the underlying low level software components along with the properties that positively influence the overall security of the product is proposed in [Zafar 2013]. This work provides explicit guidelines on how to build quality into a product. Furthermore, to incorporate security, the authors introduce several new model components and model construction guidelines. This new model is based on Dromey's quality model that addresses the core issue of quality as described in [Dromey 1995]. This model is linked to the [ISO/IEC 9126] standard.

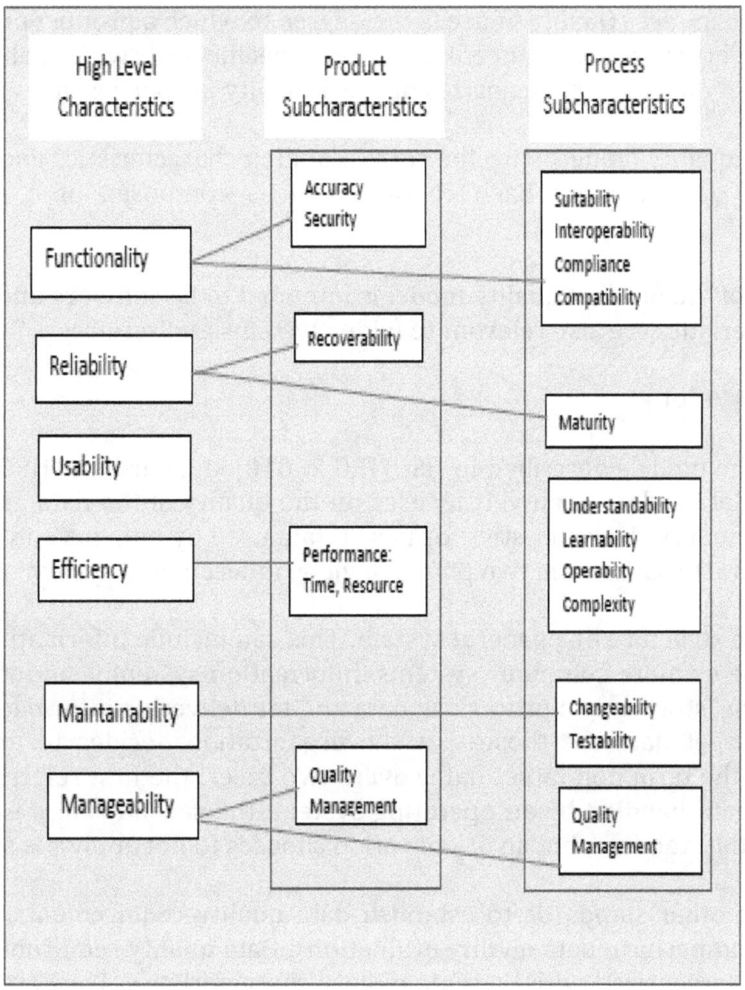

Figure 3.2 A view of quality model for COTS based systems. Adapted from [Adnan 2006]. Licensed under CC BY 4.0.

Therefore, the software industry and researchers should base their work on this new standard [ISO/IEC 25010]. Compared to [ISO/IEC 9126], from the perspective of security it is strongly recommended to use [ISO/IEC 25010] because it provides better definitions and guidance for the modern systems such as:

- The scope of the quality models has been extended to include computer systems, and quality in use from a system perspective.
- Security has been redefined and added as a characteristic, rather than a subcharacteristic of functionality.
- Characteristics and sub characteristics have been added or removed (e.g., compliance has been removed, as compliance with laws and regulations is part of overall system requirements, rather than specifically part of quality).
- When appropriate, generic definitions have been adopted, rather than using software specific definitions.
- Several characteristics and sub-characteristics have been given more accurate names and revised definitions.

In summary, the [ISO/IEC 25010] standard describes two models:

- Quality in use model - Quality in use is the degree to which a product or system can be used by specific users to meet their needs to achieve specific goals with quality characteristics
- Product quality model – it categorizes product quality properties into eight characteristics.

Table 3.1 shows the quality models with the corresponding characteristics and sub-characteristics as provided in [ISO/IEC 25010]. Each characteristic is composed of a set of related sub-characteristics.

Although the scope of the product quality model is intended to be software and computer systems, many of the characteristics are also relevant to wider systems and services.

3.3.2 Data Quality Model

Complimentary to the models described in [ISO/IEC 25010] standard, the [ISO/IEC 25012] standard describes a model for data quality. It focuses on the quality of the data as part of a computer system and defines quality characteristics for target data used by humans and systems. The standard defines data characteristics from two points of view: inherent and system.

Figure 3.3 shows the structure of a general system. This can include information systems which in turn can include one or more computer systems. Information system is an integrated set of components for collecting, storing, and processing data and for delivering information, knowledge, and digital products. Target data are those that the organization decides to analyze and validate through the model; the term non target data covers two cases: the first refers to data that are not persistent such as data handled by an operating system; the second refers to data that could be within the scope of the standard, but an organization chooses to not apply the standard to them.

It can be used with other standards to establish data quality requirements, define data quality measures, or plan and perform data quality evaluations. Data quality requirements and data quality measures can be categorized with the data quality characteristics. They are used by an evaluation process to analyze data independently from other computer system's components. It aims to support the implementation of system's life cycle processes, such as those defined in [ISO/IEC/IEEE 15288]. The standard takes into account all data types (e.g. character strings, texts, dates, numbers, images, sounds, etc.), assigned data values and relationships between data (e.g. consistency between data in the same or in different entities); the scope does not include data produced by embedded devices or real time sensors that are not retained for further processing or historical purposes.

Table 3.1 Quality models and characteristics.

Model	Characteristic	Sub-characteristics
Quality in Use Model	Effectiveness	
	Efficiency	
	Freedom from risk	Economic risk mitigation Health and safety risk mitigation Environmental risk mitigation
	Satisfaction	Usefulness Trust Pleasure Comfort
	Context coverage	Context completeness Flexibility
Product Quality Model	Functional suitability	Functional completeness Functional correctness Functional appropriateness
	Performance efficiency	Time behavior Resource utilization Capacity
	Compatibility	Coexistence Interoperability
	Usability	Appropriateness Recognizability Learnability Operability User error protection User interface aesthetics Accessibility
	Reliability	Maturity Availability Fault Tolerance Recoverability
	Security	Confidentiality Availability Integrity Non repudiation Accountability Authenticity
	Maintainability	Modularity Reusability Analysability Modifiability Testability
	Portability	Adaptability Installability Replaceability

Source: Based on [ISO/IEC25010].

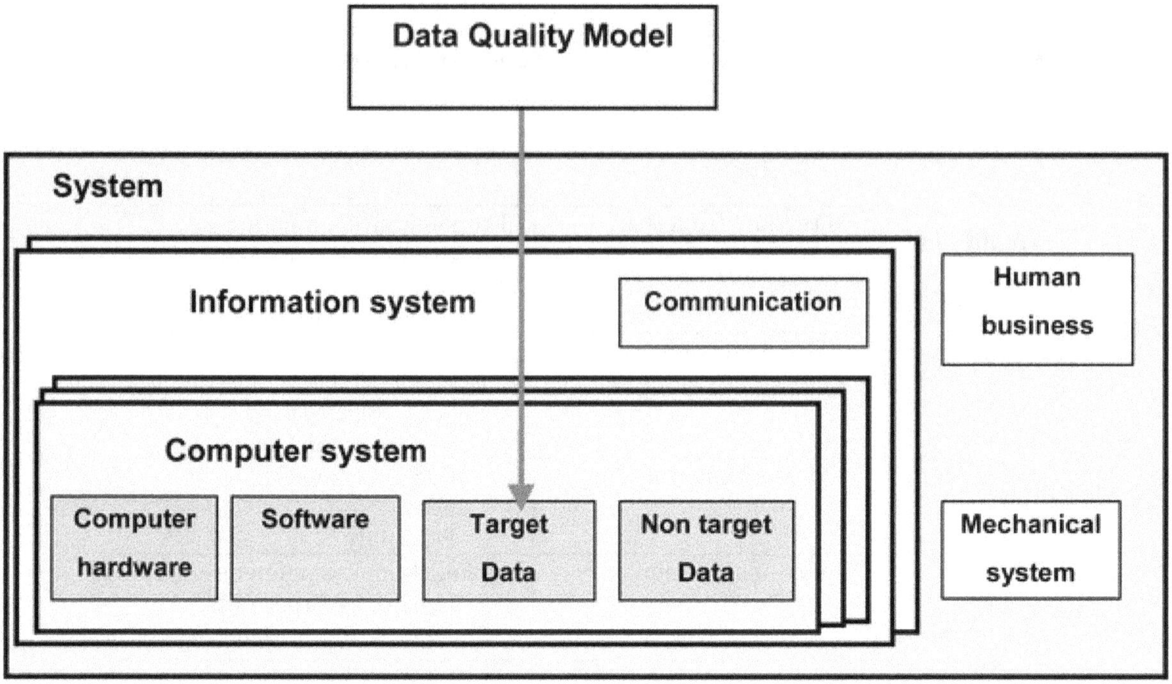

Figure 3.3 Target domain of the data quality model. Based on [ISO/IEC 25012].

Also, this standard [ISO/IEC/IEEE 24765] provides a common vocabulary applicable to all systems and software engineering work. It was prepared to collect and standardize terminology. This is intended to serve as a useful reference for those in the information technology field, and to encourage the use of systems and software engineering standards prepared by ISO and liaison organizations IEEE Computer Society and Project Management Institute. It includes references to the active source standards for each definition so that the use of the term can be further explored.

3.3.3 Quality of Security Service

A security service can provide security functionality such as confidentiality, integrity, availability, non-repudiation, privacy, authentication, etc. Quality of Security Service (QoSS) refers to the ability to provide security services according to user and system preferences and policies. An information system is a system that provides information to a user at request.

The premise of QoSS is that system and network management functions can be more effective if variable levels of security services and requirements can be presented to users or network tasks. In this approach, the level of service must be within an acceptable range, and it can indicate degrees of security with respect to various aspects of assurance, robustness, administrative diligence, etc. [Levin 2006]. This work underlines the QoSS vision is to transform security from a performance obstacle into an adaptive, constructive network management tool.

One issue that has to be considered during analysis is the planning and management of QoSS and QoS services. Both services are vital and affect the entire network infrastructure or system. While both services are necessary for safe and adequate network operations, in many organizations separate groups are responsible for security and quality of service (QoS). However, security and QoS implementations can have an impact on each other. Without information about QoS requirements, a poor choice of a security mechanism (e.g. encryption endpoints) may reduce the effectiveness of

QoS performance queuing. Without information on security requirements, a poor assignment of QoS performance levels may lead to denial of service for vital but low bandwidth data.

Therefore, QoS and security requirements must be considered together and a team has to be established with expertise on these areas. The team needs to establish a network policy framework and identify conflicts in security and QoS requirements. Security and QoS requirements can be entered in the policy framework through a single organization policy, or the security policy and QoS policy can be entered separately. In either case, if the policy framework has sufficient information about the network system, security requirements, and QoS requirements, the framework can resolve or at least identify conflicting requirements.

Enforcing both security requirements and QoS requirements can be viewed as resource allocation problems. When the policy framework is the single point that is solving both resource allocation problems, conflicts can be found or allocations can be altered to deal with the global set of requirements. When security or QoS requirements are considered separately some resource allocation decisions can be arbitrary. For example, when considering encryption requirements, two routers in the network may satisfy the security requirements equally well, but when QoS requirements are also considered, the choice may not be so arbitrary. In addition, critical process parameters and critical quality attributes have to be identified. More precisely, these have a direct and significant impact on actual or perceived quality. Therefore, a current policy framework system can adequately deal with static resolution of requirements for QoSS and QoS. The policy framework system should evolve to deal with interactions between administrative domains, more dynamic system requirements, and new system and security services.

3.4 Dependability and Security

Nowadays, people, organizations, and government are affected by information systems and have become dependent on their continued proper functioning. For example, the increased use of information systems has brought fundamental changes in internal organizational procedures and has altered the way that organizations interact. In the event of an information system failure, it may not be possible to continue business interactions as planned. Consider, for example, the effect of information system failure on the functioning and efficiency of power grid. Dependence on information systems is growing. In the same time, there is a mounting need for confidence that they will continue to be available and to operate in the expected manner. Thus, utilities recognize dependence on information systems and information.

3.4.1 Dependability Definitions

Dependability is ability to perform as and when required [IEC 62508]. This is a term used for time related quality characteristics of a product or service, and it can also be expressed as a grade, degree, confidence or probability of fulfilling a defined set of characteristics. Dependability characteristics include availability and its inherent or external influencing factors, such as reliability, fault tolerance, recoverability, integrity, security, maintainability, durability and maintenance support [IEC TC56]. The terms are generic and are applicable to all fields of dependability methodology. The International Electrotechnical Commission (IEC) develops and maintains international standards that provide systematic methods and tools for dependability assessment and management of equipment, services, and systems throughout their life cycles.

Another vocabulary [Electropedia] includes these characteristics: availability, reliability, recoverability, maintainability, and maintenance support performance, and, in some cases, other characteristics such as durability, safety, and security. The [Electropedia] is also known as International

Electrotechnical Vocabulary (IEV). The IEV (IEC 60050 series) is a general purpose multilingual vocabulary covering the field of electrotechnology, electronics and telecommunication (available at www.electropedia.org). The IEC publications encourage the use of a common dependability terminology through [IEC 60050-192].

Another IEC TC (TC57) defines dependability as probability that a system will perform correctly when required to in the presence of faults. A supporting note indicates the definition is aligned with [IEC 60834-1]. The IEC TC57 committee develops and maintains International Standards for power systems control equipment and systems including EMS (Energy Management Systems), SCADA, distribution automation, teleprotection, and associated information exchange for real time and non-real time information, used in the planning, operation and maintenance of power systems. However, the application of this definition refers to all information exchanges within power systems described above.

This [IEC 60834-1] standard aims at establishing performance requirements to teleprotection command systems used to convey command information, generally in conjunction with protection equipment. In this standard, dependability relates to the ability to issue and receive a valid command in the presence of interference and/or noise. For practical reasons the probability of a missing command (Pmc) is normally measured, dependability (D) is then given by equation (1).

$$D = 1 - Pmc \tag{1}$$

While this dependability characteristic is based on a measurement, it is valid for a specific type of command (tele protection equipment). Arguably, it is not plausible to design all systems that include different types of message exchanges (real time and non real time) as described in this standard. Therefore, complementary and improvement solutions are needed. In addition, the Smart Grid and DER systems have to be analyzed in detail to identify specific requirements for the communication technologies and networks, services, systems, and software. Also, definitions and concepts have to be identified to include:

- Network dependability is defined as ability to perform as and when required to meet specified communication and operational requirements [IEC 62673].
- Software dependability is ability of the software item to perform as and when required when integrated in system operation [IEC 62628].
- System dependability is the ability of a system to perform as and when required to meet specific objectives under given conditions of use. The dependability of a system infers that the system is trustworthy and capable of performing the desired service upon demand to satisfy user needs [AS IEC 62628].

Computing systems are characterized by five fundamental properties: functionality, usability, performance, cost, and dependability. Dependability of a computing system is the ability to deliver service that can justifiably be trusted.

Traditionally, dependability for a system incorporates availability, reliability, maintainability but since the 1980s, safety and security have been added to measures of dependability.

Maintainability is distinct from other dimensions of dependability because it is a static and not a dynamic system attribute. It is a system attribute that is concerned with the ease of repairing the system after a failure has been discovered or changing the system to include new features. This distinction is very important for critical systems as faults are often introduced into a system because of maintenance problems.

For critical systems, it is usually the case that the most important system property is the dependability of the system. The dependability of a system reflects the user's degree of trust in that system. It reflects the extent of the user's confidence that it will operate as users expect and that it will not fail in normal use.

In software engineering, dependability is the ability to provide services that can defensibly be trusted within a time period [Kumar 2015]; the ability to avoid service failures that are more frequent and more severe than acceptable. This may also encompass mechanisms and means designed to maintain the dependability of a system or software [Avizienis 2004]. Dependability is a collective set of time related performance characteristics that coexist with other requirements such as efficiency, quality, safety, and integrity. Dependability applies to any asset such as a system, product, process or service and may involve hardware, software and human aspects.

Dependability means are intended to reduce the number of failures presented to the user of a system. Failures are traditionally recorded over time and it is useful to understand how their frequency is measured so that the effectiveness of means can be assessed. Dependability can be broken down into three elements:

- Attributes provide a way to assess the dependability of a system
- Threats analysis enables an understanding of the things that can affect the dependability of a system
- Means are ways to increase a system's dependability.

However, standardization efforts provided by different communities and specific specialization (e.g., telecommunications, avionics, rail transportation, nuclear plant control, etc.) did not produce a dependability framework that is applicable to computer systems. The task of Security Engineer is to identify gaps and inconsistencies because definitions provided by standards are not complete or updated. For example, an ISO standard definition is based on the old definition for telephony and centered on availability. Meanwhile the computer and communications approaches for dependability were not considered and recognized by security communities.

Given the societies dependence on computer systems, the need for a dependability framework was more than justified when it was proposed in [Laprie 1995].

Based on the intent application of a system, dependability is viewed according to different, but complementary attributes that include [Laprie 1995]:

- Availability - readiness for use
- Reliability - continuity of service
- Maintainability - ability to undergo repairs and evolutions
- Safety - non-occurrence of catastrophic consequences to people
- Confidentiality - unauthorized access and disclosure of information
- Integrity - non-occurrence of improper alterations to information.

Associating integrity and availability with respect to authorized actions, together with confidentiality, leads to security [Laprie 1995]. Dependability and security overlap in the sense that some types of impairments fall under both properties. The undesired events to dependability are called impairments and include [Laprie 1995]:

- A system failure occurs when the delivered service deviates from fulfilling the system function.

- An error affecting the service is an indication that a failure occurs or has occurred; it can lead to subsequent failure.
- A fault is the adjudged or hypothesized cause of an error.

In order to avoid the occurrence of these impairments, a combination of methods (means) can be utilized in the development of a dependable computing system. The means provide the ability to deliver a service on which reliance can be placed and to reach confidence in this ability. These means are classified as follows [Laprie 1995]:

- Fault prevention means to prevent the occurrence or introduction of faults
- Fault tolerance means to avoid service failures in the presence of faults
- Fault removal means to reduce the number and severity of faults
- Fault forecasting means to estimate the present number, the future incidence, and the likely consequences of faults.

Dependability means are intended to reduce the number of failures presented to the user of a system. Failures are traditionally recorded over time and it is useful to understand how their frequency is measured so that the effectiveness of means can be assessed. Figure 3.4 is a graphical representation of fault activation and chain of propagation.

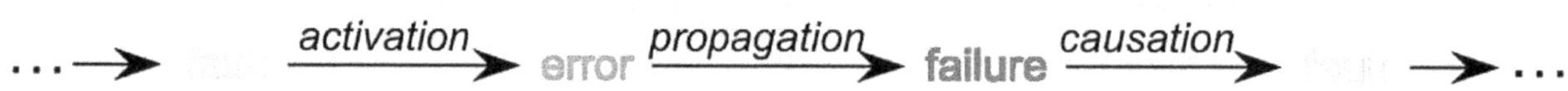

Figure 3.4 Fault-Error-Failure chain. Based on [Avizienis 2001].

The relationships between these concepts (attributes, means, impairments) are structured as dependability tree depicted in Figure 3.5. Fault prevention and fault tolerance aim to provide the ability to deliver a service that can be trusted, while fault removal and fault forecasting aim to reach confidence in that ability by justifying that the functional and the dependability and security specifications are adequate and that the system is likely to meet them.

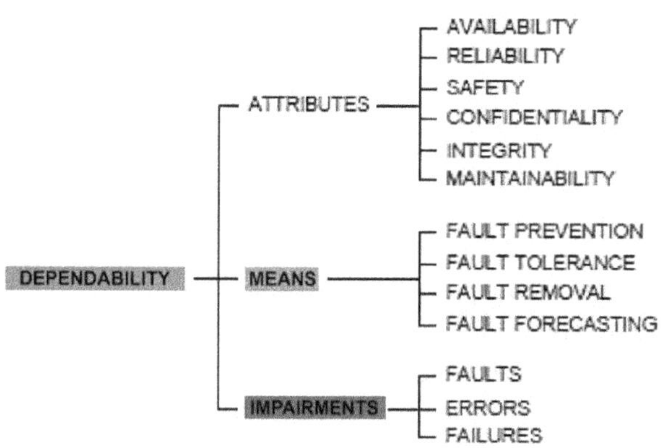

Figure 3.5 The dependability tree. Based on [Avizienis 2001].

3.4.2 Dependability and Security

Searching for unified measures of dependability and security is a continuous effort in research community. The concerns of dependability are the parameters by which the dependability of a system are judged. A dependability centric view of the world subsumes the usual attributes of reliability, availability, safety, and security (confidentiality and integrity). Each of the attributes examined has evolved within its own community. This has resulted in inconsistencies between the various points of view. Depending on the particular application of interest, different attributes are emphasized. In other words, in differing circumstances, the focus will be on differing properties of services, either, the average real time response achieved, the likelihood of producing the required results, the ability to avoid failures that could be catastrophic to the system's environment, or the degree to which deliberate intrusions can be prevented.

Computer security, communications security, information security, and information assurance are terms that have had a long development and use in the community of security researchers and practitioners, mostly without direct reference to dependability (see recommendations about information assurance implementation as promoted by the U.S. Department of Defense almost two decades ago in [DoD]). Security has not been characterized as a single attribute of dependability. However, usual definitions of security, that view it as a composite notion. Security attributes can be understood in terms of the three primary security attributes of confidentiality, integrity, availability.

Subsequent published works revised the definition and refined the relationship between attributes and dependability and suggested methods for the development of dependable computing systems [Avizienis 2001]. An alternate definition of dependability is the ability of a system to avoid service failures that are more frequent or more severe than is acceptable. Therefore, it is important to understand the propagation of the effects of a fault, from its activation to create an error in the system state, to propagation to a failure in a provided service, to causing a fault in a user system.

Computing and communication systems are characterized by fundamental properties: functionality, performance, dependability and security, and cost [Laprie 1995]. Security is sometimes called as an attribute but the view is to aggregate it together with dependability and treat dependability as a composite term called Dependability and Security in [Avizienis 2004]. This work summarizes the relationship between dependability and security in terms of their principal attributes (see Figure 3.6).

The dependability and security specification of a system must include the requirements for the attributes in terms of the acceptable frequency and severity of service failures for specified classes of faults and a given use environment. One or more attributes may not be required at all for a given system. Practically, applying security measures to the appliances of a system generally improves the dependability by limiting the number of externally originated errors.

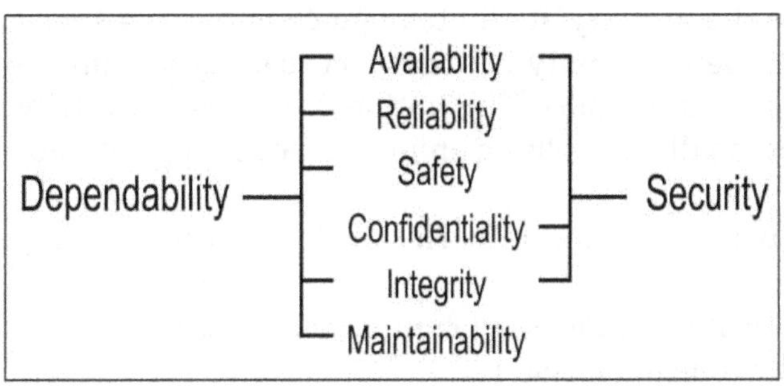

Figure 3.6 Dependability and Security Attributes. *Source:* [Avizienis 2004]. © 2004, IEEE.

A unified definition for security and the relationship between dependability and security is discussed in [Avizienis 2004]. The authors argue that the simultaneous consideration of dependability and security provides a very convenient means of subsuming these various concerns within a single conceptual framework.

3.5 Means for Dependability and Security

The fault-error-failure model is central to the understanding and mastering of the various threats that may affect a system. The model provided for the means for achieving dependability and security is extremely useful. Figure 3.7 illustrates the groupings of the means for dependability and security.

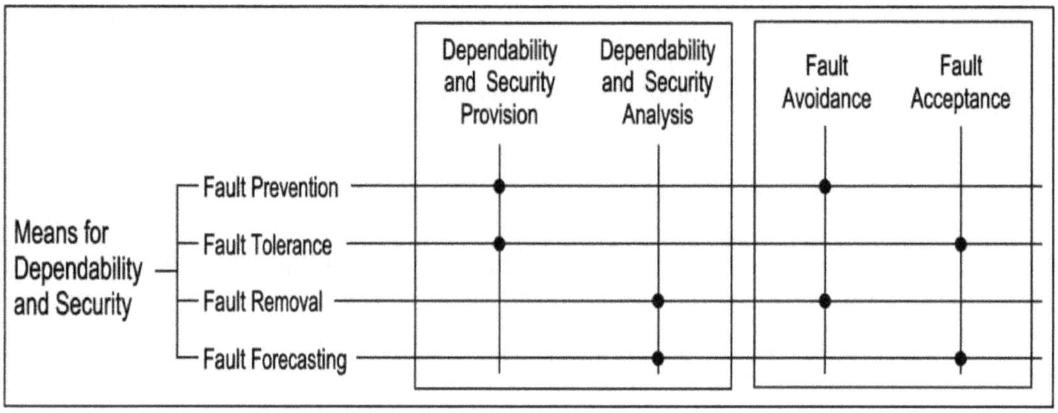

Figure 3.7 Means for dependability and security. *Source:* [Avizienis 2004]. © 2004, IEEE.

3.5.1 Taxonomy of Dependability and Security

The mix of attributes make dependability to be characterized with qualitative and quantitative measures. As these attribute definitions suggest, only availability and reliability are quantifiable by direct measurements while others are more subjective. For instance, safety cannot be measured directly via metrics but is a subjective assessment that requires judgmental information to be applied to give a level of confidence. Figure 3.8 depicts a complete taxonomy of dependability and security tree as defined in [Avizienis 2004].

Besides the attributes defined and discussed above, other, secondary, attributes can be defined, which refine or specialize the primary attributes. An example of a specialized secondary attribute is robustness, for example, dependability with respect to external faults, which characterizes a system reaction to a specific class of faults. The notion of secondary attributes is especially relevant for security and it is based on distinguishing among various types of information. Examples of such secondary attributes are [Avizienis 2004]:

- Accountability - availability and integrity of the identity of the person who performed an operation.
- Authenticity - integrity of a message content and origin, and possibly of some other information, such as the time of emission.
- Non-repudiability - availability and integrity of the identity of the sender of a message (nonrepudiation of the origin), or of the receiver (nonrepudiation of reception).

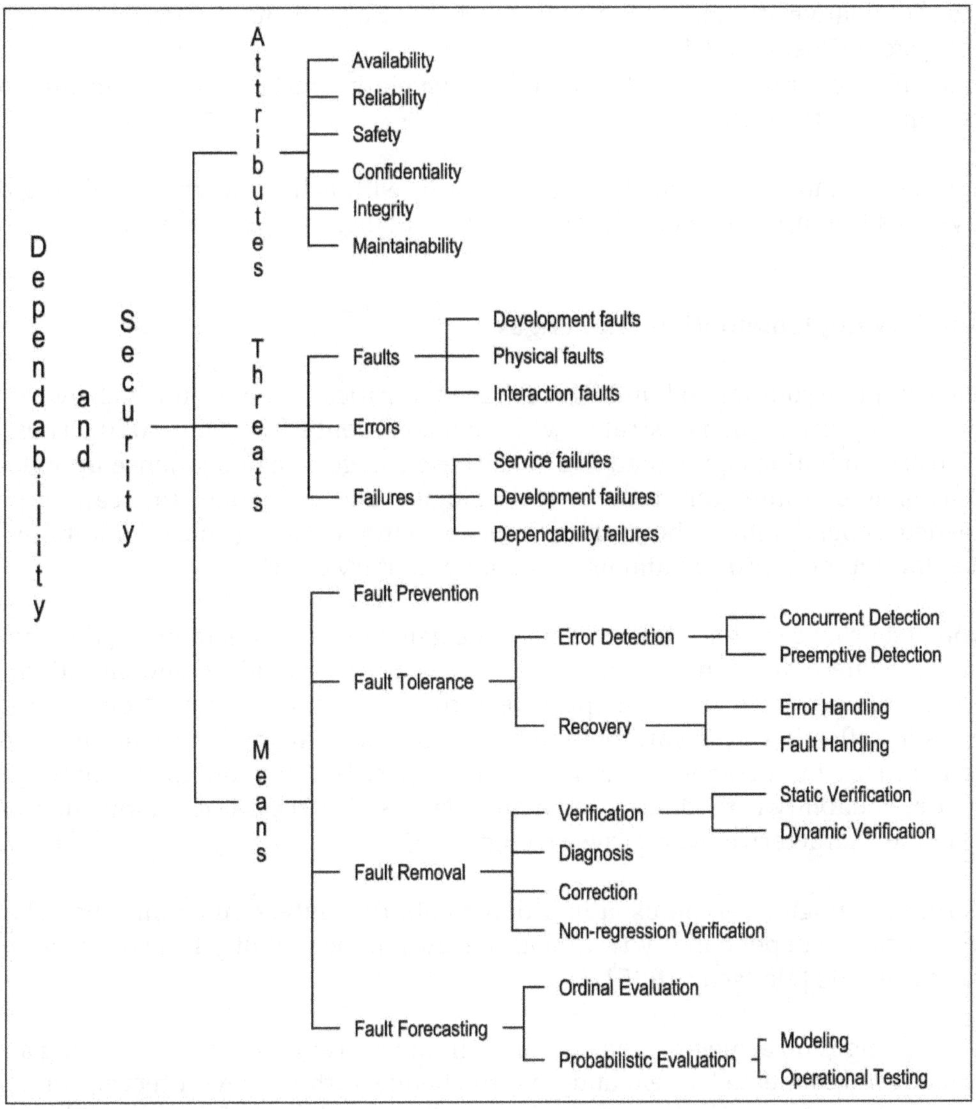

Figure 3.8 A refined dependability and security tree. *Source:* [Avizienis 2004]. © 2004, IEEE.

Examples of guidelines for development of systems with dependability features include [Avizienis 2004], [Avizienis 2001]:

- The dependability and security specification of a system must include the requirements for the attributes in terms of the acceptable frequency and severity of service failures for specified classes of faults and a given use environment.
- One or more attributes may not be required at all for a given system.
- The attributes of dependability and security should be considered in a relative, probabilistic, sense, and not in an absolute, deterministic sense to the extent to which a system possesses certain attributes.
- Depending on the applications intended for the system, different emphasis may be put on different attributes; the attributes of dependability may be emphasized to a greater or lesser extent depending on the application: availability is always required, although to a varying degree, whereas reliability, safety, confidentiality may or may not be required.
- A critical first step in understanding the nature of information system dependability is to identify the major classes of success critical information system stakeholders, and to char-

acterize the relative strengths of their dependencies on various attributes of a given information system [Boehm 2004].
- There should be strategies for dependability development to achieve it in the most cost-effective manner [Madachy 2005].

The practices of dependable design deal reasonably well with achieving and predicting dependability in ICT systems that are relatively closed and unchanging, but there are still issues that have to be addressed.

3.5.2 Dependability Implementation Challenges

There is still no complete solution to building software for critical systems in which every aspect of software dependability can be demonstrated with high confidence [Graydon 2010]. These authors argue that there are limitations embodied in any of these categories: dependence upon fallible human beings; incomplete or immature tools or technologies; and techniques that cannot practically ensure the needed dependability. The authors believe that addressing these limitations requires structuring the development process and use of complementary efforts.

There is no universal rule to define stakeholder/value gained from implementing dependability in a system because value dependencies vary significantly by stakeholders and situations [Boehm 2004]. Therefore, the authors argue developers need to understand the definitions of the dependability attributes in reflecting their variability not only due to changes in operational profiles but also due to differences in stakeholder value dependencies. Understanding the nature of information system dependability is to identify the major classes of success critical information system stakeholders, and to characterize the relative strength.

Although different paradigms can be used in a dependability context, modeling the relationships between ways to achieve dependability is difficult because dependability has so many interpretations in different contexts [Madachy 2005].

Also, there are challenges on achieving dependability in the current tendency of making all kinds of ICT systems more interconnected, open, and able to change without new intervention by designers. The existing techniques are inadequate to deliver the same levels of dependability to ICT systems [Strigini 2009]. The author argues issues such as:
- New components introduce system design faults or vulnerabilities by feature interaction or by triggering preexisting bugs in existing components.
- New patterns of use and new interconnections open the system to attack by new potential adversaries.
- Social and human factors are responsible for most dependability problems in ICT systems; for instance, influencing through design methods and constraints, or through the maintenance or use of technical systems; attackers can circumvent technical security mechanisms in ICT via social engineering.

Issues that need to be considered in the analysis and development of a dependable system include:
- Provision of measures that can be used to assess the qualities of a system; these can be assessed to determine its overall dependability using Qualitative and Quantitative measures
- Selection of the techniques to be employed on supporting each sub-characteristic should be based on the variations in the emphasis placed on the sub-characteristics that may be of varying importance depending on the application intended for the given system [Graydon 2010].

- Performing analysis via risk assessment to identify the sub-characteristics priorities help to balance techniques that need to be employed; conflicting issues (e.g., availability and safety, availability and confidentiality) may need to be resolved via tradeoffs.
- Use of an approach to the construction of critical computing systems in which the system and an argument that it meets its assurance goals are developed simultaneously [Graydon 2007].

3.5.2.1 Costs

Dependability costs tend to increase exponentially as increasing levels of dependability are required. The costs are high because of reasons such as [Sommerville 2016]:

- Extra costs for system design, implementation, and validation.
- Verification and validation costs are particularly high for systems that must be ultra-dependable such as safety-critical control system.
- Use of more expensive development techniques and hardware that are required to achieve higher levels of dependability.
- Increased testing and system validation that is required to convince the system client that the required levels of dependability have been achieved.

Figure 3.9 shows that dependability is increasing costs as dependability is increasing to a higher degree.

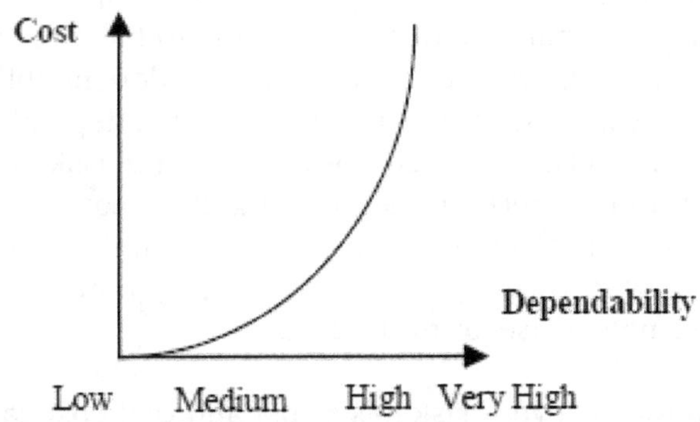

Figure 3.9 Cost vs. Dependability. *Source:* [Tsiakis 2009]. © 2009, IEEE.

System dependability is important because failure of critical computer systems can lead to large economic losses, serious information loss, physical damage or threats to human life. The dependability of a computer system is a system property that reflects the user's degree of trust in the system. According to [Sommervillle 2016], the most important dimensions of dependability are availability, reliability, safety, security, and resilience. As shown in the diagram of Figure 3.9, the cost versus dependability curve is exponential, so it is impossible to achieve 100% dependability [Tsiakis 2009].

Because of very high costs of dependability achievement, it may be more cost effective to accept untrustworthy systems and pay for failure costs. However, this depends on social and political factors. Also, a critical system failure can have severe human or economic consequence. Critical systems failure (e.g., power grid outages) can result in significant economic losses, physical damage or threats to threats to human life. More, a reputation for products that cannot be trusted may lose future business. It also depends on system type - for business systems in particular, modest levels of dependability may be adequate. The costs of critical system failure can be so high that development methods may be used that are not cost effective for other types of system. Examples of development methods to achieve dependability include:

- Formal methods of software development
- Static analysis
- External quality assurance.

Also, increasing the dependability of a system can be achieved using dependable development processes [Sommerville 2016]. Dependable processes make use of redundancy and diversity to achieve reliability. They often include different activities that have the same aim.

However, there are also benefits and needs for dependability. Today, many systems in different application domains need to be dependable. Adequate dependability is a key requirement for many different systems, such as safety critical systems, telecommunication systems, and mission critical software systems [Tsiakis 2009]. The authors argue that dependability is a measure of system trustworthiness defined in the context of the stakeholders' economic needs. Dependability is defined as the trustworthiness of a computing system that allows reliance to be justifiably placed on the services it delivers. Dependability is a system property that includes runtime behavior aspects or attributes (availability, reliability, safety, security, performance, survivability, maintainability). Although measuring dependability using quantitative methods is of great importance, it is needed to consider that dependability is contextually subjective and reflects the particular stakeholders' needs.

Also, it is observed that the focus of Security Engineering has shifted from what is technically possible to what is economically optimal. The question is how to balance the economic consequences of lowering dependability levels and the costs to implement dependability. These authors [Tsiakis 2009] argue that dependability is relative economic measure of dependability attribute risk factors since failures result to economic losses. The authors indicate the risk process can be followed from the Risk Management literature, therefore addressing the economic quantitative models and methods that could simply assign monetary values to dependability and so to measure it. The consequences of not complying with the required dependability imply a risk to be managed: either accept, ignore, transfer to someone else, or mitigate it.

As acknowledged in the literature, every risk has a cost and that cost can be (more or less precisely) quantified. The cost of a particular risk during a period of time is the probability of an adverse event occurring during the time period multiplied by the downside consequence of the adverse event. The interrelationship between risk and cost has to be understood because cost is positively influenced by risk [Tsiakis 2009]. This work illustrates that quantitative measures of dependability are challenging because of difficulty to address general quantitative evaluation methods of the system dependability that it is contextually subjective and reflects the particular needs of the stakeholders. Also, the majority of the published works refer to the general framework of risk management (risk analysis, risk assessment, risk treatment). The interrelationship between risk and cost is shown in Figure 3.10, where risk is denoted by R and cost by C. If risk R is decreased from R1 to R2, the cost C will decrease from C1 to C2, so it can be said that C is positively influenced by R (downward movement from point A to point B shows this).

This work [Tsiakis 2009] illustrates that quantitative methods are needed in order to assess the cost of the various measures which can be taken to reduce the dependability risk. These methods aim in providing estimations for the economic consequences of lowering dependability levels and the costs to implement dependability.

Methods are needed in order to assess the cost of the various measures which can be taken to reduce the dependability risk. These methods aim in providing estimations for the economic consequences of lowering dependability levels and the costs to implement dependability. In reality, most risks are part of complicated risk scenarios. In order to model such risk scenarios in a way that potential cost and time impacts are realistic, correlations and/or dependencies need to be defined and applied appropriately.

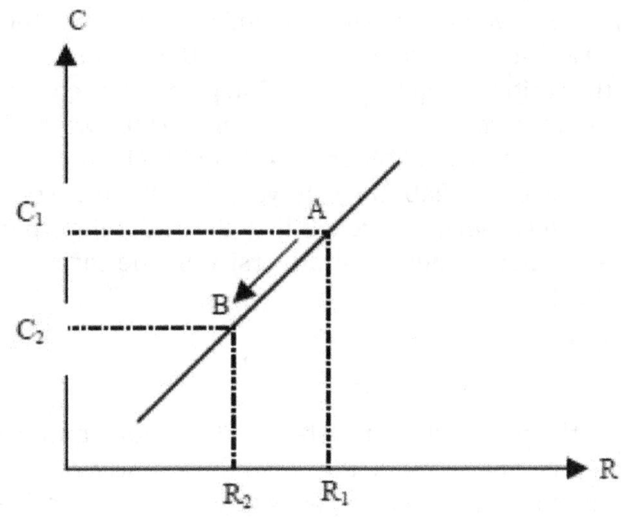

Figure 3.10 Correlating cost and risk. *Source:* [Tsiakis 2009]. © 2009, IEEE.

3.5.2.2 Other Characteristics of Dependability

Other important system attributes that affect dependability and security include usability, manageability, and adaptability that should be considered in analysis and design. This work [Hardeveld 2012] describes the main dependability characteristics of a system that include availability, reliability, maintainability, and supportability (provision of maintenance support and logistics to perform maintenance tasks). Other concepts similar to dependability include high confidence, survivability, and trustworthiness. Table 3.2 shows a comparison of these concepts that are essentially equivalent in their goals and address similar threats.

Table 3.2 Comparison of concepts.

Concept	Dependability	High Confidence	Survivability	Trustworthiness
Goal	Ability to deliver services that can be justifiably trusted; Ability of a system to avoid service failures that are more frequent or more severe than is acceptable	Consequences of the system behavior are well understood and predictable	Capability of a system to fulfill its mission in a timely manner	Assurance that a system will perform as expected

Source: Adapted from [Avizienis 2004]. © 2004, IEEE.

Trustworthiness is the attribute of an entity that provides confidence to others of the qualifications, capabilities, and reliability of that entity to perform specific tasks and fulfill assigned responsibilities [CNSSI 4009]. It is the degree to which the security behavior of a component is demonstrably compliant with its stated functionality. In the context of Systems Engineering, trustworthiness means being trusted to fulfill whatever critical requirements may be needed for a particular component, subsystem, system, network, application, mission, enterprise, or other entity [NIST SP800-160v1]. The need for trustworthy secure systems stems from a variety of stakeholder needs that are driven by mission, business, and other objectives and concerns. From a security perspective, a trustworthy system is a system that meets specific security requirements in addition to meeting other critical requirements. The principles, concepts, and practices for engineering trustworthy secure systems can be expressed in various ways, depending on which aspect of trustworthiness of concern is to stakeholders [NIST SP800-160v1]. In addition to security, other aspects of trustworthiness include reliability, safety, and resilience. Trustworthiness requirements can include attributes of safety, security, reliability, dependability, performance, resilience, and survivability under a wide range of potential adversity in the form of disruptions, hazards, and threats [NIST SP 800-160 v1].

3.6 Survivability

Survivability is defined as the property of a system, subsystem, equipment, process, or procedure that provides a defined degree of assurance that the named entity will continue to function during and after a natural or man-made disturbance [Telecom 1996]. A survivable system has the ability to continue to provide service (possibly degraded or different) in a given operating environment when various events cause major damage to the system or its operating environment. It is argued that current facets of dependability, such as reliability and availability, do not address the needs of critical information systems because they do not include the notion of degraded service as an explicit requirement.

For a given application, survivability must be qualified by specifying the range of conditions over which the entity will survive, the minimum acceptable level or post disturbance functionality, and the maximum acceptable outage duration. The notion of survivability has been used in several engineering disciplines outside of critical information systems. Survivability is defined as the capability of a system to fulfill its mission, in a timely manner, in the presence of attacks, failures, or accidents [Ellison 1999]. In this work, the term system is in the broadest possible sense, including networks and large-scale systems of systems; information survivability entails resistance, recognition and reaction to potential threats to an enterprise information infrastructure.

3.6.1 Survivable Systems

The key properties of survivable systems include resistance, recognition, recovery, and adaptation as follows:

- Resistance to attacks by providing strategies for repelling attacks; these strategies require user authentication and stochastic diversity of programs.
- Recognition of attacks and the extent of damage strategies for detecting attacks (including intrusions) and understanding the current state of the system, including evaluating the extent of damage; these strategies require recognition of intrusion usage patterns and internal integrity checking.
- Recovery of full and essential services after attack strategies for restoring compromised information or functionality, limiting the extent of damage, maintaining or, if necessary, restoring essential services within the time constraints of the mission, restoring full service as conditions permit; these actions require replication and reinitialization of data.

- Adaptation and evolution to reduce effectiveness of future attacks strategies for improving system survivability based on knowledge gained from intrusions; it requires incorporation of new patterns for intrusion.

Most computer systems are based on a narrow view of defense against computer intrusions and focus almost exclusively on hardening a system to prevent a break in or other malicious attack. This view does little about what to do once an intrusion has occurred or is under way. The shift in perspective of security to survivability is discussed during the research workshop in 2006 [CSIIRW06]. Figure 3.11 depicts a diagram that shows that survivability is a tridimensional capability that requires security, functionality, and performance.

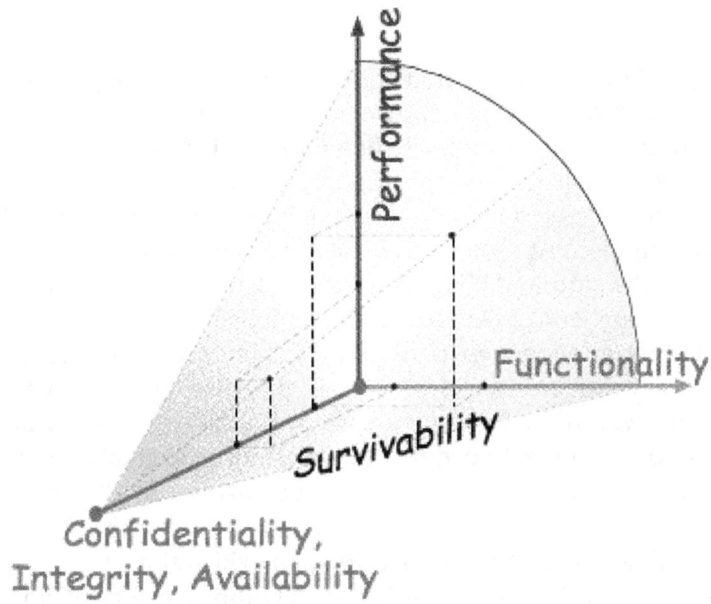

Figure 3.11 Survivability dimensions. *Source:* [CSIIRW06]. Public Domain.

The survivability problem goes beyond security and needs to include other views such as functionality and performance. Thus, the traditional view of information systems security must be expanded to encompass the specification and design of system behavior that helps the system survive in spite of active intrusions. Only then can systems be created that are robust in the presence of attack and are able to survive attacks that cannot be completely repelled.

The [CSIIRW06] workshop call, Cyber Security: Beyond the Maginot Line, is a shift from technologic IT based solutions to enterprise wide, risk management perspective that requires involvement of the whole organization and focuses on the survival of the mission rather than a particular component. Increasing dependency on complex and large scale network systems demands changes in practices and approaches. Also, an overview of power systems needs could help on identifying the adequate solutions for the Smart Grid and DER systems.

3.6.2 Power Systems and Survivability Issues

The electric utility industry faces unique information survivability challenges in the face of deregulation and increased reliance on computer networking [Hale 1998]. In addition, the fragmented nature of the industry into cooperative but independent units makes resistance, recognition and reaction to threats more difficult. This makes information survivability a challenging cooperative venture. For federated mission-critical enterprises such as electric utilities, this can only be realized with cooperative efforts, that raise awareness at all levels, promote and enforce coordinated information survival policies, and embrace new technologies that help developers deploy high assurance SCADA software. The electric industry needs to embrace trusted software development methodologies, design self-healing infrastructure systems based on a particular scheme that could answer to some possible contingencies, effective and sustainable security for the entire system life cycle, support hazard analysis and safety requirements, design systems that are predictably dependable in the face of all sorts of malice, from bombers to botnets.

Ensuring the continued delivery of electric power to the nation, electric utilities regard information survivability as critical [Moore 1996]. This need requires survivable network system and systems to collect and monitor the status of equipment and process the information about loads and demand.

In this work [Bakken 2000], the authors argue that the survivability of the grid must not be implemented and assessed with these two networks (power and communication) in isolation, as is the current practice. Rather, a combined analysis must be conducted. Also, a managed service of the power grid's communications service is needed to support status information to satisfy much more pervasive needs and optimizations being envisioned requiring more quantity, timeliness, and topological diversity of status information. At a given local level much status information must be collected. A substation must track the status of all its generators, breakers, etc. At the other geographic extreme, some status information from a remote grid can be useful to other grids. For example, if a grid knows other grids' current impedances it can optimize its operations. Also, if it can receive failure notifications from low level or mid-level devices in a remote grid, a given grid can proactively protect itself rather than waiting to become aware of the failures by cascading failures at a higher level and often too late to do much. In order to build a more survivable electric power infrastructure, the authors recommend essential research be performed in the following synergistic areas [Bakken 2000]:

- Quantifying the effect of degraded communications on grid stability.
- Developing new ways to exploit new technologies in computer networking and distributed systems to provide better control of the grid, both in normal operations and with degraded communications due to failures or attacks.
- Devising new ways to disseminate status information to a much wider set of parties and in a more survivable manner.
- Devising new ways to organize, classify, query, and use existing status information.

Therefore, survivability in power systems may include these attributes:

- Availability is readiness for correct service
- Reliability is continuity of correct service
- Safety is absence of catastrophic consequences
- Integrity is absence of improper system alterations
- Maintainability is ability to undergo modifications and repairs.

3.6.3 Survivability Approaches

As an emerging discipline, survivability builds on related fields of study, including security, fault tolerance, safety, reliability, reuse, performance, verification, and testing. Research community has been debating on definitions, approaches, and what is needed more than proposing efficient and effective automated methods including consistent definitions.

An integrated engineering framework to survivability practice is discussed in [Ellison 1999]. It includes strategies for achieving survivability, and techniques and processes for analyzing survivability of the information. In order to deliver essential services a system needs to have the capability to maintain essential properties (e.g., specified levels of integrity, confidentiality, performance, and other quality attributes) in the presence of attack, failure, or accident.

This report [Ellison 1999] describes the survivability approach to helping assure that a system that must operate in an unbounded network is robust in the presence of attack and will survive attacks that result in successful intrusions. Unbounded networks, such as the Internet, have no central administrative control and no unified security policy. Survivability focuses on preserving essential services in unbounded environments, even when systems are penetrated and compromised. Essential services are defined as the functions of the system that must be maintained when the environment is hostile or failures or accidents are detected that threaten the system.

Despite the best efforts of security practitioners, no amount of hardening can assure that a system connected to an unbounded network will not be vulnerable to attack. The flexibility of current frameworks encourages system architects to enable reconfiguration mechanisms that refocus the available, safe resources to support the most critical services rather than over provisioning to build failure proof system.

Thus, it is important to define minimum levels of quality attributes that must be associated with essential services. These quality attributes are so important that definitions of survivability are often expressed in terms of maintaining a balance among multiple quality attributes such as performance, security, reliability, availability, fault tolerance, modifiability, and affordability [Ellison 1999].

A new model for integrating a business perspective, risk management practices, and embracing disruptive innovation and technology shifts is proposed in [Allen 2002]. This work shows that survivability problem goes beyond security – security is only part of the problem; to ensure the survivability of critical information and systems, there is a need to address shifts in perspective that are essential:

- Central to Global Systems that are centrally networked under organizational control with full visibility are currently shifting to systems that are globally networked with no well-defined boundaries, with limited (if any) visibility and no centralized management or control.
- Bounded to Unbounded Systems that have well-defined geographic, political, cultural, and legal or jurisdictional boundaries are shifting to systems characterized by the absence of these boundaries. Centralized administrative control with trustworthy, known, inside users evolves to systems with distributed administrative control without central authority and unknown users.
- Insular to Networked Viewing systems as insular and fortress like, to viewing systems as being networked and interdependent; the ability to distinguish between insiders and outsiders decreases. Outsider roles go from being well defined to the realization that an outsider can be a customer, collaborator, partner, contractor, or vendor; outsider access to the network changes based on that role.
- Predictable to Asynchronous describes the shift wherein processing events that happen in predictable, prescribed sequences and patterns with predictable loads, to one where events often occur asynchronously, independent of time sequence with unpredictable loads.

- Single Responsibility to Shared Responsibility Progress from single responsibility to shared organizational responsibility to distributed responsibility. This is a shift from having a single point of known responsibility to correct failures, to having shared sometimes unknown responsibility.
- Overhead to Essential from viewing security as an overhead activity and expense, to viewing survivability as an investment that is essential to the organization, along with ensuring that there is always a contingency plan. Instead of security being IT's responsibility, with IT and the CIO constantly having to justify their budget for security, survivability is regularly reviewed and discussed in senior level management meetings and it is accepted by all as part of being in business.
- Security to Survivability from technological IT based solutions to enterprise wide based on risk management solutions. Instead of viewing security as a narrow, technical specialty accessible only to experts and focusing on the protection of specific components, survivability is embraced as a risk management perspective that requires involvement of the whole organization and focuses on the survival of the mission rather than a particular component.

An analysis of published material on the topic of survivability reveals many issues on the quality of the publications. While the abundance of publications on the topic of survivability is acknowledged in [Westmark 2004], this work identifies gaps and inconsistencies in research and practice, lack of a computational model for survivability, different interpretations of concepts published in standards, and no agreed definitions in the survivability community. It is recommended to continue research in the areas most needed to develop taxonomy of survivability.

In practice, a designer has to identify forms of degraded service that are acceptable to users, under what circumstances each form is most useful, and the fraction of time such degraded service levels are acceptable [Knight 2003]. The recommended approach for survivability is organization of system requirements into essential services and non-essential services, perhaps in terms of user categories or business criticality. Essential services must be maintained even during successful intrusions; non-essential services are to be recovered after intrusions have been dealt with.

The survivability analysis framework (SAF) described in [Ellison 2010] helps to examine the elements of an operational process and evaluate the survivability and effectiveness of the linkage among roles, dependencies, constraints, and risks to achieve critical operational capabilities. The Cybersecurity Engineering Response (CERT) team addresses security and survivability throughout the development and acquisition lifecycles, especially in the early stages. Figure 3.12 shows the phases and activities [CERT Cyber]. Security, software assurance, and survivability are addressed throughout the development and acquisition lifecycles by creating methods and solutions that can be integrated into organization's existing practices. The framework is designed to address the following [CERT SAF]:

- Identify potential problems with existing or near-term interoperations among components within today's network environments.
- Highlight the impact on successful execution as constrained interoperation moves to more dynamic connectivity.
- Increase assurance that operational mission and critical work processes can be successfully executed in the presence of stress and possible failure.

Also, SAF provides a mechanism for assembling the broad range of information that influences a business process in order to analyze it for quality and business survivability. Execution of a business process requires an extensive list of components working in harmony:

- Hardware - servers, data storage devices, PCs, PDAs, routers, telephone switches, satellite relays, physical access controls, and similar devices.

- Software - operating systems for each hardware platform, configuration management, databases, firewalls, network protocols, packet switches, authentication packages, Web applications, local and remote procedures, and others.
- People - organizational roles for data entry, inquiry, verification, audit, synthesis among multiple information sources, administration for technology components, authentication and authorization authorities, and similar roles.
- Policies and practices - certification and accreditation, third party access management, outsourcing contracts, governance controls, and the like.

This framework, a structured view of process, people, and technology, can be used by organizations to characterize the complexity of multi system and multi organizational business processes, survivability gaps, and conflicts.

In this work [Heegaard 2009], the authors argue the need for continuous provision of critical services in a telecommunication network even when undesirable events like sabotage, natural disasters, or network failures happen. The approach is to design and implement requirements to enable survivable services and network infrastructures by providing virtual connections between peering nodes with certain performance guarantees such as minimum throughput, maximum delay or loss. A methodology to estimate the security risk is described in [Youn 2011]. The authors recognize the need for quantitative evaluation of survivability and approach it by analyzing failures from the perspective of security attacks. The authors estimate the survival failure probability with regard to attack and combine it with the severity of the failure consequence obtained using the Functional Failure Analysis method. The methodology is applied for the development of an e-business application. However, this approach is limited from the perspective of security threats and a more generic computational model would be more useful for power systems and applications.

A survivability approach combines risk management and contingency planning with computer security to protect highly distributed information services and assets in order to sustain mission-critical functions. Survivability expands the view of security from a narrow, technical specialty understood only by security experts to a risk management perspective with participation by the entire organization and stakeholders. A survivable system must ensure that desired survivability properties emerge from the interactions among the components in the construction of reliable systems from unreliable components. A summary of survivability features includes:

- A broader, more enterprise wide point of view looking at solutions that are more pervasive than point solution oriented (e.g., security).
- Requirements vary substantially depending on system scope, criticality, and consequences of failure and interruption of service.
- Design of a survivable system includes management operations and administration.
- Immediate control require expanding the administrative services and providing a monitoring function as part of the infrastructure .
- Survivability places stringent requirements on system development and testing practices.
- System testing must demonstrate the performance and survivability of essential services during intrusion (intrusion usage scenarios are necessary).
- Successful implementation of survivability capabilities relies on risk analysis prevention techniques.
- Development and maintenance of survivability consumes resources in system development, operation, and evolution.
- Survivability requirements for a system should be based on costs and risks to an organization associated with loss of essential services.

- Complex infrastructure and use of required services outside an organization's bounded environment.

While it is argued that the survivability is not a quality model since it uses various subcategories of four specific main quality models (reliability, functionality, reliability, maintainability, and portability) [Westmark 2004], the refined notion of survivability that is suggested in [Knight 2000] is a valuable solution to resolve real world problems. The refined definition helps designers to view the opportunity to achieve the realization of a survivable system for critical applications. A survivable system is to address all anticipated types of fault including deliberate faults since it is the maintenance of customer value that is sought in survivability. A survivable system is expected to continue to provide one of the forms of tolerable service after many different forms of damage have occurred. Therefore, it is needed to identify some issues as suggested in [Knight 2000]. These include forms of degraded service that are acceptable to the application, information under what circumstances each form is most useful, and the fraction of time degraded service levels are acceptable.

Thus, a redefinition of survivability states that a system is survivable if it complies with its survivability specification [Knight 2003]. A specification should consider users' needs based on a balance of tolerance to degraded service and a service value factor that affects the relative value of an acceptable service to the user.

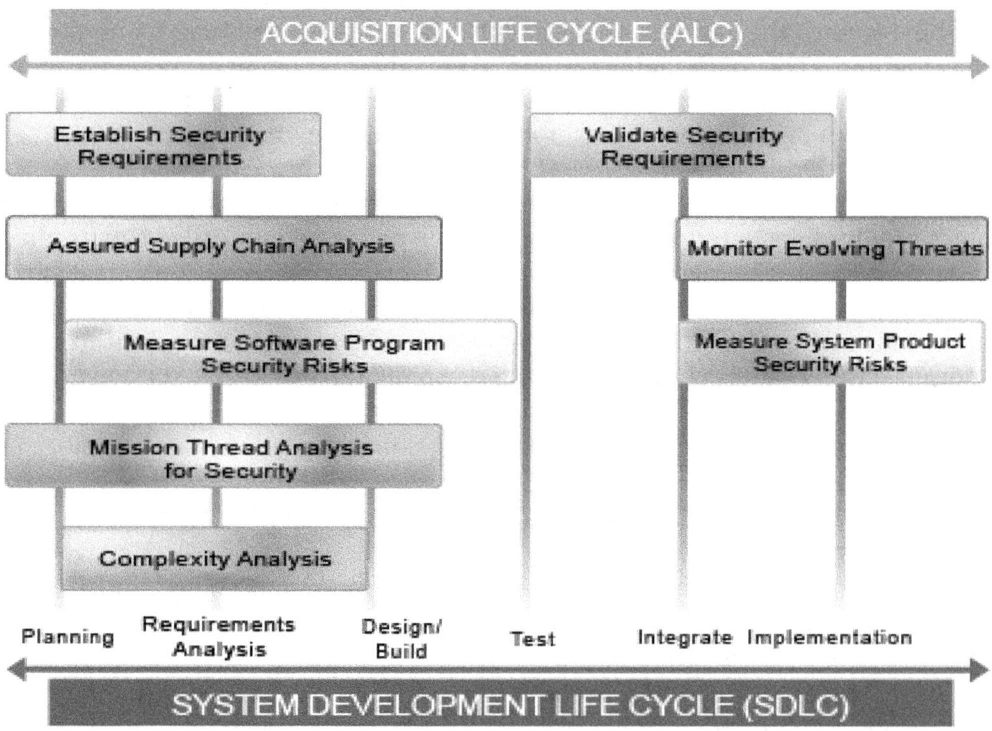

Figure 3.12 CERT system development life cycle. *Source:* [CERT Cyber]. Public Domain.

It is observed that the major organizations change the focus from security focused and IT focused views to survivability solutions embraced as a risk management perspective that requires involvement of the whole organization and focuses on the survival of the mission rather than a particular component. Using a mathematical description for survivability analysis to show its similarities and differences to reliability analysis, the authors of this work [Bai 2007] conclude that:

- A survivable system is a reliable system which has reconfiguration capability.
- System survivability is dependent on the system design and ability to identify threats.

Another defines survivability models as close cousins of performability models with the focus on transient performance immediately after an undesired event [Heegaard 2008].

Electric utility companies regard information survivability as critical to ensuring the continued delivery of electric power to the nation. While reliability is a goal of the past for the performance of the power system, a shift to survivability helps on defining more robust security. Therefore, Systems Engineers and Security Engineers need to focus on developing survivable systems for the Smart Grid.

3.7 Reliability vs. Availability

Different from the reliability that focuses on a period when the system is free of failures, the availability is concerned with a point in time at which the system does not stay at the failed state.
Simply described, availability is the proportion of time a system is in a functioning condition. It can be viewed as the percentage of time the system is available to perform its expected task. This is often described as a mission capable rate. Mathematically, this is expressed as 1 minus unavailability which is the probability that an item will not operate correctly at a given time and under specified conditions.

In telecommunication, an unavailability means that a system or component is not operable and not in a committable state at the start of a mission, when the mission is called for at an unknown time (e.g., random time). The conditions determining operability and commitment ability must be specified [Wiki Unavailability].

Availability is indicative of service duration; reliability is suggestive of failure frequency. A system can have low reliability but have high availability. Therefore, some applications require a higher degree of reliability than other types of applications which require a higher availability.

3.7.1 System Reliability

The reliability of a component or a system is the probability that the component or the system will be operating correctly without failure in a time interval. The correlation between reliability and security is a recurring theme; it has been shown that investment in software quality reduces the incidence of computer security problems, regardless of whether security was a target of the quality program or not.

3.7.1.1 Reliability Expressions

Reliability is defined as the probability that a system will not fail and that it will perform its intended function for a specified time interval. One agreed definition is specified as follows [Billinton 1992]:

Reliability is the probability of a device performing its purpose adequately for the period of time intended under the operating conditions encountered.

The reliability definitions introduce basic concepts such as probability (determined using probability theory) and engineering parameters (adequate performance, time, and operating conditions). There is no single all purpose reliability formula or technique to calculate the reliability index, it depends on assumptions made about engineering parameters. In practical applications, assumptions are made when using probability and statistical theories.

Probability provides the numerical input for the assessment of reliability, and also the first index of system adequacy (the most significant index in many instances). There are also other indices calculated that are dependent on the system and its requirements. More precisely, reliability is the probability of the system being found in the operating state at some time t in the future given that the system started in the operating state at time t = 0; failures and down states but maintenance and repair actions always return the system to an operating state. Main concepts of reliability and methods available for the analysis of general system problems need to be identified when performing reliability analysis. Based on specific needs, the computed indices for computer systems may include [Stamper 2003]:

- Expected number of failures that will occur in a specified period of time.
- Mean time to failure (MTTF) is the average time expected for component to fail when it has been running.
- Mean time between failures (MTBF) is average time between failures.
- Mean time to repair (MTTR) is average time to repair.
- Average outage duration or down time of a device.
- Expected loss in revenue due to failure.
- Expected loss of output due to failure.

Thus, the calculation of these metrics is subject to definitions and measurements used for calculation. One can simply compute MTBF as a sum of MTTF and MTTR.

The reliability can be defined by the reliability function, which is the probability that the system will not fail during a given period, is given by

$$R(t) = e^{-bt} \tag{2}$$

where, e is the natural logarithm; b = 1/MTBF or the inverse of the MTBF (Mean Time Between Failures); t is the given period of time, the time units used for MTBF calculation and t must be the same.

Reliability indicates that the required function can be performed by the entity based on stated condition for a stated period successfully. As described in [Mekkanen 2012], observations and analysis of available test data related to performance under either actual or simulated condition are the parameters that must be used to assess reliability. The data are governed by parametric probability distribution which represents the probability that the given system whether or not fail within a certain specified time [Billinton 2003]. Examples of ways to measure reliability include:

- Counting failures in periodic intervals and observing the trend of cumulative failure count - $\mu(\tau)$, or
- Counting failure intensity and observing the trend of number of failures per unit time – $\lambda(\tau)$.

  where,
  $\mu(\tau)$ denotes the total number of failures observed until execution time τ from the beginning of system execution
  $\lambda(\tau)$ denotes the number of failures observed per unit time after τ time units of executing the system from the beginning. This is also called the failure intensity at time τ.

Generally, three kinds of measurements are used when assessing the reliability of a system:

- Number of failures over number of requests
- Time between system failures
- Repair or restart time after a failure

Reliability is the ability of a system to continue operating in the expected way over time. A reliable system is expected to be operating correctly without failure in a time interval. If several components must be linked to make a system, the reliability of the system is the product of the reliability of its components [Stamper 2003].

Computer systems and quality in use from a system perspective were included in the scope of this ISO standard [ISO/IEC 9126-1]. Quality is the degree to which a set of inherent characteristics of an object fulfils requirements [ISO 9000]. Reliability of the software is a quality characteristic that is included in the ISO Quality Model (described in this standard [ISO 9216-1] among other five characteristics, namely: functionality, usability, efficiency, maintainability, portability. Once a software system is functioning as specified and delivered the reliability characteristic defines the capability of the system to maintain its service provision under defined conditions for defined periods of time [ISO 9126-1]. Software reliability is defined as the probability of failure free operation of a software system for a specified time in a specified environment. Key elements of the above definition include:

- Probability of failure free operation
- Length of time of failure free operation
- A given execution environment

Reliability assessment of a product or system must be performed during and as an integral part of the design process.

3.7.1.2 Reliability Assessment

Reliability assessment using probability methods provides a quantitative prediction of the system performance and a consistent way of evaluating alternative proposals.

In some areas of Engineering, signals are the common terms. A signal is a set of data or information. A system is an entity that processes a set of signals (inputs) to yield another set of signals (outputs) [Lathi 2009]. Examples of signals include a telephone or switch signal, the monthly sales, the daily closing prices of a stock energy market. In the power grid, we deal mostly with signals that are functions of time. In all of these applications, the signal is a function of the independent variable time. Information can be transmitted on wires by varying some physical property such as voltage or current. By representing the value of this voltage or current as a single valued function of time, it is possible to model the behavior of the signal and analyze it mathematically.

However, it should be understood that there is no single reliability evaluation technique that can be used for all problems [Bellinton 1992]. Evaluation of a system can use network modeling techniques. A system is frequently represented as a network in which the system components are connected together, either in series, parallel, meshed, or combinations of these. The validity of reliability analysis of a system is directly related to the validity of the model used for representing the system. The series and parallel systems are defined as follows:

- The components in a set are said in series from a reliability point of view if they must all work for system success or only one needs to fail for system failure.

- The components in a set are said to be in parallel from a reliability stand point of view if only one needs to be working for system success or all must fail for system failure.

For a series system of n components and each component having its own reliability index, then the reliability of the system Rs is the product of all reliability indexes.

$$Rs = R_1 \times R_2 \times ... Rn \quad (4)$$

For a parallel system with n components and each component having its own reliability index, then the reliability of the system can be calculated using this formula

$$R_S = 1 - (1 - R_1)(1 - R_2)... (1 - Rn) \quad (5)$$

System reliability is considered a complex concept which must always be considered at the system rather than the individual component level [Sommerville 2007]. A system is a collection of interrelated components that all influence the overall reliability of a system. Failures in one component can be propagated through the system affecting the operation of other components. The closely related basic components include hardware, software, and operator reliability.

So the reliability of a system (R_{SYS}) can be calculated by using equation (4), where R_h, is the product of reliability of hardware; similarly, R_s is the reliability of software elements, and R_o is the reliability of operator, (reliability of a system depends on the context in which the system is used). A system can modeled to include these components. Reliability of the system is then calculated as:

$$R_{SYS} = R_h \times R_s \times R_o \quad (6)$$

This formula of equation (6) is based on series system. The key issues for reliability include:

- The system crashes or becomes unresponsive due failures
- Output is inconsistent
- The system fails due to unavailability of other externalities such as systems, networks, and databases, etc.
- The system fails because of the software.

Therefore, a design has to account for these issues and provide techniques to overcome any reliability challenges. The correlation between reliability and security is a recurring theme; it has been shown that investment in software quality reduces the incidence of computer security problems, regardless of whether security was a target of the quality program or not. More information on the design of systems to meet the new business trends is provided in [Sommerville 2016]. Also, business needs dictate which attributes need to be satisfied. For example, financial applications typically require high reliability of networked systems while SCADA systems in power grid typically require high availability.

3.7.2 System Availability

The availability of a system is defined as the probability that the system is available at a specific time. Different from the reliability that focuses on a period when the system is free of failures, the availability is concerned with a point in time at which the system does not stay at the failed state. It can be viewed as the percentage of time the system is available to perform its expected task. Availability is indicative of service duration; reliability is suggestive of failure frequency. A system can have low reliability but have high availability. Therefore, some applications require a higher degree of reliability than other types of applications requiring a higher availability.

3.7.2.1 Availability Expressions

Availability is usually estimated based on two separated events - failure and repair - as time to failure and time to repair, also can be considered as reliability representing components.

The simple definition of availability is based on time to repair as the probability that the system will be available based on demand, or based time to failure as the capability of the system being used over a total time. As described in this work [Mekkanen 2012], determining the availability of the system is depending on the interconnections between the parts of the system. Two conditions can be considered. First, series interconnection aspect if a failure in one part causes a failure to another part. Second, parallel interconnection when a failure in one part leads to other parts operate over the fielded one in the system.

Availability is the probability that the system will be up and running to deliver useful services at the needed time. The availability can be defined by the probability function

$$A(t) = a/ (a + b) + b/(a+b) \times e^{-(a+b)t} \qquad (7)$$

where a = 1/MTTR, b= 1/MTBF, e is the natural logarithm, and t is a time interval. This equation gives the probability that a system or component will be available when required by a user. Because the exponential term approaches zero and becomes insignificant as the time interval increases, availability in such cases becomes

$$A = MTBF / (MTBF + MTTR) \qquad (8)$$

The availability of a system is defined as the probability that the system is available at a specific time. If several components (e.g., hardware, software) must be linked together to make the system available to the user, then the system availability is given by the product of the availabilities of the components [Stamper 2003] as shown in the equation (6).

$$A_{SYS} = A_h \times A_s \qquad (9)$$

where

A_h and A_s is the product of the availabilities of the hardware and software components (this is availability for series systems). Availability for parallel systems (based on a simple formula) is defined by

$$Ap = A_1 + A_2 - A_1 \times A_2 \qquad (10)$$

Reliability and availability block diagram method based on its simplicity are used [Mekkanen 2012] (see equations (9) and (10)).

Therefore, Availability is relevant for non-stop continuously running systems (e.g. AMI signal, controller signal, etc.) where users expect the system to deliver a continuous service. It measures the fraction of time system is really available for use by taking repair and restart times into account.

Availability is the probability that the system will still be operating to requirements at a given time. Availability is a function not only of how rarely a system fails (reliability) but also of how quickly it can be repaired (time to repair).

Another definition of availability is the ratio of the total time a system or component is capable of being used during a given interval to the length of the interval. However, typical availability values are specified in decimal (such as 0.9998). If each component has an availability of 0.999, the user will see a system availability (A_h x A_s) of 0.998001. Statistically, the user may find the system 2 times unavailable at every 1,000 attempts, or once every 500 attempts. However, the above formula is applied if components are linked in series. If components are interconnected in parallel (e.g., hardware) then another more complicated formula applies as described in [SYSTEM].

In applications with high availability requirements, a metric known as nines, corresponding to the number of nines following the decimal point, is used. In this system, five nines equals 0.99999 (or 99.999%) availability. Internet and global e-business application requirements demand that companies increasingly implement computing infrastructures specifically designed for at least 99.999 percent availability. High availability of the environment, at least 99.999 is the equivalent of less than 5.3 minutes of downtime a year. This is also a requirement for the SCADA networks. In response to these trends, government and SCADA owners need to address increased security and support for high availability [Hentea 2008].

3.7.2.2 Availability Assessment

The availability property for a system or component has to be analyzed in correlation with scenarios of unavailability. The analysis should also include aspects of correlation between security and availability. A system can behave erratically when software component contains errors or a network is congested causing delays or loss of data.

Availability is affected by system errors, infrastructure problems, malicious attacks, and system load. A designer has to plan for use of different techniques to maximize availability for a system in situations such as the following:

- A database server or application server fail or become unresponsive, causing the entire system to fail.
- Denial of Service (DoS) attacks, which prevent authorized users from accessing the system, can interrupt operations if the system cannot handle massive loads in a timely manner, often due to the processing time required, or network configuration and congestion.
- Inappropriate use of resources can reduce availability. For example, resources acquired too early and held for too long cause resource starvation and an inability to handle additional concurrent user requests.
- Bugs or faults in the application can cause a system wide failure.
- Frequent updates, such as security patches and user application upgrades, can reduce the availability of the system.
- A network fault can cause the system to be unavailable.

High availability system design approach and associated service implementation ensures a prearranged level of operational performance will be met during a contractual measurement period [Wiki HighAvail]. Principles of high availability engineering include:

- No single points of failure achieved by adding redundancy to the system so that failure of a component does not mean failure of the entire system.
- Reliable crossover achieved in multithreaded systems is the crossover point itself, which tends to become a single point of failure. High availability engineering must provide for reliable crossover.
- Detection of failures as they occur.

Therefore, security is an important criteria for systems to operate correctly. Without reasonable security, the reliability and availability of a system may be compromised if cyber attacks occur. It must be understood that all methods for assuring reliability and availability rely on the fact that the operational system is the same as the system that was originally installed. If a system is insecure, cyber attacks may occur causing damages such as:

- Unavailable services due to denial of service attack
- Corruption of programs and data due to unauthorized access
- Disclosure of confidential and private information.

Therefore, establishing security as a protection control objective for the system has to be addressed.

3.7.3 Power System Reliability - A Complex Problem

Power system reliability is the field within power engineering that is concerned with the ability of the power system to perform its intended function, which can have different interpretations. In practice, power system reliability concerns the development of methods to stochastically predict the number and duration of interruptions of the electricity.

The power system is a complex system with its main function to produce, transfer and provide consumers with electrical energy. Combinations of failures of components in the system can result in a failure of power delivery to certain load points and in some cases in a full blackout of power system. The power system reliability directly affects safe and reliable operation on many industries. A system is a combination of generation, transmission, and distribution components [NERC Glossary-16].

With no single definition for the reliability, we need to understand the use of this concept from the perspective of the power industry and how it relates to the reliability of information systems. According to NERC [Hirst 1999], reliability is defined as the degree to which the performance of the elements of the system results in power being delivered to consumers within accepted standards and the amount desired. It refers to two concepts: adequacy and security, defined as follows:

- Adequacy is the ability of the system to supply the aggregate electrical demand and energy requirements of the end-use customers at all times, taking into account scheduled and reasonably expected unscheduled outages of system elements [NERC Glossary-16].
- Security is the ability of the system to withstand sudden disturbances; it deals with the short run operations, ensuring that the system will remain intact after equipment outages, changes in demands, or other disturbances occur [Osborn 2001].

This standard [IEEE 1366] defines reliability as the ability of the power system components to deliver electricity to all points of consumption, in the quantity and with the quality demanded by the customer. All components of the bulk power system, including generation, transmission, and distribution, contribute to reliability.

Reliability is often measured by the frequency, duration and extent of power system abnormal states called disturbances and outages that are defined in [Osborn 2001]:

- A disturbance is any unplanned event, including an outage, that produces an abnormal system condition.
- An outage can be described in terms of frequency, duration, and amount of load (or numbers of customers) affected.

- A momentary outage is defined as an outage that lasts less than five minutes, corresponding to the time allowed for automatic reclosing schemes to try to restore a circuit if the fault was temporary.
- A sustained outage lasts longer than five minutes.

However, these definitions have to be understood in the context, perspective, and other constraints as discussed in [Osborn 2001]:

- Definitions related to reliability derive from transmission and distribution applications.
- From the consumer's perspective, transmission and distribution related outages are most important to real time reliability (system security).
- Generation and other system component outages are typically most significant to system planners, because they tend to affect the adequacy of the electricity system as a whole.
- An outage or disturbance of a system component may or may not cause an interruption of service to customers, depending on the system's configuration, and may affect power quality even if service is not interrupted.

Reliability is often measured by outage indices defined by the [IEEE 1366] standard. These indices relate to customer satisfaction and are based on both the total length of each service interruption and the frequency of interruptions. The indices are intended to apply to distribution systems, substations, circuits, and defined regions. These indices capture the effects of the number of outages, both momentary and sustained, as well as the duration of each outage, and are usually computed from the past year's or several years' utility data.

Reduced reliability affects the adequacy, security, and/or quality of the power supply. Although related to reliability, power quality is another concept that describes the characteristics, in terms of continuity and voltage, of the supplied electricity as delivered to customers at supply terminals under normal operating conditions according to Renner and Fickert [Osborn 2001].

According to [Bobba 2012], past reliability events (such as blackouts) have highlighted the need for better situational awareness and advanced applications to improve planning, operations, and maintenance. The deployment of a wide area measurement system (WAMS) is an important part of the solution to these complex problems, but it faces challenges with respect to communications and security. Although a high bandwidth and highly available networked communication system is needed to make the grid more efficient and reliable, specifications for such a communication system have been difficult to develop [Anderson 2012].

3.7.3.1 Reliability Issues and IT Technologies

The Energy Independence and Security Act of 2007 [EISA 2007] policy supports the modernization of power grid to maintain a reliable and secure electricity infrastructure by

- Increased use of digital information and controls technology.
- Dynamic optimization of grid operations and resources, with full cybersecurity.

In the power industry, the focus has been on implementation of equipment that could improve power system reliability. Until recently, communications and IT equipment were typically seen as supporting power system reliability. However, these sectors are becoming more critical to the reliability requirements of the power system. In addition, safety and cybersecurity are of paramount importance in electric power systems.

The increasing number of SCADA/EMS systems operating in the electric industry, the growing number of market participants, and the development of complex market models relying more on information technologies contribute to raising the interdependency between the operation of the power grid (including generation, transmission, and distribution) and the operation of the whole sale electric market. The electric market and the power system are every day more closely tight. The operation of one depends on the continuous and reliable operation of the other. Also, the integration of Smart Grid with DER such as renewables (e.g., solar, wind), demand/response, storage may impact the reliability of the Smart Grid due to their volatility.

3.7.3.2 Addressing Reliability and Availability Issues

Specific components of a system need to be analyzed for reliability. Besides reliable power grid, there is a need to design reliable system, reliable network, reliable protocol, reliable equipment, reliable software so requirements should be identified for different components.

In addition, reliability of communications networks is a concern for Power Engineers. However, the modern communication technologies have capabilities (e.g. faster transmission of messages, high bandwidth, low delay, near real time response) that support advanced control operations and render a more reliable power grid. DER systems need new control operations that cannot be supported by the existing communication infrastructure. Managing energy usage more efficiently through advanced metering infrastructure (AMI) and home energy management system (HEMS) require new communications technologies and a new communication architecture that is reliable, scalable, and extendable that enable future Smart Grid services and applications [Bouhafs 2012].

Specifically, there are strict availability, security, performance, and reliability requirements that are needed by power system operations such as:

- Operation of the power system must continue 24×7 with high availability (e.g., currently, availability is around 99.99% for SCADA although is recommended to be around 99.999%, and higher than 99.99% for protective relaying) regardless of any compromise in security or the implementation of security measures that hinder normal or emergency power system operations.
- Power system operations must be able to continue during any security attack or compromise (as much as possible).
- Power system operations must recover quickly after a security attack or the compromise of an information system.
- Testing of security measures cannot be allowed to impact power system operations.
- Power system management, monitoring, and control will increasingly extend away from the power entities' traditional physical and security environments into external environments that the power entity has little or no influence and control over.

Meeting these reliability challenges while effectively integrating the above resources requires an IT infrastructure (communication and information technologies) that provides [Moslehi 2010]:

- Grid wide coordinated monitoring and control capabilities.
- Fail-proof and nearly instantaneous bidirectional communications among all devices ranging from individual loads to the grid wide control centers including all important equipment at the distribution and transmission levels.
- Processing vast number of data transactions for analysis and automation.
- Fast intelligent local sub-second responses coordinated with a higher-level global analysis in order to prevent or contain rapidly evolving adverse events.
- Better telemetry for faster, time stamped, higher accuracy, sub-second scanning to

enable timely grid-wide situational awareness.
- Faster control devices based on power electronics to enable fast automated control actions, for voltage and power flow management at both transmission and distribution levels.

A high bandwidth and highly available networked communication system is required to collect and transmit phasor measurement data that is utilized to initiate fast controllers, including flexible alternating current transmission system (FACTS) devices. This system enables the control and protection of data and make the power grid more efficient and more reliable [Anderson 2012].

It is envisioned that a wide area measurement system is an important part of the solution to the reliability of the power grid and avoidance of undesirable events (e.g., blackouts) [NISTIR 7628]. WAMS provides visibility that is based on real-time data streamed from Phasor Measurement Units (PMU). However, WAMS system faces challenges with respect to reliability, availability, and security of communications networks and services [Bobba 2012]. Therefore, these concerns have to be addressed. Attacks (such as attacks upon the power system, attacks by the power system, and attacks through power system) to the Smart Grid infrastructures could bring huge damages on the economy and public safety.

Dependency of WAMC (Wide Area Monitoring and Control) on ICT infrastructure to collect measurements and further control the power equipment is another research topic. A study that includes analysis and analytical experiments to measure the maximum delay and input signal's sensitivity toward delay is described in [Zhu 2011]. This work shows that on a typical WAMC application, where the control of static var compensation (SVC) is coordinated with generator excitations using phasor measurements, there is a need for a new ICT design. An ICT architecture is recommended and requires communications networks to be available, reliable, secure, and interoperable.

Any cybersecurity measures in these systems must not impede safe, reliable power system operations. It has been shown that investment in software quality reduces the incidence of computer security problems, regardless of whether security was a target of the quality program or not.

Examples of security needs that affect reliability are described in [NISTIR 7628r1] to include:

- Availability is the most important security objective for power system reliability. The time latency associated with availability can vary such as:
  - Latency ≤ 4 ms for protective relaying.
  - Latency in the order of sub-seconds for transmission wide area situational awareness monitoring.
  - Latency in the order of seconds for substation and feeder SCADA data.
  - Latency in the order of minutes for monitoring noncritical equipment and some market pricing information.
  - Latency in the order of hours/days for meter reading and longer term market pricing information.
  - Latency in the order of days/weeks/months for collecting long term data such as power quality information.

Thus, availability of the data is crucial for some Smart Grid applications and the time latency associated with availability can vary. For example, a delay of more than 4 ms is not acceptable for protective relaying; other applications for transmission wide area situational awareness monitoring may tolerate a delay in sub-seconds.

Also, the delay is not tolerated for certain types of data because the unavailability of the data may cause application failures. The use of the Web and Internet services for certain applications has to be chosen only after a detailed analysis of the business needs because they are different from the power system operations environment.

Addressing cybersecurity is critical to enhancing the reliability of a Nation's electric grid. The vulnerability of the power system is not mainly a matter of electric system or physical system. When dealing with cybersecurity issues for the power system, it is required to find the most current terms defined by NERC organization that frequently changes terms related to cybersecurity. According to [NERC Glossary-16], a cybersecurity incident has a definition that requires understanding of other terms.

Another issue is that many security terms and standards may refer to BES, which does not include distribution systems. Another issue is the recommendation of ignoring the impact on redundancy equipment and facilities. In fact, this issue results on weak systems, because the role of redundancy is to increase the dependence on systems. Therefore, these NERC definitions for cybersecurity terms could be qualified as anachronisms that could create confusion to any security professional.

3.8 Standards, Guidelines, and Recommendations

In this chapter, we discussed many topics and pointed out the various standards, guidelines, and recommendations that could be used for the design of cybersecurity to protect the Smart Grid and DER systems. However, there are many resources that can be used for the development of secure systems for the Smart Grid. The following includes additional guidance.

3.8.1 Systems Engineering

This guide [NDIA 2008] is intended to supplement the knowledge of systems (and software) engineers who have responsibility for systems for which there are assurance concerns. Examples of related organizations for Systems Engineering Practices, Recommendations and Standards include:

- American Electronics Association (AeA)
- American Institute of Aeronautics and Astronautics (AIAA)
- American National Standards Institute (ANSI)
- Association for Computing Machinery (ACM)
- Australian Department of Defense
- Defense Science and Technology Organisation (DSTO)
- Electronic Industries Alliance (EIA)
- European Cooperation for Space Standardization (ECSS)
- Information Systems Audit and Control Association (ISACA)
- Institute of Electrical and Electronics Engineers (IEEE)
- International Council on Systems Engineering (INCOSE)
- International Function Point Users Group (IFPUG)
- National Institute of Standards and Technology (NIST)
- Object Management Group (OMG)
- Office of Government Commerce (OGC)
- Software Engineering Institute (SEI), Carnegie Mellon
- Systems Engineering Center of Excellence (SECOE)
- Systems Engineering Society of Australia
- U.S. Department of Defense (DoD)
- U.S. Department of Energy (DoE).

3.8.2 Quality Models

Given that the quantity of data and information handled by computer systems is increasing world-wide, a common prerequisite to all information technology projects is the quality of the data which are exchanged, processed and used between the computer systems and users and among computer systems themselves. Most business processes depend on the information systems and quality of data.

The data quality model defined in this international standard [ISO/IEC 25010] is intended to be used in conjunction with other Systems and software Quality Requirements and Evaluation (SQuaRE) series. The divisions within the SQuaRE series are:

- Quality Management Division (ISO/IEC 2500n)
- Quality Model Division (ISO/IEC 2501n)
- Quality Measurement Division (ISO/IEC 2502n)
- Quality Requirements Division (ISO/IEC 2503n)
- Quality Evaluation Division (ISO/IEC 2504n).

Chapter 4

Applying Security Engineering Principles to Evolving Smart Grid

4.1 Resiliency

The vision of a Smart Grid implementation requires to resist attacks, as one seven principal characteristics of the future Smart Grid [NETL 2007a] as well as to operate resiliently against physical and cyber attacks and natural disasters [EAC 2008]. These requirements imply a solution that meets the following [NETL 2007a]:

- System wide security solution
- Reduce physical and cyber vulnerabilities
- Recover rapidly from disruptions
- Demonstrate resilience to attack, even from those who are determined and well equipped
- Design and its operation should minimize the consequences of an attack and speed service restoration
- Tolerate simultaneous attacks against several parts of the electric system and the possibility of multiple, coordinated attacks over a span of time
- Include modern security protocols that contain elements of deterrence, prevention, detection, response, and mitigation to minimize impact on the grid and the economy
- Deters or withstands physical or cyber attack.

Addressing these issues require investments in technologies and ensuring the resiliency of the overall infrastructure which is dependent on both cyber and physical components. The infrastructure has to resist both physical attacks and cyber attacks as well as combined cyber-physical attacks to avoid consequences. Examples of attacks include aiming to destabilize the system and inflict some lasting damage; or misleading the operator to the true nature of the problem by modifying the information; inhibiting defensive responses and spreading the extent of an outage.

In general terms, resilience or resiliency is the ability to recover from or adjust easily to misfortune or change [Dictionary] or recover from some shock or disturbance [Wiki Res]. There are different definitions for resilience used in different contexts and domains such as characterizing a system, network, material, or human being. According to Westrum [Strigini 2009], resilience is a family of related ideas, not a single thing. It is observed that system resilience has received increased attention due to research efforts in several system domains and several more recent definitions depend on the application's domain [Meyer 2009].

4.1.1 Definitions

Examples of definitions that are still challenged include:

Resilience of a system is the ability of a system to recover from a catastrophic failure [Wiki Res].

Resilience of a network is the ability of the network to provide and maintain an acceptable level of service in the face of various faults and challenges to normal operation.

Resilience of a service is the persistence of service delivery that can justifiably be trusted, when facing changes.

Another defines resilience as the persistence of dependability when facing changes [Laprie 2008]. This definition is bounded with dependability and trustworthiness. The changes can be classified according to three viewpoints [Laprie 2008]:

- Nature - functional, environmental, or technological, either or both hardware and software
- Prospect – foreseen (versioning), foreseeable (new hardware), unforeseen (changes in service requests or new types of threats)
- Timing – short term (seconds to hours as in adhoc networks, networks of mobile nodes and sensors), medium term (hours to months as in new versioning or reconfigurations)
- Long term (months to years as in reorganizations resulting from merging of systems).

Figure 4.1 depicts resilience as a superset of survivability, disruption tolerance, traffic tolerance, trustworthiness, dependability, security, and performability.

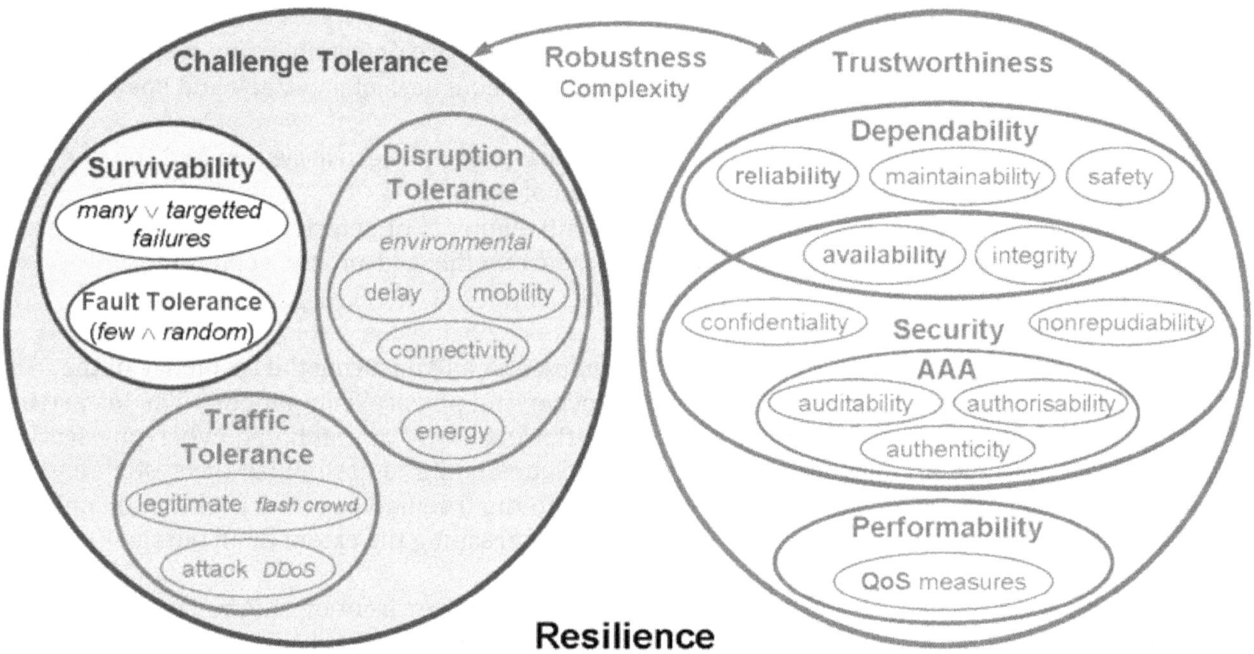

Figure 4.1 Resilience. *Source:* [Wiki Res]. Public Domain.

It is observed that the various definitions are coming from different communities and scholars still continue to provide more definitions such as the following:

- Computer networking community defines it as the combination of trustworthiness (dependability, security, performability) and tolerance (survivability, disruption tolerance, and traffic tolerance). Resilience is an important service primitive for various computer systems and networks [Trivedi 2009].
- Dependable computing community defines resilience as the persistence of service delivery that can justifiably be trusted, when facing changes [Trivedi 2009].
- Resilience is the persistence of performability when facing changes [Meyer 2009]; a performability measure quantifies a system's ability to perform in the presence of faults.

NIST definitions are based on different standards: [NIST SP800-30r1], [NIST SP800-34r1], [NIST SP800-137] and glossary [NIST 7298r3]. They are more alike with a slight variation that emphasizes the ability to adapt in this standard [NIST SP800-34r1]. However, a designer could be confused about which standard to use, then which definition to use for design compliance and guidance. An understanding of the business needs and context could help on applying a design based on the most suitable definition. Also, there should be an understanding that control systems are defined under the category of information systems, but a distinction has to be made due to their needs. Control systems require more stringent requirements for resiliency (e.g., time, diversity, fault tolerance).

Design of robust, resilient systems is a costly effort in terms of human and financial resources that are much higher compared to traditional systems. The feasibility studies for the development of dependable systems have to justify the cost and value. One argues that two criteria to judge feasibility are cost required and value to be attained [Young 1970].

4.1.2 Resiliency Evolution

The resiliency concept evolved from resilient networks. A resilient network has the ability to operate and maintain acceptable level of service under the presence of adverse conditions [Westmark 2004]. There is a substantial effort in the research community to advance the dependability and survivability to resiliency as the ability to successfully accommodate unforeseen environmental perturbations or disturbances.

Besides the multitude of definitions, research community is aware of the needs for measurements and metrics to characterize resilience by quantitative means. While resilience definitions are ambiguous and evolving, it is still recognized the need for measures. However, the identification of the measures is challenging [Strigini 2009]. As discussed in this work, since resilience is a superset of many broad concepts (e.g., dependability, survivability, fault tolerance) encompassing multiple attributes, there are multiple possible measures which are difficult to define. In addition, recognizing the multi dimensionality of resilience, the author introduces categories of systems and types of resilience as well as common problems that may recommend importing insights from some areas of study to others. A summary characterization of categories of measures related to fault tolerance and resilience, with some discussion of their uses and difficulties in measurement and prediction are proposed in this work. Measures of resilience thus quantify the persistence of such ability in the presence of changes (including faults). Characteristics of resilience measures include:

- Permits summarization of an entire history of service quality variations caused by changes that occur over a lengthy, yet bounded period of time.
- Copes with unanticipated system and environmental conditions that might otherwise cause a loss of acceptable service (failure); closer to a strict dependability interpretation.
- Relies on an underlying complementary concept of failure, basically it is success oriented.

Facets of resilience behavior and impact on other characteristics such as stability are concerns that have to be resolved. Examples include [Laprie 2008]:

- A system can be very resilient and still fluctuate greatly (low stability may induce high resilience).
- Diversity influences both stability (decreasing it) and resilience (increasing it).

It is argued that evaluating resilience involves consideration of system and environment dynamics [Meyer 2009]. Therefore, accounting for these dynamics requires evaluation methods, resilience models, and means to validate the results. This work describes a model and method to measure resilience, which stated informally, is a performability measure which quantifies a system's ability to perform in the presence of faults.

Recognizing the similarities of performability with survivability, this work [Heegaard 2008] argues that the QoS and resilience requirements should still be met even when the network suffers from overload or network element failures, irrespectively of whether this is caused by physical or logical bugs, sabotage or a natural disaster. Others propose measures that are focused on some aspects of the system's properties.

In this work [Mohammad 2006], the authors propose a model of resilience measure as a function of properties that affect the performance and resilience of networks. These properties are classified in categories such as density, mobility, channel, network traffic, node resources, and derived properties. The proposed model is based on analytical terms, although weak and incomplete, that would require the knowledge of quantitative performance of the network in all states that it may visit under normal or adverse conditions. Also, the knowledge assumption is ignoring qualitative knowledge as a component of knowledge (e.g., safety, maintainability could be characterized with quality metrics based on perception of one user or developer). In addition, the dynamics of the network are ignored; the current state is compared with predefined values of state space categorized in areas: acceptable, impaired, and unacceptable. The value ranges were assigned during the development stage, which could be incomplete or subjective. However, the initial value ranges may be different and may not be appropriate when the system is used. The current state should be computed as a function of prior state and predicted variables of current state because it is not possible to compute resiliency at time t with performance metrics measured at time t.

In [Salles 2012], the authors describe an approach to calculate the resilience factor to measure the network level of robustness and protection against targeted attacks. The resilience factor is computed based on graph metrics such as K-connectivity, average inverse shortest path length, closeness centrality, etc. This factor can be computed in advance to evaluate if topology alterations would improve the resilience of a given network.

An emerging specialty called Cyber Resiliency Engineering is focused on to develop more survivable, trustworthy systems. This specialty is a Systems Engineering discipline applied in conjunction with Resilience Engineering and Systems Security Engineering. This recent document [NIST SP800-160v2] of 2021 is focused on applying cyber resiliency concept and cyber resiliency constructs to sustain the systems' trustworthiness. Cyber resiliency is the ability to anticipate, withstand, recover from, and adapt to adverse conditions, stresses, attacks, or compromises on systems that use or are enabled by cyber resources [NIST SP800-160v2]. Cyber resiliency is intended to enable mission or business objectives that depend on cyber resources to be achieved in a contested cyber environment. Cyber resiliency engineering intends to architect, design, develop, maintain, and sustain the trustworthiness of systems with the capability to anticipate, withstand, recover from, and adapt to adverse conditions, stresses, attacks, or compromises that use or are enabled by cyber resources. From a risk management perspective, cyber resiliency is intended to reduce the mission, business, organizational, or sector risk of depending on cyber resources.

4.1.3 Resilient Smart Grid

The Presidential Policy Directive (PPD) on Critical Infrastructure Security and Resilience [PPD-21] advances a national unity of effort to strengthen and maintain secure, functioning, and resilient critical infrastructure. Proactive and coordinated efforts are necessary to strengthen and maintain secure, functioning, and resilient critical infrastructure – including assets, networks, and systems – that are vital to public confidence and the Nation's safety, prosperity, and well being. Another report [EAC 2008] calls for a transformation of the Nation's electric power grid into one that is more intelligent, resilient, reliable, self balancing, and interactive in order to enhance economic growth, foster environmental stewardship, and promote operational efficiencies, energy security, and consumer choice.

Ensuring a resilient electric grid is particularly important since it is the most complex and critical system that other sectors depend upon to deliver essential services as specified in DOE's Energy Sector Specific Plan [DOE 2007]. Over the past two decades, the roles of electricity sector stakeholders have shifted: generation, transmission, and delivery functions have been separated into distinct markets; customers have become generators using distributed generation technologies; and vendors have assumed new responsibilities to provide advanced technologies and improve security. These changes have created new responsibilities for all stakeholders in ensuring the continued security and resilience of the electric power grid.

4.1.3.1 Related Concepts

Important and beneficial properties for power grids, such as resilience, robustness, reliability have been defined by standards and research communities. While these are related concepts, they are different due to the complexity of power systems and various approaches that evaluate them. Table 4.1 shows typical definitions for these related concepts.

Table 4.1 Summary of related concepts in power grids.

Concept	Definition
Reliability	Probability that an electric power grid can perform a required function under given conditions for a given time interval (IEC definition).
Robustness	Degree to which a network is able to withstand an unexpected event without degradation in performance. It quantifies how much damage occurs as a consequence of such unexpected perturbation.
Resilience	The ability of a power system to recover quickly after a disaster or, more generally, the ability of anticipating to extraordinary, high impact, low probability events, quickly recovering from these disruptive events, and adapting its operation and structure for preventing or mitigating the impact of similar events in the future.
Stability	The ability to maintain or to recover a state of equilibrium after disturbances or contingencies.
Vulnerability	The lack of robustness. Vulnerability is often used to score low reliability of power grids.
Disturbance	An unexpected event that produces an anomalous system condition.

Adapted from [Cuadra 2015].

In this work [Arghandeh 2016], the authors argue the damage of a system following the disturbances is a function of four different parameters: probability of the disturbance, severity of the disturbance, system vulnerability, and system capacity to absorb the disturbance. This work discusses the differences between resilience and robustness, two terms that are sometimes used interchangeably (see Table 4.2). Table 4.2 compares robustness and resilience against different criteria. The authors emphasize that resilience is related to survivability and flexibility in the face of unexpected events, while robustness implies resistance to change without losing stability.

This report [Cuadra 2015] describes additional differences or similarities between these concepts including other related concepts:

- A power grid that lacks robustness will often collapse before recovery, having thus small or even no resilience.
- Resilience encompasses not only robustness but also redundancy, adaptive self organization, and rapidity.

- Resilience is the ability to recover quickly after high impact, low probability disruptive events, and is related to the potential to adapt its structure and operation for mitigating or even preventing the impact of similar events in the future.
- Resilience is related to low probability and high impact events while reliability is related to high probability and low impact events.
- Resilience is a dynamic concept while reliability is a static concept.
- Robustness is the ability to remain constant in parameters in spite of system internal and external conditions [Ross 2008]; robust systems should deliver their intended functionality under varying operating conditions without being changed.

Table 4.2 Robustness vs. Resilience in power systems.

Criteria	Robustness	Resilience
Application *	Network hardening	Network flexibility
Enterprise focus *	Utility assets	Utility services
Value proposition *	Design	Operations
Security approach *	Passive	Active
Network preference *	Isolated	Interdependent
Development *	Embedded in the system's design like control systems	Typically integrated into the system's operational components
Network coupling *	Loose	Tight
Related properties	Performance, stability	Survivability, flexibility, durability, recoverability, adaptability, rapidity
Time variance	Static	Dynamic, time dependent
Perspective	Narrow	Broader
Knowledge of networking correlation degree	Known	Unknown
Typical measure	How much damage occurs as a consequence of an unexpected perturbation	How quickly the network can retrieve from such damage

Adapted from [Arghandeh 2016]. © 2016, Elsevier.
Note: * Source [Arghandeh]

Given the synergy of power systems with IT technologies, the assessment of the power system vulnerability should be approached from both, physical and cyber, and their interdependencies perspective.

4.1.3.2 Power System Cyber-Physical Resilience

Although there are multiple definitions of electric grid resilience in the current literature on power systems, they are often conflicting and vague or focused on one perspective [Arghandeh 2016]. Therefore, this author proposes a definition of cyber-physical resilience to include all perspectives as follows:

Power system cyber-physical resilience is the system's ability to maintain continuous electricity flow to customers given a certain load prioritization scheme. A resilient power system responds to cyber-physical disturbances in real time or semi real time, avoiding service interruptions. A resilient power system alters its structure, loads, and resources in an agile way.

This definition implies the following:

- Resilience in power systems depends on the reaction time following a disturbance that maintains service availability.
- Cyber-physical system's ability to recognize, adapt to, and absorb disturbances in a timely manner.

Further, an understanding of these issues should be the first step in designing and operating resilient power systems [Arghandeh 2016]. Resilience is especially critical immediately following an event that challenges system performance and functionality. Although reliability is one goal that power industry is mostly concerned, resilience is a property as the ultimate goal to build modern power systems to cope with risks and recover from outages quickly. Therefore, addressing this problem requires control systems that are resilient by nature and remain resilient in spite of complex interactions [Rieger 2008]. When implementing a Smart Grid, it is required to deploy control systems that manage grid inputs from both constant and intermittent sources and manage grid outputs to achieve efficiency and economy by intelligently interacting with end user devices. One challenge is that the multiple competing demands to cope with increase its complexity.

With the implementation of the Smart Grid, utility companies expect full visibility and pervasive control over their assets and services. In order to achieve this capability, the Smart Grid has to be self healing and resilient to system anomalies. In addition, Smart or Intelligent Grid is expected to address the major shortcomings of the existing grid [Farhangi 2010]. While understanding the risks of disturbances and consequences to power system is important, the key issue is how to build a modern power system that copes with risks and quickly recover from failure is the key [Arghandeh 2016]. More specifically, these systems - SCADA, AMI, and DER control systems - must be resilient against disturbing events and maintain grid performance under any circumstances. These systems play a major role in power system reliability services.

Resilience is an active approach with real time reactions to disturbances that can result in disasters [Arghandeh 2016] or collapse [Theron 2016]. The response actions can include a set of real time switching and islanding actions, which can involve explicitly partitioning the grid into different areas to resist the cyber attacks and respond quickly to ensure information flow needed for decisions supporting the energy flow.

Considering the multitude of organizations that support DER systems and the Smart Grid, there is need for both organizational and systems resilience to adverse events. The multitude of numerous technologies and distribution of DER systems require a rigorous analysis and design based on resilience principles and controls applied based on the needs of each application (e.g., demand/response, wide awareness, distributed electric storage, distributed generation).

This work [Arghandeh 2014] proposes a resilience assessment framework for transmission and distribution (T&D) networks (see Figure 4.2). The authors observe resilience in power systems depends on the reaction time following a disturbance that maintains service availability.

As shown in this Figure 4.2, a resilience assessment framework starts with system identification, continues with vulnerability assessment, then evaluates the resilience of the operation of the system. Vulnerability analysis builds knowledge before, during, and after disturbance. Resilience depends on the reaction time following a disturbance that maintains service availability. The resilience operation control defines new settings for the distributed resources. These two components, grid recovery and grid absorbing potential, can absorb disturbances, adapt itself to disturbing events, then recover by mitigating disturbing events consequences.

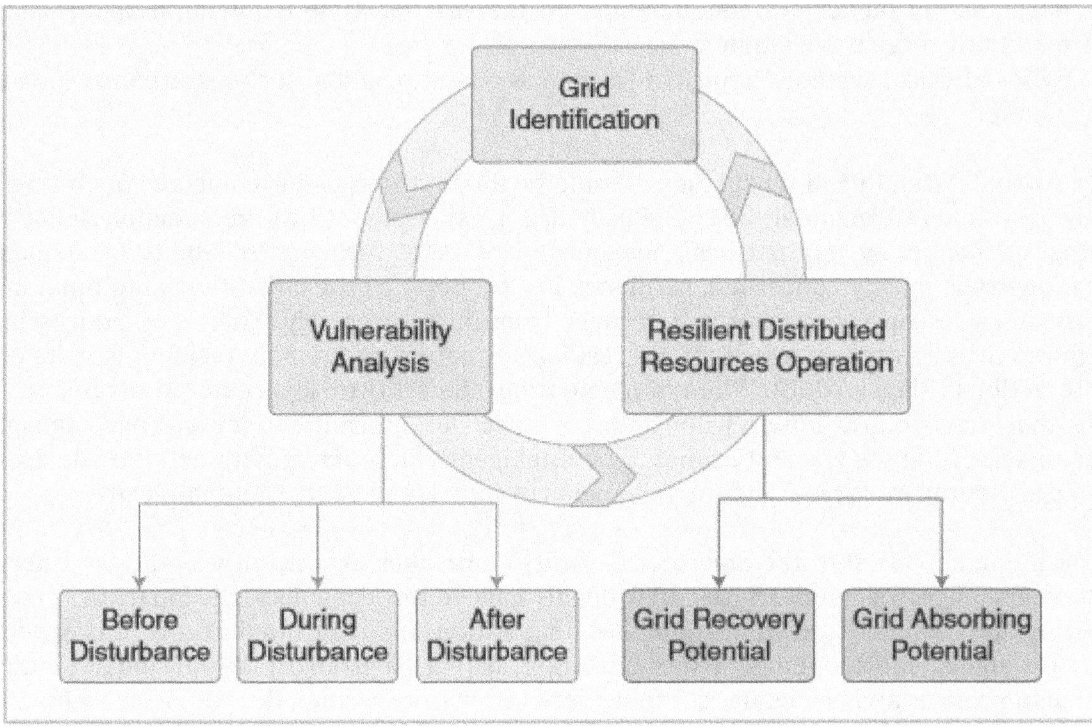

Figure 4.2 Resilience assessment framework for T&D networks. *Source*: [Arghandeh 2014]. © 2014, IEEE.

Improvements to overall service reliability are reported by diverse projects described in [DOE 2014]. For example, utilities used outage management approaches and deployed Smart Grid technologies to accelerate the service restoration following major storm events, reduce the total number of affected customers, and improve overall service reliability by reducing customer losses from power disruptions. This practice generally involves starting out with relatively small-scale deployments and emphasizes testing and evaluation before making commitments to larger scale investments.

In addition, a cyber infrastructure is required to support real time control based on multi layer control loops for multiple domains and intra domain [Sanders 2012]. The resiliency of the Smart Grid is dependent on two interdependent infrastructures (physical and cyber). Both infrastructures should be designed to meet the properties of resiliency. There is a need for data to be provided by a modern power infrastructure equipped with sensors and self-protective equipment.

Also, the analysis of the Smart Grid and DER systems has to include infrastructure needs (Smart Grid is a critical infrastructure) and address concerns mandated in US by directives such as [PPD-8], [PPD-21], [EO 13636]. Examples of other issues to be considered include:

- All businesses are dependent on critical infrastructure in order to conduct their day-to-day operations.
- Businesses can improve their overall resilience and enhance their protection against all threats and hazards by promoting preparedness, response, and recovery capabilities within their organizations.
- Impact to critical infrastructure that a business is dependent on can have devastating effects to both the business and community.
- There are a variety of resources available that can provide information regarding critical infrastructure and critical infrastructure protection.

Often, the resiliency of the Smart Grid is discussed from the perspective of providing trustworthy Smart Grid operation in possibly hostile environments, and trustworthy infrastructure dependent on both cyber and physical components [Sanders 2012]. A trustworthy system does what is supposed to do, and nothing else. However, trustworthiness is not achieved with security only, it requires to include dependability and performability properties. Also, resilience encompasses other attributes that need to be achieved. Thus, other aspects such as organizational resilience should be a design criteria for achieving the resiliency of the Smart Grid.

The dependence on control systems for the monitoring and control of critical infrastructures raises issues regarding the location and integrity of these systems and their ability to provide owner/operators a high degree of state awareness for a hostile cyber environment, uncertainties in complex networks, and human interactions. Therefore, designs of control systems have to include these concerns besides criteria for functionality and reliability. The interdependencies inherent in critical infrastructure systems can be addressed by implementing resilience in these systems.

This work [Rieger 2013] recommends that the larger issue to be addressed and understood is the mechanisms by which resilience can be achieved in the interconnected systems. Resilience is primarily achieved through dynamic coordination of different means and control strategies that enable systems to detect and respond to operating environment changes. Such changes include system component or interface failures or malicious threats from outside the system.

4.1.4 Resilient Control Systems

A research review of the resilience finds the following [Rieger 2009]:

- Two focused areas of resilience: organizational resilience and IT resilience
  - Organizational resilience is the ability of an organization to survive in the face of threats, including the prevention or mitigation of unsafe, hazardous or detrimental conditions that threaten its very existence
  - Information technology resilience considers stability and quality of service in the face of threats to the computing and networking infrastructure.
- Control systems categorized as:
  - A subset of the IT resilience because they utilize typical information technology components
  - A different category of resilience because of these differences: control systems exhibit a whole layer of complexity not adequately encompassed within the objectives of IT; control systems have to ensure proper operation of critical infrastructures, including energy and manufacturing facilities.

Arguing different definitions of resilient control systems, a revised definition for a resilient control system is [Rieger 2013]:

One that maintains state awareness and an accepted level of operational normalcy in response to disturbances, including threats of an unexpected and malicious nature.

Besides dependability and reliability qualities, the authors discuss that other areas are characterizing resilience in regard to the particular control system including design. All areas provide a level of understanding and assurance in the safe and secure operation of a plant or facility. These areas include [Rieger 2013]:

- Human systems – interactions of control systems with humans have to be understood and methods need to be implemented to consider because they introduce a level of complexity.

- Complex networks – as control systems become more decentralized, the ability to characterize interactions, performance and security of complex networks has to be provided and methods to achieve stability need to be designed; it is suggested that a true global optimization coupled with a local interaction can achieve both the assurance of a global minima, and an acceptable response when designing control system architecture.
- Cyber awareness – cyber awareness has to be included in design and architecture; dynamic mechanisms of probabilistic risk analysis that can link human reliability with the system state are promising to improve the resiliency of control systems; traditional methods of achieving reliability cannot be used to characterize cyber awareness and resilience because of limitation in predicting future behavior of the adversary and occurrence of security attacks.
- Data fusion – design and architecture of resilient control systems need to consider issues for fusion of data generated and collected by different means that include diversity of data (many data types associated with proper operation or performance of critical infrastructure, including cyber and physical security, process efficiency and stability, and process compliancy); use of this data to generate information and knowledge whether by automated and/or human mechanisms; combining the diverse types of data with security data, representation of security data within the measures of performance.

It is recommended that design objectives and priorities have to be revised to include [Rieger 2008]:

- State awareness as the primary design objective; state awareness enables the transformation from reactive to proactive control of power generation, transmission, and distribution systems.
- Defense in depth against malicious actors and actions, and support multiple, sometimes overlapping and competing regulatory agencies.

Limitations of control approaches including lacking an approach to systematically consider faults resulting from the cyber environment associated with a system pave the way to a new approach to design resilient control systems. A consistent approach to recognition and response of cyber and physical degradation is necessary to ensure resilience in a system. The design of a resilient control system is a new control design paradigm that encompasses cybersecurity, physical security, economic efficiency, dynamic stability, and process compliancy in large scale, complex systems [Rieger 2013]. Figure 4.3 shows the reference architecture of a resilient control system. As shown in this diagram, RCS includes both the cyber and the physical aspects of system design and operation, thereby overcoming the limitations of the reliable computing and fault tolerant control perspective [Rieger 2013].

As shown in this diagram, the basic element of the framework is the intelligent control that is underlying feedback control, which may be hosted on many communicating platforms, including transmitters, converters, logic solvers, and operator displays. This element is holistically integrated with cyber awareness and decision making. Data fusion is a method to combine data to yield information and knowledge, which provides state awareness and the basis for decisions.

However, the costly, numerous, and different definitions of survivability and resiliency among researchers as well as no agreed approaches with the community of practitioners (software and system developers) contribute to a slow adoption of survivability and resiliency principles by Software Engineers and developers. However, the proposed cyber resiliency engineering framework in [NIST SP800-160v2] could help in understanding and applying cyber resiliency in the system life cycle. The framework constructs include goals, objectives, techniques, implementation approaches, and design principles. Organizations can select, adapt, and use some or all of the cyber resiliency constructs described in the document by applying the constructs to the technical, operational, and threat environments for which systems need to be engineered.

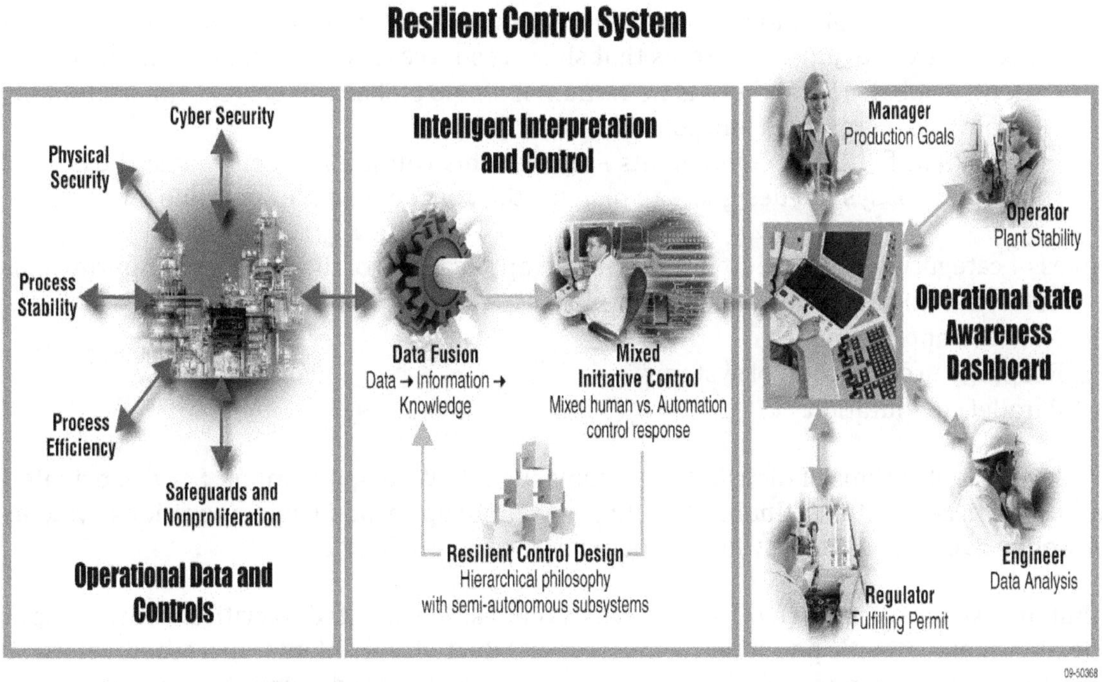

Figure 4.3 Resilient control system framework. *Source:* [Rieger 2012]. © 2012, IEEE.

4.2 Distributed Systems

A distributed system is a collection of independent computers that appear to the users of the system as a single coherent system [Tanenbaum 2003]. The computers are linked by a computer network and equipped with distributed system software which enables computers to coordinate their activities and to share the resources of the system that include hardware, software, and data [McGraw-Hill 2005].

4.2.1 Purpose, Benefits, and Issues

The components interact with each other to achieve a common goal and each computer may have its own user with individual needs. The purpose of the distributed system is to coordinate the use of shared resources or provide communication services to the users. A distributed system can be built on any networking technology including the Internet.

Although hardware for distributed systems is important, software determines what a distributed system actually looks like. So, a distributed system is viewed from this software perspective. It is defined as software system in which components located on networked computers communicate and coordinate their actions by passing messages [Coulouris 2011]. The motivation for building and using distributed systems is based on the need to share resources such as CPUs, memories, peripheral devices, networks, and data of all kinds. Also, distributed systems attempt to hide the heterogeneous hardware components by providing a virtual machine on which applications can be easily executed. While there is no single definition for distributed systems, the following defining properties are commonly used:

- Concurrency of components – in a network of computers, concurrent program execution is the norm; although adding more computers to the network is easy, the coordination of concurrently executing programs that share resources is an important and recurring topic.
- Lack of a global clock – there is no notion of single global and correct time, so timing problems have to be resolved and solutions to be provided.
- Independent failure of components – all systems can fail in many ways; thus, it is the responsibility of system designers to plan for the consequences of possible failures.

Examples of categories of distributed systems and applications of distributed computing include:

- Telecommunication networks
- Network applications
- Real time process control systems
- Parallel computation.

Other specific applications of distributed systems include Web applications, Service oriented architecture (SOA) based systems, financial trading systems, supply and chain operations systems, multi player online games to control systems, and peer-to-peer applications.

Distributed systems offer more reliable services that can be adapted to critical control applications in a flexible manner with potential to increase productivity in plants, increased safety, and decreased energy consumption. Technologies such as higher speed computation, data communications, and object-oriented software allow the integration of distributed control systems into plant wide and enterprise systems.

One of the key advantages of distributed systems is that there are enough resources to achieve fault tolerance, at least with respect to expected errors. The system can be made to tolerate both hardware and software errors, although it should be emphasized that in both cases it is the software, not the hardware, that plays a key role to handle the errors. A distributed system is one in which distribution is combined with replication to achieve fault tolerance as well as performance. With a distributed system, a high degree of fault tolerance is often, at least, an implicit goal.

However, there are issues that must be addressed for the distributed systems. Examples include:

- Security is a big challenge in a distributed environment, especially when using public networks.
- Fault tolerance could be tough when the distributed model is built based on unreliable components.
- Coordination and resource sharing can be difficult if proper protocols or policies are not in place.
- Process knowledge is required for the administrators and users of the distributed model.

4.2.2 Operating Systems as Key Component

Understanding the nature of distributed systems requires identification of the operating systems in relation to distributed computers. The operating systems can be divided in two categories [Tanenbaum 2003]:

- Tightly coupled systems referred as a distributed operating system where operating system tries to maintain a single, global view of resources it manages.

- Loosely coupled referred as network operating system which is a collection of computers each running their own operating system, but the operating systems work together to make their own services and resources available to others. The goal is to offer remote services.

A distributed operating system is a model where distributed applications are running on multiple computers linked by communications. A distributed operating system is an extension of the network operating system that supports higher levels of communication and integration of the machines on the network.

There are two types of distributed operating systems:

- Multiprocessor operating system manages the resources of a multiprocessor.
- Multicomputer operating system is developed for homogeneous multi computers.

Distributed systems can be developed using different models. Distributed computing refers to the use of distributed systems to solve computational problems. In distributed computing, a problem is divided into many tasks, each of which is solved by one or more computers, which communicate with each other by message passing.

The network services provided by a network operating system transform a user's computer into a distributed system. Network operating systems are characterized by the distribution of control that they provide. Modern operating systems include networking services such as network software protocols, augmentation of the file system to support the transfer and use of files from other locations, remote login capability, and additional utilities and tools [Englander 2003].

Many distributed systems make direct use of the programming interface offered by network operating system. Communication is expressed through operations on sockets, which allow processes on different machines to pass each other messages [Stevens 1998]. Enhancements to the services of network operating systems for a better support of distribution transparency led to what is known as middleware. Middleware is a software layer situated between applications and operating systems. Middleware lies at the heart of modern distributed systems and offers a higher level of abstractions and transparency. Middleware is similar to an operating system because it can support other application programs, provide controlled interaction, prevent interference between computations and facilitate interaction between computations on different computers via network communication services.

A typical operating system provides an application programming interface (API) for programs to utilize underlying hardware features. Middleware, however, provides an API for utilizing underlying operating system features [Technopedia]. Middleware simplifies software development by facilitating the following:

- Hides the intricacies of distributed applications.
- Hides the heterogeneity of hardware, operating systems and protocols.
- Provides uniform and high level interfaces used to make interoperable, reusable and portable applications.
- Provides a set of common services that minimizes duplication of efforts and enhances collaboration between applications.

A comparison of essential characteristics of distributed systems based on different distributed software concepts described in [Tanenbaum 2003] is shown in Table 4.3. The table shows a comparison between multiprocessor operating systems, multicomputer operating systems, network operating systems, and middleware based distributed systems. A more detailed comparison of various operating systems from different perspectives is provided at [Wiki OS]. However, more information about basic concepts of modern operating systems that can be useful to Security Professionals and Engineers is provided in [Tanenbaum 2015]. Practices on building distributed systems including information on software concepts, methods, and fundamentals are provided in [Tanenbaum 2007].

Table 4.3 Comparison of essential characteristics.

Essential Characteristics Criteria	Distributed OS-Multiprocessing	Distributed OS Multicomputing	Network OS	Middleware based DS
Degree of transparency	Very high	High	Low	High
Same OS in all nodes?	Yes	Yes	No	No
Number of OS copies	1	1 to N	N	N
Basis for communication	Shared memory	Messages	Files	Model specific
Resource management	Global, central	Global, distributed	Per node	Per node
Scalability	No	Moderately	Yes	Varies
Openness	Closed	Closed	Open	Open

Based on the control functions performed, a distributed control system can be architected into these hardware components: operator interfaces, I/O subsystem, connection buses, and field control units; and into these software modules: history modules, control modules, I/O modules, and network modules.

In advanced industrial controls, the control units in distributed systems are digital, intelligent controllers or computers, containing microprocessors. From the computer point of view, both distributed control systems and distributed computer systems have similar hardware and software architectures. Therefore, distributed operating systems are required by both distributed control and computer systems. The term distributed system is therefore used here for both distributed control systems and distributed computer systems.

4.2.3 Smart Distribution Systems

The future Smart Grid is envisioned as a large scale cyber-physical system encompassing advanced power, communications, control, and computing technologies. As Smart Grid capabilities, such as Advanced Metering Infrastructure (AMI) and demand/response, are developed, and as large numbers of distributed energy resources (DERs) and plug-in hybrid electric vehicles (PHEVs) are deployed, the automation of distribution systems becomes increasingly more important to the efficient and reliable operation of the overall power system. A technical perspective for achieving the changes to the existing distribution system into a smart distribution system is a topic of discussion by researchers and engineers [Brown 2008], [Brown 2009].

In order to help elucidate expectations of the Smart Grid from the perspective of emerging distribution systems, an imperative requirement is a unique framework for the smart distribution system that takes into account the following [Brown 2010]:

- Levels and locations of increasing intelligence in distribution systems
- Reconfiguration of distribution system architecture
- Utilization of sensors in distribution systems that aid both supervised and fully automated controls
- Enabling strategies and configurations for interconnecting renewable energy sources to distribution systems.

The results of surveys and research described in [Brown 2009] and [Brown 2010]) identify eight basic attributes of a smart distribution system as shown in the diagram of Figure 4.4. Also, the participants of the survey detailed functionality and technological applications for each system shown in this diagram [Brown 2010]. These systems have to interact and cooperate to accomplish information exchanges and intelligent tasks to maintain an efficient, reliable, and secure the envisioned Smart Distribution System, which is a system of systems. Each system shown in the diagram has to meet the requirements of the business. This can be achieved by a reliable design and implementation of the goals and functions of the distributed system as discussed in this section. More security has to be approached from the perspective of each distributed system and from the perspective of collaboration with a central repository that coordinates various tasks to accomplish the requirements of the Smart Distribution System.

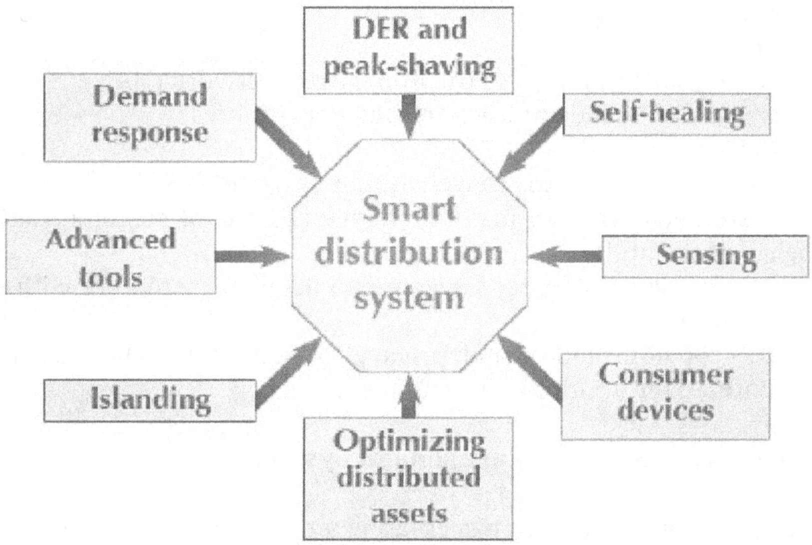

Figure 4.4 Eight attributes of a smart distribution system. *Source:* [Brown 2009]. © 2009, IEEE.

The following is an overview of different distributed systems that are key applications for the success of the Smart Grid.

4.2.3.1 Distribution Grid Management

Distribution grid management system (DGM) focuses on maximizing performance of feeders, transformers, and other components of networked distribution systems and integrating with transmission systems and customer operations. The anticipated benefits of distribution grid management include increased reliability, reductions in peak loads and improved capabilities for managing distributed sources of renewable energy. To achieve these benefits, the system should be designed with consideration for the user, the infrastructure, and the business goals. In addition, important system quality attributes (e.g., reliability, scalability, security) and measurements that measure these attributes have to be defined and planned for implementation.

4.2.3.2 Demand Response Management Systems

Power grid transformation to a Smart Grid requires more and more applications that support the markets and dynamics of changes in the operations of many electric utilities for peak load management, economic control and distribution system operations and optimization. Demand Response (DR) is critical to the ability of customers to make choices to potentially reduce their energy consumption. It requires a platform capable of adapting to ever changing business needs while remaining focused on providing the infrastructure for both command and control as well as customer empowerment. Demand Response Management Systems (DRMS) is a new wave of software that links the utility back office to its customers. It has the ability to control, operate and monitor remote assets.

Like the meter data management system (MDMS), DRMS is an information management system that connects the flow of information to the DR devices to/from the utility systems (Customer Information System and the Distribution Management System), leveraging the Smart Grid/AMI network [Vos 2009]. Therefore, there are critical functional elements that must be supported:

- All customer classes
- Utility operations
- Device agnostic
- System of Record (SoR) (billing) for the Home Area Network (HAN)
- Expandable system to include new emerging opportunities
- Scalable
- Ability to manage legacy systems as well as new systems
- Connect to customer portals to maintain active records of customer devices and manage them on behalf of the utility
- Ability to integrate with customer's systems to provide customers with valuable, decision making, information
- Ensure security of information and privacy of customers. Thus, security requirements should be identified for utilities and customers.

4.2.3.3 Distributed Energy Resource Management Systems

The proliferation of distributed generation is creating a real effect on the utility industry and jeopardizing the grid stability. A group of DERs that are geographically disperse (e.g., associated with different utility meters, residing on different feeders, or not having clearly defined electrical boundaries) is aggregated and coordinated to act as a single entity. Implementing these DERs and aggregated DERs effectively requires a good understanding of the physical topology and connectivity of the various systems where the resources are located and broad visibility to observe the state of various components and assets. In order to overcome these challenges, regions with higher distributed generation and renewables tend to adopt the new Distributed Energy Resource Management Systems (DERMS) technology.

DERMS is a software based solution that increases an operator's real time visibility into the status of distributed energy resources and allows distribution utilities to have the heightened level of control and flexibility necessary to more effectively manage the technical challenges posed by an increasingly distributed grid [Saadeh 2014]. Additionally, the control of the aggregated resources is accomplished through a mix of strategies and signals that can involve markets. Vendors approach this technology with a variety of perspectives and diverse experiences.

4.2.4 Security Challenges in Distributed Systems

There are security concerns with the middleware software because it cannot rely on the underlying local operating systems to adequately support security. Thus, security has to be partly implemented in the middleware software.

Another security challenge of distributed systems is the use of firewalls for access control. Firewalls can impede legitimate access to services when resource sharing between internal and external users is required (e.g., supply chain applications). Therefore, firewall configurations have to be designed and must be complemented by more fine grained mechanisms and policies.

The distributed systems whose resources are QoS managed benefit by having more predictable resource utilization and more efficient resource allocation. The concept of security as a dimension of Quality of Service in distributed system (called Quality of Security Service (QoSS) is proposed in [Irvine 2000]. The enabling technology for both QoSS and a security adaptable infrastructure is variant security, or the ability of security mechanisms and services to allow the amount, kind or degree of security to vary, within predefined ranges. For example, QoSS for the network has the potential to provide administrators and users with more flexibility and potentially better service, without compromise of network and system security policies. A security administrator or user may need to specify a range of security as a function of cost or other variable. All security requirements define a range of permissible behavior; a range may be unitary, or degenerate, in which case it represents no choice. Where a range represents a choice, the requirement is called security variant and can be implemented without compromising existing security policies.
Also, many security concerns arise from the use of emerging trends such as modern ubiquitous computing and new paradigm described as pervasive computing, ambient intelligence or everywhere. However, each term emphasizes slightly different aspects.

4.2.5 Ubiquitous Computing

Ubiquitous computing (UbiCom) is a concept where computing is made to appear anytime and everywhere. In contrast to desktop computing, UbiCom can occur using any device, in any location, and in any format. Ubiquitous computing is a paradigm in which the processing of information is linked with each activity or object as encountered [Technopedia]. It involves connecting electronic devices, including embedding microprocessors to communicate information.

Devices that use UbiCom have constant availability and are completely connected. Ubiquitous computing focuses on learning by removing the complexity of computing and increases efficiency while using computing for different daily activities. A user interacts with the computer, which can exist in many different forms, including laptop computers, tablets and terminals in everyday objects such as a refrigerator or a washing machine.

Ubiquitous computing (UbiCom) postulates a world where people are surrounded by computing devices and a computing infrastructure that supports us in everything we do [Poslad 2009]. Therefore, the features that distinguish UbiCom systems from distributed systems include [Poslad 2009]:

- Situated in human centered personalized environments, interacting less obtrusively with humans
- UbiCom systems are part of, and used in, physical environments, sensing more of the physical environment.

The main properties of UbiqCom are classified into five main types or groups of system properties to support the five main requirements for ubiquitous computing [Poslad 2009]:

- Distributed system properties - distributed system, middleware, set of generic services (e.g., universal, seamless, heterogeneous; mobile, nomadic)
- Implicit Human–Device Interaction (iHCI) system properties (e.g., Non intrusive, hidden, invisible, calm computing; user aware)
- Context aware system properties (e.g., unique, situated, localized; ICT awareness)

- Autonomous system properties (e.g., automatic; emergent, self organizing)
- Intelligent system properties (e.g., reactive, reflex; learning, adaptive).

These groups of properties act to provide a higher level of abstraction of the important characteristics for analyzing and designing ubiquitous systems. Security would be also needed and it should be designed and adapted for ubiquitous computing use.

A taxonomy of properties for ubiquitous computing and a framework are described in [Poslad 2009]. The author mentions some intricacies such as: the group of properties are not exclusive; different notions and visions for ubiquitous computing overlap; some properties are similar but are referred to by different terms. Figure 4.5 shows a UbiCom system model boundary, components, and properties. The dotted line indicates the UbiCom system boundary.

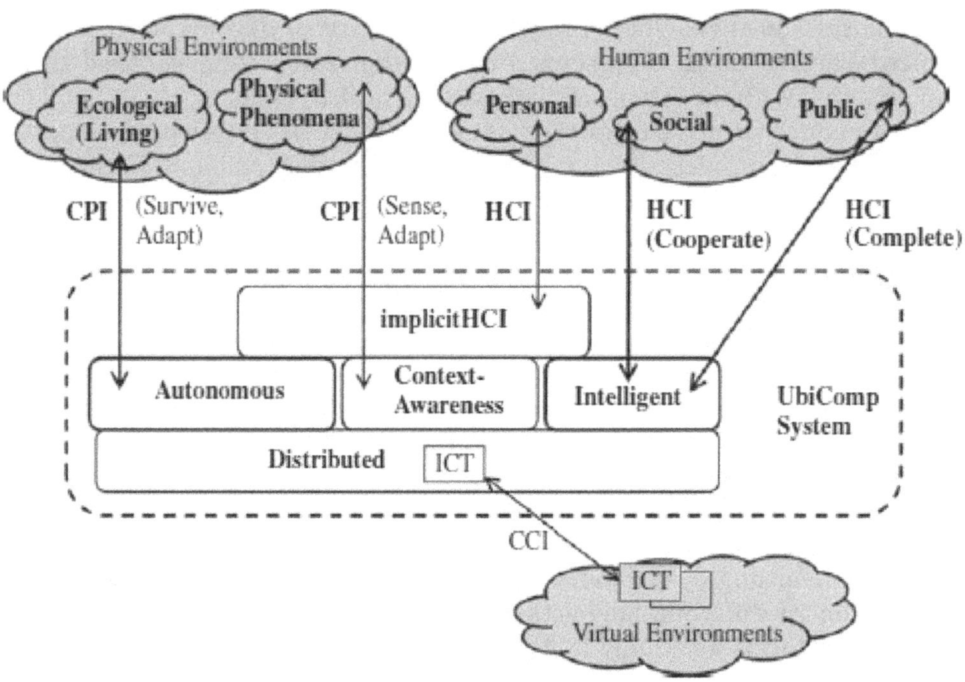

Figure 4.5 A UbiCom system model. *Source:* [Poslad 2009]. © 2009, Wiley.

Ambient intelligence is closely related to the long term vision of an intelligent service system in which technologies are able to automate a platform embedding the required devices for powering context aware, personalized, adaptive and anticipatory services. A typical context of ambient intelligence environment is a home environment. Design of such system requires considering security issues on several levels as described in [Bieliková 2001].

The successful implementation of emerging DER applications requires use of these new technologies and paradigms. Demand/response applications adjust energy use are based on response to dynamic price signals and policies. These applications can be designed using a ubiquitous system based on multi agent system and autonomic systems models. Also, the ubiquitous home environment is designed to regulate the consumption of a utility (such as energy or heating) and to improve usage efficiency [Poslad 2009].

An example of Ubicom application is the utility regulation for the ubiquitous home environment designed to regulate the consumption of a utility such as water, energy or heating and to improve usage efficiency. As described in [Poslad 2009], multiple devices can self manage themselves and cooperate to adhere to users' policies such as minimizing energy expenses. For example, if several overlapping devices are deemed to be redundant, the system will decide which individual one to switch on. Energy usage costs will depend upon multiple factors, not just the time a device is switched on, but also upon the energy rating which varies across devices and the tariff, for example, the cost of energy usage varies according to the time of day. Advanced utility consumption meters can be used to present the consumption per unit time and per device and can empower customers to see how they are using energy and to manage its use more efficiently. Demand response designs can adjust energy use in response to dynamic price signals and policies. For example, during peak periods, when prices are higher, energy consuming devices could be operated more frugally to save money. A direct load control system, a form of demand response system, can also be used, in which certain customer energy consuming devices are controlled remotely by the electricity provider or a third party during peak demand periods.

It is clear that several underlying technologies support ubiquitous computing and each technology has its own security challenges. Therefore, a security engineer needs to identify and analyze each technology for security gaps and assess the risks besides other issues of design, modeling, and interfacing with users. For example, a domestic ubiquitous computing environment might interconnect lighting, appliances, smart meter, and environmental controls with personal information. Privacy is easily the most often cited criticism of ubiquitous computing (ubicomp) and may be the greatest barrier to its long term success [Hong 2005]. This work describes a novel tool kit that is designed to provide support for building privacy sensitive ubicomp applications that are optimistic, pessimistic, and mixed initiative. The kit provides reusable mechanisms for both application developers and for end-users in managing personal information, as well as mechanisms and abstractions for developers designing privacy sensitive ubicomp systems.

4.3 Fault Tolerance

Fault tolerance and failure recovery are important aspects of Systems Engineering and Security Engineering. Fault tolerance is the capability of a system to perform in accordance with design specifications even when undesired changes in the internal structure or external environment occur [McGraw-Hill 2003].

Fault tolerance is also called graceful degradation because it enables a system (often a computing system) to continue operating properly in the event of the failure caused by one or more faults within some of its components [FreeDictionary]. The operating quality is decreasing proportional with the severity of the failure.

The goal of fault tolerance is to ensure that system faults do not result in system failure. Fault tolerance capability is needed in situations where system failure could cause catastrophic accidents, threats to human life, or economic losses. The basic characteristics of fault tolerance require:

- No single point of failure
- Fault isolation of the failing component
- Fault containment to prevent propagation of the failure
- Availability of reversion modes.

Fault tolerance techniques may be applied to both hardware and software. Hardware fault tolerance is a well understood problem. It consists of including two or more identical copies of every working part, constructed in such a way that the backup can take over for any system that breaks. It also requires that broken parts can be swapped out with new ones while the system is still operational.

4.3.1 Fault Tolerance and Security

Fault tolerance is one approach to develop dependable software. Computing systems such as high availability systems, life critical systems, control systems, and security critical systems need to be designed with fault tolerance capability. Fault tolerance may also characterize the rules by which the entities interact. For example, the Transmission Control Protocol (TCP) is designed to allow reliable two-way communication in a packet switched network, even in the presence of communications links which are imperfect or overloaded. The end points of the communication are designed to expect packet loss, duplication, reordering and corruption, so that these conditions do not damage data integrity, and only reduce throughput by a proportional amount.

When we talk about fault tolerance and security, we have to distinguish between ensuring security to a fault tolerant system and designing fault tolerance properties to security mechanisms or security architectures. Fault prevention and fault tolerance aim to provide the ability to deliver a service that can be trusted, while fault removal and fault forecasting aim to reach confidence in that ability by justifying that the functional and the dependability and security specifications are adequate and that the system is likely to meet them.

One method to achieve fault tolerance property is by using spare components in three different ways:

- Replication - provides multiple identical instances of the same system or subsystem, directing tasks or requests to all of them in parallel, and choosing the result on the basis of a quorum; replication could be with double or triple elements.
- Redundancy – provides multiple identical instances of the same system and switching to one of the remaining instances in case of a failure (fail over).
- Diversity – provides multiple different implementations of the same specification and using them like replicated systems to cope with errors in a specific implementation.

In Smart Grid and DER systems, the needs for the availability of the information or applications have to be identified and prioritized in order to provide the appropriate and cost-effective solutions for security. Also, it is needed to understand when system failures are most likely to happen and assess the reliability requirements to include fault tolerance techniques.

The design of fault tolerance capability requires effort in time and resources. Fault tolerance is among the most expensive possible solutions and should be justified. Usually, fault tolerance is used for the most critical information of mission-critical systems [Harris 2013] and to improve the reliability of a system [Sommerville 2007]. Solutions are chosen to balance the value of having the information available against the cost of keeping that information available. All the concepts of fault tolerance are hard enough and the proper way to do is to consider use cases and abuse cases of a system. These issues are still not well established in the practice by community of security professionals and research is needed to resolve the interaction between fault tolerance and security. Interactions of fault tolerance with security can happen in different ways such as failure model, the nature of resilience, the location of redundancy used, and defenses against service denial attacks [Anderson 2008]. Therefore, needs for fault tolerance capability in a system or application have to be assessed before starting any design.

Fault tolerance property may be required for the infrastructure (networks, computing devices) supporting control functions. In networks, fault tolerance is often achieved by a combination of hardware and software techniques that improve the reliability of a system [Stamper 2003]. For example, backup devices help ensure network tolerance (resistance to network failures and crashes).

Besides fault tolerance architecture where capabilities are implemented in hardware and software, a complementary approach is defensive programming [Sommerville 2007].
Defensive programming is an approach based on assumptions that there may be undetected faults and consistencies. So, redundant code is included to check the system state after modifications to ensure the changes are consistent.

4.3.2 Fault Tolerance Actions

Fault tolerance actions are triggered by fault detection capabilities. In addition, fault recovery and fault repair capabilities have to be available. Examples of fault detection requirements include:

- System must detect that a fault (an incorrect system state) has occurred
- Damage must be assessed
- Components of the system state affected by the fault must be detected.

Fault recovery capability must restore system after a failure to a known safe state. Failure recovery is a procedure that allows for restart of a failed system in a way that either eliminates or minimizes the amount of incorrect system results [FailRecovery]. Appropriate methods need to be applied for different system states (privileged or user) to ensure that recovery does not cause compromises that could be exploited [Harris 2005]. Fault repair implies that the system may be modified to prevent recurrence of the fault. As many software faults are transitory, this is often unnecessary.

4.3.3 Security vs. Fault Tolerance Issues

From the security perspective, there are issues that have be understood how to use and when to use fault tolerance methods. The following is an overview of these issues.

Classical fault tolerance is usually based on specific mechanisms that are greatly complicated when it must withstand malicious attacks on these mechanisms [Anderson 2008]. The author argues that fault tolerance interacts with security in a number of ways such as the failure model, the nature of resilience, the location of redundancy, and the defence against denial service of attacks.

Availability is one security service that depends on network and resource availability. A network needs to be properly maintained and its resources are always available when they are needed. Another example includes control systems that are based on temporal considerations (real time) rather than on architectural characteristics. Availability of such systems is crucial and different fault tolerance mechanisms such as redundancy, replication, fault tolerant systems, and recovery mechanisms are used to cope with system failures.

Failure access is a type of incident in which unauthorized access to data results from hardware or software failure. While fault tolerance capabilities are provided with some networking technologies, other availability solutions are chosen to balance the value of having information available against the cost of keeping that information available. Examples include [Harris 2013]:

- Redundant hardware ready for hot swapping allows replacement of the failed component while the system continues to run and information remains available; it may cause degraded performance, but unplanned time is avoided.
- Fault tolerance technologies keep information available against not only individual device or software faults but even against whole system failures.
- Service level agreements (SLAs) can be designed based on the type of availability is needed.

One potential risk to the availability of a system and information is a single point of failure. Examples of devices that could be single points of failure are firewalls, routers, network access servers, communication lines, switches, bridges, hubs, and authentication servers. Best defenses against single points of failure vulnerability include regular backups, redundancy (e.g. redundant array of independent disks (RAID), storage area networks (SAN), and fault tolerance technologies (e.g., clustering).

Another aspect is replication versus confidentiality. If the replication of data is chosen to make a system fault tolerant there is a chance to increase the risk of a compromise of confidentiality.
From the cybersecurity perspective, there is a need for failover, fail safe, fail secure, and fail soft capabilities.

Failover capability allows to switch over automatically (typically without human intervention or warning) to a redundant or standby information system upon the failure or abnormal termination of the previously active system [CNSSI 4009]

Fail safe is the automatic protection of programs and/or processing systems when hardware or software failure is detected [CNSSI 4009]

Fail secure is the mode of termination of system functions that prevents loss of secure state when a failure occurs or is detected in the system (but the failure still might cause damage to some system resource or system entity) [RFC 4949]

Fail soft is the selective termination of affected, non-essential system functions when a failure occurs or is detected in the system [RFC 4949].

Therefore, it should be understood that fail safe is a mode of termination of system functions that prevents damage to specified system resources and system entities (e.g., specified data, property, and life) when a failure occurs or is detected in the system, but the failure still might cause a security compromise, moving the system to an unsecure state.

If a tolerance technology has a failover capability is still needed to plan for switching to a working system when there is a failure that cannot be handled through normal means. Therefore, failure control is a methodology that is used to detect imminent hardware or software failure and provide fail safe, fail secure, or fail soft recovery solutions.

Fault management is one area of the ISO network management model. Fault management is the process of identifying, tracking, diagnosing, and resolving faults in the systems. Because information security involves systems based on both technology and people, the management needs to focus on finding faults and remediation on operations and systems.

4.3.4 Fault Resilience

From a grid perspective, the power grid is designed and operated with built in redundancy to ensure a certain degree of reliability and resilience. Industry planning criteria assume a local grid area can be operated even if one asset is out of service. In addition, during unforeseen events, the industry provides mutual aid to assist in emergency response and prompt restoration.

Engineering concepts that are related to resilience therefore include for instance fault tolerance, redundancy, stability, feedback control, and safety. However, the task of assessing that a system is adequate secure, is more challenging than proving that it is safe. Often, the goals of safety and security may create contrasting requirements, so that designers need to perform detailed analysis to decide tradeoffs. The difficulty lies in the multi dimensionality of resilience, which is a family of related issues, not a single thing.

One definition of resilience for current and future ICT is ability to deliver, maintain, improve service when facing threats and evolutionary changes [Strigini 2009]. In comparison with the concept of fault tolerance, this extension to the definition emphasizes the fact that the perturbations that current and future systems have to tolerate include change. There is shift from fault tolerance in the architecture of a system component (e.g., the computer) to fault tolerance in the architecture of the system. In addition, as ICT becomes more pervasive and interwoven with human activities, the dependability of the technical components in isolation may not be the only part necessary in the study of dependability and thus of resilience. The boundaries between strict technical ICT systems and socio-technical systems are fuzzy, and for many applications the recognition of social components in determining meaningful assessment of resilience and dependability is required [Strigini 2009].

One difficulty is the prediction of the fault resilience such that adequate design is provided. It is recognized the modeling methods based on analytical models do not always provide useful results. One reason is assessing and measuring the fault resilience only based on quantitative measurements, thus quality attributes are needed in the resilience model. However, there are challenging issues on selecting the appropriate qualitative variables that could contribute to a better description of the fault resilience model. It is suggested the degradation degree of a service may be measured based on decision mechanisms for actions which are part of risk strategy - either security risk or organizational risk [Theron 2013].

Although no clear path to design fault resilience, this is required for many distributed systems and cloud applications that are emerging in electrical utilities. Many issues are discussed and solutions are proposed to build fault resilient distributed systems and cloud management stacks in [Ju 2013]. For example, OpenStack - a popular open source cloud management stack, is considered fault resilient during the processing of an external request, if it maintains correct and consistent states and behaviors, even in case of occurrence of faults.

When introducing resilience mechanisms into a system, it is needed to understand the purpose and domain to be applied. Economical alternative to fault tolerant systems, fault resilient systems duplicate only a few critical components instead of duplicating all (the entire system) [FaultRes].

According to this work [Theron 2016], resilience is a struggle against collapse. Resilience stems from empirical engineering and requires different plans and mechanisms. These include mechanisms for prevention of threats (P3) and coping with threats (R3). Therefore, it is needed to understand the mechanisms that are provided by Resilience Engineering discipline to defend against threats. In addition, different levels of collapse require different plans. In ICT systems, resilience is a dynamic, engineered aptitude that allows to survive critical incidents that occur despite defense barriers crafted in those systems. Figure 4.6 is a representation of the related mechanisms and threats and their relationship.

According to [Dondossola 2013], resilience is recognized as an opportunity to develop new strengths in business and critical infrastructures and to grow organizations to a higher level of performance. Resilience has also been equated to critical software that is resistant to cyber attacks and extraordinary events may occur in parallel. This is a concern because traditional risk management measures are not efficient in this case [Dondossola 2013]. Risk assessment centers around the probability of hitting a system's weak, while resilience assessment depends on the temporal dimension of potential disturbances and mitigating actions [Arghandeh 2016].

An analysis of abstract principles for resilience as needed for the ICT domain is suggested in [Jackson 2013]. The author argues that concrete solutions cannot be developed from abstract principles individually but must be implemented in combination with other specific principles. The abstract principles are only a starting point for determining the characteristics of a resilient system. Therefore, the Security Engineer needs to consider the interactions, vulnerabilities, limitations, and conflicts associated with the principles as they appear in the concrete solutions.

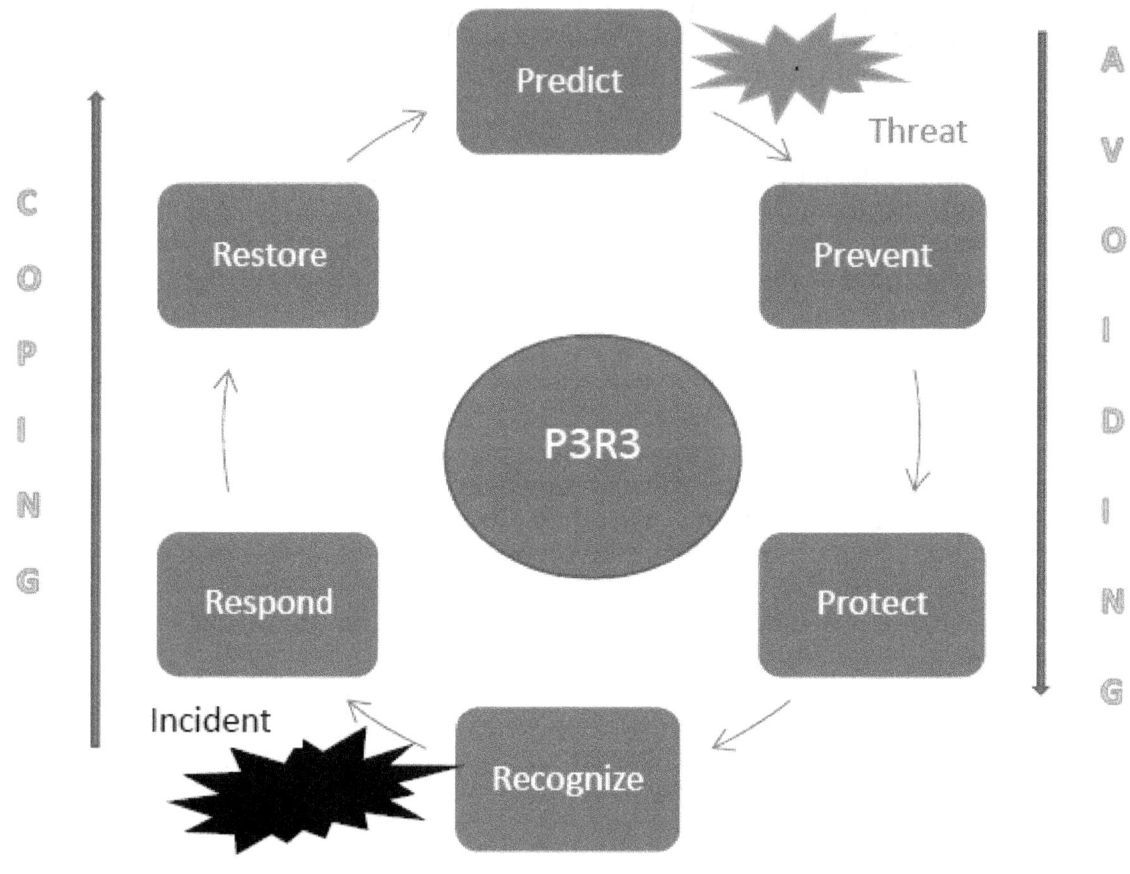

Figure 4.6 Examples of Resilience Engineering mechanisms.

4.3.5 Fault Tolerant Systems

A fault tolerant system is a system that can continue its operation after some system faults have caused a failure within a subsystem or component. An important distinction should be made between systems that are fault tolerant and those that are fault intolerant. A fault tolerant system is one that can continue functioning (perhaps in a degraded form) even if something goes wrong. A fault intolerant system collapses as soon as any error occurs.

In critical situations, software systems must be fault tolerant. Fault tolerance is required where there are high availability requirements or where system failure costs are very high. Fault tolerance means that the system can continue in operation in spite of software failure. Even if the system has been proved to conform to its specification, it must also be fault tolerant as there may be specification errors or the validation may be incorrect.

The basic functions that have to be implemented in a fault tolerant system include [Sommerville 2004]:

- Fault detection – detect fault that occurred and could lead to a system failure
- Damage assessment – identification of failed component
- Fault recovery – restoring system to a safe state by correcting the damaged state (forward error recovery) or by restoring the system to a known state (backward recovery)
- Fault repair – modifying the system such that the fault does not recur.

A fault tolerant system must identify that a failure has occurred and then take corrective action to ensure that the continuity of operations is maintained with the least possible delay [Hansche 2004]. Examples of mechanisms available to support fault tolerant systems include network data mirroring, hot spares, redundant servers, and clustering.

In addition, fault tolerant systems are characterized in terms of both planned service outages and unplanned service outages. These are usually measured at the application level and not the hardware level. One metric used is availability expressed as a percentage. It is defined as the ratio of the total time a functional unit is capable of being used during a given interval to the length of the interval. In high availability systems, the metric is known as 5 nines (0.99999) which is 99.999% availability.

4.4 DER System Characteristics

As many countries around the globe, the United States is transitioning from a centralized power system towards a more decentralized system. As greater amounts of distributed energy resources are available, it is vital to understand the characteristics of these systems in order to design their basic security solutions. The goal is to derive the common security needs and help on their integration with the Smart Grid thus minimizing the impacts to the power system.

A Smart Grid is a modernized electricity network which is being utilized as a way of addressing energy independence, global warming and emergency resilience issues. Components such as renewable, distributed generation, distributed energy storage systems, plug-in electric vehicles identified as DERs are becoming also part of Smart Microgrids. Figure 4.7 shows a model of a possible architecture for a DER system within a Smart Grid network, distinguishing the electrical infrastructure and the communications infrastructure.

4.4.1 Security

The vision of a Smart Grid is enhanced security that includes resistance to attacks as one of seven principal characteristics of the future power grid [NETL 2007b], [NETL 2009]. These unique characteristics are the reasons why cybersecurity of the Smart Grid is imperative [EAC 2008]. However, many technologies that leverage the Smart Grid can increase the vulnerabilities and exposure to threats. Therefore, we need to understand the security characteristics and challenges with smart meters, sensor networks, communication technologies, DERs deployment leverage the benefits of increased threat awareness while mitigating the risks.

Figure 4.8 shows the functions of various components of Smart Grid architecture and how they interact to make the Smart Grid works. Components named as generators, central power plant, renewable distributed resources, isolated microgrid are being integrated using different technologies and controls to support Smart Grid goals [Ghansah 2012].

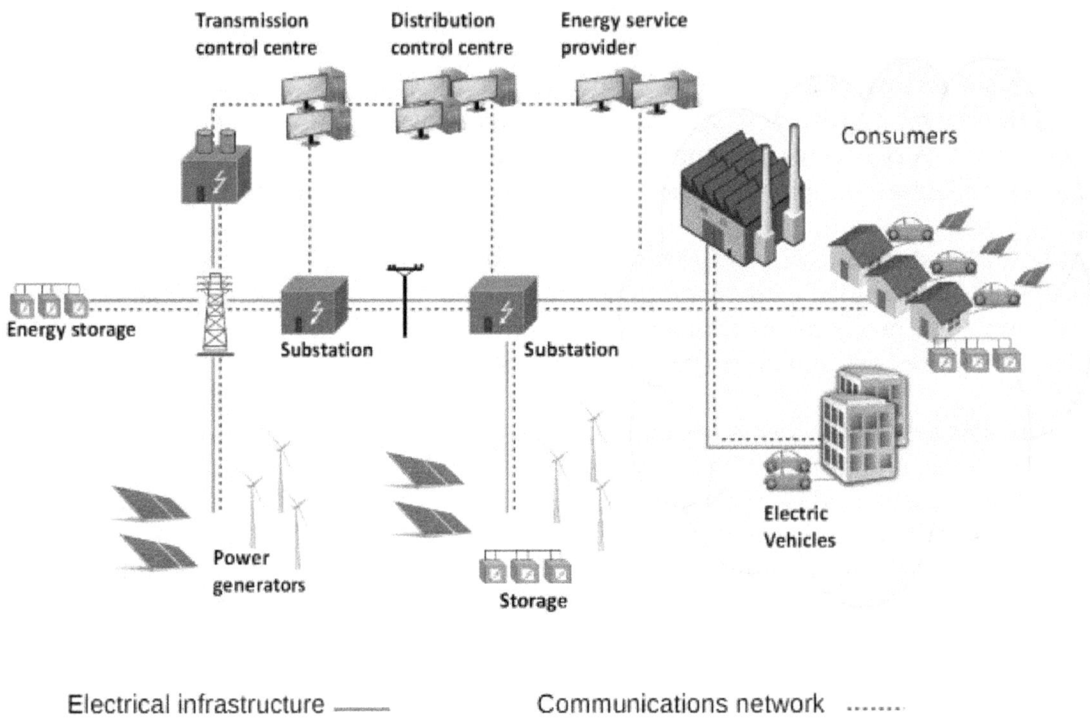

Figure 4.7 Distributed Energy Resource (DER) diagram. *Source*: [ENISA 2015]. © 2015, ENISA.

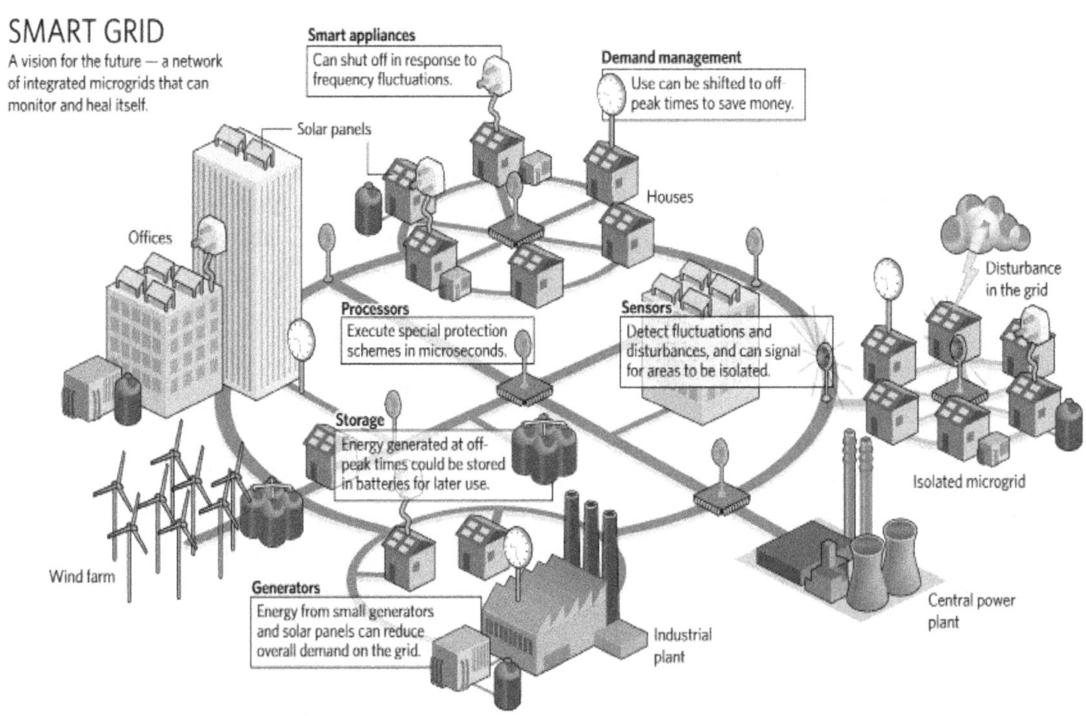

Figure 4.8 SMART GRID A vision for the future — a network of integrated microgrids that can monitor and heal itself. *Source:* [Marris 2008]. © 2008, Springer Nature.

Security issues for many important Smart Grid components are discussed in several reports. Other documents describe DER issues. For example, threats and failure scenarios are described in [Ghansah 2012], [NESCOR 2013a], [NESCOR 2013b]. The document [NESCOR 2013b] includes categories (called generic) of failure scenarios due to malicious and non malicious cybersecurity events that impact the delivery of electricity, the business of running a utility and/or the interests of the customers of a utility. The analysis includes possible vulnerabilities, impacts, and mitigations (suggested cybersecurity solutions) for each system of a number of systems (called domains), but without considering issues such as:

- Resilience capabilities already built in a system so the impacts may result in other scales requiring maybe different mitigations.
- Top down approach, both bottom up and top down approaches should be used to provide more robust assessments.
- Scenario failures may create a ripple effect failure due interdependencies between systems.
- Evaluating impacts of possible threats based on quantitative measures is the key on providing mitigations; scoring based only on empirical data and qualitative attributes (high, low, medium) can create an inadequate cybersecurity posture.
- Impacts to other critical infrastructures and Nation's economy and safety of people.

Another document [EPRI 2013] is focused in cybersecurity requirements for DER systems defined to include: generation and storage systems, both renewable (photovoltaic systems, wind turbines, bio-fuel systems, fuel cells, battery storage systems, electric and thermal storage systems, co-generation systems and small hydro plants) and non-renewable (diesel generators, gas turbine generators, etc.)

4.4.2 System Organization

DER systems are active participants in grid operations and must be coordinated with other DER systems and distribution grid devices. Given the high number of DER systems, potentially millions of DER systems in the field, utilities cannot exercise a direct control of DER systems, so a hierarchical approach is proposed such that utilities can interact with the widely dispersed cyber-physical systems. DER functions in the Smart Grid and proposed DER architectures are described in [EPRI 2013]. A hierarchy and the various devices structured in levels is proposed in this document. The reason for choosing hierarchical model is that a direct control by utilities is not feasible for the thousands and potentially millions of DER systems in the field [EPRI 2013]. Figure 4.9 illustrates this hierarchy and the various devices that may be included in a DER system.

At the local level, DER systems manage their own generation and storage activities autonomously, based on local conditions and DER owner preferences. DER systems are active participants in grid operations and must be coordinated with other DER systems and distribution grid devices. In addition, the distribution utilities must interact with other Regional Transmission Organizations (RTOs) and/or Independent System Operator (ISOs) for reliability and market purposes. This hierarchy is organized in five levels:

- Level 1 is the lowest level and includes autonomous DERs.
- Level 2 is DER energy management facility and manages the operation of the Level 1 DER systems.
- Level 3 is DER communications that extends beyond the local site.
- Level 4 is distribution utility operational analysis to determine which requests or commands should be issued to specific DER systems.
- Level 5 is the highest level and involves the larger utility environment (e.g., transmission and market operations).

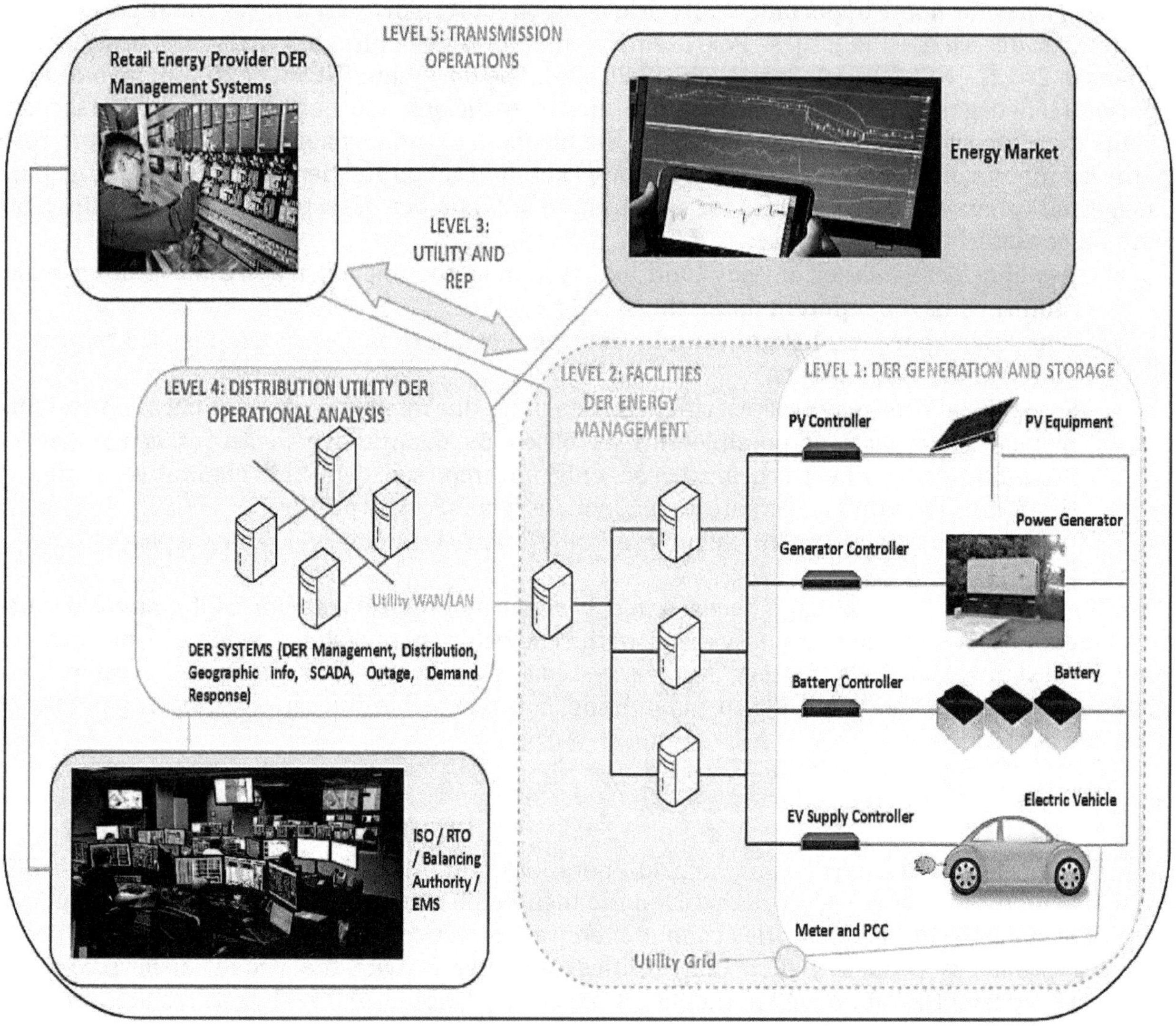

Figure 4.9 Hierarchical DER system architecture. *Source:* [ENISA 2012]. © 2012, ENISA.

At the local level, DER systems must manage their own generation and storage activities autonomously, based on local conditions, preestablished settings, and DER owner preferences. The lowest level includes the actual cyber-physical DER systems operated autonomously. This autonomous operation can be modified by DER owner preferences and/or by settings and commands issued by utilities.

The configurations will include the security profiles as defined by standard profiles. However, the incorrect settings or modification of settings by an intruder are critical to ensure accuracy of data and quality DER functionality.

There are no universally recognized DER size boundaries, but DER systems are usually defined as being interconnected to the distribution power system or possibly to the sub transmission power system. From a security perspective, the multiple combinations of architectures introduce complexity and hidden interactions and dependencies that can be unknown and difficult to assess.

At level 4, distribution utilities implement DER management software applications to analyze and coordinate DER generation and storage as part of the broader DMS functions that manage the safe, efficient, and reliable planning and operation of the power grid. As part of this analysis, the DER applications handle DER registration, determine the optimal settings for DER autonomous functions, establish pricing signals for DER energy and ancillary services, initiate control commands when necessary, and transmit this information to the appropriate facilities within the utility jurisdiction. Therefore, DER software applications are highly critical to safe and reliable power system operations and their commands can affect large number of DER systems.

Figure 4.10 shows a view of DER operational analysis architecture comprising six systems that have interactions as illustrated. Therefore, each interface has to be assessed such that appropriate security controls are applied to protect the sensitive information and all the other assets.

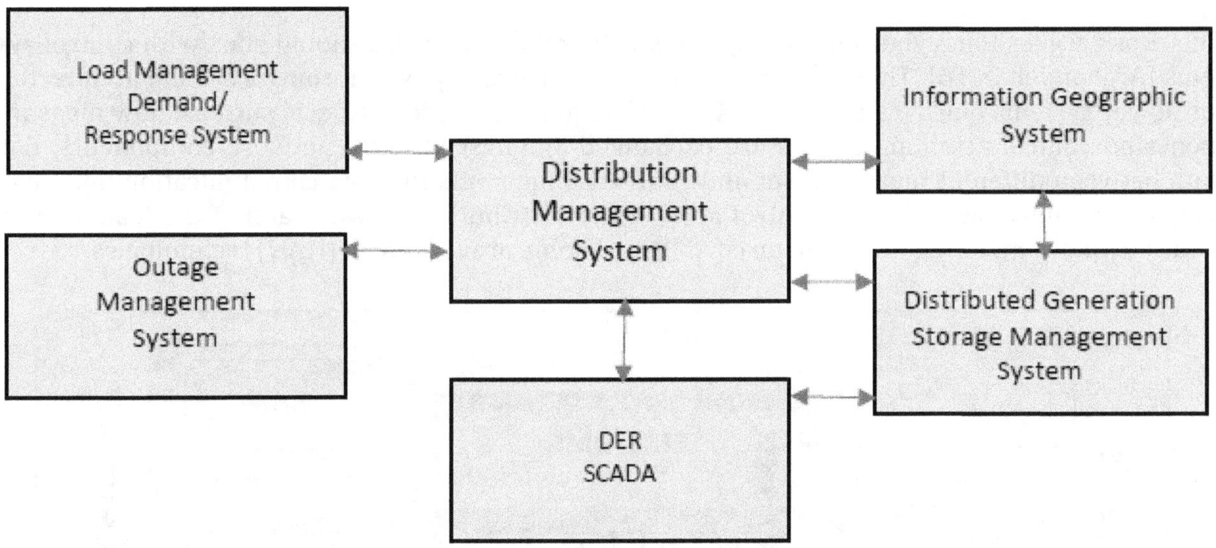

Figure 4.10 Level 4 distribution utility DER operations architecture view.

DER characteristics can be described from different perspectives such as cyber-physical systems, unique DER system, location, interconnections, management, uses, environment, etc. Therefore, these characteristics can provide the basis for understanding and analysis of security needs.

4.4.3 Cyber-Physical System

A cyber-physical system (CPS) is a system of collaborating computational elements controlling physical entities. Cyber-physical systems are hybrid networked cyber and engineered physical elements codesigned to create adaptive and predictive systems for enhanced performance. These smart systems present a key opportunity to create a competitive advantage for U.S. industrial innovation and to improve the performance and reliability of new and existing systems. From smart manufacturing and Smart Grid to smart structures and smart transportation systems, CPS will pervasively impact the economy and society.

DER system as cyber-physical systems consist of physical/electrical components (e.g., the electric generator and storage components) and cyber components (e.g., DER controllers, information), thus making them cyber-physical systems. They are generally designed to operate autonomously according to settings in the DER controller. Each type of DER system has its own unique characteristics, but in general, each DER system can be treated as a small to medium sized source of electric power. Electric Vehicles (EVs), microgrids, and virtual power plants act as DER systems too.

Many traditional parts of the Smart Grid are increasingly CPS dominated, so the real time response of Smart Grid applications depends on the cyber-physical systems' response. Also, cybersecurity is a critical cross cutting discipline that provides confidence that cyber-physical systems, their information, and supporting communications and information infrastructures are adequately safeguarded. CPS are increasingly being utilized in critical infrastructures and other settings.

It is predicted that Smart Grid is a DER CPS dominated infrastructure (other examples of CPS are smart meters, control systems), which is another challenge to a Smart Grid due to increased amount of information generated. In [Vellaithurai 2015], the authors argue that Smart Grid should be treated of being a CPS and security of the grid should be assessed with this view.

This work argues that cyber-physical systems vulnerability analysis should start with control systems [Arghandeh 2016]. The authors describe a typical cyber-physical control system architecture for power systems illustrated in Figure 4.11. The figure shows key aspects such as: how measurement and control actuation signals are exchanged amongst physical network components; data path between different measurement and control components through communication lines, control and communication levels. Control level of the distribution network and at the load level include advanced metering infrastructure (AMI) and home area network (HAN) technologies.

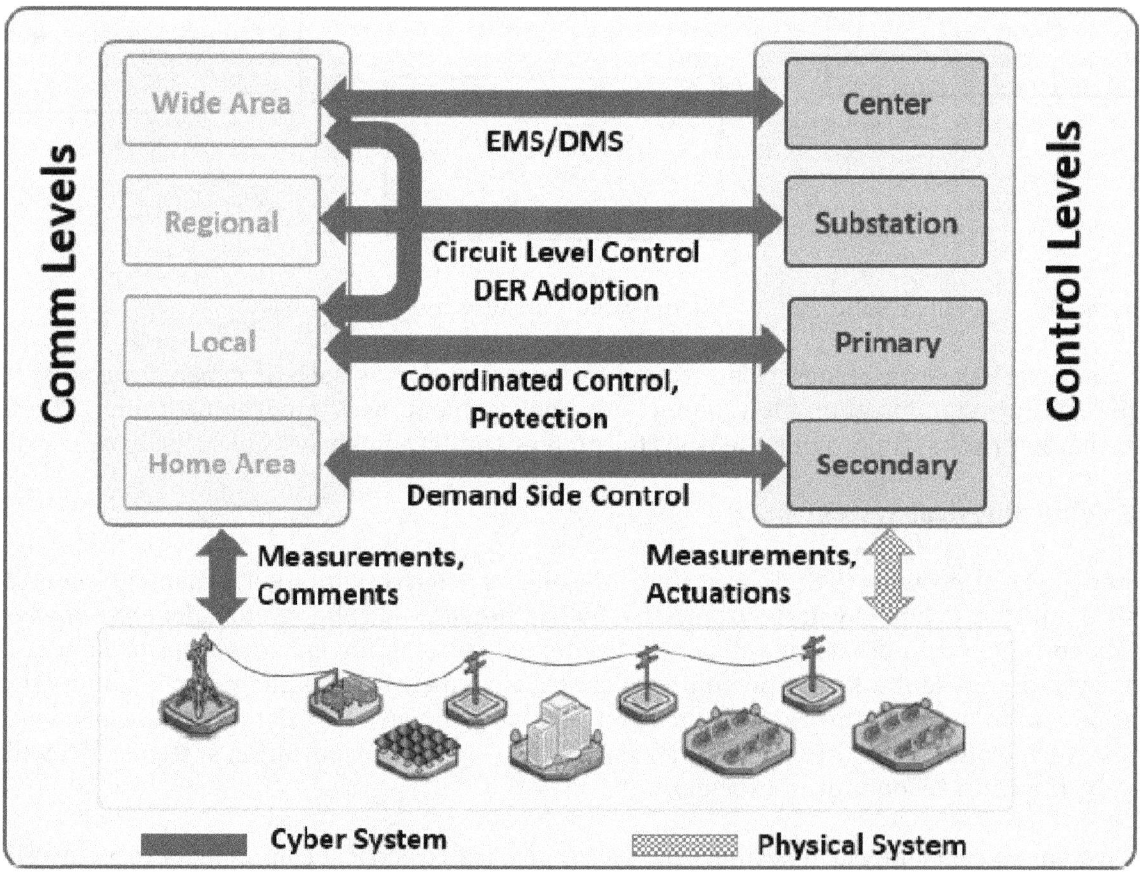

Figure 4.11 Cyber-Physical Control System for an Electricity Grid. *Source:* [Arghandeh 2016]. © 2016, Elsevier.

DERs are penetrating every segment from generation to homes. Customer side (homes, commercial/industrial buildings) use DERs such as electric mobility (plug-in vehicles) and distributed installation of small scale generators (e.g. solar panels) such that a customer is producer and consumer depend on CPS for initiating and monitoring network connections as well as the information exchange among the various entities.

Smart Grid interactions with DERs and among its different actors is expected to be automatic via intelligent user configured agents and smart CPS that allow trading of energy and interaction with value added services (e.g., billing, roaming, inter utility exchange, etc.) including monitoring customer's context (e.g., preferences, current location, capabilities, existing contracts, etc.) and taking automated decisions on his behalf.

Common applications of CPS typically fall under sensor based communication enabled autonomous systems. For example, many wireless sensor networks monitor some aspect of the environment and relay the processed information to a central node.

While integration and cooperation of CPS systems are major goals, security and interoperability have to be addressed from the start [Karnouskos 2011]. So, assessing the information and modeling the DER systems are very challenging tasks. A concept map of cyber-physical systems is available (see [CPS 2012]). Although not a complete diagram in terms of cybersecurity, the diagram illustrates basic elements involved in the description of the cyber-physical systems.

In general, cyber-physical systems (e.g., DER systems and the power system) should be designed for resiliency, to cope with equipment failures and inadvertent cybersecurity events through fault tolerant equipment design, redundancy of equipment, and software applications that model the physical systems using the laws of physics (e.g., power flow based applications). These design protections may counter potential cybersecurity events. In addition, existing cyber-physical reactions may be invoked to mitigate the impact of an attack. These designs and intrinsic mitigations may require improvements to address new cybersecurity events, attacks and threats.

The CPS characteristics may be supported by DERs, but some unique characteristics including real time response, reliability, and cybersecurity are being defined for a DER system by the manufacturer and system designer. In addition, various technologies and applications impact the DER system characteristics including cybersecurity attributes. Therefore, security requirements for DER system should be based on DER characteristics instead of following the approach of tailoring several security requirements from [NISTIR 7628r1] as suggested in [EPRI 2013]. An analysis of these security requirements included in [EPRI 2013] reveals drawbacks such as incompleteness and no feasibility of some security requirements to be realized for the real world systems.

A cyber-physical attack is using both cyber and physical means to attack a target [GAO 2011]. Therefore, various impacts have to be analyzed such as:

- Physical impacts such as power outages and damaged equipment caused by compromising the cyber information and/or control commands.
- Cyber impacts to processes, information, functions, operations, and CPS components including cybersecurity controls; if the charging and discharging processes are compromised by cyber incidents, then the reliance on DER systems is affected.

Therefore, it is needed to protect against coordinated cyber-physical attacks as the potential consequences may be severe [Vellaithurai 2015]. The authors argue that CPS security systems must be able to differentiate between physical and cyber attacks and respond accordingly. The Smart

Grid should be treated to its true nature of being a cyber-physical system and security of the grid should be assessed with this view.

4.4.4 System as Critical System

One issue that requires planning in the early stages of any system is the identification of the criticality of the system and sensitivity of the information. As systems and applications are developed and integrated, the identification of critical systems should be a priority. Critical systems can be defined as those in which a certain class of failure is to be avoided if it all possible.

In addition, understanding vulnerabilities and the associated attack vectors to exploit the critical systems is essential to building effective security mitigation strategies. An evaluation tool is provided by ICS-CERT [Cyber Tool]. Depending on the class of failure, critical systems may be:

- Safety-critical – can lead to loss of life, serious personal injuries, or damage to the environment (e.g., fault operation of electrochemical batteries may result in spills that cause harm).
- Mission-critical – can lead to an inability of the overall system, such as loss of critical infrastructure or data; for example, loss of data or unavailability of utility DER Management System (DERMS) for distribution (level 4) due to a threat may impact the reliability of power grid.
- Business-critical – can lead to significant economic costs such as loss of business or reputation; for example, any threat that causes modification or loss of DR signals during real time communication can cause disruption to DR applications and impact businesses as well as the reliability of load and demand response functions.
- Security-critical – can lead to loss or modification of sensitive data; the continuous and most recent updates of [NERC CIP] standards provide guidelines for the identification of critical cyber assets to support reliable operation of the for electrical power grid (see CIP updates); the standards require the identification and documentation of the critical cyber assets associated with the critical assets that support the reliable operation of a system using a risk based assessment approach. One example of security critical type is the DER SCADA system (level 3 or level 4) defined as a control point that provides monitoring and control, automatic generation control, real time power system modeling, and real time inter utility data exchange.

4.4.5 System Location

DER systems are located at residential, commercial, and industrial customer sites and are usually owned and managed by the utility customers located at those sites. Utility owned DER systems may be located at utility sites, such as substations, or may be located by mutual agreement at customer sites (e.g., rent-a-roof contracts). Also, multiple DER systems may be colocated in physical power plants, combined in aggregations as virtual power plants, deployed as utility owned units for ancillary services support, or included in market driven microgrids.

These sites have little or no security and with owners who have minimal or no cybersecurity expertise. In addition, unlike utility owned smart meters, the customers must be permitted to interact with the DER systems that they own, since they often use these systems to meet their specific requirements. These factors may increase the potential cybersecurity vulnerabilities to the interfaces between DER systems and utilities.

Figure 4.12 is a high level view of a Smart Microgrid model with its elements. This microgrid includes the renewable electricity generation systems that have been complemented with an energy battery, a power conditioning system, and an advanced energy management system to help manage renewables variability and uncertainty. The storage system serves to not only support flexibility of the grid system but also to provide initial start up power to the wind turbine. In addition, the automated power management system, which tracks electricity on a real time basis, helps to optimize grid operation. From a criticality point of view, the microgrid is a business critical system to provide energy to consumers.

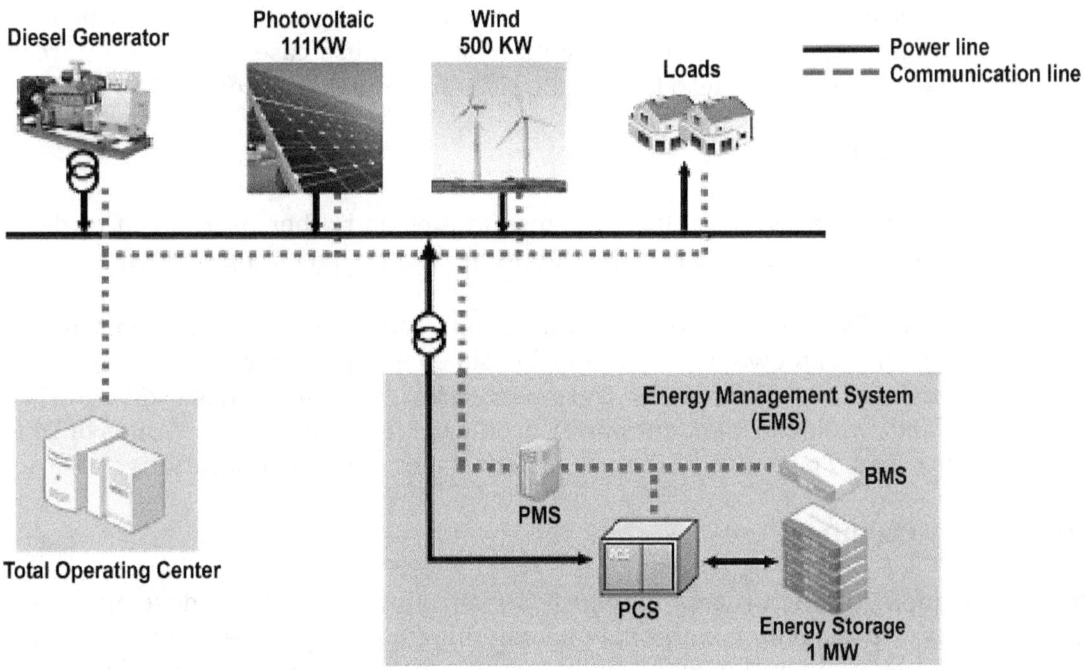

Figure 4.12 Schematic illustration of the Gapa Island microgrid model. *Source:* [Speer 2015]. Public Domain.

As described in this report [Speer 2015], wind and solar PV generation systems of 500 kW and 111 kW, respectively were installed on Gapa Island to replace a 450-kW diesel generator. The renewable electricity generation systems have been complemented with a 1-MW Lithium-ion battery, a power conditioning system, and an advanced energy management system to help manage renewables variability and uncertainty. The storage system serves to not only support flexibility of the grid system but also to provide initial start up power to the wind turbine. In addition, the automated power management system, which tracks electricity on a real time basis, helps to optimize grid operation.

Based on this proven model, which generates electricity from renewables and storage and power system management technologies, Korea plans to roll out the self sufficient Smart Grid concept on other small islands that have traditionally depended on instable fossil fuel supplies [IRENA 2013].

4.4.6 System Interconnections

The complex DER domain is still evolving. The increasing numbers of DER systems that are being interconnected, more stakeholders managing these interconnected systems, and updates to regulations are being debated [EPRI 2013]. Since most of the focus is on getting the DER functions understood, the DER systems installed and operational, and the interconnection requirements standardized, not too much attention is on cybersecurity requirements [EPRI 2013]. In addition, many different communication protocols and standards are currently being used that often have inadequate cybersecurity capabilities.

DER interactions cross many different domains and many different organizations. Although several issues related to the security of DER systems and interconnections to the Smart Grid are described in [EPRI 2013], there should be more focus on defining security and resilience requirements for the interfaces to prevent events that lead to disasters. More security requirements implemented in the DER systems do not necessarily will make them more secure because of increased complexity that may not be well handled by the manufacturers and designers. Also, inherent conflicting requirements and security mechanisms could impact the overall performance and interoperability with other Smart Grid systems. Therefore, depth analysis of use cases and DER interconnections could highlight security gaps and lead to mitigate risks to avoid disruptions.

4.4.7 Managing Systems

Given the autonomous behavior of DER systems, utilities and other users need to develop their own policies and requirements for the DER systems integration including security.

Utilities do not typically have direct organizational control over these DER systems and often need to operate through the DER owners, commercial retail energy providers, aggregators, virtual power plant managers, or other third parties. Third party providers may manage some DER systems as virtual power plants, while independent power producers may manage the groups of DER systems that form power plants. To resolve DER management issues, several applications can be used.

4.5 Contingency Planning

Contingency planning is the process by which organizations position to develop responses to a contingent event or possibility conditional on something uncertain. The main goal is the restoration to normal modes of operation with minimum cost and disruption to normal business activities after an unexpected event [Whitman 2011].

Contingency planning is the family of security and privacy controls, called CP, in this standard [NIST SP800-82r2]. Contingency planning policy and procedures address the controls in the CP family that can be implemented at organization, mission/business process, or system level. Contingency planning policy and procedures address the controls in the CP family that are implemented within organizations. Security and privacy programs collaborate on the development of contingency planning policy and procedures. Security and privacy program policies and procedures at the organization level are preferable, in general, and may obviate the need for mission- or system-specific policies and procedures. The policy can be included as part of the general security and privacy policy or be represented by multiple policies that reflect the complex nature of organizations. Procedures can be established for security and privacy programs, for mission or business processes, and for systems, if needed. Procedures describe how the policies or controls are implemented. However, an update to contingency planning policy and procedures may be triggered by events that include assessment or audit findings, security incidents or breaches, or changes in laws, executive orders, directives, regulations, policies, standards, and guidelines.

As described in [NIST SP800-34r1], an organization must have the ability to withstand all hazards and sustain its mission through environmental changes (gradual or sudden due to a disaster event). Organizations need to work toward building a resilient infrastructure, minimizing the impact of any disruption on mission essential functions. The goal of a resilient organization is to continue mission essential functions at all times during any type of disruption. Risk management, contingency, and continuity planning are individual security and emergency management activities that can also be implemented in a holistic manner across an organization as components of a resiliency program.

4.5.1 Developing Contingency Plans

Contingency planners define a contingency plan devised for an outcome other than in the usual (expected) plan. A contingency plan is a course of action designed to help an organization respond effectively to uncertainty or a significant future event or situation that may or may not happen. A contingency plan is sometimes referred to as Plan B because it can be also used as an alternative for action if expected results fail to materialize. Contingency management defines what should take place during and after an event. Although negative events are first considered, a good contingency plan should also address positive events that might disrupt operations - such as a very large order.

Effective contingency planning begins with the development of an organization contingency planning policy and subjection of each information system to a business impact analysis (BIA). This document [NIST SP800-34r1] defines various types of IT contingency plans - including business continuity plans (BCPs) and disaster recovery plans (DRPs) - and outlines a six step planning process for creating contingency plans:

- Develop the contingency planning policy statement
- Conduct the business impact analysis (BIA)
- Identify preventative controls
- Create contingency strategies
- Plan testing, training and exercises
- Plan maintenance.

Contingency planning is developing responses in advance for various situations that might impact business. Contingency plans address how to deal with small incidents that do not qualify as disasters, as in power outages, server failures, failure of a communication link to the Internet, or the corruption of software [Harris 2013]. Contingency plan is the program developed to anticipate, react to, and recover from events that threaten the security of information and information assets in the organization, and, subsequently, to restore the organization to normal modes of business operations [Whitman 2011].

Although in the security industry contingency planning and business continuity planning (BCP) are commonly used interchangeably, it is important to understand the difference. BCP addresses how to keep the organization in business after a disaster occurs. Contingency planning is the process by which organizations position to detect, react, and recover from events that threaten the security of information resources and assets. Therefore, a contingency plan should be developed for events that might disrupt a business. Plans may include threats such as:

- Natural, intentional, unintentional events
- Failure of firewall to protect the network
- Undisciplined personnel (e.g., employees and contractors find easy ways to access protected data)
- Failure of security profiles on the configured devices
- Personnel, such as death, or security technicians going on strike
- Data loss, such as loss due to natural disasters, sabotage, or other criminal action (such as an attack on a website)
- Mismanagement, such as theft, neglect of critical duties, or accidental destruction
- Congested networks
- Supply issues
- Production issues, such as a huge order that requires reallocation of plant resources.

Contingency planning should ensure continuous information systems availability. It consists of three major components or plans that can be established after business impact analysis is completed [Whitman 2014]:

- Incident response plan (IRP)
- Disaster recovery plan (DRP)
- Business continuity plan (BCP).

Regardless, if any of these components is one plan method or multiple plan method, it must be addressed and developed on its entirety. Contingency plans are designed to maintain or restore business operations including computer operations.

The contingency planning process includes steps such as:

- Analyze possible risks for each event from the list of the possible events that could disrupt operations
- Determine the likelihood and impact of risks, including only the events with the highest impacts
- Develop a process for each event that was selected
- Evaluate alternative solutions and select the most efficient and effective ones.

Contingency plans are part of continuity business strategy. Therefore, the development and implementation of contingency plans should include both proactive and reactive strategies [Benson 2000]. For example:

- Proactive strategy – a set of steps that help to minimize existing security policy vulnerabilities and develop contingency plans.
- Reactive strategy - helps security personnel to assess the damage caused by the event, to repair the damage or implement the contingency plan developed in the proactive strategy.

The contingency plans should be implemented, tested, and refined continuously. The process shares certain characteristics with risk management and system development. It is recommended to distribute a copy of the contingency plan to the local site; a copy should be stored on site but in a protective mode, fireproof safe; and a copy at an offsite location [Harris 2013].

4.5.2 Contingency Plans for Smart Grid

The electric industry is facing the trend of IT systems integration with control systems. Given that the main objective of these plans is support for an organization's business and mission, it is recommended to follow a unified approach and generate unified contingency plans.

A unified analysis of both IT systems and industrial control systems to generate and manage these contingency plans would benefit the electric sector industry to take advantage of accumulated expertise in security problems and eliminate the gap on disparate views on managing the security problems of these systems during contingent events. For example, in the event of a failure of the communications, the concept of a contingency mode of operation in which power is safely delivered to customers of the system is highly required. During the process for defining data flow requirements, it is recommended to perform an analysis and design a system to be able to ride through certain failures in contingency mode of operation [P2030 2011]. The existence of this contingency mode at all levels of the Smart Grid could mitigate the need to spend exorbitant amounts of money on extremely reliable interfaces.

As critical infrastructures, organizations, and public depend on telecommunications infrastructures in order to assess the resilience of any national critical infrastructure, there is a need to consider the failure modes and resilience capabilities of telecommunications infrastructures. A consequence of this principle is that the failure of telecommunications infrastructures must be considered in all contingency plans, in drills and exercises, as well as the recovery strategies that are used to mitigate the consequences of an adverse event [Johnson 2013]. Although a range of measures have been introduced to encourage and support contingency planning for the critical infrastructures in Europe [ECD/114], telecommunications infrastructures are not in the list of critical sectors. The telecoms should be included as a basis for contingency planning across critical information infrastructures. Therefore, techniques are needed for planning multi party drills that include telecommunications failures during wider contingencies that affect national critical infrastructures including energy system.

This standard [NIST SP800-82r2] describes contingency planning as family category of the security controls that can be implemented for an organization, mission-business critical system or other system (e.g., SCADA, microgrid, storage system). Each family contains security controls related to the general security topic of the family. Along with planning, controls also exist for contingency training, testing, and plan update, and for backup information processing and storage sites. The security controls that fall within the contingency planning family of the standard provide policies and procedures to implement a contingency plan by specifying roles and responsibilities, and assigning personnel and activities associated with restoring the information system after a disruption or failure. Supplemental guidance for the contingency plan controls is provided in the following documents:

- [NIST SP800-34r1] document provides guidance on contingency planning (more on recovery strategies).
- [NIST SP800-100] provides guidance on information security governance and planning.

However, contingency plans have to be developed with some discretion. According to [NIST SP800-53r5], contingency plans for high impact information systems may contain significant amounts of implementation detail and be quite lengthy. In contrast, contingency plans for low impact systems may contain considerably less detail and be quite succinct. Organizations need to give consideration to the scalability factors in particular operational environments. Scaling security controls to the appropriate system impact level facilitates a more cost effective, risk based approach to security control implementation, expending only the level of resources necessary to achieve sufficient risk mitigation and adequate security.

Because business continuity and disaster recovery plans are particularly important for ICS, they are described in more detail in this document [NIST SP800-82r2].

4.5.3 Incident Reporting and Recovery

Usually, security management is concerned with occurrence of a security incident, which is defined as a security event that involves a security violation [RFC 4949]. Sometimes these terms, incident and event, are used interchangeably. However, there are subtle differences between these two terms. An event is any observable occurrence, especially one of some importance. Events may occur within an organization, network, or information system. Events sometimes provide indication that an incident is occurring affecting the security posture. In an organization, many incidents may occur (e,g., a broken patch to a system), so the identification of the incidents that are managed by security professionals is a challenge for incident response team.

4.5.3.1 Incident Security Management

Generally, Incident Management (IcM) is an IT service management process to restore a normal service operation as quickly as possible and to minimize the impact on business operations, thus ensuring that the best possible levels of service quality and availability are maintained. It is one process area within the broader Information Technology Infrastructure Library (ITIL) and [ISO/IEC 20000] environment. This [ISO/IEC 20000] is the first international standard for managing and delivering IT services. Part 1 of this standard specifies requirements for an organization to establish, implement, maintain and continually improve a service management system.

Related to IT service management is information security incident management defined in [ISO/IEC 27035]. This standard covers the processes for managing information security events, incidents and vulnerabilities. For better management of processes and practices, it is recommended to review definitions of these concepts as being used by an organization or industry. The recommendations are generic and intended to be applicable to all organizations, regardless of type, size or nature. The updated [ISO/IEC 27035] of 2016 aims to include a cross reference between these standards. The standard expands on the information security incident management section of [ISO/IEC 27002]. It includes a section on management of information security incidents and improvements. The section covers responsibilities and procedures to manage (report, assess, respond to and learn from) information security events, incidents and weaknesses consistently and effectively, and to collect forensic evidence.

Information security incident management includes proactive and reactive processes. Proactive measures need to be put in place, so that incidents can actually be detected in a controllable manner, and reactive measures need to be put into place, so those incidents are then dealt with properly [Harris 2013].

Most organizations choose only the reactive approach called incident response. Other organizations use an approach called SIEM (security information and event management) which seeks to provide a holistic view of organization's information technology security. SIEM is a term for software products and services combining security information management (SIM) and security event management (SEM). SIEM technology provides real time analysis of security alerts generated by network hardware and applications. SIEM is sold as software, appliances or managed services, and are also used to log security data and generate reports for compliance purposes.

Some organizations may have a structured incident handling program, called Computer Incident Response Team (CIRT) to perform the main tasks [Krutz 2004]:

- Analysis of an event notification
- Response to an incident if the analysis warrants it
- Escalation path procedures.

Resolution, post incident follow up, and reporting to the appropriate parties. Figure 4.13 shows the stages in time – initial failure at t0, failure propagation time from t_0 to t_1, response is the time between t_0 and t_2, recovery time is the time between t_2 and t_3.

Response time includes detection time and decision time. The response time is dependent on the detection time and decision on how to recover from the failure. This diagram shows the system performance decreases from t_0 to t_1 and it comes back to the performance before the failure at time t_3. The time sequence of these specific events (e.g., detection time, decision time) can be used to characterize the performance of a dynamic system after a disruption. This work [Ouyang 2012] argues that performance response process of an infrastructure system after a disruptive event is taking three stages, reflecting system resistant, system absorptive, and restorative capacities. These restorative capacities can be used to characterize the system level resilience in a quantitative fashion.

Also, recovery can start after a decision (manual or automated) has been made. In real world practice, the response time depends on several factors, such as the type of incident, the criticality of the resources and data that are affected, the severity of the incident, existing service level agreements (SLAs) for affected resources, the time and day of the week, and other incidents that the team is handling. Generally, the highest priority is handling incidents that are likely to cause the most damage to the organization or to other organizations.

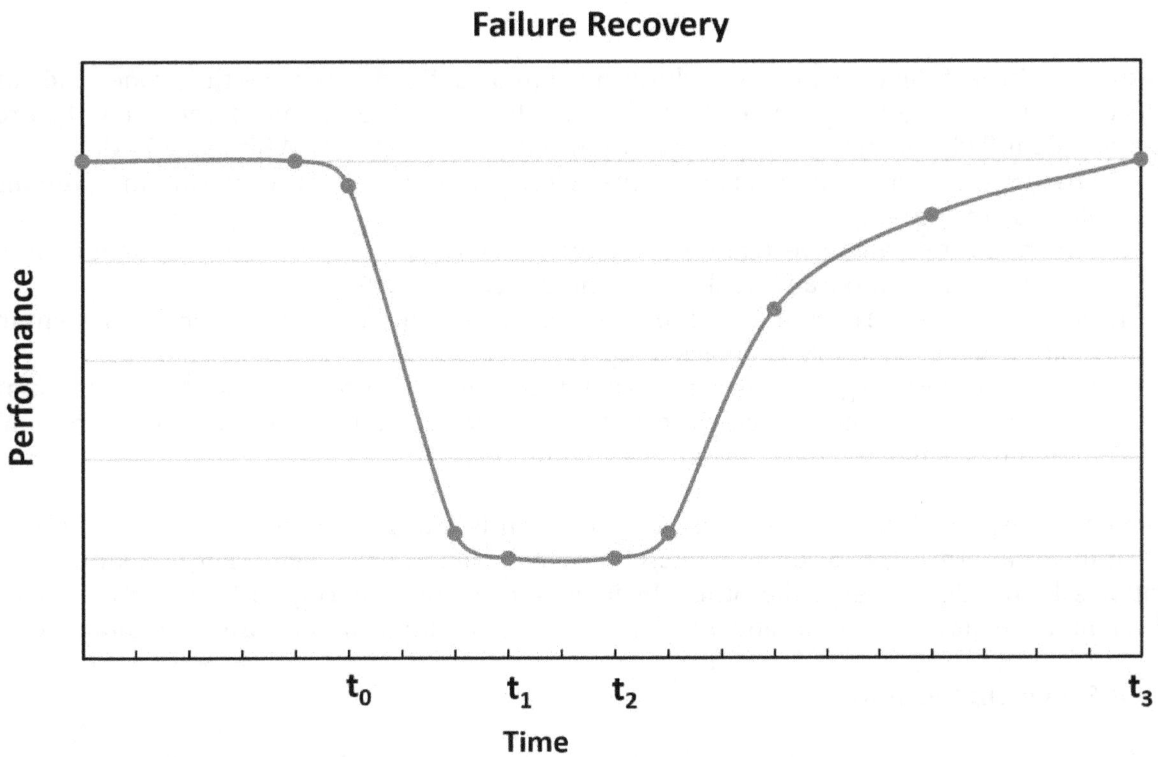

Figure 4.13 Stages to recovery.

4.5.3.2 Incident Response Plan

Commonly, the Incident Response Plan (IRP) is the documentation of a predetermined set of instructions or procedures to detect, respond to, and limit consequences of an incident against an organization's IT systems(s) [CNSSI 4009]. The IRP contains a detailed set of processes and procedures that anticipate, detect, and mitigate the effects of an unexpected incident that might compromise information resources and assets. An incident is an assessed occurrence that actually or potentially jeopardizes the security objectives of an information system including a violation or imminent threat of violation of security policies, security procedures, or acceptable use policies.

The concept of computer security incident response has become widely accepted and implemented. One of the benefits of having an incident response capability is that it supports responding to incidents systematically (e.g., following a consistent incident handling methodology) so that the appropriate actions are taken. Incident response helps personnel to minimize loss or theft of information and disruption of services caused by incidents. Another benefit of incident response is the ability to use information gained during incident handling to better prepare for handling.

Guidelines and assistance to incident response and disaster recovery are provided via many published materials and courses (e,g., [Whitman 2007], [Harris 2013]) including [NIST SP800-61r2]. This document [NIST SP800-61r2] assists organizations in establishing computer security incident response capabilities and handling incidents, but it addresses only adverse events that are computer security related, not those caused by natural disasters, power failures, etc. Adverse events are events with a negative consequence, such as system crashes, packet floods, unauthorized use of system privileges, unauthorized access to sensitive data, and execution of malware that destroys data.

The responsibilities of the team and procedures need to be defined to include the proper and current documentation. Regulations by industry and standards need to be considered when the procedures are identified. Commonly, the procedures include [Harris 2013], [Whitman 2014]:

- Preparing and documenting the plan (strategies, procedures, team responsibilities, refining to keep up to date)
- Incident detection and investigation (e.g., analysis, collecting information, incident identification and classification, and tracking potential security events)
- Incident response (e.g., notification of key personnel, containment, preserving system information and events, logs, and backup ups)
- Incident recovery (e.g., damage assessment, actions to restore the systems and information, and many other activities to finally restore the confidence of the organization's communities of interest including an after action review).

One important aspect of incident response coordination is information sharing, where different organizations share threat, attack, and vulnerability information with each other so that each organization's knowledge benefits the other. Incident information sharing is frequently mutually beneficial because the same threats and attacks often affect multiple organizations simultaneously.

4.5.3.3 ICS Incident Response

Recently, there is more guidance for industrial control systems. In USA, ICS-CERT advises how to create an incident response plan [CERT IRP] and recommends the guidelines and practices provided by the US Homeland Security Department [DHS 2009]. Although there are common IT tasks that can be performed for ICS incident response, there should be also specific tasks on procedures that provide resolutions to accommodate the needs for industrial control systems and differences between these systems. In fact, ICS-CERT recommended practices are adapted practices for IT systems described in [DHS 2009]. ICS-CERT recommends similar activities for the industrial control systems. Figure 4.14 illustrates an IRP sequence of key activities recommended for control systems.

As shown in this diagram, these activities - planning, incident prevention, and post incident analysis/forensics – are proactive in nature to prevent an incident or better allow the organization to respond when one occurs. Other activities are reactive in nature and are typically carried out under severe time constraints and great visibility. These activities include detection, containment, remediation, and recovery and restoration.

A proper collection of evidence is required for reporting the incident to law enforcement and for identifying the possible gaps in security mechanisms or security policies. Systematic collection of safety incident/accident data has been common in many industries for decades. Similarly, organizations can benefit from capture data about incidents, determining the cause of incidents, and deriving knowledge on how to handle more efficiently the security incidents.

An information security incident reporting system (ISIRS) is proposed in [Reed-Mohn 2007]. This is a system enabling collection and analysis of incident data (over time), building knowledge to minimize or avoid future risk through organizational awareness and improved business processes. Also, this work assumes that a system model of a general incident reporting system has similar characteristics to the model used in Safety Science.

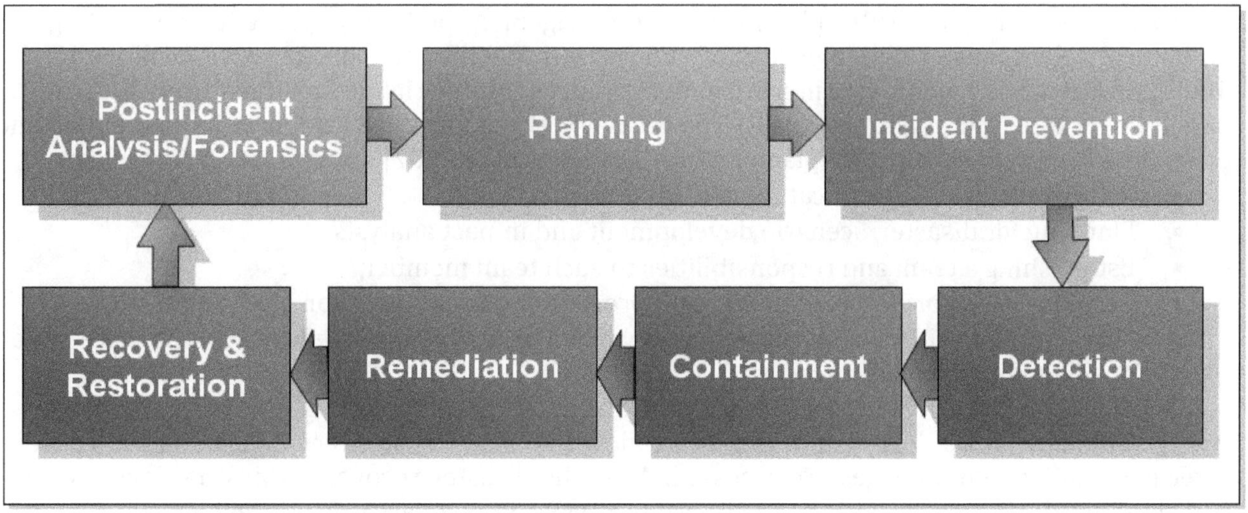

Figure 4.14 IRP key elements. *Source:* [DHS 2009]. Public Domain.

This standard [ISO/IEC 21827] defines activities to be used for ICS in depth defense strategy that include:

- Identify based on continuous identifying, evaluating and managing of cyber risks using best practice risk assessment and management methods
- Protect based on structured and robust built in security architecture, network protection, host protection, interface protection, and secure remote connection
- Detect based on capabilities to detect viruses and other cyber annoyances, as well as sophisticated cyber attacks, both on the network and inside of the system and each host
- Respond using well established and efficient processes to handle cyber attacks
- Recover to a quickly return to normal or degraded operation after an attack.

The measures that organizations should adopt, implement, and improve upon to have in place include: Incident Response policy, plan, and team that is capable to minimize the effects of an incident, not if it occurs, but when it occurs. Also, it is important to develop an effective defense because is not possible to prevent every cyber attack so containment is equally important as prevention [Wolf Davis 2015]. Utility managers need to focus just as much on post breach damage control measures as they do on active defense of power systems. The incident response team should include security professionals, process control engineers, IT cybersecurity experts, systems administrators, network administrators, and managers that support the activities and management of activities on a permanent basis, although some members could be called when it is needed.

Sometimes an incident may escalate to a disaster and the IRP may no longer be able to handle the effective and efficient recovery from loss. The incident is qualified as a disaster and the Disaster Recovery Plan (DRP) is needed by the enterprise risk managers.

4.5.4 Disaster Recovery Plan

A disaster is an unplanned event that causes the general system or major application to be inoperable for an unacceptable length of time (e.g., minor or extended power outage, extended unavailable network, or equipment or facility damage or destruction). Disaster recovery planning is the preparation for and recovery from a disaster, whether natural (e.g., fire, flood, earthquake, etc,) or human made (e.g., acts of terrorism, war, cyber terrorism, hacktivism, sabotage, etc.).

A Disaster Recovery Plan (DRP) plan includes a management policy and procedures used to guide an enterprise response to a major loss of enterprise capability or damage to its facilities [CNSSI 4009]. A DRP is used when the enterprise must recover (at its original facilities) from a loss of capability over a period of hours or days. The recovery procedures are designed, implemented, and tested in advance. Basically, DRP planning includes [Whitman 2007], [Harris 2013]:

- Identification and classification of disasters
- Planning for disaster, scenario development and impact analysis
- Establishing a team and responsibilities to each team member
- Establishing priorities, communication procedures, documentation, and action steps
- Responding to the disaster based on advanced restoration procedures, which were developed and tested in advance.

A DRP should include a degree of flexibility to allow the team to use alternative actions after a disaster is classified and damages are assessed. After the disaster recovery, a post review of events and activities during a disaster is key to improve a DRP and recovery procedures.

Security must be included in disaster recovery planning to ensure sensitive data is protected. Often, organizations overlook data protection provisions in their disaster recovery/business continuity plans so there is the risk of dealing with a double crisis: a security breach on top of a recovery situation.

When a disaster threatens the viability of an organization at the primary site, the disaster recovery process becomes a business continuity process. Disaster recovery/business continuity plans must ensure that an organization's information security policies are maintained in a recovery situation. That means making sure the recovery site has proper security including updated antivirus and firewall protection. It also means conducting proper due diligence of any disaster recovery provider and taking proper precautions in a shared recovery facility. Transmission of data for backup purposes must also be secured.

4.6 Business Continuity Management

Business continuity management is a framework to review the way an organization provides its services while increasing its resilience to disruption, interruption or loss. Business continuity management includes aspects such as:

- Project management and initiation
- Business impact analysis
- Recovery strategy
- Plan, design, and development
- Business continuity planning
- Testing, maintenance awareness and training.

4.6.1 Business Continuity Planning

Business continuity planning (BCP) prepares an organization to reestablish critical business operations during a disaster that affects operations at the primary site [Whitman 2014]. The business continuity domain addresses the preservation and recovery of the business in the event of outages to normal business operations. In the power grid, a disaster is any event that creates an inability on any organization or plant to support critical power generation, transmission, distribution, and business functions for some predetermined period of time.

This document [NIST SP800-82r2] describes that business continuity planning for ICS addresses the overall issue of maintaining or reestablishing production in the case of an interruption. These interruptions may take the form of a natural disaster (e.g., hurricane, tornado, earthquake, flood), an unintentional man made event (e.g., accidental equipment damage, fire or explosion, operator error), an intentional man made event (e.g., attack by bomb, firearm or vandalism, attacker or virus), or an equipment failure.

The major tasks for business continuity planning include [Whitman 2007], [Harris 2013]:

- Establishing continuity strategies
- Plan for continuity of operation
- Continuity management.

Although there are no unique or simple steps to follow on setting the business continuity planning and straight guidelines for the content of contingency plans, each organization has to provide methods and resources for dealing with longer term objectives. The timing and sequence of contingency plans need to be planned in advance. A timing and sequence of contingency plans is suggested in [Whitman 2007].

Although there is overlapping of tasks performed concurrently, each plan is activated at a specific time in the life of an incident. For example, the BCP occurs concurrently with DRP when the damage is major or long term, requiring more than simple restoration of information and information resources. A BCP takes a broader approach to the problem. In addition, BCP for Smart Grid control systems should be a distinct plan given the need for satisfying the specific requirements for real time and availability constraints. Figure 4.15 illustrates major tasks in contingency planning for common IT systems.

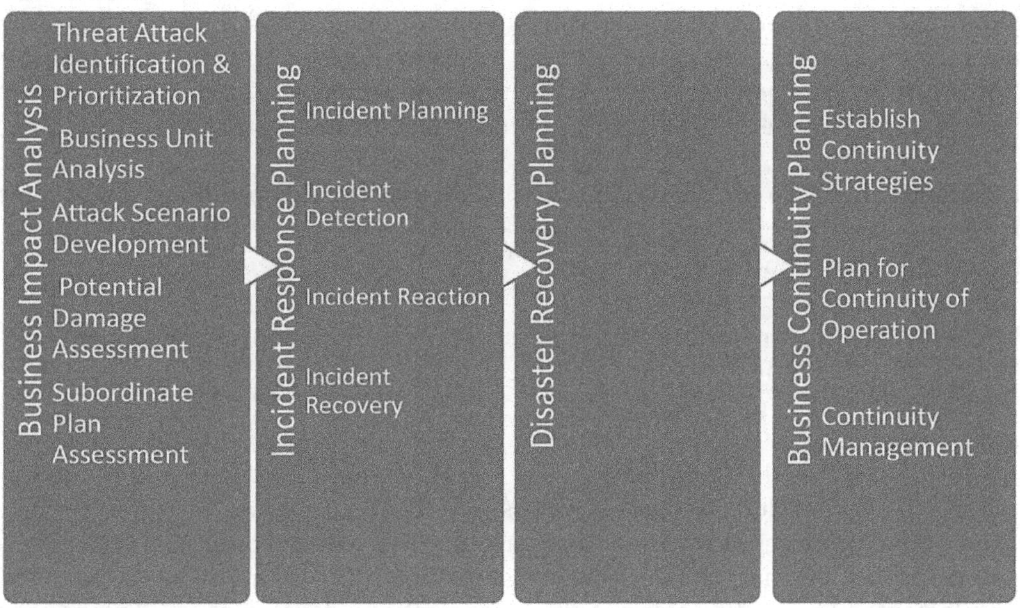

Figure 4.15 Major tasks in contingency planning.

However, each plan should be adapted to Smart Grid systems that include both IT and control systems. Business continuity planning is often developed based on best practices and recommendations provided by standards organizations and professional associations. While the processes and plans are important for an organization, a business continuity management process has the objective to allow organization to continue to perform business operations under various conditions [Harris 2013].

Essentially, continuity plans address every critical function of an enterprise. The BCP plan should cover all aspects of an organization, such as:

- Personnel
- Facilities
- Infrastructure
- Support systems
- Information systems.

The BCP framework involves the preparation, testing, and maintenance of specific actions to protect critical business processes from the effect of extended data services outages [Hansche 2004]. This document [NIST SP800-34] provides how various types of plans revolve around the BCP.

Business continuity should not be a standalone entity, but it should be integrated into the security program and be part of business decisions. If a disaster rendered the current business location unusable, there must be a plan to allow the business to continue its function. Business continuity planning is the process that ensures that critical business functions can continue if a disaster occurs. The result of this process is the business continuity plan.

4.6.2 Business Continuity Plan

The BCP is the documentation of a predetermined set of instructions or procedures that describe how an organization's business functions will be sustained during and after a significant disruption [CNSSI 4009]. A BCP is activated and executed in parallel with the DRP when a major or long term disaster occurs and recovery requires more complex and full restoration of information and IT resources. While DRP is focused on the reestablishment of the technical infrastructure and business operations after events (e.g., security incidents, disasters) occur, the BCP reestablishes critical business functions, often at an alternate site when they cannot be restored at the actual site. Usually, the plan is managed by the chief enterprise officer (CEO) of an organization.

The business continuity plan is a set of advanced arrangements and procedures that enable an organization to:

- Ensure the safety of people.
- Minimize the amount of loss.
- Facilitate the recovery of business operations to reduce the overall impact of an event, while at the same time resuming the critical business functions within a predetermined period of time.

The key aspect of a BCP is the identification of critical business functions and the resources that support them. This can be done using a business impact analysis that identifies attack scenarios and assess potential damages. The determining factor is based on continuity strategies and cost. Besides reducing risks, business impact analysis should be focused on preventative measures that could be implemented to avoid disasters. Some organizations develop specific plans for each task (e.g., continuity of operation, security incident response plan, disaster recovery plan, etc.). Because

DRP and BCP are related, many organizations combine DRP and BCP into a single document, but distinct from IRP.

4.6.3 Testing Contingency Plans

Testing contingency plans should be done regularly at least once a year. It can be done following these methods:

- Desk check – reviewing the plan and creating a list of correct and incorrect components.
- Structured walkthrough – all members of the team walk through the steps they would take during a response to an event.
- Simulation – each task is simulated and then the performance is evaluated.
- Parallel testing – testing of the plan without interfering with normal business operations.
- Full interruption – testing is done after the business operations are interrupted, for example, after normal business hours.

Besides testing, recovery drills and exercises are used too because the team faces more situations that are closer to the assumed event and expected problems. Before any tests are performed, there should be a clear indication of what is being tested, how success and mistakes will be determined, and how to deal with problems. In addition, testing contingency plans that involve control systems will require careful planning since these systems require continuous operation and high availability such that they cannot be easily interrupted to allow testing. One approach can be based on lab testing with a simulated environment. Technologies and proposed applications including algorithms can be tested and validated using simulation and modeling techniques.

4.7 Simulation and Modeling Techniques

Modeling and simulation is the use of models, including emulators, prototypes, simulators, and stimulators, either statically or over time, to develop data as a basis for making managerial or technical decisions. It includes models, tools including the interface of real world systems with models or simulations, standards and specifications, and modeling system descriptions, interfaces, and data communication methods.

In order to study a system or process, it is necessary to make a set of assumptions about how system works. These assumptions, which usually take the form of mathematical or logical relationships, constitute a model. Since most real world systems are too complex to allow realistic models to be evaluated analytically (by means of algebra, calculus, or probability theory), there is a need to study these models by means of simulation.

4.7.1 System Modeling

System modeling is an important method used during engineering processes. System Engineering encompasses a collection of top down and bottom up methods to navigate the system hierarchy. The system engineering process usually begins with a world view (business domain, e.g., renewable integration), refined to a domain view (domain of interest, e.g., electrical storage systems), and further the targeted system is analyzed at the element level, and finally detailed views are defined for specific types of battery (e.g., chemical batteries). The steps to create models include [Pressman 2005]:

- Define processes serving the needs of the view under consideration
- Represent the behavior of the processes and the assumptions on which the behavior is based
- Define inputs and outputs of the model

- Define functions and constraints.

The resultant system model (at any view) can be a completely automated solution, semi-automated solution, or a manual approach that allows System Engineer to modify the variables to determine the desired output. The most important aspect of a system model is that it leaves out detail; system model is an abstraction of the system being studied [Sommerville 2007]. The author describes different types of system model are based on different approaches to abstraction. Examples of these types of system model that can be produced during the analysis process include:

- A data process model shows how data is processed at different stages in the system.
- A composition model shows the composition and relationships between entities of the system.
- Architectural model shows the subsystems that make up a system.
- Classification model shows the common characteristics of object/class inheritance diagrams.
- Stimulus response model shows state transition diagrams and how the system reacts to internal and external events.

4.7.2 Simulation

Simulation is a tool for the analysis of complex systems. It provides the engineer with the ability to predict the behavior of a real time system before it is built. Simulation is used for building systems that interact with real world in a reactive fashion (e.g., real time control systems, embedded systems) and processes (e.g., COTS acquisition process) and must operate with a high degree of reliability. Simulation is a problem solving methodology for the solution of many real world problems.

Simulation is also used to describe and analyze the behavior of a system, ask what if questions about the real system, and aid in the design of real systems. Both existing and conceptual systems can be modeled with simulation. The simulation tools can be applied during engineering process, while the roles of some components (e.g., hardware, software, people) are being specified. The execution of simulations allows to gain insight into the function, operation, performance, and response prior to actual implementation. Simulation requires to build a model.

Simulation is effective and economical when adequate tools are used. Simulation tools ca be built internally or can be gathered from commercial manufacturers. The four modeling structures that are used in simulation include [Banks 1998]:

- Process interaction method is a computer program emulates the flow of an object through a system (for example, the movement of an entity for identity identification and authentication takes a sequence of steps; it can enter next activity, be halted, delayed, or exit from the system).
- Event scheduling is to advance time when something next happens (e.g., when a battery is discharged, the event then reallocates available batteries by scheduling activities where they can participate).
- Activity scanning, known as two phase approach, produces a simulation program composed of independent modules waiting for execution (e.g., if a reconnaissance event occurs, the system state is updated and logs are written).
- Three phase method where the time is advanced in the first phase; resources are released in the second phase, and starting activities when resources are available.

Simulation has a number of distinct functions in the design of complex systems. The major expectations from simulation are viewed as an aid in system synthesis and analysis. By using simulation, an evolutionary process of system design can be achieved. Among the benefits of using the simulation include [Reitman 1984]:

- The problem definition becomes more clear.
- Design uncertainties are reduced.
- Problem definition is more complete.
- A dynamic representation is produced.

One important step in simulation is verification and validation of simulation results to be sure that the simulation model is an accurate representation of the real world system. Many study reports describe the application of modeling and simulation techniques for a wide range of problems.

4.7.3 Applications

System modeling and simulation tools are used for building systems that interact with real world in a reactive fashion [Pressman 2005]. Simulation is the imitation of the operation of a real world process or system over time [Banks 1998]. In a simulation, a computer is used to evaluate a model numerically, and data is gathered in order to estimate the desired true characteristics of the model [Law 2007]. However, a system can be studied in different ways and from different perspectives.

Assuring the survivability of mission-critical systems developed under this new paradigm of distributed systems is an effort for software engineering research. This effort requires that traditional computer security measures be augmented by new and comprehensive system survivability strategies. One key issue is how to measure survivability which falls into one of three major categories connectivity, network performance, or a function of other quality or cost measures [Westmark 2004]. This work recommends expansion of the survivability function and defines a simulation model for managing survivability of networked information systems including reliability, latency, and cost benefit model.

Modeling methods of survivability networks are described in [Heegaard 2009]. This work provides an assessment of the survivability of a network with virtual connections exposed to failures using both simulation and analytic models to cross validate the assumptions.

Approaches to simulating complex systems described in [Law 2007] can be applied to simulate security issues for the Smart Grid and DER systems. Often, optimization techniques can be used to design a system to yield optimal expected outcome [Andradottir 1998].

Research is also focused on models and simulation for different aspects of information security such as security risk analysis, security incidents prediction and detection, security risk management, etc. A survey of models targeted at information security and simulations of incident reporting system is provided in [Reed-Mohn 2007].

A risk propagation model based on the Markov process, which can be applied to diverse threats to information systems is overcoming the limitations of the existing risk propagation models specified only on one specific threat, such as a single virus or worm [Kim 2007]. Usually, a threat is related to other threats. Using the model based on Markov process, the occurrence probability and frequency for each threat to information systems can be predicted globally and applied to establish effective countermeasures.

A model based simulation of the system for risk assessment based on the security property (e.g., integrity, confidentiality, authenticity, non repudiation) approach aims to help assessing risks in a process oriented or service level view of a system thus deriving a more detailed estimation on a technical level is proposed in [Winkelvos 2011]. Also, this work proposes a simulation based risk analysis tool that enables a security property oriented view of risk. The authors argue that their approach is superior to current automatic security analysis tools that are mostly confined to a technical point of view and thereby missing implications on an application or process level. The tool facilitates probabilistic simulation, providing information about the probability distribution of satisfaction or violation of specified properties. The work contributes to risk analysis methods by facilitating a broader view including the relation between actions within the IT system and their external influence.

4.7.4 Smart Grid Applications Based on Wireless Technologies

The future Smart Grid is envisioned as a large scale cyber-physical system encompassing advanced power, communications, control, and computing technologies. Key tools and methods to assist Smart Grid system designers in making informed decisions about existing and emerging wireless technologies are provided in [NISTIR 7761r1].

Wireless technologies for technical and business communications have been used in the power industry for system monitoring, metering and data collection. The advanced applications and widespread use now foreseen for the Smart Grid require highly reliable, secure, well designed, and managed communication networks. However, a complete technology assessment with respect to capacity and coverage including technology attributes such as latency and security must also be taken into account for a full Smart Grid suitability assessment. Wireless is one of several communications options for the Smart Grid that must be approached with technical rigor to ensure communication systems investments are well suited to meet the needs of the Smart Grid. Also, the utilities need to implement systems that satisfy their specific needs.

The decision to apply wireless technologies for any given set of applications is a decision that must take into account several important elements including both technical and business considerations. Smart Grid applications requirements must be defined with enough specification to quantitatively define communications traffic loads, levels of performance, and quality of service. Applications requirements must be combined with a complete set of management and security requirements for the life cycle of the system. These requirements can then be used to assess the suitability of various wireless technologies to meet the requirements in the particular applications environment.

The general modeling framework described in [NISTIR 7761r1] document includes simulation and modeling studies to allow a broad assessment of many wireless technologies that may be used for Smart Grid applications. The modeling framework helps to develop an analytical structure that is flexible enough to enable users to employ a variety of modeling techniques that can be used with virtually any proposed wireless technology. Thus, the network designers can use the modeling and simulation models to project and predict the real temporal (and spatial) requirements of the data flows (for the utility, application, or operating mode in question) and to then select and implement technologies and topologies that will provide the needed capacity, reliability, security, cost effectiveness, etc. The overall structure of the framework model is shown in Figure 4.16.

The framework's main components include a Medium Access Control (MAC) sublayer model, a Physical layer (PHY) model, a module that performs coverage analysis, a channel propagation model, and a model for multiple links (multi link). The studies include many technologies anything from radio, satellite, Bluetooth, to WiMAX. The wireless technologies chosen in the study encompass different capabilities, cost, and ability to meet different requirements for advanced power systems applications.

Another work reports the use of simulation techniques to verify the results of an algorithm. This is a strategy based on game theory that is implemented in an autonomous and distributed demand side energy management system to minimize the energy costs for consumers [Mohsenian-Rad 2010].

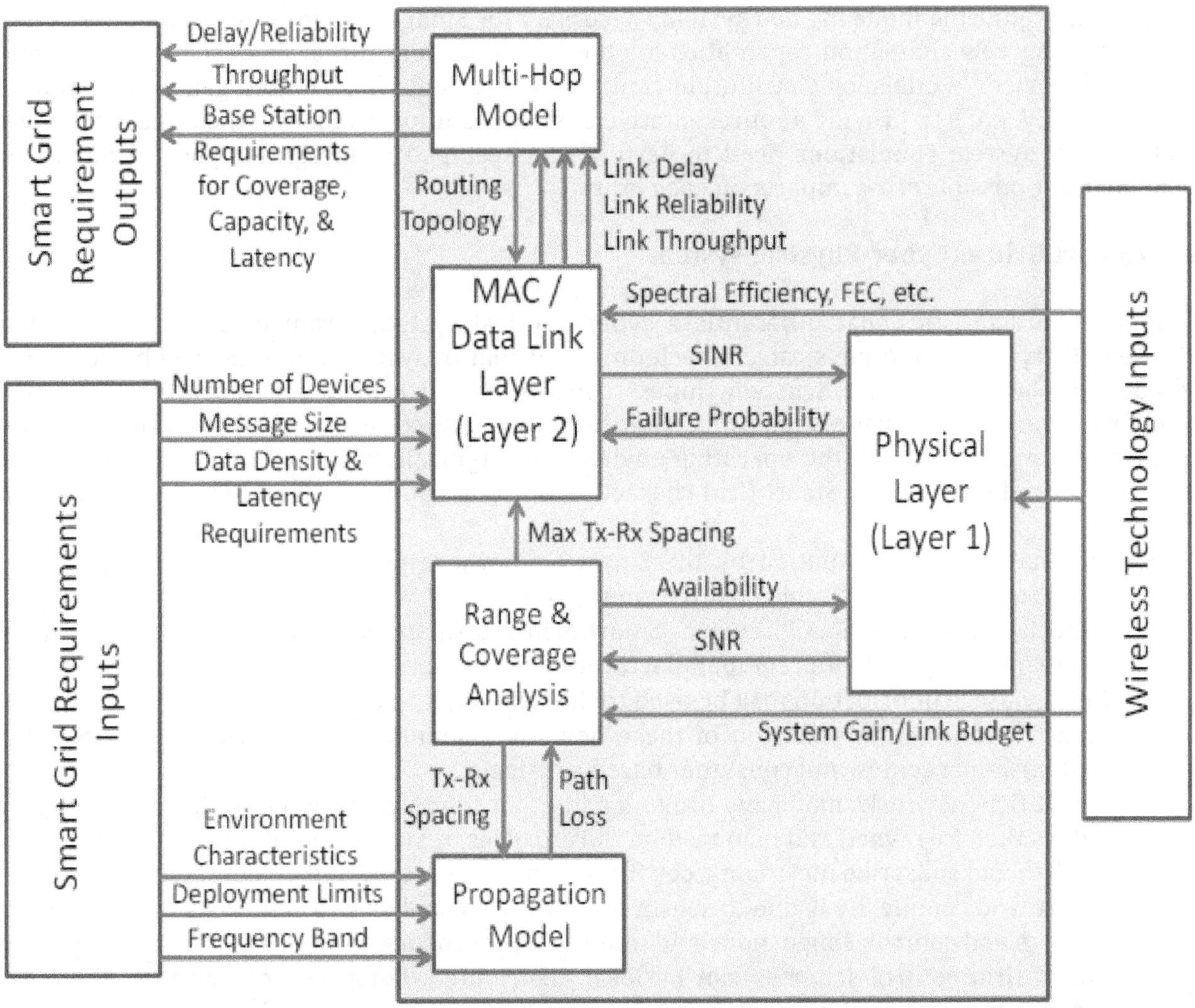

Figure 4.16 Wireless modeling framework building blocks. *Source:* [NISTIR 7761r1]. Public Domain.

Guidelines for modeling and simulation of cyber-physical systems are described in [Vellaithurai 2015]. The authors focus on the modeling the power system, data measurement services, data collection services, and communication networks used to deliver data to the control center. The cyber-physical characteristics of the Smart Grid are analyzed by using modeling and simulation for different layers represented by simulators, hardware sensors, and controllers to realize a real time test bed for power applications. The authors conclude that in order to develop cyber-physical analysis of the Smart Grid, it is necessary to use modeling and simulation methodologies with sufficient detail. Also, there are challenges on integrating different simulators.

Simulation can help predeployment decision making process, but simulating system models for complex CPS is becoming impractical for the evaluation of specialized design aspects [Akkaya 2015]. The authors argue that the existing modeling techniques are inadequate for capturing the heterogeneous nature of CPS, so they propose the aspect modeling paradigm.

However, more research is needed for developing advanced methods and tools to simulate the intricacies of CPS systems. Also, more research is needed to improve the current modeling, simulation, and emulation technologies for future networks using IPv6 protocol [NISTIR 7628r1]. Also, research is needed to ensure that the IPv6 based network will be stable, reliable, and secure. To develop and refine the modeling and systems necessary for Smart Grid applications, there is a need for developing new simulation capabilities for the distribution grid that incorporate communications with devices/models for distribution control, DER systems (distributed generation, storage, EV/PEV/PHEV, etc.), to provide a representative environment for evaluating the impact of various events. Also, system simulations need to be further developed to integrate cyber elements and evaluate their possible cross impacts on each other.

4.7.5 Smart Grid as Cyber-Physical System

Common characteristics that differentiate cyber-physical systems from other systems include [Hentea 2021]: cyber and physical, connectedness, system of systems, emergent behavior, methodology, application enabler, trustworthiness concern, broad range of platform and algorithm complexity, variety of modes of communication, heterogeneity, co-design, typically a time-sensitive component, interaction with the operating environment, typically a human environment. The following is a brief overview of Smart Grid characteristics that can be categorized as CPS aspects or concerns:

- Optimal energy distribution by the Smart Grid where power consumers and producers work together is a desirable positive emergent effect.
- Especially in large scale CPS such as Smart Grid and smart city, many of the subsystems are the responsibility of different manufacturers.
- Energy usage information may be used to diagnose equipment faults.
- Smart Grid is the intersection of these domains: manufacturing and energy distribution systems, smart cities, and consumer based sensing.
- Smart Grid networks may have brokers and other infrastructure based devices and aggregators that are owned and managed by third parties, resulting in potential trust issues, e.g., publish and subscribe messaging, certificate authorities, type and object registries.
- Smart Grid complexity is due to sensors that range from basic to smart; static and adaptive sensors and control; single mode and multi faceted sensors.
- Smart Grid control schemes can be local, distributed, federated, or centralized; control loops that rely on a single data source and those that fuse inputs; and so on.
- Interactions in Smart Grid can be loosely coupled, as in repurposing of distributed sensing that is part of an existing CPS, as well as tightly coupled as in Smart Grid operations.
- Coupling is both an opportunity to fulfill the vision of CPS and a challenge to CPS assurance.
- Emergent behaviors can become part of the intent of new services or may be unwanted.
- To mitigate complexity, SmartGrids may be a product of co-design; in codesign, the design of the hardware and the software are considered jointly to inform tradeoffs between the cyber and physical components of the system.
- Smart Grid applications are time sensitive, e.g., synchrophasor measurements.
- Accurate timestamps in Smart Grid are typically needed to facilitate better root cause analysis, and sometimes also for legal or regulatory reason; accurate time intervals are useful for coordinating actions in CPS of large spatial extent; the accuracy of event timestamps is a constraint on a time value, in this case between the actual time of the event and the value of the timestamp.

A complete methodology is proposed with common vocabulary, structure, and analysis methodology is introduced in [NIST SP1500-201] document.

4.7.6 NIST Cyber-Physical System Framework

The analysis methodology described in [NIST SP1500-201] is addressing concerns that are specific to CPS systems and those that are applicable to any device or system. As a process method, the CPS analysis methodology enables and facilitates CPS systems engineering using a range of development approaches, such as waterfall, agile, spiral, and iterative. Although there are many well documented system engineering process documents and flows (e.g., CMMI), the framework focuses on the detailed scope of CPS and the specific concerns implementers and analysts have in designing and understanding them.

The CPS Framework captures the generic functionalities that CPS provide, and the activities and artifacts needed to support conceptualization, realization, and assurance of CPS. This analysis methodology includes addressing concerns that are specific to CPS and those that are applicable to any device or system. The key elements of the Framework are shown in Figure 4.17 and can be summarized as follows:

- Domains represent the different application areas of CPS.
- Concerns, as expressed by many different stakeholders in their unique and collective viewpoints, are a fundamental concept that drives the CPS Framework methodology; they are addressed throughout the CPS development cycle; concerns that are conceptually equivalent or related are grouped into Aspects, which are addressed by activities within the facets.
- Properties are the concrete assertions that address the concerns; they include requirements, design elements, tests, and judgments.
- Aspects consist of groupings of conceptually equivalent or related concerns; aspects and concerns are not considered orthogonal; there are nine defined aspects: functional, business, human, trustworthiness, timing, data, boundaries, composition, and lifecycle.
- Facets are views on CPS encompassing identified responsibilities in the Systems Engineering process; each facet contains a set of well defined activities and artifacts (outputs) for addressing concerns; there are three identified facets: conceptualization, realization, and assurance.

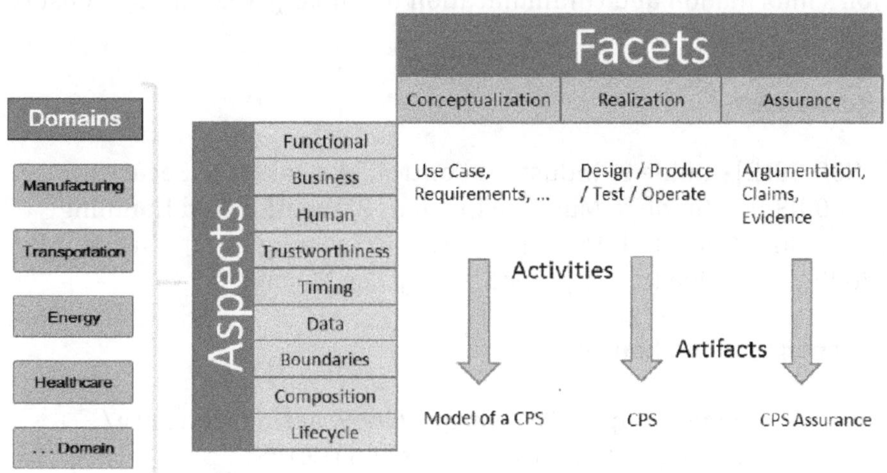

Figure 4.17 CPS Framework – domains, facets, aspects. *Source:* [NIST SP1500-201]. Public Domain.

The CPS framework includes the identification of foundational goals, characteristics, common roles, and features across CPS domains, while considering cybersecurity, privacy, and other cross cutting concerns. The CPS Framework is an abstract framework, or metamodel, for understanding and deriving application domain specific CPS architectures.

4.8 Standards, Guidelines, and Recommendations

There are many resources that can be used for the development of secure systems for the Smart Grid. The following sections include additional guidance.

4.8.1 Business Continuity Planning

Practices, recommendations, and guidelines for business continuity planning are provided by international standards and governments. The British Standards Institute standard [BS 25999] for business continuity management (BCM) include:

- [BS EN ISO 22313] standard that provides guidance on the use of [ISO 22301].
- [BS EN ISO 22301] specifies generic requirements that are intended to be applicable to all organizations around the world, or parts thereof, regardless of type, size and nature of the organization.
- [BS EN ISO 22313:2020] supersedes previous version of 2014; it includes some technical changes concerning BS EN ISO 22313:2014 standard; the changes include: structural and content alterations to align this document with the latest edition of ISO 22301 standard; additional guidance to explain key concepts and terms.

The International standards includes developed by ISO include:

- [ISO 22301] provides a framework to plan, establish, implement, operate, monitor, review, maintain and continually improve a business continuity management system (BCMS). It is expected to help organizations protect against, prepare for, respond to, and recover when disruptive incidents arise.
- [ISO 22313] provides guidance and recommendations for applying the requirements of the business continuity management system (BCMS) specified in [ISO 22301] standard. The guidance and recommendations are based on good international practice.
- [ISO/IEC 27031] provides a framework of methods and processes to identify and specify all aspects (such as performance criteria, design, and implementation) for improving an organization's information and communication technology readiness to ensure business continuity.

4.8.2 Incident Response

- [NIST SP800-82r2] - Guide to Industrial Control Systems (ICS) Security
- [NIST SP800-83r1] - Guide to Malware Incident Prevention and Handling
- [NIST SP800-86] - Guide to Integrating Forensic Techniques into Incident Response
- [NIST SP800-92] - Guide to Computer Security Log Management.

4.8.3 Incident Response Organizations

- Anti-Phishing Working Group (APWG) http://www.antiphishing.org/
- Computer Crime and Intellectual Property Section (CCIPS)
- U.S. Department of Justice http://www.cybercrime.gov/

- CERT® Coordination Center, Carnegie Mellon University (CERT®/CC) http://www.cert.org/
- European Network and Information Security Agency (ENISA) http://www.enisa.europa.eu/activities/cert
- Forum of Incident Response and Security Teams (FIRST) http://www.first.org/
- Government Forum of Incident Response and Security Teams (GFIRST) http://www.us-cert.gov/federal/gfirst.html
- High Technology Crime Investigation Association (HTCIA) http://www.htcia.org/
- InfraGard http://www.infragard.net/
- Internet Storm Center (ISC) http://isc.sans.edu/
- National Council of ISACs http://www.isaccouncil.org/
- United States Computer Emergency Response Team (US-CERT) http://www.us-cert.gov/

Chapter 5

Security Engineering Core Practices

5.1 Applying Systems Security Engineering

Security Engineering evolved to Systems Security Engineering (SSE) as an element of System Engineering (SE) that applies scientific and engineering principles to identify a system's security vulnerabilities and minimize or contain risks associated with these vulnerabilities. It uses mathematical, physical, and related scientific disciplines, and the principles and methods of engineering design and analysis to specify, predict, and evaluate the vulnerability of the system to security threats [DoD 1995]. More often, security is associated with information technology and software, a perception which has led to gaps in systems science and engineering principles and inappropriate guidance and solutions for SSE [Baldwin 2009]. This association and practical needs led to the definition Information Systems Security Engineering (ISSE) as an engineering process that captures and refines information protection requirements and ensures their integration into Information Technology (IT) acquisition processes through purposeful security design or configuration [Defense 2003]. Not integrating ISSE approaches leads to weak security implementations and results in system users accepting an unknown level of risk when using the system.

Recognizing that security became a system level risk due to increasing threats, the United States Department of Defense (DoD) included systems Security Engineering as a critical element of SE. The guidebook [NDIA 2008] describes two perspectives for systems security engineering. First, it describes that SSE needs to be an integral to the technical and management processes of Systems Engineering. The second perspective is the overlay of security throughout the lifecycle of a system. Obviously, the publication of this guide was the first step toward motivating the community in the defense industry to adopt SSE as a recognized consideration in SE.

SSE applies scientific, mathematical, engineering, and measurement principles, concepts, and methods to coordinate, orchestrate, and direct the activities of various security engineering and other contributing engineering specialties. The SSE discipline provides the security perspective to systems engineering processes, activities, tasks, products, and artifacts. These processes, activities, and tasks are conducted in consideration of all system elements [NIST SP800-160v1].

5.1.1 System Definitions

A system is composed of interrelated parts or components (structures) that cooperate in processes (behavior). Although no provision for a unique definition of a system, in the field of SE, definitions from prominent standards are recognized:

- An aggregation of end products and enabling products to achieve a given purpose [ANSI/EIA 632].
- A set or arrangement of elements and processes that are related and whose behavior satisfies customer/operational needs and provides for lifecycle sustainment of the products [IEEE 1220].
- A combination of interacting elements organized to achieve one or more stated purposes [ISO/IEC/IEEE 15288].
- An organized hierarchy of components (hardware, software, data, personnel, and communications, for example) having a specified purpose and performance requirements [Vacca 2012].

- An assembly of computer and/or communications hardware, software, and firmware configured for the purpose of classifying, sorting, calculating, computing, summarizing, transmitting and receiving, storing, and retrieving data with a minimum of human intervention [CDSE 2012].
- A construct or collection of different elements that together produce results not obtainable by the elements alone. The elements, or parts, can include people, hardware, software, facilities, policies, and documents; that is, all things required to produce systems level results. The results include system level qualities, properties, characteristics, functions, behavior and performance. The value added by the system as a whole, beyond that contributed independently by the parts, is primarily created by the relationship among the parts; that is, how they are interconnected [INCOSE 2015].
- System is synonym for information system where information system is an organized assembly of computing and communication resources and procedures - e.g., equipment and services, together with their supporting infrastructure, facilities, and personnel – that create, collect, record, process, store, transport, retrieve, display, disseminate, control, or dispose of information to accomplish a specified set of functions [RFC 4949].
- A discrete set of information resources organized for the collection, processing, maintenance, use, sharing, dissemination, or disposition of information; Information systems also include specialized systems such as industrial/process controls systems, telephone switching and private branch exchange (PBX) systems, and environmental control systems [CNSSI 4009].

5.1.2 Evolving Systems Engineering Perspectives

SE is an interdisciplinary, collaborative approach that derives, evolves, and verifies a life cycle balanced system solution which satisfies customer expectations and meets public acceptability [IEEE 1220].

According to INCOSE (the International Council on Systems Engineering), SE means to enable the realization of successful systems. It focuses on defining customer needs and required functionality early in the development cycle, documenting requirements, and then proceeding with design synthesis and system validation while considering the complete problem including activities such as operations, performance, test, manufacturing, cost and schedule, training and support, disposal.

SE consists of two significant disciplines: the technical knowledge domain in which the Systems Engineer operates, and Systems Engineering Management (SEM) [DoD 2001].

SEM is an element of program management that ensures system security tasks are completed. These tasks include developing security requirements and objectives; planning, organizing, identifying, and controlling the efforts that help achieve maximum security and survivability of the system during its lifecycle; and interfacing with other program elements to make sure security functions are effectively integrated into the total system engineering effort.

The origin of structured SE lies in the development of large, complex, operational systems. Complexity incorporates not only engineering systems, but also the logical human organization of data. At the same time, a system can become more complex due to an increase in size as well as with an increase in the amount of data, variables, or the number of fields that are involved in the design. An understanding of the basic principles and processes is necessary in order to benefit from the knowledge embodied in SE processes. However, the basic principles are not either unique or diligently applied by the developers.

SE focuses on how to design and manage engineering projects over their lifecycles to ensure that all likely aspects of a project or system are considered and integrated into a whole system. It deals with processes, optimization methods, and risk management tools. Tools for graphic representations of a system are used to communicate a system's functional and data requirements. Examples of common tools include [Sommerville 2007]:

- Functional flow block diagram
- Model based design
- Data flow diagram
- N2 Chart
- IDEF0 diagram
- Use case diagram
- Sequence diagram.

A graphical representation relates the various subsystems or parts of a system through functions, data, or interfaces. Any or each of the above methods are used in an industry based on its requirements. For instance, the N2 chart may be used where interfaces between systems is important. Part of the design phase is to create structural and behavioral models of the system.

5.2 Why Security Engineering Methods?

Traditionally, Information Security is focused on applying information protection techniques to previously established system architectures and design specifications. Also, it is often applied without using the available methods provided by SE specialty (e.g., analysis methods, design methods, simulation, measurement methods, verification/validation methods, logistics methods, acceptance methods, maintenance methods, sanitization, etc.).

5.2.1 Systems Engineering Management

Systems engineering provides the foundation for a disciplined and structured approach to engineering trustworthy secure systems. Trustworthiness, in this context, means simply worthy of being trusted to fulfill whatever critical requirements may be needed for a particular component, subsystem, system, network, application, mission, enterprise, or other entity [Neumann 2004]. Trustworthiness requirements can include, for example, attributes of safety, security, reliability, dependability, performance, resilience, and survivability under a wide range of potential adversity in the form of disruptions, hazards, and threats [NIST SP800-160v1]. Effective measures of trustworthiness are meaningful only to the extent that the requirements are sufficiently complete and well defined and can be accurately assessed. The security considerations and the planning to address those considerations are captured as part of the Systems Engineering Management plan. SEM is accomplished by integrating three major activities [DoD 2001]:

- Development phasing that controls the design process and provides baselines that coordinate design efforts.
- A Systems Engineering process that provides a structure for solving design problems and development tracking requirements flow through the design effort.
- Lifecycle integration that involves customers in the design process and ensures that the system developed is viable throughout its life.

Figure 5.1 depicts the scope of SEM that includes process, planning, integration, development, and management.

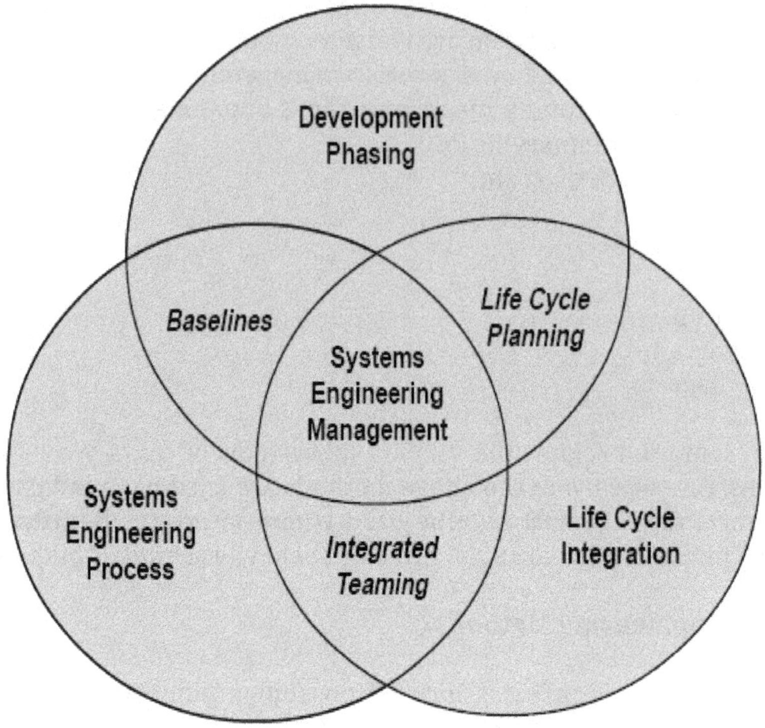

Figure 5.1 The Systems Engineering management scope. *Source:* [DoD 2001]. Public Domain.

Increasing the trustworthiness of systems is a significant undertaking and a fundamental cultural change to the current approaches. Introducing a disciplined, structured, and standards-based set of SSE activities and tasks provides an important starting point and forcing function to initiate needed change [Ross 2018].

SE involves technical and human centered disciplines such as control engineering, industrial engineering, organizational studies, and project management. SE encourages the use of modeling and simulation to validate assumptions or theories on systems and the interactions within them. For example, a recommended methodology for the planning and realization of projects is provided by the V-Model.

5.2.2 The V-Model

The V-Model is a term applied to a range of models, from a conceptual model designed to produce a simplified understanding of the complexity associated with systems development to detailed, rigorous development lifecycle models and project management models. Figure 5.2 shows activities to achieve objectives such as:

- Minimization of project risks - improves project control by specifying standardized approaches and describes the corresponding results and responsible roles; it recognizes risks and improves process management, thus reducing the project risk.
- Improvements and guarantees of quality - the standardized process model, V-Model, ensures that results to be provided are complete and have the desired quality.
- Reducing total project cost over system lifecycle - the effort for the development, production, operation and maintenance of a system can be calculated, estimated and controlled in a transparent manner by applying a standardized process model.

- Satisfaction of stakeholders - all relevant elements and terms are the basis for the mutual understanding between all stakeholders.

The Project Management Body of Knowledge (PMBOK) guide, an IEEE standard, redefines the verification and validation phases [IEEE 1490] as follows:

- Validation - assurance that a product, service, or system meets the needs of the customer and other identified stakeholders, acceptance and suitability with external customers.
- Verification - evaluation of a product, service, or system complies with a regulation, requirement, specification, or imposed condition; it is often an internal process.

The V-Model demonstrates the relationships between each phase of the development lifecycle and its associated phase of testing (see more in [IEEE 1490] guide).

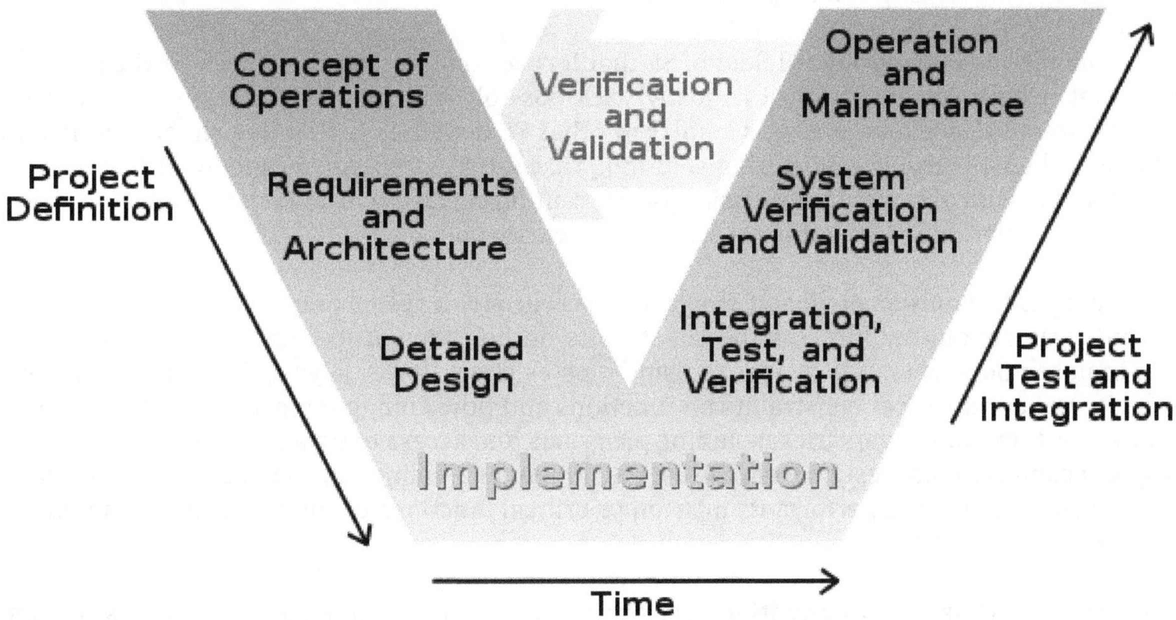

Figure 5.2 The V-model of the Systems Engineering process. *Source:* [DoT 2005]. Public Domain.

PMBOK is the [IEEE 1490] guide to the project management body of knowledge (see [IEEE 1490] standard). The sixth edition of PMBOK was released in 2017. The seventh edition was released in 2021 because the body of knowledge evolves over time. Also, this post [V-Diagrams] is focused on teaching the basics of engineering project management. Specifically, it describes the V-Diagrams – a tool that helps plan, execute, monitor, and control the procurements [Warner 2019] of an engineering project.

5.2.3 Recommended System Engineering Methods

The development of smarter control algorithms, microprocessor design, and analysis of environmental systems also come within the purview of Systems Engineering. SE encourages the use of tools and methods to better comprehend and manage the development of systems. Examples of methods that are described in [DoD 2001] and guidelines:

- Requirements Engineering
- Systems analysis
- Modeling and simulation
- System architecture

- Optimization
- System dynamics
- Risk management
- Statistical analysis
- Reliability analysis
- Decision making
- Knowledge engineering.

Development of systems using Systems Engineering concepts and methods is also described in [Wasson 2015]. Methods and practices for engineering complex systems are also described in [Kossiakoff 2011].

5.2.4 Systems Security Engineering Constraints

Security Engineering is a specialized field of SE that focuses on the security aspects in the design of systems that need to be able to deal robustly with possible sources of disruption, ranging from natural disasters to malicious acts. It is similar to other SE activities in that its primary motivation is to support the delivery of engineering solutions that satisfy predefined functional and user requirements, but with the added dimension of preventing misuse and malicious behavior. These constraints and restrictions are often asserted as a security policy.

Security policy is a definition of what it means to be secure for a system, organization or other entity. For an organization, it addresses the constraints on behavior of its members as well as constraints imposed on adversaries by mechanisms such as doors, locks, keys and walls. For systems, the security policy addresses constraints on functions and flow among them, constraints on access by external systems and adversaries including programs and access to data by people.
Security is a condition that results from the establishment and maintenance of protective measures that enable an enterprise to perform its mission or critical functions despite risks posed by threats to its use of information systems.

The term system, as used in Information Security, refers to those systems that process, store, and transmit data to provide a service or function, or to monitor and control physical devices, other systems, or capabilities. Examples of these systems include general and special purpose information systems; command, control, and communication systems; crypto modules; central processing unit and graphics processor boards; process control systems; flight control systems; weapons, targeting, and fire control systems; medical devices and treatment systems; financial, banking, and merchandising transactions systems; and social networking systems [CNSSI 4009].

Protection measures may involve a combination of deterrence, avoidance, prevention, detection, recovery, and correction that should form part of the enterprise's risk management approach [CNSSI 4009]. This definition focuses on security as an organizational enterprise objective.

SE focuses on all types of systems, including information systems (e.g., a type of system). Information systems security engineering is the process that captures and refines information security requirements and ensures their integration into information technology component products and information systems through purposeful security design or configuration. In this book, when the term information system is used, the intended meaning is any type of system. ISSE is a specialization of SE [McAllister 2002]. Therefore, SSE must have that same focus. Security principles applied to phases of system's lifecycle are described in literature (books, articles, custom handbooks), [NIST SP800-27revA], [NIST SP800-160v1] documents, and standards (e.g., [ISO/IEC/IEEE 15288]). Commonly known are the five lifecycle phases to include initiation, development/acquisition, implementation/assessment, operation/maintenance, and disposal for the system acquisition/development process as depicted in Figure 5.3.

This document [NIST SP800-160v1] describes SSE activities and tasks aligned with and developed as security extensions to the system life cycle processes in [ISO/IEC/IEEE 15288] standard. The standard defines a plethora of processes for the system life cycle that are organized and grouped into four families:

- Agreement processes
- Organization project enabling processes
- Technical management processes
- Technical processes.

The processes are applied across all stages of the system life cycle that includes these stages: concept, development, production, utilization, support, and retirement. The system life cycle processes are not prescriptive in their execution and do not map explicitly to specific stages in the system life cycle. By design, the processes can be applied concurrently, iteratively, or recursively at any level in the structural hierarchy of a system at any stage in the system life cycle, in accordance with acquisition, SE, or other process models. The processes can be customized so providing flexibility for optimized and efficient use for different stakeholders of interest and sectors, system types, technologies, and trustworthiness objectives.

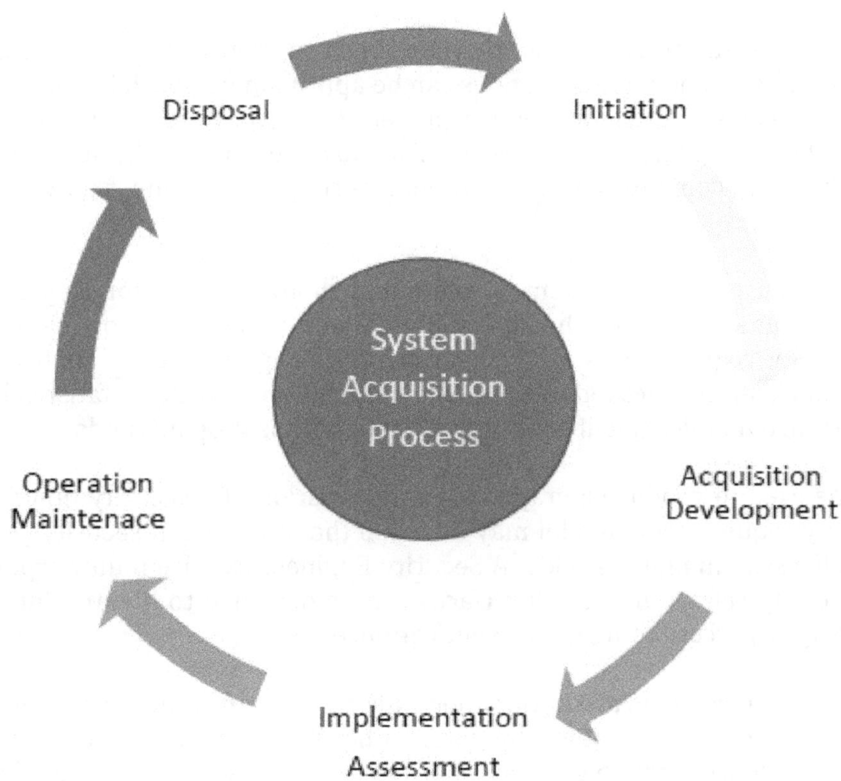

Figure 5.3 System acquisition/development process.

This work [Higginbotham 1998] describes the use of a SE tool to identify security relevant threads and to manage and provide discipline for the design process. This information is further applied to the security architecture development and the technical documentation needed for risk assessment. SE tools support requirement analysis, behavioral analysis, architectural design; verification; document generation, and requirements traceability.

5.2.5 Applying Engineering Methods to Security

Although Security Engineering has been recognized as a discipline, the security industry is still not using consistently sound engineering methods to defend and respond to attacks including problems such as [Kadrich 2007]:

- Security of end point (e.g., laptop, phone device, industrial controller) is not ensured.
- Vendors use their marketing tools to sell products as security solutions for a regulation such as HIPAA (The Health Insurance Portability and Accountability Act of 1996), SOX (The Sarbanes–Oxley Act of 2002), and GLBA (The Gramm–Leach–Bliley Act (GLBA), also known as the Financial Services Modernization Act of 1999 instead of focusing on real security issues.
- Security professionals are using the vendor's template without searching for a good solution to security problem.
- Methods based on the perspective of eliminating vulnerabilities (a reactive method) failed because eliminating the vulnerabilities only reduces the number of attack vectors, but threat agents still find security flaws in the software.
- Relying solely on vulnerability management is not any more practical because there could be other threats. For example, trusted systems can do untrustworthy things.

Traditionally, Engineering tries to solve the non-linear system problem while accepting that for small perturbations, most non linear systems can be approximated with linear systems, significantly simplifying the analysis. Linear systems represent the main class of systems for which general techniques for stability control and analysis exist. However, many physical systems (e.g., power system) are inherently complex systems and engineering practice must now include elements of complex systems.

Generally, methods used for the system development should be used for security and privacy too. These methods include analysis methods, design methods, implementation methods, testing, etc. Analysis of security requirements can be performed using structured analysis, object-oriented analysis, agile, and adaptive methods. Common techniques may include statistical analysis, reliability analysis, system dynamics (feedback control), and optimization methods.

Modeling can be used to produce a graphical representation of a security function or system. For example, security requirements model may describe the information security protection and performance, which a system must provide. A Security Engineer can use multipurpose graphical tools to represent security related models. Interfaces can be designed to identify internal and external security features (e.g., access controls, authentication, etc.).

Once the security requirements are understood, it is the responsibility of a Systems Security Engineer to refine them, and to determine, along with other engineers, the best alternative for security. SSE adds value and contributes to general SE processes in these areas: protection needs; security relevance; trustworthiness and assurance; and security risk management. SSE, as part of the requirements analysis process, transforms the stakeholder security requirements into the system security requirements and refines the security concept of operations to achieve the operational objectives of the organization. Therefore, it has to be understood that Security Engineering for Smart Grid systems is about developing and implementing systems that are predictably dependable in the face of all types of threats. The systems and applications must also be resilient to error or threats (intentional or unintentional). In addition, the responsibility of Security Engineering is to ensure protection of systems that evolve constantly and face a changing spectrum of threats [P2030 2011]. Considering the needs of Smart Grid and DER systems, there is a need to apply the methods that support security objectives and the best interests of an organization.

5.2.6 Approach to Systems Security Engineering

One common approach used by Security Professionals takes the form of threat model - security policy - security mechanisms [Anderson 2008a]. This approach has its limitations when requirements engineering is not done right or completely avoided. Security requirements engineering could be difficult if the available approaches are not properly used.

A challenge is that most Security Engineers don't have the Systems Engineering background required to approach a security problem holistically [Bayuk 2011a]. Security Engineering involves aspects of SE, safety engineering, social science, psychology, economics, computer engineering, computer science, electronics, physics, chemistry, mathematics, and architecture. Innovative methods, tools, and procedures can be used in a SSE approach that has to address systemic security requirements. While approaches to SSE are not unique, one suggested sequence of steps is described in [Bayuk 2011a] (see Figure 5.4).

As illustrated in the diagram, the sequence is defined as the following: analyze system mission and purpose in operational context, define security, design secure architecture, devise security metrics, then devise SSE methods. By extracting the definition of security from the context in which a system operates, the security architecture is integrated into systems architecture, customizing it rather than bolting it on. Next, after devising the architecture metrics it is possible to verify whether security features meet security functional requirements.

When requirements gathering techniques are being employed in the service of security, they present functional requirements at the systems level. In this approach, security is defined in the context of a given system mission. This allows design of alternative security architectures, as well as metrics. By applying the metrics to those architectures, it is possible to determine their effectiveness in maintaining system security. This approach enables system owners and operators to:

- Identify security features requiring system level functions.
- Evaluate the extent to which security features protect systems from deliberate damage that would cause system failure.
- Devise verification and validation metrics at the system level that show that security requirements are met.

However, this method is a sequential approach that is costly inefficient, takes more time, and leaves room for gaps and makes. In practice, steps are performed in parallel (see SE steps depicted in Figure 5.5).

Others consider different principles. For example, the principles used for the development of systems include [Anderson 2008a]:

- Risk management that is decomposed into seven actionable elements: risk identification, risk quantification, risk sensitivity, risk mitigation, factoring risk in decisions, document risk mitigation, tracking effectiveness of mitigation plans.
- Types of risk issues to deal include technical, cost, schedule risk, programmatic risk.
- Technology maturity and maturation rate.

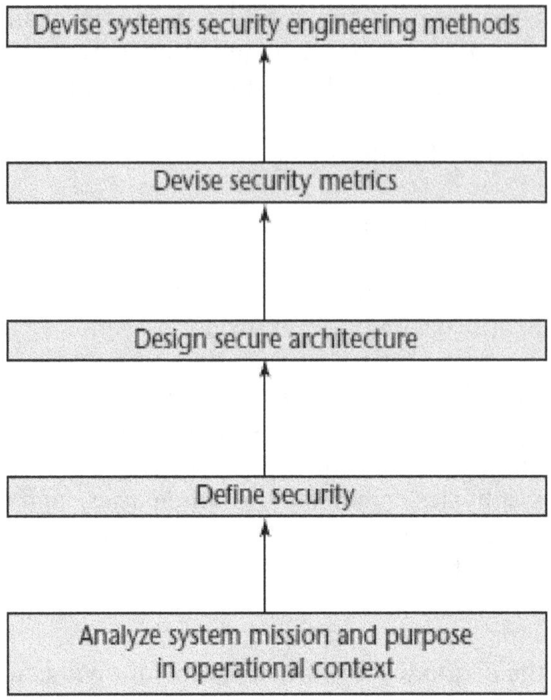

Figure 5.4 An approach to Systems Security Engineering. *Source:* [Bayuk 2011a]. © 2011, IEEE.

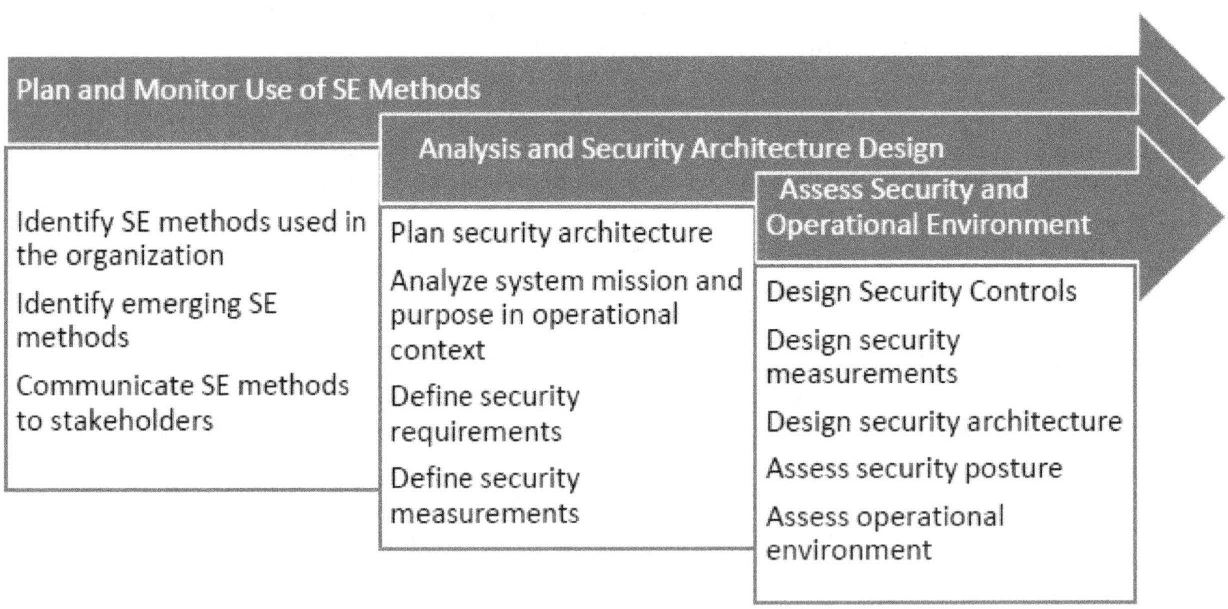

Figure 5.5 An approach to Systems Security Engineering activities.

Different than the above diagram of Figure 5.5, this work [Cantor 2006] summarizes the following principles for system development:

- Decompose systems, not requirements (technical).
- Enable both separation and integration of key systems development concerns (technical).
- Specifications flow up and down the architecture (technical).
- Systems and components collaborate, so should development teams.

- Development organizations should reflect product architectures.
- Base the development lifecycle on removing risk and adding value.

To be effective in protecting their information infrastructure, enterprises need to take a Systemic Security Management (SSM) approach to build in security [Kiely 2008]. The authors argue that security issues have been studied too simplistically, in terms of - technology, process, people - as a somewhat static collection of three independent issues. By adding the new concept of organizational design, the authors emphasize the need of understanding how people, process, technology, and organizational design all interact among themselves to create that complex mix of elements and issues that the question of security really is. This approach to security can serve the extended enterprise, going well beyond the boundaries of the company to include not just people, process, technology and organization, but also partners, suppliers, customers, and communities. A refined systemic approach for Internet of Things (IoT) security is described in [Riahi 2013]. The security model is made up of four nodes: person, technological ecosystem, process and intelligent object. Therefore, Systems Security Engineers need to adopt the strategy of building systems based on robust models including and not limited to methods and objectives as suggested in [CSIIRW06], [Sheldon 2006]:

- Knowledge discovery and management
- Critical infrastructure protection
- Deobfuscating tools for the validation and verification of tamper proofed software
- Computer network defense technologies
- Scalable information assurance strategies
- Assessment driven design for trust
- Security metrics and testing methodologies
- Validation of security and survivability properties
- Threat assessment and risk analysis
- Early accurate detection of the insider threat
- Security hardened sensor networks and ubiquitous computing environments
- Mobile software authentication protocols
- A new model of the threat to replace the Maginot line model.

Ideally, the most efficient approaches cover all phases of information security management and system lifecycle processes including tools and techniques to support methods for decomposing systems to identify critical components, architectural approaches to neutralize threats, verification and validation, and ways to optimize lifecycle security costs. The challenge is the availability of commercial tools with these capabilities. Many advanced methods are being demonstrated by researchers as a prototype and the implementations are lagging behind.

Also, these authors [Bugeja 2016] argue that smart home security and privacy issues might go unnoticed or poorly addressed by researchers as the commercial side of this paradigm is evolving at high pace. Certain security methods while they increase privacy or security may carry side effects, so further investigation is needed.

5.2.7 System of Systems Approach

System of Systems (SoS) is a collection of task oriented or dedicated systems that pool their resources and capabilities together to create a new, more complex system which offers more functionality and performance than simply the sum of the constituent systems [Wiki SoS]. [Tzanev 2013] defines SoS consisting of a mix of intensively interacting independent systems with common goals during task execution. SoS are large scale concurrent and distributed systems with the components of which are complex systems themselves. Smart Grid is an example of SoS [P2030 2011]. However, many physical systems (e.g., Smart Grid and DER systems) are inherently complex systems in terms of the definition above, and engineering practice must now include elements of complex systems research.

The methodology for defining, abstracting, modeling, and analyzing system of systems problems is typically referred to as System of Systems Engineering (SoSE). While the individual systems constituting a system of systems can be very different and operate independently, their interactions typically expose and deliver important emergent properties that must be recognized, analyzed, and understood. The SoS approach does not advocate particular tools, methods, or practices; instead, it promotes a new way of thinking for solving grand challenges where the interactions of technology, policy, and economics are the primary drivers.

SoS typically exhibit the behaviors of complex systems. But not all complex problems fall in the realm of SoS. Inherent to SoS problems are various combinations of behavior due to:

- Interactions
- Interoperability
- Geographical distribution
- Heterogeneity
- Networks.

Design for SoS problems is performed under some level of uncertainty in the requirements and the constituent systems, and it involves considerations in multiple levels and domains [Curry 2012]. Whereas SE addresses the development and operations of monolithic products, SoSE addresses the development and operations of evolving programs. In other words, traditional systems engineering seeks to optimize an individual system (e.g., the product), while SoSE seeks to optimize a network of various interacting legacy and new systems brought together to satisfy multiple objectives of the program. SoSE should enable the decision makers to understand the implications of various choices on technical performance, costs, extensibility and flexibility over time; thus, effective SoSE methodology should prepare decision makers to design informed architectural solutions for SoS problems.

Due to varied methodology and domains of applications in existing literature, there is no single unified consensus for processes involved in SoSE. More information on methods can be found in this survey [Tzanev 2013]. A proposed SoSE framework [DeLaurentis 2006] recommends a three-phase method where a SoS problem is defined, abstracted, modeled and analyzed for behavioral patterns. This review [Axelsson 2019] finds game theory can be applied to SoS in a wide range of application areas, specifically problems related to acquisition, design, and operations. In particular, the operational formation of SoS is feasible and it often requires the use of simulation techniques.

Figure 5.6 depicts a visual diagram of several discipline areas that contribute to the understanding and development of complex systems. The organizational map is broken into even subgroup areas. The three circles on the left (Nonlinear Dynamics, Systems Theory, and Game Theory) are the historical roots of complex systems science, while the other four (Pattern Formation, Evolution and Adaptation, Networks, and Collective Behavior) are the more recently studied topical areas. The key problems of complex systems are difficulties with their formal modeling and simulation. For systems that are less usefully represented with equations, various other kinds of narratives and methods for identifying, exploring, designing and interacting with complex systems are used.

Complex systems science is a rapidly growing scientific research area that fills the huge gap between the two traditional views that consider systems made of either completely independent or completely coupled components. Complex systems science develops conceptual, mathematical, and computational tools to describe systems made of interdependent components. It studies the structural and dynamical properties of various systems to obtain general, cross disciplinary implications and applications.

Although not easy task, Security Engineer and the team have to evaluate existing approaches and tools within an organization and then evaluate and adopt other new approaches based on some criteria that guarantees security protection and meets the business needs.

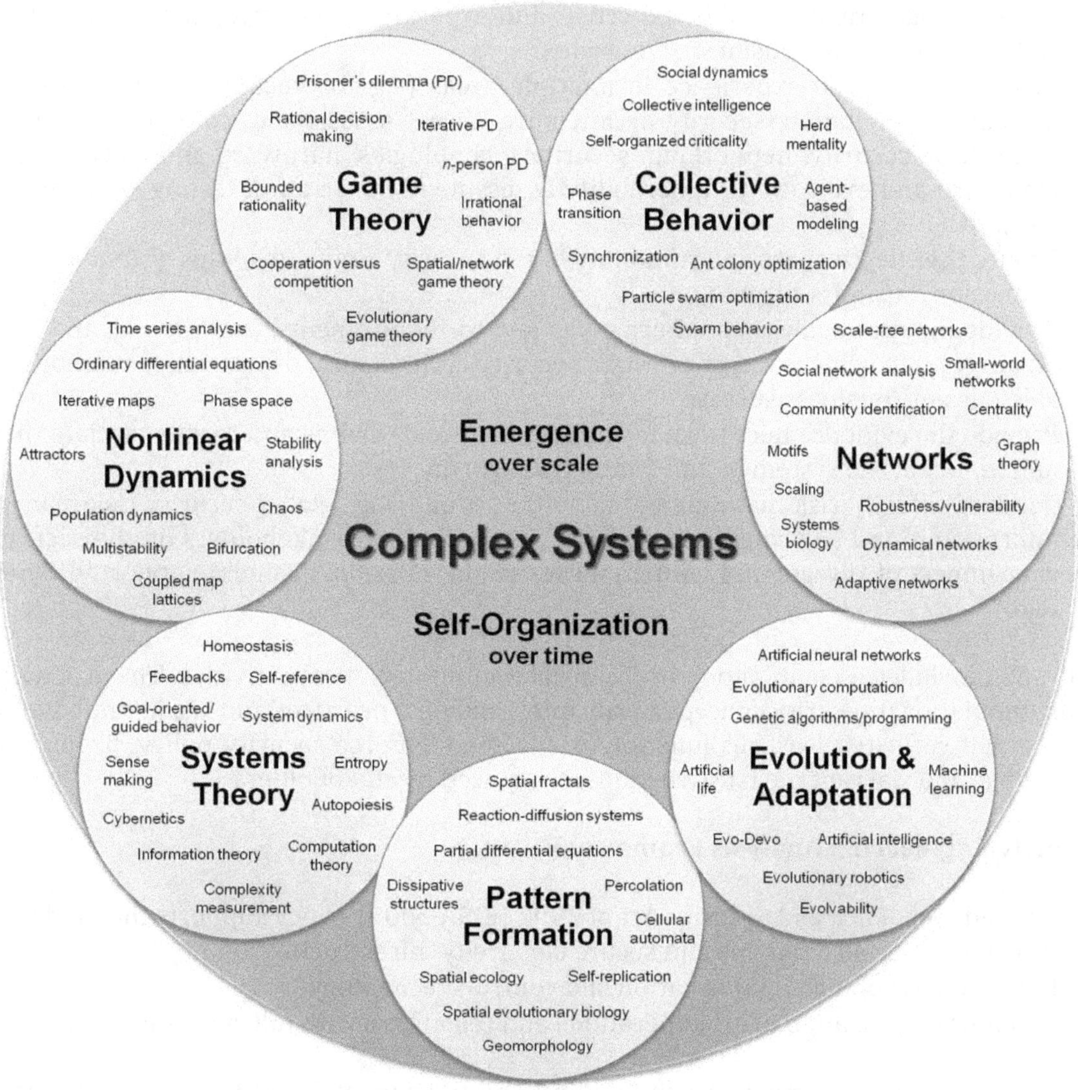

Figure 5.6 Organizational map of complex systems. *Source:* [Sayama 2010]. Licensed under CC BY 3.0.

5.2.8 Security Engineer Responsibilities and Skills

Security professionals should use engineering methods in describing and finding solutions for security problems [Kadrich 2007]. One example is closed loop process control method that can be used for a network and end points. A demonstration of the process control techniques applied to security problems is introduced in this work [Kadrich 2007]. Because modern systems cut across many areas of human endeavor, Security Engineers not only need to consider the mathematical and physical properties of systems, they also need to consider attacks on the people who use and form parts of those systems using social engineering attacks. Secure systems have to resist not only technical attacks, but also social engineering attacks.

The Security Engineer has a foundational understanding of SE, to include the processes and roles for which the Systems Engineer is responsible. A summary of Systems Security Engineer role and responsibilities are described in [NIST SP800-160]:

- Participate in a multidisciplinary Systems Engineering team, applying fundamental systems security understanding, skills, expertise, and experience to develop a system that satisfies stakeholder mission/business requirements.
- Have expertise and experience in multiple areas (e.g., protection needs assessment, requirements elicitation, security architecture, threat assessment, computer security, communication security, networking, security technologies, hardware and software development, test and evaluation, vulnerability assessment, penetration testing, and supply chain risk).
- Ensure the effectiveness and suitability of the security elements of the system as an enabler to mission/business success.
- Provide input to analyses of alternatives and to requirements, engineering, and risk trade off analyses to achieve a cost-effective security architectural design for protections that enable mission/business success.
- Provide the evidence necessary to support assurance claims and to substantiate the determination that the system is sufficiently trustworthy.
- Conduct security risk management activities, producing related security risk management information, and advising the engineering team and key stakeholders on the security relevant impact of threats and vulnerabilities to the mission/business supported by the system.

Basically, SSE includes a combination of technical and nontechnical activities. These activities are based on foundational security concepts, principles, and best practices and are intended to provide confidence that security measures function as specified, enforce security policy, produce the desired outcome, and warrant the trustworthiness required by stakeholders.

5.3 Security Engineering Analysis Framework

The Energy Independence and Security Act of 2007 [EISA 2007] policy supports the modernization of power grid to maintain a reliable and secure electricity infrastructure by

- Increased use of digital information and controls technology
- Dynamic optimization of grid operations and resources, with full cybersecurity.

More precisely, in order to meet security and reliability goals of Smart Grid it is required a good security engineering approach based on the framework (see Figure 5.7) of four elements to work together as described in [Anderson 2008a]:

- Policy – a succinct statement of a system's security strategy
- Incentives – motives of people operating a system and of attackers trying to defeat a policy
- Mechanism – the countermeasures to defeat a security event that may cause a cyber attack
- Assurance – the amount of reliance on a particular mechanism.

Figure 5.7 shows the framework comprising these elements which interact with each other. There is a policy that has to be achieved and a mechanism implements the policy so enabling an assurance of protection given that there are incentives for users to keep the system and attackers to defeat the policy. For example, utilities need a Demand/response Management System (DRMS) to enable market operators and utilities to efficiently, reliably and securely manage Demand Response (DR) processes end-to-end [Vojdani 2006], [Neumann 2007]. D/R refers to changes in electric usage by end-use customers from their normal consumption patterns in response to changes in the price of electricity over time, or to incentive payments designed to induce lower electricity use at times of high wholesale market prices or when system reliability is jeopardized. There are benefits (incentives) to make this product secure and operate correctly.

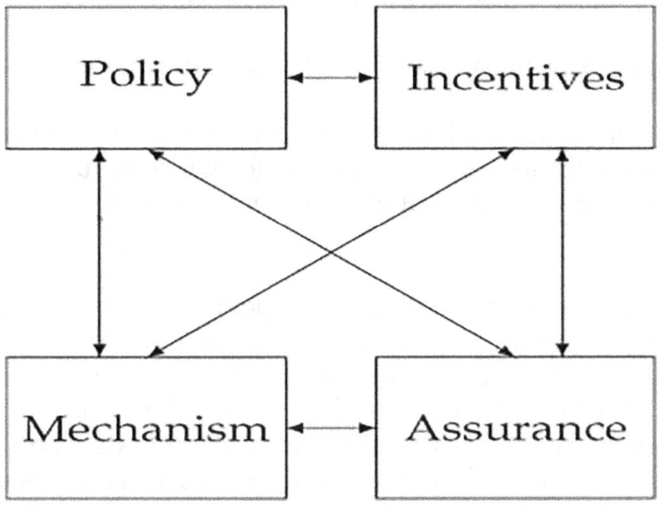

Figure 5.7 Security Engineering analysis framework. *Source*: [Anderson 2008a]. © 2008, Wiley.

When using the security engineering analysis framework to implement the necessary protection for the information, DRMS system and customers' privacy, there are elements that are needed. There is a policy across energy sector to manage the consumer demand for electricity. Also, there is a security policy that mandates the implementation of proper security controls (mechanisms) to protect customer and utilities information. Example of a security policy for this Smart Grid system is one that addresses the constraints on use of the DRMS and behavior of its users (utility, customers, service providers) as well as constraints imposed on functions and flow among them. Also, it includes constraints on access by external systems and adversaries including programs and access to data by people as well as protection of customers' privacy.

There should be an assurance that security controls are effective and cannot be defeated by intruders. Finally, there is assurance, a confidence that the security mechanisms enforce the security policy (security assurance) and protect the information and the system. Security assurance is a property of DRMS that the utility and customers have confidence that the system operates such that the system's security policy is enforced [RFC 4949]. There are benefits (incentives) to make this product secure and operate correctly. Therefore, the Security Engineer has to understand that all these aspects interact and have to be addressed separately and as a whole framework for the DRMS system. The following provides an overview of the analysis framework.

5.3.1 Policy

Security policy is a definite goal, course, or method of action to guide and determine present and future decisions concerning security in a system [RFC 4949].

A security policy is a (partial) system specification, lack of adherence to which will be regarded as a security failure. In practice, there may be a hierarchy of such security policies, relating to a hierarchy of systems —for example, an entire organization, its information systems department, and the individuals and systems in that department. However, the term security policy is used in several ways, including:

- Security policy objectives include a statement of intent to protect identified assets within the specific scope of stakeholder responsibility and security risk concerns.
- Organizational security policy is the set of laws, rules, and practices that regulate how an organization manages, protects, and distributes assets to achieve specified security policy (protection) objectives.
- System security policy specifies what a trusted system is trusted to do.

There is a hierarchical relationship established among the uses of the term security policy (e.g., security policy objectives subsume organizational security policy which in turn, subsumes system security policy). Figure 5.8 shows the hierarchical relationship among the different uses of the term security policy.

From the systems Security Engineering perspective, security policy objectives and all subsequent security policy forms are derived from protection needs and inputs which are assessed and transformed into the security requirements that specify the security capability needs that are to be satisfied. Example of security objectives for Smart Grid and DER systems include:

- Develop more penetration resistant, trustworthy, and resilient systems that are capable of supporting critical missions and business operations with a level of assurance or confidence that is consistent with the risk tolerance of the organization.
- Provide assurance evidence to substantiate the trustworthiness of protective measures.

According to [EPRI 2013]., every aspect of a Smart Grid must be secure. Cybersecurity technologies are not enough to achieve secure operations without policies, ongoing risk assessment, and training. The development of these human focused procedures takes time and needs to take time to ensure that they are done correctly. System security policy may be reflected in semiformal or formal models and specifications. Mathematical methods and techniques (e.g., formal methods) may be used to provide assurance in the correctness of security policy models. The models and specifications are used as the basis for design and implementation of the mechanisms that enforce the policy. Verification activities demonstrate that the mechanism is a correct implementation of the security policy model. Proper planning and help for writing information security policies including sample policies are available in [Barman 2002].

Figure 5.8 Security Policy Hierarchy. *Source:* [NIST SP800-160]. Public Domain.

According to [NIST SP800-160v1], security policy distinguishes the security relevant aspects of integrity protections and availability protections from the integrity and availability protections that are addressed by other specialty disciplines. Figure 5.9 illustrates the scope of security policy properties.

Security policy addresses all aspects of confidentiality but only the security relevant aspects of integrity and availability. Whereas confidentiality protections against unauthorized access and unauthorized disclosure of resources are fully addressed by the security discipline, the scope of integrity protections and availability protections are addressed by the combination of security and other specialty disciplines.

The integrity and availability scope overlap may lead to confusion when the basis for the protection need is not properly allocated or understood. The confusion is often the source of design conflicts and contradictions that are best resolved through informed trade space analysis among all impacted and contributing disciplines.

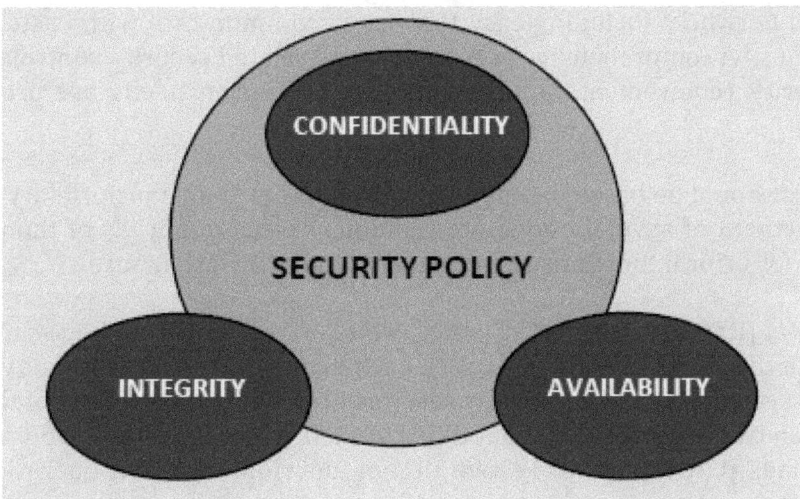

Figure 5.9 Security policies properties scope. *Source:* [NIST SP800-160]. Public Domain.

Confidentiality concerns are fully within the scope of security disciplines and therefore, security policy. Integration and availability concerns span security and non security disciplines such as performance, fault tolerance, reliability, safety, and functional correctness. Security policy identifies the security relevant aspects of integrity and availability.

The confidential nature of personal identifiable information (PII) suggests that identity related data should be the subject of administrative policies. These include rules for application access, company policies about the use of personal or corporate data, legal and regulatory requirements. The quality and reliability of user data and confirmation of user or provider consent regarding its use may also be the subject of administrative policy.

The issue of availability is a difficult one for any quality model that seeks to minimize the overlap of quality factors in its taxonomy. When systems and software engineers think of availability requirements, they think in terms of non malicious causes of lack of availability. However, a denial of service (DoS) attack is a security problem. To maintain the disjointed nature of the taxonomy of quality factors, the scope of availability will remain non-malicious and DoS should be dealt as a violation of a type of security requirements [Firesmith 2003].

5.3.2 Mechanism

A mechanism is a method, tool, or procedure used in a system to implement a security service that is provided by or within the system [RFC 4949]. A security mechanism is a security control of these categories: management, operational, technical. Examples of security services include access control, data confidentiality, data integrity, availability, nonrepudiation, authentication, etc. Security services give a specific kind of protection to system resources [RFC 4949]. Usually, security services are rated in terms of strength of service and assurance of the design [CNSSI 4009].

Security controls are applied based on decisions that balance risks and costs and include protections for authorized access to functions and users based on needs; integrity of data, commands, and system; confidentiality of business information and intellectual property (technologies, business decisions and strategies); guarantees for availability of information and system; guarantees that response and demand inputs are not repudiated.

In addition, the security controls should address the absence of security features in legacy systems, security gaps of emerging technologies, and a heterogeneous environment of the multitude of devices, protocols, and networks including easy tool set to communicate with customers and allow input of their decisions. A comprehensive set of assurance related security controls to help organizations satisfy security requirements and stakeholder protection needs are provided in [NIST SP800-53r5].

The security controls should be based on open standards rather than a proprietary one and should consider the management of several endpoints (millions versus hundreds of thousands) and differentiated controls for critical functions and communications infrastructure.

According to a tutorial [RFC 4949], security policy is one of four layers of the Security Engineering process (as shown in the diagram of Figure 5.10). Each layer provides a different view of security, ranging from what services are needed to how services are implemented. Each of these four layers is a mapping between two points of view that differ in their degree of abstraction, according to the perspectives of various participants in system design, development, and operation activities, as follows:

- Mission functions view is the perspective of a user of system resources.
- Domain practices view is the perspective of an enterprise manager who sets protection standards for resources.
- Enclave services view is the perspective of a system designer who allocates security functions to major components.
- Agent mechanisms view is the perspective of a System Engineer who specifies security mechanisms to implement security services.
- Platform devices view is the perspective of an as built description of the system in operation.

The security policy models and specifications are used as the basis for design and implementation of the mechanisms that enforce the policy. Verification activities demonstrate that the mechanism is a correct implementation of the security policy model.

5.3.3 System Assurance

Security assurance is an attribute of an information system that provides grounds for having confidence that the system operates such that the system's security policy is enforced [RFC 4949]. Related to security assurance are the concept of trust, trust level, trusted system, and trustworthy system. Security assurance offers grounds to a trusted system, that the system does what it claims to do and does not perform unwanted functions and will not fail with some certainty based on inconclusive evidence [RFC 4949]. One area in which SSE adds value and contributes to general SE processes includes trustworthiness and assurance. This area guides SSE activities, tasks, and outcomes.

```
What Security Services
Should Be Provided?          +- - - - - - - - - - - - - -+
^   +- - - - - - - - - - - - -| Mission Functions View   |
|   | Security Policy         |- - - - - - - - - - - - - -+
|   +- - - - - - - - - - - - -| Domain Practices View    |
|   | Security Model          |- - - - - - - - - - - - - -+
|   +- - - - - - - - - - - - -| Enclave Services View    |
|   | Security Architecture   |- - - - - - - - - - - - - -+
|   +- - - - - - - - - - - - -| Agent Mechanisms View    |
|   | Security Mechanism      |- - - - - - - - - - - - - -+
v   +- - - - - - - - - - - - -| Platform Devices View    |
How Are Security             +- - - - - - - - - - - - - -+
Services Implemented?
```

Figure 5.10 Security Engineering views. *Source:* [RFC 4949]. Public Domain.

System assurance is a structured and systematic approach to ensuring the requirements related to reliability, availability, maintainability and safety [Scarbo 2013]. Assurance requirements should be included during a System Engineering lifecycle. However, ensuring information security goes beyond advanced technological solutions as it involves a complex interaction process between human actors and technological solutions. System assurance ensures confidence that the system functions as intended and is free of exploitable vulnerabilities, either intentionally or unintentionally designed or inserted as part of the system at any time during the lifecycle [NDIA 2008]. The guide provides two perspectives for SSE:

- First, process perspective explains how criticality analysis and Security Engineering are integral to the technical and management processes of SE as defined in [ISO/IEC/IEEE 15288]; it recognizes security relationships with these processes that lead to the maturing of guidance for SE teams faced with security as a design requirement.
- Second perspective is the overlay of security throughout the lifecycle; it is critical for program and SE teams to address security requirements while the largest possible trade space exists and ensure the technical maturity of the security solution throughout the acquisition lifecycle.

A sufficiently trustworthy system results from adherence to the security concepts, principles, and practices in the application of security specific methods, techniques, and analyses, some of which stand alone and others that supplement existing SE techniques, methods, and analyses. Safety, reliability, and fault tolerance each have constructs and techniques that are shared with security. However, the application of those constructs and techniques to achieve safety, reliability, and fault tolerance does not necessarily achieve security.

5.3.4 Incentives

The descriptions of the SSE processes assume that sufficient time, funding, and human/material resources are available to ensure a complete application of the processes within a comprehensive SE effort.

Experts from government and industry explore market incentives that may or may not currently exist to stimulate greater corporate investment in cybersecurity protection measures [News 2011]. The discussion is about market incentives and business drivers that may or may not currently exist to stimulate greater corporate investment in cybersecurity protection measures. Such business drivers may include the consideration of relating proactive cybersecurity activity to a favorable valuation of company performance, liability protection measures such as insurance coverage for cyber vulnerabilities, or statutory liability protection. The experts explore the likelihood that these or other market incentives may emerge across industry and government to enhance our collective defense against sophisticated cyber threats. Also, establishing a solid cyber environment requires incentives to develop proactive cybersecurity controls than reactive response.

In the case of DRMS product, increased revenue for a manufacturer of DRMS system and increased productivity and efficiency of utility for operations is predicted by recent study [News 2014]. Market incentives and reliability imperatives have created a wave of new capital investment and technology advancement in support of these new demand response market dynamics. For example, the DRMS system, can benefit from the use of the security analysis framework by applying it to the system and to each component of the system. Instead of focusing on reactive measures, more proactive approaches should be provided for preventing security breaches from taking place.

5.4 Decision Engineering Framework

The need for a unified methodology of decision making is driven by a number of factors that organizations face as they make difficult decisions in a complex internal and external environment. Engineering design is increasingly recognized as a decision making process. This recognition brings with it the richness of many well developed theories and methods from economics, operations research, decision sciences, and other disciplines. Done correctly, it forces the process of engineering design into a total systems context and demands that design decisions account for a product's total lifecycle. It also provides a theory of design that is based on a rigorous set of axioms that underlie value theory. But the rigor of decision based design also places stringent conditions on the process of engineering design.

Decision Engineering seeks to use a number of engineering practices to the process of creating a decision. These include requirements analysis, specification, scenario planning, quality assurance, security, and the use of design principles as described above. During the decision execution phase, outputs produced during the design phase can be used in a number of ways; monitoring approaches like business dashboards and assumption based planning are used to track the outcome of a decision and to trigger another evaluation and planning as appropriate as new information is available. Decision Engineering has potential benefits such as:

- Improve the quality of decisions made, the ability to make them more quickly.
- Enables the ability to align organizational resources more effectively around a change in decisions, and lowers the risks associated with decisions.
- Improves the ability to reason about complex systems as well as to enhance collaboration.

Decision Engineering practices may alleviate the management of a unified information infrastructure of Smart Grid and the interactions between information infrastructure and power infrastructure. Also, control engineers and security engineers could more efficiently find solutions for problems. However, an Information Assurance (IA) framework is needed to address the complexity of information security systems for the Smart Grid including DER systems.

5.4.1 Information Assurance Framework

Many different perspectives and definitions have been constructed on Information Assurance (IA). One common definition refers to IA as the set of measures intended to protect and defend information and information systems by ensuring their availability, integrity, authentication, confidentiality, and non-repudiation [CNSSI 4009]. Another perspective is that IA is a part of risk management. The goal of IA is to assure the quality of information, which results in inspiring confidence on information. In practice, an appropriate level of information assurance confidence is designed based on the worthiness of the information. In doing so, information assurance enables the following features:

- Confidence on the quality of information.
- Performance assessment of the underlying information security systems from policy to implementation.
- Auditing the current performance level of the information security protection in the production system.

Major properties that the IA framework is concerned with include the following:

- Specifies the guiding principle for the information security policy; by tracking the costs of measures and total value of information as deployment ratio there could be assurance that resources are justly deployed without spending more on protection than the value of the information.
- Supports the information lifecycle; information protection must be presented throughout the information lifecycle (creation, consumption, and disposition of the information). When information is created, IA mostly concerns the completeness and quality of information creation and the proper operations of the states within the computing systems. At the creation stage, IA is concerned with ownership and confidentiality. At the consumption stage, IA is concerned with the integrity, confidentiality, and availability of the information. In the disposition stage, IA concerns the proper storage, accessibility and integrity of the information.

Techniques can be used to assure that the quality of the information at each stage is within the specification of the IA framework. A piece of information should be stopped from passing through to the next stage if the information does not meet the requirements of the IA specification. This could be the result of irregularities of data being exposed or vulnerabilities causing data to be in an unprotected state.

5.4.2 Using Information Assurance Framework to Develop Proactive Protection

By reviewing the performance of techniques and how effective they are in stopping the exposure of confidential data, the IA framework can be used as a guide for developing a proactive approach to information privacy protection. With the proper operations of these techniques, the IA framework would help prevent the occurrence of human error.

Techniques can be used to assure that the quality of the information at each stage (creation, consumption, and disposition) is within the specification of the Information Assurance framework.

By reviewing the performance of techniques and how effective they are in stopping the exposure of confidential data, the IA framework can be used as a guide for developing a proactive approach to information privacy protection. When products are evaluated for the level of trust (the system does what it claims to do and does not perform unwanted functions) and assurance they provide, operational assurance is part of the evaluation process. Operational assurance enables a customer to continually obtain the necessary level of protection when using the product. Examples of operational assurances are access controls, separation of privileged from user code, auditing and monitoring capabilities, covert channel analysis, and trusted recovery in case of failure.

5.5 Core Practices

The success of security depends on the use of core practices of Security Engineering that include:

- Security requirements analysis
- Secure coding
- Security architecture
- Security testing
- Security management
- Engineering product lifecycle
- Economics of security.

Each core practice relies on methods, some are based on known techniques and practices used in SE and adapted for security problems, other methods are emerging to include more formal methods, advanced simulation, modeling methods, measurements, and verification and validation of results. As threats evolve, so advancements in the field of SSE are needed. Methods for different core areas and systems are more investigated and several publications describe useful methods [Khan 2012]). SSE views security as a:

- Protection capability that satisfies security requirements and is delivered as part of the system, and
- Quality property of the entire delivered system when it is properly configured and used to enforce the security policy associated with mission/business operations.

Although the V-Model is no longer used as much as would be desirable [BD 2004], we view the basis of core practices illustrated by the V-Model (see Figure 5.2) as a useful Systems Engineering process. The left side of the V-Model represents the creation of system design specifications and the side represents the integration of parts and their validation. This model is also adopted by International Council on Systems Engineering (INCOSE) as a common graphical representation of the system engineering lifecycle [SEbook 2010]. The phases of V-model process are summarized to include [SE cycle]:

- Concept development
- Requirements engineering
- System architecture
- System design and development
- Systems integration
- Test and evaluation

More about characteristics, benefits, and limitations of the V-model can be found in [Forsberg 2005].

5.5.1 Security Engineering Activities

SSE activities are intended to be performed throughout every stage of the system lifecycle. Security must be included in all phases of the system development lifecycle, from design phase through implementation, maintenance, and disposition/sunset [NISTIR 7628r1]. Security must be managed over the lifecycle of the systems deployed. Fundamentally, security is about managing risk which is commensurate with the vulnerabilities and exposures for any given application. Security should be considered at the time the application requirements are being developed since the domain experts are in the best position to understand what is at stake.

Some organization may prefer to organize all security activities and create a variation of the process model, called security systems development process (e.g., Security Systems Development Lifecycle (SecSDLC) [Whitman 2014]).

A system is either built inhouse or the system is purchased from a vendor. A system has its own development lifecycle (SDLC) comprised of the following phases with security activities [Harris 2013], [ISO/IEC 27002] as follows:

- Initiation – determine the need for a specific system, security objectives, preliminary risk assessment
- Acquisition/development – requirements analysis, risk assessment, security assurance, risk management, design, purchase or develop
- Implementation – system testing, installation, certification, accreditation
- Operations/maintenance - system is operated and maintained to perform its services, security monitoring and control
- Disposal – information can be preserved, system is disposed (hardware, software, media).

Each phase includes its own subphases and processes. There are different SDLC models and methodologies, but each follows the same or similar phases. The management can choose a framework of activities which are described by a process model such as waterfall, incremental, evolutionary [Pressman 2005]).

To address system security at every stage of the system lifecycle, SSE processes are also defined in [NIST SP800-160v1r1] extensions to the SE processes defined in [ISO/IEC/IEEE 15288], an International standard for systems and software engineering. The recommended processes contain a specific set of security enhancements that augment/extend the process outcomes, activities, and tasks. There are several methods (formal or informal) that do not include specific security activities. For example, key Security Engineering activities and where they should be integrated into the application or system lifecycle development are described in [Meier 2005], but some activities are missing specific security features.

Examples of distinct security related activities that should be an integral part of an application lifecycle are summarized as follows:

- Security objectives - define security objectives and requirements early in the process; security objectives are goals and constraints that affect the security attributes such as confidentiality, integrity, and availability, non repudiation of information and data.
- Design guidelines for security - to avoid many of the vulnerabilities introduced by poor design choices, design activity should use proven design practices, patterns, and principles. By organizing these design patterns and practices into common vulnerability categories, it is possible to focus on those areas where security mistakes are most often made.
- Threat modeling - threat modeling helps on understanding and identifying the threats and vulnerabilities relevant to a specific application scenario.
- Architecture and design for security - the architecture and design review process analyses the architecture and design from a security perspective. It examines a number of aspects including deployment and infrastructure, overall application architecture and design, and each tier in the application.
- Code review for security - all code should be subject to code inspections where the emphasis is on identifying security vulnerabilities. This should be a continuous activity during the

development and test phases of the application lifecycle.

- Security testing - Use of a risk-based approach and use of the output from the threat modeling activity helps establish the scope of testing activities and define the test plans.
- Deployment review for security - when an application is deployed, there is a need to verify and ensure that weak or inappropriate configuration settings do not introduce security vulnerabilities.

This guidebook [NDIA 2008] highlights the need for assurance case and provides process and technology guidance to increase the level of system assurance for security. This is the enabling and critical mechanism to show that the system will meet its prioritized requirements, and that it will operate as intended in the operational environment, minimizing the risk of being exploited through weaknesses and vulnerabilities. Thus, assurance for security must be integrated into the SE processes. Also, the modern view of core activities is more supportive of Security by Design concept.

A recent survey [Surakhi 2017] discusses known secure software development (SSD) methods used for developing secure software. SSD methods include, among others, security specification languages, security requirements engineering processes, and software security assurance methods. Also, this work provides a list of the most widely used specification languages with the advantages and disadvantages for each. A comparison of the main security features supported by SSD methods finds gaps that require refining the processes and activities.
Figure 5.11 depicts a summary of the secure activities that can be added to the software development life cycle as suggested in [Surakhi 2017].

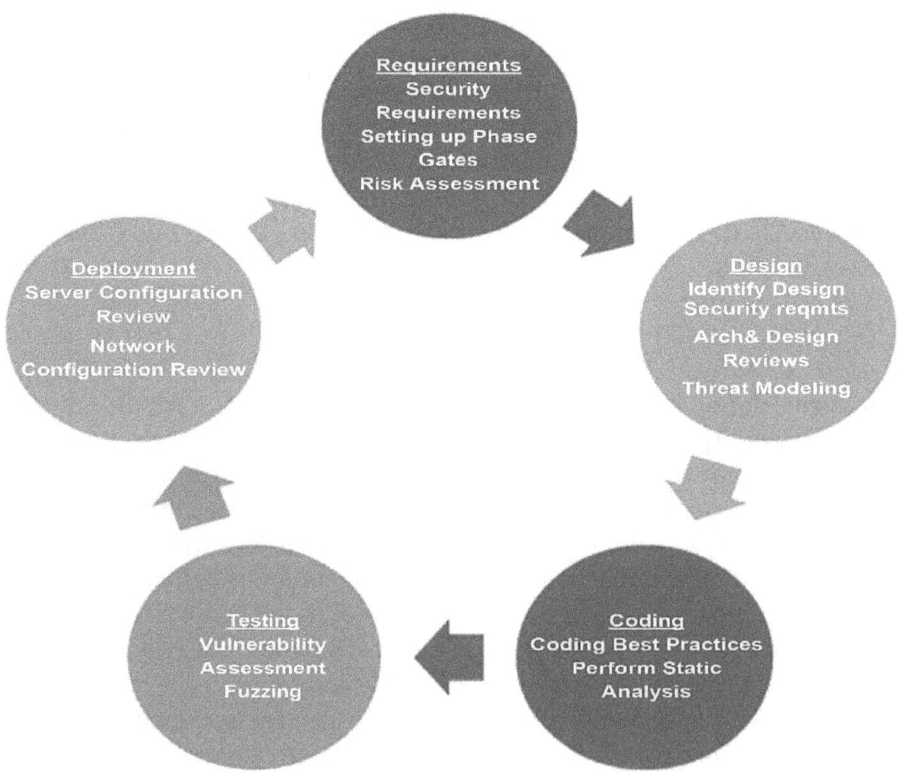

Figure 5.11 Recommendations to secure software development cycle. *Source:* [Surakhi 2017]. Licensed under CC BY 4.0.

The results of the comparison and the suggested recommendations aim to enable software developers with a guideline that allows selecting the best secure process that includes security activities in each phase of the software development.

5.5.2 Is Security by Design a Core Issue?

Security by Design, called also secure design, in the enterprise context was first positioned by Gartner in a whitepaper [Kreizman 2006]. Security by Design, in Software Engineering, means that the software has been designed from the ground up to be secure. Aspects of Security by Design include:

- Software security assurance (designing software systems that are secure from the ground up and minimizing the impact of system breach when a security vulnerability is discovered).
- Preserving privacy in the enterprise environment.
- Ensuring identity across heterogeneous vendors.

According to [Devanbu 2000], security concerns must inform every phase of software development, from requirements engineering to design, implementation, testing, and deployment]. As illustrated in Figure 5.12, Security by Design or Secure by Design with Capability Maturity Model Integration (CMMI) is a structured framework for secure product development. It defines security characteristics for the development process that include process characteristics and specific implementation characteristics.

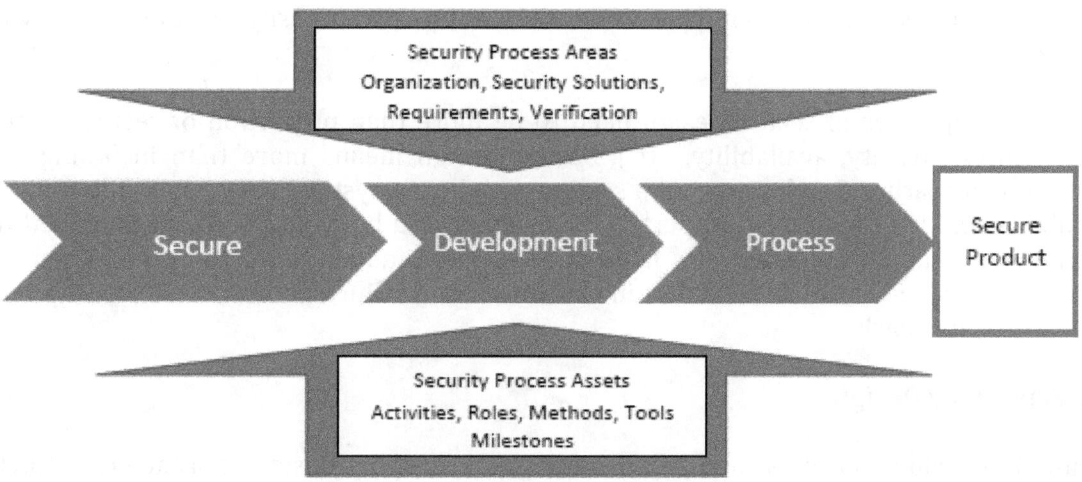

Figure 5.12 Security by Design with CMMI for development.

5.5.2.1 Characteristics

The process characteristics include the organization's preparedness for secure development, the security management of the project, the requirements and security technical solutions, and the verification and validation procedures. The specific implementation characteristics include security process assets and others such as activities, roles, milestones, etc.

As specified in [CMMI 2013] guide, standard processes and other organizational process assets that expressly address these needs are specifically designed for the development of secure products. An organizational process asset library holds these standard processes, methods, procedures, templates, measurements, and training material. Examples of processes for which standard processes and process assets are created include the following:

- Product security risk management
- Security requirements engineering
- Security reviews (e.g., architecture reviews, code reviews, chip design reviews)
- Security testing
- Defect management
- Security validation
- Secure coding.

Other techniques for Security by Design approach include [Wiki CompSEC]:

- The principle of least privilege, where each part of the system has only the privileges that are needed for its function.
- Automated theorem proving to prove the correctness of crucial software subsystems.
- Defense in depth, where the design is such that more than one subsystem needs to be violated to compromise the integrity of the system and the information it holds.
- Default secure settings, and design to fail secure rather than fail insecure.
- Audit trails tracking system activity, so that when a security breach occurs, the mechanism and extent of the breach can be determined.
- Full vulnerability disclosure policy (versus nondisclosure or coordinated disclosure policies) as early as possible so potential victims are as knowledgeable as attackers.

Also, a more extended list of techniques to ensure enterprise security system are provided in [McGraw 2013].

However, the objectives of secure design need to be more than protection of security attributes (confidentiality, integrity, availability, etc.). Secure design means more than including security principles during earlier development of a system and during system lifecycle. It means also more dependability and fault tolerance properties to be supported by a design. There is a need to provide confidence that a system survives failures (e.g., due to adversarial attacks) and can recover from failures. The needs for survivability in case of system failure because of a security incident, requires a new approach.

5.5.2.2 Privacy by Design

The ability to provide security seamlessly by taking a Security by Design approach is of considerable importance to any organization. Equally important is to ensure that privacy is embedded into businesses practices and policies, by default. Therefore, the protection of privacy has taken a similar path, transitioning from a strict compliance issue to an entirely new regulatory paradigm called Privacy by Design. The standard [ISO/IEC 21827] identifies a list of subdisciplines that contribute to secure design:

- Operations security
- Information security
- Network security
- Physical security
- Personnel security
- Administrative security
- Communications security
- Emanation security
- Computer security.

While companies are investing in emerging technologies in the new Internet of Things (IoT) industry, the U.S. Federal Trade Commission (FTC) recommends that companies should also be making appropriate investments in privacy and security [Corbin 2015]. Although there are differences between security and privacy concepts, technically, security and privacy have commonalities and a cost effective, practical trade off must be found [Shannon 2015].

The concept of Privacy by Design emphasizes the need to adopt a proactive rather than a reactive compliance approach to the protection of privacy. As technologies evolve, the expectations and concepts of privacy also evolve. In the past, privacy was viewed as a personal good rather than a societal one. As such, privacy was regarded as a matter of individual responsibility [Cavoukian 2011]. Currently, the practices are changing as general data protection regulation [GDPR 2016] takes effect in Europe (compliance as of May 2018) and globally. The General Data Protection Regulation 2016/679 is a regulation in European Union (EU) law on data protection and privacy in the European Union and the European Economic Area. The regulation's rules include seven principles: lawfulness, fairness and transparency; purpose limitation; data minimization; accuracy; storage limitation; integrity and confidentiality (security); and accountability.

Like Security by Design, Privacy by Design is focused on processes rather than a singular focus directing technical outcomes. This approach reflects the reality that it is difficult in practice to favorably impact both consumer and user behavior after the fact. Privacy is best interwoven proactively into business processes and operational practices. To achieve this, privacy principles should be introduced early on – during architecture planning, system design, and the development of operational procedures. These principles, where possible, should be rooted into actual code, with defaults aligning both privacy and business imperatives.

Privacy by Design prescribes that privacy be built directly into the design and operation, not only of technologies, but also how a system is operationalized (e.g., work processes, operational practices, management structures, and networked infrastructure). The concept of Privacy by Design extends to IT systems, accountable business practices, and networked infrastructure. Privacy by Design should be based on requirements and software engineers need to provide solutions that are both holistic, robust, and are embedded by default.

Also, Privacy by Design can be used to approach operational and management issues. It assists in creating a particular culture of privacy within the enterprise. This culture of privacy is what emerges when organizations approach privacy as a business issue – not a compliance issue. The management of an organization needs to envision that the implementation of positive privacy controls creates – rather than constrains – business opportunities.

Although the concept of Privacy by Design was developed back in the 90's, only a few years ago it was recognized that the use of Privacy Enhancing Technologies (PETs) was not a seen as the solution and a more substantial approach is needed. Extending the use of PETs to PETS Plus, a full functionality approach advances the view of privacy. The objectives of Privacy by Design can be accomplished by practicing the seven foundational principles [Principles]:

- Proactive not reactive; preventative not remedial.
- Privacy as the default setting; it is built into the system, by default.
- Privacy embedded into design and architecture of IT systems and business practices.
- Full functionality - Positive-Sum, not Zero-Sum.
- End-to-End security - full lifecycle protection; strong security measures are essential to privacy, during lifecycle management of information, end-to-end.
- Visibility and transparency - keep it Open; operating according to the stated promises and objectives, subject to independent verification.

- Respect for user privacy - keep it user centric; architects and operators are required to protect the interests of the individual uppermost by offering such measures as strong privacy defaults, appropriate notice, and empowering user friendly options.

Therefore, Privacy and Security by Design continue to evolve into an essential component of information technologies and operational practices of organizations, as well as becoming an integral part of entire systems of data governance and privacy protection [Cavoukian 2013].

5.6 ISO/OSI and GWAC Stack Reference Model

One requirement in Smart Grid is interoperability. Interoperability refers to the ability of diverse systems and organizations to work together (interoperate). In the context of the electricity system, interoperability refers to the seamless, end-to-end connectivity of hardware and software from end-use devices through the transmission and distribution system to the power source, enhancing the coordination of energy flows with real time information and analysis.

5.6.1 Interoperability Principles

Smart Grid interoperability is defined in [GridWise 2005] as:

The capability of two or more networks, systems, devices, applications, or components to exchange and readily use information - securely, effectively, and with little or no inconvenience to the user. The Smart Grid will be a system of interoperable systems. That is, different systems will be able to exchange meaningful, actionable information. The systems will share a common meaning of the exchanged information, and this information will elicit agreed-upon types of response. The reliability, fidelity, and security of information exchanges between and among Smart Grid systems must achieve requisite performance levels.

This definition is adopted in NIST documents of earlier release in 2010 [NIST SP1108r1] and later release as of 2021 [NIST SP1108r4]. These documents define a framework and road map for Smart Grid interoperability standards.

Characteristics of interoperability as defined in [GridWise] include a requisite quality of service, reliability, fidelity, and security. The GridWise Alliance underscores that with sound planning, thorough design, and coordinated execution, a safe, secure, and reliable Smart Grid can be achieved [GridPrinciples]. One initiative [www.gridwise.org] is advancement of grid visualization technology to provide wide area grid awareness, integrating real time sensor data, weather information, and grid modeling with geographical information. Also, interoperability and cybersecurity policies and legislation, and the resulting guidelines and standards play a critical role in realizing a transformed electric grid. The benefits of interoperability are broadly recognized and have been demonstrated in other industries, and interoperability is a cornerstone for realizing future benefits from the Smart Grid.

The vision of advancement of technologies and successful integration of DERs can be done by improving the interoperability between the automation systems needed to enable the Smart Grid applications. DER is an emerging domain (e.g., the types, use, and deployment of distributed energy resources are evolving). DERs may be directly controlled by the customer (e.g., in situations where they elect to participate or opt out of DER signals); by operations for safety and contracted to microgrids. Customer is the end user of the electrical process. End-users may consume and produce electricity for premise use or resale. As an electricity producer, they are also members of the DER domain when the electricity produced is sold. Customers include residential, commercial, and industrial facilities.

Examples of technologies that are used for DERs integration in distribution include:

- Smart inverter
- Solid state transformer
- Distributed energy storage
- Solar assisted PV charging station.

Examples of technologies needed at the consumer location include:

- Integrated inverter
- EMS portal and panel
- In-home displays
- Grid ready appliances and devices
- Vehicle to grid two way power converters
- Residential storage for back up
- Industrial and commercial storage for power quality
- Commercial building automation.

These technologies need to work together in several applications and significant messages containing data or commands have to be shared so incompatibilities barriers (e.g., semantics, format, timeliness) have to be resolved during design.

5.6.2 Different Layers of Interoperability

Large, integrated, complex systems require different layers of interoperability, from a plug or wireless connection to compatible processes and procedures for participating in distributed business transactions. Referred to as, the GWAC stack, the GWAC Interoperability Context-Setting Framework, the eight layers comprise a vertical cross section of the degrees of interoperation necessary to enable various interactions and transactions on the Smart Grid ([GWAC 2008]).

Very simple functionality, such as the physical equipment layer and software for encoding and transmitting data, is confined to the lowest layers. Communication protocols and applications reside on higher levels, with the top levels reserved for business functionality. As functions and capabilities increase in complexity and sophistication, more layers of the GWAC stack are required to interoperate to achieve the desired results. Each layer typically depends upon - and is enabled by - the layers below it.

The GWAC stack eight layers (see Figure 5.13) are classified into three categories called drivers (Organizational, Informational, Technical) and each layer requiring a different level of interoperability. Cross cutting issues are areas that need to be addressed and agreed upon to achieve interoperation. They usually are relevant to more than one interoperability category of the framework. The most important feature of the GWAC stack is that the layers define well known interfaces which are loosely coupled and establishing interoperability at one layer enables flexibility at other layers.

The GWAC stack structure could be useful for the elicitation and management of Smart Grid requirements since it addresses the facets of interoperability from regulatory policy down to the physical connection between components. GWAC model covers basic connectivity needed to establish communication between distributed assets and applications within its lowest layer (layer 1).

Correlating with ISO/OSI reference model, this layer addresses physical and link layer of this model. The most important feature of the GWAC stack and the OSI model which preceded it is that layering defines well known interfaces: establishing interoperability at one layer can enable flexibility at other layers.

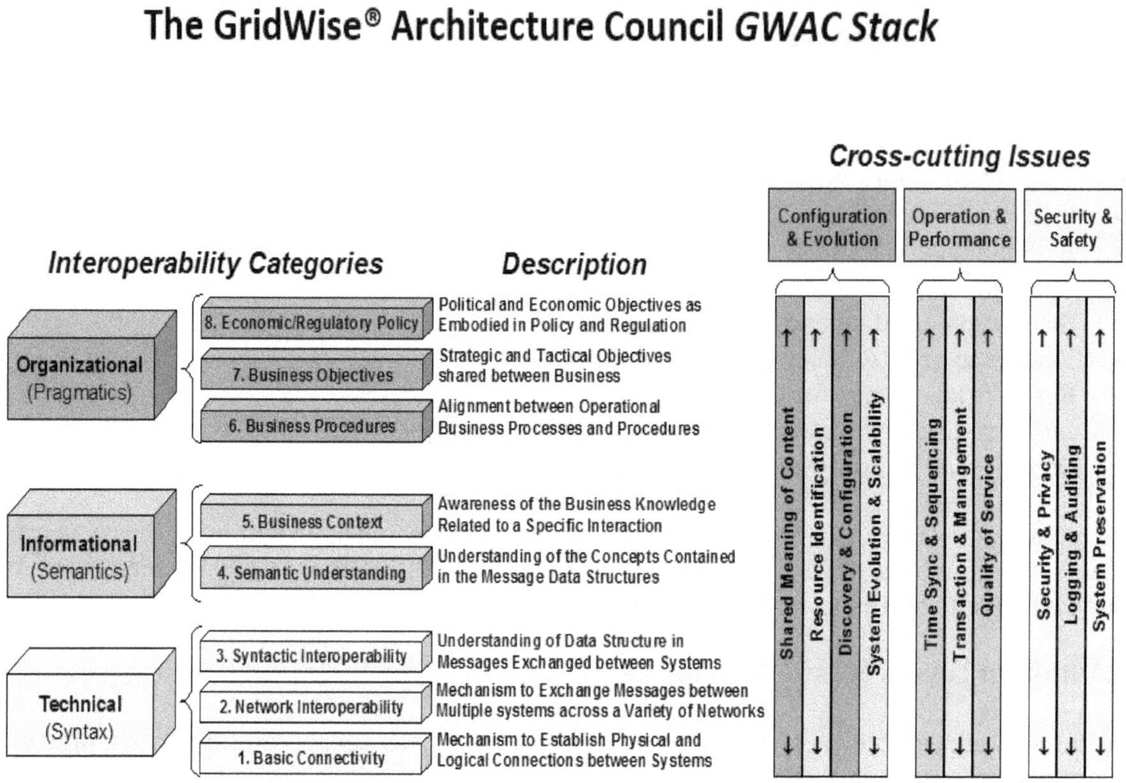

Figure 5.13 GWAC Stack Reference Model. *Source:* [GWAC 2008]. © 2008, GridWise Architecture Council.

However, correlating cybersecurity with specific information exchange standards, including functional requirements standards, object modeling standards, and communication standards, is needed although it is very complex [OASIS 2006].

To simplify and provide a high level view of this landscape, the cross cutting issues are proposed to be collected into three cross-cutting issue areas with specific issues for each area as shown in Figure 5.13.

Therefore, cybersecurity should be viewed as a stack of different security technologies and communication standards and procedures, woven together to meet the security requirements of a particular implementation of a stack of policy, procedural, and communication standards, all designed to provide specific security services to protect the organization's business. For example, when cybersecurity is applied to the information exchange standards, it is described as profiles of technologies and procedures which can include both power system methods (e.g., redundant equipment, analysis of power system data, and validation of power system states), policies, and information security technology controls (e.g., encryption, role based access control, and intrusion detection).

This study [Schütz 2021] finds that no standard or key concept alone can guarantee interoperability. Also, interoperability solutions based on CIM model or IEC 61850 standard describe suitable information models that are able to represent the relevant use cases, but they do not define a context and therefore do not guarantee pragmatic interoperability.

5.6.3 European Modified GWAC Stack Layers

The European Smart Grid framework adopts a simplified GWAC stack for the interoperability based on five layers instead of eight layers. The layers include: business, function, information, communication, and component. As shown in the diagram of Figure 5.14, the eight layers of the GWAC stack (on the left) can be mapped onto the five distinctive layers like depicted on the right.

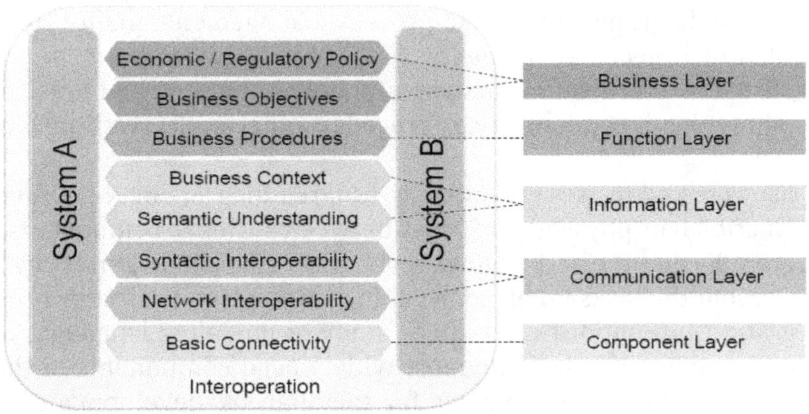

Figure 5.14 Grouping into interoperability layers. *Source:* [CENELEC 2012]. © 2012, CENELEC.

5.7 Analysis Perspectives

We discuss the analysis phase from the perspective of Requirements Engineering and Security Requirements Engineering.

5.7.1 Analysis

Analysis is the phase of the SDLC (Software Development Lifecycle) that assesses the organization's readiness, its current system status, and its capability to implement and then support the prosed systems, including development of preliminary system requirements. Unique steps to Security Software Development Lifecycle (SecSDLC) include analysis of existing security programs, current threats and controls, and risk analysis including definition of security requirements.

5.7.2 Requirements Engineering Overview

Requirements Engineering (RE) deals with analysis and topics such as stakeholder identification, measurable goals, and use cases. A fundamental activity in RE is modeling which deals with the construction of abstract descriptions that are amenable to interpretation. A number of RE textbooks focus almost entirely on modeling methods and their associated analysis techniques. Models can be used to represent a whole range of products of the RE process. Moreover, many modeling approaches are used as elicitation tools, where the modeling notation and partial models produced are used as drivers to prompt further information gathering [Kaur 2010].

The analysis model is viewed as a bridge between the system description and the design model. In addition, tools are needed to support the methods to be used in various engineering activities. Examples of tools include graphical tools and support methodologies for system analysis, design, modeling, simulation, testing, and documentation. Tools for graphic representations help a Systems Engineer to comprehend a complex problem. For example, graphic representations of a system are used to communicate a system's functional and data requirements. Examples of common graphical tools are described in [Dennis 2012], [Ashrafi 2009]:

- Functional flow block diagram is a classic business process modeling of a system's functional flow.
- Data flow diagram is a classic business process modeling of a system's functional flow.
- Flow chart may represent a solution for an algorithm, workflow or process.
- Control flow diagram describes the control flow of a business process, process or review; different types of diagrams are used in process management, quality control, configuration management, and project management.
- Gantt chart is used for scheduling activities, showing the start and finish dates of the activities and to show the dependency (e.g., precedence network) relationships between activities.
- N2 chart (also referred to as N^2 diagram, N-squared diagram or N-squared chart) is used to represent functional or physical interfaces between system's components.
- PERT (Program Evaluation and Review Technique) diagram is a statistical tool used to analyze and represent the tasks involved in completing a given project or program.
- IDEF (Integration DEFinition) refers to a family of modeling languages in the field of systems and software engineering; the most widely used components of the IDEF family are IDEF0 (a functional modeling language for the analysis, development, reengineering, and integration of information systems; business processes; or software engineering analysis), and IDEF1X, which addresses information models and database design issues.
- Model based design is a mathematical and visual method of addressing problems associated with designing complex control, signal processing, and communication systems; control systems can be defined with models with advanced functional characteristics using continuous time and discrete time building blocks; also models can be used with simulation tools for design, rapid prototyping, software testing, and verification.
- Use case is a list of steps, typically defining interactions between a role or actor (as known in Unified Modeling Language (UML)) and a system, to achieve a goal; the actor can be a human, an external system, or time.
- Sequence diagram, called event diagram or event scenario, is an interaction diagram that shows how processes and objects operate in time with one another and what is their order.
- State diagram describes the behavior of a system or a reasonable abstraction of a system; many forms of state diagrams exist (e.g., UML state diagrams), which differ slightly and have different semantics.

Different analysis methods that can be used for control systems may include [Shaw 2001]:

- Data and system diagrams provide a useful visualization which can be easily developed and evolved by a team.
- Tabular languages provide a way to express and communicate requirements in a comprehensive and complete process.
- Formal methods are sophisticated mathematical techniques that can provide powerful description and verification of systems; examples include regular expressions, propositional logic, predicates, and temporal logic.
- Finite state machines (shortly, state machines) have proven to be a language of choice for modeling and design projects; a number of adaptations have been proposed and used effectively in real time system designs for automation, engineering applications, communication protocol design, etc.

Critical to requirements engineering is requirements analysis.

5.7.3 Why Requirements Analysis?

Proper analysis of system requirements is very important to the success of real time systems [Lawrence 1987]. Similarly, a security engineer performs tasks focused on security issues during the analysis phase. Analysis is the process of defining the requirements for a solution to a problem [Martin 1985]. During this phase, the needs of users are examined, the system properties are defined, and system constraints and functional and non-functional requirements are identified. The major output is the requirements document.

Requirements analysis is the process of identifying the functions needed by the organization to be carried out by the desired system. Figure 5.15 depicts a SE perspective on requirements analysis [DoD 2001]. It shows a cycle of activities to be performed during the process. These include functional requirements analysis, refining, and defining requirements. However, this diagram does not include the notion of RE to support description and modeling properties of the environment. It is argued that only by describing the environment and expressing what the new system must achieve in that environment, it is possible to capture the system's purpose, and reason about whether a given design will meet that purpose [Kaur 2010].

A requirement is a statement that identifies an operational, functional, or design characteristic or constraint of a product or process. A requirement is defined as a condition or capability needed by a user to solve a problem or achieve an objective [IEEE Std 610.12].

Ideally, a requirement should be unambiguous, testable or measurable, and necessary to the acceptability of the processor product (by consumers or those responsible for verifying the product/process conformance to internal quality assurance guidelines) [Goertzel 2008]. The requirements should be documented, actionable, measurable, testable, traceable, related to identified business needs or opportunities, and defined to a level of detail sufficient for system design. Specific activities include [Laplante 2014]:

- Requirements elicitation/discovery
- Requirements analysis and reconciliation
- Requirements representation/modeling
- Requirements verification and validation
- Requirements management.

5.7.3.1 System Analysis

Related to RE is system analysis, which is a problem solving technique. It decomposes a system into its component pieces for the purpose of the studying how well those component parts work and interact to accomplish their purpose [Bentley 2007]. Analysis and synthesis, as scientific methods, always go hand in hand; they complement one another. Every synthesis is built upon the results of a preceding analysis, and every analysis requires a subsequent synthesis in order to verify and correct its results. Further, systems analysis tools include structured analysis and object oriented analysis.

Use cases are widely used systems analysis modeling tools for identifying and expressing the functional requirements of a system. Each use case is a business scenario or event for which the system must provide a defined response. Use cases evolved from object oriented analysis.

Others argue that RE is not a standalone approach [Danekas 2013], [Uslar 2013], [Kotonya 1998]. This work [Danekas 2013] uses RE as the main term to refer to the activities of elicitation, analysis, negotiation, documentation, validation, and management of requirements. RE within engineering methodologies relates requirements to the systems stakeholders, project constraints, regarding time and costs, implementation artifacts and test results; it relates to processes such as risk management, process improvement, etc. A perspective of relationships of RE activities with other disciplines and processes for Smart Grid is depicted in Figure 5.16.

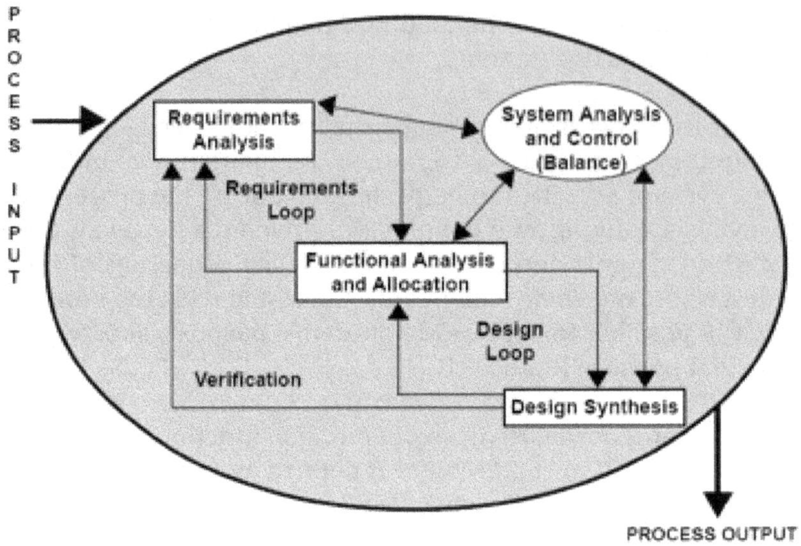

Figure 5.15 Systems Engineering perspective on requirements analysis. *Source:* [DoD 2001]. Public Domain.

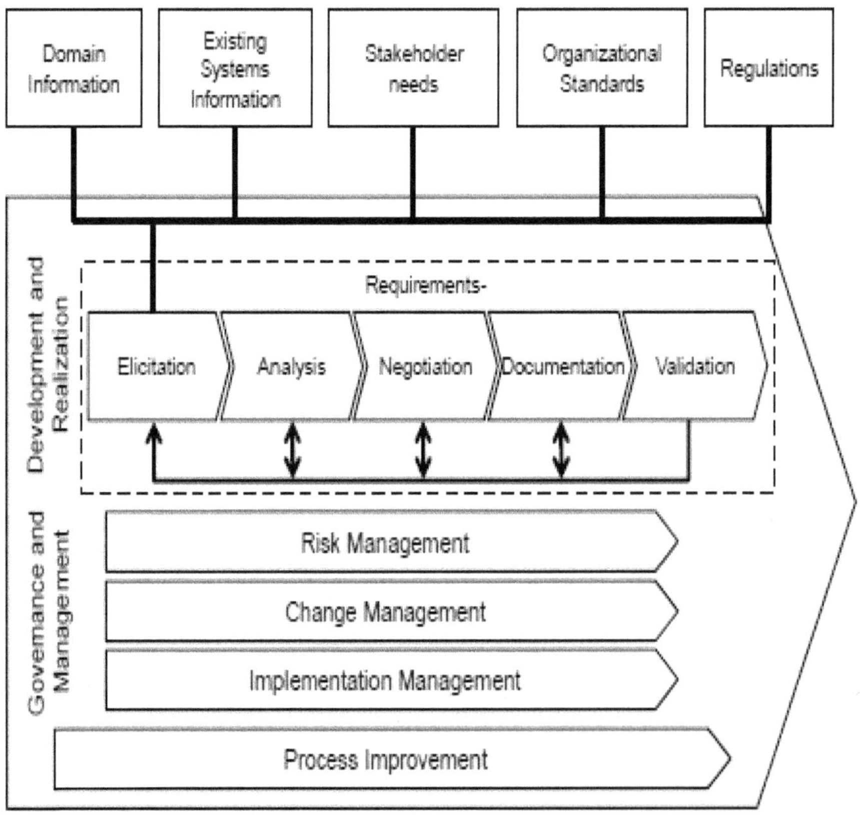

Figure 5.16 Requirements engineering activities and process inputs. *Source:* [Danekas 2013]. © 2012, Springer-Verlag.

5.7.3.2 System Security Policy

If the security policy of the system was not defined in the initiation phase, during the analysis phase, Security Engineer has to analyze existing security program and policy for the organization and define a security policy for the system (collaboration with the management and all team members is needed) and perform security requirements analysis. Steps include analysis of current organization's security posture, threats, legal issues, operating environment of the system, interaction of the system with other systems (internal and external), security requirements identification, writing, and documentation. During the analysis phase, an understanding of the goals and needs for the system is crucial because security requirements need to support user needs, business functions, organization's security program objectives, and security principles to be implemented into the system.

5.7.3.3 Security Requirements Analysis

Although more security standards and guidelines are published by different organizations, there is a need for understanding how to perform the analysis to derive security requirements, how to assess the security risks due to design, implementation, platforms, and standards.

The analysis for security results in a security policy for the system, resolving conflicting and missing requirements and security requirements documentation. Analysis phase is an important activity of the SDLC either a separate activity in the initiation phase, or integrated with activities such as risk assessment, SRE, and risk management. Analysis should address requirements to achieve security objectives: availability, integrity, confidentiality, non-repudiation, accountability, assurance [NIST SP800-33]. SRE is often not performed during the requirements engineering process, or when performed, the security requirements are not properly integrated with the main system requirements [Zafar 2013]. SRE identifies security risks and mitigation strategies in the early stages of the system development cycle.

In many organizations, the analysis phase is skipped to expedite the development of a system or security issues are to be resolved after a product or system was developed. Even when security requirements are provided, there is a tendency to provide solutions instead of making declarative propositions regarding the level of protection required [Mellado 2006]. Therefore, result is the failure of the project and/or security.

5.7.4 SCADA Analysis

Several issues related to SCADA systems have to be investigated during analysis. These issues may include devices, networks, protocols, operations, and many other. SCADA networks traditionally used dedicated telephone lines to send control messages to field devices from the control station and get the current status of the field equipment.

However, the modern SCADA networks, integrated with corporate networks and the Internet, have become far more vulnerable to unauthorized cyber attacks putting the National infrastructures at risk and easy targets of attacks by terrorists. Control systems evolved from static to dynamic systems with access to Internet. By sending a false control message from a computer connected to the Internet, an unauthorized intruder can manipulate traffic signals, electric power switching stations, chemical process control systems, or sewage-water valves, creating major concerns to public safety and health. Control systems challenges are numerous and more issues are described elsewhere (e.g., [Hentea 2008a]). For example, the electric power industry is using World Wide Web (WEB) for information exchange, analytical simulations, and actual control of the electric power grid [Wildberger 1999]. Thus, the use of the WEB technologies raises several serious issues such as [Hentea 2008a]:

- Can the security of communications be assured?
- Can time critical operations be carried out reliably?
- Can delays be accurately predicted and securely compensated for?

In addition, standardization and use of open market technologies are current requirements in control systems. Modern products are often based on component architectures using commercial off-the-shelf products (COTS) elements as units. This architecture leads to control systems that are becoming very complex software applications with the following characteristics [Sanz 2003]:

- Time critical
- Embedded
- Fault Tolerant
- Distributed
- Intelligent
- Large
- Open
- Heterogeneous.

Hans-Jurgen Weidemann [Conference 2006] stressed the complexity of control systems and greater need to compliance for safety, quality of service, and security of systems and data. SCADA security design and information security management can be improved by applying a wide range of control principles and methods as well productivity control, involving decision making under uncertainty with increased levels of decision support. Therefore, the improvements for SCADA security have to be broad - at the systems level, and detailed - at the component level [Hentea 2008a].

5.7.5 Analysis Activities and Methods

SRE is often the most critical task of managing secure system development [Anderson 2008a]. Security requirements are often viewed as a subset of non functional requirements. Besides processes, requirements engineering needs analysis methods and tools to produce security requirements at the system level rather than the security technology level and the trend should be to escape from best practices checklists and return to core systems-engineering methods, processes, and tools [Bayuk 2011a]. The analysis activities illustrated in Figure 5.4 can be applied to SRE. The specific analysis activities include two aspects:

- Security functional requirements analysis identifies the protection levels that must be provided by the system to meet all business, regulatory, legal, and policy compliance needs.
- Security assurance requirements analysis identifies the assurance level the system must provide; the activities that need to be carried out to ensure the desired level of confidence in the system are determined, these are usually specific types of tests and evaluations.

However, in the subfield of security requirements analysis, there is an abundance of methods that can used for security requirements design. Surveys and comparison of different methods are published in recent works [Fabian 2010], [Mead 2007]. Approaches to SRE are classified in four categories in [Nhlabatsi 2010]:

- Goal based approaches - security requirements engineering views the system to be and its environment as a collection of active components called agents.
- Model based approaches –requirements engineering and development of the information system is by means of models, e.g., system model, security model, assurance model, security risk model, etc.

- Problem oriented approaches –focused on decomposing the problem into simple subproblems; uses a lower abstraction level that describes the entire system to be; e.g., driven by environment in which a system will operate.
- Process oriented approaches – focused on activities driven by a process; e.g., Security Requirements Engineering Process (SREP) is a standard centered process and a reuse based approach which deals with the security requirements at the earlier stages of software development in a systematic and intuitive way by providing a security resources repository.

However, a classification can be useful when adopting an approach, but not all of these approaches are bound to specific modeling languages and tools that support them. It is argued that there are benefits when combining several approaches, so additional capabilities may be needed. It is argued the need to extend the tool based process of engineering by guiding each activity of the process with knowledge acquired in other activities [Roudier 2014]. This work recommends use of meta modeling, which can help increase the expressivity of security requirements and automate verifications through the introduction of meta level reasoning capabilities.

A knowledge based meta model, supported through the definition of security ontologies is integrated with SysML-Sec (a model driven environment for developing safety and security requirements for embedded systems). The ontologies are used to reason about the security concepts defined in SysML-Sec models. This integrated approach enables the security engineers to have access to various ontological concepts and their relationships, and to reason over semantic graphs. The integration of these capabilities relies on an engine to bridge the two approaches as depicted in Figure 5.17.

Security metadata produced at different phases of the security requirements engineering process can be easily shared and reused. The meta model also makes it possible to automate verifications through the introduction of meta-level reasoning capabilities. A knowledge centric meta model that is integrated into SysML-Sec platform is described in [Roudier 2014]. As described in this work, capabilities of the knowledge centric meta model include:

- Agreement on definitions
- Elicitation of requirements
- Consistency checking
- Prioritization of requirements.

A meta model, supported through the definition of security ontologies, can be combined with other approaches to support the need for documentation of security requirements including input and visualization. Numerous uses of ontologies in various contexts of security engineering including finding discrepancies and incompletenesses in security standards are described in [Roudier 2014].

Integrating knowledge into system model enables engineers to define more precise relationships in the course of a typical model based development process. Capabilities of ontologies are used to reason about the security concepts defined in SysM-Sec models and enable a security engineer to have access to various ontological concepts and their relationships, and to reason over such semantic graphs.

A SysM-Sec framework is described in [Apvrille 2013]. This is a SysML (Systems Modeling Language) environment aimed at making security experts collaborate with system designers at all methodological stages of the design and development of an embedded system. SysM-Sec is also meant to support the assessment of the impact of security over safety. Security and safety concerns are captured in extended SysML diagrams elaborated according to an iterative process centered around the software/hardware partitioning of the architecture. SysML is defined as an extension of a subset of the Unified Modeling Language (UML). It is a general purpose modeling language for systems engineering applications. It supports the specification, analysis, design, verification and validation of a broad range of systems and systems of systems methods.

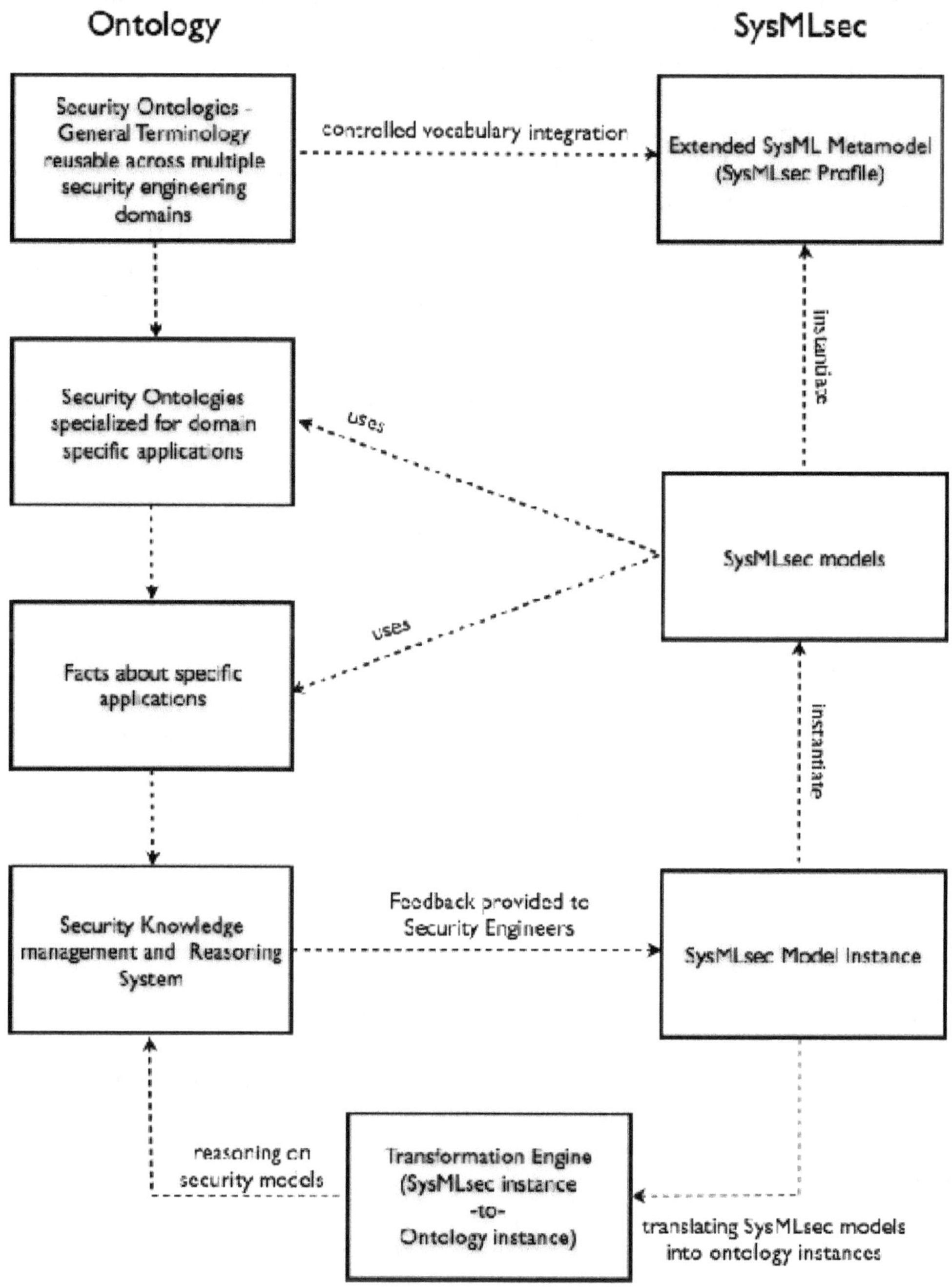

Figure 5.17 Integration of security ontologies and reasoning into SysML. *Source:* [Roudier 2014]. Public Domain.

To overcome the shortcomings of current approaches (e.g., perimeter security design), a methodology to identify classes of new reusable system security solutions and an architectural framework based on reuse of the patterns of solutions are introduced in [Bayuk 2011b]. This work also introduces a new methodology for security metrics intended to stimulate critical solution design trade off analyses as part of security design reuse considerations. This work suggests that requirements may include a security feature as a system capability with a clear and measurable goal, which is customized in context. In order to show that security requirements are met, it is imperative to devise verification and validation metrics at the system level.

This work [Firesmith 2004] argues that security requirements can be reusable across multiple systems and has proposed the use of parameterized templates to model reusable security requirements. Unlike typical functional requirements, security requirements can potentially be highly reusable, especially if specified as instances of reusable templates.

This work [Slankas 2013] is focused on extracting security requirements from existing functional requirements and requirements like documents and not considering issues related to regulatory compliance or policy specification. This is a similar approach to [Firesmith 2004] approach, but security requirements are generated from a set of context specific templates. Also, this work emphasizes that security objectives of software systems involve not only technical aspects from system development perspective but also operational and management aspects. Security requirements are functional and non-functional requirements that formalize security objectives without specifying how to achieve those objectives. By identifying the security objectives expressed or implied by a particular sentence within a document there is ground on understanding the underlying intent of the statement as well as possible controls and mechanisms to establish that intent. Figure 5.18 shows a hierarchy of security objectives as identified by the authors.

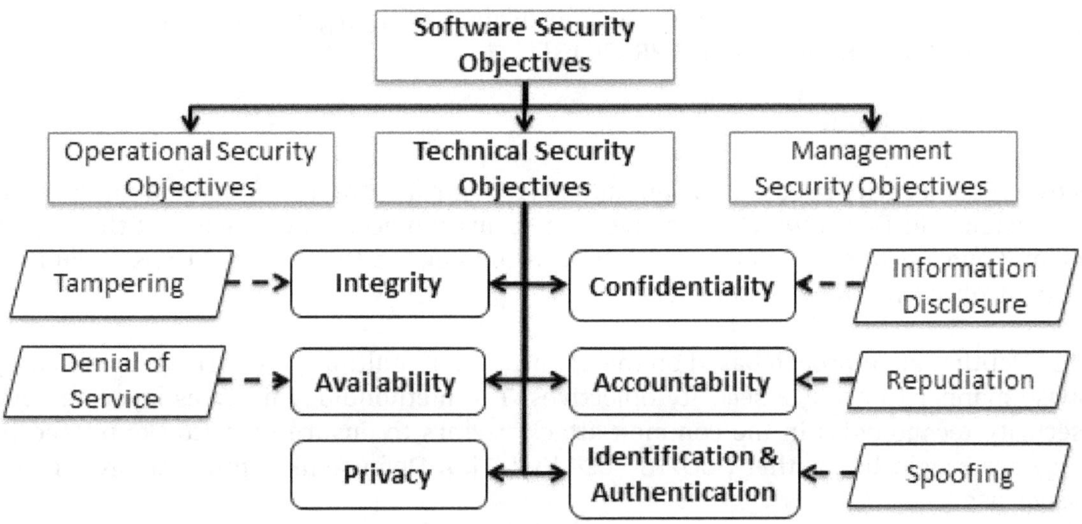

Figure 5.18 Security objectives hierarchy. *Source:* [Slankas 2013]. Public Domain.

5.7.6 Depth Analysis for Smart Grid

RE for the Smart Grid should be based on models that depict user scenarios, functional activities, problem classes and their relationships, system and class behavior, power flows and associated measurements, and the flow of data as it is transformed. Developing security requirements for Smart Grid applications requires understanding and analysis of existing security posture to recommending future security development. This may require investigation of legacy approach, but the possible use of security by obscurity should be limited or avoided when planning security requirements for DER systems.

Although a prescriptive list of specific steps or issues is not feasible due to dynamics of the Smart Grid evolution and technology developments, it is worth to mention a few issues that require attention. Examples of issues include security policies, risk assessment methods and processes, risk management, evaluation and selection of security mechanisms, monitoring and control, forensics data collection, critical infrastructure protection issues.

A broader vision of security in the context of the Smart Grid and DER systems implies more effort and requires a multitude of skills from Systems Engineers and Security Engineers. Security of the computer and device includes aspects of processes as well as physical and technical security measures applied to the hardware, software, and all the components that make the computer or device. Security features provided by vendors of the SCADA devices and computers need to be implemented. In addition, analysis has to be performed to identify the levels of security of SCADA protocols, vendor interfaces to SCADA systems, and myth of obscurity has to be removed.

The first effort in this expanded scope would be to investigate the most current documentation (standards, recommendations, guidelines, regulations) and near future planned documents. Several National and International key players are involved in the development of the Smart Grid systems. These include standard organizations, electrical sector, computing and security industries, and many others. For example, NIST conceptual model [NIST SP1108r4] provides a good framework for discussing the characteristics, uses, behavior, interfaces, requirements and standards of the Smart Grid. Also, high level security requirements are included in Volume 1 of [NISTIR 7628r1] document. These are generic requirements and need to be adapted to the results of analysis and security requirements engineering processes performed by an organization. These requirements were also adapted for DER systems in [EPRI 2013].

5.7.7 ICS Defense in Depth

An effective cybersecurity program for an industrial control system (ICS) should apply a strategy known as Defense in Depth which uses layers of security mechanisms such that the impact of a failure in any one mechanism is minimized. It is recommended that organizations should not rely on security by obscurity [NIST SP800-82r2].

Defense in Depth is an approach based on the application of multiple countermeasures in a layered or stepwise manner to achieve security objectives. The methodology involves layering heterogeneous security technologies in the common attack vectors to ensure that attacks missed by one technology are caught by another [ISO/IEC 62443-1-1]. A Defense in Depth strategy includes objectives such as:

- Developing security policies, training and educational material that applies specifically to the ICS.
- Addressing security by design – security throughout the lifecycle of the ICS from architecture design to procurement to installation to maintenance to decommissioning.
- Implementing network topology for the ICS that has multiple layers, with the most critical communications occurring in the most secure and reliable layer.
- Providing logical separation between the corporate and ICS networks.
- Separating information and infrastructure based on security requirements.
- Designing critical systems for graceful degradation (fault tolerant) to prevent catastrophic cascading events.

Failure protection strategies that employ replication of policy enforcement mechanisms can allow the system to continue in a secure state even when one mechanism has failed to protect the system.

5.7.8 Security Pattern

The phenomenon of security pattern can help Security Engineers to decrease the time for the requirements specification. Security requirements are the major reasons for secure software developments. Security pattern approach is one of many methods, model and approaches that have been designed to ensure a correct definition of security requirements. The approach aims to help requirement engineers who are not security experts to better identifying and understanding security concerns and leads to a correct implementation [Yahya 2019]. This work evaluates various security patterns by classifying and structuring the security requirements patterns.

Security patterns enable best practices for avoiding security related design flaws in software [Behrens 2018]. A security pattern describes a particular recurring security problem that arises in a specific security context and presents a well proven generic scheme for a security solution. A security pattern occurrence is observed in different situations:

- When defining organization security policy and responsibilities related with different security roles.
- Security patterns at the architecture level.
- Security patterns in security application code.

This survey [Yoshioka 2008] identifies security patterns from the published works to use them for each phase of software development, but they found that some security patterns are difficult to apply. Also, the authors describe requirements for the use of security patterns such as security patterns are intended to be secure and to use security functions efficiently. This means that a security pattern aims to use security functions efficiently with requests from several clients. Also, there are requirements for security pattern effectiveness. For the effectiveness of security, a security metric, such as the frequency of security violations or the inconsistency of security requirements and a function.

Also, by using the knowledge of domain that a product belongs to, developers can customize the security pattern for particular domains [Wu 2007]. This work identifies security functional requirement patterns for different domains of security products and it describes a method of evaluating the effectiveness of patterns. Then, developers can choose the applicable patterns and customize them for the use on a specific security application. The solutions of the patterns are illustrated using class diagram and sequence diagram [Yoshioka 2008]. The authors argue that while understanding the pattern may be easy, its actual implementation may be challenging because the problem and context are not clear. So, it is required vulnerabilities to be described concretely. Security context is a combination of the communication protection proxy, security context and subject descriptor pattern. Figure 5.19 shows the class diagram of the secure logger pattern with the secure log strategy for a secure logger pattern.

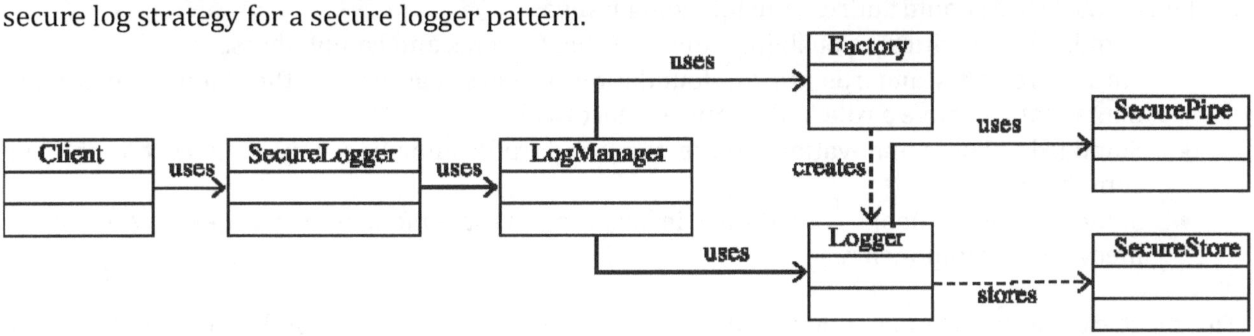

Figure 5.19 Secure logger pattern with secure log store strategy. *Source:* [Yoshioka 2008]. © 2008, National Institute of Informatics.

Commonly, security pattern types may be classified as: IT service security patterns; IT application level patterns; IT infrastructure; (central) security services (e.g., identification, authentication). One example is the authenticator pattern also known as the pluggable authentication modules or Java authentication and authorization service (JAAS).

5.8 Standards, Guidelines, and Recommendations

There are many standards, guidelines, and recommendations that support engineering systems and security of the systems. Included are examples of key standards focused on Systems Engineering, Systems Security Engineering, and Smart Grid interoperability. Also, the vast majority of organizations have adopted a security framework (e.g., [NIST 2018]), or more often an amalgam of frameworks - often with very different productive results.

5.8.1 Standards Focused on Systems Engineering

This International standard [ISO/IEC/IEEE 15288] provides a common process framework for describing the life cycle of systems created by humans, adopting a SE approach.

This publication [NIST SP800-160v1] is intended to be used in conjunction with and as a supplement to ISO/IEC/IEEE 15288, Systems and software engineering — System life cycle processes. Organizations using this publication are advised to obtain the standard [ISO/IEC/IEEE 15288] in order to fully understand the context of the security related activities and tasks in each of the system life cycle processes. Appendixes of [NIST SP800-160v1] include summaries that include: roles, responsibilities, and skills associated with SSE design principles for security; and engineering and security fundamentals.

Also, these standards [ISO/IEC 15026], [ISO/IEC 12207], [ISO/IEC 27036], [ISO/IEC TR 24748-1] are supporting this document [NIST SP800-160v1]. For example, this standard [ISO/IEC 27036] is a multi-part standard offering guidance on the evaluation and treatment of information risks involved in the acquisition of goods and services from suppliers. The implied context is business-to-business relationships, rather than retailing, and information-related products. In addition, these handbooks, INCOSE Systems Engineering Handbook [INCOSE 2015] and MITRE Cyber Resiliency Engineering Framework [Bodeau 2012], provide additional guidance on how to tailor the SE processes.

5.8.2 Standards Focused on Smart Grid Interoperability

Efforts directed toward enabling interoperability among the many diverse components of the evolving Smart Grid should address the following issues:

- Standards are critical to enabling interoperable systems and components.
- Mature, robust standards are the foundation of mass markets for the millions of components that will have a role in the future Smart Grid.
- Standards enable innovation where thousands of companies may construct individual components.
- Standards also enable consistency in systems management and maintenance over the lifecycles of components.

The most frequently recommended guidelines are based on [ISO/IEC 27000] series and [NISTIR 7628r1], [NIST SP1108r4], [NIST 2018]. These guidelines enable organizations to identify and prioritize risks, gauge the maturity of their cybersecurity practices and better communicate internally and externally.

Chapter 6

Risk Management

6.1 Risk Management Methods

An effective security program starts with a strategy and a foundation based on risk management strategy. Risk is an expectation of loss expressed as the probability that a particular threat will exploit a particular vulnerability with a particular harmful result [RFC 4949]. In an organization, risk management is done in a much broader framework of managing all the risks including information security risks. Risk is a measure of the extent to which an entity is threatened by a potential circumstance or event, and typically a function of [CNSSI 4009]:

- Adverse impacts that would arise if the circumstance or event occurs, and
- Likelihood of occurrence.

Types of risk include, but are not limited to, mission/business, safety, security, programmatic, monetary, schedule, training, maintenance, and sustainment.

6.1.1 Risk Assessment and Risk Management Overview

Risk management includes a set of coordinated activities to direct and control an organization with regard to risk [ISO 73 Guide]. Risk management is performed following a process which includes systematic application of management policies, procedures and practices to the activities of communicating, consulting, establishing the context, and identifying, analyzing, evaluating, treating, monitoring and reviewing risk [ISO 73 Guide].

The discipline of risk management reduces risks by defining and controlling vulnerabilities and threats. The focus of risk management is how much to spend on protection against what. It should be understood that it is not possible to identify all risks during risk assessment phase, nor it is possible to eliminate all risks. The remaining risk is called residual risk. The residual risk, after risk treatment decision have been taken, should be estimated to ensure that sufficient protection is achieved. If the residual risk is unacceptable, the risk treatment process should be iterated.

Major types of impacts may include physical damage, human interaction, equipment malfunction, misuse of data, loss of data, system and application failures and errors [Harris 2013]. When performing risk assessment for ICS, an adverse event can have multiple consequences and different types of impact, at different levels, and in different time frames. As described in [NIST SP800-82r2], organizations can experience the consequences/impact of adverse events at the individual ICS system level (e.g., failing to perform as required), at the mission/business process level (e.g., failing to fully meet mission/business objectives), and at the organizational level (e.g., failing to comply with legal or regulatory requirements, damaging reputation or relationships, or undermining long term viability). Therefore, the guidelines included in this document recommend that organizations should assess the impacts to organizational operations (e.g., mission, functions, image, and reputation), organizational assets, individuals, other organizations, and the Nation (e.g., using [FIPS 199] or a more granular approach).

The risk assessment helps identify any weaknesses that contribute to information security risks and mitigation approaches to reduce the risks, but risk assessments should be conducted multiple times during a system's life cycle. The focus and level of detail varies according to the system's maturity.

Guidelines (e.g., [Harris 2013], [Whitman 2014], [NIST SP800-53r5], [NIST SP800-82r2]) and best practices can be used to select the security measures to reduce the risk, but is up to the single organization to choose the most appropriate one according to its business strategy, constraints of the environment and circumstances. The choice should be rational and documented. The importance of accepting a risk that is too costly to reduce is very high and led to the fact that risk acceptance is considered a separate process. These documents [NIST SP800-53r5] and [NIST SP800-82r2] incorporate baseline security controls that derive from the determination of the impacts.

As security industry matured, so definitions of risk management and risk assessment including other related terms continued to evolve from different perspectives. Definitions of risk management and risk assessment are also provided in known standards such as [ISO 13335-2], [NIST SP800-39], [NIST SP800-30r1], [NIST SP800-37r1], [ISO/IEC 27000], and others. Also, the European Union Agency for Network and Information Security (ENISA) is a center of expertise for cybersecurity in Europe and maintains an updated list of research projects and inventory of methods [ENISA 2016].

Recognizing that definitions evolve, there is an abundance of definitions and concepts. Thus, it is very important that a Security Engineer or a Process Control Engineer understands the concepts since all definitions provide a perspective on the process and often different steps (phases) for a process. An understanding of the concepts and a unified use of definitions in an organization is necessary for decision making and collaboration. Basically, risk management is an element of managerial science concerned with the identification, measurement, control, and minimization of uncertain events [NIST ManagContr].

Although the basic concepts of risk management are fairly described in several publications, there is still a need to check the most current standards, guidelines, and recommendations for security risk management provided by different forums and associations such as ISO, ISO/IEC, NIST, DOE, ISA, ENISA, etc.

6.1.2 Understanding Security Risk

Risk in the context of information security, also called information security risk, is the likelihood that something bad will happen that causes harm to an informational asset (or the loss of the asset) or the possibility of damage to information and systems. Security risk or security relevant risk is that risk associated with the loss or degradation of confidentiality, integrity, or availability of assets [CNSSI 4009]. In the context of information security, the impact of a risk could be possibly other losses (lost income, loss of life, loss of real property, loss of reputation).

Risks can be categorized and viewed on from or to a perspective. As defined in [NIST SP800-30r1] standard, information security risk is the risk to organizational operations (including mission, functions, image, reputation), organizational assets, individuals, other organizations, and the Nation due to the potential for unauthorized access, use, disclosure, disruption, modification, or destruction of information and/or information systems. Therefore, information security risk is never a matter purely for the information system as human and organizational factors are just as important as having the right hardware and software. Security risk related to the operation and use of information systems is just one of many components of organizational risk that senior leaders/executives address as part of their ongoing risk management responsibilities [NIST SP800-39]. Some publications may refer to IT risk while others explicitly refer security risk. These terms are used in the evaluation of methodologies, methods, and tools for risk management and risk assessment, called information security risk management (as process, framework, approach, or activities).

6.2 Information Security Risk Management

Information security risk management is the process of identifying, measuring, and controlling (e.g, mitigating) risks in information systems so, as to reduce the risks to a level commensurate with the value of the assets protected [RFC 4949]. As described in this standard, one activity of risk management is risk analysis which is an assessment process that systematically (a) identifies valuable system resources and threats to those resources, (b) quantifies loss exposures (e.g., loss potential) based on estimated frequencies and costs of occurrence, and (c) (optionally) recommends how to allocate available resources to countermeasures so as to minimize total exposure. Risk analysis provides the basis for risk evaluation and decisions about risk treatment.

Others define risk management as the process of discovering and assessing the risks to an organization's operations and determining how those risks can be controlled or mitigated [Whitman 2014]. The authors define risk assessment as the identification of risk for all assets and assignment of a comparative risk rating to each specific information asset based on the value of the asset, possible threats, and vulnerabilities. This rating or score represents the relative risk impact for a specific vulnerability of a specific information asset. The result is a list of documented vulnerabilities, ranked by criticality of impact. The assessment activity is viewed as part of the risk management process.

Another defines information risk management as the process of identifying and assessing risk, reducing it to an acceptable level, and implementing the right mechanism to maintain that level [Harris 2013]. This process requires an information risk management policy that should be mapped to the organizational security policies. Risk assessment is viewed as a tool for risk management; it is a method of identifying vulnerabilities and threats and assessing the possible impacts to determine where to implement security controls. Then the results are analyzed to determine that security is cost effective and relevant.

In [NIST SP800-39] standard, information risk management is described from the organizational, mission/business process, and systems perspective. This publication places information security into the broader organizational context of achieving mission/business success. Recognizing that risk management is a complex activity and managing risk requires the involvement of the entire organization to address risk from the strategic level to the operational level, this standard defines risk management as a comprehensive process that includes activities:

- Frame risk (e.g., establish the context for risk based decisions)
- Assess risk
- Respond to risk once determined
- Monitor risk on an ongoing basis using effective organizational communications and a feedback loop for continuous improvement in the risk related activities of organizations.

Further, this standard [NIST SP800-39] requires support of an information security management system and a risk management process similar to those described in [ISO/IEC 27000] standards, but it recommends of being used as part of a more comprehensive Enterprise Risk Management (ERM) program.

Risk treatment is the process to modify risk [ISO Guide 73]. However, risk treatment options are not necessarily mutually exclusive or appropriate in all circumstances. Generally, the risk treatment process aims at:

- Selecting security measures to reduce, retain, avoid, or transfer risk and
- Produce a risk treatment plan, that is the output of the process with the residual risks subject to the acceptance of management.

Risk management is considered to be the umbrella under which several processes/activities concerning the identification, mitigation, management and control of risks take place. The risk treatment options may include:

- Avoiding the risk by deciding not to start or continue with the activity that gives rise to the risk
- Taking or increasing the risk in order to pursue an opportunity
- Removing the risk source
- Changing the likelihood
- Changing the consequences
- Sharing the risk with another party or parties (including contracts and risk financing)
- Retaining the risk by informed decision.

Risk transfer applies where the risk has a very high impact but is not easy to reduce significantly the likelihood by means of security controls. The insurance premium should be compared against the mitigation costs, eventually evaluating some mixed strategy to partially treat the risk. Another option is to outsource the risk to somebody more efficient to manage the risk.
Risk avoidance describes any action where ways of conducting business are changed to avoid any risk occurrence. For example, the choice of not storing sensitive information about customers can be an avoidance for the risk that customer data can be stolen.

The [ISO/IEC 27005] standard assists in implementation of information security based on a risk management approach. In broad terms, the risk management process consists of:

- Identification of assets and estimating their value; assets include people, buildings, hardware, software, data (electronic, print, other), and supplies.
- Conduct a threat assessment that include acts of nature, acts of war, accidents, malicious acts originating from inside or outside the organization.
- Conduct a vulnerability assessment, and for each vulnerability, calculate the probability that it will be exploited; evaluate policies, procedures, standards, training, physical security, quality control, technical security.
- Calculate the impact that each threat would have on each asset by using qualitative analysis or quantitative analysis.
- Identify, select and implement appropriate controls; provide a proportional response and consider productivity, cost effectiveness, and value of the asset.
- Evaluate the effectiveness of the control measures; ensure the controls provide the required cost effective protection without discernible loss of productivity.

Due to the probabilistic nature and the need of cost benefit analysis, the IT risks are managed following a process that accordingly can be divided in the following steps [NIST SP800-30r1]:

- Risk assessment
- Risk mitigation
- Evaluation and assessment.

Also, this standard recommends that effective risk management must be totally integrated into the Systems Development Life Cycle. Information risk analysis conducted on applications, computer installations, networks and systems under development should be undertaken using structured methodologies.

Basically, risk management process includes two major subprocesses [ISO 31000]:

- Risk assessment (risk identification, risk analysis, risk evaluation)
- Risk treatment.

Therefore, different publications combine risk assessment and risk management in a process or define it in two separate processes.

6.3 Risk Assessment Process

Risk assessment, the first component of risk management, is often considered synonymous with risk analysis. It is defined as the process of identifying, estimating, and prioritizing risks to organizational operations (including mission, functions, image, reputation), organizational assets, individuals, other organizations, and the Nation, resulting from the operation of an information system [NIST SP800-39]. Distinct from this standard, another NIST document [NIST SP800-30r1] provides guidelines on how to implement risk assessment process including communicating and maintaining the assessed results current.

In case of ICS, the organization may perform a detailed risk assessment for the highest impact systems and assessments for lower impact systems as deemed prudent and as resources allow. The risk assessment helps identify any weaknesses that contribute to information security risks and mitigation approaches to reduce the risks. Risk assessments are conducted multiple times during a system's life cycle. The focus and level of detail varies according to the system's maturity.
Risk assessment is a study of the vulnerabilities, threats, likelihood, loss or impact, and theoretical effectiveness of security measures. The results of a risk assessment are used to develop security requirements and specifications [NIST RiskManag].

6.3.1 Risk Assessment and Analysis: Overview

Basically, risk assessment is the process of gathering data such as identifying assets, value of information and assets, identifying vulnerabilities and potential threats that could affect these assets. Risk assessment identifies the areas of risk to an organization including circumstances that may require better controls. Risk assessment is the process of identifying, estimating, and prioritizing information security risks. Risk assessment characterizes both the process and the result of analyzing and assessing risk. Assessing risk requires analysis of threat and vulnerability information to determine the extent to which circumstances or events could adversely impact an organization and the likelihood that such circumstances or events will occur. Generally, risk is calculated as a function of the threat, vulnerability, likelihood, and consequence/impact:

$$\text{Risk} = f(\text{threat, vulnerability, likelihood, consequence/impact}) \quad (1)$$

The purpose of the risk assessment component is to identify [NIST SP800-39]:

- Threats to organizations (e.g., operations, assets, or individuals) or threats directed through organizations against other organizations or the nation.
- Vulnerabilities internal and external to organizations.
- Harm (e.g., consequences/impact) to organizations that may occur given the potential for threats exploiting vulnerabilities and likelihood that harm will occur.

The end result is a determination of risk (e.g., the degree of harm and likelihood of harm occurring). To support the risk assessment component, organizations need to identify:

- Tools, techniques, and methodologies that are used to assess risk.
- Assumptions related to risk assessments.
- Constraints that may affect risk assessments.
- Roles and responsibilities.
- How risk assessment information is collected, processed, and communicated throughout organizations.
- How risk assessments are conducted within organizations.
- Frequency of risk assessments.
- How threat information is obtained (e.g., sources and methods).

Risk assessment and analysis are performed by a team using a methodology that includes process, risk model, assessment approach, and analysis approach [NIST SP800-30r1]. The risk assessment methodology typically includes [NIST SP800-39]:

- A risk assessment process.
- An explicit risk model, defining key terms and assessable risk factors and the relationships among the factors.
- An assessment approach (e.g., quantitative, qualitative, or semi qualitative), specifying the range of values those risk factors can assume during the risk assessment and how combinations of risk factors are identified/analyzed so that values of those factors can be functionally combined to evaluate risk.
- An analysis approach (e.g., threat oriented, asset/impact oriented, or vulnerability oriented), describing how combinations of risk factors are identified/analyzed to ensure adequate coverage of the problem space at a consistent level of detail.

Governance structures provide direction and oversight for risk management activities conducted by an organization. Organizations can use a single risk assessment methodology or multiple assessment methodologies. Selection of a specific methodology depends on some factors such as time, budget, complexity and maturity of organizational mission/business processes, the phase of the information systems in the system development life cycle, the criticality/sensitivity of the information and information systems supporting the core organizational missions/business functions. In addition, organizations need to define the risk models, identify threats and vulnerabilities including methods as described in current standards and publications.

6.3.2 Risk Measures

Quantitative risk is the use of measurable, objective data to determine asset value, probability of loss, and associated risks. Security industry practices for computing an IT asset risk are described in [Harris 2013]. However, computing quantitative risk for an ICS asset should be different than computing risk for and IT asset. Qualitative risk is the measure of risk or asset value based on rank or separation into categories such as low, moderate, high (see more in [Hentea 2008b]). One key issue is that organizations in the electricity industry need to apply the process that may be used by more organizations that share information. In this respect, the proposed risk management process by US Department of Energy (DOE) could provide common grounds.

6.4 U.S. Department of Energy Guidance on Risk Management

The electricity subsector provides the cybersecurity Risk Management Process guideline [DOE Process] that was developed by DOE, in collaboration with National Institute of Standards and Technology (NIST) and North American Electric Reliability Corporation (NERC).

6.4.1 DOE Risk Management Process

As described in [DOE Process], the risk management includes risk assessment, cost benefit analysis, selection, implementation, test, and security evaluation of safeguards. The overall system security evaluation considers both effectiveness and efficiency, including impact on the mission and constraints due to policy, regulations, and laws. The risk management process is a continuous and iterative process which is based on a comprehensive structure, called cycle, that requires organizations to [DOE Process]:

- Frame risk - establish the context for risk based decisions and security strategies.
- Assess risk – calculate and evaluate risk.
- Respond to risk – risk mitigation.
- Monitor risk - monitor and evaluate effectiveness of security controls for continuous improvement.

Each element of risk management cycle includes inputs, activities, and outputs as described in the following summary.

6.4.2 DOE Risk Management Summary

Risk framing describes the environment in which risk-based decisions are made such as:

- Assumptions about threats, vulnerabilities, impacts, and likelihood of occurrence.
- Legislation, regulation, and resource constraints (time, money, and people).
- Risk tolerance, which identifies the level of acceptable risk.
- Trust relationships, such as physical interconnections, third party service providers, supply chain, financial and investment partners, or device vendors.

Risk assessment identifies, prioritizes, and estimates risk to an organization's information, Information Technology (IT) and Industrial Control Systems (ICS) systems, operations, assets, individuals, and other interconnected electricity subsector organizations. Activities include:

- Defining assumptions and constraints.
- Identification of tools, methodologies, and resources.
- Gathering and processing data.
- Identification of threats, vulnerabilities, impacts, and likelihood or probability that an event will occur.
- Calculation and evaluation of risks.
- Communication and sharing of results.

Risk response addresses how an organization handles a potential security risk. It includes organization wide and consistent activities such as:

- Mitigate risk (accept, avoid, mitigate, share, or transfer).
- Develop and evaluate alternative actions for responding to risk.
- Determine appropriate security controls and actions.
- Implement the courses of action.
- Document and report the results to management.

Risk monitoring addresses how risks are monitored and communicated over time within organization. For example, a utility may need to know that it has a good handle on its risk assessment and mitigation strategy. The frequency of monitoring is driven by changes in IT and ICS systems, operational environments, and regulations. Examples of activities include:

- Verify if risk response measures are implemented and how security requirements are satisfied.
- Evaluate the ongoing effectiveness of risk response measures.
- Provide feedback for improvement of security controls at the system level or other tiers.
- Identify changes that may impact risk to an organization's IT and ICS and the operational Environment.

It is important to understand that effectiveness monitoring is different than compliance monitoring. Failure to achieve desired levels of effectiveness (e.g., risk increase) may be an indication that risk response measures have been implemented incorrectly or are not operating as intended. Determining the effectiveness of risk response measures is generally more challenging than determining whether the measures have been implemented correctly and are operating as intended (e.g., meeting identified compliance requirements). Risk response measures implemented correctly and operating as intended do not guarantee an effective reduction of risk, so other means such as measures and criteria to assess effectiveness have to be implemented.

6.4.3 DOE Risk Management Process and Integration Perspectives

The need to incorporate cybersecurity risk management into the organizational enterprise risk management program resulted in a model that represents a fairly standard strategic approach to breaking down business responsibilities for enterprise risk management. Responsibility for risk management is divided into three different business layers and their associated areas of business functionality.

To integrate the risk management process, a three-tiered approach is used to address risk from different perspectives and levels (tiers). Figure 6.1 shows the diagram of adapted risk management model for electric subsector [DOE Process]. This model is based on the risk management model described in [NIST SP800-39r1].

- Organization (tier 1)
- Mission/business processes (tier 2)
- Information systems (tier 3).

Tier 1 at the organization level produces an initial cybersecurity risk management strategy (if one does not already exist) that includes a risk assessment methodology, a risk monitoring strategy, and a cybersecurity governance program. This strategy is iteratively informed and revised based on outputs from Tiers 2 and 3.

The risk management process is carried out seamlessly across all tiers with the overall objective of continuous improvement in the organization's risk related activities and effective inter-tier and intra-tier communication among all stakeholders.

Use of such common roles in the risk management process facilitates the integration of cybersecurity risk management into existing enterprise risk management programs as does the focus on aligning technical security needs with business operations and requirements. This model allows organizing and presenting the strategic tasks necessary to either integrating cybersecurity risk into existing risk management processes or creating a risk management function specific to cybersecurity.

Tier 1 addresses risk from an organizational perspective by establishing and implementing a governance structure consistent with the strategic goals and objectives of the electricity subsector organization. Cybersecurity specific activities at Tier 1 may include:

- Establishing the organization's cybersecurity risk management strategy.
- Planning recovery for critical mission and business processes.
- Defining techniques and methodologies for assessing cybersecurity risk.
- Defining risk management constraints and requirements.

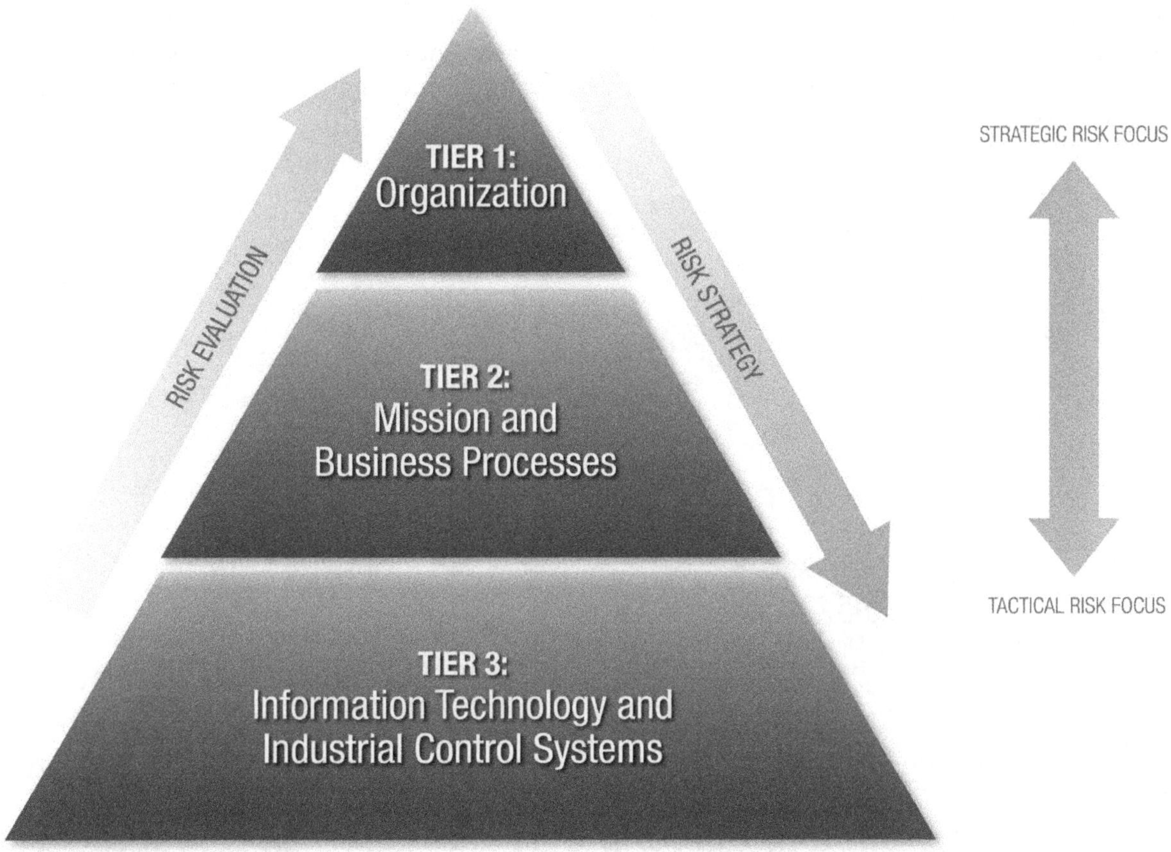

Figure 6.1 Risk management model for electric sector. *Source:* [DOE Process]. Public Domain.

Tier 2 addresses risk from a mission and business process perspective including system architecture (IT and ICS technical architecture). Tier 2 activities are inputs to activities in Tier 3, and provide feedback to Tier 1. Specific security activities may include:

- Identifying and prioritizing assets necessary to support the mission and business processes of an organization defined in Tier 1.
- Identifying cybersecurity processes needed to successfully execute mission and business Processes.
- Mapping cybersecurity requirements against mission and business processes.
- Developing a disciplined and structured approach for managing IT and ICS assets that support mission and business processes.
- Ensuring implementation, maintenance, and monitoring of security controls and protection.
- Developing cost effective, efficient, and resilient IT and ICS systems.

Tier 3 addresses system risk from an IT and ICS perspective and it is guided by the risk context, risk decisions and activities at Tiers 1 and 2. Activities at this level provide feedback to Tier 2 and Tier 1 on the organization's risk posture. Specific security related activities may include:

- Categorizing risk categorization for IT and ICS systems.
- Allocating security controls to systems and the environments in which they operate.
- Establishing a process to routinely reassess a system's security posture.

Figure 6.2 depicts the integration of risk management cycle within each tier as defined in [DOE Process]. Risk management activities can be integrated into each system development life cycle with the outputs at each phase having an effect on subsequent phases. All systems, including operational systems, systems under development, and systems undergoing modification, are in some phase of the system development life cycle.

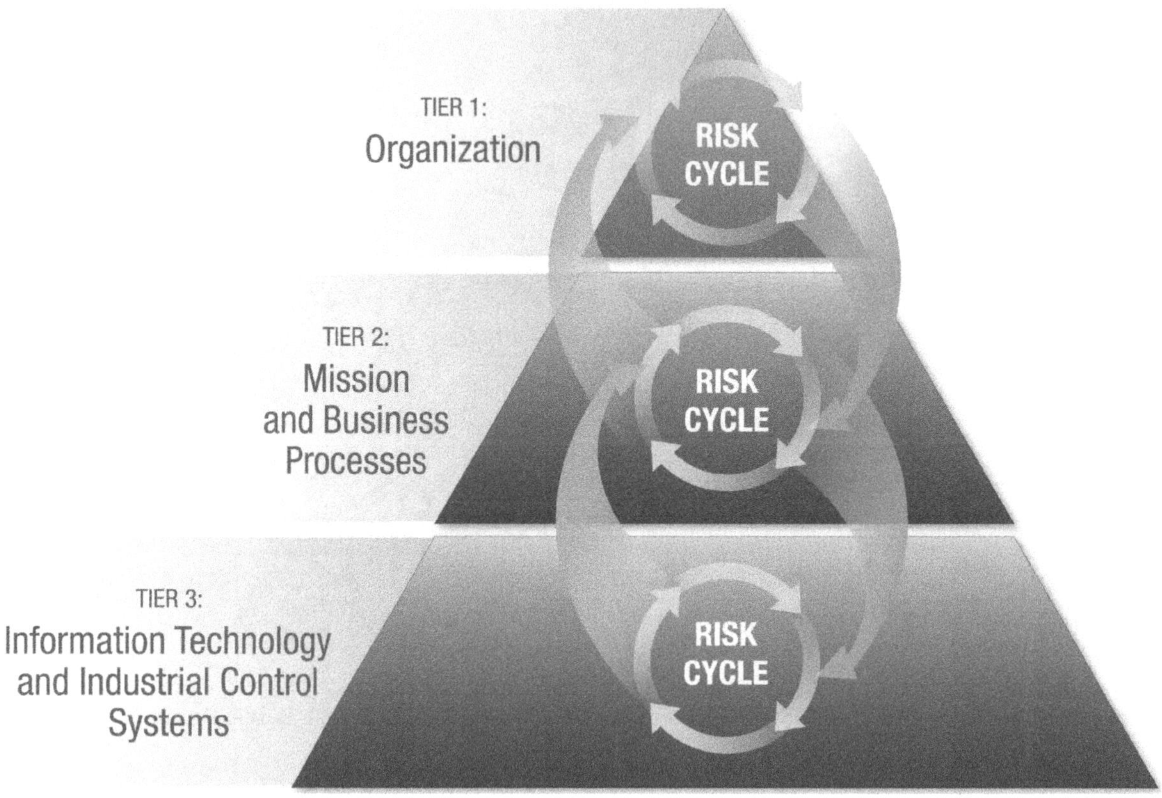

Figure 6.2 Risk Management Process. *Source:* [DOE Process]. Public Domain.

6.4.4 Applying DOE Guidelines

The use of consistent workflow descriptions (e.g, the inputs, activities, and outputs), throughout the life cycle sections allows for consistency throughout risk management planning and implementation [ICS-CERT 2012a].

There are some benefits of using the DOE risk management model. It could be applied more to the security program (information security management model) to be part of the enterprise risk management. However, this model may not be very practical for a system. Also, DOE model requires organizations to have specific organizational structures. In reality, structures are not unique across electrical sector and third party providers. Another issue is that many concepts and risk management process are more based on practices for Federal Information Systems.

Therefore, it is necessary to understand that utilities, in terms of business and organizational structures, vary widely, and not all frameworks, maturity models or metrics are appropriate. Some customization may be required and individual needs and processes may be adopted. In addition, many organizations already have established processes for risk management cycle. More information about these processes is included in other standards and practices (e.g., [IEC 62443], [IEC 62351], [ENISA 2016]) that provide more specific recommendations suitable to the Smart Grid and power systems.

6.4.5 Risk Models

Risk models define the risk factors to be assessed and the relationships among those factors. Risk factors are characteristics used in risk models as inputs to determine levels of risk in risk assessments.

Risk factors are also used extensively in risk communications to highlight what strongly affects the levels of risk in specific situations, circumstances, or contexts. Typical risk factors include threat, vulnerability, impact, likelihood, and predisposing condition. Risk factors can be decomposed into more detailed characteristics (e.g., threats decomposed into threat sources and threat events). Therefore, these definitions are important for organizations to document prior to conducting risk assessments because the assessments rely upon well defined attributes of threats, vulnerabilities, impact, and other risk factors to effectively determine risk.

Risk models differ in the degree of detail and complexity with which threat events are identified. When threat events are identified with great specificity, threat scenarios can be modeled, developed, and analyzed. Threat events for cyber or physical attacks are characterized by the tactics, techniques, and procedures employed by adversaries. Understanding adversary based threat events gives organizations insights into the capabilities associated with certain threat sources.

6.4.6 Information Security Risk Management Challenges

Although many security experts agree on the need to implement both risk management and risk assessment, there are issues that need to be addressed. Since information security management is focused on protection of systems, applications, devices, and networks, it is required that security professionals understand the risks and map them to business drivers [Harris 2013] and this management activities must be done within a broader framework of managing the non IT risks too [Anderson 2008a]. Information technology risk, or IT risk, IT related risk, is any risk related to information technology. However, there are many challenges on implementing these processes. For example, some organizations implementing risk management often tend to neglect risk assessment. Other organizations rudimentarily implement risk management and risk assessment as part of their Information Security Management System. In addition, many small organizations do not implement them at all (e.g., SMEs, large organizations in the area of commerce).

Generally accepted by Information Security experts, risk assessment and risk management are major components of information security management. However, the communication among stakeholders and experts in the areas of risk management and risk assessment has to be based on recognized and agreed definitions. Often, the communication is impacted by the use of different terms for relevant terms [ENISA 2006]. Also, the use of differentiated terminology in the relevant literature (e.g., National or International standards, tools, good practices, etc.) aggravates the comparability of the achieved results, used methods and tools. With no single definition for these terms, the information security industry is more concerned on making money than focusing on issues such as consolidating definitions and advancing the fundamental practices. Security industry is a big business today, while the loss caused by security breaches is growing too.

While information security risk management can be carried out separately, the emerging threats and business needs drive integration of information security risk management in the organization's wider risk management process. Also, risk management should integrate with other operational and security related functions such as: product evaluations (e.g., Common Criteria), information security management systems, security controls deployment, incident handling, business continuity planning, operational processes, etc.

An attempt to consolidate different definitions in one view is depicted in Figure 6.3. Also, these definitions are refined as follows [ENISA 2006]:

Risk management is the process, distinct from risk assessment, of weighing policy alternatives in consultation with interested parties, considering risk assessment and other legitimate factors, and selecting appropriate prevention and control options.

Risk assessment is a scientific and technologically based process consisting of three steps, risk identification, risk analysis, and risk evaluation.

Figure 6.3 shows that risk assessment is part of the risk management process and the relationship is adopted from [OCTAVE]. After initialization, risk management is a recurrent activity that deals with the analysis, planning, implementation, control and monitoring of implemented measurements and the enforced security policy. On the contrary, risk assessment is executed at discrete time points (e.g., once a year, on demand, etc.) and – until the performance of the next assessment - provides a temporary view of assessed risks and while parameterizing the entire risk management process.

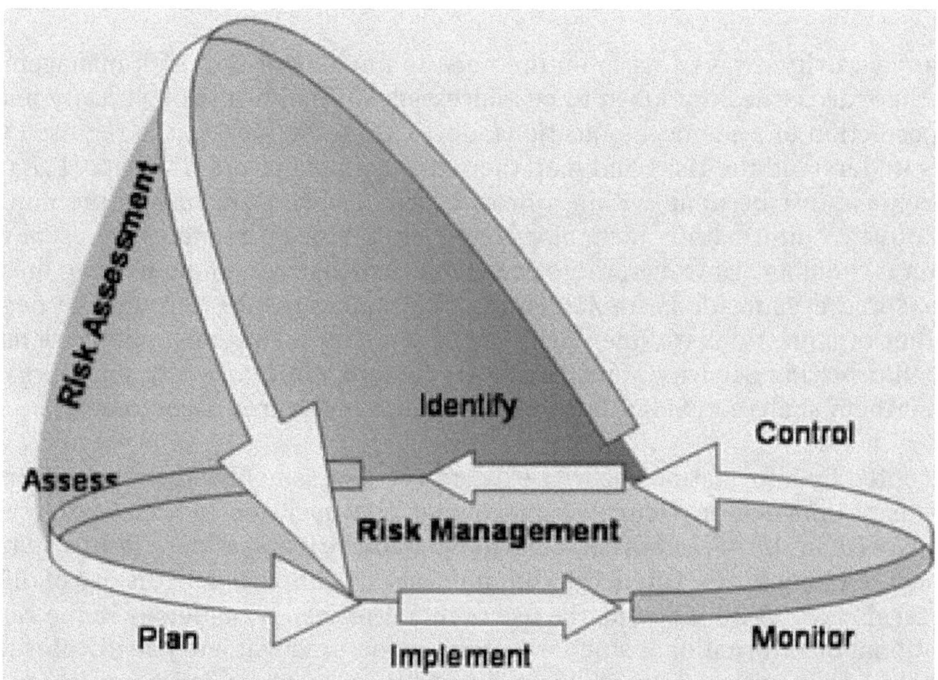

Figure 6.3 ENISA risk management. *Source:* [ENISA 2006]. © 2006, ENISA.

Risk assessment and risk management processes comprise the heart of the Information Security Management System (ISMS) because these processes transform on one hand the rules and guidelines of security policy and the targets; and on the other to transform objectives of ISMS into specific plans for the implementation of controls and mechanisms that aim at minimizing threats and vulnerabilities.

6.5 Customized Risk Management for Smart Grid

The U.S. Department of Energy recommends organizations should be provided with tailored guidance on implementing risk management for the electricity sector. Therefore, this document [NISTIR 7628r1] and other security standards and guidelines can be used for defining a methodology as part of an organization wide risk management process and methodology to reduce the cyber vulnerabilities and limit the impacts of cyber-physical attacks. Because the Smart Grid includes systems from the IT, telecommunications, and electric sectors, the risk assessment process is applied to all three sectors as they interact in the Smart Grid.

6.5.1 Cybersecurity Recommendations

The overall strategy used in the development of this document [NISTIR 7628r1] examines both domain specific and common requirements when developing a risk mitigation approach to ensure cybersecurity and interoperability of solutions across different parts of the infrastructure. The cybersecurity strategy addresses prevention, detection, response, and recovery.

The NIST Cybersecurity Framework [NIST 2018] and Roadmap [NIST SP1108r1] document defines Smart Grid Conceptual Reference model (see updated model of [NIST SP1108r4] in Appendix B). The model identifies seven domains to include: Generation, Transmission, Distribution, Operations, Markets, Customer, and Service Provider. A Smart Grid domain is a high level grouping of organizations, buildings, individuals, systems, devices, or other actors with similar objectives and relying on - or participating in - similar types of applications.
By expanding the high level Smart Grid conceptual reference model to include logical interfaces between actors and domains, a logical reference model for the Smart Grid is presented in [NISTIR 7628r1]. The logical reference model illustrates the logical interfaces which are grouped in about 22 categories based on criteria of common level of risk impact (low, medium, high) to security objectives (confidentiality, integrity, availability). However, using these levels of impact is problematic and may not fit well in each organization.

An organization may use or customize the security requirements defined in [NISTIR 7628r1] document to meet its own business and operational needs. Since this document is based on subjective definition of the risk impact levels that are based on expert's evaluation, an organization can establish an objective method to determine the cyber attack consequences and further identify the risk impact levels.

6.5.2 Building Risk Models

Among the challenges associated with applying security risk management is that it does not mean a single task [Hentea 2008b]. Rather, information security management, particularly risk management, comprises a collection of activities whose ultimate purpose is to produce a model meeting business needs and user requirements. These activities should be based on system identification concepts.

System identification is described as the method of developing empirical mathematical models of systems based on excitation and response measurements [Fassois 2007]. Given the nature of communications networks and Internet, the traffic behavior cannot be modeled using only principles derived strictly from physical principles. SCADA systems are no exceptions to these difficulties on building the model for performing the design of security. A model is essential to analyze system properties and risks, perform simulation, prediction, state estimation, monitoring, fault diagnosis and control. System identification model is a collection of activities to produce a model that meets user requirements. These activities include experiment design, model structure selection, parameter estimation, and model validation.

System identification is an interdisciplinary approach that uses systems theory, signal processing, optimization, and statistics methods. Considering the nature of high speed malware spread in a network including the damages and losses that cause to the attacked device or network, there are needs such as:

- Identify the security risks and model the dynamics of SCADA systems to find an optimal solution to monitor for cyber attacks.
- Control the ongoing cyber attacks using prediction models based on parameter estimation using data collected by SCADA devices and specialized software. Statistical and machine learning approach are techniques that can be used to estimate and derive optimum parameters to control the ongoing cyber attacks as well as fault propagation to several devices connected to SCADA systems. In other words, we can also use the principle of locality to minimize the damages of SCADA and stop propagation of cyber attacks to a larger scale.
- Support identification of nonlinear models for short term network load forecasting. An effective load forecasting model needs to recognize the effects of overloading or non available services at multiple levels and must be able to explain irregular events.
- Use SCADA data and enterprise data to achieve experimental designs that are informative and yet display time domain characteristics such that design solutions are acceptable for operations.

To address the needs, nonlinear optimization techniques can be applied to obtain coefficients for multivariable inputs and relationships that result in data leading to models with improved closed loop performance over traditional designs. Nonlinearity and uncertainty are some of the solution features. An estimation of vulnerabilities and threats models provides a characterization of the risk management performance. For example, frequency domain identification method allows to address specific challenges such as performance measures and compensation for errors and better countermeasures. Use of special techniques for the subspace algorithm could identify missing data points that could lead to the attack detection and prevention.

Identifying needs and security requirements for distributed energy resources (DER) systems provide the basis of security and privacy protection. Since many DER systems include functions on real time monitoring and control, the choice of models is a key decision. There is a need to describe a system unambiguously and prove that it can meet desired criteria of complexity and scalability before starting the design. Modeling and design of complex systems require notations for describing concurrency, asynchronous event handling, communication between independent machines, and timing properties. The complexity and the scale of any DER system can cause other problems. For example, race conditions due to concurrency or deadlock and starvation due to mutual exclusions need to be identified to allow design of optimal solutions. Analysis and design of such system become difficult when relying only one approach. A combination of methods may result on more models that provide a more complete and consistent coverage of the system's requirements. Therefore, a synergy of various techniques can be used to define an efficient model for security risks.

One component of this analysis is to identify the security constraints and issues associated with each network interface and the impact level of a security compromise of confidentiality, integrity, and availability, non-repudiation, and authenticity.

6.5.3 Process Control Model

Security problem is about to understand what the problem is and how to manage it. The application of process control model is recommended in [Hentea 2008b] and discussed on other works such as [Kadrich 2007], [Hentea 2005]. The SCADA network and the endpoints that populate it can be expressed as a closed loop process control problem. When assessing the risks, it is required to evaluate the threats and potential attack methods including the target endpoints and the probability, or, risk, that a cyber attack will succeed. By using process control model, we can identify metrics that have greater impact on security. We can identify inputs, outputs, and the feedback in a network as well as for endpoints. For example, time constraints can be added to associate it with the control process. The model is then used to identify risk to be introduced in a network and setting a low limit to endpoints for security criteria. When approaching the problem of security using closed loop process control model, there is a need to identify those components (nodes) in SCADA network that can assist with the basic modes of control: proportional, integral, and derivative controls.

A device can be a control node – a place used to enforce a condition or extract data for the purpose of managing the process. A control node can have the capability to decide about the traffic or the status of the device. In other words, the devices report data that enables a human to make decisions in support of either derivative or integral control functions. Most of the devices in a SCADA network, with a few exceptions, can be categorized as performing derivative controls. By using log data collected by a device, it is possible to enable an integral control. Following these activities, we get a mapping of devices (hardware, software, logs, alerts, probes, and correlation) to control modes. Next, we identify the control signals using metrics that are security relevant such as failed login attempts, firewall rule violations, access attempts, alerts, new software requests, new protocol connections, network access requests, etc.
By using this model, it is possible to set a more accurate and acceptable limit of risk for SCADA networks to be protected more effectively.

6.5.4 Threat Model

A threat model is essentially a structured representation of all the information that affects the security of an application [OWASP Threat]. In fact, it is a view of the application and its environment through security glasses.
Threat modeling is a process for capturing, organizing, and analyzing all of this information. Threat modeling enables informed decision-making about application security risk. In addition to producing a model, typical threat modeling efforts also produce a prioritized list of security improvements to the concept, requirements, design, or implementation.

A threat is any circumstance or event with the potential to adversely impact organizational operations and assets, individuals, other organizations, or the Nation through an information system via unauthorized access, destruction, disclosure, or modification of information, and/or denial of service. Threat events are caused by threat sources. In general, types of threat sources may include:

- Hostile cyber or physical attacks
- Human errors of omission or commission
- Structural failures of organization-controlled resources (e.g., hardware, software, environmental controls)

- Natural and man-made disasters, accidents, and failures beyond the control of the organization.

Various taxonomies of threat sources have been developed. Some taxonomies of threat sources use the type of adverse impacts as an organizing principle. Multiple threat sources can initiate or cause the same threat event—for example, a provisioning server can be taken off line by a denial of service (DOS) attack, a deliberate act by a malicious system administrator, an administrative error, a hardware fault, or a power failure. Risk assessment must include emerging threats such as new technology, culture and change in culture, unauthorized use of technology (e.g., wireless technologies, rogue modems, PDAs, unlicensed software).

Threat assessment is a very detailed and, in some cases, difficult task. Threats are classified as human (intentional or unintentional) and natural disasters. Many types of threat agents can take advantage of several types of vulnerabilities, resulting in a variety of specific threats. Once the threats are identified for an entity (application, system, business unit, organization), the vulnerabilities must be investigated.

Therefore, it is crucial to understand the behavior of attacks against cyber-physical Smart Grid, which can be classified into three broad subsets [NISTIR 7628r1]:

- Physical attacks informed by cyber could cause a much greater impact than the physical attack of a random substation; the use of information gathered by cyber means allows an adversary to plan and execute an improved or enhanced physical attack.
- Cyber attacks enhancing physical attacks increase the impacts of a physical attack by either making the attack more successful (e.g., greater consequences) or interfering with restoration efforts (thereby increasing the duration of the attack.
- Intentional use of a cyber system (e.g., changing the settings or control commands of a control system) to cause physical harm/damage to equipment or safety hazards.

Cyber-physical attacks can greatly enhance the overall impact and/or consequences of an attack or increase the duration of those consequences by delaying or interfering with responses. However, good cyber, physical, and operational security planning and implementations can minimize these impacts. Defensive measures that can be used to minimize the likelihood of successful cyber attacks and physical attacks contribute to reducing the impacts of a cyber-physical attack. Security operators need to consider both types of attacks and how they may be used together in order to better develop systems that are resilient to cyber-physical attacks.

In the later releases, [NIST SP1108r3] and [NIST SP1108r4], Smart Grids are viewed from the perspective of cyber-physical systems (CPS) - hybridized systems that combine computer based communication, control, and command with physical equipment to yield improved performance, reliability, resiliency, and user and producer awareness.

6.5.5 Vulnerabilities and Predisposing Conditions

Risk assessments are a key cornerstone in defining, understanding, and planning remediation efforts against specific threats and vulnerabilities. A vulnerability is a weakness in an information system, system security procedures, internal controls, or implementation that could be exploited by a threat source. Besides software vulnerabilities, it is necessary to recognize the possible vulnerabilities in the systems hardware, firmware, and integration [Baldwin 2009].

Often, information system vulnerabilities can be associated with security controls that either have not been applied (either intentionally or unintentionally), or have been applied, but retain some weakness. However, it is also important to allow for the possibility of emergent vulnerabilities that can arise naturally over time as organizational missions/business functions evolve, environments of operation change, new technologies proliferate, and new threats emerge. In the context of such change, existing security controls may become inadequate and may need to be reassessed for effectiveness. The tendency for security controls to potentially degrade in effectiveness over time reinforces the need to maintain risk assessments during the entire system development life cycle and the importance of continuous monitoring programs to obtain ongoing situational awareness of the organizational security posture.

Vulnerabilities are not identified only within information systems. Viewing information systems in a broader context, vulnerabilities can be found in organizational governance structures (e.g., the lack of effective risk management strategies and adequate risk framing, poor intra-agency communications, inconsistent decisions about relative priorities of missions/business functions, or misalignment of enterprise architecture to support mission/business activities). Vulnerabilities can also be found in external relationships (e.g., dependencies on particular energy sources, supply chains, information technologies, and telecommunications providers), mission/business processes (e.g., poorly defined processes or processes that are not risk aware), and enterprise/information security architectures (e.g., poor architectural decisions resulting in lack of diversity or resiliency in organizational information systems).

Formal risk assessment identifies vulnerabilities and threats in the proposed system and the potential risk levels as they pertain to security attributes (e.g., CIA). This builds on the preliminary risk assessment carried out in the initiation phase of a system development/acquisition cycle. The results of this assessment help the team build the system's security plan and develop an initial description of the CIA requirements of the system. The assessment should define the environment in which the system will operate within and any identified vulnerabilities.

6.5.5.1 Analysis of Vulnerabilities

A vulnerability is a potential avenue attack. A vulnerability is a property of a system or its environment which, in conjunction with an internal or external threat agent, can lead to a security failure. A vulnerability is characterized by the difficulty and the level of skill that is required to exploit it. When identifying specific vulnerabilities, it is necessary to locate all the entry points to the organization and to find all access points to information (in both electronic and physical form) such as:

- Internet connections
- Remote access points
- Connections to other organizations
- Physical access to facilities
- User access points
- Wireless access points.

During analysis, it is required to understand both electricity infrastructure and communication networks carrying information for delivery and control of power. There is also a significant risk of overloading beyond system or network specification. Different situations may occur such as [Sommer 2011]:

- A network cannot carry the traffic due to a cyber attack or overloaded communication path.
- Some network designs may be vulnerable to a large scale cascade triggered by the disabling of a single key node.

- Most systems and networks are designed for expected normal levels of activity, with an allowance for expected growth and a further allowance to cope with —unexpected demand.
- In a catastrophe the networks can reach a tipping point where the design specifications are exceeded.
- Failure cascade due to interdependencies.

In resilience planning, these scenarios have to be identified and requirements should be defined to anticipate accordingly. One alternative could be passing the load to another similar system or network, which is put at risk of collapse. This can affect other systems that may rely for their functioning on receiving information from the first system. Therefore, resilience planning could benefit by developing models for modeling vulnerability cascading and tipping points. In addition, electricity infrastructure can have a significant impact on the systemic consequences of a system or network failure when a power blackout occurs. Thus, interdependencies, backups, and redundancies have to be carefully investigated when defining security requirements. So, protection measures have to be applied to backups and redundancies too.

Another example is the use of risk assessments for design of distribution networks that pose challenges for information and communication technology. It is argued that existing analysis approaches focus on analyzing the power networks components separately [Koziolek 2012]. For example, communication simulation provides failure data for communication links, while power analysis makes predictions about the stability of the traditional power grid. Therefore, these aspects have to be combined and design of these networks should be based on a holistic assessment of traditional power system topology and ICT architecture. Cybersecurity protection is also correlated with electricity operations.

6.5.5.2 Fault Analysis

Generally, the meaning of a fault is associated with a mistake or defect [Fault]. Thus, standards and dictionaries define it more specifically for a domain. Since Smart Grid is system of systems, a fault may refer to a power system, technology, or computing system as defined in [FaultTech]:

- Fault (technology), an abnormal condition or defect at the component, equipment, or subsystem level which may lead to a failure.
- Fault (power engineering), an unintended electrical connection, a short circuit of several kinds.
- Fault (computing), a type of interrupt in software or operating systems.

A technology fault or combinations of faults may lead to a failure. Failure is defined in as the lack of ability of a component, equipment, sub system, or system to perform its intended function as designed [Failure]. Technology faults may occur in hardware or software used in power systems or computing systems.

6.5.5.3 Power System Faults

Electric systems are exposed to different types of faults. A fault is any abnormal electric current [FaultPower]. Many faults in the power lines are transient in nature and a protection device isolates the area of fault. A transient fault is a fault that is no longer present if power is disconnected for a short time and then restored. For example, a short circuit is a fault in which current bypasses the normal load. The prospective short circuit current of a fault can be calculated and protective devices are used to detect fault conditions to limit the loss of service due to a failure. Power system protection is the process of making the production, transmission, and consumption of electrical energy as safe as possible from the effects of failures and events that place the power system at risk. One objective of systems design and power system protection is to detect and interrupt power system faults. Power system protection must determine from measurements of currents and/or voltages whether the power system is operating correctly. Measurements, data processing, and control are the critical elements for protective relays to be effective.

However, the centralized power plant paradigm is increasingly becoming outdated. A new generation of sensors is providing grid operators with more information about what is happening across the grid at any given moment. Since the grid of the future needs to be flexible, transparent, and dependable, more attention should be directed to the analysis and design of the infrastructure needed by power systems protection.

The electric grid operates as a series of networks that are defined by geography and many dependencies exist in the system. Using the network theory, a recent mathematical study of spatial networks finds that spatial networks are necessarily dependent on any number of critical nodes whose failure can lead to abrupt—and unpredictable—collapse [Tollefson 2013a], [Tollefson 2013b].

Recognizing the cascading effect of a failure causing instability in the power grid, other researchers and practitioners argue that network theory is not sufficient and it must be complemented with more complex models that attempt to represent both the physical realities and the responsiveness of the modern electric grid. It is argued that this study could be applied entirely to US blackout of 2003. An analysis of the blackout (about 50 million people without power) determined that the failure of service was caused by a combination of bad vegetation management and a series of monitoring and communications breakdowns [Blackout 2004], [ELCON 2004], [Pourbeik 2006].

The identification of power faults, failure causes, and relationships to technologies and networks could help in design of communications and security of networks carrying real time data about the equipment, loads, and environment conditions. In addition, a Security Engineer has to understand that hidden failures of monitoring and control devices can be exploited by a suite of threat agents.

Therefore, analysis has to include the security risk assessment of the sensors, communication technologies, and new protection systems equipped with Intelligent Electronic Devices (IEDs). IEDs collect and transmit more power system data, enhance operations and maintenance, and permit the use of adaptive system configuration for protection and control systems. For example, one area of concern is the security of IEDs transfer data to central control facilities and Human Machine Interfaces (HMIs) using electronic communications.
The integrity of the data and consistency of the data are essential in the analysis of power systems behavior and failures. A systematic method for the analysis of system failures and a prototype system capable of mitigating the risk of wide area disturbance and full scale applications is described in [Qiu 2003].

Although power faults analysis is mainly a power engineer task, the design team of a protection system should include systems engineers, security engineers, and other domain experts.

6.5.5.4 Computing System Faults

A fault in a computing system may cause an error, which is an incorrect state. The error may lead to a failure which is a deviation from the system's specified behavior [Anderson 2008a]. Therefore, the Systems Engineer and Security Engineer need to understand that different categories of faults may be associated with a specific domain of the Smart Grid conceptual model. Also, in order to control the impact of faults, there is a need to apply the appropriate methodology to detect the category and interdependence between systems because a fault in one system may create faults in another system. For example, a fault in a network connecting an electrical storage provider with the utility may cause a disturbance to utility when tries to use the stored energy to satisfy an increasing load and demand. Also, an understanding the failure effect is key. Failure is defined as the state or condition of not meeting a desirable or intended objective. The criterion for failure depends on the context of use and it may be particular to an observer and system behavior.

In the context of computing systems, failures are associated with faults occurring in hardware and software, environment, and human errors. A failure access to data may be caused by an incident in which access to data is not possible because of a hardware or software failure. Also, the failures may be caused by a threat agent exploiting a vulnerability in the system. A vulnerability may be the result of a flaw, which is an error in the design, implementation, or operation of an information system [RFC 4949]. Security bugs or defects introduce security vulnerabilities by compromising one or more of:

- Authentication of users and other entities
- Authorization of access rights and privileges
- Data confidentiality
- Data integrity
- Privacy.

Security defects may be also traced to security mechanisms and protocols implementation. The most recent bug of April 2014, called Heartbleed, is a security bug in the OpenSSL cryptography library, which is a widely used implementation of the Transport Layer Security (TLS) protocol [Heartbleed]. This vulnerability is classified as a buffer over read, a situation where software allows more data to be read than should be allowed. Also, errors in design, architecture, configuration management may expose systems to vulnerabilities that can be exploited.

Although not uniform, different taxonomies of security bugs are provided by security professionals and researchers to be used by programmers. In addition, government sponsored tools (e.g., [MITRE 2011], [CERT Response]), provide information about discovered security bugs, vulnerabilities, recommendations, and other tools.

Failure analysis is the process of collecting and analyzing data to determine the cause of a failure. Thus, a methodology to provide failure control should not be limited only to the methodology used to detect hardware and software failure, it requires to provide fail safe, fail secure or fail soft termination and recovery of system functions [RFC 4949]. Thus, an application or system has to return to a safe and more secure state. Often fail safe is used as a synonym for fail secure [Security], although there are subtle differences.

Fail safe is a mode of termination of system functions that prevents damage to specified system resources and system entities (e.g., specified data, property, and life) when a failure occurs or is detected in the system (but the failure still might cause a security compromise) [RFC 4949]. Fail safe assumes automatic protection of programs and/or processing systems when hardware or software failure is detected [CNSSI 4009]. After a failure, software by default to no access instead of allowing full control.

Fail secure is a mode of termination of system functions that prevents loss of secure state when a failure occurs or is detected in the system (but the failure still might cause damage to some system resource or system entity) [RFC 4949].

Fail soft is the selective termination of affected, non essential system functions when a failure occurs or is detected in the system.

However, fail safe and fail secure definitions are opposing design alternatives. Designers have to identify the entities that are concerned with safety and the principle of fail safe defaults to be applied (which states that a subject should be denied access to an object, unless a subject is given explicit access) [Bishop 2005].

6.5.5.5 Fault Management

Fault management (an area of ISO model) involves identifying, tracking, diagnosing, and resolving faults in the system. Information security fault management involves identifying faults in the applied information security profile and then addressing them [Whitman 2014]. Analysis techniques such as flaw hypothesis and fault tree are often recommended to be used.

A flaw hypothesis methodology is often applied during analysis or design of penetration tests. This method is an evaluation or attack technique in which specifications and documentation for a system are analyzed to hypothesize flaws in the system. Then, the list of hypothetical flaws is prioritized on the basis of the estimated probability that a flaw exists and, assuming it does, on the ease of exploiting it and the extent of control or compromise it would provide.

Fault tree analysis is a technique used to calculate the probability of an event. This method is based on a structure called fault tree (or threat tree, as it is often called in Security Engineering) in which an undesired state of a system is specified and the system is studied in the context of its environment and operation to find all credible ways in which the event could occur. The specified fault event is represented as the root of the tree. The remainder of the tree represents AND or OR combinations of subevents, and sequential combinations of subevents, that could cause the root event to occur.

The analysis for the safety-critical systems is based on hazard analysis. This can be done using the Failure Modes and Effects Analysis (FMEA) method, which is a bottom up rather than top down analysis. This involves tracing the consequences of a failure of each of the system's components all the way up to the effect of the mission. Understanding the consequences of a failure of any of the protection mechanism is often useful in Security Engineering. A real thorough analysis of failure modes may combine top down and bottom up approaches. However, many other techniques can be used for critical systems. Critical systems can be defined as those in which a certain class off failures is to be avoided if it all possible. Depending on the class of failure, they may be safety-critical, business-critical, security-critical, critical to the environment [Anderson 2008a], [Anderson 2005].

In addition, it is important to consider failover capabilities for critical systems. The failover capability enables switching over automatically (typically without human intervention or warning) to a redundant or standby information system upon the failure or abnormal termination of the previously active system. Many more analysis challenges arise when performing a system of systems because of limitations in analysis mechanisms such as [Ellison 2008]:

- Means to focus on challenges arising from integrating multiple systems
- Support architecture trade offs carrying impacts beyond a single system
- Link technology to critical organizational functions.

To overcome these inconveniences, an analysis framework, called Survivability Analysis Framework, can be used to [Ellison 2008]:

- Identify potential problems with existing or near term interoperations among components within today's network environments.
- Highlight the impact on survivability as constrained interoperation moves to more dynamic connectivity.
- Increase assurance that mission threads can survive in the presence of stress and possible failure.

Adopted by Computer Emergency Response Team (CERT), this framework help organizations analyze and understand stresses and gaps to survivability for operational and proposed business processes. Figure 6.4 depicts a broad view of the SAF adopted by CERT [SAF].

The framework can be used to construct a view of an operational process that documents the interrelationships of people, actions, and technology. This view must be shared by stakeholders and should identify critical steps and the ways in which a step failure could lead to an operational process failure. The framework can be used characterize the specific actions of each step in an operational process and the linkages between each step (preconditions and post conditions). Therefore, the SAF provides a structure for gathering and visually displaying the operational process information that can be useful to management, users, technology architects, system engineers, and software engineers.

As defined in this framework, preconditions have the role to establish the resources provided to the step. They may trigger the execution of the step actions (for example, data or a human command), or the actions may be continually executed (such as a sensor). Each step has outcomes (post conditions) that may interact with subsequent steps. It is stipulated that the contract with prior and subsequent steps is not necessarily static and may have to be negotiated during execution to reflect changing conditions. In addition, the identity of prior and subsequent steps may vary across executions of an operational process.

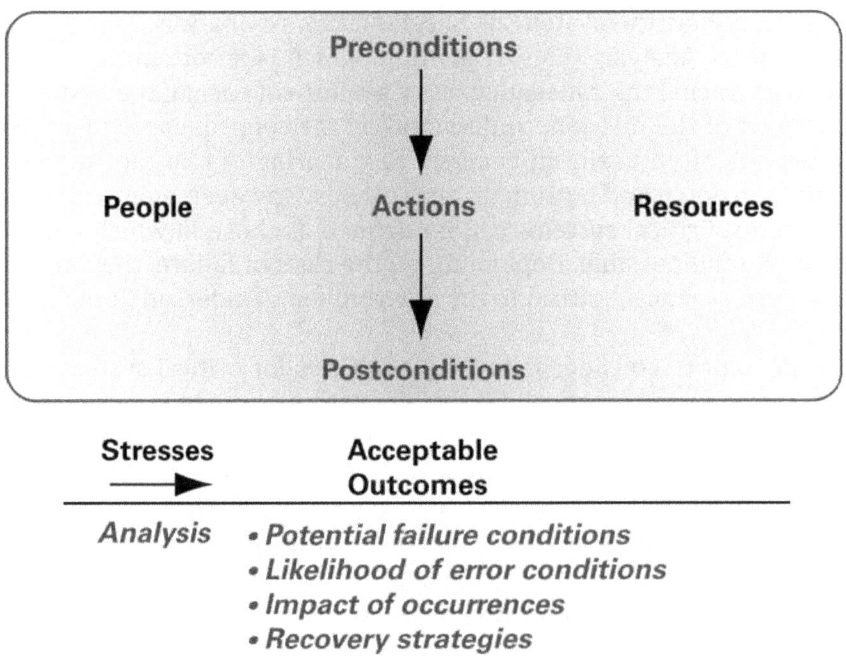

Figure 6.4 Survivability Analysis Framework. *Source:* [SAF]. Public Domain.

6.5.5.6 Vulnerability Analysis Based on Proactive, Discovery, and Adaptation Solutions

A changing vulnerability analysis and threat landscape and continuing requirements for compliance are the main drivers for vulnerability management programs to expand towards proactive, discovery, and adaptation solutions. In addition, a strategy to deal with cyber attacks against the nation's critical infrastructure requires first understanding the full nature of the threat.

As described in [Hentea 2008b], vulnerability analysis must focus on identifying (and reducing) the vulnerability of engineered systems to both natural (e.g., weather related) and man made (e.g., sabotage, terrorism) disruptions. This implies new tasks related to conceptual and methodological development of risk and vulnerability research in vulnerability modeling, cause mitigation analysis and process component definition including risk and vulnerability assessments for SCADA networks. Also, the focus should be in the development of tools that can provide discovery and training on vulnerability and adaptation. Currently, several vulnerabilities are modeled based on heuristics. In protecting against an attack and maintaining continuous operation, research must focus in vulnerability management.

Vulnerability management consists of a combination of technologies and processes to improve security posture. Targeted threats drive the need for more effective and proactive infrastructure protection solutions. For example, a control system should monitor for cyber attack activities and automatically generate patches to protect an application source code and identify new vulnerabilities [Keromythis 2004]. This assumes that an analysis engine can identify the potential vector attack from the information collected in real time and discover new vulnerabilities.

Vulnerabilities in software (bugs and flaws) can be grouped together by central characteristics and give rise to particular attack patterns [Hoglund 2004]. This is based on the premise that related programming errors give rise to similar exploit techniques. A particular exploit usually amounts to the extension of a standard attack pattern to a new target. An attack pattern is a blueprint for exploiting software vulnerability. Solutions such as identifying attack patterns to provide general descriptions of vulnerabilities are discussed in [Jiwnami 2002], [Gegick 2005].

6.5.6 Risk Management for SCADA Systems

Information security management is a continual cyclical process comprised of key phases such as assessment (risk management approach), policy, implementation, training, and auditing. Risk management characterizes an overall process to identify, measure, control, and minimize losses associated with uncertain events or risks. The first phase called risk assessment includes analyzing assets, identifying vulnerabilities and potential risks due to threats, risk reducing measures, and decisions related to the acceptance, avoidance, or transfer of risk. Risk assessment characterizes both the process and the result of analyzing and assessing risk. The second phase includes the process of assigning priority to, budgeting, implementing, and maintaining appropriate risk reducing measures.

On the other hand, risk management for SCADA control systems is not a common practice and it is not based on a unique standard or method to follow. Several methods for risk assessment are emerging. Risk management challenges include limitations of these methods, untrained and unskilled developers in business management and organizational areas, various classifications and nonunique taxonomies of vulnerabilities [Hentea 2008b]. Figure 6.5 illustrates a typical ICS risk management process that includes risk identification, risk mitigation, and applying security controls to protect the assets.

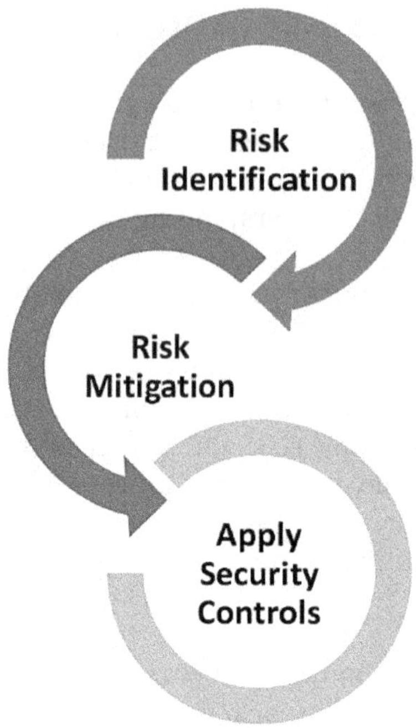

Figure 6.5 Typical phases of ICS risk management.

6.5.6.1 SCADA Analysis Challenges

Risk management challenges include limitations of methods, untrained and unskilled developers in business management and organizational areas, various classifications and non unique taxonomies of vulnerabilities. It is necessary to address not only the individual vulnerabilities, but the breadth of risks that can interfere with critical operations and safety of people, and public in large [Hentea 2008b]. In addition, cyber attacks exploit vulnerabilities previously not modeled or unknown to a system. SCADA networks were initially designed with little attention to security threats. The use of insecure IT protocols in SCADA systems can be high risk because attackers are aware of their known vulnerabilities and may have access to automated tools created to exploit these vulnerabilities. For example, network protocol analysis tools can intercept and decode most common protocols. Similarly, password cracking tools can decode messages and passwords that have been encrypted with weak or improperly implemented encryption schemes. Insecure protocols and services connected to the SCADA system hosts can create a high risk access path into the system. This introduces a significant vulnerability because it can provide remote access to SCADA hosts along with access to the functionality allowed by the privilege level of the remote user whose authentication credentials were stolen.

Also, protocols with many optional fields and features are less secure because no two implementations will agree on what is optional, which invites incorrect assumptions. Also, if protocols contain seldom used or never used components, then those components tend to contain more vulnerabilities than well used components because they will be tested to a lesser degree.

Another consideration in analysis of SCADA protocols is the encryption feature. Encryption can pose a risk to network throughput, bandwidth, availability, and Intrusion Detection System (IDS) capabilities, as well as CPU availability on endpoint systems. Encryption can also make system monitoring difficult and can add undesired latency to communications. However, encryption may not fix the vulnerabilities in a SCADA system, but it can be used as a layer of defense. SCADA designers and administrators need to assess the priorities of each communication channel when implementing encryption.

The primary cause of the US power blackout of 2003 was a software bug in the alarm system at a control room of an energy company, considered one of the largest utilities in US [Blackout 2003]. A lack of alarm left operators unaware of the need to redistribute power after overloaded transmission lines hit unpruned foliage, which triggered a race condition in the control software. What would have been a manageable local blackout cascaded into widespread distress on the electric grid.

Therefore, a combination of many issues (malfunctioning of a device and SCADA system, management, operator procedures, etc.) caused this blackout affecting millions of customers, industries, and economy of two countries (US and Canada). Another report of 2008 [Minkel 2008] still reveals issues with power grid and the need for Smart Grid to address the issues. For example, the centralized power plant paradigm is increasingly becoming outdated. A new generation of sensors is providing grid operators with more information about what is happening across the grid at any given moment.

Since the grid of the future needs to be flexible, transparent, and dependable, more attention has to be directed to the analysis and design. The identification of power faults causes and relationships to technologies and networks could help design of communications and security of the SCADA networks carrying real time data about the equipment, loads, and environment conditions.

Therefore, security analysis should be approached from a broader perspective that includes these aspects for survivability too. The analysis of security issues needs to be performed in collaboration with analysis for reliability and safety needs for Smart Grid and particularly for each system.

6.5.6.2 Protocol Analysis

There are issues with the performance and security of various protocols used for Smart Grid and DER systems. Therefore, a detailed analysis of different communication protocols including [IEC 61850] and [IEC 62351] suite of protocols is strongly recommended before making any device selection and implementation for a Smart Grid application. Aspects to be investigated range from interoperability, compatibility of security features, suitability to application, operation and control needs.

It is reported that security controls defined in [IEC 62351] standard could impact communications performance for critical messages defined in [IEC 61850] standard (e.g., GOOSE or SMV) due to processing time and overhead required by the encryption algorithm or authentication function [Yang 2012]. Therefore, improvements to encryption algorithms may be needed because the performance of each service is important for the stability of the power system. The overhead time due to applied security mechanisms may not have an impact on the normal Internet traffic, but in some situations and for certain applications, the security function delay could have a critical impact on successful operation of a substation.

Constraints for performances such as communication delay for each message, service, and signal have to be identified and compared before any security function is added or any device is planned to communicate with another device. Different situations may occur. For example, a sender device may include an authentication security feature, but if the receiving device does not have it implemented or configured, then the receiving device will accept the sender's message as unauthenticated message, so the danger of accepting messages from intruders and a possible venue for attack.

In another situation, it could be an inconsistency of timing requirements for the suite of protocols used in the communication stack. The [IEC 61850] standard may specify certain timing requirements for a signal or service. However, work around may be necessary if any discrepancies are relevant.
Therefore, analysis of the specifications has to be considered before starting any design and be investigated with appropriate modeling and simulation tools to ensure the expected functionality and performances of a system based on IEDs. Related work to these topics is documented in [Juárez 2011], [Hou 2010], [Hohlbaum, 2009].

6.5.6.3 Devices

Since real time applications alone may never be capable of addressing all security requirements, incorporating security features into a device can further enhance system security. Individual devices erect their own security perimeters and defend their own critical resources, such as a network link or storage media [Hentea 2008a]. Therefore, analysis needs to include aspects such as device functions, vulnerabilities, threats, and security controls. In addition, there is a need to identify the security constraints and issues associated with each device, application, system, network, and network interface including the impact of risk when security attributes (confidentiality, integrity, availability, and non-repudiation, authentication) are compromised.

Another need is the analysis of interoperability and security features built in home network devices. The Security Engineer needs to analyze the threats and identify the vulnerabilities, risks, and then decide where and how devices support the security needs in the home network. As attractive as this concept is, achieving plug-and-play is not easy and in many, complex situations it is not practical to specify standard interfaces to this level of detail.

6.5.6.4 Hazard Analysis and Safety Requirements

One way of doing the hazard analysis is failure modes and effects analysis (FMEA), pioneered by NASA, which is a bottom up rather than top down approach. This involves tracing the consequences of a failure of each of the system's or domain's components all the way up to the effect on the mission. Also, safety must be built in as a system is developed. For example, correct and secure information related to emergency lighting, fire protection, temperature and humidity controls, water damage, power equipment and power cabling, and lockout/tagout is a requirement for safety.

6.5.7 Innovative Risk Management Approaches

The need for tailored risk management for SCADA systems is recognized widely. In addition, SCADA and ICS are key components of critical infrastructure. Therefore, innovative approaches are investigated.

6.5.7.1 SCADA Systems

General models for managing risk through its various phases are available for IT systems [Craft 1998], [Glavin 2003]. However, methods for risk management that are based on automated tools [Ozier 2002] and intelligent techniques are more beneficial to SCADA systems because they require minimum or no human intervention in controlling the processes [Hentea 2006]. In assessing the risk for SCADA systems, use of general methods for risk analysis including specific conditions and characteristics of a control system need to be applied.

Another important issue is applying vulnerability management life cycle that offers guidance on design and operational processes and technologies needed to find and remediate security weaknesses before they are exploited. It is imperative to analyze risk as a function of asset value, threat and vulnerability. New concepts to analyze the threats and vulnerabilities have to be applied regularly and uniformly. Methods for SCADA security risk analysis based on combining concepts of vulnerability tree analysis, fault tree analysis, attack tree analysis, and the cause consequence are discussed in [Hentea 2008b], [Patel 2006].

6.5.7.2 Calculate Risk as Impact to Security and Safety

Effective risk analysis for SCADA systems requires a unified definition for mishap and identification of potential harm to safety. As computer systems are more integrated, the distinction between security and safety is beginning to disappear. In bridging the gap between these domains, a unified risk framework is proposed in [Hentea 2008b]. This work suggests a new definition of mishap with an expanded definition of hazard to include the security event. The unified definition for mishap described in [Stoneburner 2006] is enhanced by including the hazard due to security event in a model depicted in Figure 6.6. The figure shows the unified security/safety risk framework. Also, the figure suggests the risk should be calculated as impact of hazard multiplied by likelihood of mishap event.

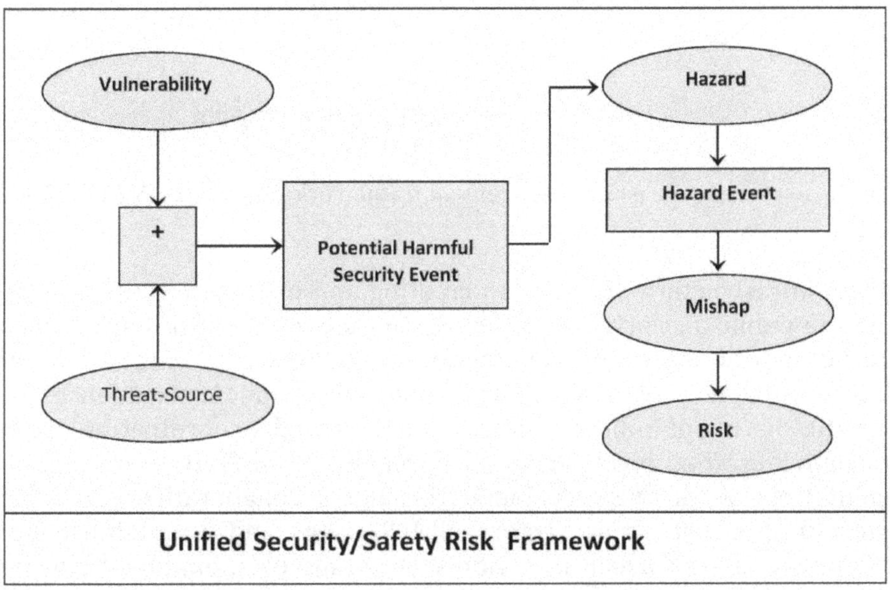

Figure 6.6 SCADA Unified Security/Safety Risk Framework: *Source:* [Hentea 2008b]. © 2008, ISCA.

6.5.7.3 Risk Management for Critical Infrastructure

Risk management is considered as the foundation of critical infrastructure security and resilience of critical infrastructure plan and it is relevant at the National, regional, State, and local levels [DHS 2013a].

Risk management is one of the most important components of a security program. Risk management contributes to the achievement of resilience by identifying opportunities to build resilience into planning and resourcing to achieve risk reduction in advance of a hazard, as well as enabling the mitigation of consequences of any disasters that do occur [DHS 2011].

To manage risk management function, an organization may need a team or just one individual responsible for information risk management that works in a coordinated manner with the management of the security program. The overall goal of the team is to ensure the organization is protected in the most cost effective manner.

Figure 6.7 depicts the components and the steps of the National infrastructure protection plan (NIPP) risk management framework. Specifically, the three elements of critical infrastructure (physical, cyber, and human) are integrated throughout the steps of the framework. Also, the framework depicts the importance of information sharing throughout the entire risk management process. Information is shared through each step of the framework, to include the measure effectiveness step, facilitating feedback and enabling continuous improvement of critical infrastructure security and resilience efforts.

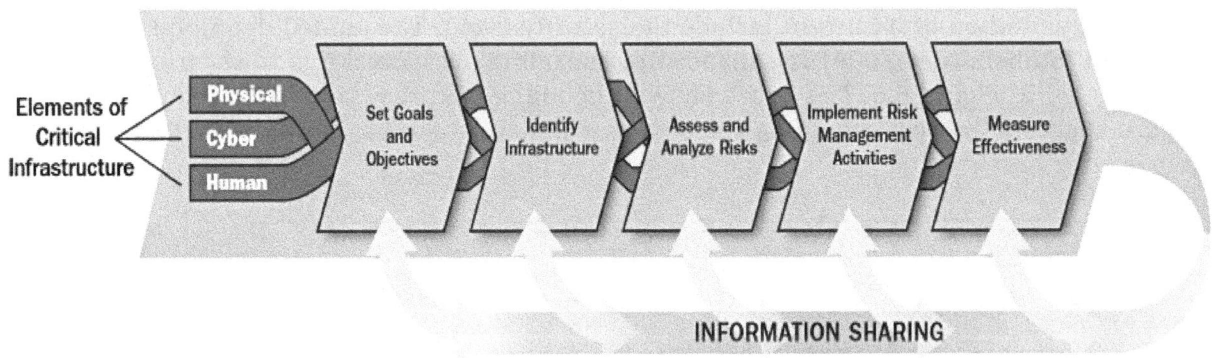

Figure 6.7 NIPP 2013 critical infrastructure risk management framework. *Source:* [DHS 2013b]. Public Domain.

The resilience of critical infrastructure depends on creating and maintaining sustainable, trusted partnerships between the public and private sector. While individual entities are responsible for managing risk to their organization, partnerships between public and private sectors are necessary. They help to improve understanding of threats, vulnerabilities, and consequences and how to manage them through the sharing of indicators and practices and the coordination of policies, response, and recovery activities at all levels, from local, regional, and State to National level. The NIPP risk management framework was developed through a collaborative process involving stakeholders from all critical infrastructure sectors, all U.S. states, and from all levels of government and industry. It provides a clear call to action to leverage partnerships, innovate for risk management, and focus on outcomes.

6.6 Unified Risk Management for Smart Grid and DER Systems

Given the trend of convergence of IT systems and ICS systems, a unified risk management approach is essential to support efficient and effective security controls for both IT and ICS systems within an organization of electrical sector. Unified and interoperable security solutions are essential to assure communication and information security among these systems and assets.

Generally, DER systems employ both IT and ICS assets, which could be dedicated to one system, shared by both systems, or interconnect to exchange information. Therefore, a unified risk management approach helps identify best practices on the basis of meaningful benchmarks and demonstrate various regulatory and business compliances for both systems, although these systems could have different performance and security requirements. This approach needs to consider the interconnected vulnerabilities and common threats to evaluate and aggregate risks in order to provide better risk response. By using a unified risk management approach, an organization can evaluate risks and protection measures across both IT and ICS domains (which often can be highly correlated) for all its assets using a common principle. An attempt to characterize and correlate different elements of the risk in both IT and OT domains is described in [Ray 2010].

By the end of 2015, the energy sector visions major improvements for control systems as described in [DOE 2010]:

- Energy asset owners are able to perform fully automated security state monitoring of their control system networks with real time remediation.
- Next generation control system components and architectures that offer built in, end-to-end security will replace older legacy systems.
- Control system networks will automatically provide contingency and remedial actions in response to attempted intrusions.

Energy asset owners and operators are working collaboratively with government and sector stakeholders to accelerate security advances. A Risk Management Methodology (SRMM) is developed by an association of organizations and projects for Smart Grids [CORDIS 2016]. The methodology is targeted to be used by organizations to assess risk in their Smart Grid developments. It is providing advanced capabilities such as: guidance for its use, risk estimation with notion of dependencies and responsibilities between systems and/or stakeholders in a system of systems environment, threat actor capability and motivation, impact assessment including an assessment of societal impact. The SRMM methodology is distinct from other approaches by taking into account relevant aspects, such as society impact and stakeholder interests in the Smart Grid area. This methodology helps evaluate, further develop and enhance risk management approach specifically suited for the future Smart Grids that can be characterized by multiple administrative domains and high technical complexity.

6.6.1 DER Security Requirements

Key to the success of the analysis and defining the security requirements is the application and efficiency of the security requirements process. To avoid the consequences of design failures, it is crucial to include processes and activities for Security Requirements Engineering to support these DER goals:

- Identifying security requirements for control systems requires broad and detailed analysis. The task of identifying the requirements that define measures of normalcy for cybersecurity, physical security, process efficiency, process stability, and process compliancy has to be performed in the context of a process application. These measures provide the operating requirements that are monitored for state awareness and definition of the state space. Within each of the state awareness task areas, data is retrieved, analyzed, fused and tailored to the user or consumer of the data, whether operator, manager or engineer.
- The requirements engineering for DER systems have to identify the interdependencies between DER systems and other systems. For example, distributed generation, storage and load control would be unable to operate within the limitations of the AMI networks in terms of bandwidth, throughput, and inadequate security protection.
- Cybersecurity and engineering strategies have to be included for designing and installing DER systems to provide DER resilience to anomalous power system events and cyber attacks.

6.6.2 Security Requirements Process

The Security Requirements Engineering Process (SREP) is an asset based and risk driven method for the establishment of security requirements in the development of systems. The process describes how to integrate the generic requirements (described in Common Criteria [ISO/IEC 15408]) into the software lifecycle model together with the resources repository to support reuse of security requirements, assets, threats and countermeasures. Figure 6.8 shows activities of SREP that are integrated with the process for system development. This integrated process could provide a better organization of the multiple activities to achieve the planned results.

The unified process lifecycle is divided into a sequence of phases, called inception, elaboration, construction and transition. Each phase may include many iterations and each iteration may contain all the core workflows (analysis, requirements, design, implementation, and test), but with different emphasis depending on where the iteration is in the lifecycle. The core of SREP is a microprocess, made up of nine activities which are repeatedly performed at each iteration throughout the iterative and incremental development, but also with different emphasis depending on what phase of the lifecycle the iteration is in.

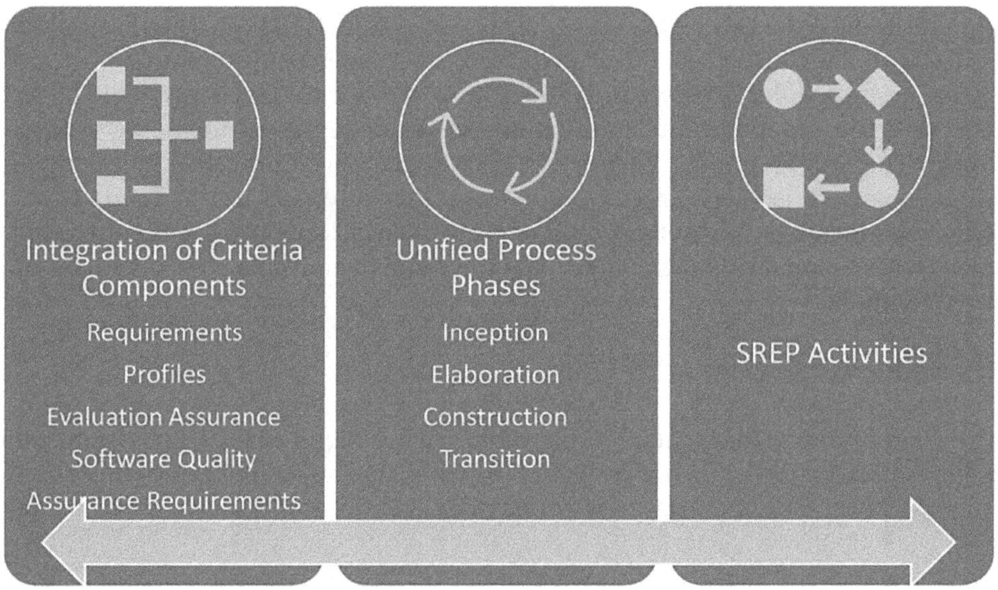

Figure 6.8 SREP main components.

The SREP micro process includes the following activities:

- Agree on definitions
- Identify vulnerable and/or critical asset
- Identify security objectives and dependencies
- Identify threats and develop artifacts
- Risk assessment
- Elicit security requirements
- Categorize and prioritize requirements
- Requirement inspection
- Repository improvement.

The SREP model is iterative and incremental, and the security requirements and their associated security elements (threats, security objectives, etc.) evolve along the life cycle. SREP provides a way for identifying what need to be protected against what. SREP also integrates security requirements defined in CC [Common Criteria] and SSE-CMM (Systems Security Engineering Capability Maturity Model) in a software development cycle using Unified Process [Jacobson 1999], UMLSec, misuse cases, security use cases, and treat/attack trees. A number of documents are created and updated in every iteration respectively.

Based on a comparison of SREP process with Security Engineering processes defined by approaches such as Security Quality Requirements Engineering (SQUARE), Agile, modeling, reuse, and others, SREP serves the purpose of defining security requirements from the beginning of the software development [Deshmukh 2009]. Obviously, the dynamics of security threats and vulnerabilities cannot be dealt with a process based on waterfall model. The vulnerabilities and threats are changing so analysis and changes to requirements have to be performed too. Also, an iterative process for systems development aligns better with the continuous process for information security management based on standards such as [IEC 62443] and [IEC TS 62351-1]. As indicated in [NISTIR 7628r1], standards have different degrees for expressing requirements, and the security requirements must match these degrees of mandatory, recommended, or avoided.

6.7 Issues on Developing Security Requirements for Smart Grid

Each organization needs to use engineering techniques to analyze business objectives and determine security requirements for each system [Anderson 2008a] and perform risk assessment and risk impact analysis prior to taking countermeasures [ICS-CERT]. For example the successful implementation of automation and optimization of Smart Grid distribution systems (e.,g., DER systems) requires to implement security requirements and maintain a security process. A continuous process is summarized as being to assess the vulnerabilities, identify and address changes in regulations and standards, establish metrics to assess performance and vulnerabilities [Abi-Samra 2012]. A set of minimum security requirements for an organization are provided in Appendix B of [NIST SP800-55r1] document.

Information security requirements are a subset of the requirements levied on information systems and are incorporated into the system development life cycle simultaneously with the other requirements. For example, DER system security requirements define the protection capabilities provided by the system, the performance and behavioral characteristics exhibited by the system, and the evidence used to substantiate claims that the system security requirements have been satisfied. The security requirement and threat analysis are mostly regarded as a preliminary document to risk analysis. Risk analysis generally decides whether to introduce security countermeasures into the DER system. The system security requirements and security concept of operations provide the basis for the architectural design, implementation, verification, and validation of the DER system.

6.7.1 Approaches

The system security requirements are developed by performing a requirements analysis of the stakeholder security needs that includes engineering trade space analyses of alternative means to provide the required protections and to implement risk treatment decisions. They are concurrently developed with the system requirements and in collaboration with systems engineers, security engineers, stakeholders, and management. Therefore, security requirements are defined from the perspective of different communities of interest [NIST SP800-53r5]. The trend should be to escape from best practices checklists and return to core Systems Engineering (SE) methods, processes, and tools.

Although more standards and guidelines are published by different organizations, there is a need for understanding how to perform the analysis to define security requirements, how to assess the security risks due design, implementation, platforms, and standards. Security requirements can be defined at a high level of abstraction (e.g., policies, standards, mission/business needs) or at lower detail such as security requirements for contracting purposes, operations and management.

Systems engineers, security engineers, system developers, and systems integrators develop the security design requirements for the information system, develop the system security architecture and the architecture specific derived security requirements, and subsequently implement specific security functions at the hardware, software, and firmware component level. It is important to define the context for each use of the term security requirement so the respective communities (including individuals responsible for policy, architecture, acquisition, engineering, and mission/business protection) can clearly communicate their intent. When defining system security requirements, Security Engineer needs to identify the security functional boundary, security domains, trust domains, and context of the system in terms of the security behavior and properties to be provided [NIST SP800-160v1].

The security functional boundary identifies the set of system elements that compose the system to be delivered and it may be physical and/or virtual. Security domains and trust domains reflect a security focused partitioning of the system that group trust relationships between security enforcing and security supporting functions.

The system context consists of the environment in which the system operates including constraints established by the stakeholders; interconnections between the system and other system elements; services provided by the system or provided to the system by the other system elements; data and information flows and control flows between the system and other system elements; and assignment of responsibility for implementation of system security requirements individually to the system or to other system elements. The system context is defined before refining the security concept of operations and developing the system security requirements.

Besides stakeholder business needs for protection, the requirements analysis for Smart Grid systems may use regulations, directives, policies, guidance, and standard documents; industry best practices and other proven means and methods; and requirements from similar systems. Often, a system needs to demonstrate standards compliance (e.g., NERC standards) or standards can be used for establishing interoperability across the functions and services that compose the system and among interconnected systems [NISTIR 7628r1]. In order to manage complexity, there is a need to impose a security policy as an abstraction or even a rule of thumb that enables a Security Engineer to focus on exceptions [Anderson 2008a].

Security requirements can be formulated on different abstraction levels. For example, at the highest abstraction level they may reflect security objectives (e.g., availability, integrity, confidentiality, etc.). In general, information security requirements describe functional and non functional requirements that need to be satisfied in order to achieve the security attributes of a system.

The need for documentation and developing quality requirements is stressed in several published materials. Like other requirements, all security requirements should satisfy the SMART attributes, meaning that should follow a few basic properties as described in [OWASP 2014], [Mannion 1995]:

- Specific - detailed as necessary so that there are no ambiguities in the requirement; it requires consistent terminology among requirements.
- Measurable - to determine whether the requirement has been met, through analysis, testing, or both.
- Appropriate – requirement should be validated, thereby ensuring that it not only derives from a real need or demand.
- Reasonable - some validation is needed to determine whether meeting the requirement is physically possible, and possible given other likely project constraints.
- Traceable – requirement can be isolated to make it easy to track/validate throughout the development lifecycle.

Security shall be an important part of the system life cycle process. In addition, all information technology services and systems must address the security implications of any changes made to a particular service or system [DHHS 2005]. As stressed by DHS policy, documentation should provide complete information on the system's security aspects and shall include:

- Application design and development
- Security controls
- Test plans and results
- Operations and maintenance procedures.

A sufficiently trustworthy system results from adherence to the security concepts, principles, and practices in the application of security specific methods, techniques, and analyses, some of which stand alone and others that supplement existing SE techniques, methods, and analyses.

6.7.2 Security Requirements Classification

No unique classification criteria have been defined. Therefore, various categories of security requirements are used by different stakeholders. The [ISO/IEC 15408], just called the Common Criteria (CC), categorizes the security requirements in two groups of requirements:

- Security Functional Requirements (SFR) - a translation of the security objectives for the target of evaluation (TOE) into a standardized language.
- Security assurance requirements (SARs) - a description of how assurance is to be gained that the TOE meets the SFRs.

The SFRs are usually at a more detailed level of abstraction, but they have to be a complete translation (the security objectives must be completely addressed) and be independent of any specific technical solution (implementation). Also, [ISO/IEC 15408] standard requires this translation into a standardized language for several reasons:

- To provide an exact description of what is to be evaluated; as security objectives for the TOE are usually formulated in natural language, translation into a standardized language enforces a more exact description of the functionality of the TOE.
- To allow comparison between two security terms. As different security professionals may use different terminology in describing their security objectives, the standardized language enforces using the same terminology and concepts. This allows easy comparison.

Use of Common Criteria Protection Profiles (PP) and Security Targets (ST) greatly helps the development of products (and to some extent systems) that have IT security functions.

The protection profile construct allows creation of generalized reusable sets of the security requirements. The PP can be used by prospective consumers for specification and identification of products with IT security features which will meet their needs. The security target expresses the security requirements and specifies the security functions for a particular product or system to be evaluated, called the target of evaluation (TOE). The ST is used by evaluators as the basis for evaluations conducted in accordance with CC and [ISO/IEC 15408] standard. The CC was standardized as [ISO/IEC 15408] in 1999, last revision is from 2005 [ISO/IEC 15408]. As of September 2014, several countries around the globe agree on reform to improve cybersecurity certification through international public-private collaboration [CC News]. One objective of the agreement [CC Arrangement] is to ensure that Evaluations of Information Technology (IT) Products and Protection Profiles are performed to high and consistent standards; these are seen to contribute significantly to confidence in the security of those products and profiles.

Security functional requirements may:

- Determine a security service (confidentiality service, integrity service, availability service, authentication service, etc.)
- Determine logical & physical data architecture
- Define user roles
- Determine access needs
- Map users to data access
- Define criteria to data access
- Identify data owner
- Identify security policies
- Define security requirements for a particular business solution (e.g., Portal, control, marketing, remote access, etc.).

The Common Criteria includes a set of possible assurance requirements that can be selected. Assurance requirements may include:

- Determine level of concern (data protection level)
- Determine probability of data insecurity
- Determine business impact from data insecurity.

The Open Security Architecture (OSA) organization suggests specific requirements that are more useful to a system designer or software architect. These requirements are classified [OSA IT]:

- Secure functional requirements - define security related description that says what shall not happen and it is integrated into each functional requirement
- Functional security requirements - define security services that have to be achieved by the system under inspection. Examples include authentication, authorization, accountability, data access, etc.
- Non-functional security requirements – define related architectural requirements, like resilience, dependability, or minimal performance and scalability
- Secure development requirements – describe required activities during system development which assure that the outcome is not subject to vulnerabilities. Examples include data classification, coding guidelines, test methodology.

Security functional requirements, security assurance requirements, and security policy combine as the means that establish the basis for trustworthiness which is different than assurance. Figure 6.9 depicts a taxonomy of security requirements. Security requirements can be derived from applicable laws, Executive Orders, directives, policies, standards, instructions, regulations, procedures, user needs, or organizational mission/business needs to ensure the confidentiality, integrity, availability, and non-repudiation of the information.

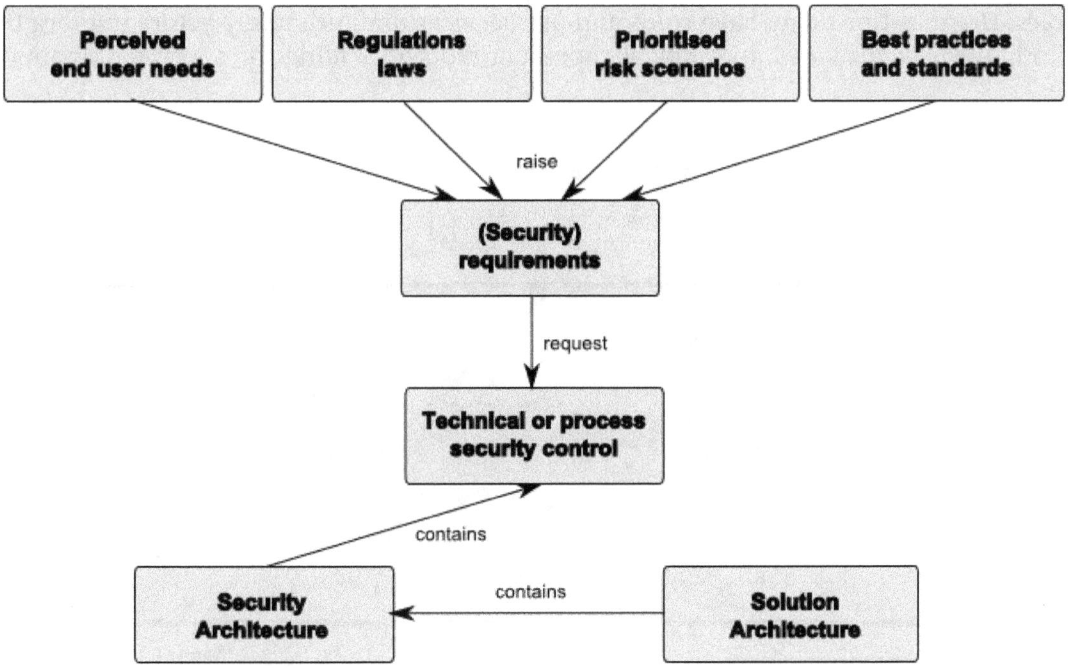

Figure 6.9 IT Security Requirements Taxonomy. *Source:* [OSA Taxonomy]. Licensed under CC BY 3.0.

6.7.3 Information Path

To fully realize the Smart Grid goals of vastly improving the control and management of power generation, distribution, and consumption, the current state of information network interconnectivity must be improved so that information can flow securely between the various actors in the Smart Grid.

The domain diagram (or the conceptual reference model) of the Smart Grid is not limited to a single domain, single application, or single use case. The conceptual model assumes that Smart Grid includes a wide variety of use cases and applications, especially (but not limited to) functional priorities and cross cutting requirements identified by the Federal Energy Regulatory Commission (FERC). The conceptual reference model of Smart Grid [NIST SP1108r4] shows domains, interactions and overall scope such as power flow and information communication among domains (generation, transmission, distribution, markets, customer, operations, service providers) (see more information in Appendix B).

The levels of conceptual model, as shown in Figure 6.10, down to a specific domain provide information about actors and roles that need to be identified. A domain is a high level grouping of physical organizations, buildings, individuals, systems, devices, or other actors that have similar objectives and that rely on—or participate in—similar types of services. Communications among roles and services in the same domain may have similar characteristics and requirements. Domains contain subdomains. Moreover, domains have many overlapping functionalities, as in the case of the Transmission and Distribution domains. Because transmission and distribution often share networks, they are represented as overlapping domains.

Actors exist primarily in the physical and implementation architectures. They include devices, computer systems, software applications, or the individuals within an organization that participates in the Smart Grid. Actors perform roles to make decisions and to exchange information with other roles. Organizations may have roles and use services and ultimately actors in more than one domain. Figure 6.11 provides a graphical representation of drilling for specific domains, actors, and roles.

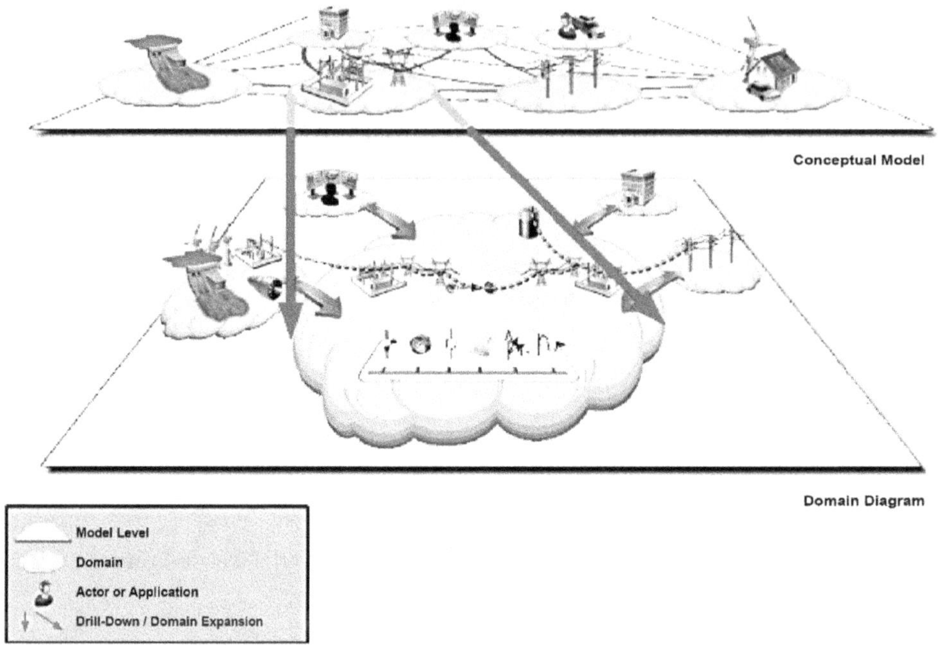

Figure 6.10 Levels of NIST conceptual model. *Source:* [IETF SGIP]. Public Domain.

To design security protection, the information path has to be understood since information crosses domains. There are needs for security (e.g., availability, integrity, nonrepudiation, privacy, etc.) characteristics including performance, and assurance that all require identification of the security requirements. Although generic cybersecurity requirements are described in [NISTIR 7628r1], each organization needs to adapt the recommended requirements to its specific context and business.

More practical recommendations on how to develop security functional requirements are provided in [OWASP 2014]. Systematic security requirements analysis for the Smart Grid can add significant value to a system, especially when it is incorporated in the early stages of system development.

By applying the SQUARE method as a representative system security requirements analysis method from the system of systems perspective, security requirements are identified for the customer domain including specification of secure Smart Grid intersystem interactions [Zafar 2013].

6.7.4 Use Cases - Examples

Use cases are helpful for scoping out interoperability requirements at lower levels of architectural focus for specific areas of functionality - such as on premises energy management or predictive maintenance for grid equipment. Use cases are viewed from a variety of stakeholder perspectives and application domains.
By combining roles and interactions from multiple use cases using existing roles permits the Smart Grid to be rendered as a collection of transactional relationships that allow possible reuse of legacy

applications, within and across domains. One goal is to provide stakeholders with a normalized use case repository of architecturally significant use cases and assist in the development of requirements. Detailed use cases can be found at [NIST IKB]. Use cases can be used on designing security requirements for interfaces, networks, and services.

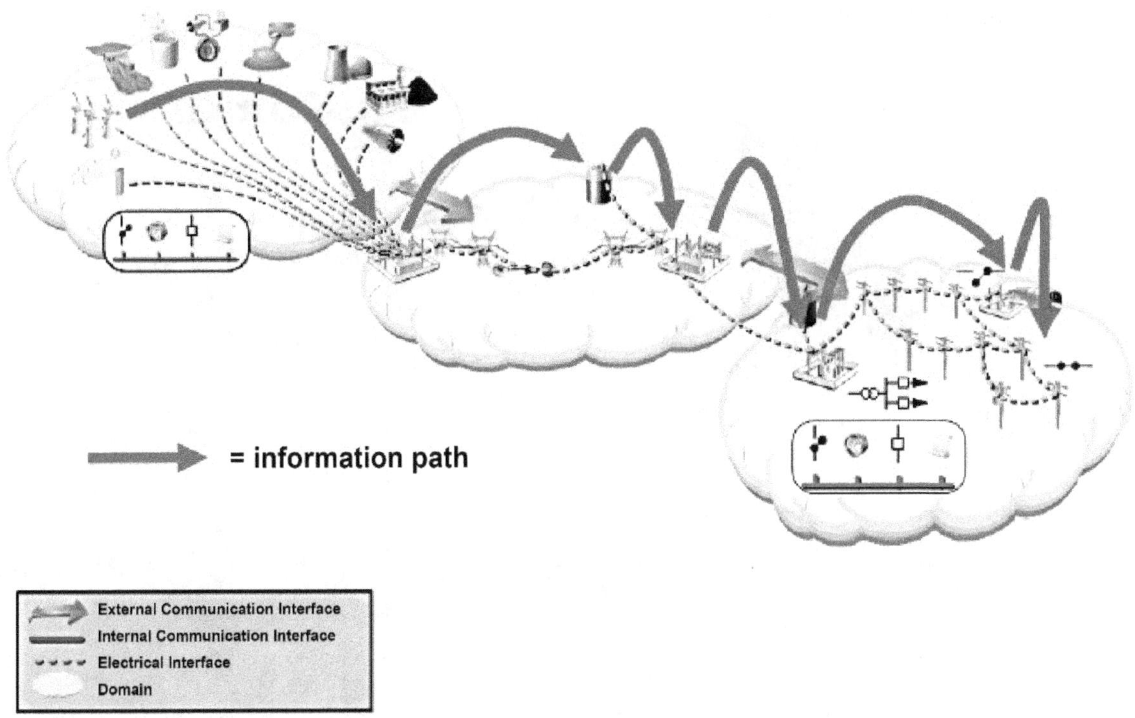

Figure 6.11 Information path through the model. *Source:* [IETF SGIP]. Public Domain.

Based on the parlance of the Smart Grid concepts, a new type of networks, called active distribution networks, are moving from the conventional and passive operation approach (without any monitoring and control in lower distribution) for the integration of DER. These networks are capable of handling bidirectional power and information flows based on the latest automation, information and communication technologies, as well as on corresponding metering services. The following is in overview of the pictorial interactions of each Smart Grid domain with other domains.

6.7.4.1 Customer Domain

Defining requirements for customer domain should consider several objectives that include applications, devices, communication networks, interfaces, regulations, cybersecurity and safety for consumers.

Figure 6.12 provides a graphical representation of key actors, entities, applications associated with customer domain including interactions with other domains. A customer is the end user of the electrical process. End users may consume and produce electricity for premise use or resale. As an electricity producer, they are also members of the DER domain when the electricity produced is sold. Customers include residential, commercial, and industrial facilities. DER systems at the customer site interact with service provider, market, operations, and distribution domains. Therefore, use cases have to be developed for each application and information path.

Defining specific DER requirements should be based on the analysis of DER communications network including:

- Layered networks architecture
- Typical types of traffic that it may carry and associated quality of service requirements
- Technologies both wired and wireless medium including protocols
- Networking issues (topologies, services, operations, etc.)
- Quality of security services.

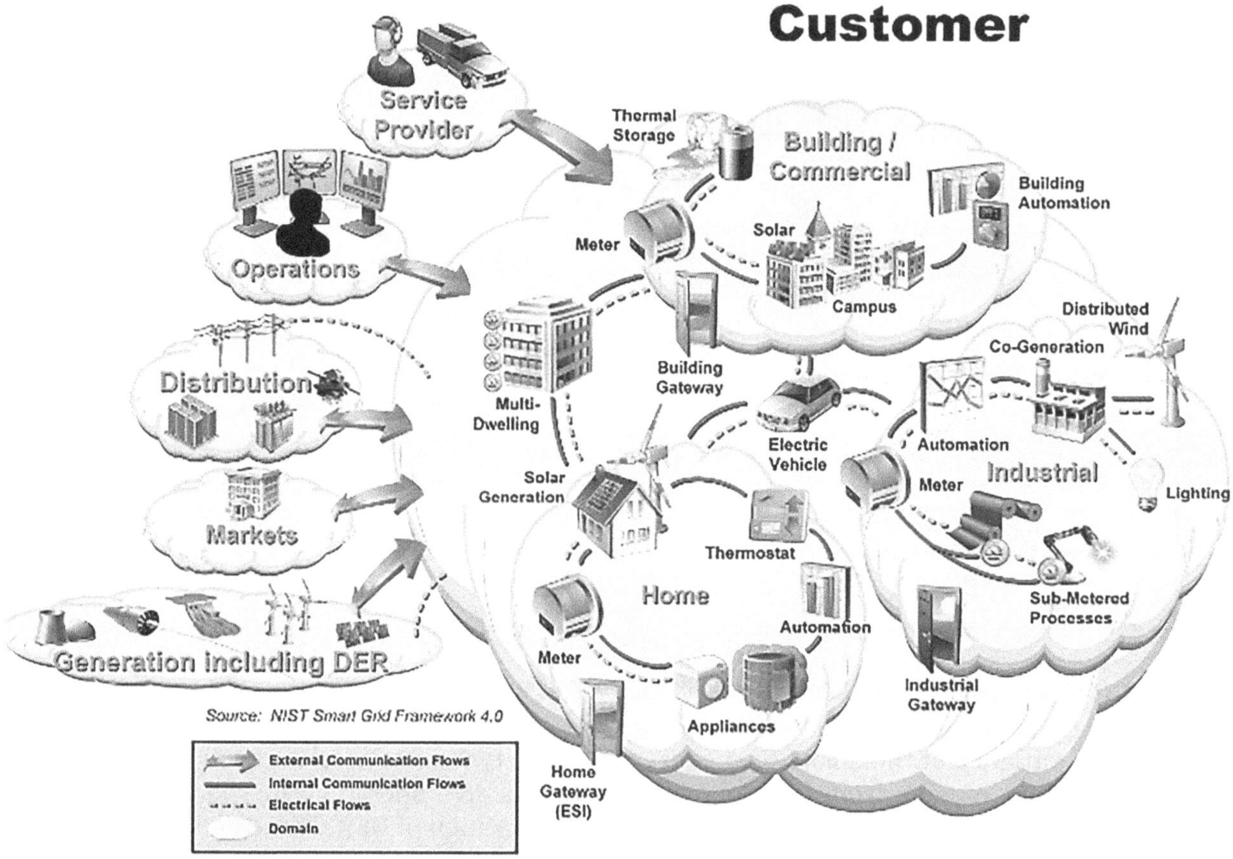

Figure 6.12 Customer domain – actors and interactions. *Source:* [NIST SP1108r4]. Public Domain.

6.7.4.2 Operations

Figure 6.13 provides an overview of operations domain and interactions with other domains. DER systems need specific applications in the operations domains for monitoring and control of DER resources. All these needs require security protection and obviously analysis of needs and requirements analysis is mandatory.

6.7.4.3 Markets

Figure 6.14 provides an overview of markets domain where DERs are part of DER aggregation application. Aggregation is the process of organizing small groups of industrial, commercial or residential customers into a larger, more effective bargaining unit with a view to strengthening their participation in electricity trading.

Other characteristics include:

- Aggregation can link demand (demand response, DR), and/or generation and, if applicable, storage units.
- An aggregator is a legal organization that consolidates or aggregates a number of individual customers and/or small generators into a coherent group of business players aiming at optimizing energy supply and consumption both technically and economically.
- Process optimization improves the coherent operation of DER to support system balancing; an aggregator can design and offer energy services downstream to energy customers and upstream to several key players (system operators, electricity traders, etc.).
- Aggregators can be considered as power plants that use distributed resources together with other generation solutions to supply their aggregated customers (a group of consumers together to buy electricity).
- Aggregation process may comprise a portfolio of flexible energy resources.

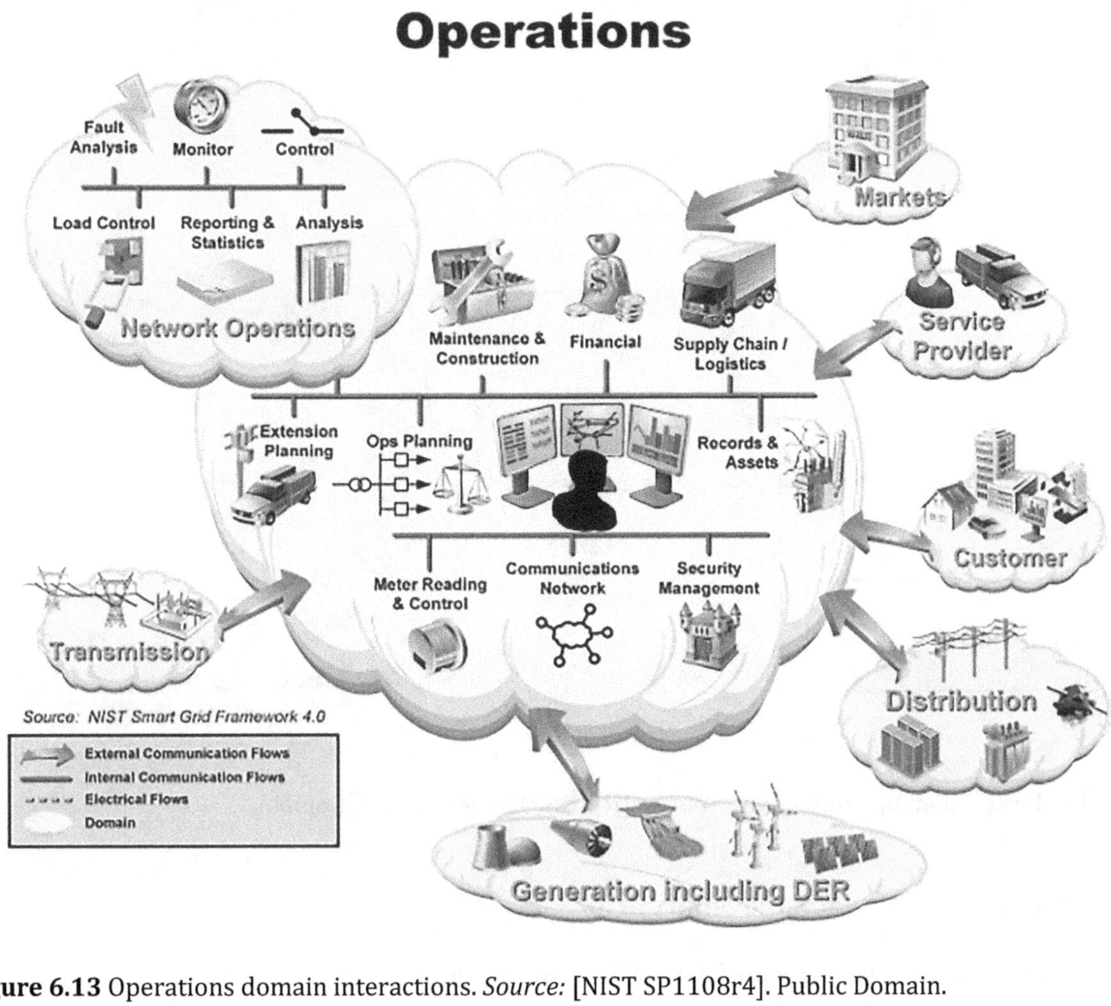

Figure 6.13 Operations domain interactions. *Source:* [NIST SP1108r4]. Public Domain.

Flexible DER is defined as a DER with these key characteristics:

- Controllability through an existing communication and control infrastructure - DER is able to react to remote signals, such as schedules and set points sent through communication channels.
- Adaptability - DER output is variable and DER is able to adapt (up/down) its generation/load profile according to incentives, instructions, or price signals.

One remark is that renewable resources generation (even if connected to the required information and communication infrastructure) may only be controllable (reduce its output) if not operated in a curtailed mode. Thus, this unit is not fully adaptable according to the above definition. Examples of renewable energy resources include any of the following: wind, solar, geothermal heat, wave, tidal, hydro (falling water), ocean energy, biomass, landfill gas, sewage treatment plant gas and biogases.

Therefore, the aggregation application requires security and privacy protection. For example, this issue implies access control - registration and authentication of different key players including consumers, monitoring and control of DERs, operations, etc.

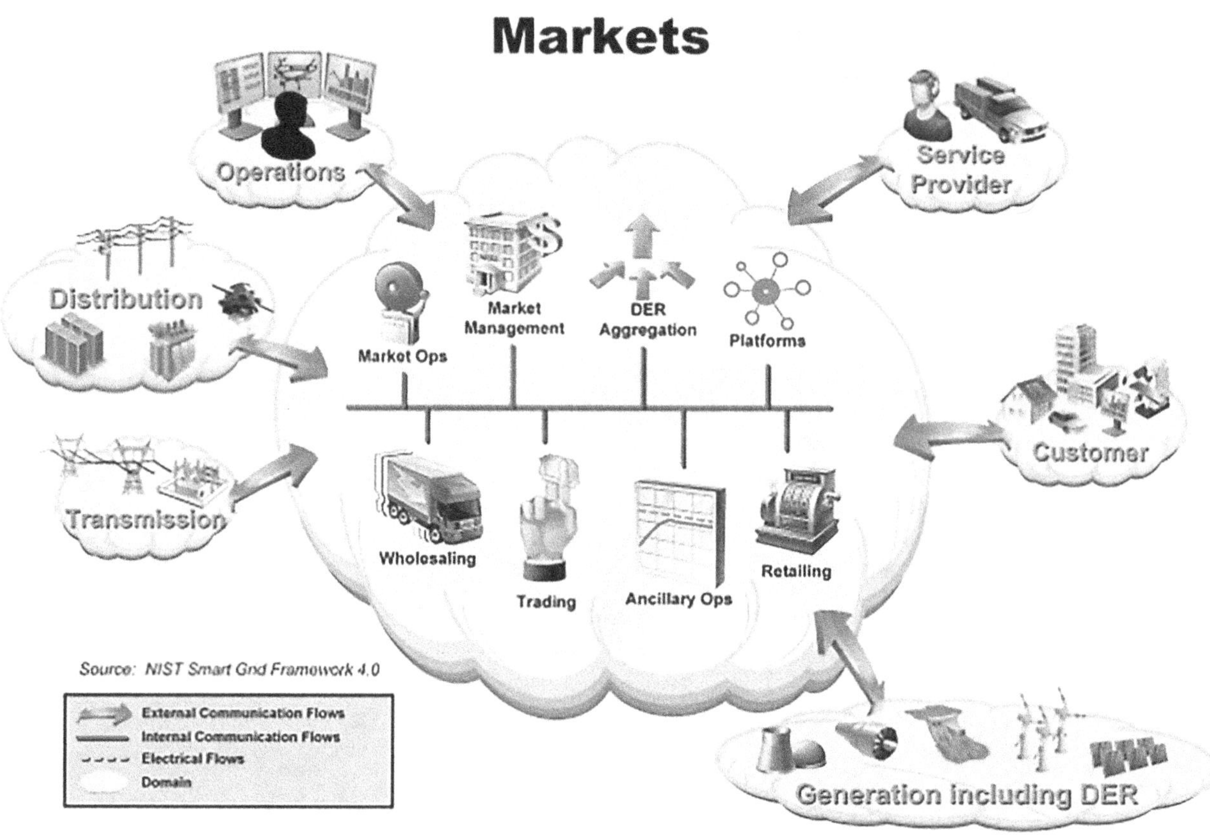

Figure 6.14 Markets domain interactions. *Source:* [NIST SP1108r4]. Public Domain.

6.7.5 A Perspective on Privacy and Smart Grid

Generally, privacy is discussed as consisting of four dimensions:

- Privacy of personal information relating to an individual, who can be identified, directly or indirectly, by that information and in particular by reference to an identification number or to one or more factors specific to his or her physical, physiological, mental, economic, cultural, locational or social identity. Privacy of personal information involves the right to control when, where, how, to whom, and to what extent an individual shares their own personal information, as well as the right to access personal information given to others, to correct it, and to ensure it is safeguarded and disposed of appropriately.

- Privacy of the person is the right to control the integrity of one's own body; it covers such things as physical requirements, health problems, and required medical devices.
- Privacy of personal behavior is the right of individuals to keep any knowledge of their activities, and their choices, from being shared with others.
- Privacy of personal communications is the right to communicate without undue surveillance, monitoring, or censorship.

As described in Volume 2 (Privacy and the Smart Grid) of [NISTIR 7628r1], most Smart Grid entities directly address the first dimension, because most data protection laws and regulations cover privacy of personal information. Also, these guidelines recommend that the other dimensions should also be considered in the Smart Grid context because new types of energy use data may be created and communicated.

For instance, unique electric signatures for consumer electronics and appliances could be compared against some common appliance usage profiles to develop detailed, time stamped activity reports within personal dwellings. Also, charging station information might reveal the detailed whereabouts of an electric vehicle/plug-in electric vehicle. Other privacy issues associated with advanced metering infrastructure are related to the behavioral inferences that can be drawn from the energy usage data collected by the meter at more granular frequencies and collected intervals. Additionally, smart meter data also raises potential surveillance issues relating to the methods by which the data is collected and transmitted (electronic collection transmittal rather than manual meter reading and compilation).

The privacy concerns increase due to smart appliances and devices on the consumer side of the meter if data is transmitted outside of the HAN or energy management system (EMS) and do not have documented security requirements (e.g., a smart appliance being able to send data back to the manufacturer via telematics), thereby effectively extending the reach of the system beyond the walls of the premises. Also, third party entities may also seek to collect, access, and use energy usage data directly from customers, rather than from the utility (e.g., vendors creating energy efficiency or demand response applications and services specifically for smart appliances, smart meters, and other building based solutions). The ability of the customer to understand these risks may require customers to be better educated and informed on the privacy consequences of decisions regarding these third party services.

While the NIST guidelines [NISTIR 7628r1] focus on the customer privacy risks, it is equally important to focus on privacy of employees and personnel carrying activities in the power grid engineering, business, and operation. Table 6.1 shows potential data elements within the Smart Grid that could impact privacy if not properly safeguarded.

6.7.6 Cyber-Physical Systems Security Requirements

Key to Security Engineering and analysis task is to consider that Smart Grid is dependent on several cyber-physical systems (CPS) technology trends and business needs. Therefore, these emerging trends require different solutions. CPS need to support security for both computational as well as physical components. Protection features may be provided by both CPS manufacturers and service providers. Consequently, given the basic requirement for power system resiliency, it is important to identify the cybersecurity requirements that utilities should consider before permitting these potentially vulnerable DER systems to be interconnected with other systems.

Traditional analysis tools are unable to cope with the full complexity of CPS or adequately predict system behavior. For example, the current electric power grid may experience blackouts over large regions that are triggered by minor events that may escalate with surprising speed into widespread power failures. In addition, traditional real time performance guarantees are insufficient for CPS when systems are large, and spatially and temporally distributed in arrangements that may rapidly change. With the greater autonomy and cooperation possible with CPS, greater assurances of safety, security, scalability, and reliability are necessary.

Table 6.1 Information potentially available through Smart Grid.

Data Element	Description
Name	Party responsible for the account
Address	Location where service is being provided
Account Number	Unique identifier for the account
Meter reading	kWh energy consumption recorded between 15 to 60 minute intervals and once daily intervals during the current billing cycle
Financial Information	Current or past meter reads, bills, and balances available, including history of late payments/failure to pay, if any
Lifestyle	When the home is occupied and unoccupied, when occupants are awake and asleep, how much various appliances are used
Distributed Energy Resources	The presence of on-site generation and/or storage devices, operational status, net supply to or consumption from the grid, usage patterns
Meter Unique Identifier	The Internet Protocol (IP) address, media access control (MAC) address, or other network identifiers for the meter, if applicable

Source: [NISTIR 7628r1]. Public Domain.

Managing risk associated with CPS cybersecurity, therefore, requires consideration of these properties along with classic IT security concerns. The properties of safety, reliability, privacy, cybersecurity, and resilience have, for the most part, evolved within distinct silos (see Figure 6.15). Historically, systems design has occurred within disparate disciplines.

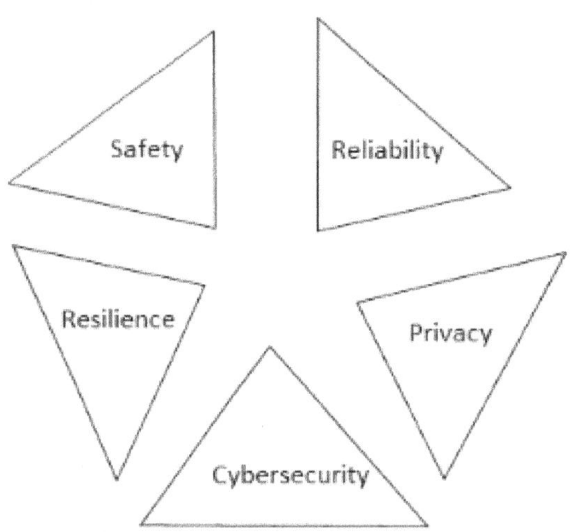

Figure 6.15 Evolution of systems design property silos. *Source:* [NIST SP1500-202]. Public Domain.

Also, risk management activities have been conducted within each silo, rather than across them. Industry trends suggest that these SE disciplines are converging toward increased interdependency as illustrated in Figure 6.16.

This document [NIST SP1500-202] highlights that the future of CPS design, integration, and risk management is evolving toward a multi-disciplinary approach where systems designers and integrators will increasingly be required to work across properties, with cybersecurity becoming a common requirement for all. Also, systems engineers responsible for each property need to consider the interdependencies among all five properties throughout the system lifecycle.

Figure 6.16 Interdisciplinary approach to CPS Engineering. *Source:* [NIST SP1500-202]. Public Domain.

A key aspect of CPS design will be the application of controls that balance the achievement of positive outcomes among the five properties. In addition, developing effective CPS cyber protection and mitigation actions requires understanding of functions and interactions of all three types of components of CPS: cyber, analog, and physical (see CPS representation in Figure 3 of [NIST SP1500-202]). Further, CPS designers and integrators should consider both the intended and unintended effects resulting from the combination of properties where the goals of each may contradict or be complimentary to their counterparts. This may result in challenges such as trade-off decisions that have to be considered in support of the SoS objective; implementation of cybersecurity measures can create risks to privacy (e.g., log retention, system monitoring). Figure 6.17 shows the CPS risk properties to include safety, resilience, reliability, privacy, and cybersecurity. Because Smart Grid is a large cyber-physical system, governance and risk management processes should address all risks, not just cybersecurity.

There are concerns related to the ability of the Smart Grid to prevent entities (people, machines) from gaining access to data stored in, created by, or transiting a CPS or its components such that individuals or groups cannot seclude themselves or information about themselves from others. An additional privacy concern is the confidentiality of customer data, including data of commercial and industrial customers. For example, the availability of high frequency energy usage data collected for the purpose of energy monitoring may facilitate the unintentional release of private, confidential information of residential users. By collecting pieces of information from various sources and then using algorithms or machine learning to analyze this information, the adversaries may combine privacy protected data from many sources to create unsafe results that reveal confidential information about individuals, commercial, and industrial consumers and prosumers. Therefore, methods and algorithms should be developed for removing personally identifiable information and sensitive personal information from monitored electric usage data.

It is recommended a SoS design or integration approach for CPS based on risk model analysis that considers both the impact to each system objective individually and the SoS objective as a whole [NIST SP1500-202]. For example, a SoS with the highest priority goal being to deliver resilience should have a risk model that favors resilience. Risk models could help on placing emphasis on the most appropriate type of component—physical, analog, or cyber. System risk analysis may provide context when considering how best to apply desired CPS risk related properties. Despite of differences on priorities, it is recommended that CPS owners and operators should use a similar process when evaluating risk in operational situations. The challenge is analysis and understating of the strengths and weaknesses of the system in place, the role of each architectural layer, and the interactions among the layers.

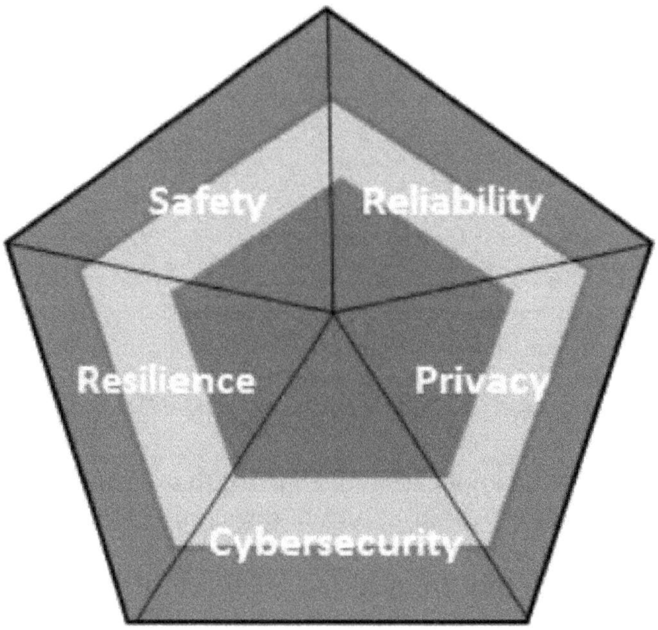

Figure 6.17 CPS risk properties. *Source:* [NIST SP1500-202]. Public Domain.

Design of applications is more challenging because of multi domain interactions and constraints that need to be satisfied at several layers. In addition, a change on the security policy may lead to redesign of the interactions among different entities. This may have impacts on resources use and the QoS of the delivered service. Since CPS add more complexity to Smart Grid, new tools and design methodologies including standardization and research are needed in the areas of integration of control for networked embedded systems, large scale monitoring, simulation, high performance analysis, and cross layer modeling.

6.8 Standards, Guidelines, and Recommendations

Smart Grid standards harmonization efforts are under way with (but are not limited to) the following groups:

- The Institute of Electrical and Electronic Engineers (IEEE) P2030 has developed a view of the Smart Grid organized into three major areas: physical, communications, and information.

- The European Telecommunications Standards Institute (ETSI), together with the European Committee for Standardization (Comité Européen Normalisation - CEN) and the European Committee for Electrotechnical Standardization (CENELEC), have started the development of a Smart Grid architecture.
- The Chinese Electrical Power Research Institute (CEPRI) adapting the architecture with some very specific changes that support the difference in the Chinese market.
- The Korean Smart Grid Association (KSGA) has not published an architecture document yet, but pieces of the architecture have been released, including IT, physical field devices, and interfaces.
- The Japanese Federal Government architecture has been focused, to a large extent, on the customer domain with strong links to the other six NIST Conceptual Reference Domains.
- IEC TC 57 and TC8 working groups have architecture development artifacts under development and have published initial versions for standards integration across the IEC.
- NIST published cybersecurity guidelines in three volumes that can be used to expedite the development of a system [NISTIR 7628r1]. However, each organization needs to assess the security and privacy risks and manage these risks based on its business goals and mission. However, the documents include security issues (threats, vulnerabilities, technologies) that may not be current. Threats, vulnerabilities, and technologies are emerging. In addition, many security requirements are generic, risk impacts are not quantified using a methodology, and many security controls are mostly based on reactive strategies. Also, the Security Engineer needs to consider evolving regulations, standards, and applications.
- DOE Guidance is focused on key technologies, processes, and cybersecurity. Therefore, DOE in collaboration with the National Institute of Standards and Technology (NIST), the North American Electric Reliability Corporation (NERC), members of industry, and utility specific trade organizations developed guidance on implementing risk management process described in [DOE Process] for the electricity subsector, to include organizations responsible for the generation, transmission, distribution, and marketing of electric power, as well as supporting organizations such as vendors.

Chapter 7

Developing Requirements Using Frameworks and Guidelines

7.1 Cybersecurity Framework

A cybersecurity framework to support the cybersecurity needs of power asset owners and operators including strategies for building security awareness and increasing the resilience of energy delivery systems was first described in [NIST 7628] of 2010, then updated in document [NIST 7628r1] of 2014. These needs can be satisfied by requirements to implement means such as:

- Sharing information and practices within energy sector, vendors, people, stakeholders, public, and other industries as described in [ICS-CERT 2012a], [ICS-CERT 2012b] and recommended in [WH 2011].
- Assessing and monitoring risk continuously which allow an organization to understand the current security posture, vulnerabilities and threats of all energy delivery system architecture levels and across cyber-physical domains.
- Develop and implement new protective measures to reduce system risks.
- Use of countermeasures based on effectiveness and efficiency (costs) of a system architecture that provides proactive detection of cyber intrusion attempts, defense in depth and employs components that are interoperable, extensible, and able to continue operating in a degraded condition during a failure caused by a cyber incident.
- Manage cybersecurity incidents by use of remediation and recovery to minimize the impact. Each system should have an incident reporting and recovery plan that is developed in correlation with the organization's plan for incident management.

Based on the cybersecurity framework [NIST 2018], NIST created a cybersecurity risk profile, called Smart Grid Profile, described in [NIST 2051TN]. The profile examines the cybersecurity functions and activities of the framework through the lens of power system owner/operators, and cybersecurity outcomes are evaluated for relevance against the following four high level business objectives: maintain safety, maintain power system reliability, maintain power system resilience, and support grid modernization. The profile also provides a list of considerations relevant to the challenges power system owners/operators may experience as they implement these cybersecurity activities in infrastructures with high concentrations of distributed energy resources (DERs). In this Smart Grid Profile, power system owners/operators are primarily distribution grid owners rather than customers and the assets that they own behind the meter. The power system owners/operators may use the Smart Grid Profile to:

- Prioritize organizational cybersecurity activities to align with available resources.
- Enable better informed decision making about cybersecurity activities by referencing the considerations listed for each cybersecurity activity.
- Convey cybersecurity requirements to an external entity such as a service provider.
- Verify their organizational cybersecurity posture against the prioritized cybersecurity activities.

However, the profile is focused on cybersecurity needs of Smart Grid owners/operators and therefore is not sufficiently useful by itself to all the diverse parties in the Smart Grid.

7.1.1 NIST Security Requirements for Smart Grid

In this document, Volume 1 (Smart Grid Cybersecurity Strategy, Architecture, and High Level Requirements) [NISTIR 7628r1], the cybersecurity requirements are grouped into three categories:

- Governance, risk, and compliance (GRC) - applicable to all Smart Grid information systems within an organization and are typically implemented at the organization level and augmented, as required, for specific Smart Grid information systems.
- Common technical - applicable to all Smart Grid information systems within an organization.
- Unique technical – applicable to a specific asset or situation.

The document identifies logical interfaces (see Reference Logical Model) between actors, assigns those logical interfaces to one of 22 logical interface categories (LICs), and then allocates the high level security requirements to each of those LICs. The GRC security requirements apply to most of the interfaces since they cover policies and procedures. The common technical requirements address confidentiality, integrity, and availability common to all LICs while the unique technical requirements address specific LICs. The high level security requirements are also organized into categories primarily based on standards as follows:

- Access Control
- Awareness and Training
- Audit and Accountability
- Security Assessment and Authorization
- Configuration Management
- Continuity of Operations
- Identification and Authentication
- Information and Document Management
- Incident Response
- Smart Grid System Development and Maintenance
- Media Protection
- Physical and Environmental Security
- Strategic Planning
- Security Program Management
- Personnel Security
- Risk Management and Assessment
- Smart Grid System and Services Acquisition
- Smart Grid System and Communication Protection
- Smart Grid System and Information Integrity.

This user guide [NIST 2014] is intended to provide an easy to understand approach to navigate the content of [NISTIR 7628r1]. Organizations - from electric utilities to providers of energy management services to manufacturers of electric vehicles and charging stations - can use the approach and activities described in this guide for assessing the risk, and then identifying and applying the security requirements to mitigate the risk. However, the cybersecurity requirements are recommended, but they are not mandatory. In addition, the document does not include controls such as contingency planning, personally identifiable information processing and transparency, supply chain risk management, and others that are security controls recommended in [NIST SP800-53r5].

The NIST document [NISTIR 7628r1] includes guidelines for organizations as they develop their cybersecurity strategy, perform risk assessments, and select and modify security requirements for the Smart Grid information system implementations. However, additional criteria must be used in determining the cybersecurity requirements before selecting and implementing the security solutions. These additional criteria must take into account the characteristics of the interfaces, including the constraints and issues posed by device and network technologies, the existence of legacy components/devices, varying organizational structures, regulatory and legal policies, and cost criteria. Each organization needs to perform a risk assessment to determine the applicability of the requirements to their specific business needs. Additional work and customization have to be performed before a design is finalized.

In addition, the cybersecurity requirements of the Smart Grid systems have to be correlated with the Catalog of Standards [SGIP CoS], which include information exchange standards, functional requirements standards, object modeling standards, and communication standards. There is rarely a one-to-one correlation, with more often a one-to-many or many-to-one correspondence. Since, cybersecurity is a cross cutting issue, it should be reflected in requirements at all levels of the ISO reference model and GWAC stack. Therefore, cybersecurity requirements should be mapped to address the cybersecurity layers of both reference models. Also, cybersecurity requirements should be added or modified to address the missing security features and gaps in the standards. Cybersecurity must also address how to cope during an attack, recover from it afterwards, and create a trail of forensic information to be used in post attack analysis. In addition, the cybersecurity requirements must reflect the environment where a standard is implemented rather than the standard itself - how and where a standard is used must establish the levels and types of cybersecurity needed.

According to [NIST SP11208r3], cybersecurity must be viewed as a stack or profile of different security technologies and procedures, woven together to meet the security requirements of a particular implementation of policy, procedural, and communication standards designed to provide specific services. Cybersecurity, as applied to the information exchange standards, should be described as profiles of technologies and procedures which can include both power system methods (e.g., redundant equipment, analysis of power system data, and validation of power system states) and information technology (IT) methods (e.g., encryption, role based access control, and intrusion detection).

Also, distributed energy resources (DERs) such as solar photovoltaics are connected to Industrial Internet of Things (IIoT) including sensors, data transfer and communications systems, instruments, and other commercially available devices. DERs introduce information exchanges between a utility's distribution control system and the DERs to manage the flow of energy in the distribution grid.

The primary goal of Information Security is to guarantee safety of the information stipulated by the security attributes such as confidentiality, integrity, availability, non repudiation, authenticity, accountability and other aspects such as reliability, privacy and anonymity. Safety can also be defined to be the control of recognized hazards to achieve an acceptable level of risk. Therefore, a broader risk analysis is the approach when designing security controls to manage the risk.

7.1.2 Broader View of Security Requirements

Following a different approach than one used in NIST document [NISTIR 7628] of 2010, the OpenHAN Task Force (TF), in conjunction with UCAIug SG Security Working Group, evaluated the security risks associated with the home area network (HAN) and developed security requirements using the GWAC stack. Specific security requirements for HAN system are described in [OpenHAN 2010]. The requirements define HAN to enable successful functionality for consumers, utilities, service providers, and vendors. In creating the requirements, a number of categories are considered such as:

- HAN applications
- Communications

- Security
- Performance
- Operations, maintenance, logistics.

In addition to the high level security requirements of [NISTIR 7628r1] document, more specific security requirements are described in this guide [P2030 2011]. Examples include customer requirements, operational requirements, functional requirements, nonfunctional requirements, performance requirements, design requirements, derived requirements, allocated requirements, security requirements [P2030 2011].

Another categorization of requirements includes functional requirements, data requirements, and system quality requirements that include security requirements [Consulting 2003].
The emerging interactions between end-use devices and the power grid mean that physical interoperability must evolve to include functional aspects and be much more than the design of the physical connections (e.g., socket and plug design). Interoperability is a principal enabler of system control schemes that can manage and rely on the active participation of DERs while empowering customers to provide solutions across numerous scales. Operational trustworthiness emerges as a requirement for grid modernization. Interoperability allows utilities, system operators, and other grid participants to select and implement high priority capabilities from a menu of available operational strategies. Customer focused and utility centric control strategies are therefore no longer mutually exclusive.

In support of interoperability, there are requirements such as:

- Requirements for metrology, observability, and controllability help optimizing any cyber physical system — the electric grid included. The measurement parameters must be properly designed to capture the physical phenomena of interest, which becomes more challenging as the physical interactions of devices with the grid become more dynamic and complex.
- Requirements for trustworthiness. Trustworthiness in the grid is often viewed from a cybersecurity perspective. It is generally understood to be whether devices or systems can be expected to preserve the confidentiality, integrity, and availability of the information being processed, stored, or transmitted by the system across the full range of threats. Besides this information driven understanding of trustworthiness, it is necessary to support the operational trustworthiness because edge devices, or field level nodes, continue to gain energy production and management capabilities. Aspects of a modern electrical grid highlight that trustworthiness for a cyber-physical system (CPS), like Smart Grid, includes physical security, reliability, and resilience in addition to the typical informational trustworthiness concerns.
- Requirements for customer empowerment due to technical innovations that impact the customer and enabling capabilities to realize substantial benefits from new interactions with the grid.

7.1.3 DER Cybersecurity Goals and Strategies

Besides understanding the functions of a DER system, the Security Engineer needs to identify cybersecurity related goals and strategies. Also, analysis of any DER system must be focused on information cycles and flows, communications, users, environment, interactions with power grid and other DER systems (or applications), technologies, and interactions among technologies.

Distributed energy technologies are therefore not a special class of generation assets, but a typical asset that can use power generation or demand response to participate in balancing supply and demand in the electrical grid. Accordingly, DER assets can reside in numerous domains with numerous operational strategies. Some examples include:

- Utility owned DER assets - reside in the Generation domain and prioritize supporting conventional markets and grid infrastructure.
- Customer sited DER assets - reside in the Customer domain and may prioritize local or non utility services or operational strategies over utility priorities. Although sited at the customer premises, these assets may be controlled by customers, third party aggregators, or utilities.

Achieving resilient Smart Grid systems requires more than a focus on compliance, instead a culture focused on achieving comprehensive and effective cybersecurity based on analysis and robust models [NSF 2011]. Security requirements are also the result of a process called risk analysis. This implies the implementation of a system to be secure by design that includes processes, development, and assets as described in this guideline [CMMI 2013].

Risk analysis team for Smart Grid systems must include personnel from key areas of an organization such as management, engineering staff, security experts, process control engineers, IT architects, application analysts, programmers, network administrators, and users. The team members must be at the right levels of each area in terms of understanding the processes, business objectives, engineering principles, technologies, and regulations. The team must address issues related to selection of proper process, methods, risk models, and standards.

7.1.4 Defining DER Security Requirements

This [NISTIR 7628r1] document provides generic security recommendations for the Smart Grid, but these requirements must be tailored into DER specific security requirements for utilities, manufacturers, integrators, and owners/operators. Based on this document, the security requirements for DER systems are adapted in [EPRI 2013]. However, the included DER requirements follow the same organization with a focus on compliance to standards and often suggesting specific security controls and implementation solutions instead of strategies and approaches to be used. Some suggested requirements in [NIST 7628r1] are incomplete, not implementable, and focused on solutions. Utilities, DER manufacturers, and DER integrators should review the cybersecurity requirements in this document and ensure they are complete, implementable, and coordinated with the DER functional capabilities. Also, there should be security requirements on policies, agreements, and service level agreements (SLAs). The updated roadmap to Smart Grid [NIST SP1108r4] recognizes the need to address the challenges of DERs technologies.

7.1.4.1 Security Issues for Interface Categories

The Smart Grid is more dependent on information exchange interactions between domains, systems, and applications than the legacy power grid due to DERs high penetration and integration. A critical aspect of information exchange is information security to ensure power systems operations and energy delivery to consumers. This document [NIST SP1108r4] describes new DER interfaces and communication paths that can be associated in the existing categories and security requirements defined in [NIST 7628r1] document or they need to have new requirements identified. The focus NIST's Smart Grid program is on interoperability and securing associated information exchanges.

The diagram of 7. 1 represents new power system entities and logical interfaces for a High-DER architecture, where Uxx-labeled blue interface arrows are the same as those originally shown in [NISTIR 7628r1] and Dxx-labeled red interface arrows are newly introduced interfaces for the example described in [NIST SP1108r4]. This diagram is based on the Logical Interface Reference Model of [NISTIR 7628r1], called Spaghetti Diagram. For example, the High-DER scenario depicts a power system that relies on increasing communications between, and coordinated management of, diverse assets to provide essential grid services. Figure 7.1 shows new types of communications interfaces such as:

- New interfaces for new entities (e.g., Distributed Energy Resource Management System (DERMS) into the grid operations domain).
- New interfaces between subsystems (e.g., the customer-sited DER asset, electric vehicle asset, and the utility-scale DER or cogeneration asset have been split to reflect the different logical interface requirements between asset controllers (Figure 7.1, boxes 4a, 4c, and 6a) and the equipment (Figure 7.1, boxes 4b, 4d, and 6b) connected to the grid physically consuming or supplying electrons).
- New interfaces for legacy systems (e.g., both the utility-scale DER or cogeneration asset and the facility energy management system interface directly Supervisory Control And Data Acquisition (SCADA) system via a new logical interface (Figure 7.1, red lines D03 and D04); aggregators interact with new actors in the High-DER scenario and the logical interfaces increase accordingly (Figure 7.1, red lines D08, D52, and D92).

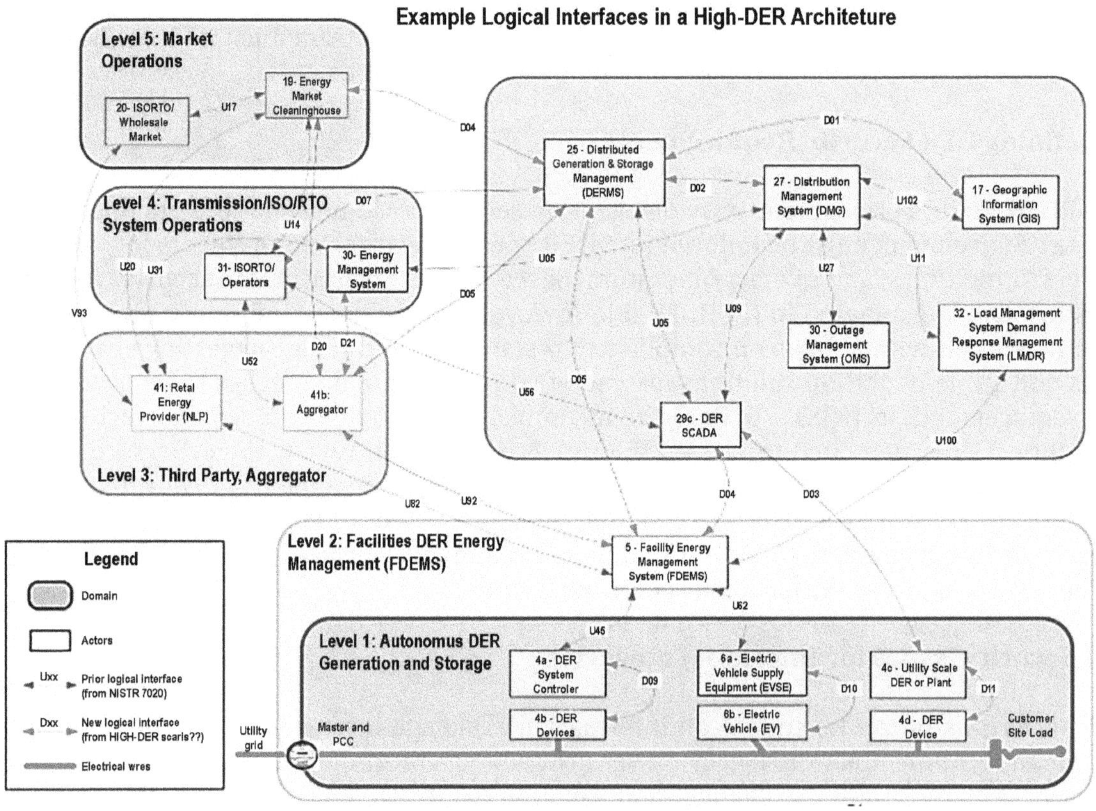

Figure 7.1 Example logical interfaces in a High-DER architecture. *Source:* [NIST SP1108r4]. Public Domain.

This figure includes only those entities requiring new logical interfaces for this high-DER example. The complete evaluation of each information exchange in Figure 7.1 is provided in Appendix I of [NIST SP1108r4] document, and the LIC interface mapping is shown graphically in Figure 7.2. However, the document emphasizes that not all conventional approaches for responding to and coping with cybersecurity attacks that were described in [NISTIR 7628r1] will map directly to evolving Smart Grid architectures. As the criticality of DERs to system operations increases, it may not be plausible to abort communications and potentially cease device operations in response to cyberattacks as was described in [NISTIR 7628r1] for several LICs in traditional power system architecture. Therefore, mapping emerging interfaces to existing LICs and the associated [NISTIR 7628r1] cybersecurity guidance provides a foundation upon which the cybersecurity of emerging grid architectures can be built.

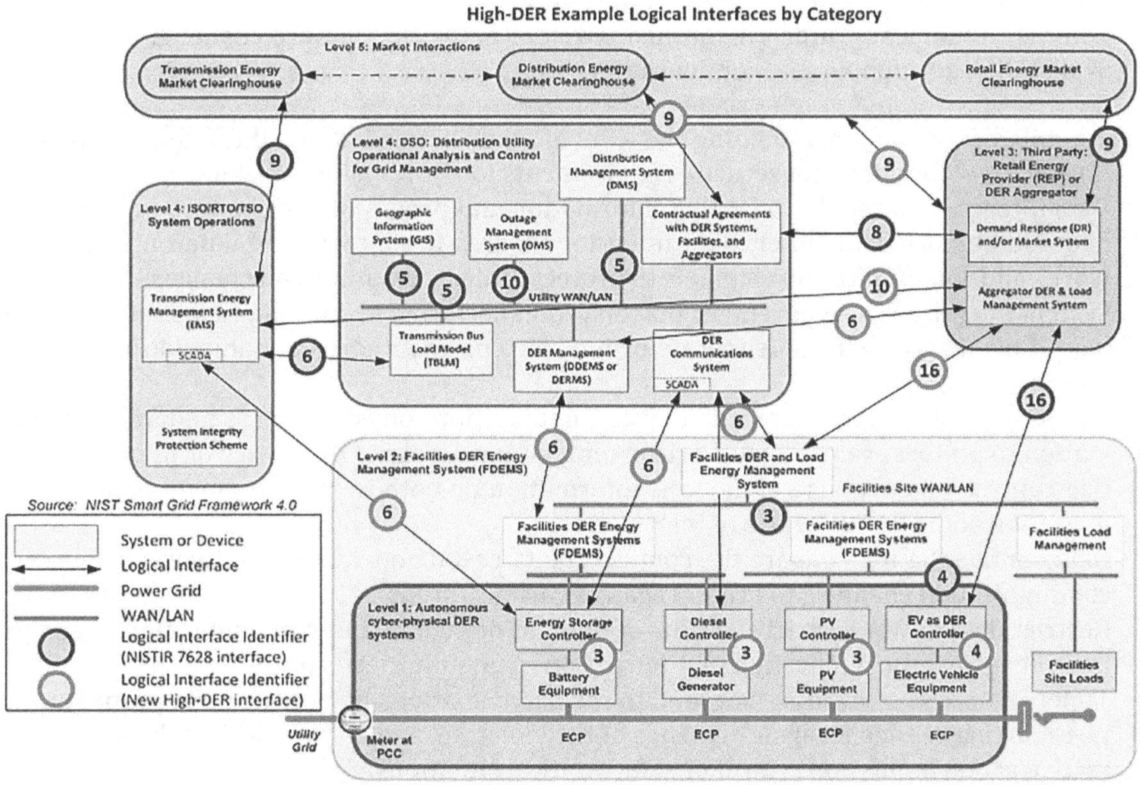

Figure 7.2 Logical interface categories for the High-DER example. *Source:* [NIST SP1108r4]. Public Domain.

There is no single set of cybersecurity requirements that addresses each of the DERs logical interface categories. Additional criteria must take into account the characteristics of the DER interface, including the constraints and issues posed by device and network technologies, the existence of legacy components/devices, varying organizational structures, regulatory and legal policies, and cost criteria.

Figure 7.3 shows logical interfaces of category 14 and DER actor as described in [NISTIR 7628r1]. In this diagram, customer DER is identified as actor number 4 that interacts directly with other actors (actor 5 – Customer Energy Management System, actor 7 – Home Area Network Gateway, actor 10 – Submeter energy usage). Logical interface category 14 covers the interfaces between systems that use the AMI network with high availability:

- Between LMDR/DRMS (Load Management System/Demand Response Management System) and customer DER
- Between DMS (Distributed Management System) applications and customer DER, and
- Between DMS applications and distribution automation (DA) field equipment.

Examples of security issues for this interface category include:

- Privacy of information for customer DER.

- Integrity of data is very important, lack of protection can affect the reliability and/or efficiency of the power system.
- Availability is a higher requirement for those parts of the AMI networks that will be used for real time interactions and/or rapid request response requirements.
- Physical security of DER equipment and DA equipment varies (DER could be located in unsecured locations that are not under utility control)
- Communications are often via meshed wireless networks that are challenging in terms of availability, throughput and configurations.

However, implementing and maintaining the level of security needed for DER interactions is challenging. Examples of foundational security requirements for DER systems include:

- Data integrity is a requirement that information and programs are changed only in a specified and authorized manner; data integrity ensures the integrity of data on selected networks and communication channels to protect against unauthorized changes.
- System integrity is a requirement that a system performs its intended function in an unimpaired manner, free from deliberate or inadvertent unauthorized manipulation of the system.
- Access control - control access to devices, implies both physical access and access to information to protect against unauthorized temper and access of the device or information.
- Use control – control use of devices, information or both to protect against unauthorized operation of the device or use of information.
- Data confidentiality – ensure the confidentiality of data on selected networks and selected communication channels to protect against various threats.
- Restrict data flow – restrict the flow of data on networks and communication channels to protect against the publication of information to unauthorized sources.
- Timely response to events – respond to security violations by notifying the proper authority, reporting needed forensic evidence of the violation, and automatically taking timely corrective action in mission critical or safety critical situations.
- Network resource availability - ensure the availability of all network resources to protect against denial of service attacks.
- Business continuity and survivability of critical services due to intentional and unintentional threats.

This work [Uslar 2014] describes a methodology to use both the modeling approach of the European Smart Grid Architecture Model (SGAM) model (see Figure 7.4) and the security guidelines of [NISTIR 7628] for a secure architecture development of the Smart Grid as critical infrastructure. The logical interfaces described in [NISTIR 7628] framework are mapped onto the SGAM and its planes, domains and zones, bridging the gap for the experts for security analyses. The resulting architecture is described using appropriate descriptors. By showing the strengths and drawbacks of these approaches and underlining the necessity to make them interoperable, the authors pave the way for the foundation of a methodology of secure design. The methodology supports an approach of integrating security into the development process of a smart grid architecture from the very beginning of the envisioned new business and use case.

As illustrated in Figure 7.4, the SGAM comprises five so called core viewpoint layers. The domains, and zones support a holistic view on architecture including business processes which usually are out of scope for standardization.

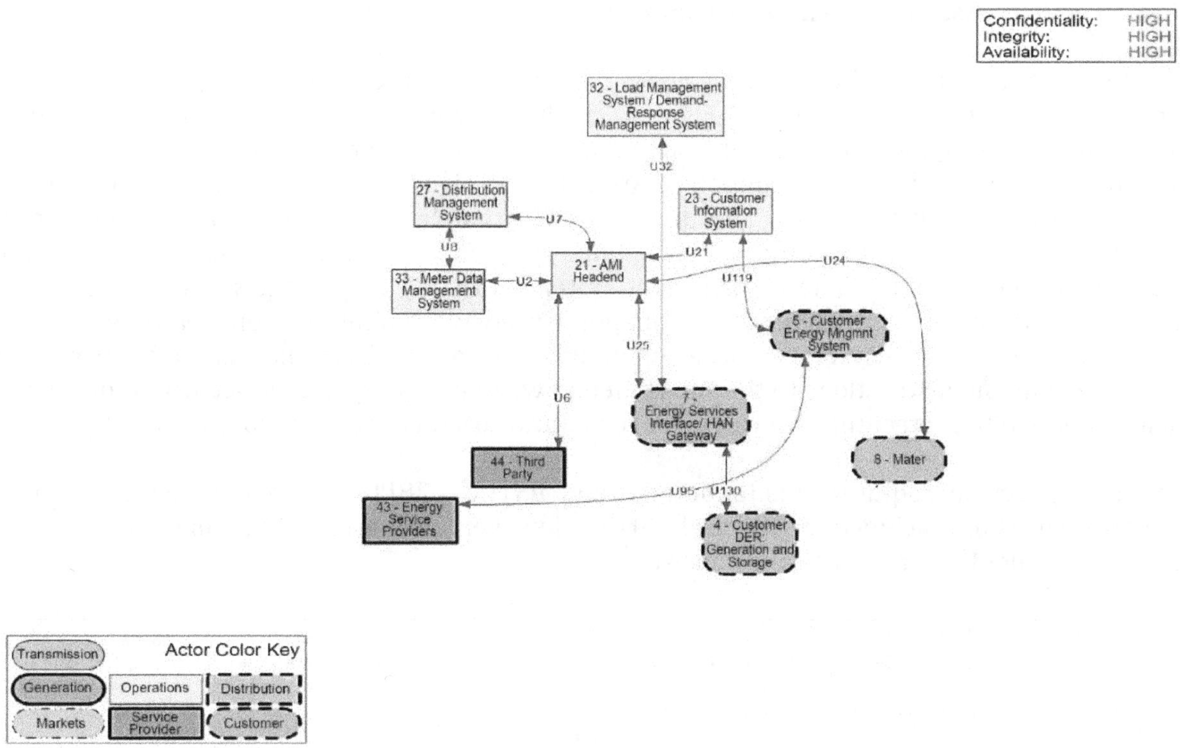

Figure 7.3 Logical Interface Category 14. *Source:* [NISTIR 7628r1]. Public Domain.

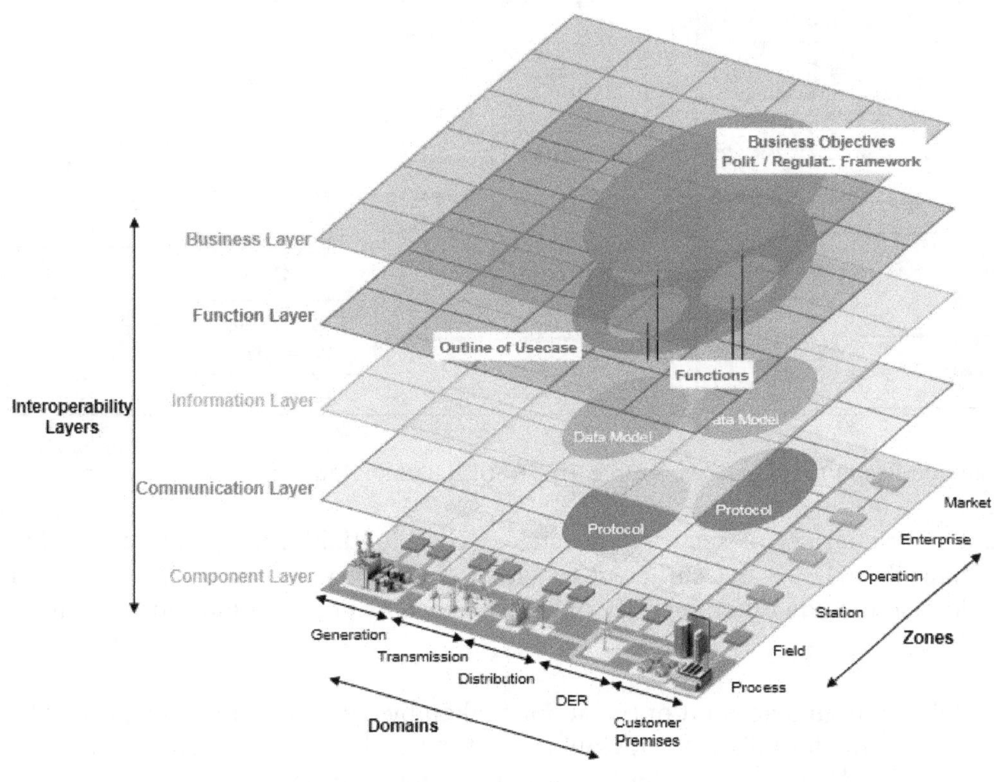

Figure 7.4 The Smart Grid Architecture Model (SGAM). *Source:* [CENELEC 2012]. © 2012, CENELEC.

7.1.4.2 Security Issues for Demand Response Interfaces

Security issues are analyzed using scenarios and by looking at interfaces of components that affect Demand/Response (DR) applications. Figures 7.5 shows basic DR components and interfaces for typical DER as being part of DR applications that relate to NIST domains and other components including the advanced metering infrastructure network. Figure 7.6 shows the diagram of an AMI infrastructure and its components. This diagram is an example of an underlying network that provides communication between different smart meters it controls, the intermediate data concentrators and the central Meter Data Management systems. Existing technologies, such as power line, radio frequency or wireless networks are commonly used as a communication means between customer premises and AMI systems. These communications are bidirectional, as the control system can send command instructions to the smart meters when necessary. Again, security requirements must be defined to protect these devices from cyberattacks and unauthorized access.

Examples of security requirements (adopted from [NISTIR 7628] for DR that are described for different scenarios are included in [Ghansah 2012]. However, each organization may need to revise these requirements for accuracy and consistency.

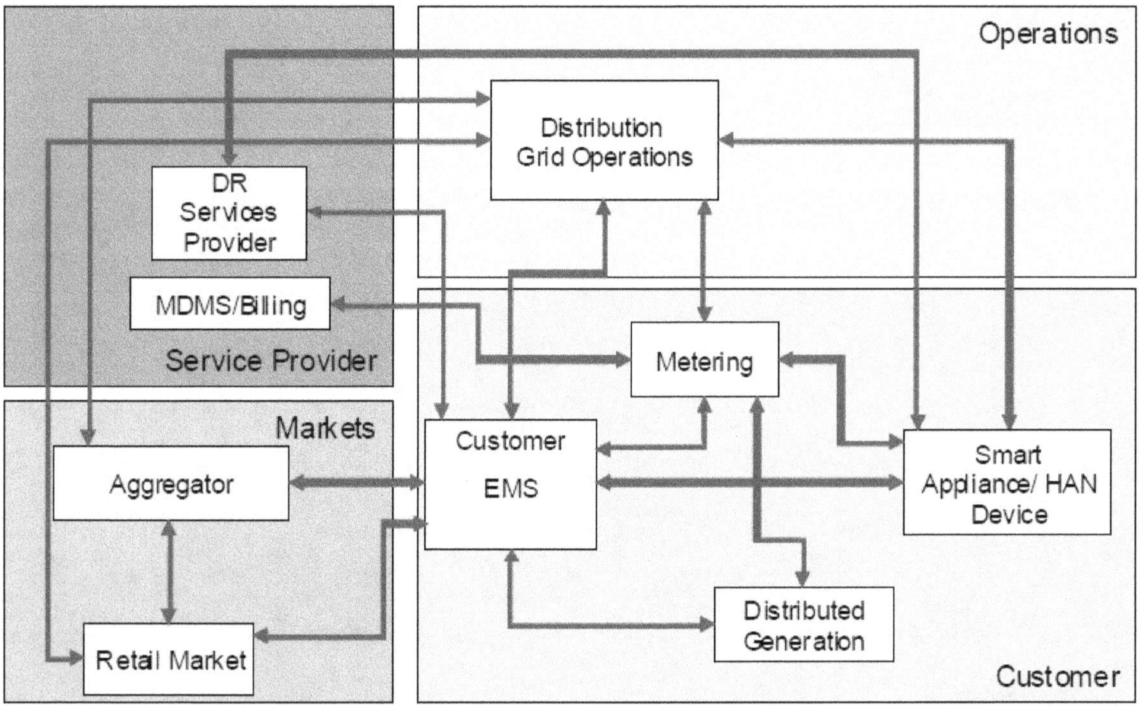

Figure 7.5 Demand Response interfaces. Source: [IETF SGIP]. Public Domain.

In addition, a restricted set of security objectives (confidentiality, integrity, availability) as described in [Ghansah 2012] is not sufficient to satisfy the information security needs of DER systems. It should also include other security objectives such as non-repudiation, authentication, accountability, etc.

For example, failure to hold account of the actions taken by DR communicating parties because of the invalid service provider information would result in the dispute between parties and decrease customer confidence. Also, non-repudiation protection for transactions is very important for DR customers and DER services when dealing with responses to market offers or utilities' commands to reduce load. Many security requirements suggested in this document are conflicting with the needs described in the same document.

The integrity of the pricing signal is critical because if it can be manipulated, it could lead to financial impacts on the organization or customers. Thus, most of the DR functions in the Smart Grid, such as load shedding, time-of-use pricing (ToU), dynamic pricing, etc. require data integrity and/or confidentiality to maintain the reliability of the grid and prevent adversaries to manipulate the information in the system. Failure to provide data integrity and/or data confidentiality could result in the exposure of customer's information, unauthorized modification and manipulation of the information [Ghansah 2012]. These needs require more robust security protection than the suggested protections for many scenarios described in the same document.

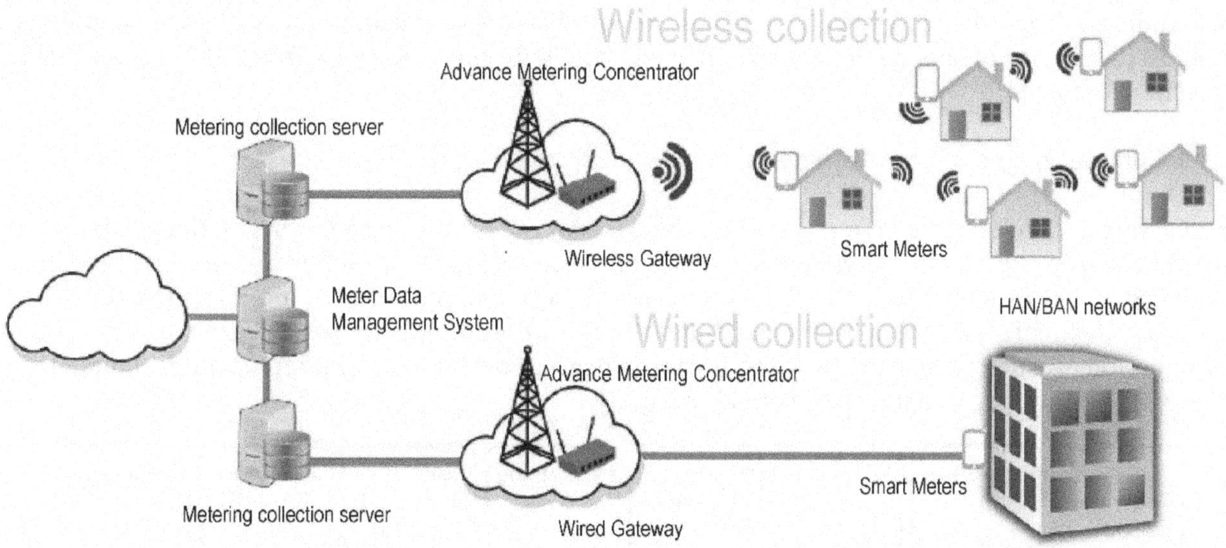

Figure 7.6 AMI infrastructure components. *Source:* [ENISA 2015]. © 2015, ENISA.

Also, there is need for consistency of security requirements, if only one function implemented without security protection, then it could expose the data and systems to cyber threats regardless of other functions' security protection. Therefore, security requirements for Smart Grid and DER systems should be defined based on rigorous analysis of the stakeholders' needs and compliance to standards mandated by energy sector.

7.1.4.3 DER Resilience

Resilience of DER systems to cyber threats has benefits on DER systems and Smart Grid. DER interconnections to grid and operating the grid with significantly large numbers and capacities of DER systems can impact grid reliability and security. It is not just the utilities who must take responsibility for achieving this resilience goal. Many stakeholders need to be involved in the design, implementation, and operation of DER systems, including manufacturers, integrator/installers, users, information and communication technology (ICT) providers, security managers, testing and maintenance personnel, and ultimately utility regulators.

Thus, more guidance on these topics can be found in the proposed report [IEC TR 62351-12]. This technical report provides cybersecurity recommendations and engineering/operational strategies for improving the resilience of power systems with interconnected DER systems. It covers the resilience and security requirements for the many different stakeholders of these dispersed cyber-physical generation and storage devices, with the goal of enhancing the safety, reliability, power quality, and other operational aspects of power systems, particularly those with high penetrations of DER systems. The focus is describing the impact of DER systems on power system resilience, and this covers cybersecurity and engineering strategies to improve power system resilience with high penetrations of DER systems. However, this report only addresses the additional reliability and resilience issues caused by 3rd party managed DER systems which may not be secured or operated with the same reliability as the utility managed power system. Additional DER security requirements for the information exchange and networks are provided in these documents [IEC 62351-9] and [IEC 62443-3-3].

The resilience issues for cyber-physical DER systems interconnected with the power grid are being discussed by SGIP groups. Resilience of the grid is often associated with making the grid able to withstand and recover from severe weather and other physical events, but resilience should also include the ability of the cyber-physical grid to withstand and recover from malicious and inadvertent cyber events.

7.1.5 European Smart Grid Approach

As ICT technologies emerged and provided the basis of the Smart Grid roadmaps, their roles were adopted within the functional viewpoint of the European Smart Grid Reference Architecture [Bruinenberg 2012] as depicted in Figure 7.7. As shown in this diagram, ICTs are allocated to the domains of the Smart Grid. Because the roadmap relates different viewpoints towards the Smart Grid within a long term engineering process, it can be used to derive and evaluate requirements at the strategic level within an enterprise [Danekas 2013].

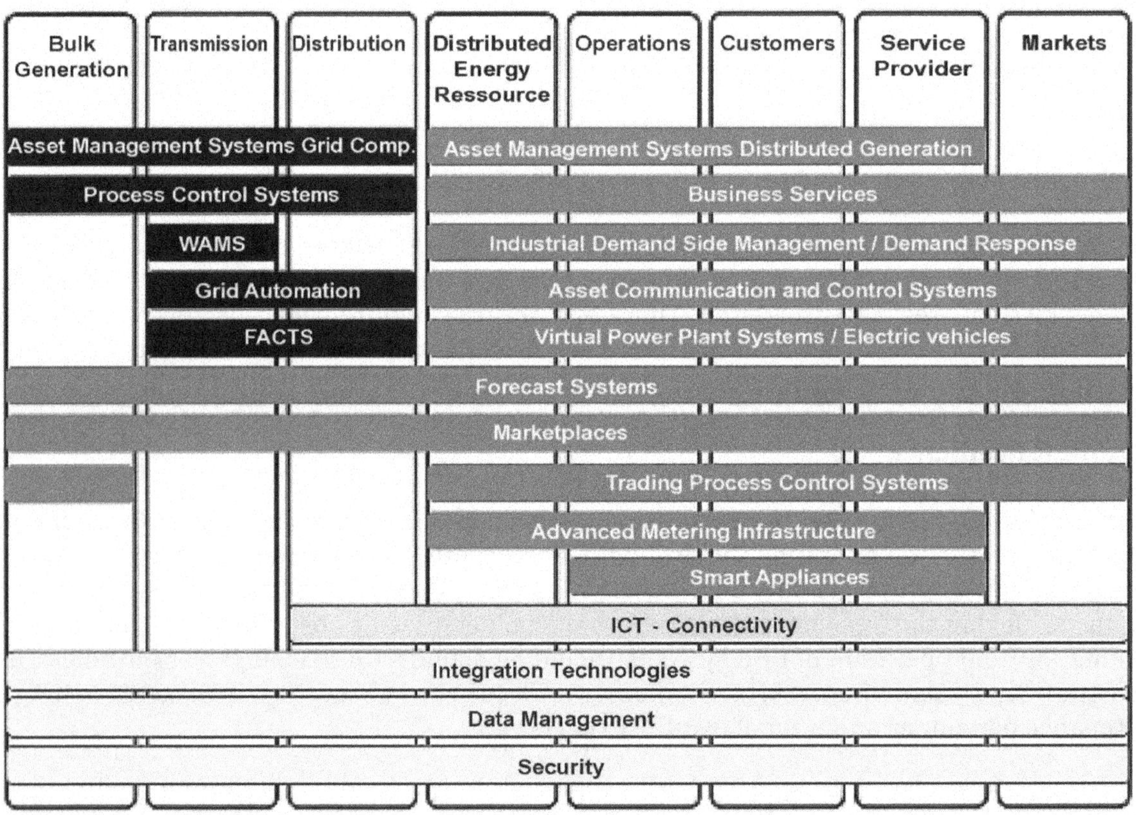

Figure 7.7 Functional viewpoint of the European Smart Grid reference architecture. *Source:* [CENELEC 2012]. © 2012, CENELEC.

As depicted in Figure 7.7, security is a cross function requirement across different dimensions (domains, zones, and layers). Methods and processes for requirements analysis and engineering concerning the system's layers are described in [Rohjans 2012], [Danekas 2013].

7.2 Critical Issues for the Security Requirements of Power Systems

Smart Grid development should rely on established standards and well defined requirements to ensure a secure and interoperable power system. Therefore, a Security Engineer needs to address critical issues and be involved in many activities such as:

- Collect and develop security requirements to eliminate redundancy and consolidate energy industry needs.
- Select existing standards that best satisfy the power system's security requirements and continue to adopt new standards.
- Develop security requirements to address possible security gaps and to facilitate interoperability.
- Create testing and validation tools (e.g., compliance with standards, automated testing tools, evaluation of security features for the products under development).
- Develop security requirements for plug-in interfaces and guidelines to test the security features of products under development and to test defects in the products or specifications.
- Develop requirements for interfaces to ensure the exchange of information between participating actors and domains.

7.2.1 Challenges and Approaches

The approaches for the Security Requirements Engineering (SRE)are diverse and in a continuous change. One survey reveals that many known approaches are based on various techniques and activities with no common agreement on a definition of security requirement, requiring different levels of expert knowledge, and suggesting artifacts that are too complex for regular developers [Tondel 2008]. This work identifies common activities such as misuse or threat identification, generation of artifacts, and support of security objectives. The authors recommend an approach that should include the following steps:

- Define security objectives which are high level requirements or goals that are most important to stakeholders and the requirements that must be met to comply with relevant legislation, policies, and standards.
- Identify assets and prioritize security requirements based on the value of the assets that results from the user, system owner, and attacker viewpoints; also, it is necessary to include different stakeholders' priorities for the asset's confidentiality, integrity, and availability protection. For example, confidentiality of a piece of information or data may be very important for a market application and may not be important for a process control application. So, the protection mechanisms have to be discretely provided in the context of needs.
- Assets have interrelated characteristics that include value, criticality, and the degree to which they are relied upon to achieve organizational mission/business objectives. From these characteristics, appropriate protections are to be engineered into solutions employed by the organization.
- Threat analysis to be focused on the most important assets and using predefined threat categories such as STRIDE (Spoofing, Tampering, Repudiation, Information disclosure, Denial of service, and Elevation of privilege) threat model; further elaboration should be based on the most important threats identified using techniques such as attack trees.
- Documentation of the security requirements must be comprehensive and understood by the developers and requirements should focus on what should be achieved and not how.

Based on the above approach, security objectives and asset identification typically result in policy level requirements. From these requirements, other more detailed requirements can be derived. One high level requirement may be defined such as:

The privacy of the service's users must be protected.

Then, this might result in a number of additional security requirements such as:

The personal profile of users must be protected in transit.

This can be detailed further, for example:

A protection mechanism should be effective and costly efficient.

Then, the mechanism is selected (e.g., encryption algorithm, key management) during design phase. Further, effort has to be dedicated to documenting the security requirements. Once the requirements are understood, it is now the responsibility of a Systems Engineer to refine them, and to determine, along with other engineers, the best technology for a job. At this point starting with a trade study, Systems Engineering (SE) encourages the use of weighted choices to determine the best option. The trade study in turn informs the design, which again can affect the architecture representation of the system (without changing the requirements).

In a SE process, this stage represents the iterative step that is carried out until a feasible solution is found. A decision matrix is a list of values in rows and columns that allows an analyst to systematically identify, analyze, and rate the performance of alternative solutions. A decision matrix is often populated using techniques such as statistical analysis, reliability analysis, system dynamics (feedback control), and optimization methods. For example, when using encryption techniques, it is needed to ensure that the sender domain does not pass encrypted data to a receiver domain that expects it to be unencrypted. Therefore, it is needed to know the performance of the encryption algorithm and all the dependencies including where the data is stored and used. Another issue is the expected delay. Processing overhead of encrypted data has to be counted and when processing large amounts of data, the mathematical algorithms used in the encryption and decryption are computationally intensive and may take longer time than the expected delay of the data transfer.

Security Engineer needs to understand the layers, characteristics, and disturbances of a power system when designing the cybersecurity requirements to avoid interactions between power system functions and cybersecurity functions.

Based on DOE Smart Grid definition, Figure 7.8 shows the proposed concept of Smart Grid layers with security in another dimension that covers all layers as described in [Gao 2012]. At the bottom layer, security is concerned with physical energy infrastructure that distributes energy. Communication infrastructure is defined on the top of the physical energy infrastructure to entire supply chain. Computing and control systems/information technology is above the communication infrastructure for timely decision making. Smart Grid applications are on the top layer to create electrical system/societal values.

In conclusion, defining security requirements for the Smart Grid could be a tedious task and many sources could help on handling it. For example, one source is [SecReq], which is providing information resources and guidance to build security in. In addition, process optimization tools such as metrics/management dashboards, simulation and forecasting, bottleneck detection, and what if analysis. A tool set that supports the analysis and evaluation of various Smart Grid scenarios with respect to their security relevance is presented in [Chinnow 2013].

Research is in progress for identifying the methodological relation between requirements engineering and architectural design in the energy sector [Trefke 2012]. Real time constraints are still an issue for critical events to meet the latency requirements.

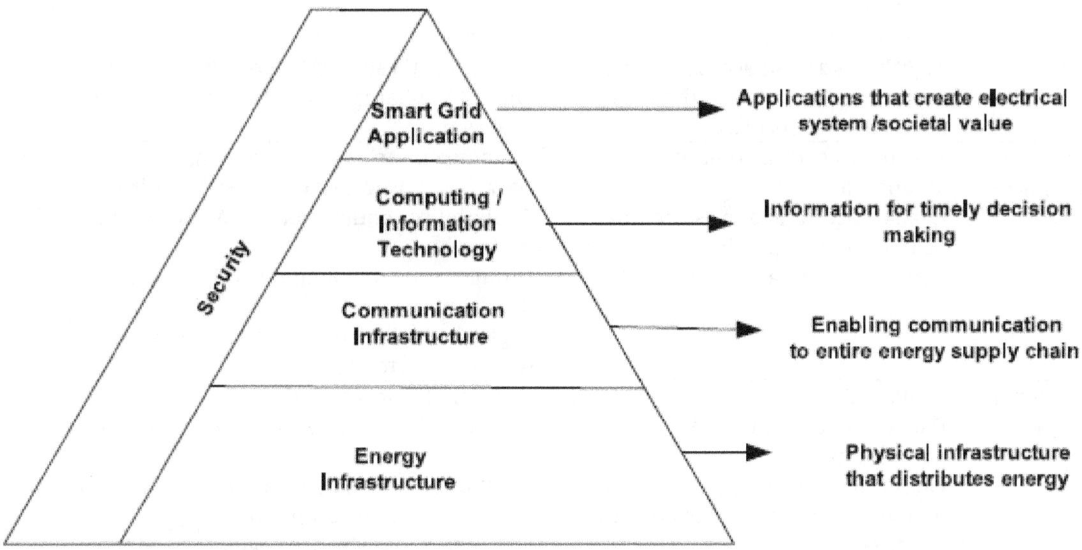

Figure 7.8 Smart Grid view. *Source*: [Gao 2012]. © 2012, Elsevier.

Although several protocols may include security features, no single protocol is capable of meeting all security requirements needed by applications. Examples of relevant protocols include: DNP3, TLS, TCP/IP, SIP, RTP, and XML. One action is to include additional security requirements such as:

- Data must be available with high probability, both for storage and transmission, even in the presence of failures, and
- Adversary must not be able to read the data or interfere with it.

One solution to ensure data availability is distributing data via coding among several nodes.

7.2.2 IT Systems Requirements vs Control Systems Requirements

When comparing Industrial Control Systems (ICS) and Information Technology (IT) systems to determine the security requirements, there is a need to identify differences and similarities for different aspects (see typical differences for category examples in Table 7.1).

Table 7.2 summarizes a set of key differences between IT and ICS systems based on various criteria that need to be considered when defining the security requirements.

Although some of these differences refer to either components, functions, or performance, they also imply differences in security requirements and security controls.

More specific differences focused on security characteristics of IT and OT systems are illustrated in Table 7.3. It shows differences in security environments, communication performances, use and type of security controls.

Table 7.1 Summary of IT system and ICS differences.

Category	Information Technology System	Industrial Control System
Performance Requirements	Non real time Response must be consistent High throughput is demanded High delay and jitter may be acceptable Less critical emergency interaction Tightly restricted access control can be implemented to the degree necessary for security	Real time Response is time critical Modest throughput is acceptable High delay and/or jitter is not acceptable Response to human and other emergency interaction is critical Access to ICS should be strictly controlled, but should not hamper or interfere with human-machine interaction
Availability (Reliability) Requirements	Responses such as rebooting are acceptable Availability deficiencies can often be tolerated, depending on the system's operational requirements	Responses such as rebooting may not be acceptable because of process availability requirements Availability requirements may necessitate redundant systems Outages must be planned and scheduled days/weeks in advance High availability requires exhaustive pre-deployment testing
Risk Management Requirements	Manage data Data confidentiality and integrity is paramount Fault tolerance is less important – momentary downtime is not a major risk Major risk impact is delay of business operations	Control physical world Human safety is paramount, followed by protection of the process Fault tolerance is essential, even momentary downtime may not be acceptable Major risk impacts are regulatory noncompliance, environmental impacts, loss of life, equipment, or production
System Operation	Systems are designed for use with typical operating systems Upgrades are straightforward with the availability of automated deployment tools	Differing and possibly proprietary operating systems, often without security capabilities built in Software changes must be carefully made, usually by software vendors, because of the specialized control algorithms and perhaps modified hardware and software involved
Resource Constraints	Systems are specified with enough resources to support the addition of third party applications such as security solutions	Systems are designed to support the intended industrial process and may not have enough memory and computing resources to support the addition of security capabilities
Communications	Standard communications protocols Primarily wired networks with some localized wireless capabilities Typical IT networking practices	Many proprietary and standard communication protocols Several types of communications media used including dedicated wire and wireless (radio and satellite) Networks are complex and sometimes require the expertise of control engineers
Change Management	Software changes are applied in a timely fashion in the presence of good security policy and procedures. The procedures are often automated	Software changes must be thoroughly tested and deployed incrementally throughout a system to ensure that the integrity of the control system is maintained. ICS outages often must be planned and scheduled days/weeks in advance. ICS may use OSs that are no longer supported
Managed Support	Allow for diversified support styles	Service support is usually via a single vendor
Component Lifetime	Lifetime on the order of 3 to 5 years	Lifetime on the order of 10 to 15 years
Components Location	Components are usually local and easy to access	Components can be isolated, remote, and require extensive physical effort to gain access to them

Source: [NIST SP800-82r2]. Public Domain

Table 7.2 Examples of key differences: ICS vs. IT systems.

Criteria	Variable	ICS	IT
Performance	Response	Real time, time critical	Non real time, reliable
	Throughput	Low	High
Reliability	Operations	Continuous	Scheduled
	Failure Tolerance	Not tolerated	Occasional tolerated
Testing	Beta field testing	Not allowed	Accepted
	Penetration testing	Not allowed	Accepted
Risk management	Goals	Human safety	CIA
	Impact	Loss of life, equipment, environmental damage	Loss of data, loss of business
	Recovery	Fault tolerance	Rebooting
	Security perimeter	Well defined	Not defined

7.2.3 Developing Secure SCADA Systems

Many of today's SCADA systems employ an enterprise wide area network to facilitate the exchange of data between the field sites and the central location. Also, the enterprise network is usually connected to the Internet. Figure 7.9 depicts a refined reference model [Knowles 2015] for the industrial control systems of third generation (networked), which frequently use open technologies, while connecting to and communicating over other (potentially non-industrial control system) networks. This openness increases the exposure to threat agents compared to first generation (monolithic using security through obscurity) and second generation (distributed) industrial control systems (using proprietary and closed source components and standards, with limited connectivity to IT systems). So, the need for communications between enterprise zone (layers 4 and 5) and manufacturing zone (layers 0 to 3) justifies appropriate security requirements to be defined.

The reference model highlights the multitude of interconnections between industrial control systems and information technologies both within and between layers.
Layers 0, 1, and 2 include information and functions related to physical processes and devices (e.g., sensors and actuators). Contemporary industrial control systems contain many information technologies at layer 3, and this layer frequently communicates with business applications at layers 4 and 5. The devices include historians (e.g., databases of time stamped industrial control system events such as process outputs and alarms), and authentication, authorization and accounting (AAA) services. Layer 4 relies on standard information technologies and provides business administration services, such as enterprise applications (e.g., e-mail servers) and non-critical industrial control functions (e.g., inventory management). Layer 5 consists of the majority of centralized information technology services (e.g., business to customer services).

The strategy of SRE for ICS requires balancing real time requirements with performance, reliability, flexibility, safety, and security requirements. Although security expertise and domain expertise are needed, they are often separated. Also, there are difficulties on specifying the requirements for ICS operating in complex environments. Examples of issues that have to be considered include [NIST SP800-82r2]:

- Timeliness and performance requirements
  ICS are generally time critical, require reliable, deterministic responses; in contrast, IT systems typically require high throughput, and they can typically withstand some level of delay and jitter. Some ICS are built on real time operating systems (RTOS), where real time

Table 7.3 Security characteristics for OT and IT systems.

Security Characteristic	OT Systems	IT Systems
Physical Security	Often unmanned, dispersed locations, less secure when far located	Secure Facility, Computer room
Connection speed	Slow serial link and dial up not uncommon	Relatively fast communications
Computing power	Low	High
Availability	Timely authorized access (24x7x365)	Can tolerate some delay in access and availability
Integrity	Less assured	Assured
Confidentiality	Less salient	Always or more often first priority
Authenticity	Data source without deception and provenience are critical	Connection and information sharing could be adhoc and could draw a variety of data sources
Encryption (key, digital certificate, signature)	Uses are limited to access constraints, computing speed, and memory limitations	Extensively used
Message authentication control	Resource limitations make it to be used less	A variety of authentication schemes make it relevant and useful
Non-repudiation	Commonly, not used; more prevalent for Smart Grid transactions	Common, often mandatory
Attack signature	Sparsely available	Well characterized
Pattern and rules	Limited and not used	Diverse and more used
Communication protocols	Proprietary protocols such as DNP3, Modbus,	Open TCP/IP reference model
Characteristics of communication protocols	A numerous set of communication protocols, several communication and security gaps; harmonization is needed	A more limited number and better defined
Use of unsecure communication services (e.g., telnet, ftp)	Frequently used	Rarely
Access controls (e.g., passwords, firewalls, authentication, etc.)	No mandatory, usually not enforced	Mandatory
Differentiation of access privileges	Generally, no distinction between administrators and end-users or not implemented	Always distinct
Unneeded OS services running by default	No rigorous control	More control
Security features (e.g., authentication, integrity, confidentiality) in protocols and applications	Limited development, often not used	More robust development and disciplined use
Access paths	Common back door access for maintenance by vendors, users, and administrators	Controlled and supervised
Control software (e.g., firmware)	Usually not protected and changes are permitted	Protected, no alteration
Changes to configuration settings	Adhoc changes are permitted, no rigorous control	Not allowed in adhoc manner
Security risk assessment	Not a routine activity	Periodically performed
Security policies	Not in the norms of their operations	Standard procedures and operations to implement the security policies

Adapted from [Ray 2010]. © 2010, IEEE.

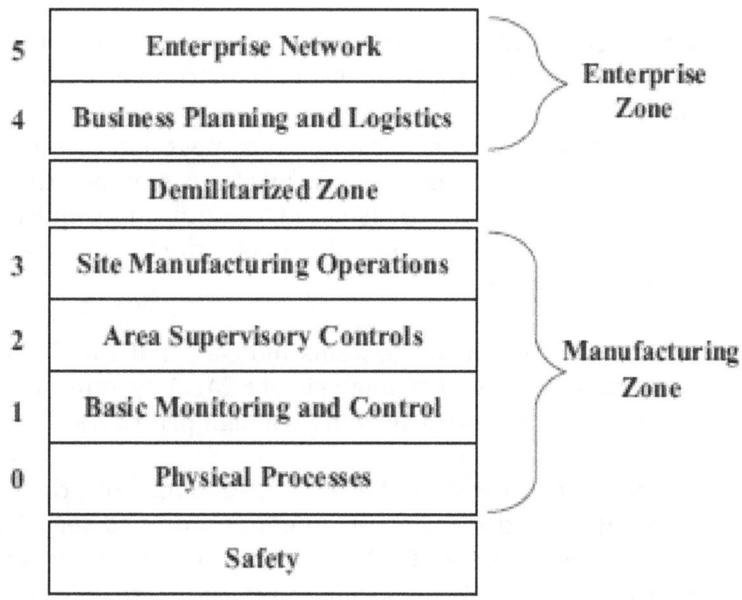

Figure 7.9 Industrial control system reference model. *Source:* [Knowles 2015]. © 2015, Elsevier.

refers to timeliness requirements. IT security technology can impact timing, inhibit performance.

- Availability and reliability requirements
  Many ICS processes are continuous in nature, unexpected outages of systems that control processes are not acceptable; outages often must be planned and scheduled days or weeks in advance; in contrast, IT systems use procedures such as rebooting a component, usually not acceptable solutions due to the adverse impact on the requirements for high availability, reliability and maintainability of the ICS. Some ICS employ redundant components, often running in parallel, to provide continuity when primary components are unavailable.
- Risk management requirements
  In a typical IT system, data confidentiality and integrity are typically the primary concerns; for an ICS, human safety and fault tolerance to prevent loss of life or endangerment of public health or confidence, regulatory compliance, loss of equipment, loss of intellectual property, or lost or damaged products are the primary concerns. The personnel responsible for operating, securing, and maintaining ICS must understand the important link between safety and security. Any security measure that impairs safety is unacceptable. These differences create huge differences in acceptable security practice.
- Physical effects
  ICS field devices (e.g., PLC, operator station, DCS controller) are directly responsible for controlling physical processes. ICS can have very complex interactions with physical processes and consequences in the ICS domain that can manifest in physical events. Understanding these potential physical effects often requires communication between experts in control systems and in the particular physical domain.
- System operation
  ICS operating systems (OS) and control networks are often quite different from IT counterparts, requiring different skill sets, experience, and levels of expertise.
- Resource constraints
  ICS and their real time OSs are often resource constrained systems that do not include typical contemporary IT security capabilities.

- Communications
  Protocols and media used by ICS environments for field device control and intra-processor communication are typically different from most IT environments, and may be proprietary.
- Component location
  Most IT components and some ICS are located in business and commercial facilities physically accessible by local transportation. Remote locations may be utilized for backup facilities. Distributed ICS components may be isolated, remote, and require extensive transportation effort to reach. Component location also needs to consider necessary physical and environmental security measures.

The increasing threats against SCADA control systems indicate that there should be more directions in the development of these systems. The analysis of SCADA systems suggests that achieving better quality and more secure SCADA control systems is a high priority.

Key requirements and enhancing features should be provided to improve the security and safety of SCADA systems [Hentea 2008a]. Also, this work recommends a unified security/safety risk framework for defining security requirements for SCADA systems. In addition, it is emphasized that the trend should be to escape from best practices checklists and return to core SE methods, processes, and tools. Addressing critical issues for SCADA systems require actions such as the following [Hentea 2008a]:

- Critical path protection
  It is mandatory to protect the critical components from cyber attacks. An approach to security evaluation for the identification of critical components is discussed in [Rae 2006]. In control systems, a component failure greatly increases the likelihood of multiple simultaneous failures. Also, the high speed of SCADA networks facilitates the quick propagation of malicious code.
- Stronger safety policies and procedures
  Solutions for preventing the attacks are becoming more important to enterprises and software vendors. The best first steps recommended for preventing attacks against control systems is increased awareness of potential vulnerabilities and solutions, as well as implementing stronger safety policies and procedures. It is essential that every enterprise is taking security seriously when developing software, or managing the control systems.
- Enhanced security for device
  Since real time applications alone may never be capable of addressing all security requirements, incorporating security features into a device can further enhance system security. Individual devices erect their own security perimeters and defend their own critical resources, such as a network link or storage media [Cummings 2002]. Although protecting code in embedded systems is sometimes a business decision, a company must weigh the cost of implementation of protection versus the potential loss of service revenue when the vulnerabilities are discovered [Fisher 2000].
- Sensor networks solutions
  Security and privacy challenges will require new technological solutions for sensor networks. Sensor networks implementations include monitoring factory instrumentation, pollution levels, etc. Sensor networks consist of hundreds or thousands of sensor nodes. Each node represents a potential point of attack. However securing each node is impractical. Sensor detectors offer one defense against attacks and have the capability to differentiate between the transmissions of authorized and unauthorized sensor networks and other devices [Chan 2003].
- Security requirements early in the software development cycle
  Saydjari [Saydjari 2002] recommends developing a top down architecture and engineered approach to the problem because current technology is insufficient to defend against cyber

warfare. Developing requirements for control systems with security features and use of simulation model based on a framework could improve the definition of requirements and reveal problems early in the software development cycle [Menzies 2006].

- Ensure authentication, confidentiality, integrity, availability, and non-repudiation
  Security of SCADA control systems must ensure far more than the confidentiality of information in transit. It must also ensure that only authorized parties have access to such information, a task that will require abuse resistant method for identifying such parties. In the information security context, the SCADA control systems require features that support key security concepts such as authentication, authorization, confidentiality, integrity, availability, and non-repudiation. Also, control protocols should be improved to include security features.

7.3 Developing Integrated Systems

A Smart Grid system may consist of:

- IT which is a discrete system of electronic information resources organized for the collection, processing, maintenance, use, sharing, dissemination, or disposition of information.
- Operational technologies (OT) or industrial control systems (ICS), which is a general term that encompasses several types of operational and control systems, including SCADA systems, distributed control systems (DCS), and other control system configurations such as skid mounted Programmable Logic Controllers (PLC) often found in the industrial sectors and critical infrastructures.
- Integrated system is a synergy of IT and ICS functions; utilities heavily rely on the information to operate the power grid; so they need a robust business information environment that requires interconnections among the control and business information system domains, the external internet, and other peer organizations.

Examples of Smart Grid systems include Energy Management System (EMS), Distribution Management System (DMS), Supervisory Control and Data Acquisition (SCADA), Market Management System (MMS), Advanced Metering Infrastructure (AMI), Customer Information System (CIS), Outage Management System (OMS), Meter Data Analysis (MDA). All these systems require use of a secure automated information exchange, analysis and intelligent decision making throughout the Smart Grid entities.

A unified risk management approach for Smart Grid systems is proposed in [Ray 2010]. This approach is a blended approach (not a sum of two approaches) for both ICS (e,g., OT) and IT systems with the objective to identify not only specific security risks associated with ICS and IT systems, but to identify the security risks due to interconnections and interfaces between ICS and IT systems.

Although the integration of IT systems with SCADA data collection and monitoring are beneficial to run the Smart Grid more efficiently and effectively, the design approach has to provide a security model that accommodates the differences between the security architecture and priorities of SCADA systems versus IT traditional systems. Issues such as modems connected to one system compromising the other, merging the disparate culture of SCADA and IT practices, integrating security measures between the SCADA system and the enterprise communications network, and multitude of specific regulations and requirements for each system have to be considered in order to design appropriate security solutions [Hentea 2008a].

The automation and control systems for power system operations have many differences from most business or corporate systems. Some particularly critical issues related to security requirements include:

- Operation of the power system must continue 24x7 with high availability (e.g., currently 99.99 % for SCADA and higher for protective relaying for the current systems) regardless of any compromise in security or the implementation of security measures which hinder normal or emergency power system operations.
- Power system operations must be able to continue during any security attack or compromise (as much as possible).
- Power system operations must recover quickly after a security attack or compromised information system.
- The complex and many fold interfaces and interactions across the power system makes security particularly difficult since it is not easy to separate the automation and control systems into distinct security domains, and yet end-to-end security is critical.
- There is not a one size fits all set of security practices for any particular system or for any particular power system environment.
- Testing of security measures cannot be allowed to impact power system operations.
- Balance is needed between security measures and power system operational requirements. Absolute security is never perfectly achievable, so the costs and impacts on functionality of implementing security measures must be weighed against the possible impacts from security breaches.
- Balance is also needed between risk and the cost of implementing the security measures.

Applications of real time systems are ubiquitous and proliferating in DER systems. Examples include:

- Process control for renewable, energy storage systems, microgrids
- Telephone, radio, and satellite communications
- Interfaces based on multimedia systems (text, graphics, sound, video)
- Building controls (heat, lights, doors, air conditioning)
- Protection devices (relays, circuit breakers, etc.)
- Syncrophasor systems
- Distribution automation.

When defining the security requirements for Smart Grid and DER systems, it is recommended to include tasks such as:

- Balancing and prioritizing requirements that exhibit real time constraints; IT technologies can impact timing of control systems.
- Specifying requirements and testing scenarios for complex systems in operational Environments.
- Balancing requirements for performance, reliability, flexibility, and safety with security requirements.
- Providing security expertise and domain expertise to successfully define the security requirements.

7.3.1 Characteristics of Integrated Systems

Integrating operational information like equipment status, phasor measurements, distributed generation and storage status with business level information such as consumer preferences and energy usage and market pricing allows organizations to achieve higher end-to-end business and operational efficiency, reduce grid stress and improve overall enterprise productivity.

Therefore, engineering security requirements for these integrated systems requires a revision of traditional approaches and improvements toward more advanced techniques that support analysis and planning for similar characteristics as well as differences between IT systems and control systems. Thus, security requirements for control systems need to ensure:

- System availability which is a high priority in critical systems that requires addressing several key issues (e.g., adequate network infrastructure with high bandwidth and throughput, secure communications and operating platforms).
- System efficiency in its use of computation and communication resources that have capabilities such as error management for proper handling of failures, fail safe to avoid resource exhaustion even in the face of adversarial action, adequate redundancy built into it so the entire system does not collapse, auxiliary security functions to detect to and respond to cyber attacks.
- System scalability is important to be defined in key dimensions such as controlled devices, aggregated DERs, distributed communication networks, devices connected to AMI networks or PKI and cryptographic systems in which large amount of data from hundreds of thousands of smart meters are collected and processed; sound techniques for optimal solutions need to be used when elaborating scalability goals and security requirements. Examples include design of information paths flow to be simple and uniform, limited private information flow for bottom up data flows, detection of security threats and incidents that could destabilize demand patterns and load management applications to avoid potential failures (e.g., blackouts).
- Although some characteristics are similar to IT systems, AMI systems also have characteristics that differ from traditional information systems. Many of these differences stem from the fact that AMI systems are integrated into the physical power grid. In some cases, adversely impacting an AMI system can pose significant risk to the health and safety of human lives and serious damage to the environment, as well as serious financial issues such as production losses. Furthermore, the goals of safety and efficiency sometimes conflict with security in the design and operation of AMI systems.

To address the challenge of building a secure AMI, the National Institute for Standards and Technology (NIST) and users' groups, such as the Open Smart Grid [OpenSG] and [OSGP], have been producing reports and requirements to enable technology and policy makers to include security from the very beginning of the development process. These documents range from risk assessment [FERC 2006], [FERC 2012] to security requirements [Berthier 2010], [UCAIUG 2008], [ISORTO 2010], attack testing methodology [Carpenter 2009] or testing tools [Higgins 2012]. However, the requirements that are suggested in these documents are subject to more precise determination by the utility organization to apply a unified risk management approach.

7.3.2 Addressing DERs Differences

Most of security requirements described in [NISTIR 7628r1], [EPRI 2013] documents do not give a full perspective in the future Smart Grid that integrates more DERs (e.g., renewable sources) and some scenarios are missing (e.g., communication between utilities and customer owned DER systems).

While common security requirements were defined in [EPRI 2013], there are more differences due to technologies that are specific to each type of DER. Therefore, more analysis is necessary such that security needs have been identified and further security requirements are defined accordingly. Although both, wind power and solar power are renewable technologies in the category of DER systems, it is necessary to look at the operating characteristics and different type of data associated with each technology. Also, storage systems operate in different modes, have interactions with different actors, and generate specific data that have a different set of data to be collected and shared among systems and often utilities, service providers, and markets. More functional and security differences are also based on different types of storage technologies [Connolly 2010].

7.3.3 Energy Storage Systems

As renewable take a bigger share of generation capacity and where electrical systems cannot keep up with the increasing demand, energy storage systems could provide services and improvements to power grid systems, so storage may one day be ubiquitous in power systems [EPRI-DOE 2003]. It is believed that energy storage will be a key asset in the evolving Smart Grid [Rastler 2012]. As more storage options become available and nations around the globe continue to enrich their portfolios of renewable energy, the use of energy storage is increasing such that security is taking new dimensions. For example, increased deployment of energy storage devices in the distribution grid could make this process more effective and could improve system performance [Roberts 2012] if exchange of information is timely and secure. Mainly, energy storage mediates between variable sources and variable loads; it works by moving energy through time [SANDIA 2013].

However, it is very difficult to compare the various types of energy storage technologies to one another as they are individually suited for certain applications, because no technology is perfect for everything [Connolly 2010]. According to this author, a storage technology can be analyzed under the following key criteria: performance characteristics, operation and advantages; applications; costs; disadvantages; and future potential. Although security should be explicitly listed, it can be assumed under criteria such as operation, application, or future potential.

Energy storage applications differ from other DER options, such as distributed generation or energy efficiency and have unique characteristics such as [Rastler 2010]:

- No typical operating profile or load shape that can be applied prospectively
- Limited energy resources with a narrow band of dispatch and operation
- Support participation in multiple wholesale markets
- Provide several benefits simultaneously to the wholesale power system, electric distribution companies, and customers
- Deficient in monetizing multiple stakeholder benefits.

These characteristics imply security needs that have to considered require confidentiality and integrity of the energy prices and transactions, privacy of the customers, etc. Besides performance characteristics specific to a storage technology, there are more requirements for the interconnection and communications infrastructure. So, security requirements have to be designed to include all these aspects. For example, some storage technologies for most applications must receive and respond to appropriate control signals. In some cases, storage may have to respond to a dispatch control signal. In other cases, the signal may be driven by a price or prices. Storage response to a control signal may be a simple ramp up or ramp down of power output in proportion to the control signal. A more sophisticated response, requiring one or more control algorithms, may be needed. In addition, a storage system needs to respond to price signals or to accommodate more than one application. Messages may have different priority needs and distinct requirements for protection.

Therefore, interconnection is another key area that a Security Engineer needs to investigate. If storage is to be charged with energy from the grid or to inject energy into the grid, it must meet applicable interconnection requirements. At the distribution level, an important point of reference is the [IEEE 1547] standard. This standard describes a limited set of security features so additional security controls may be needed for some states. Also, utilities may have more specific interconnection rules and requirements, so security requirements have to be provided for these use case scenarios.

Another issue is safety. The energy storage systems should be able to protect themselves from internal failures and utility grid disturbances. This means that systems should be self-protecting for ac or dc component system failures. In addition, systems should be able to protect themselves from various types of grid faults and other abnormal conditions on the grid. Systems should reflect the safety standards of the utility. For example, utility and local fire personnel should be notified of particular safety issues and the appropriate response in case of an emergency. Energy storage systems need to timely notify users about degradation of functional parameters of chemical storage batteries. High temperatures or harsh environments could cause insulators to become conductors and self-discharge the battery thus a serious safety concern.

Depending on the utility requirements, energy storage systems provide energy discharge for a defined period of time and as needed by different applications such as voltage and frequency regulation, load and resource shifting, ramping, and dispatchability. Security of the information and equipment is critical during this discharge time. The energy storage systems and security features can be designed for the needs of distribution and/or transmission systems and for single purpose or multi-purpose operation (see Figure 7.10).

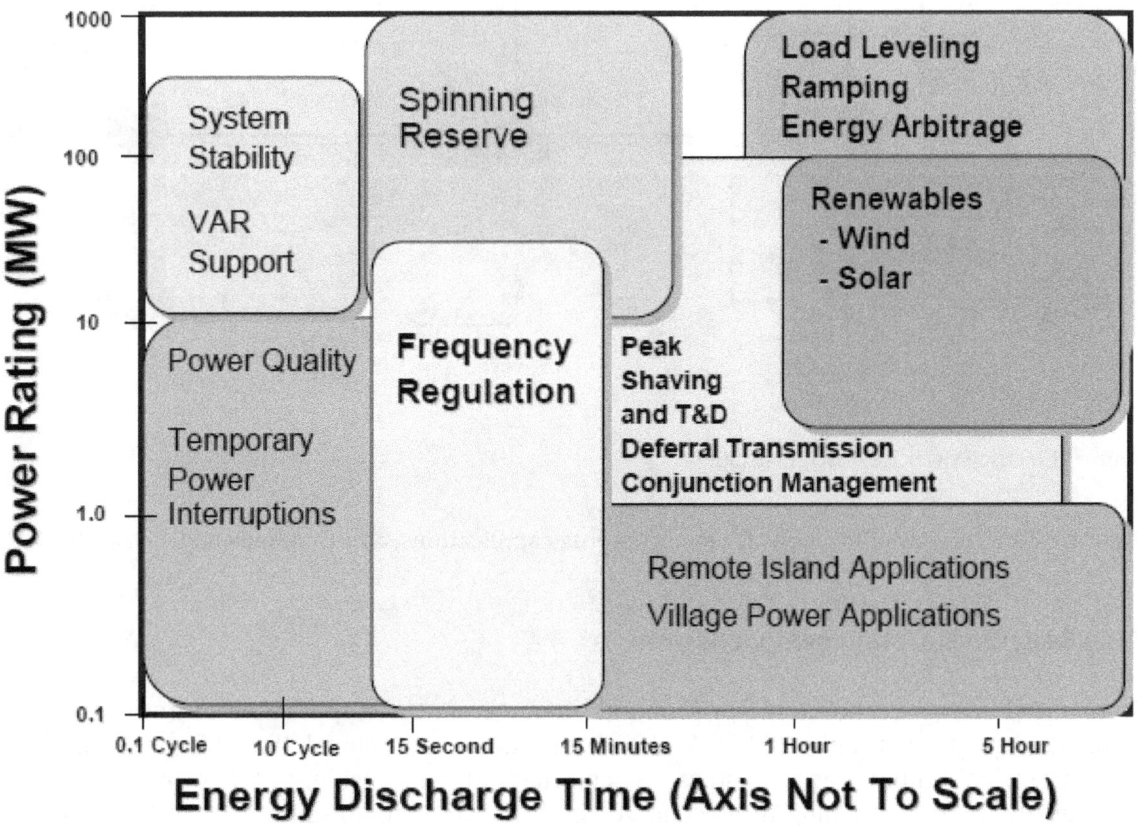

Source: Electric Power Research Institute

Figure 7.10 Overview of Energy Storage Use Cases. *Source:* [Steeley 2011]. © 2011, EPRI.

The three essential energy storage applications include substation based storage, distributed energy storage systems (DESS), and renewables integration. These serve different functions on the electric grid. They vary in categories such as location on the electric grid, capacity, operating modes and use cases, and the required duration of energy storage. Figure 7.11 illustrates possible location options for energy storage applications.

These systems generate data and events that have to be monitored and collected for different applications. Therefore, security requirements need to be detailed to satisfy the protection needs for each particular system because generated data and events have different characteristics such as volume, real time constraints, reporting periods, processing and communications infrastructures [Steeley 2011], [Rastler 2010]. Also, impacts of cyber intrusions need to be estimated in order to mitigate the security risks and recommend the appropriate security controls.

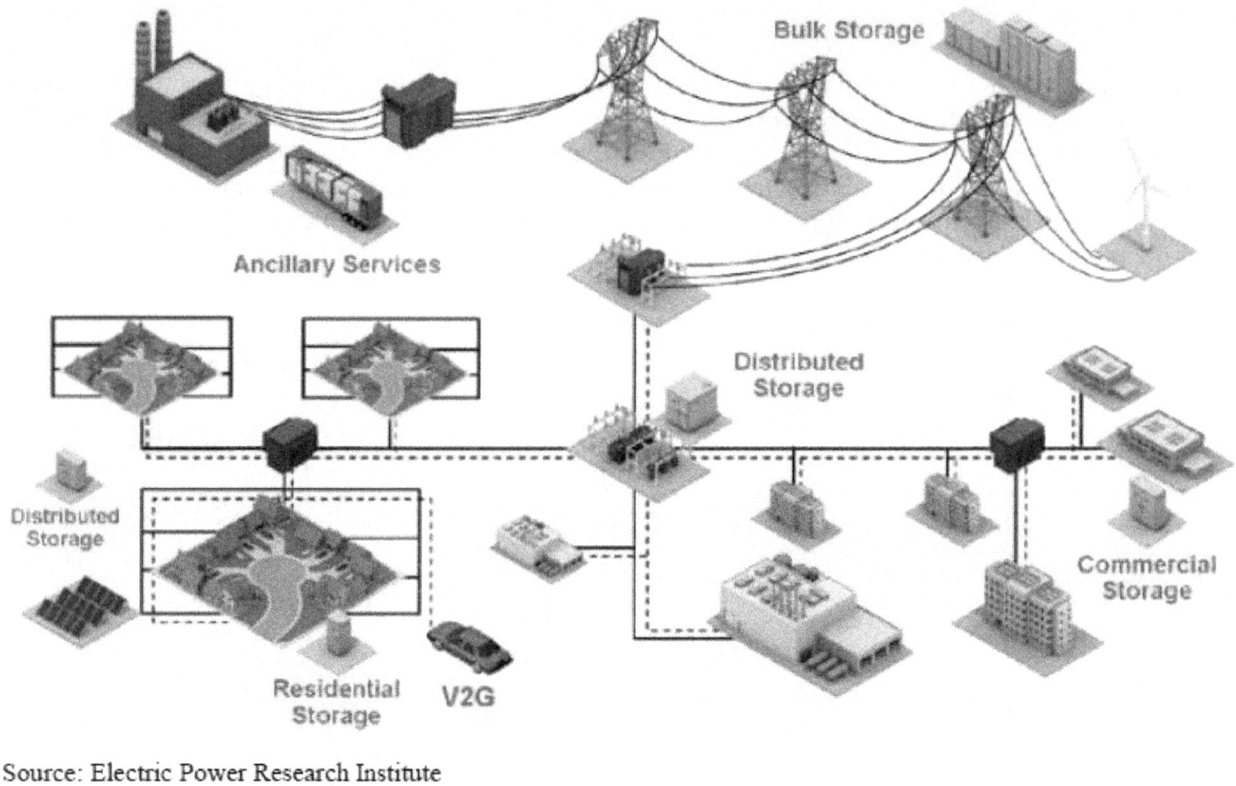

Figure 7.11 Overview of locations for energy storage applications: *Source:* [Steeley 2011]. © 2011, EPRI.

7.3.4 Distribution Management System

Distributed energy resource integration and management are applicable to both distribution and transmission systems [Feng 2012]. At the distribution system level, the DERs are small scale wind generators, community energy storages, commercial and residential photovoltaic cells, microturbines, fuel cells, and plug-in electric vehicles. At the transmission system level, DERs are the mega wind farms, solar farms, geothermal power plants, etc.

DR and DER applications manage the dispatch of D/R and DERs to reduce the power demand during peak hours and/or shift some of the demand to hours of the day when the demand is low, thus improving the load factor (utilization efficiency) of transmission and distribution assets.

The basic benefits of distribution management system (DMS) are the management and optimization of near real time power flows. The foundation to perform its task is the data model containing both static and operational data. Static system data describes the distribution system over an extended time frame. Updates to this data tend to be long term in nature and are generally considered permanent. Operational data is short term and generally considered to be temporary changes. Operational data frequently deals with changes to connectivity or equipment settings in response to system conditions or fieldwork. Operational data is generally entered directly into the DMS by distribution system operators or updated automatically by the outage management system. So intentional or unintentional errors may expose the systems to cyber attacks. Therefore, security requirements should be provided, and security protections should be installed to avoid compromises of the systems.

These data are combined with actual, telemetered data from SCADA devices for error correction, to perform near real time power flow calculations and analysis that reflect the current conditions on the system. These power flow calculations then enable the optimizations and advanced functions [Russell 2013].

7.3.5 Demand Response Management System

A Demand Response Management System (DRMS) is an example of system that integrates requirements for both IT systems requirements and control systems. DRMS enables the management of load, optimization of distribution system operations, and maintain economic control. However, advancements in Smart Grid technology and renewable energy pose new challenges that a DRMS must be able to handle. From an enterprise systems point of view, the DRMS falls into a category of an information management system much like the Meter Data Management System and connects the flow of information to the DR devices to/from the utility, leveraging the Smart Grid AMI network. Therefore, critical functional elements include:

- DR functions for all customers (residential, commercial, and industrial).
- Support of utility operations; DRMS is a link from utility systems such as the Customer Information System and the Distribution Management System.
- Device agnostic breaking the current use of proprietary systems and control 3rd party devices.
- System of Record (SoR) for the Home Area Network (HAN) devices besides the system of record for billing systems.
- Platform approach viewed for the next generation of systems and not a single solution; an expandable system as new opportunities emerge.
- Ability to manage heterogeneous environments (multiple devices, several protocols, and different networks.
- Support strategy for migrating from legacy technologies and legacy systems to new system.
- Link to customer portals and information systems and ability to maintain active records of customer devices and manage them on behalf of the utility.
- Integration of Web portals with customer facing systems provides customers with valuable, decision making, and information for DR application.
- Integration of data without duplication among systems such as Automatic Meter Reading (AMR), Advanced Metering Infrastructure (AMI), remote AMR systems, Meter Data Management System.
- Designs of plug-and-play services into the Service Oriented Architectures (SOA) of back office utility' service and seamless utilization of the data.
- Provides a complete DR transaction log, process visibility, and audit trail.
- DR process optimization tools such as metrics/management dashboards, simulation & forecasting, bottleneck detection, and what if analysis.

- Configurable, standards based interfaces with other customer systems.

In order to support these functions, electrical industry needs to increase the design efforts, adopt appropriate standards, new techniques for designs, implement security, and perform testing at a larger scale to ensure the availability of applications and information [Vos 2009]. In addition, enabling technology (e.g., real time metering) and standardized business protocols are key for the development of DRMS [Neumann 2006]. Specific requirements for successful results and cost effective DRMS implementation include [Neumann 2007]:

- Fast, reliable, automated and secure communications between multiple players in the DR domain in real time.
- Standardized protocols for customer enrollment, DR event notification, and customer participation followed by timely and accurate business processes for invoicing and settlement.

Other examples of security needs include:

- Management of DR processes to support end-to-end security
- Secure customer signup and program registration processes
- Customer identification and access verification
- Automated and secure notifications to customers, aggregators, and distribution/grid operators
- Automated triggering of any type of intelligent load control for registered devices over any network
- Tracking event participation and actual demand reduction for settlement
- Seamlessly interfacing with other corporate applications
- DR workflow management with features such as secure portal inboxes, deadline monitoring, escalation, substitution, and e-mail notifications
- Secure, real time communications with DR participants and resources
- Vulnerability management and protection against cyber threats
- Reporting of cyber incidents and cybersecurity attempts to back office utilities and to customer's residential gateway or smart phone
- Providing logging of cyber incidents and failures.

Also, it is necessary to develop security requirements for any of these services supported by a DRMS to include:

- Managing customer signup and program registration processes
- Real time visibility of available DR resources and interruptible demand
- Automated notifications to customers, aggregators, and distribution/grid operators
- Automatically triggering any type of intelligent load control devices over any network
- Interfaces with other corporate applications
- Customers making choices to potentially reduce their energy consumption (critical capability)
- Operations of electric utilities for peak load management, economic control and distribution system operations and optimization
- Two way, real time communication link from utility to customers (residential, commercial, or industrial customers)
- Critical linkage between the DER resources and the utility back offic
- Upgrades for security improvements
- Knowledge acquisition and evaluation of the DR program performance
- Logging significant cybersecurity events (e.g., logins, compromised data notifications, etc.).

7.4 Addressing Differences

When developing Smart Grid or DER systems, the differences between IT and ICS systems require mitigations and optimizations.

7.4.1 Recommendations

Since the demarcation of boundaries between IT systems and control systems is disappearing, the SRE task is to identify the constraints of each system and blend into new system to optimize the security functions.

Security Engineering requires consideration of security objectives in each type of requirements already discussed in previous chapters. For example, functional requirements describe what outcome is needed by identifying the necessary input, task, action, or activity that must be accomplished. Functional requirements analysis results in defining user roles and access needs, mapping users to data access, identifying data owner and information security policies, etc.

7.4.2 Summarizing Issues for DER Systems

Although there could be only specific requirements for DER systems, there is a need to relate security issues to Smart Grid in a broader view context. Smart Grid is defined a system of systems and implies also a network of information networks. Smart Grid security requires a system of systems approach in engineering design and operations such that security responsibility is extended beyond the Smart Grid.

While security requirements for the broader Smart Grid may or may not be within the scope of a single utility's responsibility, there should be imposing requirements through agreements and/or regulatory mandates upon cooperating interconnecting systems and corresponding capabilities to meet and/or support some aspects of the Smart Grid security objectives.

AMI systems use smart metering technology that is the base of any Smart Grid strategy, as it provides the platform for offering digital era services to consumers. To incorporate intermittent energy resources, a category which renewable energy falls into, electricity networks will have to become smarter grids with integrated communication systems and real time balancing between supply, demand, and storage [Crossley 2010]. One of the greatest challenges for future electricity grids lay on the demand side response and creating a system that can shift peak demand, at the same time as being socially acceptable.

Consumer needs such as privacy, technical efficiency of gadgets, lifestyles, and socio-cultural factors influence the use of energy demand response applications. The abundance of various consumer goods facilitates an easy access to these applications. However, this demand for greater consumption on electronic gadgets turns complex when consumer privacy and security of Smart Grid applications have to be implemented and be user friendly. This technological transfer phenomenon needs to be viewed from the perspective of daily life too. More, any requirement for manual intervention (applications require users to manually program the components in their homes) is prone to human error and threats from adversarial actions. So, these applications and future grid are expected to detect, analyze and respond automatically to changes where human intervention may not be required or the action is too critical to wait for human input.

A Smart Grid can improve energy usage levels via energy feedback to users coupled with real time pricing information and from users with energy consumption status to reduce energy usage, and real time demand response and management strategies for lowering peak demand and overall load via appliance control and energy storage mechanisms. This implies a factor of scalability of control and management of DERs that need to be analyzed and included when designing security requirements for the Smart Grid operation.

Correct control of DER processes is required because an inappropriate control action or management decision could affect the operation of the grid. Grid technology provides the chance of a simple and transparent access to different information sources. For network operation purposes, more accurate real time state estimation of the whole network gives information on voltages, loads, losses and stressing of components, and also makes it possible to optimize, e.g. network topology, voltage control, and load control actions [Järventausta 2010]. A data grid can be interpreted as the consolidation of different data managing systems furnishing the user with data, information, and knowledge.

Obviously, the control and management of the Smart Grid operation is an interdisciplinary field which crosses the boundaries of communication, optimization, control, dynamic optimization techniques, advanced intelligent control, and even social and environmental constraints [Momoh 2009].

A Smart Grid is already leveraging benefits of networking. The networking trends are transforming from its traditional flow to an advanced Smart Grid style of flow, spurring a positive change of networking trends towards a more distributed, digitized and multiple source of supply. Its technologies may even integrate with social networking sites, as well as associated security risks. Security requirements need to cover network security, trust, and privacy concerns with regard to networking sites among consumers. Personal data from profiles may also leak outside the network when search engines index them [Shin 2010].

Therefore, the Smart Grid network is required to connect the magnitude of electric devices in distributed locations and exchange their status information and control instructions. The system wide intelligence is feasible only if the information exchange among the various functional units is expedient, reliable and trustable. The current communication capabilities of the existing power systems are limited to small scale local regions that implement basic functionalities for system monitoring and control, such as power line communications and the SCADA systems, which do not yet meet the demanding communication requirements for the automated and intelligent management in the next generation electric power systems.

Therefore, the goals of DER resources integration require an open computing system and the establishment of a shared information model, based on the data integration system, and optimization of network operation and management. Resources can be identified and employed in the most effective manner when the environment is fully connected with suitable mechanisms for sharing information, ensuing great positive transformation of electricity service quality and value for both utility provider and end user.

The underpinning of situational awareness for DER systems is to identify adversaries, estimate impact of attacks, evaluate risks, understand situations and make accurate decisions on how to protect these systems promptly and efficiently. More, the significant relationship of the Smart Grid security to consumer requirement has to be taken into consideration as how it would impact consumer trust and satisfaction. One example are consumer requirements described in [Ai Ling 2011a]. In this work, sixteen functional requirements are grouped in five categories (philosophy, human behavior, rule base social system, strategic system, and hardware) (see Table 7.4).

In the process of identifying these requirements, the background of the Smart Grid information infrastructure and the needs for the Smart Grid information security are described based on hermeneutic circle methodology [Ai Ling 2011a]. The authors identify information technology and security as components of the hermeneutic circle methodology in reference to Smart Grid security. It refers to the idea that one understanding of the text (e.g., Smart Grid) as a whole is established by reference to the individual parts (e.g., information technology, security) and one understands each part by reference to the whole (Figure 7.12).

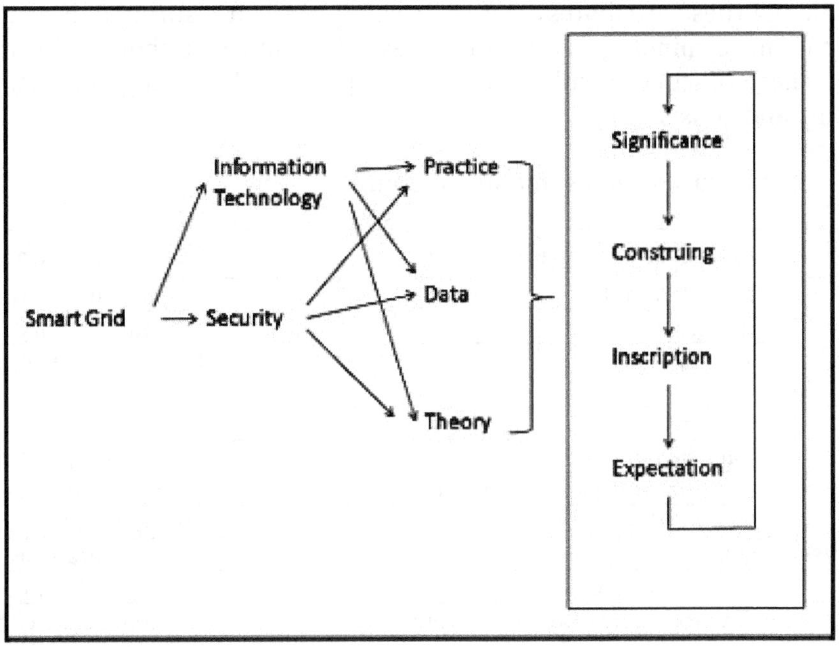

Figure 7.12 Using the hermeneutic circle methodology. *Source*: [Ai Ling 2011a]. Courtesy of the authors.

However, one needs to understand at least three components – information technology, security, and Smart Grid (as an electricity system), in order to understand the security of the Smart Grid. As depicted in Figure 7.12, by applying the method described in [Ai Ling 2011a], the approach starts with the construing and explaining the meaning of IT, Security, and Smart Grid. This is followed by the inscription of the need for Smart Grid security. Then the functional requirements (see also [Ai Ling 2011b]) serve to achieve the expected consumer assumption. The efforts are concluded with the significance study as the stages of this process are supported by these main elements: theory, practice, data.

This research methodology (based on the interpretive theory) has its own merit, but the approach of identifying requirements only from published documentation did not produce a complete and correct set of requirements. In fact, some items listed in [Ai Ling2011a] are security controls, basically, these are solutions (e.g., data backup and recovery, cryptography and key management, high bandwidth of communications channels) as they are found in published materials. Normally, security controls are selected using a criteria that includes impact to security risk and other issues (e.g., costs and economic benefits).

7.4.3 Addressing Safety

Although hazards are considered related to safety and threats are related to security, they are also obviously related in that both involve incidents that cause harm to valuable assets. A new concept, called danger, is created to act as an abstract superclass for both hazard and threat [Firesmith 2003]. Hazard is a danger that may result in one or more related accidents. Threat is a danger that may result in one or more related attacks. As such, dangers are ways of organizing related categories of incidents. Fact is that security is defined as freedom from danger, safety in this dictionary [Merriam].

Another issue that has to be considered is the relationship between security, safety, and survivability attributes of a system and how to design requirements that do not conflict or miss similarities and differences among these attributes. Typically, safety, security, survivability, and requirements engineers use different terminologies and processes that emphasize their differences and obscure their similarities. These practices result in requirements typically lacking necessary characteristics such as verifiability and consistency.

Table 7.4 Information Security requirements mapping.

Category	IS Consumer Requirements	IS Functional Requirements
Philosophy	Confidentiality	Limited Information Access
	Integrity	Data authenticity
	Availability	Data collection and storage management
	Privacy	Compliance to privacy laws
Human Behavior	Tactical oversight monitoring system	User security policies
	Facilities misuse prevention	Encryption policies
Rule base Social System	Networking Issues	Interoperability & security
	Quality assurance	Reliable data storage system
	Mature or proprietary protocols	Cryptographic protocols
Strategic System	Cryptography & key management	Personal key exchange
	Reliable system	Trusted network
	Strategic support	Gap analysis
	Security of wireless communications	Secret protection
Hardware	Reliable device	Device and system configuration management
	High bandwidth for communication channels	Enhanced wireless technology
	Microprocessor performance, memory availability, and computing capabilities	Controlled power consumption

Source: [Ai Ling 2011a]. Courtesy of the author.

A standard terminology that is supported by a set of information models and concepts that explain the similarities between the asset based, risk driven methods for identifying and analyzing safety, security, and survivability requirements is introduced in [Firesmith 2003]. In this work, these attributes are defined as follows:

- Safety is the degree to which accidental harm is prevented, detected, and reacted to.
- Security is the degree to which malicious harm is prevented, detected, and reacted to.
- Survivability is the degree to which both accidental and malicious harm to essential services is prevented, detected, and reacted to; essential service is any mission-critical service that must continue to be provided in spite of either accident or attack.

According to [Firesmith 2003], survivability engineering is approximately the combination of safety and security engineering, the difference is the type of asset to be protected (critical services that need to be maintained). This work recommends defining safety, security, and survivability requirements as quality requirements for the quality factors of safety, security, and survivability, respectively. By using the information models, the similarities between the foundational concepts of these three disciplines (safety, security, and survivability) can be identified. This allows engineers to develop relatively uniform development processes to avoid incompleteness and ambiguity of requirements. Also, the similarities of requirements can derive common architectural mechanisms and control mechanisms (safeguards, security practices and countermeasures, survivability mechanisms) that are costly efficient and effective. All three disciplines require:

- Prevention or reduction of risks associated with hazards and/or threats.
- Recognition and response to associated accidents and attacks require to prevent, detect, or react to harm that may occur to some asset.
- Addressing potential vulnerabilities of assets to harm.

Also, a new abstract quality factor, called defensibility, is defined. This is a subclass of dependability and which, in turn, is subclassed into the three quality factors of safety, security, and survivability (see Figure 7.13). So, defensibility can act as a focal point for common concepts and related processes (e.g., asset based risk analysis) that can be inherited by its three subclasses. By combining all these attributes in a common attribute, defensibility, a standard set of quality subfactors are identified. These quality subfactors are decomposed into a standard set of lower level quality subfactors that are essentially common to safety, security, and survivability.

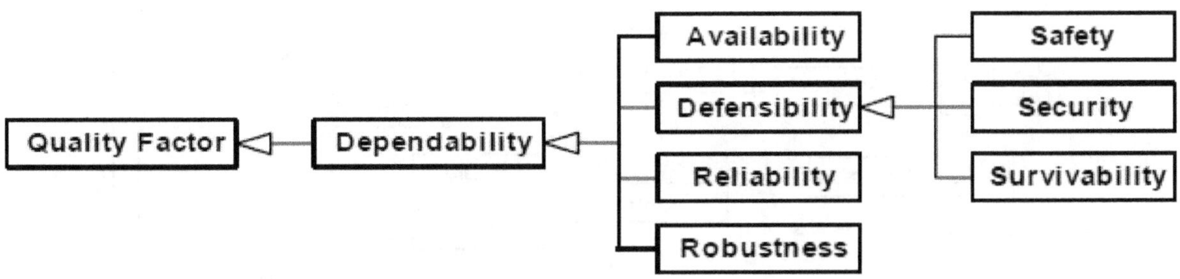

Figure 7.13 Defensibility as a kind of dependability. *Source:* [Firesmith 2003]. © 2003, CMU.

Using a common process, the defensibility requirements can be identified, analyzed, and defined (see the information model in Figure 7.14).

The defensibility requirements should be initiated in parallel with the other requirements. If the defensibility is defined after the architecture exists and components or architectural mechanisms have been determined, the resulting defensibility requirements will not have driven the architecture and will not have been checked for consistency with the other requirements. In addition, costs and time may increase when trying to add safety, security, and survivability to an existing architecture.

7.5 Secure Smart Grid Architecture

Architectural goal is one of the key goals of a Smart Grid development. Since 2009, NIST has worked cooperatively with industry to develop and refine a framework to advance Smart Grid architectures and cybersecurity. The need to understand how a Smart Grid architecture relates to other architectures is compelling discussion about concepts, methods, tools, and standards.

7.5.1 Enterprise Architecture

An enterprise architecture is a conceptual blueprint that defines the structure and operation of an organization. Enterprise architecture is the analysis and documentation of an enterprise in its current and future states from an integrated strategy, business, and technology perspective [iEAi]. Although there are many definitions for the enterprise architecture, all definitions have a reference to structure and relationships combined with a reference to a set of governing principles that provide guidance and support for directions and decisions. One known definition [Gartner 2009] states:

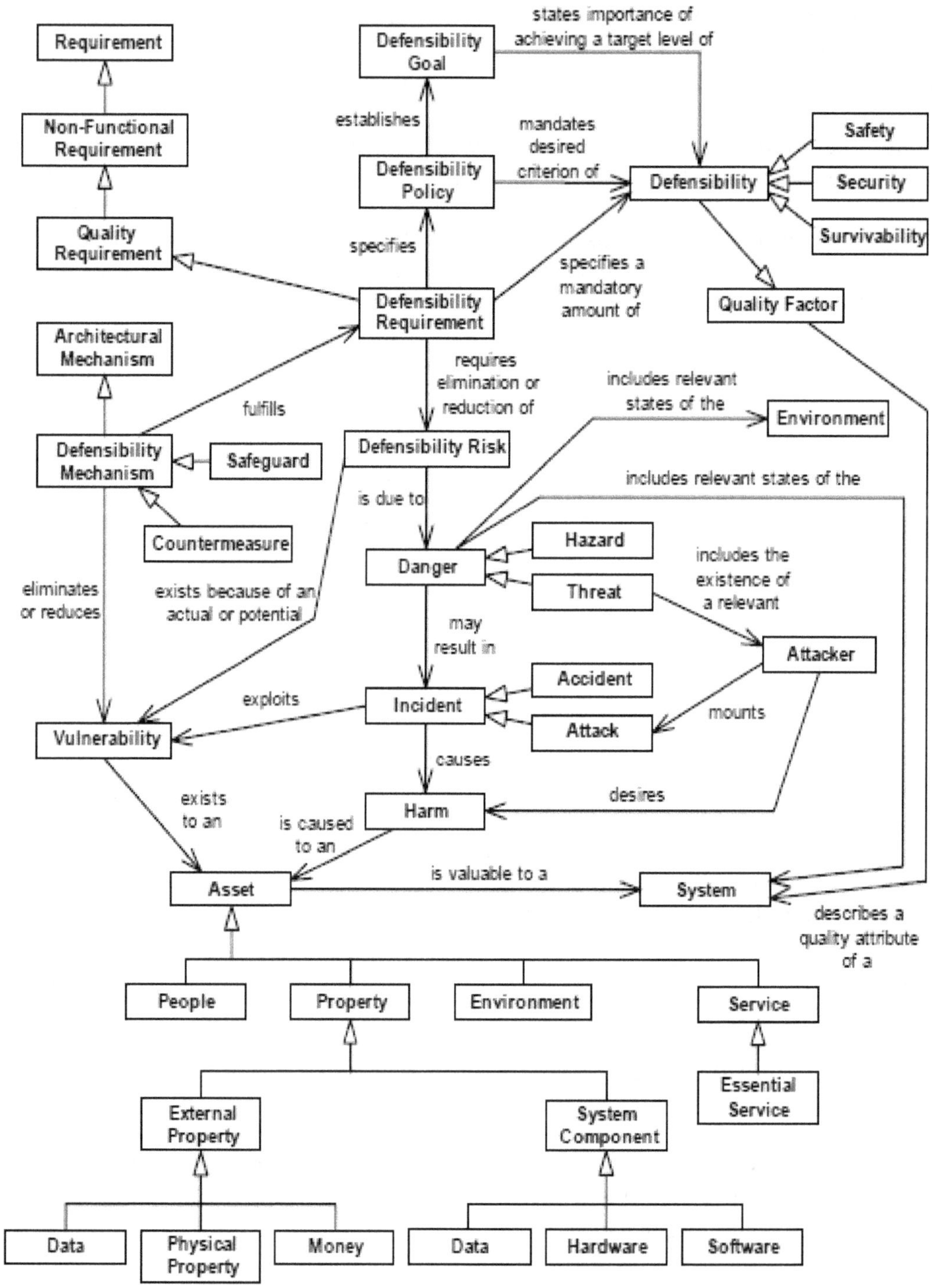

Figure 7.14 Information model for defensibility engineering. *Source:* [Firesmith 2003]. © 2003, CMU.

Enterprise architecture (EA) is the process of translating business vision and strategy into effective enterprise change by creating, communicating, and improving the key principles and models that describe the enterprise's future state and enable its evolution.

This definition is largely accepted in industry and adopted by practitioners [Clements 2010]. The benefits of the enterprise architecture can be summarized as follows:

- Defines how to organize the structure and views associated with an enterprise architecture.
- Focuses on shaping and governing the design of the future enterprise using principles to stipulate future direction and models to underpin and visualize future states. The intent of an enterprise architecture is to determine how an organization can most effectively achieve its current and future objectives [TechTarget].
- Brings together business models (e.g., process models, organizational charts, etc.) and technical models (e.g., systems architectures, data models, state diagrams) making possible to trace the impact of business changes on the systems. A key benefit to be gained from enterprise architecture is the support for decision making in changing business Enterprise architecture is a practice for conducting enterprise analysis, design, planning, and implementation, using a holistic approach at all times, for the successful development and execution of strategy [FEAPO 2018]. There are some challenges facing the users of enterprise architectures, and practical approaches for developing them incrementally over time to provide just in time utility to decision making [Martinez 2014].
- Analyzes the gaps between the future state and the currentstate, and provides the road maps and pragmatic artifacts such as requirements, specifications, guiding principles, and conceptual models that describe and support the next major stage of evolution and closing the gaps [EABOK], [FEAPO 2018].
- Encompasses the essential and unifying components of an organization [Harris 2013] which enables the ability to support decision making in changing businesses. It expresses the structure (form) and behavior (function). Enterprise architecture brings together business models (e.g., process models, organizational charts, etc.) and technical models (e.g., systems architectures, data models, state diagrams, etc.) making possible to trace the impact of organizational change on the systems, and also the business impact of changes to the systems.
- Regards the enterprise as a large and complex system or system of systems (SoS) [Manthorpe 1996]. Hardware, software, network, and application systems are merging into integrated information systems. As a result, the variety of feasible integrated system architectures and their complexity is rapidly increasing. Such systems become very large and very complex systems which are, in fact, Systems of Systems as their components are themselves complex systems [Kotov 1997].

The concept of architecture structure is applied to information, application, network, enterprise, information security, and system. In practice, an enterprise architecture covers several views that blend together to form this architecture. Other views may include Information Architecture, Business Architecture, and Governance Architecture, Security Architecture [Op't Land 2009]:

- Information Architecture or IT architecture is the structure and relationships of IT systems including the way IT supports the enterprise to achieve its business goals. The design artifacts describe the structure of components, their inter relationships, and the principles and guidelines governing their design and evolution over time. It also defines the sources and destinations of information, its flow through the enterprise, as well as the rules for persistence, security and ownership.
- Business Architecture is the integrated structure of the overall business itself (in terms of organization, people and processes, and resources). Business architecture supports busi-

ness change with a more holistic perspective. This approach is becoming more important with the move toward service oriented architecture at the business level.

- Governance Architecture addresses the full range of governance, from business governance (how to manage overall business processes, both formal and informal) to organizational and systems governance and also IT systems management capabilities.
- Security Architecture addresses the full range of security, from business and information security to IT security. It also addresses the required security for organizational and business related services. It is often linked to governance aspects to address security management.

Another view of the enterprise architecture is as a service oriented architecture composed of:

- Business Architecture
- Information Architecture
- Technology Architecture
- Security Architecture.

Another definition of enterprise architecture consists of four types of subset architectures:

- Business Architecture (similar to Business Model in the Zachman Framework)
- Data Architecture (IT has, since the very beginning of computing, always separated process and data)
- Applications Architecture (a blueprint for individual application systems, their interactions, and their relationships to the core business processes of the organization)
- Technology Architecture (enables the logical and physical application and data components and the architecture vision, addressing the request for architecture work and stakeholder concerns)

Figure 7.15 shows one dimensional view of enterprise architecture as a service oriented architecture, called BITS (Business Architecture, Information Architecture, Technology Architecture, Security Architecture).

BITS is one dimensional view of Enterprise Architecture as a Service orientated architecture. It also reflects the addition of Security to the enterprise architecture family.

Because the discipline of enterprise architecture is so broad, and because enterprises can be large and complex, the models associated with the discipline also tend to be large and complex. To manage this scale and complexity, an architecture framework provides tools and methods that can bring the task into focus and allow valuable artifacts to be produced when they are most needed. Also, in order to manage the scale and complexity of the power systems (described as a System of Systems (SoS) [P2030 2011]), there is a need for an architectural framework that provides tools and approaches to help architects abstract from the level of detail that builders work at to bring enterprise design tasks into focus and produce valuable architecture description documentation.

An architectural framework includes conventions, principles and practices for the description of architectures established within a specific domain of application and/or community of stakeholders [ISO/IEC/IEEE 42010]. It is a tool for assisting in the production of organization specific architectures and it consists of a technical reference model, a method for architecture development and a list of component standards, specifications, products and their interrelationships which can be used to build up architectures [TOGAF].

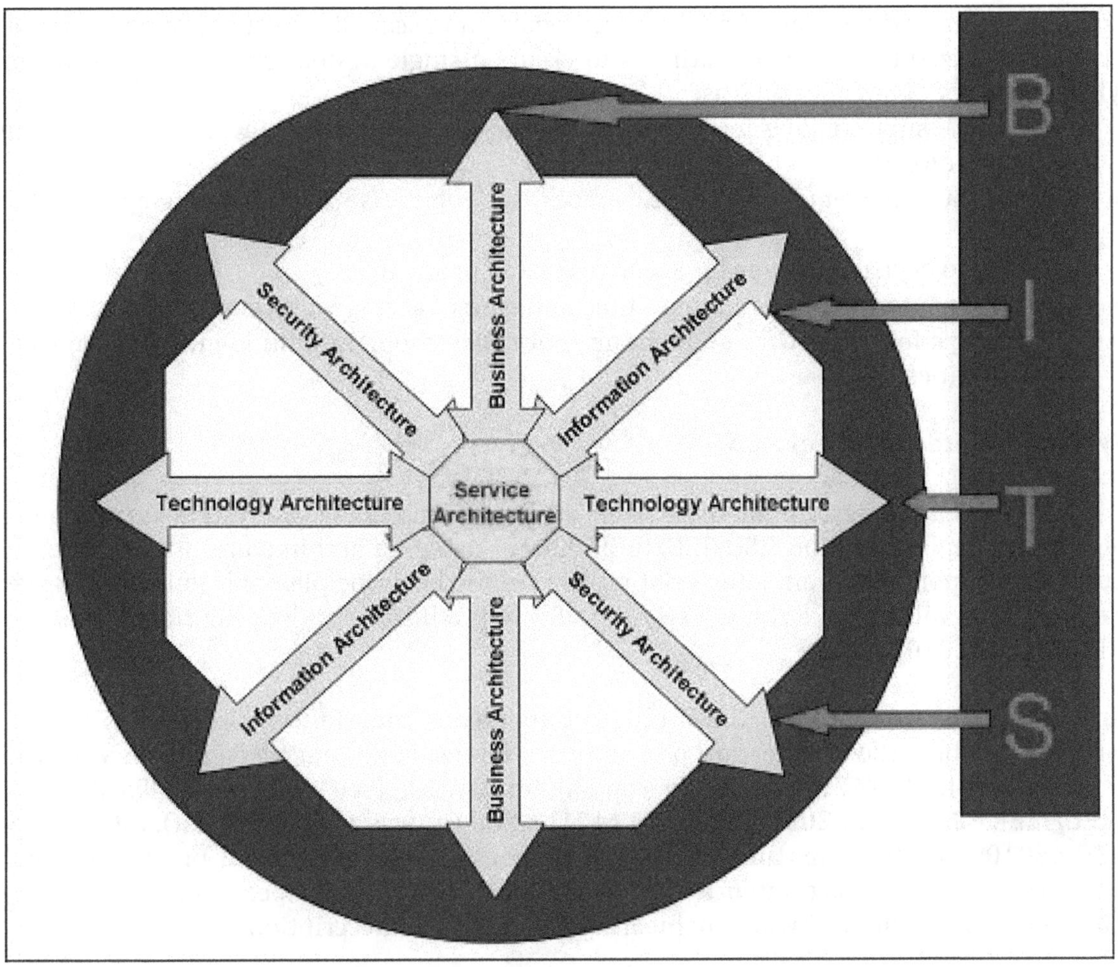

Figure 7.15 BITS view of enterprise architecture. *Source:* [Huxham 2006a]. Public Domain.

Examples of enterprise architecture frameworks include Zachman and Defense frameworks (DODAF, TOGAF, MODAF, FEAF, TEAF, MDA), AGATE (adopted a standard meta model which defines the critical architectural elements and the dependencies between them); open source frameworks (IDEAS Group supports a common ontology for architecture interoperability); RM-ODP (The Reference Model of Open Distributed Processing based on [ITU-T X.901-X.904] and [ISO/IEC 10746-3] defines an enterprise architecture framework for structuring the specifications of open distributed systems). The analyses of comparing modeling or enterprise architecture frameworks is challenging since the scope of the different frameworks varies [Bouwens 2013]. The author argues that some frameworks focus on providing a complete taxonomy where others focus on process or present a methodology.

7.5.2 Electrical Utility Architecture

As utilities converge their OT and IT operations, there is an increase of data collected from different devices and sensors. In addition to smart meters, utilities have to manage increasing amounts of data from devices, technologies and platforms such as workforce management systems, substation management systems, RTU upgrades, GIS, outage management systems, distribution automation, smart homes, and PMU sensors on power lines. Utilities need to establish the infrastructure to support not only processing huge amounts of data, but also processing data in real time and sharing the data with customers and other energy organizations. Therefore, it is critical to have an overarching architecture in place that is secure, reliable and scalable. A robust data and application integration architecture will enable utilities to create a single system or record, and this in turn will yield a number of benefits such as:

- Strategic – support long term view of technical direction, not invalidated by changes in technical direction.
- Consistent management of IT risk across the enterprise while leveraging industry best practices.
- Support to decision making for security related problems.
- Promoting interoperability, integration, and ease of access while managing the risk.
- Providing a focal point for eliminating root cause problems and identifying improvement opportunities.

7.6 System Architecture Overview

Architecture is a formal description of a system, or a detailed plan of the system at component level, to guide its implementation [ISO/IEC/IEEE 42010] using an architectural framework. It is the structure of components, their inter relationships, and the principles and guidelines governing their design and evolution over time [TOGAF3]. There is a history of the evolution of this standard [ISO/IEC/IEEE 42010].

In 2011, the standard [ISO/IEC/IEEE 42010] supersedes the standard [IEEE 1471], which is known for describing the architecture of a software intensive system, also known as software architecture. The [IEEE 1471] uses a conceptual framework (see more in [Hilliard 2007] and [ISO/IEC/IEEE 42010]). In 2007, the [IEEE 1471] standard was adopted by ISO/IEC JTC1/SC7 as ISO/IEC 42010:2007 (Systems and Software Engineering -- Recommended practice for architectural description of software intensive systems). In 2011, it was superseded by ISO/IEC/IEEE 42010:2011 (Systems and software engineering - Architecture description).

A system is a type of enterprise architecture component that is comprised of hardware, and software, and activities that has inputs and outputs [iEAi]. System architecture is the conceptual model that defines the structure, behavior, and more views of a system. Associated with this definition is the concept of architecture description which is a formal description and representation of a system using a language called architecture description language (ADL). Architecture description is governed by a framework which is an encapsulation of a minimum set of practices and requirements for artifacts that describe a system's architecture. Models are representations of how objects in a system fit structurally in and behave as part of the system. Views are a partial expression of the system from a particular perspective (e.g., security). A viewpoint is a set of representations (views and models) of an architecture that covers a stakeholder's issues. Concerns drive viewpoint selection and each concern is addressed by an architectural view. Each viewpoint can be specified by modeling techniques, or analytical methods. A viewpoint may also include:

- Any consistency or completeness checks associated with the underlying method to be applied to models within the view.
- Any evaluation or analysis techniques to be applied to models within the view.
- Any heuristics, patterns, or other guidelines which aid in the synthesis of an associated view or its models.

The standard [ISO/IEC/IEEE 42010] is organized around the terms and concepts of the diagram shown in Figure 7.16. It captures terms and concepts of systems and their architectures, as a context for understanding architecture description, architecture frameworks, and views. It depicts the contents of an architecture description (AD) and the relations between those content items when applying the standard to produce an AD to express an architecture for some system of interest.

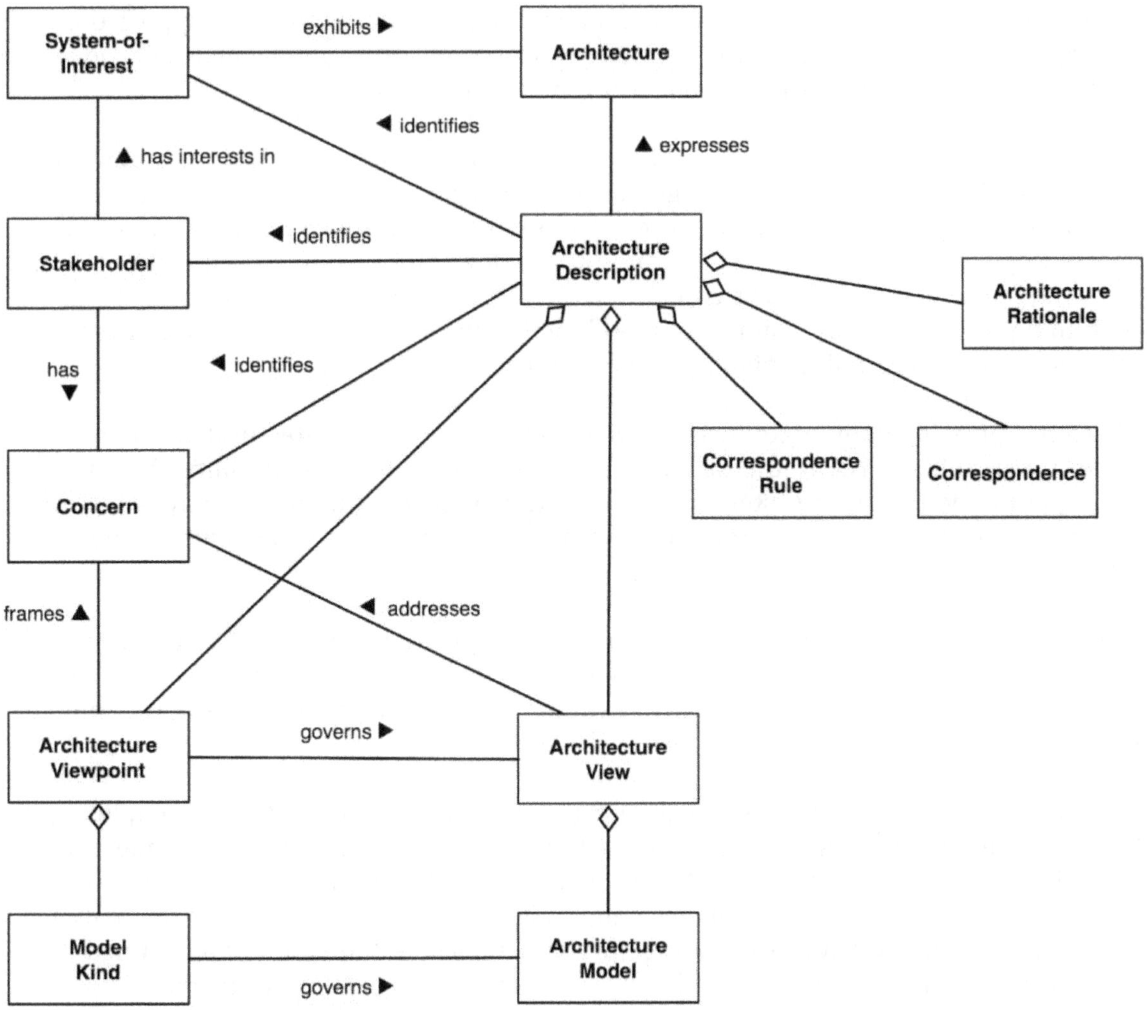

Figure 7.16 Conceptual model of an architecture description based on [ISO/IEC/IEEE 42010].

The architecture of the system describes the fundamental concepts or properties of a system in its environment embodied in its elements, relationships, and in the principles of its design and evolution [ISO/IEC/IEEE 42010]. The standard is based upon a conceptual model – a meta model of the terms and concepts pertaining to AD. An AD is an artifact that expresses an architecture. The conceptual model is presented in the standard using UML class diagrams to represent classes of entities and their relationships. Architects and other system stakeholders use ADs to understand, analyze and compare architectures, and often as blueprints for planning and construction. ADs are the primary subject of ISO/IEC/IEEE 42010 and the focus of this diagram.

Also, the standard [ISO/IEC/IEEE 42010] defines architecture framework and specifies requirements on architecture frameworks. Architecture frameworks and ADLs are two mechanisms widely used in architecting. Instances of each can be specified by building on the concepts of Architecture Description (see the contents of an Architecture Framework (AF) in Chapter 2).

In this work [Hilliard 2012], the author argues that majority of software architects must operate within the confines of a prescribed architecture framework or ADL as dictated by their organization or client. Issues related to the rigor of the levels of tool support with AFs and ADLs are making difficult for the architect to provide a representational solution to the specific challenges of the project. Therefore, the authors propose an automated infrastructure to support the architecture description related activities of the architect is proposed. The approach has been evaluated in the context of a complex, real world, public transportation system.

The environment of a system includes developmental, technological, business, operational, organizational, political, economic, legal, regulatory, ecological and social influences. As shown in Figure 7.17, concepts, principles and procedures of architecting are applied to help manage the complexity faced by stakeholders of systems. Conceptualization of a system's architecture, as expressed in an architecture description, assists the understanding of the system's essence and key properties pertaining to its behavior, composition and evolution, which in turn affect concerns such as the feasibility, utility and maintainability of the system.

There is a growing body of knowledge in the application of architectural concepts to systems to attain the benefits of reduced costs and increased quality, such as usability, flexibility, reliability, interoperability and other system qualities. However, there is no universally agreed definition of which aspects constitute a system architecture, and various organizations define it in different ways.

One describes system architecture as a generic discipline to handle systems, in a way that supports reasoning about the structural properties of these systems. Also, system architecture is the response to the conceptual and practical difficulties of the description and design of complex systems [Golden].

Also, a system architecture is more than a model or a method to design complex systems, it is based on thinking with a systemic approach and architectural principles, while overcoming the difficulties related to complexities (technical, social, and cognitive) of the system and its design. Commonly, system architecture involves multiple views, with each view as an analytical description of the system. However, the architecture of complex systems cannot be defined by an analytical decomposition and there is a need to use best practices from various fields.

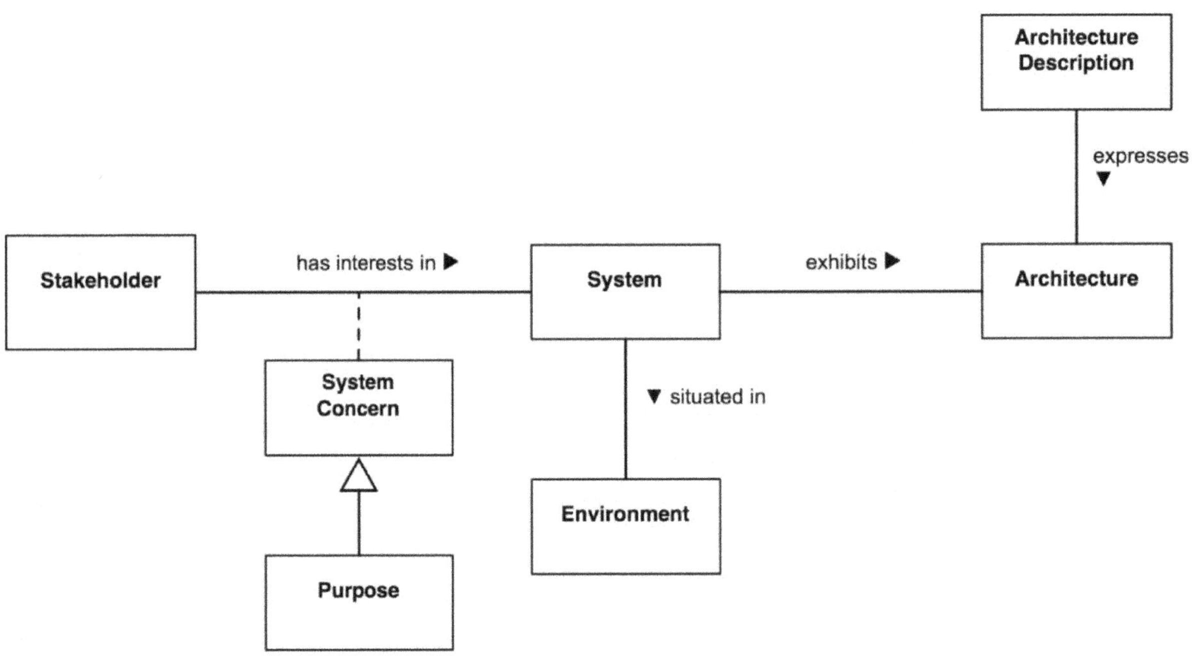

Figure 7.17 Concepts of system and architecture based on [ISO/IEC/IEEE 42010].

A practical, heuristic approach to systems architecting including methods for integrating business strategy with technical architectural decision making are described in [Maier 2009]. This book includes also practical guidelines for avoiding common systematic failures while implementing new systems. System architecture can be contrasted with system architecture engineering, which is the method and discipline for effectively implementing the architecture of a system [Firesmith 2008]. The importance of an architecture can be summarized in these domains of importance as a way to understand complex systems, to design them, to manage them, and to provide long term rationality by means of standards [Crawley 2004]. Complex systems have behaviors and properties that no subset of their elements have. These behaviors are products of the interactions between the elements. They may be anticipated and designed in, or they may be unanticipated, in which case they are called emergent. Both anticipated and emergent properties may be desirable or undesirable.

The system architects are attempting to address the enterprise architecture goals such as effectiveness, efficiency, agility, and durability [Bass 2012]. Although many new models and frameworks were developed, there is no single comprehensive model and no single universal approach from which all essential characteristics of a system can be derived [Grady 2009]. The author compares different frameworks and methods and concludes that:

- All models attempt to support several views using different modeling artifacts and rules.
- Each model has useful unique features that are helpful in analyzing problem space.
- No model is comprehensive such that a single model can be used as the basis for analyzing problem space to define all of the appropriate requirements for a given problem space.
- This situation causes a very ineffective approach in integrating and optimizing across these available models resulting in some difficulty in hardware and software integration.
- A solution is to use a universal architecture framework that would reorder the software development pattern to encourage improved hardware software integration and management visibility while simultaneously improving the software development process.
- A universal architecture description framework can be employed on any program no matter the problem space or the intended methods of implementation.

Examples of architecture frameworks include MODAF, TOGAF, RM-ODP, SABSA, OSS, etc. (see [ISO Survey]). In the long list of examples, these two architecture frameworks – SABSA and OSSAF – are focused on security. Also, they do not resemble the [ISO/IEC/IEEE 42010] architecture framework, the following is a brief description:

- SABSA (Sherwood Applied Business Security Architecture) is a model and a methodology for developing risk driven enterprise information security architectures and for delivering security infrastructure solutions that support critical business initiatives. The model comprises six layers: the Business View: Contextual Security Architecture; the Architect's View: Conceptual Security Architecture; the Designer's View: Logical Security Architecture; the Builder's View: Physical Security Architecture; the Tradesman's View: Component Security Architecture; the Facilities Manager's View: Operational Security Architecture.
- OSSAF (Open Safety & Security Architecture Framework) aims to provide a simple, open and prescriptive approach to coordinate the perspectives of different types of stakeholders within an organization's security from strategic, operational, functional and technical view.

A system architecture consists of a comprehensive modeling set that includes coverage of the problem and solution spaces plus a means to capture the requirements derived from the models [Grady 2009]. However, typical SoS (e.g., Smart Grid) have to satisfy many strict requirements, among which are cost effectiveness (SoS are often unique and very expensive), responsiveness, throughput, scalability and extensibility, availability, maintainability, reliability, fault tolerance, recoverability, data and application integrity, security. Many of these requirements are conflicting. Therefore, modeling can help to find the best compromise solutions. With advances in technology, architectural frameworks are evolving to fill the need for characterizing and developing complex system and SoS environments.

It is recommended that selection of a systems framework approach for any class of SoS problems begins with an exploration of the requirements of the community developing the system of interest and the goal of implementing the framework whether it is to capture a view of the system or to perform specific analysis of system characteristics [Bouwens 2013].

7.6.1 System Architecture Types

Types of system architectures include [Bass 2012]:

- Hardware architecture refers to the identification of a system's physical components and their interrelationships.
- Software architecture is the set of structures needed to reason about the system, which comprise software elements, relations among them, and properties of both including documentation [Clements 2010], [Bass 2012].

In general, a system consists of a mix of ingredients that include not just software, but also hardware, people, processes, and services. According to [Schultz 2007], the architecture is a complex effort that must balance the needs of the business with the reality of how technology can support it. Every architecture requires principles that allow measurement and control attributes that then enforce consistency of usage.

7.6.2 Architecture Principles

Principles are rules that are fundamental to making decisions related to the design and construction of the architected system. Architecture principles focus on two main areas:

- Developing the architecture
- Implementing the architecture

Most common principle groupings that are applied to architecture design include the following:

- General architecture
- Business goals
- Technology
- Information/data
- Management
- Industry trends
- Security
- Testing.

Principles are created from business objectives and architecture drivers (business and design) and are used as the foundation for governance over the development of the architecture and the implementation of the architecture. Architecture principles, required capabilities, and requirements are all used to make architectural decisions, which reflect the created architecture. In [Muller 2011], the author provides systems architects insight into the business process processes, and especially into the processes to which they actively contribute. Also, this work discusses the relationship of the systems architect with the enterprise and management as well as methods and techniques that facilitate the architect's contribution to the business strategy process.

7.7 Enterprise Security Architecture

The security architecture is an architecture view (work product) expressing the architecture of a system from the perspective of specific system security concerns. The security architecture addresses the full range of security, from business and information security to IT security. It also addresses the required security for organizational and business related services. It is often linked to governance aspects to address security management.

Enterprise security architecture is a part of the enterprise architecture focusing on information security throughout the enterprise. Although often associated strictly with information security technology, it relates more broadly to the security practice of business optimization in that it addresses business security architecture, performance management and security process architecture as well. While some organizations prefer to deploy adhoc security solutions to protect their information assets against a multitude of threats resulting in an elevated unmet needs, other organizations pursue a more strategic approach by understanding the security requirements and developing an enterprise security architecture and solutions to ensure business needs are met.

An enterprise security architecture includes the information security strategy consisting of policies, standards, and procedures and the way they are connected across an enterprise strategically, tactically, and operationally. Besides security, the benefits of enterprise security architecture include better interoperability and integration.

Although the enterprise security architecture can follow the same model as the security program discussed in the previous chapters, it provides more granularity because it works its way from the policy level to the component level. Therefore, when developing an enterprise security architecture, it is required to support the following:

- Strategic alignment - business drivers and the regulatory and legal requirements are met by the security architecture.
- Process enhancement - refinement and integration in system life cycles and day-to-day operations.
- Business enablement - core business processes are integrated into the security operating model.
- Security effectiveness - how security solutions perform in meeting the defined requirements.

The practice of enterprise security architecture involves developing an architecture security framework to describe a series of operational, tactical, and strategical reference architectures.

One recommends developing two types of Security Architecture designs: logical and physical [Farah 2004]:

- A logical security architecture design is needed to organize the logical structure and implement security in all identified architectures. The logical structure includes information, processes, technology, and people.
- Physical architecture design includes network diagrams illustrating security controls (e.g. firewalls, virtual private networks, etc.), internal and external connections and devices used, and diagrams of other architectures in relation to security architecture.

The security architecture component of the enterprise architecture (see BITS View) is designed so that the A-I-C (Availability-Integrity-Confidentiality) goals of information security can meet the business and security needs of an organization. It is standards based, and business driven blueprints can be used to develop and implement an enterprise security program, core security architecture and operating model architecture. It is important to think of security architecture as including several underlying architectures. The key components of the security architecture include:

- Management processes that leverage foundations for design, business processes, development life cycles and day-to-day operations, evolutionary integration of security measures across the enterprise.
- Compliance and monitoring to report security performance and effectiveness by measuring key performance metrics
- Security models
- Security zones and perimeters
- Information classification (e.g., confidential, private, sensitive, public)
- Solution implementation which defines, develops, and implements security architecture consistent with the organization's defined risk profile to ensure end-to-end security services (availability, integrity, confidentiality, and non-repudiation).

Security architecture is a high level design used to satisfy a system's security requirements as defined in an organization's security policy. Besides enterprise security architecture and system security requirements, developers design a system security architecture.

7.7.1 System Security Architecture

Security architecture is a set of physical and logical security relevant representations (e.g., views) of system architecture that conveys information about how the system is partitioned into security domains and makes use of security relevant elements to enforce security policies within and between security domains based on how data and information must be protected.

The security architecture reflects security domains, the placement of security relevant elements within the security domains, the interconnections and trust relationships between the security relevant elements, and the behavior and interactions between the security relevant elements. The security architecture, like the system architecture, may be expressed at different levels of abstraction and with different scopes.

The security architecture of a system is very important and comprises many aspects such as:

- Security policy is the foundation for the specifications of security for a system; it provides the baseline for evaluating a system after it has been built.
- Core tenets include trusted computing base, security perimeter, reference monitor, and the security kernel are requirements described initially in Trusted Computer System Evaluation Criteria TCSEC, known as Orange Book, which it has been replaced by the Common Criteria, originally published in 2005 [Common Criteria]. Common Criteria is an international standard used to assess the effectiveness of the security controls built into a system from functional and assurance perspectives. Common Criteria also replaces the European Information Technology Security Evaluation Criteria (ITSEC) first published in 1990.
- Security models are important concepts in the design and analysis of secure systems; security model is a symbolic representation of a security policy.
- Security modes of operation describe the security conditions under which the system actually functions. Security modes include dedicated security mode, system high security mode, compartmented security mode, and multilevel security mode [Harris 2013]. The multilevel security mode permits two or more classification levels of information to be processed at the same time. All users can access some data, based on their need to know, clearance and formal access approval.

System security architecture must describe the system's components, their connections, their interactions, security controls, and the nature of the interactions between the system and its environment [Barbacci 1999]. Evaluating a system design before it is built is good engineering practice. The architecture should include the factors or parameters of interest for each attribute model. Parameters that are common to more than one attribute model influence multiple attributes and can be used to tradeoff between attributes. For each of the attribute models, we must identify those parameters that have a major effect on the results for that model. A sensitive parameter is one that has a great impact on the model. In the past, methods used for the development of critical computer systems were focused on a quality attribute [Barbacci 1995]:

- Performance - from the tradition of hard real time systems and capacity planning.
- Dependability - from the tradition of ultra reliable, fault tolerant systems.
- Security - from the traditions of the government, banking, and academic communities.
- Safety - from the tradition of hazard analysis and system safety engineering.

Systems often fail to meet user needs (e.g., lack quality) when designers narrowly focus on meeting some requirements without considering the effect on other requirements or by taking them into account too late in the development process.

Building security into information systems is not a matter of referring to a checklist of technical and procedural controls and applying the appropriate security measures on the list [Sherwood 2009]. The authors argue that it requires a model for security architecture such as SABSA layered model that includes the layers (views) as business, logical architecture, design, physical architecture, component, service.

At the top layer is the business requirements definition stage. With each lower layer a new level of abstraction is developed until the selection of technologies and products (component architecture at the lowest layer (in this respect, SABSA aligns closely with ITIL v3). There is a growing body of knowledge in the application of architectural concepts to systems to attain the benefits of reduced costs and increased quality, such as usability, flexibility, reliability, interoperability and other system qualities. When systems exhibit similar architecture patterns, they may use common security architecture models. The existence of such models allows to develop tools to guide future engineering efforts toward more secure solutions. Security patterns are reusable solutions to security problems. A survey of approaches to their integration in the system development is discussed in [Yoshioka 2008].

Therefore, achieving an architecture that ensures information and interoperability in a secure, reliable, and timely manner requires several key elements such as repeatable methodology, standardized output models, formal validation, governance, collaboration guidelines, configuration management, tools, and patterns.

We need to understand that the system security architecture is a component of enterprise security architecture, which includes the overall physical and logical components that make up security architecture in the enterprise. Security Architecture is simply a view of an overall system architecture from a security perspective. It provides some insight into the security services, mechanisms, technologies, and features that can be used to satisfy system security requirements. The security architecture is only one aspect of the enterprise or system architecture, which may also include network architecture or physical connectivity architecture [Hansche 2004]. Security architecture is more of a design overview, describing at abstract level the relationships between key elements of the hardware, operating systems, applications, network, and other required components to protect the organization's interests. Recommendations for the development include [Hansche 2004]:

- Security architecture should also describe how the functions in the system development process follow the security requirements.
- During the beginning stages of system development processes, the security architecture should outline high level security issues, such as the system security policy, the level of assurance required, and any potential impacts security could have on the design process.
- As the system is developed, the security architecture should evolve in parallel, and may even need to be slightly ahead of the development process so that the security requirements will guide the development process.

With many definitions about architecture and security that may have some differences, we need to apply some rules based on principles.

7.7.2 Security Architecture Principles

Architecture principles are needed to define the underlying general rules and guidelines for the use and deployment of all security mechanisms, resources, and assets across the enterprise. They reflect a level of consensus among the various elements of the enterprise and form the basis for making future security decisions. Given the broad range of topics spanned, security architecture is often defined within a common framework that describes all of the components and provides a structure for process and governance. The primary purpose of creating an enterprise security architecture is to ensure that business strategy and IT security are aligned. As such, enterprise security architecture allows traceability from the business strategy down to the underlying technology. Security by design and enterprise information security architecture were first formally positioned by Gartner in their whitepaper [Kreizman 2006]. The authors argue that by including security requirements in the EA process and security professionals in the EA team, enterprises can ensure that security requirements are incorporated into priority investments and solutions. Since this publication, security architecture has moved from being a silo based architecture to an enterprise focused solution that incorporates business, information, and technology.

Security principles are value statements that the business requires for the delivery of security. Examples of security principles are identified in [Schultz 2007] as follows:

- Defense in depth obtained by layering defenses.
- Managed risk such as security risk and security controls should be balanced according to business objectives - security controls should be proportionate to risk.
- Security by design such that security considerations should begin with the requirements phase of development and be treated as an integral part of the overall system design.
- Requirements based access such that a user (human or computer) should only be given enough privileges to do those tasks needed to perform a specified job activity, function, or task; no more, no less.
- Transparency such that security should be user transparent and not cause users undue extra effort; the administration and configuration of security components should not be overly complex or obscure.
- Resilience in response by designing and operating systems to limit vulnerabilities and to be resilient in response.
- Enforced policy by implementing processes, procedures, and systems that promote enforcement of organizational security policies.

7.8 Standards, Guidelines, and Recommendations

Additional guidance for designing cybersecurity for Smart Grid and DER systems may be required.

7.8.1 NIST Cybersecurity Framework Profile for Smart Grid

In Smart Grid environments, power system owners/operators rely on and interact with a larger community of diverse third parties than in traditional grid environments. These third parties include but are not limited to vendors, suppliers, contractors, distributed generation owners/operators, and consumers. Furthermore, supply chain risk management considerations are relevant in these relationships especially when Smart Grid devices and systems are interconnecting with third parties.

In addition to using the cybersecurity framework [NIST 2018] and Smart Grid Profile [NIST 2051TN], power system owner/operators may need to use more specific guidance when assessing and managing risks associated with third parties. This guidance include cybersecurity procurement language for energy delivery systems [ESCSWG 2014] and utilities telecom council (UTC) white paper [UTC 2015].

Energy delivery systems are used to monitor and control the production, transfer, and distribution of energy. These systems include SCADA systems, Energy Management Systems (EMSs), Distribution Management Systems (DMSs), and Distributed Control Systems (DCSs). Energy delivery systems comprise components such as:

- Sensors and actuators used for monitoring and controlling energy delivery processes.
- Computer-based systems that analyze and store data.
- Communication pathways and networks that interconnect the various computer systems.

Also, this standard [ISO/IEC 27019] provides requirements and guiding principles by extending code of practice for information security controls described in [ISO/IEC 27002:2013] to process control systems and automation technology as used in the energy utility industry. The standard describes requirements for a general and overall information security framework for the process control domain of the energy utility industry including security requirements for controlling and monitoring the production or generation, transmission, storage and distribution of electric power, gas, oil and heat, and for the control of associated supporting processes. Energy utility organizations should consider the continuity of the general energy supply as one of the key elements of business continuity management while ensuring safety of the general public and the security of assets. This standard [ISO/IEC 27031] provides guidance for information and communication technology readiness for business continuity.

7.8.2 Interoperability Standards

As an attribute of the electric grid, interoperability is challenging to achieve. Interoperability is foundational to ensuring that new technologies can be cost-effectively integrated with the legacy power system. It is also foundational to ensuring that diverse, distributed, and decentralized stakeholder groups can realize the anticipated benefits of Smart Grid. The Appendix J of [NIST SP1108r4] includes a subset of 169 standards from a set of about 240 standards (drawn from multiple sources that are relevant to the Smart Grid). From this subset relevant to interoperability a small percentage of standards were assessed for the testing and certification program. The reviews and methodology for the standards analysis, with detailed descriptions of each of the 240 standards reviewed are documented in [NIST 2042TN].

7.8.3 Data/Information Models

Within the field of Software Engineering, data models and information models are used to develop software for information systems. A summary of existing and emergent canonical data models in Smart Grid standards include [NIST SP1108r3]:

- CIM (Common Information Model) - aims to allow applications to exchange information about an electrical network; it defines a common vocabulary and basic ontology for aspects of the electric sector; CIM was adopted by IEC (see IEC 61968 and IEC 61970 standards); IEC 62325 series of standards is a CIM and framework for energy market communications encompassing European style and North American style markets [Uslar 2012].
- MultiSpeak – is a standard for software interoperability in the power domain mainly applied in the US; the mapping is performed on profile level. A harmonization of CIM and Multispeak means that each CIM profile is mapped onto an appropriate MultiSpeak element; however gaps have been identified.
- IEC 61850 –a standard for the design of electrical substation automation; the abstract data models defined in IEC 61850 can be mapped to a number of protocols (e.g., MMS (Manufacturing Message Specification), GOOSE (Generic Object Oriented Substation Event), SMV (Sampled Measured Values), and Web Services (a software system designed to support interoperable machine-to-machine interaction over a network). A mapping of IEC 61850 objects is provided in IEC 61400-25 standard; IEC 61400 is a set of standards for designing wind turbines.
- BACnet - a data communication protocol for building automation and control networks; it describes a method for representing any type of building automation equipment in a standard way; it defines messages that can be sent across a computer network to monitor and control such equipment.
- SEP (Smart Energy Profile) 2.0 aims to create a standard and interoperable protocol that connects smart energy devices in the home to the Smart Grid; it is an emerging standard for enabling informed consumer participation and implementation of Demand Response Programs by the Utilities; SEP 2.0 resource representations are built to be compatible with CIM; a coalition of Alliances had been formed (composed of Wi-Fi Alliance, ZigBee Alliance, HomePlug Alliance and HomeGrid Alliance) with the intent of developing SEP 2.0 application level interoperability testing and certification program.
- The American National Standards Institute (ANSI) suite of protocols, called Smart Grid Meter Package, provides requirements and guidance on electricity metering, device data tables, meter interfacing to data communication networks, etc.
- ANSI C12.19 standard defines a table structure for utility application data to be passed between an end device (usually, an electric meter) and a computer (usually, a handheld device carried by a meter reader, or a meter communication module which is part of an Automatic Meter Reading system).
- OASIS (Organization for the Advancement of Structured Information Standards) provides a suite of standards consisting of three parts; the first two parts define information models, expressed as an XML vocabulary for describing information in their scope; the third is an information model, architecture, and set of Web services definitions for communicating among participants in the Smart Grid.

Chapter 8

Perspectives on Smart Grid Architectures

8.1 Smart Grid Architectures and Requirements

A significant view of any Smart Grid system architecture is the architecture of industrial control networks that generally have a much deeper architecture than commercial networks [Galloway 2013]. However, there are many perspectives to look at Smart Grid resulting in different architectures, each representing a perspective. Services are performed by one or more roles within a domain (see NIST updated model in [NIST SP1108r4]). A relevant requirement is to identify services within a domain that are performed by one or more roles. Examples of services include home and building automation, distributed energy resources (DERs) and demand/response (DR) customer load control, and near real time wide area situation awareness (WASA).

With the move towards establishing the Smart Grid, regarded as the next generation power grid, there are many proposals for the major systems (e.g., [Fang 2012], [Sollecito 2009], [EPRI 2012]) and enabling technologies for the Smart Grid to address reliability, security, and privacy issues.

In addition, a modern power grid needs to become smarter in order to provide an affordable, reliable, and sustainable supply of electricity [Li 2010]. Several application domains such as Advanced Metering Infrastructure, Smart Home, Distributed Generation/Microgrids, and Wide Area Measurement and Control are also key in driving the Smart Grid's development. Therefore, these developments require an organized effort to define an architecture for the Smart Grid. This could overcome isolated development and unite power system organizations to provide a more capable, secure, and manageable energy provisioning and delivery system [Sollecito 2009].

8.1.1 Conceptual Smart Grid Architecture

This report [EPRI 2002] recommends an architecture for Smart Grid should be technology independent, based on standard common services, a common information model, and generic interfaces to connect it together. Figure 8.1 shows a conceptual view of such architecture proposed in [Sollecito 2009].

This architecture highlights the use of adapters to accommodate the heterogeneity and the use of standard information models and standardized set of abstract interfaces applicable to a variety of utility services. Key security decision areas that provide the basis for consistency across the architecture include:

- Security zones – definition of various relationships and boundaries between the Smart Grid domains and business partners; a framework to support a consistent approach for managing risk and supporting security policy.
- Information classification – consistent definition of the various levels of security sensitivity attached to data.

The organization (e.g., utility) must design and implement a process that ensures continual movement from the current state to the future state. The future state will generally be a combination of one or more closing gaps between Information Technology (IT) and Supervisory Control And Data Acquisition (SCADA) systems, present and desired future security strategy, upgrades and replacements of hardware and software on both IT and SCADA systems.

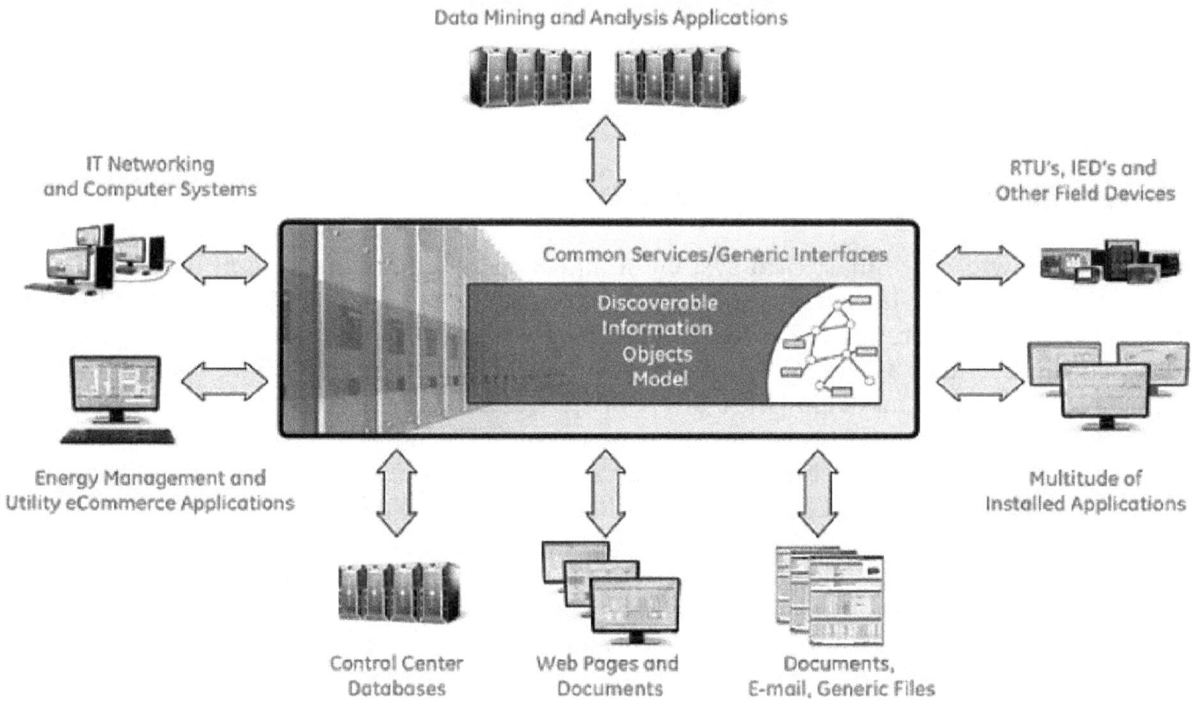

Figure 8.1 Smart Grid architecture. *Source*: [Sollecito 2009]. © 2009, GE Digital Energy.

One issue that needs to be addressed is mapping this architecture to the enterprise architecture including other specific architectures such as energy enterprise architecture, security architecture, and information architecture.

8.1.2 Smart Grid Architectural Model

A Smart Grid Architectural Model (SGAM) provides architectural guidance for achieving interoperability of such massively scaled and distributed system as the Smart Grid [NIST SP1108r3]. Architectures should support standard interfaces with other systems and manual processes if a standard exists. This includes interoperability among third party products and management and cybersecurity infrastructures.

The SGAM utilizes several information and communications technology (ICT) architecture standards. As described in [NIST SP1108r3], it allows to:

- Represent existing Smart Grid architectures in a universal way
- Represent current and future Smart Grid scenarios and also migration paths
- Validate Smart Grid use cases in respect of standards' support.

8.1.2.1 NIST Smart Grid Architectural Model (SGAM)

NIST developed a conceptual domain model to support planning, requirements development, documentation, and organization of the diverse, expanding collection of interconnected networks and equipment that will compose the Smart Grid. For this purpose, NIST adopted the approach of dividing the Smart Grid into seven domains (generation, transmission, distribution, operations, market, customer, and service provider) with specified roles and services. The conceptual domain model described in [NIST SP1108r1] in 2010 is the basis for conceptual Smart Grid architecture that evolved to SGAM.

The SGAM is a template for architects to follow while building aspects of a Smart Grid architecture, regardless of an architect's specialty (such as in areas of transmission, distribution, IT, back office, communications, asset management, and grid planning). SGAM utilizes an enterprise wide, service oriented approach to describe a Smart Grid architecture.

A definition of the architecture in the context of SGAM model is described as the structure and overall organization of the Smart Grid from the point of view of its use or design. This includes technical and business designs, demonstrations, implementations, and standards that together convey a common understanding of the Smart Grid. The architecture embodies high level principles and requirements that designs of Smart Grid applications and systems must satisfy [PNNL 2003].

Different aspects of Smart Grid such as roles, services, and requirements that enable the functionality of the Smart Grid can be described in various architectural artifacts and at lower levels of architecture by standardized business and use cases, which detail specific envisioned Smart Grid requirements. In general, roles in the same domain have similar objectives. To enable Smart Grid functionality, the roles in a particular domain often interact with roles in other domains.

Each domain encompasses Smart Grid roles and services to include types of services, interactions, and stakeholders that make decisions and exchange information necessary for performing identified goals, such as: customer management, distributed generation aggregation, and outage management.

Appendix A of [NIST SP1108r4] describes the seven Smart Grid domains in more detail including the type and scope of interactions within and across domains. However, particular domains may also contain components of other domains. For example, the Independent System Operators (ISOs) and Regional Transmission Organizations (RTOs) in North America have roles in both the markets and operations domains. Similarly, a distribution utility is not entirely contained within the distribution domain - it is likely to contain roles in the operations domain, such as a distribution management, and in the customer domain, such as monitoring. On the other hand, a vertically integrated utility may have roles in many domains.

NIST SGAM has been an evolving framework described in several releases of the Smart Grid roadmap with first release published in 2010, current fifth release published in 2021. NIST has been working through the SGIP's Smart Grid Architecture Committee (SGAC) to align this effort with the European Union Smart Grid - Coordination Group (SG-CG), the International Electrotechnical Commission (IEC) TC57 WG19 (IEC 62357) and IEC TC8 WG5 (Methodology and Tools).

8.1.2.2 IEEE 2030 Model

The IEEE 2030 Smart Grid Interoperability Reference Model (SGIRM) is a conceptual representation of the Smart Grid architecture from three interoperability architectural perspectives (IAP) that include [P2030 2011]:

- Power systems perspective (PS-IAP) emphasizes the production, delivery, and consumption of electric energy including apparatus, applications, and operational concepts; it defines seven domains common to all three perspectives: bulk generation, transmission, distribution, service providers, markets, control/operations, and customers.
- Communications technology perspective (CT-IAP) emphasizes the communications technology such as connectivity among systems, devices, and applications; it includes communication networks, media, performance, and protocols.

- Information technology perspective IAP (IT-IAP) emphasizes the control of processes and data management flow; it includes technologies that store, process, manage, and control the secure information data flow.

IEEE defines interoperability as the capability of two or more networks, systems, devices, applications, or components to externally exchange and readily use information securely and effectively [P2030 2011].

The SGIRM contains both entities and relationships within the environment of the Smart Grid and defines interfaces in a technology agnostic manner [P2030 2011]. The goal of each perspective's architecture is to address interoperability among the elements of the Smart Grid. While the technologies share this common goal, each perspective contains unique aspects addressed from its individual architectural specific technology purposes. The three IAPs primarily relate to logical and functional considerations of power systems, communications, and information technology interfaces for Smart Grid interoperability. Using IEEE SGIRM model, architectural perspectives were adapted for DERs in [P2030.2 2015]. This document includes the representation diagram of entities and relationships included in the architectural communication perspective for Energy Storage System (ESS).

8.1.2.3 European Smart Grid Architecture Model

The European Conceptual Model has been adapted from the Smart Grid Conceptual Model introduced first by NIST Roadmap for Smart Grid interoperability [NIST SP1108r1].

When comparing NIST conceptual model and European SGAM model, there are similarities and differences. The European Conceptual Model for the Smart Grid is an evolution of the NIST model. In order to take into account some specific requirements of the EU context that the NIST model did not address, this model includes Distributed Energy Resources (DER). Basically, the European model introduces two domains to create the EU Conceptual Model. The first one is the Distributed Energy Resource (DER) domain that allows addressing the very important role that DERs play in the European objectives. The second one is the Flexibility concept that groups consumption, production and storage together in a flexibility entity.

The DERs are defined in a separate domain (Distributed Energy Resources), which is (in terms of electricity and communications) is connected with the other NIST domains. The EU Conceptual Model is a top layer model (or master model) and acts as a bridge between the underlying models in the different viewpoints of the Reference Architecture. A Reference Architecture describes the structure of a system with its element types and their structures, as well as their *i*nteraction types, among each other and with their environment.

In general, power system management distinguishes between electrical process and information management viewpoints. These viewpoints can be partitioned into the physical domains of the electrical energy conversion chain and the hierarchical zones (or levels) for the management of the electrical process (refer to [IEC 62357], [IEC 62264]).

One dimension of the Smart Grid Plane covers the complete electrical energy conversion chain, partitioned into 5 domains: Bulk Generation, Transmission, Distribution, DER, and Customer Premises (see domains listed in Table 8.1). According to this concept. domains are physically related to the electrical grid (Bulk Generation, Transmission, Distribution, DER, Customer Premises) are arranged according to the electrical energy conversion chain. The conceptual domains Operations and Market are part of the information management and represent specific hierarchical zones. The conceptual domain Service Provider represents a group of actors which has universal role in the context of Smart Grid. This means that a Service Provider can be located at any segment of the smart grid plane according to the role he has in a specific case.

Applying this concept to the Smart Grid conceptual model allows the foundation of the Smart Grid Plane (see Figure 8.2). This Smart Grid plane enables the representation on which levels (hierarchical zones) of power system management interactions between domains take place.

The other dimension of the Smart Grid Plane represents the hierarchical levels of power system management, partitioned into 6 zones: Process, Field, Station, Operation, Enterprise, and Market [IEC 62357] (see zones listed in Table 8.2). These zones reflect a hierarchical model which considers the concept of aggregation and functional separation in power system management.

In order to allow a clear presentation and simple handling of the architecture model, the interoperability categories described in the GridWise Architecture Framework [GridWise] are aggregated into five abstract interoperability layers: Business, Function, Information, Communication and Component.

The SGAM framework is established by merging the concept of the interoperability layers with the previous introduced Smart Grid plane. This merge results in a model which spans three dimensions:

- Domains – generation, transmission, distribution, distributed energy resources (DER), customer (see Table 8.1)
- Interoperability layers – business, function, information, communications, component
- Zones - process, field, station, operation, enterprise, market.

Table 8.1 SGAM domains.

Domain	Description
Bulk Generation	Representing generation of electrical energy in bulk quantities, such as by fossil, nuclear and hydro power plants, off-shore wind farms, large scale solar power plant (i.e. PV, CSP)– typically connected to the transmission system
Transmission	Representing the infrastructure and organization which transports electricity over long distances
Distribution	Representing the infrastructure and organization which distributes electricity to customers
DER	Representing distributed electrical resources directly connected to the public distribution grid, applying small-scale power generation technologies (typically in the range of 3 kW to 10.000 kW). These distributed electrical resources may be directly controlled by DSO (distribution system operator)
Customer Premises	Hosting both - end users of electricity, also producers of electricity. The premises include industrial, commercial and home facilities (e.g., chemical plants, airports, harbors, shopping centers, and homes). Also, generation in form of photovoltaic generation, electric vehicles storage, batteries, and micro turbines is hosted

Source: [CENELEC 2012]. © 2012, CENELEC.

The SGAM consists of five consistent layers representing business objectives and processes, functions, information models, communication protocols and components. These five layers represent an abstract version of the interoperability categories (a modification of GWAC stack). An illustration of SGAM reference model is depicted in Figure 7.4 (chapter 4). The layers are defined as follows [CENELEC 2012]:

- Business layer defines the business viewpoint regarding strategic and tactical goals, processes as well as regulatory aspects.
- Function layer is IT oriented, technology independent description of use cases, functions and services.
- Information layer defines business objects and data models of the function layer to enable interoperability.
- Communication layer includes specification of protocols and procedures for the data exchange between components based on the information layer.
- Component layer includes physical and technical view on Smart Grids components; besides power system related infrastructure and equipment, ICT infrastructure and systems are also considered.

Each layer covers the Smart Grid plane, which is spanned by Smart Grid domains and zones. The intention of this model is to allow the presentation of the current state of implementations in the electrical grid, but furthermore to present the evolution to future Smart Grid scenarios by supporting the principles of universality, localization, consistency, flexibility and interoperability.

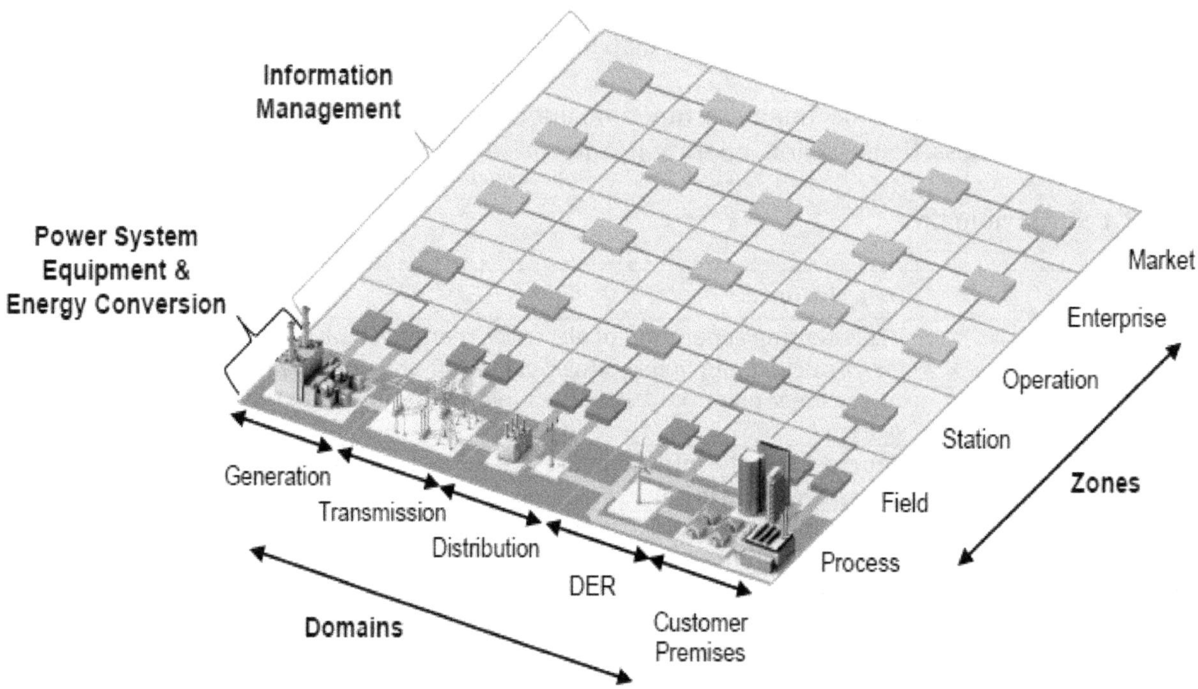

Figure 8.2 Smart Grid plane – domains and hierarchical zones. *Source:* [CENELEC 2012]. © 2012, CENELEC.

When defining the Zones, there are two important concepts that have to be understood. These include the concept of aggregation and the concept of functional separation. The concept of aggregation considers multiple aspects in power system management:

- Data aggregation –data from the field zone is usually aggregated or concentrated in the station zone in order to reduce the amount of data to be communicated and processed in the operation zone.

- Spatial aggregation –from distinct location to wider area (e.g., HV/MV power system equipment is usually arranged in bays, several bays form a substation; multiple DER form a plant station, DER meters in customer premises are aggregated by concentrators for a neighborhood).

The concept of functional separation considers the following:

- Different functions are assigned to specific zones.
- Assignment is typically the specific nature of functions (e.g., real time functions typically in the field and station zone - metering, protection, phasor measurement, automation, etc.), but also considering user philosophies (e.g., functions covering an area, multiple substations or plants, city districts usually located in operation zone such as wide area monitoring, generation scheduling, load management, balancing, area power system supervision and control, meter data management).

Table 8.2 SGAM zones.

Zone	Description
Process	Including the physical, chemical or spatial transformations of energy (electricity, solar, heat, water, wind, etc.) and the physical equipment directly involved. (e.g., generators, transformers, circuit breakers, overhead lines, cables, electrical loads any kind of sensors and actuators which are part or directly connected to the process, etc.)
Field	Including equipment to protect, control and monitor the process of the power system, e.g., protection relays, bay controller, any kind of intelligent electronic devices which acquire and use process data from the power system
Station	Representing the area aggregation level for field level, e.g., for data concentration, functional aggregation, substation automation, local SCADA systems, plant supervision, etc.
Operation	Hosting power system control operation in the respective domain, e.g. distribution management systems (DMS), energy management systems (EMS) in generation and transmission systems, microgrid management systems, virtual power plant management systems (aggregating several DERs), electric vehicle (EV) fleet charging management systems
Enterprise	Includes commercial and organizational processes, services, and infrastructures for enterprises (utilities, service providers, energy traders, etc.), e.g., asset management, logistics, work force management, staff training, customer relation management, billing and procurement, etc.
Market	Reflecting the market operations possible along the energy conversion chain, e.g., energy trading, mass market, retail market, etc.

Source: [CENELEC 2012]. © 2012, CENELEC.

In general organizations can have actors in several domains and zones. For example, according to the business area of a transmission utility it is likely that the utility covers all segments of the transmission domain, from process to market. A service provider offering weather forecast information for distribution system operators and DER operators could be located to the market zone interacting with the operation zone in the distribution and DER domain.

8.1.3 Communication Networks Architecture

As digital controllers began replacing analogue control, the movement toward digital systems resulted in new communication protocols (called fieldbus protocols) to the field as well as between controllers. More recently, digital control systems started to incorporate networking at all levels of the industrial control, as well as the inter networking of business and industrial equipment using Ethernet standards. This has resulted in a networking environment that appears similar to conventional networks at the physical level, but which has significantly different requirements. While an enterprise may have one or more office Local Area Networks (LANs) connected by a backbone network or Wide Area Network (WAN), small industrial networks tend to have a hierarchy three or four levels deep. Figure 8.3 shows communication infrastructures in use for the European SmartGrids at different domains: end-user premises, transmission grids, distribution grids and generation.

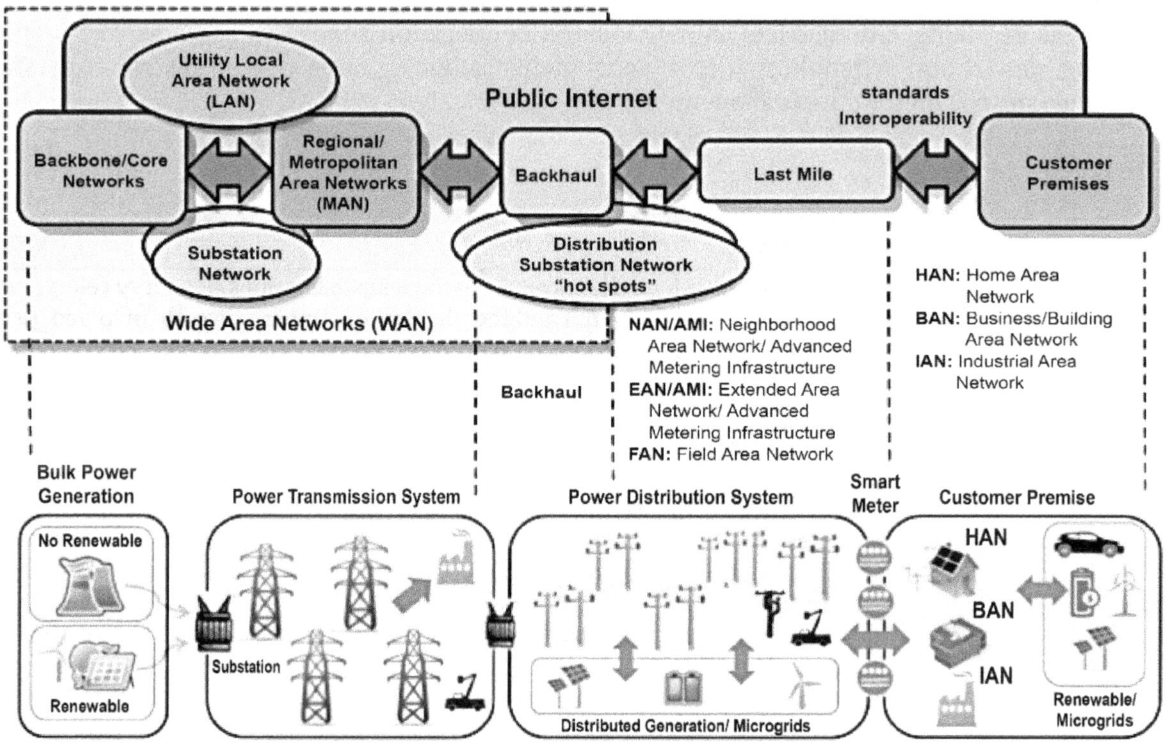

Figure 8.3 End-to-End Smart Grid communications architecture. *Source:* [ENISA 2015]. © 2015, ENISA.

8.1.4 Information Architecture

Within the SGAM, one particular aspect of the layer is the level of data exchanged between the various layers. The focus of the layer within the SGAM is the meaningful representation and localization of the data models, abstract communication system interfaces towards the communication layer and the functional (system) layers implementing the logical interfaces and the Smart Grid component using standards and data models.

The conceptual reference model viewpoint described in [NIST SP1108r1] provides a high level, overarching logical architecture representation of a few major relationships that existing applications have to Smart Grid domains. This model is a useful tool in identifying which existing applications may be a good candidate for a Smart Grid role, it also suggests what their possible information communications paths could be in a Smart Grid. It is also a useful way to identify potential intra and inter domain interactions between existing and new applications, along with capabilities enabled by these interactions.

8.1.4.1 Security at All Levels

In order to efficiently build Smart Grids inherently secure by design, security should be involved at all levels of the Smart Grid in order to secure Smart Grid operations and related IT operations. The security requirements vary depending on the SGAM Domain/Zone the systems are located.
European Electrical Grid stability has been chosen as reference to define SGIS Security Level (SGIS-SL) and create a bridge between electrical operations and information security. Thus, focus is made on power loss caused by ICT systems failures. Translating this fact into the SGAM means that information security should be considered in all domains, zones, and layers.

8.1.4.2 Security View per Layer

Additional layers, called Security view per Layer, have been proposed with the Reference Architecture. The Security View per Layer is inserted under each SGAM layer [CENELEC 2014]. As described in this document, security view per layer emphasizes that security is a cross functional topic, which must be obeyed in each of the SGAM layers. This can even be more underlined as security is actually to be obeyed per layer, per domain, and per zone and thus basically per SGAM cell.

8.1.5 Information Security Architecture

Information Architecture or IT Architecture is defined as a set of design artifacts that are relevant for describing an object such that it can be produced to requirements (quality) as well as maintained over the period of its useful life (change). The design artifacts describe the structure of components, their inter relationships, and the principles and guidelines governing their design and evolution over time.

Information architecture is a component of enterprise architecture that deals with the information component when describing the structure of an enterprise. Information architecture is the structural design of shared information environments; the art and science of organizing and labelling websites, intranets, online communities and software to support usability and findability; and an emerging community of practice focused on bringing principles of design and architecture to the digital landscape [IA]. Typically, it involves a model or concept of information which is used and applied to activities that require explicit details of complex information systems. These activities include library systems and database development [Wiki IA].

Consequently, the Information Security Architecture consists of design artifacts that describe how the security controls (security countermeasures) are positioned and how they relate to the overall Information Architecture. These controls serve the purpose to maintain the system's quality attributes, among them confidentiality, integrity and availability. A blueprint of enterprise information technology architecture exists in many organizations. Information includes all sources such as paper, graphics, video, speech and thought that define the enterprise.

Enterprise information security architecture is only a subset of enterprise architecture. The intermediate outcome of an architecture process is a comprehensive inventory of business security strategy, business security processes, organizational charts, technical security inventories, system and interface diagrams, and network topologies, and the explicit relationships between them. The inventories and diagrams are merely tools that support decision making. But this is not sufficient. It must be a living process.

The primary purpose of creating an enterprise information security architecture is to ensure that business strategy and IT security are aligned. As such, enterprise information security architecture allows traceability from the business strategy down to the underlying technology supported by this standard [ISO/IEC 27014].

Implementing the enterprise information security architecture generally starts with documenting the organization's strategy and other necessary details such as where and how it operates. The process then cascades down to documenting discrete core competencies, business processes, and how the organization interacts with itself and with external parties such as customers, suppliers, and government entities. Thus, the enterprise information security architecture documents the current state of the technical security components listed above including a desired ideal future state (Reference Architecture) and finally a target future state which is the result of engineering trade-offs and compromises versus the ideal state. The result is a nested and interrelated set of models, usually managed and maintained with specialized software available on the market.

Having documented the organization's strategy and structure, the architecture process then flows down into the discrete information technology components such as:

- Organization charts, activities, and process flows of how the IT Organization operates.
- Organization cycles, periods and timing.
- Suppliers of technology hardware, software, and services.
- Applications and software inventories and diagrams.
- Interfaces between applications that includes events, messages and data flows.
- Intranet, Extranet, Internet, eCommerce.
- Data classifications, databases, and supporting data models.
- Hardware, platforms, hosting servers, network components and security devices.
- Local and wide area networks, Internet connectivity diagrams.

Along with the development of Security Architecture, all other architectures, data, applications, and infrastructure (e.g., hardware, networks, operating systems) are being defined and developed by specialized architects and Systems Engineers. All these architectures are components of the enterprise architecture, which forms the umbrella architecture for all technologies used by the enterprise (organization).

8.1.6 Information Security Architecture for Smart Grid

Enterprise information security architecture is the practice of applying a comprehensive and rigorous method for describing a current and/or future structure and behavior for an organization's security processes, information security systems, personnel and organizational subunits, so that they align with the organization's core goals and strategic direction. Although often associated strictly with information security technology, it relates more broadly to the security practice of business optimization in that it addresses business security architecture, performance management and security process architecture as well.

There, we need to clarify that information security is not equivalent with computer security or information assurance concepts that are frequently used interchangeably. Although there are common security goals for ensuring the security services, there are differences in the methodologies and techniques used.

Often, information security architecture is viewed as simply the relations between different technology components in an IT system. However, technology cannot stand on its own without integration into all other elements. Architecture, as applied to information security, is the overall design or structure of a system typically described as the interconnection of hardware, software, and components that make up an organization's infrastructure. This is then complimented by the processes, policies and procedures that govern the practices.

Organizations (e.g., utilities) need to design the information security architecture and to adopt an enterprise architecture approach when attempting to secure their environment. This approach is to be the guide for standard protection when designing and implementing security solutions to ensure the security needs are met. Identification of functional issues (e.g., differences between control systems and business systems) combined with approaches for security models and security architectures are discussed briefly in several NIST documents.

Many of today's SCADA systems employ an enterprise wide area network to facilitate the exchange of data between the field sites and the central location. Also, the enterprise network is usually connected to the Internet. Although the integration of IT systems with SCADA data collection and monitoring are beneficial to run the Smart Grid more efficiently and effectively, the design approach has to provide a security model that accommodates the differences between the security architecture and priorities of SCADA systems versus IT traditional systems. Issues such as modems connected to one system compromising the other, merging the disparate culture of SCADA and IT practices, integrating security measures between the SCADA system and the enterprise communications network, and multitude of specific regulations and requirements for each system have to be considered in order to design an appropriate information architecture.

An organization must design and implement a process that ensures continual movement from the current state to the future state. The future state will generally be a combination of one or more closing gaps between IT and SCADA systems, present and desired future security strategy, upgrades and replacements of hardware and software on both IT and SCADA systems.

The dynamics of IT and Operational Technology (OT) convergence allow utilities to better leverage the move to a Smart Grid information architecture. The new Smart Grid architecture must have these characteristics [Meyers 2013]:

- Supports convergence of OT – the grid physical infrastructure assets and applications–and IT – the human interface that enables rapid and informed decision making.
- Supports the integration of existing enterprise systems within the grid modeling, monitoring, and management environment.
- Includes a utility's current and future ecosystem.
- Flexible and adaptive to meet future needs while providing the scale and security required for mission-critical aspects of the environment.

The SGIP Architecture subgroup (SGCC) has initiated the development of a conceptual Smart Grid cybersecurity architecture based on the high level requirements, standards analysis, overall Smart Grid architecture, and other cybersecurity information [NIST SP1108r3]. Also, the business viewpoint addresses the concerns of users, planners, and business management. This provides input to the subsequent phases, defining its realization through technology. The technology architecture addresses the software and hardware capabilities that are required to support deployment of business, data, and application services. However, defining the business architecture requires to identify and examine information relevant for the particular stakeholders. The Open Group Architecture Framework (TOGAF) [TOGAF] defines information systems architecture viewpoint as is further subdivided into Application Architecture and Data Architecture. Based on the set of terms defined by TOGAF standard, the ARCHI framework is based on ArchiMate standard. The ARCHI framework architecture (promoted by Open Group Archimate Forum) offers an integrated architectural approach that describes and visualizes the different architecture domains and their underlying relations and dependencies. This survey [Frameworks] includes a list of frameworks for architecture representation. With so many frameworks used by different developers, the selection of the appropriate framework is based on many factors, needs, availability, costs, knowledge, benefits, standardization, etc. This standard [ISO/IEC/IEEE 42010:2011] focuses on the specification of architecture frameworks.

Considering different aspects of information systems for the Smart Grid, the information systems can be developed after business and technology architecture viewpoints are developed. These two architectures illustrate the relations to the business architecture. The information (represented by business objects) is expressed as data and business processes or their functionality respectively is realized by applications. The order in which the relevant views (Application Architecture and Data Architecture) are developed may vary. They can be chosen depending on the context and the project's objectives to focus on only one of the domains, e.g., either data or application may be possible [Trefke, 2013].

The Application Architecture provides description of applications on a logical level and it is technology neutral. Identified applications are relatively persistent over time while technology changes more often. Typical applications in the energy sector are, e.g., Supervisory Control and Data Acquisition (SCADA) systems, billing, master data management or planning.

Depending on the functions considered, existing information models from different domains may be used as the foundation for more problem specific models [Danekas 2013]. The authors recommend to identify the information needed to support the business functions embedded in processes and procedures, then to establish a common understanding regarding the semantics within the information models. The authors argue that this semantic understanding of the information contained in the models is especially important since the Smart Grid aims at the plug and automate paradigm, which means that technical assets and applications should be able to understand each other without a high amount of configuration by domain experts. Instead, the human knowledge on the meaning and rules applying to model entities has to be codified explicitly within the model itself.

The Data Architecture should include the structure of logical and physical data required to exchange information and data management resources. It is relevant for data management, data flows, and quality management. Generally, Data Architecture is concerned with the identification of enterprise wide data, e.g., information relevant for business processes, but not with the design or development of data management systems like databases. Possible considerations regarding data management can for instance begin with the identification of components relevant for creation, saving or use of data. In the context of Smart Grid systems, the information can be customer data, metering data or billing data exchanged across several involved Smart Grid actors.

An information model has to be established because it provides formalism to the description of a problem domain without constraining how that description is mapped to an actual implementation in software. There may be many mappings of the information model. Such mappings are called data models, irrespective of whether they are object models (e.g., using UML, entity relationship models or XML schemas [Wiki Infmodel]. Definitions of concepts such as information model, data model, Common Information Model (CIM) are available in [RFC 3198]. Most of the management models standardized to date are data models. Compared to information models, data models provide these benefits [RFC 3444]:

- Define managed objects at a lower level of abstraction.
- Include implementation and protocol specific details.

Data model is basically the rendering of an information model according to a specific set of mechanisms for representing, organizing, storing and handling data. One benefit of using a data model is for evaluating a vendor software. Since a data model actually represents the infrastructure of an organization, vendor software can be evaluated against a company's data model in order to identify possible inconsistencies between the infrastructure implied by the software and the way the company actually does business.

The next building block to be analyzed includes Common Information Model (CIM) profiles. The basis of the CIM is a comprehensive data model. CIM was originally developed by the Electric Power and Research Institute (EPRI) in the midst of the 90's. It offered an internal database model for EMS and SCADA systems. Over the years, CIM has outgrown its original purpose and now contains a pretty large domain ontology.

Although the extensive size of the data model providing with almost all important objects can be seen as an advantage, this size complicates the data model's application. In order to solve this problem, it is possible to define and generate profiles for specific scenarios. A CIM profile is a real subset of the original data model. It includes classes and associations required for the scenario. Furthermore, attributes of the considered classes can be defined as optional or mandatory. Finally, additional restrictions such as cardinalities of associations can be specified. Creating profiles is an established and recommended means to cope with the CIM data model for the Smart Grid and DER applications.

Profiles can be created individually by single users in enterprise specific contexts or officially by working groups or other organizations. Figure 8.4 shows the overall relationships among domain industry standard canonical data models.

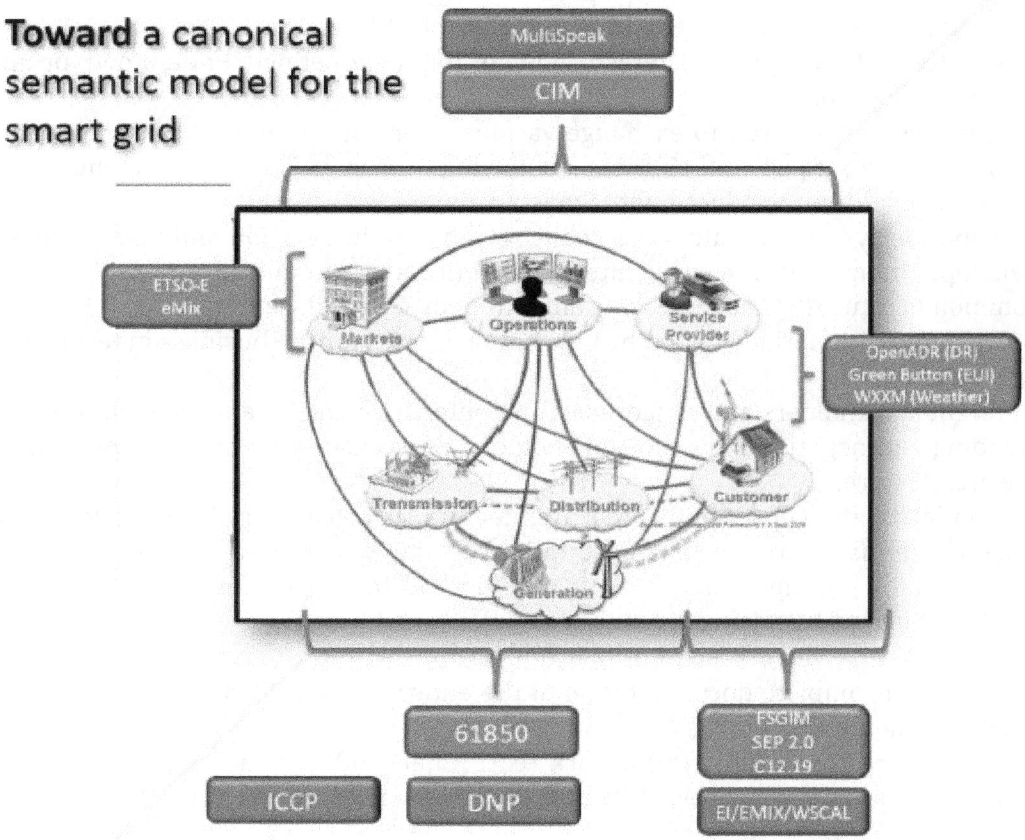

Figure 8.4 Proposal to use select models for canonical data model. *Source:* [NIST SP1108r3]. Public Domain.

Although this model provides useful information for the Smart Grid developers in US, it is important to understand that it is not a complete view, many other standards are being developed to fill the gap. More information about the information models including applying to developments for the Energy sector and specifically to Smart Grid applications is available in [Gonzalez 2013], [Specht 2013], [Uslar 2013].

8.1.7 SGAM Implementation Challenges

Costs and other constraints do not allow utilities to design and implement an architecture from scratch. As with communications infrastructure, no operating utility has the luxury of designing and implementing an architecture from scratch [Meyers 2013]. But industry proven architectural roadmaps and guidelines exist to help utilities work through the design process.

The GWAC stack can be used for the elicitation and management of Smart Grid requirements, since it addresses the facets of Smart Grid interoperability from regulatory policy down to the physical connection between assets. Each layer covers distinctive functions depending on the abstraction level or degree of formalization of the requirements connected to them. However, the SGAM model may need context specific tailoring (e.g., enterprise architecture) to fit the organization or generally, speaking the context it shall be applied to. On the other hand, a unique solution that fits all Smart Grid requirements both in terms of technical solutions and its corresponding standardization requirements is unlikely to be found. Challenges of implementing SGAM interoperability model within an enterprise may include:

- Functions of enterprises in the Smart Grid context differ from country to country, mainly depending of the type of market and regulations. Examples of such roles, which are taken on by enterprises, are network operators, suppliers, traders, electricity generators or measurement service providers, DER aggregators, utilities, etc.
- The concept of enterprise architecture is not uniquely defined, understood, or applied by each organization.
- Enterprises are required to exchange various information with others and to adapt their business processes as well as their supporting information technology due to new roles and functionalities introduced with Smart Grids.
- Distribution system operators are challenged by the integration and management of their field/operational technologies; integration efforts are required and will become more common in context of the increasing amount of distributed generation.
- Efficient operation is demanded by new opportunities for the business in terms of products or services.
- Different stakeholders and concerns within enterprise and its environment require understanding and negotiation of the stakeholders' perspectives on architecting and managing the architecture.
- The understanding of the enterprise architecture scope can vary from modeling the enterprise (in terms of architecture description) to implementing architectures within enterprises and to a professional discipline or method to manage the enterprise architecture (architecting and managing the architecture).
- Artifacts may not be consistent or may not exist.
- Missing or incomplete documentation of the enterprise architecture make difficult to consider stakeholders' concerns and requirements.
- Enterprise interoperability framework (e.g., [Chen 2008]) is a complex issue with many facets that includes interoperability approaches, barriers, and concerns [Curry 2012].

Thus, interoperability is a complex issue with many facets. Also, within many modern information systems, interoperability is not seen as a strong requirement within their design. It is argued that Smart Grid is a complex system of systems and a system of systems carries several noteworthy burdens [P2030 2011].

Generally, modern information systems are not designed with a strong interoperability requirement. This results in limited interoperability between systems and a high cost associated with system alignment. An enterprise can use multiple information systems to support its business processes including sales, marketing, management reporting, research and development, financial reporting, customer billing, and similar areas. Integrating these systems is a key challenge for any utility because the number of systems involved can be significant, and it can be time consuming and expensive to bring together the legacy systems, so they use common data and processes.

One significant technical challenge exists in terms of information interoperability that requires overcoming conceptual (syntax and semantics) and technological barriers [Curry 2012]. The author describes a concept to address interoperability issues for the Enterprise Energy Management System, which is a System of Systems (SoS). The concept uses the information interoperability based on the dataspace data management abstraction and the linked data approach to sharing information on the Web.

Another challenge is SoS interoperability within an enterprise. First, not all enterprises within electric sector could follow the same SGAM architectural model to implement interoperable SoS. Differences on mission, business objectives, processes, and technologies could be challenging for an enterprise such a utility. Second, there are different definitions for a SoS and different System of Systems classifications (virtual, collaborative, acknowledged, and directed). For example, one issue is the diversity of SoS implementation types such as:

- Directed SoS are built and centrally managed to fulfil specific purposes; component systems maintain an ability to operate independently.
- Acknowledged SoS have defined objectives and dedicated resources; constituent systems retain their independent ownership and objectives.
- Collaborative SoS component systems interact and collectively decide the means for enforcing and maintaining standards.
- Virtual SoS have no central management authority but relying on some mechanisms to maintain them.

Third, there are different approaches (federated, united, integrated) for the enterprise architecture. More issues regarding SoS interoperability are described in [Chen] [Chen 2006] [Sledge 2010].

8.2 DER Logical Hierarchical System Architecture

The hierarchical DER System architecture shown in Figure 4.9 of Chapter 4 is abstracted to the logical architecture shown in Figure 8.5. This logical architecture is a hybrid combination of five architectures.

As illustrated in this figure, the systems hierarchy of five levels includes various devices that are defined in [EPRI 2013] as follows:

- Autonomous DER Generation and Storage at level 1 includes the DER cyber-physical systems interconnected to the utility grid, usually operating autonomously according to preestablished settings.

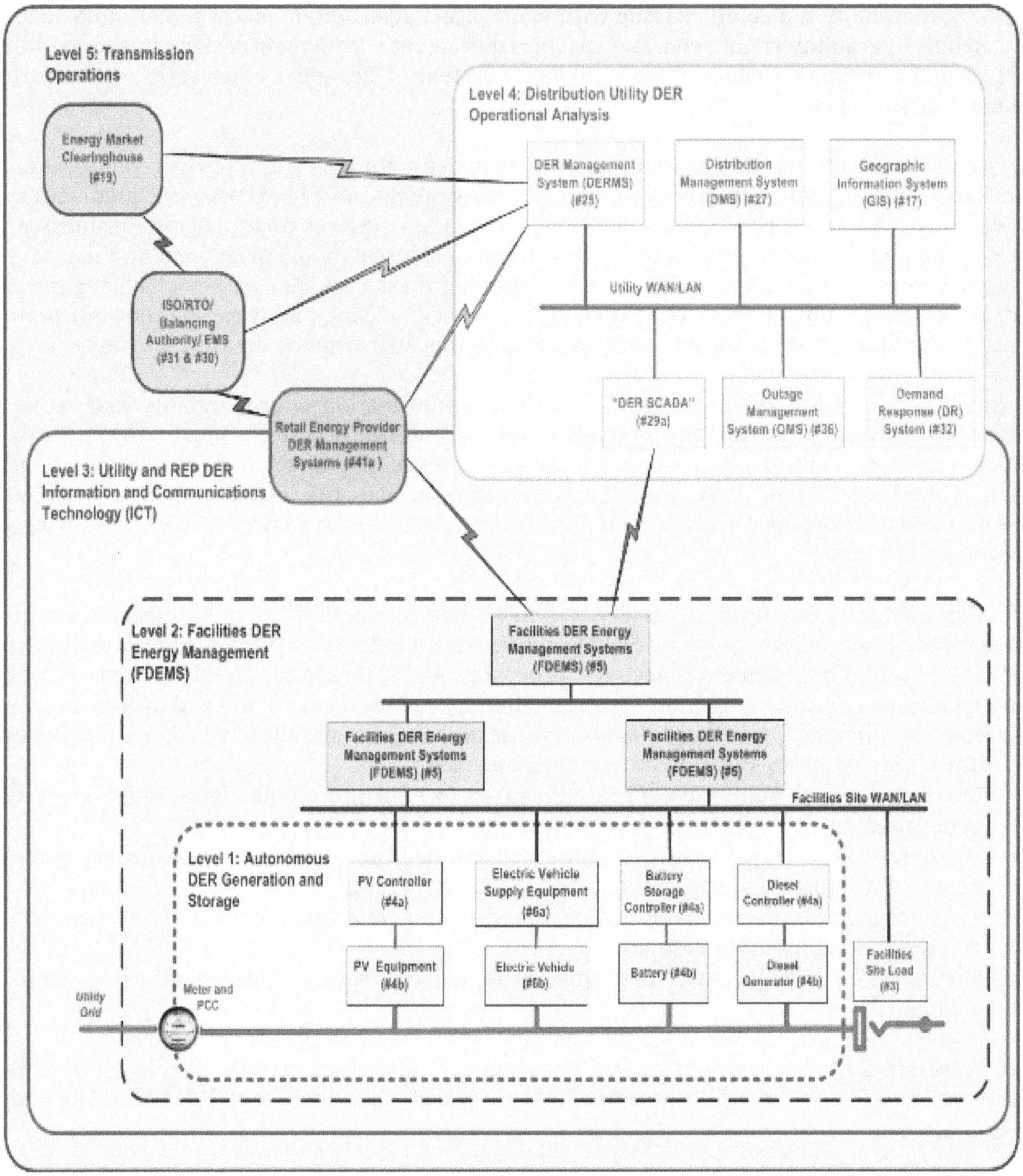

Figure 8.5 Logical hierarchical DER system architecture. *Source:* [Cleveland 2013]. © 2013, Xanthus International.

- Facilities DER Energy Management at level 2 includes a facility DER management system (FDEMS) that manages the operation of the DER systems located at lower level. This FDEMS may be managing the DER systems in a residential home, but more likely DER systems in commercial and industrial sites (e.g., university campus, shopping mall, virtual power plant).
- Utility and REP Operational Communications at level 3 extends beyond the local site to allow utilities and possibly REPs to request or require DER systems (typically through a FDEMS) to take specific actions. The settings for autonomous DER operations are modifia-

ble by utilities and REPs. Controls include turning on or off devices, setting or limiting output, providing ancillary services (e.g., volt-var control), and other grid management functions.

- Distribution Utility Operational Analysis at level 4 applies to utility applications that are needed to determine which requests or commands should be issued to specific DER systems. Utilities monitor the power system and assess if metrics such as efficiency, reliability, or market advantage can be improved by having DER systems modify their operation. This utility assessment involves many utility control center systems, including Geographic Information System (GIS), Data Management System (DMS), Outage Management System (OMS), Demand Response (DR) systems, as well as the Distributed Energy Resources Management Systems (DERMS).
- Transmission and Market Operations at level 5 is the highest level, and involves the larger utility environment. RTOs or ISOs may need to exchange information about the capabilities and operational status of larger DER systems and/or aggregated DER systems.

8.2.1 DER Cybersecurity Logical Reference Model

Figure 8.6 depicts the DER Reference Model as described in [Cleveland 2013] report. This model is a modified version (an extract/expansion) of the cybersecurity logical reference model for Smart Grid described in [NISTIR 7628r1]. The DER Logical Reference Model covers the high level actors and logical interfaces relevant to DER systems. This diagram shows relevant DER actors and logical interfaces between actors of various domains and levels (see more information in [Cleveland 2013].

Each DER system can be viewed as composed of two classes of components: physical hardware/firmware components and controller components that manage the physical components. These components are typically installed at a customer site behind the meter or in some cases within a utility substation. The DER equipment is connected to the Local Electric Power System (EPS) [IEEE 1547]. Some DER systems include a simple Human-Machine Interface (HMI) (or a port for a laptop HMI) that can be used for status information and maintenance access. This HMI provides status information and may be used during maintenance.

8.2.2 Addressing DER Architecture Gaps

The only external communications between the utility and these DER systems are the meter readings, typically measured at the point of common coupling (PCC) between the local EPS and the area EPS. Although DER systems are dependent on metering functions, they are not included in this architecture discussed in this report [Cleveland 2013].

However, we stress the need to understand both DER functions and metering functions including other aspects such as data, communications systems and protocols used for metering control, networks, and security controls. Since DER may be located outside of utility's perimeter, requirements for physical security need to be considered and recommendations for cost effective controls have to be followed. If only significant events are logged, then a criteria has to be defined to qualify a significant event.

In order to support protection against denial of service, communication paths between the utility DER SCADA systems and FDEMS need to support secure fail over capabilities. A security boundary between the utility DER SCADA and the FEDMS has to be established and allow only authorized interactions.

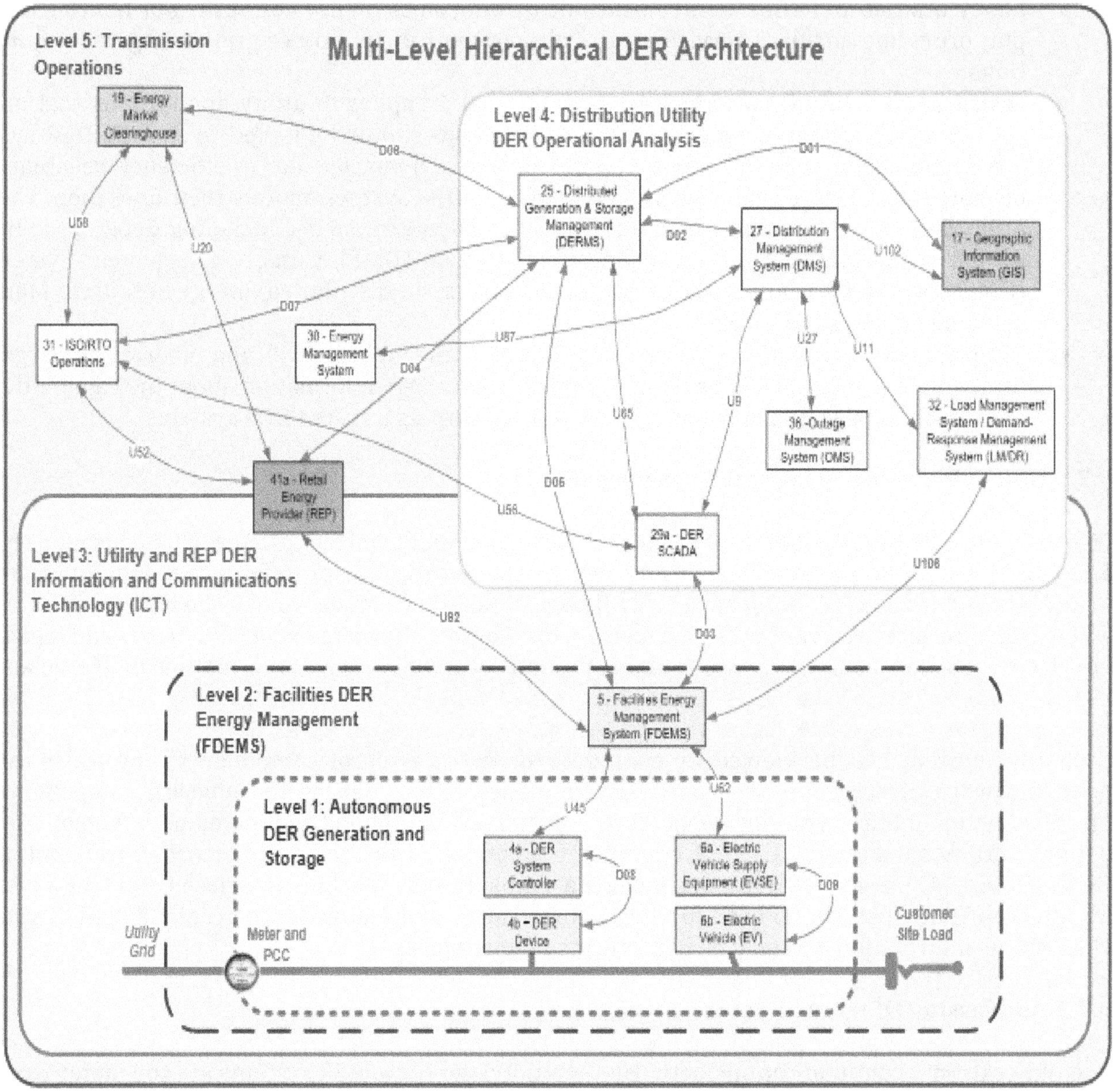

Figure 8.6 Hierarchical DER architecture mapped to the NISTIR 7628. *Source:* [Cleveland 2013]. © 2013, Xanthus International.

Besides understanding the existing architecture organization, Security Engineer needs to understand the various components (actors) and the interactions that take place between them as well as the possible vulnerabilities that may be exploited in a cybersecurity event. While common security requirements may be applied for all levels, there could be specific security requirements for a level. Therefore, at each level security concerns are addressed by identifying security risks, defining security requirements, and design appropriate security controls.

Due to the distributed nature of these DER systems, it becomes necessary to properly interconnect them with the rest of the Smart Grid network. The communication networks must cover all the devices from these systems, in order to enable remote control from the utility and operation centers and ensure their efficient interaction within the whole grid. Furthermore, as they are distributed, and on more than one occasion will make use of the Internet network, these communications will need to be secured to protect them from attacks and accesses from unauthorized users.

Advanced DER functionalities also enable new grid architectures incorporating microgrids that can separate from the grid when power is disrupted and can interact in cooperation with grid operations to form a more adaptive resilient power system. Microgrid makes a more resilient power system.

8.3 Security Architecture Frameworks

Numerous architecture frameworks have been developed and matured over the past decade and the focus of these frameworks is diverse. A list of architectural frameworks is provided in [Frameworks], but not all them support security. The Enterprise Architecture Body of Knowledge (EABOK) is a guide to Enterprise Architecture [EABOK]. The EABOK includes five knowledge areas that correspond to phases in the enterprise architecture development lifecycle: planning, managing, developing, using and measuring the enterprise architecture. Another review provides an analysis of challenges related to producing a security architecture including various issues related to security architecture frameworks, information security models, and security architect roles and skills [Ross 2014], [NIST SP160v1].

Standards and recommendations (NIST, DOE, ISO/IEC) are useful to a security architect when selecting the frameworks, methodologies, and practices. Thus, security architecture provides a balanced relationship between strongly opposing elements and the interplay of conflicting elements and serves as a device for regulating relations and interactions among an organization's governance, technology architecture and its operations [Kiely 2008]. The following include examples of frameworks that provide security recommendations or are more focused on security architecture.

8.3.1 Network Applications Consortium Framework

An earlier framework, Network Applications Consortium (NAC) framework, defines a policy driven process for security architecture that starts with a security program; then it continues with in depth focus on the three major components that make up enterprise security architecture: governance, technology architecture, and operations. Enterprise security architecture must be understood in the larger corporate context, where it is part of an overall enterprise security program. It must relate appropriately to the organization risk management, IT governance, enterprise architecture, and physical security programs of the organization. Figure 8.7 illustrates a perspective of the enterprise security program and relationships to core tenets (risk management, architecture, governance, physical security).

8.3.2 Open - Enterprise Security Architecture

The recent update of the NAC framework, called O-ESA (Open-Enterprise Security Architecture) [O-ESA 2011]) is describing a policy driven security architecture which starts with a policy framework for identifying security principles; authorizing their enforcement in specific control domains through a set of policies; and implementing the policies through technical standards, guidelines, and procedures. These frameworks place the security program in the larger context of the enterprise architecture and describe enterprise security architecture based on three components: governance, security technology architecture, and security operations.

8.3.3 Sherwood Applied Business Security Architecture

The Sherwood Applied Business Security Architecture (SABSA) is a framework and methodology for enterprise security architecture and service management. The framework supports the Enterprise Security Architecture and Service Management including Risk Management, Information Assurance, Governance, and Continuity Management.

The SABSA Model comprises six layers (see Table 8.3). It has a similar structure to Zachman framework and a methodology for developing risk driven enterprise information security architectures, although it has been adapted somewhat to a security view of the world [Sherwood 2009]. Each layer represents the view of a different player in the process of specifying, designing, constructing and using the business system. It supports design of an architecture which is derived from an analysis of the business requirements for the information security. The primary business requirements for information security are business specific and should be expressed in terms of protecting the availability, integrity, authenticity and confidentiality of business information, and providing accountability and auditability in information systems. Deriving these requirements requires a detailed analysis of the business processes and identification of data and information needed by users to support the business needs.

Figure 8.7 Enterprise security program context.

Table 8.3 SABSA Layered Architecture Views.

View	Architecture
The Business View	Contextual Security Architecture
The Architect's View	Conceptual Security Architecture
The Designer's View	Logical Security Architecture
The Builder's View	Physical Security Architecture
The Tradesman's View	Component Security Architecture
The Service Manager's View	Security Service Management Architecture

Source: [Sherwood 2009]. Public Domain.

Another configuration of these six layers which is represented in the diagram of Figure 8.8. In this diagram the security service management architecture has a meaning in the context of each of these other layers, so it has been placed vertically across the other five layers.

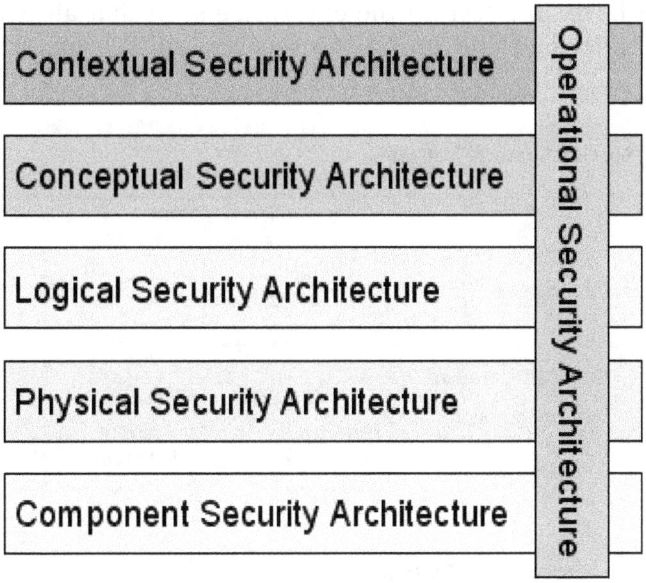

Figure 8.8 SABSA model for security architecture development. *Source:* [Sherwood 2009]. Public Domain.

The primary characteristic of the SABSA model is that everything must be derived from an analysis of the business requirements for security, especially those in which security has an enabling function through which new business opportunities can be developed and exploited. Although one may find many architectural approaches that can use, the most suitable should be driven from a clear understanding of the business requirements for the system [Sherwood 2009], [Van Haren 2013]. Therefore, SABSA is another example of a methodology that can be used for Smart Grid systems (the converged IT (information technology) and OT (operational technology) systems). However, this methodology is following a SABSA life cycle (see [Sherwood 2009]) that includes activities - manage and measure, allow to set the target for the performance metrics. Once the system is operational, it is essential to measure actual performance against targets, and to manage any observed deviations.

8.3.4 Open Security Architecture

The Open Security Architecture (OSA) focuses on controls and patterns. The security architecture (landscape) includes entities and relationships between security controls, risks, security patterns, policies, standards, and security services (see Figure 8.9).

8.3.5 Huxham Security Framework

The enterprise information security architecture within a generic framework is represented as a high level conceptual security architecture framework in this diagram [Huxham 2006b]. The elements of this framework combine Zachman architecture framework with a limited set of areas of interest (What, Who, Which, Where) and a view of security components for information and technology, all driven by business strategies and security policies. Figure 8.10 is an illustration of how Huxham's security framework can be applied to Smart Grid system.

Security architecture supports a vision aligned with information security governance. In general, it is expected to provide different types of architecture (e.g., information security architecture, technology architecture, etc.) and architecture descriptions based on views and different perspectives as identified in the chosen frameworks, standards, and regulatory documents.

Security considerations should have an effect on the enterprise such that all enterprise architecture development needs to be informed and utilize the security policy, constraints, governance, artifacts, and building blocks.

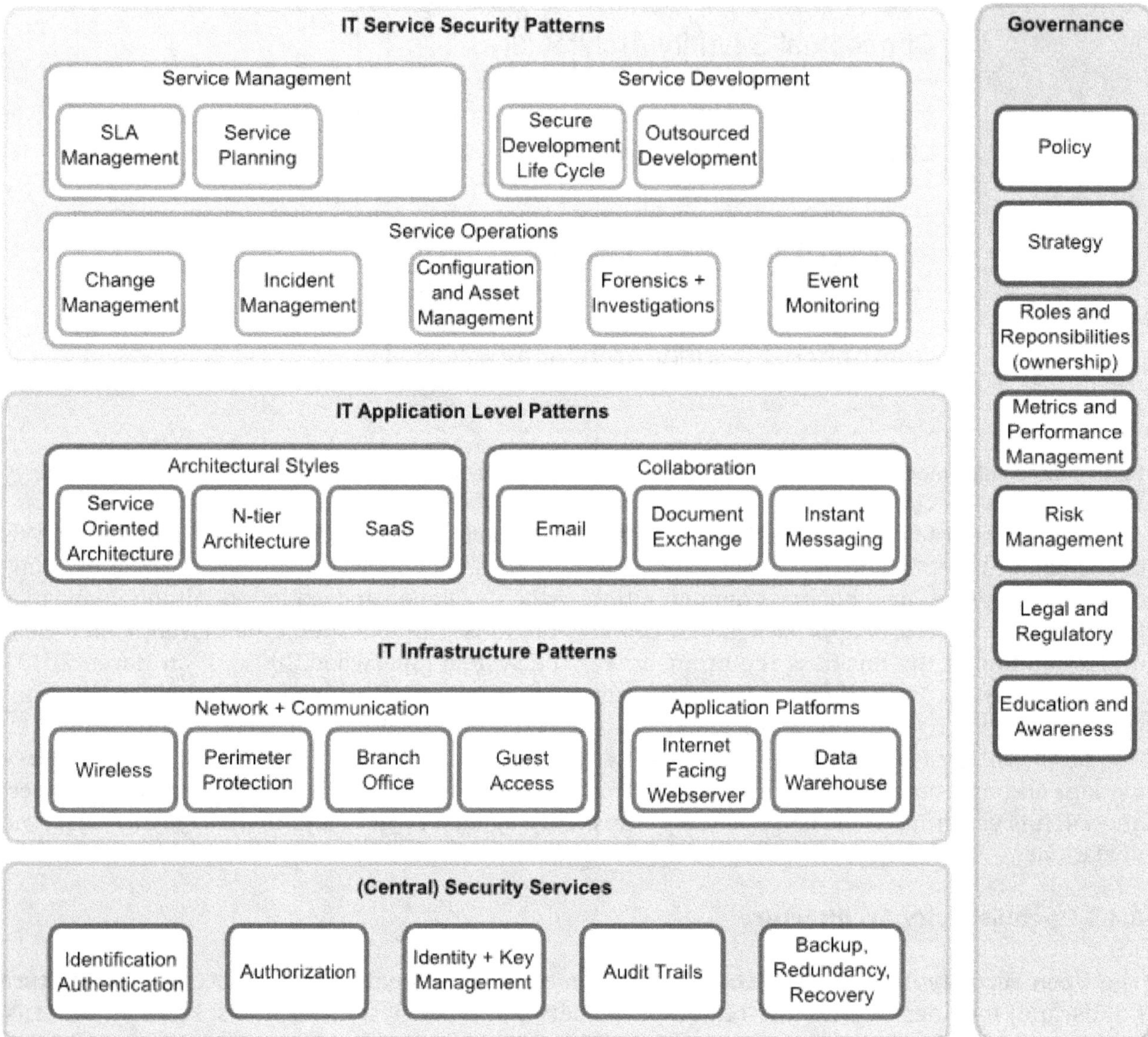

Figure 8.9 OSA framework. *Source:* [OSA Framework]. Licensed under CC BY 3.0.

8.3.6 FMLS Architecture Description Framework

The architecture frameworks should describe a system's processes and information flow, physical and logical architecture, related interaction with the environment, operational and lifecycle requirements, and system outputs [Carter 2016].

The FMLS Architecture Description Framework (FMLS-ADF) version 3.0 provides a description of the systems in a systems context. The framework supports three horizons (long term, medium term and short related and levels of detail) and six aspects of the architecture: domain (business), information system, information, infrastructure, security and governance; definition and deployment of systems.

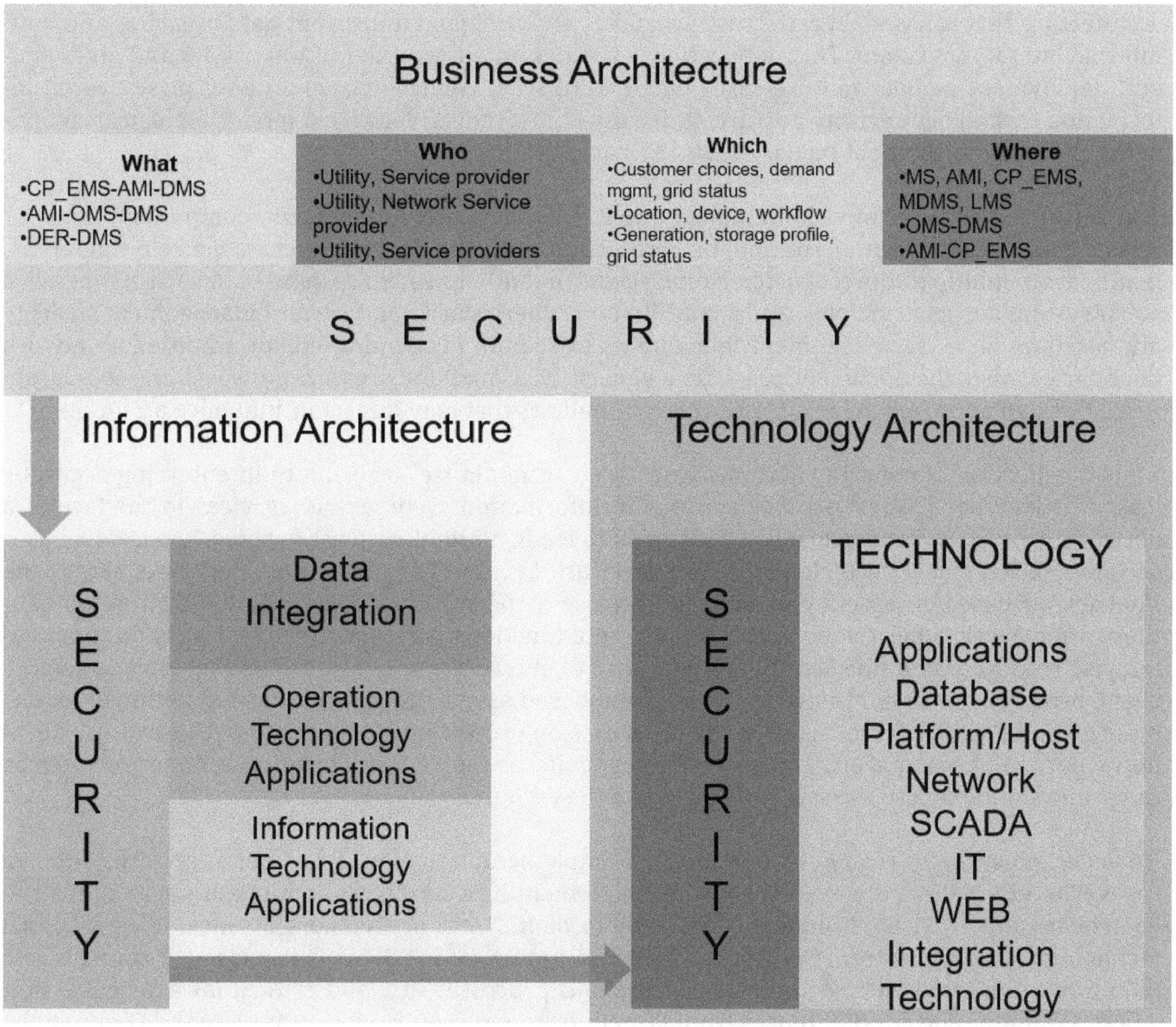

Figure 8.10 Applying Huxham's Security Framework to Smart Grid. Adapted from [Huxham 2006b]. Public Domain.

8.3.7 Rozanski Woods Software Systems Architecture

The Rozanski Woods Software Systems Architecture (RWSSA) framework provides a guide for implementing a software system architecture. It describes a set of architectural viewpoints (context, functional, information, concurrency, development, deployment, operational) and perspectives (security, performance and scalability, availability and resilience, evolution) [Rozanski 2011].

8.3.8 Zero Trust Architecture

Zero trust (ZT) is the term for an evolving set of cybersecurity paradigms that move defenses from static, network based perimeters to focus on users, assets, and resources. A zero trust architecture (ZTA) uses zero trust principles to plan industrial and enterprise infrastructure and workflows. This document [NIST SP800-207] defines zero trust architecture is an enterprise cybersecurity architecture that is based on zero trust principles and designed to prevent data breaches and limit internal lateral movement. This publication provides an abstract definition of ZTA and gives general deployment models and use cases where zero trust could improve an enterprise overall information technology security posture. It discusses ZTA tenets, zero trust model, its logical components, possible deployment scenarios, and threats.

The zero trust security model eliminates implicit trust in any one element, component, node, or service and instead requires continuous verification of the operational picture via real-time information from multiple sources to determine access and other system responses. A security model, a set of system design principles, and a coordinated cybersecurity and system management strategy are based on an acknowledgement that threats exist both inside and outside traditional network boundaries. Also, the document presents a general road map for organizations wishing to migrate to a ZT design approach and discusses relevant policies that may impact or influence a ZTA.

ZT is a collection of concepts and ideas designed to minimize uncertainty in enforcing accurate, least privilege per request access decisions in information systems and services in the face of a network viewed as compromised. ZT assumes there is no implicit trust granted to assets or user accounts based solely on their physical or network location (e.g., local area networks versus the Internet) or based on asset ownership (enterprise or personally owned). Authentication and authorization (both subject and device) are discrete functions performed before a session to an enterprise resource is established. ZT is a response to enterprise network trends that include remote users, bring your own device (BYOD), and cloud based assets that are not located within an enterprise owned network boundary. Zero trust focuses on protecting resources (assets, services, workflows, network accounts, etc.), not network segments, as the network location is no longer seen as the prime component to the security posture of the resource.

However, most topics related to ZTA focus on implementations for IT business networks and use cases. Therefore, there is a need to consider implementing a zero trust architecture in an industrial control system (ICS) environment. Components of ICS that are vulnerable include operational technology (OT) and Internet of things (IoT) and Industrial IoT. This is because OT controls and monitors industrial assets, equipment, events and processes. ICS and critical infrastructure networks exhibit important differences from IT networks and require to reformulate the IT zero trust approach. For example, the decision of where to encrypt traffic and at what point that end-to-end encryption impacts the availability of protection and control devices.

It may be possible to consider extending ZTA architecture and example solution to include dynamic access control for DERs or other grid edge devices connecting to the distribution grid.

8.3.9 Microgrid Security Architecture

Based on the microgrid communications pathway described in [NIST SP 1108r4] (see Figure 8.11), this publication [NIST SP1800-32] describes reference architecture to address the cybersecurity of the microgrid communications pathways. Figures 8.12 and 8.13 depict the reference architectures used to protect information exchanges. The reference architecture demonstrates the capabilities to protect, monitor, and audit DER information exchanges such as [NIST SP1800-32]:

- All information exchanges are by and between authenticated and authorized entities.
- The networks used to exchange information are monitored, and suspicious activity is detected and reported.
- A distributed ledger of information exchanges is maintained by a third party to allow both DER operators and the utility to independently verify the information exchanges.
- A DER operator log collection, data analysis and visualization capability provide controlled results sharing with the utility and other DER operators.

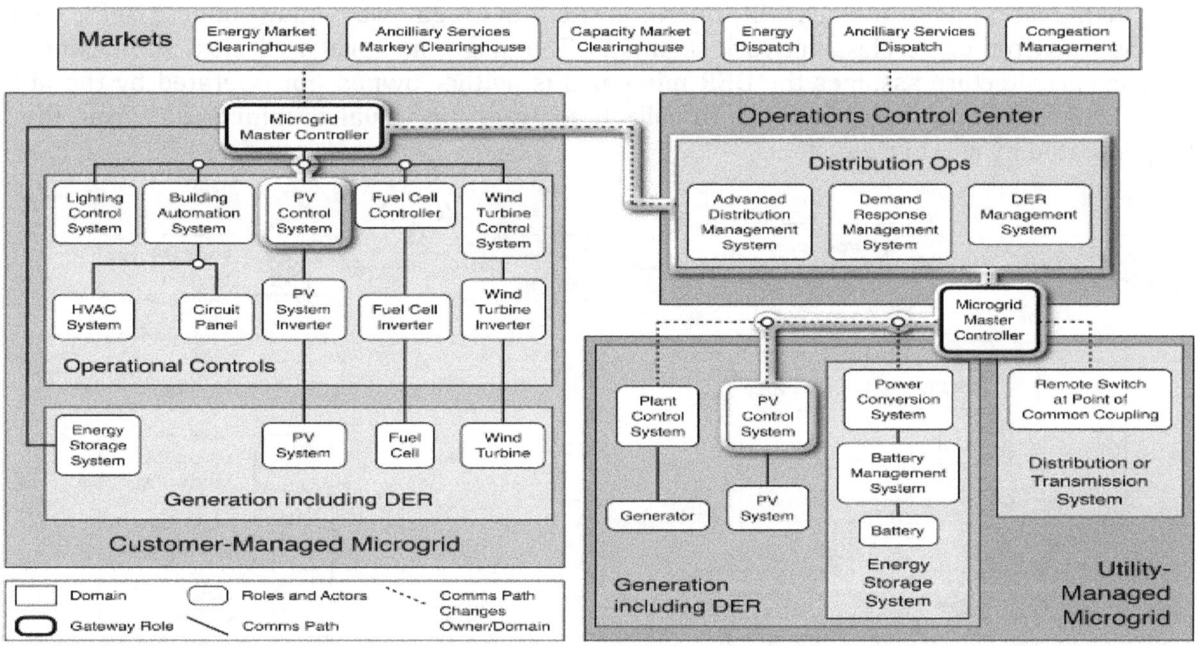

Figure 8.11 Microgrid communications pathways scenario. *Source:* [NIST SP1108r4]. Public Domain.

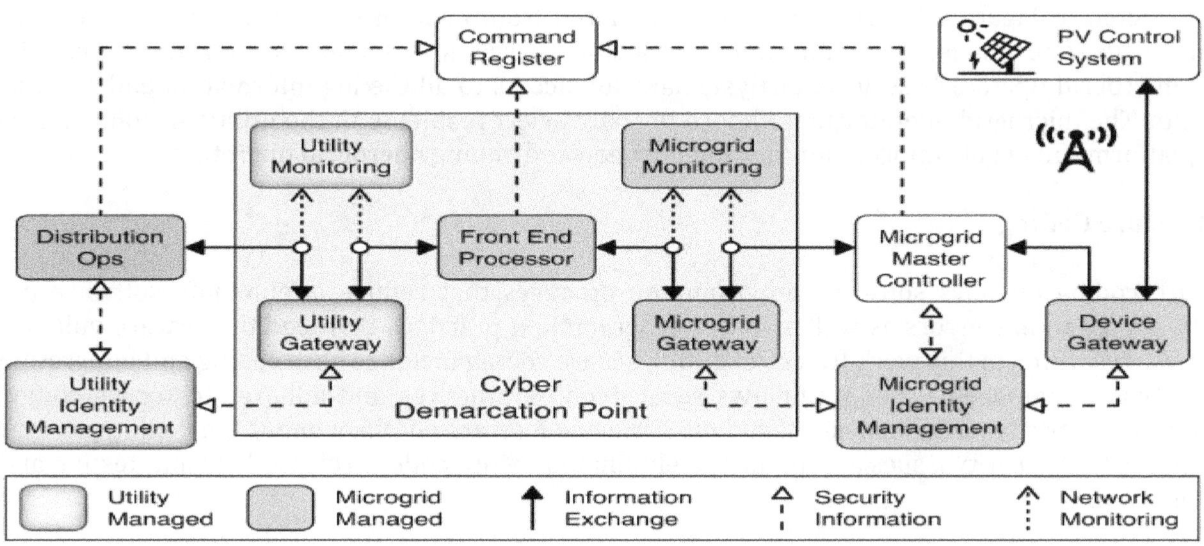

Figure 8.12 Information exchange, monitoring, and distributed ledger reference architecture. *Source:* [NIST SP1800-32]. Public Domain.

Figure 8.12 shows the elements of the reference architecture for protecting information exchanges, monitoring network traffic, and recoding information exchanges in a distributed ledger. The photovoltaic (PV) control system controls the PV DER.

The core element of this architecture is the cyber demarcation point. The cyber demarcation point separates a utility network and a microgrid network that is owned and controlled by a DER operator. The cyber demarcation point is responsible for independently enforcing two distinct security policies - the utility's security policy and the microgrid owner's security policy. There should be a cyber demarcation point at each DER operator site. Device gateways may connect to PV control systems via wired or wireless network segments. The diagram shows a wireless connection. The reference architecture assumes the DER microgrid is neither owned nor operated by the utility. Figure 8.13 illustrates the capabilities to collect, analyze, and visualize information from the log files generated by microgrid systems.

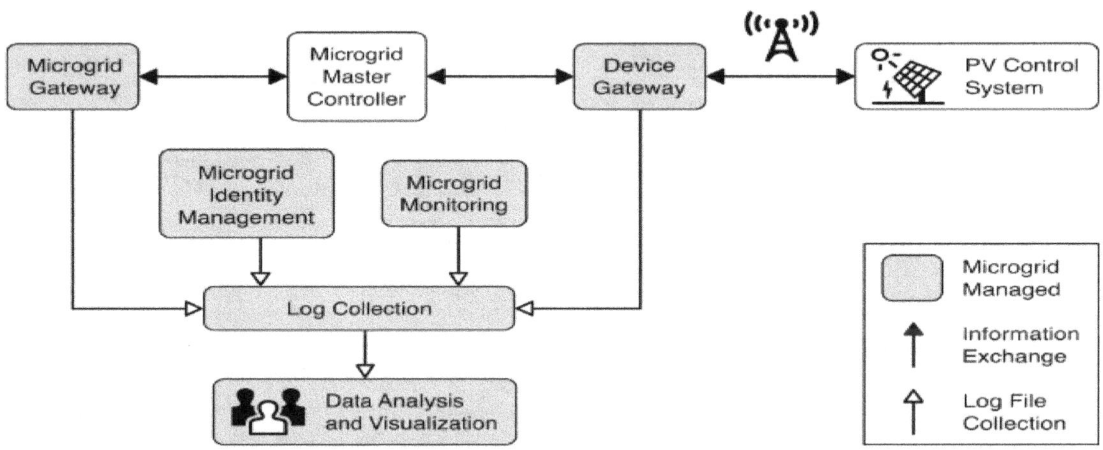

Figure 8.13 Log collection, data analysis and visualization reference architecture. *Source:* [NIST SP1800-32]. Public Domain.

As depicted in Figure 8.13, the log files are gathered from microgrid systems by a log collector which aggregates the log data and sends it to a cloud based analysis and visualization capability. The microgrid operator's security analysts have full access to all the log information and analysis results. The microgrid operator may choose to share select results with the utility. A cloud analytics platform can enable information sharing between and among microgrid operators.

8.4 Secure Coding

Secure coding employs software programming practices that reduce or eliminate software defects/programming errors as well as other programming practices that lead to software vulnerabilities. According to this work [Goertzel 2008], secure coding includes both coding and integration of software components. Coding follows secure coding practices and adheres to secure coding standards. Static security analysis of code is performed iteratively throughout the coding process, to ensure that security issues are found and eliminated before code is released for unit testing and integration.

Coding errors cause the majority of software vulnerabilities. A search of National Vulnerability Database (NVD) [NVD] for the first ten months of 2014 year shows the number of vulnerabilities is much higher than in 2013 and more than 50% of vulnerabilities were caused by programming errors. Improving the software development process and building better software are ways to improve software security, by producing software with fewer defects and vulnerabilities. Therefore, secure software has to support security and other attributes such as dependability, trustworthiness, and survivability.

8.4.1 Software Security

Software security called also secure software is concerned with developing defect free software than can function robustly in the production environment. The continuous increasing of potential costs associated with security incidents, the emergence of increasingly complex regulations, and the continued operational costs associated with staying up to date with security patches all require that organizations give careful consideration to how they address software security [Oracle]. Most software that has been released typically comes with defects – implementation bugs and design flaws.

Secure software is different than security software which performs security functions. However, security software is just as likely to contain flaws and bugs as other software and it should not be assumed secure.

Secure software is software that consistently exhibits the properties of dependability, trustworthiness, and reliability defined as follows [Goertzel 2008]:

- Dependability is the ability of a system to perform its intended functionality or deliver its intended service correctly and predictably whenever it is called upon to do so, including under hostile conditions such as when the software comes under attack or runs on a malicious host. These properties of software directly contribute to its dependability: availability, integrity, reliability, survivability, trustworthiness, safety, and fault tolerance.
- Trustworthiness is the characteristics of containing few if any vulnerabilities or weaknesses that can be intentionally exploited to subvert or sabotage the software's dependability, and containing no malicious logic that would cause the software to behave in a malicious manner.
- Reliability is the ability of a software system, including all of its individual components, to accomplish their objectives without failure, degradation, or unspecified behavior. Software that possesses the characteristic of reliability to the extent that it can be expected to consistently perform its intended functions satisfactorily. This implies a time factor in that reliable software is expected to perform correctly over a period of time. It also encompasses environmental considerations in that the software is required to perform correctly in whichever conditions it finds itself - this is sometimes termed robustness. Reliability is the probability of failure free (or otherwise satisfactory) software operation for a specified/expected period/interval of time, or for a specified/expected number of operations, in a specified/expected environment under specified/expected operating conditions.

For purposes of software assurance, secure software needs only realize these properties with a justifiably high confidence without having to guarantee absolutely the substantial set of explicit security properties and functionality, including all those required for its intended usage.
In most cases, software defects are caused by:

- Non-conformance, or a failure to satisfy requirements – require improvements through a software security assurance program reduce the defects.
- Error or omission in the software requirements – generally cannot be handled within the software discipline.

Other causes of defects include software complexity, incorrect assumptions, flawed specifications, unintended and unexpected interactions. Defects in software are commonly classified as bugs (errors in coding and implementation) and flaws (errors in the software requirements, architecting, or design). Software defects may cause vulnerabilities and weaknesses by increasing software exposure to threats.

A vulnerability is a weakness in an information system, system security procedures, internal controls, or implementation that could be exploited or triggered by a threat source. Classes of potential vulnerabilities for the Smart Grid that are identified by category such as management, operational, technical, training, and platform software/firmware are described in Volume 3 (Supportive Analyses and References) of [NISTIR 7628r1]. Using the CWE and OWASP taxonomies as a guide, classes and subclasses of vulnerabilities in platform software and firmware are described in this document. Discovery and reporting of vulnerability instances occurs continuously and the Common Vulnerability and Exposures (CVE) specification establishes a common identifier for known vulnerability instances. The Common Weakness Enumeration (CWE) and the Vulnerability Categories defined by OWASP are two taxonomies which provide descriptions of common errors or oversights that can result in vulnerability instances.

Errors or oversights in software and firmware design, development, and deployment may result in unintended functionality that allows attackers or other conditions to affect, via programmatic means, the confidentiality, integrity, and/or availability of information. These errors and oversights are discovered and reported as vulnerability instances in platform software and firmware.

8.4.2 Improving Software Security

Organizations are looking to ways to improve software security by reducing the number of vulnerabilities to a level that can be fully mitigated in operational environments. This can be achieved by preventing coding errors or discovering and eliminating security flaws during software development, implementation, and testing.

In each stage of the software development lifecycle, best practices can be used for instilling security in a software application. Software security best practices are described to include crucial development phases [SAFECode 2008]:

- Requirements phase should explicitly define functionality as this will affect subsequent programming, testing and management resources expended in the development process.
- Design and documentation phase should provide systematic specifications of each requirement for a software application; a level of detail that will adequately enable production of near final drafts of documentation has to coincide with final release of the product.
- Programming phase is where programmers translate the design and specification into actual code; effective coding requires implementers to enforce consistent coding practices and standards throughout all aspects of producing the application.
- Software security best practices include secure coding, secure source code handling, security testing, and security documentation.

In addition, achieving secure software for SCADA systems and ICS requires to address deficiencies in the software/system development and operations such as [Hentea 2008a]:

- System development skills
  Specific development issues need to be addressed by SCADA systems developers. First, the evaluation of the software is crucial. Software plays an increasingly important role in all types of controllers in SCADA systems. Developers of control systems require programming skills in addition of engineering skills to be able to understand the broad picture and detailed control problems. A common perspective and security solution to the control systems is essential in the design of these systems. New approaches and skills for software de-

velopment of distributed control systems architectures are required to ensure that software is stable and reliable to avoid hazard conditions [Heck 2003]. Control methods are essential in distributed software development and E-commerce applications [Hellerstein 2005].

- Operating system based on microkernel architecture
  A more radical method on developing operating systems software based on microkernel architecture to limit the number of lines of code and implicitly the number of bugs is recommended in [Tanenbaum 2006]. The microkernel technique was used in the past to develop dedicated operating systems for process computer control. This technique might be making a comeback to its potentially higher reliability.
- Increasing quality of software with security features
  Some organizations consider that software security is an important and growing aspect of software quality for control systems. New products are developed to normalize the events in SCADA application logs that are not understood or used in security monitoring. Current products can detect the security events and provide the pattern a Security Event Monitoring (SEM) for SCADA systems.
- Compliance to standards for software development
  Control software development can be improved by following documents such as DOE published guidelines (e.g., [SCADA Steps], NIST guidelines for control systems security (e.g. [NIST SP800-82r2], [NIST Roadmap]), general assessment methods and tools for SCADA vulnerabilities (e.g., [Permann 2005], [ISS]) and Holzman's rules [Holzmann 2006].
- Integration of different technologies
  New languages and platforms such as Java, C, C#, CORBA, and Phyton are promising increased ease of use, portability, and safety and contribute to making heterogeneous distributed control system platforms possible. Low level, real time technology needs to be combined with high level aspects, such as programming, networking, security, simulation, and control.

8.4.3 Build Security In

Build Security In approach [BuildSecurity] is a US-CERT initiative to improve security and software assurance. It provides practices, tools, guidelines, rules, principles, and other resources that software developers, architects, and security practitioners can use to build security into software in every phase of its development.

Organizations can insert secure development practices into their software life cycle process either by adopting a software development life cycle process as supported by US-CERT in the area of Build Security In [BuildSecurity], or through the evolutionary security enhancement of their current practices, as described in other documents such as [Goertzel 2008].

Addressing the SoS integration challenge from an architectural perspective requires that integration is secure [Kazman 2013]. Security is one quality attribute of integration. Examples include no alteration of the exchanged information may occur, and it must be possible to assure the source of the data.

8.4.3.1 Secure Coding Standards

The approach to security recommends that security threats to the software are anticipated and addressed during the software's design, development and testing through secure engineering practices and secure coding standards. This requires a focus on code quality and functional requirements to reduce unintentional vulnerabilities in the code [Simpson 2011a].

Faults resulting from malicious intent or other unintentional causes make the software vulnerable to threats and likely to compromise its security. According to [Schneier 2006], security is about preventing adverse consequences from the intentional and unwarranted actions of others.

8.4.3.2 Applying Coding Standards

Developers and software designers can apply coding standards during software development to create secure software and systems. Coding standards help programmers to follow a uniform set of rules and guidelines determined by the requirements of the project and organization, rather than by the programmer's familiarity or preference.

The US-CERT Secure Coding Initiative coordinates the development of secure coding standards by security researchers, language experts, and software developers using a wiki-based community process [Coding Standards]. The purpose of the secure coding standards is to promote software security. This site provides recommended practices for preventing the most critical exploits in software including support for the development of secure coding standards for commonly used programming languages.

However, because of the relationship between security and other system attributes, the coding standards provide recommendations that deal primarily with some other system attribute that also has a significant impact on security. The dual nature of these recommendations will be noted in broader scope guidelines such as Java Coding Guidelines and CERT C Secure Coding Standard [Long 2013], [Seacord 2014]. The US-CERT security coding model for C language is also supported by ISO/IEC standard [ISO/IEC TS 17961]. Also, other professional groups have expressed an interest in supporting this model.

Improving software security by implementing code that conforms to the US-CERT secure coding standards can be a significant investment for a software developer, particularly when refactoring or otherwise modernizing existing software systems [Seacord 2013]. However, a software developer does not always benefit from this investment because it is not easy to market code quality.

Security is one of many system attributes that must be considered in the selection and application of a coding standard for the Smart Grid systems including DER systems. Other attributes of interest include safety, portability, reliability, availability, maintainability, readability, and performance. Incorporating security features is one of the most important and challenging tasks in designing distributed systems like DER systems.

Many of these attributes are interrelated in interesting ways. For example, readability is an attribute of maintainability; both are important for limiting the introduction of defects during maintenance that can result in security flaws or reliability issues. Reliability and availability require proper resources management, which contributes also to the safety and security of the system. System attributes such as performance and security are often in conflict requiring trade offs to be considered.

8.4.3.3 Why Software Security Assurance?

Software assurance is defined as the level of confidence that software is free from vulnerabilities, either intentionally designed into the software or accidentally inserted at any time during its lifecycle, and that the software functions in the intended manner [CNSSI 4009]. Software assurance has emerged in response to increases in business and mission risks that are now known to be attributable to exploitable software. Software assurance is a confidence that software is free from vulnerabilities, either intentionally designed into the software or accidentally inserted at any time during its life cycle, and the software functions in the intended manner [Goertzel 2008].

As organizations worldwide increase their reliance on software controls (security software) to protect their computing environment data, the topic of software security assurance grows in importance. Software security assurance is defined as an attribute of an information system that provides grounds for having confidence that the system operates such that the system's security policy is enforced [RFC 4949]. Software security assurance is the process of ensuring that software is designed to operate at a level of security that is consistent with the potential harm that could result from the loss, inaccuracy, alteration, unavailability, or misuse of the data and resources that it uses, controls, and protects.

Once information systems are deployed, assurance related controls can help organizations to continue to have confidence in the trustworthiness of the systems. These controls include, for example, conducting integrity checks on software and firmware components, conducting penetration testing to find vulnerabilities in organizational context. Assurance is measure of confidence that the security features, practices, procedures, and architecture of an information system accurately mediates and enforces the security policy.

In practice, software vendors take action in three key overlapping areas to achieve software assurance - security, authenticity and integrity [Simpson 2011a]. Figure 8.14 depicts the key overlapping areas.

As discussed earlier, secure software is software that consistently exhibits the properties of dependability, trustworthiness, and reliability. For purposes of software assurance, secure software needs only realize these properties with a justifiably high confidence without having to guarantee absolutely the substantial set of explicit security properties and functionality, including all those required for its intended usage [Goertzel 2008].

Tools and techniques that can be used to improve software security assurance include:

- Software packages that are available to support software security assurance activities.
- Common weaknesses enumeration (CVE) allows a better understanding of the most common weaknesses that can affect software security.
- Security architecture/design analysis verifies that the software design correctly implements security requirements. Basic techniques that are used include logic analysis, data analysis, interface analysis, and constraint analysis.
- Secure code reviews, inspections, and walkthroughs verify that the software source code is written correctly, implements the desired design, and does not violate any security requirements.

Specific techniques and updated improvements to core areas of design, development and testing are described in [Simpson 2011b]. Also, it includes best practices for implementers including Common Weakness Enumeration (CWE) references and Verification guidance.

8.4.3.4 Challenges of Secure Design

According to [McGraw 2006]:

- Design flaws account for 50% of security problems
- One cannot find design defects by staring at code

- A higher level understanding is required; that is why architectural risk analysis plays an essential role in any solid software security program.

Secure design concept recommends that early design phase must identify and address potential threats to the system and ways to reduce the risks. The objectives can be achieved with threat modeling and risk mitigation and analyzing vulnerabilities and attack vectors from an adversary's perspective. During the early stages of product development, security requirements must be defined and address architecture and design, product development and programming, including requirements for assurance, testing, and metrics for each phase of development. A systematic approach to threat modeling is described in [Hernan 2006].

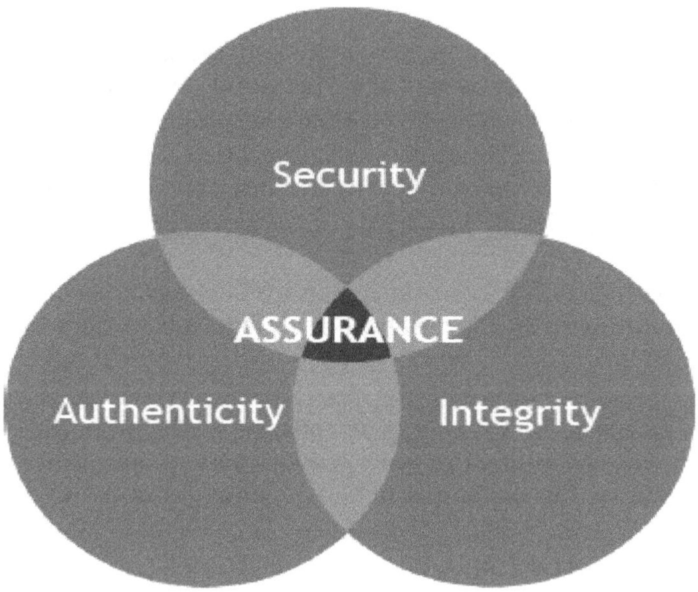

Figure 8.14 Software assurance in practice. *Source:* [Simpson 2011a]. Public Domain.

The goal of a secure design is to enable a system that supports and enforces the necessary authentication, authorization, confidentiality, availability, and non-repudiation requirements even when the system is under attack. If organizations design secure systems which avoid software defects, they can significantly reduce the number and impact of security breaches. A first order approach is to identify the critical software components that control security related functions and pay special attention to them throughout the development and testing process. This approach helps to focus scarce security resources on the most critical areas.

There has been a larger focus on finding bugs rather than on identifying design flaws. Design recommendations are provided in [IEEE Design]. For example, late discovery of errors introduced in requirements and architecture design of cyber physical systems have resulted in costly rework, making up as much as 70% of the total software system cost [Delange 2013]. Methods to overcome software and design issues for cyber-physical systems (CPSs) need to be employed. CPSs are being integrated with Smart Grid applications. However, the challenge of building secure software for Smart Grid is to prevent unwanted damages due to software errors.

Also, the development of CPSs requires rigorous practice standards to avoid any error because they perform critical functions where a fault might have catastrophic consequences (mission failure, severe injuries, etc.). An error behavior feature is added to SAE AADL (Society for Automotive Engineers (SAE) Architecture Analysis and Design Language AADL), a standard for modeling embedded software system architectures [Delange 2013]. This capability provides analysis methods to automate safety analysis and production of safety assessment reports, as requested by safety recommended practices standards (such as [ARP4761]).

To incorporate security features during the development of a distributed system requires a sound analysis of potential attacks or threats in various contexts. Threat modeling is an approach for analyzing the security of an application. It is a structured approach that allows to identify, quantify, and address the security risks associated with an application. To reduce the level of security expertise required, threat modeling can be supported by threat libraries (structured or unstructured lists of threats), which have been found particularly effective in industry scenarios; or attack taxonomies, which offer a classification scheme to help developers find relevant attacks more easily [Uzunov 2014a].

Although threat modeling is not an approach to reviewing code, it does complement the security code review process. The inclusion of threat modeling in the SDLC can help to ensure that applications are being developed with security built in from the very beginning. This, combined with the documentation produced as part of the threat modeling process, can give the reviewer a greater understanding of the system. As more software is delivered on the Internet or operates on Internet connected devices, the design of secure software is absolutely critical.

How to build better security into the design of systems, software, or services from the outset using various threat modeling approaches, how to test designs against threats, and effective ways to address threats that have been validated by software companies are discussed in [Shostack 2014]. Code reviews and penetration tests can be automated using a threat modeling tool [Howard 2006].

Although there are many guides focused on improving the state of the practice in developing secure software, others argue that security methodologies are not being applied in any significant measure, or, if they are, they are not successful in their aims [Uzunov 2012a]. Therefore, a universal framework is evolving as a collection of models [Grady 2008]. The author shows how subsets of models can be assembled into a universal set coordinated with the content of a universal specification format and a way to establish traceability across the gaps when using one particularly troublesome pair of models. In another survey [Uzunov 2012a], the authors describe a range of security methodologies that allow industry stakeholders to better qualify risks and make informed decisions in adopting a systematic security approach. A criteria for selection of methodologies for distributed systems is proposed to be based on comprehensiveness, applicability and uniqueness characteristics aimed at detailed picture of the state of the art.

However, adopting such methodologies in real life scenarios also depends on various non-technical factors that have to be addressed. Examples include:

- A code centric view of security - vulnerabilities are introduced during coding.
- A limited emphasis on software only – developing software application in isolation.
- Stubborn vision of security as an expense, thus seeing security features during development as an extra expense.

Researchers and security practitioners have shown that design level problems are no less critical; often one must consider the system as a whole, e.g., the application together with its whole environment, and organizations need to understand the value of the information about their customers.

One issue that has to be considered by Smart Grid developers is that securing DER systems requires an evaluation of security patterns and methodologies for their capabilities and appropriateness for securing distributed systems. Using security patterns show software developers how to design and implement solutions to common security problems. A survey of pattern based methodologies and list of security patterns ranked according to a set of quality characteristics are provided in [Uzunov 2012b].

A methodology described in [Uzunov 2014b] is based on construct pattern based threat taxonomies for more specific system types and/or technology contexts by specializing one or more threat patterns. It allows developers to quickly consider a range of relevant threats in various architectural contexts using a taxonomy applied to distributed systems.

8.4.3.5 Using Resources

Aligned with Build Security In approach, one valuable resource is the Common Weakness Enumeration (CWE) [CWE]. This is a free platform for public and international in scope that allows the use of software security tools and services that can find weaknesses in source code and operational systems as well as better understanding and management of software weaknesses related to architecture and design.

CWE is used for training programmers on how to eliminate all too common errors before software is delivered and put into operation. The CWE provides a standard means for understanding residual risks and thus enables more informed decision making by suppliers and consumers about the security of software. More information for understanding the most significant exploitable software constructs or many other recommendations that include best mitigation practices are provided by US-CERT (e.g., [Mitigating], [Practices], [Assurance]).

Another source is the Open Web Application Security Project (OWASP) report which captures the top ten risks associated with the use of web applications in an enterprise [OWASP TOP10].

The National Vulnerability Database (NVD) [NVD] includes security checklists, security related software flaws, misconfigurations, product names, and impact metrics. NVD is a product of the National Institute of Standards and Technology (NIST) Computer Security Division and is sponsored by the Department of Homeland Security's National Cyber Security Division. It supports the U.S. government multi agency (OSD, DHS, NSA, DISA and NIST) Information Security Automation Program. It is the U.S. government content repository for the Security Content Automation Protocol (SCAP). The Security Content Automation Protocol (SCAP) is a method for using specific standards to enable the automated vulnerability management, measurement, and policy compliance evaluation of systems deployed in an organization, including e.g., FISMA (Federal Information Security Modernization Act of 2002) compliance.

The SCAP suite of specifications standardizes the nomenclature and formats used by these automated vulnerability management, measurement, and policy compliance products. SCAP enables automation of the linkage between computer security configurations and the [NIST SP800-53r5] controls framework.

The NVD is based upon the Common Vulnerabilities and Exposures (CVE) standard vulnerability dictionary and provides Common Vulnerability Scoring System (CVSS) scores for all CVE vulnerabilities as well as expresses the applicability of these vulnerabilities using the Common Platform Enumeration (CPE). CVE is a dictionary of publicly known information security vulnerabilities and exposures [CVE]; it is International in scope and free for public use. CVE's common identifiers based on this [ITU-T X.1520] recommendation enable data exchange between security products and provide a baseline index point for evaluating coverage of tools and services.

The Department of Homeland Security's United States Computer Emergency Readiness Team (US-CERT) leads efforts to improve the nation's cybersecurity posture, coordinate cyber information sharing, and proactively manage cyber risks to the Nation while protecting the constitutional rights of US citizens. Also, US-CERT strived to be a trusted global leader in cybersecurity. US-CERT distributes vulnerability and threat information through its National Cyber Awareness System (NCAS) and operates a Vulnerability Notes Database to provide technical descriptions of system vulnerabilities. US-CERT is part of DHS National Cybersecurity and Communications Integration Center (NCCIC) [NCCIC].

In addition, the Industrial Control Systems Cyber Emergency Response Team (ICS-CERT) [ICS-CERT] provides security information (alerts, incidents, vulnerabilities, periodic monitoring reports) including assessments for critical infrastructure sectors and industrial control systems and networks. For example, alerts provide timely information about current security issues, vulnerabilities, and exploits. An enterprise strategy for secure software is described in [Steven 2006].

An overview of many common types of attacks can be found in [Howard 2014]. Other resources for building secure software include [Rosenberg 2004], [Thompson 2005]. Also, reviews of current and past security advisories from security bulletins are available at [Microsoft], [CERT], [Secfocus] and other vendors and third parties. Adherence to different guides cannot be accomplished on day one; it has to be part of an organization's strength that is infused opportunistically.

8.5 Testing Criteria

Testing is determination of one or more characteristics of an object of conformity assessment, according to a procedure [ISO/IEC 17000]. Testing is the process of exercising one or more assessment objects under specified conditions to compare actual and expected behaviors.

Security testing is specialized validation that ensures that the security requirements were met and the secure design and coding guidelines were followed. Security testing is an assessment method to determine how effectively an entity being assessed (e.g., host, system, network, procedure, person—known as the assessment object) meets specific security objectives. Security testing allows to validate the security strength of an information system component (e.g., hardware, software, or firmware). Security strength is determined by the degree to which the security functionality implemented within that component is correct, complete, resistant to direct attacks (strength of mechanism), and resistant to bypass or tampering.
Testing may include vulnerability analysis, penetration testing, or use of security testing techniques such as fuzzing or varying external inputs to identify potential buffer overflows and other errors. Some vendors not only do internal testing, but also submit their products to external testing or certification. Periodically, all security controls, mechanisms, and procedures should be tested. The testing can be focused on technology analysis implemented in a device or system and focus on security features and capabilities and/or perform penetration testing.

Penetration testing is a test methodology in which assessors, typically working under specific constraints, attempt to circumvent or defeat the security features of an information system [NIST SP800-53r5]. Penetration testing performed by independent teams can uncover vulnerabilities that would not be detectable using other means. Penetration testing is the process of simulating attacks on a network and its systems at the request of the owner, senior management. Its goal is to determine the vulnerabilities and effectiveness of their security measures. Because some tests could be intrusive and disruptive, it is recommended to use a timeframe when the productivity is not affected [Harris 2013]. Specific recommendations related to testing security of information systems are provided in NIST documents such as [NIST SP800-115], [NIST SP800-53r5], [NIST SP800-53Ar4].

The results from static/dynamic testing and code analysis can provide important evidence that the information systems (including the components that compose those systems) will be more reliable and trustworthy. Security evidence can also be generated from security testing conducted by independent, accredited, third party assessment organizations (e.g., Common Criteria Testing Laboratories, Cryptographic/Security Testing Laboratories, and other assessment activities by government and private sector organizations).

Testing recommendations such as testing criteria and activities that are described in [SAFECode, 2008] include:

- Determine attack surface
- Use appropriate testing tools
- Perform fuzz/robustness testing
- Perform penetration testing.

8.5.1 Testing Smart Grid and DER Systems

It is recommended that testers should follow a series of testing principles supporting consistent technology analysis for the AMI power devices and systems [Carpenter 2009]. Through these techniques, a team of analysts can produce a cohesive assessment of one or more target devices. For example, fields of expertise required for effective AMI security analysis include [Carpenter 2009]:

- Knowledge of the assembly language for processors used by meters and associated hardware.
- Software vulnerability identification and analysis skills.
- Experience writing software based exploits.
- Radio frequency (RF) modulation and coding analysis experience.
- Experience analyzing standards based protocols including infrared, IEEE 802.15.4/ZigBee, IEEE 802.16/WiMAX, BPL/PLC, ANSI C12.18/C12.19/C12.21/C12.22 and proprietary protocols (see more information in Chapter 9).
- Basic and advanced understanding of electronics and principles of electricity.
- Safety training and experience working with high voltage electronics.
- Experience developing embedded technology including embedded software development programming and hardware engineering.
- Analysis skills for evaluating cryptographic algorithms and associated functions.
- Ability to creatively evaluate technology for the goal of subversion.

The results of testing must confirm or negate the presence of vulnerabilities in the target systems. Identifying and evaluating weaknesses allows for these vulnerabilities to be addressed, protecting customers, utilities and vendors alike. The nature of vulnerabilities to be evaluated can be broadly classified as being the fault of design or implementation. Effective recommendations for testing Smart Grid and DER systems include:

- Utilities and vendors should perform the security testing regularly for the several areas as indicated in [Carpenter 2009].
- Disclosure of the penetration tests and tools to the public is not recommended. Some security companies may offer it to the vendor or utility while others will publish on the Internet.

Manufacturers of smart meters have to include a thorough design to support a wide range of security and communications requirements.

Understanding the overall effectiveness of the security requirements implemented in a Smart Grid system and its operational environment is essential in determining the risk to the organization's operations. The use of consistent, standardized cybersecurity evaluation criteria and methodologies contributes to the repeatability and objectivity of test results, which provide insight into the extent to which the requirements are implemented correctly, operating as intended, and producing the desired security posture for the Smart Grid system and system components.

Testing and certification of how standards are implemented in Smart Grid devices, systems, and processes are essential to ensure interoperability and security under realistic operating conditions. Industry has indicated that this is a high priority.

8.5.2 Certification, Conformance, Compliance Testing

Certification and accreditation are the result of a comprehensive technical evaluation of the security components. An evaluation is done by comparing the system to a defined criteria. Certification is the process to indicate that a system provides the functionality and security needed. Accreditation is the formal acceptance of the overall system security and functionality by the management.

System evaluation methods are used to provide an assurance evaluation for the security relevant parts of a system, access control mechanisms, reference monitor, kernel, and protection mechanisms. There are different methods (e.g., [ISO/IEC 15408], also known as Common Criteria).
The testing and certification of the Smart Grid cybersecurity requirements provide assurance that systems and system components are conformant to the requirements selected by the organization. The use of consistent, standardized cybersecurity evaluation criteria and methodologies contributes to the repeatability and objectivity of test results, which provide insight into the extent to which the requirements are implemented correctly, operating as intended, and producing the desired security posture for the Smart Grid information system and system components.

Test procedures, detailed methodologies and specific test cases are necessary to rapidly and effectively introduce test programs that help assure conformance and interoperability for the Smart Grid systems. When testing security features, it is important to understand that thorough tests should cover platform software and firmware vulnerabilities besides testing security features (e.g., identification, authentication, etc.). Known taxonomies (CWE and OWASP) provide descriptions of common errors or oversights that can result in vulnerability instances. For example, Code Quality Vulnerability (CWE-398) description is Poor code quality, which leads to unpredictable behavior (Open Web Application Security Project (OWASP)). From a user's perspective that often manifests itself as poor usability. For an attacker it provides an opportunity to stress the system in unexpected ways. In addition, it has to be understood that a taxonomy provides a way of describing the causes of vulnerabilities, which are largely independent of the operational environment, but the impact of these vulnerabilities may differ in a Smart Grid environment compared to a traditional IT enterprise (see more information in Chapter 7 of [NISTIR 7628r1] guide).

Currently, only a small number of Smart Grid standards have associated industry recognized test programs, and those programs that exist are primarily conformance based and are limited in scope relative to end to end interoperability. Some new programs have emerged in the past year and benefitted from NIST participation in their development [NIST 2016].

NIST, in consultation with industry, government, and other stakeholders, proposes to develop an overall plan for a testing and certification framework. Guidance and methodologies for cybersecurity testing of Smart Grid systems, subsystems, and components are being developed [NIST SP1108r3], [NISTIR 7628r1]. Guidance on how to develop and implement a framework for testing and certification is provided in [NIST 2010]. However, testing and certification is taking on increased urgency as industry reaches consensus on the underlying standards for the Smart Grid. Figure 8.15 illustrates the steps of a methodology for penetration testing described in [NESCOR 2013a].

The following is a brief overview of the phases if applied to Smart Grid testing.

- Penetration test scoping is deciding which system will be the target of the assessment; it could be a substation, generation plant or any other device; the scope could even be the entire system.
- Architecture review is about the information acquisition such as context of the entire system, checking the documentation of the system and analyzing the configuration of the devices part of the scope.
- Target system setup of an environment with the same configuration or closer as much as possible to the configuration of the Smart Grid production environment.
- Server OS, server application, network communication and embedded device penetration tasks are the specific penetration tests within the target system.
- End-to-end penetration testing analysis is ensuring that all possible inputs from external systems to all systems in the scope have been tested and evaluated as possible vulnerable points for attacks.
- Result interpretation and reporting should include the vulnerabilities that could be exploited, the risks associated, the remediation steps and other recommendations that could be applied.

However, there are challenges related to these activities that need to be addressed.

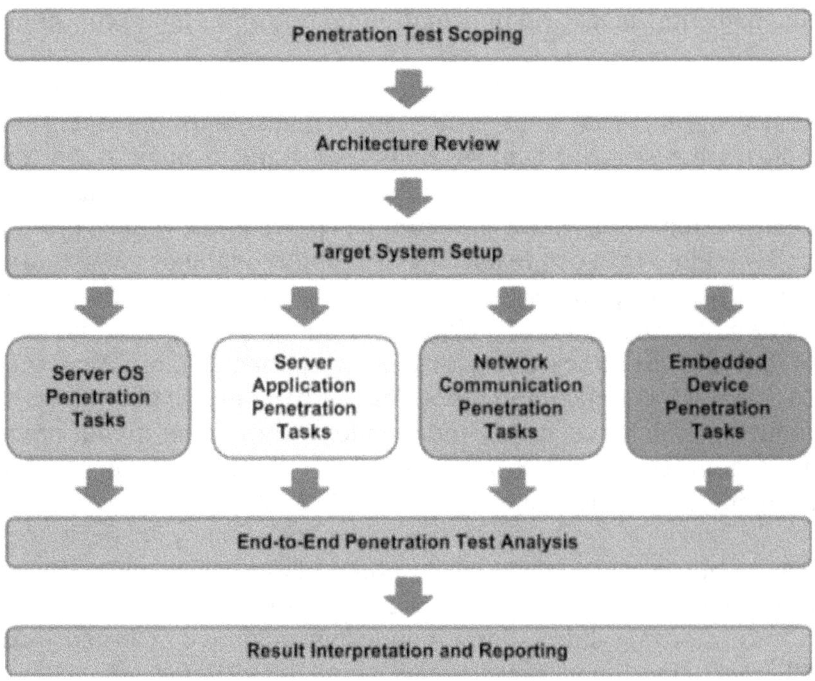

Figure 8.15 Methodology for penetration testing. *Source:* [NESCOR 2013a]. Public Domain.

8.5.3 Challenges

As many other organizations are providing recommendations for the Smart Grid developments, it is necessary to compare the recommendations and adopt those that could provide the best results. Therefore, organizations should be aware of other published documents and issues.

A conformity assessment program for security is established in [NISTIR 7628r1] document; another document details its recommendations on processes and best practices that enhance the introduction of interoperable and secure products in the marketplace [IPRM]. The recommendations build upon international standards based processes [ISO/IEC 17025] and ISO/IEC Guide 65, revised by [ISO/IEC 17065]) for interoperability testing and certification for testing laboratories and certification body management systems. Guidance on testing and certification of the Smart Grid cybersecurity requirements is also provided in [NIST Guide].

However, these documents provide recommendations for specific activities such as:

- Software security testing, which includes penetration testing, confirms the results of design and code analysis, investigates software behavior, and verifies that the software complies with security requirements. Security testing can be conducted in accordance with a security test plan and procedures to establish the compliance of the software with the security requirements. However, security testing efforts are often recommended only for the software requirements that are classified as critical items. Also, the testing methods may be limited and other methods need to be included.
- Penetration testing as a method often involves simulating or issuing real attacks on real systems and data, using common tools and techniques used by attackers. Most penetration tests involve looking for combinations of vulnerabilities on a single system or multiple systems that can be used to gain more access than could be achieved through any single vulnerability. However, the testing to determine risk and vulnerabilities for SCADA systems cannot be performed in the same manner as for any IT system because of differences such as
  - SCADA penetration testing cannot be performed in production environment; SCADA devices are sensitive to unexpected commands that could pose harmless to regular IT environments, but they could be catastrophic to the process availability in a SCADA environment.
  - SCADA devices have specific outputs for the industrial process they are controlling. The architecture and operating systems are not the same, so risks assessment approach is not performed in the same way. For electrical systems, it is necessary to address devices belonging to AMI, DR, DER, Distributed Grid Management (DGM), Electric Transportation (ET) and Wide Area Monitong, Protection and Control (WAMPAC). Table 8.4 shows the category of devices used in Smart Grid and DER systems that differ of IT conventional IT devices (servers, network devices, services, laptops, desktop computers, or mobile devices).
- Appropriate security oriented reviews and tests should be performed throughout the software/system development life cycle of Smart Grid systems. Tests plans based on scenarios may include abnormal and hostile conditions among anticipated conditions under which the software may operate, and test criteria include those that enable the tester to determine whether the software satisfies its requirements for dependability, trustworthiness, and survivability under all anticipated conditions. The review and test regime include a broad and enough variety and combination of tests to enable the tester to determine whether the software satisfies its requirements for dependability, trustworthiness, and survivability under all anticipated conditions.

- Following the suggested mitigations in [NESCOR 2013b] without a rigorous risk assessment and impacts could result in poor security practices. A manufacturer is often inclined to rush a product to market and may choose the mitigation that requires least effort in time and resources including the lowest costs. This practice may also lead to more concerns including security, interoperability, performance, reliability, and safety. CPS must be carefully designed and validated as an error potentially means mission failure or loss of life, hazards must be evaluated and eliminated. For that reason, development of such systems requires following rigorous practice standard to evaluate system safety and show evidence of absence of critical failure. Although, a new version of the failure scenarios are being described in the new version [NESCOR 2015], there are many vulnerabilities that are included in [NISTIR 7628r1], but they are omitted in the revised document. Fact is that many software vulnerabilities (e.g. software coding errors, failures, design flaws, etc.) could be exploited by threat actors. As it is now, the document [NESCOR 2015] does not include any mitigations directed to this kind of exposure to threats. In addition, the selected types of vulnerabilities and threats do not provide a path to a complete security risk assessment leading to a shallow security protection.

Table 8.4 Examples of systems and devices in the Smart Grid.

Systems	Devices
AMI	Meters
	Relays
	Aggregators
	Access points
DR	Energy resources
	Digital control unit
DER	DER managed generation and storage devices
	Customer energy management system
DGM	Automated reclosure
	Remote fault indicators
	Capacitor banks
	Automated switches
	Load monitor
	Substation breakers
WAMPAC	Phasor measurement units
	Device which includes phasor measurement unit capabilities
	Field deployed phasor data concentrator
	Field deployed phasor gateways

Note: see acronyms in Appendix C.

In addition to the development of secure coding standards, US-CERT supports the development of the Source Code Analysis Laboratory (SCALe), which supports conformance testing of systems against these coding standards [SCALe]. SCALe applies conformance assessment in accordance with [ISO/IEC 17000] defined as a demonstration that specified requirements relating to a product, process, system, person, or body are fulfilled.

A goal of conformance testing is to provide an incentive for industry to invest in developing conforming systems by performing conformance testing against US-CERT secure coding standards, verifying that a software system conforms with a secure coding standard, using the CERT seal when marketing products, and maintaining a certificate registry with the certificates of conforming systems.

For each secure coding standard, the source code is certified as probably nonconforming, conforming, or provably conforming against each guideline in the standard. Also, SCALe provides a consistent measure that can be used by power grid asset owners and operators and other industry stakeholders to assess the security of deployed software systems, specifically by determining if they are free of coding errors that lead to known vulnerabilities. This helps to reduce the risk to these systems from increasingly sophisticated hacker tools [Seacord 2010].

Once the security policy, model, and architecture have been developed, the system can be built, tested, evaluated, and rated. Customers use the rating to understand the system (product) they are buying and how they can trust the system. Consumer end users also play a role in Smart Grid product selection as they become educated about certification of devices that they handle directly. Similar to consumer expectations raised through programs such as UL safety listings, Wi-Fi enabled devices, EnergyStar rated products and the like, consumers can raise the bar on vendor devices to help assure interoperability at the level where there is direct daily interaction by the public.

Successful conformance testing of a software system indicates that the SCALe analysis does not detect violations of rules defined by a CERT secure coding standard. Successful conformance testing does not provide any guarantees that these rules are not violated nor that the software is entirely and permanently secure. Conforming software systems can be insecure, for example, if they implement an insecure design or architecture. Software that conforms to a secure coding standard is likely to be more secure than nonconforming or untested software systems. However, no study has yet been performed to prove or disprove this claim.

SCALe does not test for unknown code related vulnerabilities, high level design and architectural flaws, the code's operational environment, or the code's portability. Conformance testing is performed for a particular set of software, translated by a particular implementation, and executing in a particular execution environment [ISO/IEC 27005].

While purchasers and users of Smart Grid products and services create the demand for conforming and certified products, vendors may have some reservations about the added costs of formalized conformance and certification programs where none exist today.

Aside from the definitions provided earlier in this document, a practical matter of standards is that conformance to a specific standard, does not in and of itself assure interoperability of devices and systems when complex systems are constructed. Certainly, conformance to a standard is an essential building block on the road to interoperability. However, additional measures may be needed to assure end to end interoperability.

The current testing and certification market trends to focus on single element or technology issues. For example, there are various programs that individually address specific wireless communications technologies, but few formal programs analyzing the interaction and interoperability across those alternative approaches – occasional test events or plug fests may be held as a venue to bring technologists across those areas together, but those are often more informal and development focused. Similarly, the various power technology and data structure standards testing efforts tend to focus on specific aspects within their respective domains. Broader testing programs across technologies and domains tend to occur most often within end user lab and field trials, as well as in academic or government test beds.

Another issue to be addressed is the meaning of different terms such as product certification, conformance testing, and compliance testing. The terms conformance testing and compliance testing are loosely used across industry generally with the same intent and meaning. In both cases, the terms refer to testing a system or device against a defined set of criteria and evaluating the test results against the metrics defined within the criteria. These concepts are defined in the framework, but some limitations need to be distinguished and addressed. For example:

- Conformance or compliance to the criteria implies a passing or successful result. The term conformance is more widely used and generally associated with testing programs that are of a voluntary or market driven nature.
- The term compliance is more closely associated with mandatory or regulatory oriented programs. It should be noted that these generalizations are not absolute, but a trend, and in most cases the terms are used interchangeably.
- Product Certification is a more specific term, and typically an extension of conformance/compliance testing. In most cases, programs granting certification against a set of criteria require that the system under test meet those 100% of those criteria, or other specific definition of required results necessary towards the granting of certification. Certification can also be viewed as a snapshot in time – the product meets all the criteria required for certification for the specific units tested and on a given date. Certification programs often add an element of ongoing surveillance or quality checks to give greater value to the certification and help assure end users that products remain certified. Certification of Smart Grid applications ensures a consistent level of testing for products based on the same Smart Grid standards, as well as ensuring consistency in the implementation of test programs among different standards.

8.6 Operations Security and Maintenance

Operations security (OPSEC) is a process to identify, control, and protect evidence of the planning and execution of sensitive activities and operations, and thereby prevent potential adversaries from gaining knowledge of capabilities and intentions [RFC 4949]. Operations security is all about ensuring that people, applications, equipment, and the overall environment are properly and adequately secured. Operations security encompasses security controls to protect resources, information, and the hardware on which the resources and information reside [Harris 2013]. It comprises all activities needed to keep networks, computer, systems, and environments up and running in a secure and protected manner. More specifically, operations security includes maintenance activities, vulnerability testing, penetration testing, monitoring and control of security posture. Operations security is described as the controls over the hardware in a computing facility, over the media used in a facility, and over the operators using these resources in a facility [Krutz 2004]. Related activities are categorized as follows:

- Controls and protections to ensure security objectives
- Monitoring and auditing the controls
- Threats and vulnerabilities applicable to operations domain.

Operations take place after a system and environment is developed and implemented. This includes the continuous maintenance of the environment and activities on day to day basis. Audit and monitoring activities permit the identification of security events. They support subsequent action to identify and report the key elements of these events to the appropriate individual, group, or process. Operations security is used to identify the controls over hardware, media, and operators and administrators with access privileges to any of these resources. Operations security activities are documented in standards (e.g., [ISO/IEC 27001], [ISO/IEC 15408], [NIST SP800-53r5], [NIST SP800-12]).

The environment for operations security includes all computing resources of an organization – hardware, software, media, telecommunications equipment, supporting utilities, personnel, and facilities such as computer/server rooms and buildings. Operators, network or system administrators have elevated access rights and permissions in order to perform required activities. Operator is a person who has been authorized to direct selected functions of a system.

Another way of structuring activities is based on ISO model for network management. The ISO-based Network Management Model that includes five areas is transformed into five areas for security management [Whitman 2014]:

- Fault management
- Configuration and change management
- Accounting and auditing management
- Performance management
- Security program management

In addition, there are strict performance and reliability requirements that are needed by power system operations as documented in [NISTIR 7628r1]:

- Operation of the power system must continue 24/7 with high availability for SCADA and higher for protective relaying regardless of any compromise in security or the implementation of security measures that hinder normal or emergency power system operations.
- Power system operations must be able to continue during any security attack or compromise (as much as possible).
- Power system operations must recover quickly after a security attack or the compromise of an information system.
- Testing of security measures cannot be allowed to impact power system operations.
- Power system management, monitoring, and control will increasingly extend away from the power entities' traditional physical and security environments into external environments that the power entity has little or no influence and control over.

In several situations, operators are expected to take actions. Today's power system operators take action in the multi second to multi minute time frame, but PMU based system can make and execute decisions in the 100 millisecond time frame [Rihan 2013]. Also, during real time emergency transmission operations, the power system takes some automated actions and the operators can also take actions such as management of emergency alarms, system restorations based on system restoration plans prepared (authorized) by operation management. The operator's response has to be available within the expected constraints of time (e.g., 1 second or less in some situations) and safety of the people and equipment. The electronic commands submitted by operators have to be secure and accurate. There is an expectation that SCADA and Energy Management System (EMS) systems are protected to avoid exposure to cybersecurity threats. That means the security mechanisms continue to work as intended during emergency situations. Therefore, these real time constraints have to included in the design of security requirements, specifically aimed at system performance, security architecture, and security controls performance.

Electric grid operators utilize their EMS to run sophisticated contingency analysis programs every 5 to 10 seconds to identify the most critical components of the electric grid. The operators need to be aware of the critical components, as well as the consequences if a key component is removed from service, and to operate the system to mitigate the loss of any key components.

Basically, the future of grid operations may see smaller market intervals, more variable generation, demand and distributed resources (including integration with distribution management systems), and much more and frequent data on system state. In addition, the grid of the future requires more closed loop (computer driven) and wider area controls for voltage and power flow control than the more traditional open loop (human intervention) and local drivers for voltage and power flow controls.

8.6.1 Maintenance

A basic security program must include system management and maintenance models. Examples of models include ISO network management model (focuses on methods to manage and operate systems) and security maintenance model (focuses on methods to maintain security of systems).

One essential aspect of operations is maintenance of the security program. Objectives of maintenance include:

- Ensure systems and support activities are conducted in a secure manner.
- Maintain the security of system software and data.
- Maintain the security controls ongoing basis (e.g., checking for failures and errors, presence of unexpected services, human interventions, etc.).

In planning for ongoing operations, security maintenance models can be used. A security maintenance model is designed to focus organizational effort on maintaining information systems (see Figure 8.16). The maintenance model is based on five subject areas: external monitoring, internal monitoring, planning and risk assessment, vulnerability assessment and remediation, readiness and review. It is necessary to monitor external and internal environment.

A complete security program for Smart Grid includes a maintenance program of cybersecurity and physical security for OT and IT assets based on intelligence. Integrating these is the key to overcome the unique challenges of securing data and systems. A maintenance management program employs prevention, detection, proactive planning, response, testing, analyzing, logging, and strategy that evolves over time and with an organization's needs. Ensuring a strong defense against attacks requires a continuous effort.

A maintenance program has to be effective, manageable and predictable in terms of effort – time and costs. It requires ongoing compliance activities such as logging, monitoring, preventative scanning, and awareness training.

The objective of external monitoring is to provide early awareness of new threats, threat agents, vulnerabilities, and attacks that is needed to mount an effective defense. Also, it includes activities such as collecting intelligence from external data sources and giving that intelligence context and meaning for use to organizational decision makers.

Threats and vulnerability data is gathered and processed to generate information that can be used by decision makers. Useful information comes from three classes of sources: vendors, computer emergency response teams (CERTs), SANS, VERIZON), and public network sources. Data is analyzed in the context of organization's security environment to be useful. An organization must consider many threats, including:

- Disclosure of information – release of information to unauthorized persons or administrators can access most corporate data and permission files.
- Destruction of information – obliteration of information or destruction of hardware/equipment, accidental erasure of files.

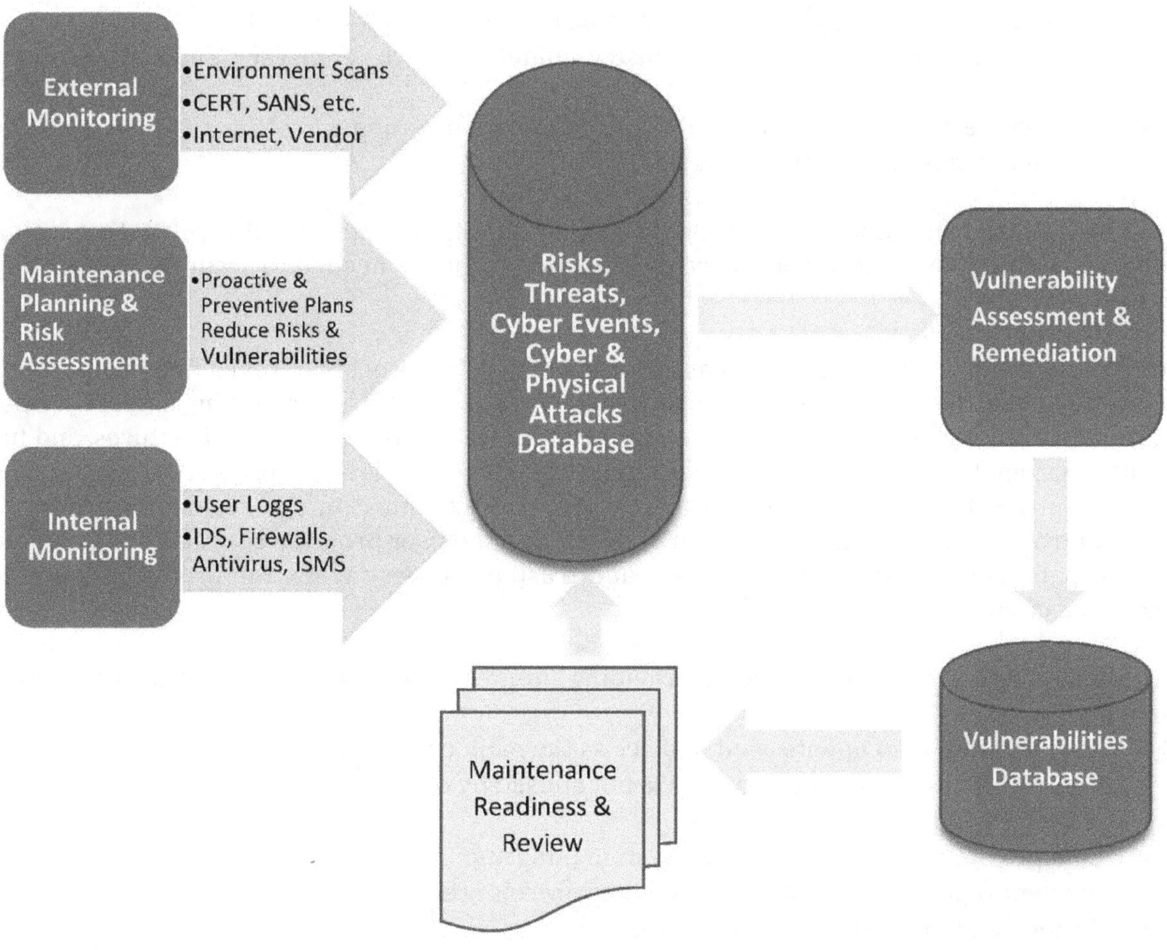

Figure 8.16 Basic components of the maintenance model.

- Interruption/non availability of information, software, hardware, equipment, telecommunications when administrators and operators may accidentally or intentionally disable or disconnect devices.
- Corruption/modification of information through alteration of information, software, information systems when administrators or operators may intentionally or by accident may change file permission and internal tables.
- Theft/removal causing loss of information or equipment.

The objective of internal monitoring is to:

- Maintain informed awareness of state of organization's networks, systems, and defenses by maintaining inventory of IT infrastructure and applications.
- Real time monitoring of IT activity using intrusion detection systems.

- Automated difference detection methods that identify variances introduced to network or system hardware and software.

Requirements of operations security include:
- Reduce the vulnerabilities and threats to availability, integrity, confidentiality, non-repudiation and privacy.
- Balance ease of use and adequate system controls by considering the value of data (intrinsic value and monetary value) and ongoing business need for the data.
- Compliance with legislation or industry guidelines such as privacy and data protection laws.
- Protect the resources, people, and assets of the organization.
- Protect modes of operation as needed.

Not to be confused with operations security, another concept is operational security that includes system capabilities, or performance of system functions that are needed to securely manage a system [RFC 4949].

Another related concept is operational assurance which is the process of reviewing an operational system to see that security controls, both automated and manual, are functioning correctly. Operational assurance concentrates on the system or product architecture, embedded features, and functionality that enable a user to continually obtain the necessary level of protection when using a system or product. Examples of operational assurances examined in the evaluation process are access control mechanisms, the separation of privileged and user program code, auditing and monitoring capabilities, covert channel analysis, and trusted recovery when the product experiences unexpected circumstances [Harris 2013].

Operational management duties involve providing support for overall operations security, such as reviewing:
- Implementation of updates and patches as the main concern to be current
- Operating logs ensuring the system is not abused by operators
- Inventory and status of resources
- Change control practices ensuring system integrity
- Incident reporting to verify how well the controls are operating
- System/audit logs ensuring their integrity
- Audits/security reviews to verify that intrusions and attempts are properly addressed.

8.6.2 DER Operations Overview

The control and management of Smart Grid operation is an interdisciplinary field which crosses the boundaries of communication, optimization, control, dynamic optimization techniques, and even social and environmental constraints. In many situations, the time available for operators to make decisions is only seconds. Thus, the modern grid needs applications and tools that enable grid operators and managers to make decisions quickly. Decision support with improved interfaces enhance human decision making this capability with near real time response and at all levels of the grid operations including operations security. Of course, all these tools require security protection to be available when they are needed.

One major issue is that many DER systems could reside at customer sites that have little or no security and with owners who have minimal or no cybersecurity expertise. In addition, unlike utility owned smart meters, the customers must be permitted to interact with the DER systems they own, since they often use these systems to meet their specific requirements. These factors increase the potential cybersecurity vulnerabilities to the interfaces between DER systems and utilities. Therefore, these modes of operation need to include adequate security controls to avoid threats and vulnerabilities exposure.

Mode of operation is a classification for systems that execute in a similar fashion and share distinctive operational characteristics (e.g., production, online, real time, interactive, etc.). Resources needed for protection range from people to software, hardware, and system utilities, logs, audits, backup files, procedures, security technologies (e.g., firewalls, intrusion detection and prevention systems, cryptography systems, etc.).

Therefore, DER technologies are increasingly being evaluated, and the operational requirements for interconnecting these DER systems are being assessed. But DER systems can have very different operational characteristics (e.g., exclusively driven by their energy source, such as wind and solar energy with more control capability, but constrained by emission limits or availability limits). These may require flexibility and appropriate management controls for the security program.

For example, at Level 4, distribution utilities implement DER management software applications to analyze and coordinate DER generation and storage as part of the broader DMS functions that manage the safe, efficient, and reliable planning and operation of the power grid.

These DMS and DER focused applications are responsible for executing power flow based situational awareness, contingency analysis, generation and load forecasts, planning studies, and other assessment functions to determine DER operational requirements. As part of this analysis, the DER applications handle DER registration, determine the optimal settings for DER autonomous functions, establish pricing signals for DER energy and ancillary services, initiate control commands when necessary, and transmit this information to the appropriate FDEMS within the utility jurisdiction.

The Distribution Utility DER Operational Analysis (DERMS) manages the requests and commands to the DER systems. The types of interactions are illustrated in Figure 8.17. For this architecture, the DERMS is also responsible for the database of interconnection authorizations and registrations of DER systems. The six main DER operational analysis systems consist of [Cleveland 2013]:

- Distribution utility DER operational analysis
- Distribution management system (DMS)
- Geographic information system (GIS)
- Load management and demand response system (LM/DR)
- Outage management system (OMS)
- DER SCADA system (logically separate from the distribution SCADA system that interacts with the DER systems within the utility's perimeter).

Utilities monitor the power system and assess if efficiency, reliability, or market advantage can be improved by having DER systems modify their operation. Level 4 utility DER software applications are highly critical to safe and reliable power system operations since their results can affect large numbers of DER systems. Cyber events on these utility applications could cause decreased power system resiliency, major outages, and/or widespread equipment damage.

Therefore, these related applications require expertise (knowledge of control systems, IT operations skills, communication networks, security operations), coordination of processes and people, security controls (e.g., access control, authentication, auditing), monitoring, collection of data and events, data management of the configuration information, etc.

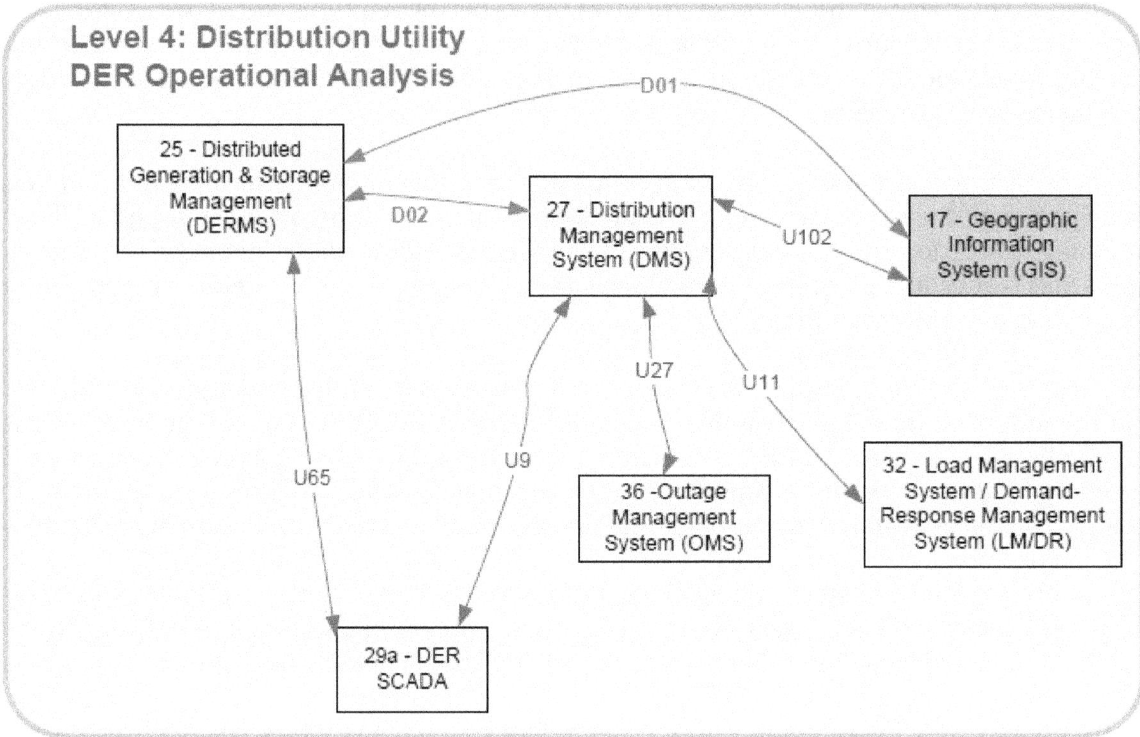

Figure 8.17 Level 4, distribution utility DER operational analysis (DERMS) for distribution operations architecture. *Source:* [Cleveland 2013]. © 2013, Xanthus International.

8.7 Economics of Security

The investment on security is a continuous issue in many organizations. The increasing need for more budgets to provide information security protection prompted economists to look at the economics of security and a new discipline emerged.

The economics of information security addresses the economic aspects of privacy and computer security. It addresses individual and organizational decisions and behaviors with respect to security and privacy as market decisions [Gordon 2002].

The birth of economics of security is attributed by many to Ross Anderson's paper [Anderson 2001] pointing the need of aligning incentives with the security technology to enable rational adoption. A security technology should enable the party at risk to invest to limit that risk.

A year before, in 2000, the authors of this paper [Camp 2000] predict the rise of the market for vulnerabilities. They introduced a mechanism for creating a market for security vulnerabilities based on vulnerability credits which can be exploited. This paper is a first step to developing a pricing scheme for vulnerabilities to increase infrastructure security.

For almost two decades, many ideas and opinions are debated on the challenges and various aspects of economics of security (see these Web sites [Anderson], [ISE]). However, this field evolves with many developments among economists and computer scientists. The most important aspects include security as a tradable good, investments, models, frameworks, users, manufacturers, etc.

8.7.1 Facets of Economics vs. Security

The use of economics proposes that security must be some kind of tradable or measurable good. However, it is argued that security is more complex than most goods in that its primary function will be subverted by its users. The nature of security as a good is complicated by the fact that it is inherently a bundled good and cannot be purchased. Security level is a public good, while defensive costs are private [Kunreuther 2003]. Others argue that security is an investment and not an expense, because investing in security measures makes possible to satisfy the changing business requirements and risks without hurting the business's viability [Avolio 2000]. Systems often fail because the organizations that defend them do not bear the full costs of failure [Moore 2011]. This survey reveals that the economic challenges facing information security include:

- Misaligned incentives between those responsible for security and those who benefit from protection are rife in IT systems.
- Information asymmetries such as bad security products drive out good ones from the marketplace.
- Externalities where an individual's actions have side effects on others; two key externalities types: network externalities and externalities of insecurity; for example, externalities occur when the target is someone other than the host computer (e.g., botnets send spam, host phishing scams and other illegal content, launch denial of service attacks, and provide anonymous cover for attackers).

Security economists have discovered that systems often fail not for some technical reason, but because the incentives were wrong [Anderson 2008b]. Others describe how to calculate the return on security investment (ROSI) (see [Geer 2001]) while others view that calculating the return on investment has to go beyond a single ROSI calculation [ENISA 2012]. Therefore, economics combined with a management, organization theory, and computer security together can address the chronic problems of economic security. The following includes attempts to resolve the challenges related to the economics of security.

8.7.2 Economic Models

Lately, several economic models have emerged to determine the optimal amount to invest to protect a given set of information. Several publications, conferences, and books propose criteria and ways to model security investments and control security risk are made known via these dedicated Web sites [Anderson], [ISE].

An analysis of models and approaches to economics of security are provided in [Thomas 2013]. Also, this work proposes a new risk assessment method based on branching activity model that unifies theoretical and empirical social research for information security, and professional practice, and identifies the consequences of information security breaches including cascading consequences in critical infrastructures. However, many assumptions in this model require testing and validation in order to be applied in practice. It is not clear how reliable is this modeling approach if much information about threats is unknown, incomplete, ambiguous, and contradictory.

A risk management framework to extend the Inoperability Input-Output Model (IIM) capability for evaluating investment options in terms of their implementation costs and loss reduction potentials is proposed in [Anderson 2007]. The authors apply the Inoperability Input-Output Model (IIM) to measure the financial and inoperability effects of the Northeast Blackout of 2003 in U.S. The authors estimate the losses due to effects of power grid inoperability and effects on other critical infrastructures. An inoperability of the power sector can result from different types of disasters (e.g., accidents, natural catastrophe, or willful attacks). The IIM estimated losses are of the same magnitude as other published reports.

While many models are available or simple formula are used, the analysis to define an economic model has to consider many aspects of security landscape (vulnerabilities, loss, threats), trends in software and threats, dynamic issues, game theoretic aspects of information security. Although such consideration is not a practice in a small retailer business, a thorough analysis approach to define an economic model for security investment is needed for Smart Grid systems and DER systems. The game theoretic aspect was first noted in [Jajodia 1993], computer security is a kind of game between two parties, the designer of a secure system, and a potential attacker. Game theory offers promising perspectives, insights, and models to address the ever changing security threats in cyber space. Game theory has the capability of examining hundreds of thousands of possible scenarios before taking the best action. A survey and taxonomies to categorize several game theoretic approaches that have been proposed to address network security issues is described in [Roy 2010]. However, the use of proposed solutions requires understanding of the capabilities, limitations, and assumptions.

Also, there is a significant gap in methods and frameworks between theoretical and empirical researchers, and also a gap between academic research and professional practice. Theoretical models of security investment assume some assumptions and focus only on some aspects of loss (e.g., asset) and often they do not take a broader view of impacts affecting all stakeholders.

8.7.3 Research

More research and regulatory frameworks are needed. Other key areas of active research in the economics of information security include [Moore 2011]:

- Modeling attack and defense
- Breaches of personal information
- Malware and botnets
- Payment system security.

The Economics of Security is a research and policy relevant initiative that was formerly managed by the German Institute for Economic Research (DIW Berlin) and is now hosted by the Stockholm International Peace Research Institute (SIPRI). The aim of the Economics of Security initiative is to understand the economic and social costs of terrorist attacks and anti terror policies, as well as to use this knowledge to produce more effective and efficient security policies. This aim is currently pursued through two separate projects:

- The first is EUSECON (European Security Economics), an EU 7th Framework Programme project that investigates the human induced drivers of insecurity in society.
- The second is NEAT (the Network for the Economic Analysis of Terrorism), which brings together Europe's leading economists to collaborate on research concerning the economic impacts of terrorist and counter terrorist actions.

8.7.4 Improvements

Improvements to modeling attack and defense are suggested in [Moore 2011]. The authors argue that the approach used by computer scientists to design information security systems starting from assumptions about the adversary's capabilities is not sufficient. It is needed to consider also their motivation. It is pointed that empirical research on privacy breaches should continue, as they are currently the best source of public data on information security incidents. The reason is that measuring the effect of breach notifications on firm behavior is one promising area, but it is also needed a better understanding of the effect on consumers.

In order to stop malware and botnets propagation, it is recommended researchers to study the effects of different policies supported by different countries and to identify what works best for different circumstances.

8.7.5 Costs

The need to justify the costs for security mechanisms drives security practitioners, business managers, and researchers on a continuous dialog about what model to use. Understanding the risk model used by an organization and establishing an economic model aimed for an optimal solution during the initial phase of a system development and acquisition cycles help decision makers on allocating the appropriate resources for information security.

Different aspects of the economics of information security investments may be available at this Web site [Anderson]. However, there is no common practice for organizations accumulating and analyzing all costs related to information security. How much to spend on security is still a debatable issue. Some argue that the security costs are too high while another argues the contrary [Schneier 2002].

Manufacturers of privacy enhancing technologies have failed in the market place, because the great majority of people pay little attention to privacy and they will give away the most sensitive information for little benefit [Anderson 2008a]. This work points out that the economists are interested in changes of public's attitude towards privacy and security economics is a growing research area that it may approach privacy from the interests of both the public and manufacturers.

However, both research and real world experience indicate that one way to decrease security costs is correcting weaknesses and vulnerabilities as early as possible in the software's life cycle. This is far more cost effective over the lifetime of the software than developing and releasing frequent security patches for deployed software.

8.7.6 Regulations

Challenging is what information security issues should be handled at the state level and what issues may require harmonization or coordination using a framework. An appropriate regulatory framework is important for protecting economic and other activity online as it is offline. Based on the analysis of cybercrime costs, it is suggested that users and countries should spend less in anticipation of cybercrime (on antivirus, firewalls, etc.) and more in response – that is, on the prosaic business of hunting down cyber criminals [Anderson 2012].

8.7.7 Trade Off Perspective

It is argued that security is always a trade off which sometimes works, sometimes doesn't work [Schneier 2007]. The aspects of trade off include [Schneier 2007]:

- The severity of the risk
- The probability of the risk
- The magnitude of the costs
- How effective the countermeasure is at mitigating the risk
- The tradeoff itself.

Another argues that economics based approach should provide when vulnerability disclosure is good for the ecosystem, and when have incentives to disclose [Swire 2007]. Others argue that while one major goal of disclosure is to eventually force vendors to develop secure software, there is no evidence that suggests that disclosure indeed creates such incentives [Telang 2005], but empirical studies provide some evidence of the value of secure software [Telang 2007].

Therefore, with no uniform agreement and conclusion on economics of security, the users continue to spend significant amount of money on security protections while security industry is increasing the profits regardless of effectiveness and efficiency of security technologies.

8.8 Standards, Guidelines, and Recommendations

Currently, there are several architectures being developed by different Smart Grid stakeholder groups in US and other countries (e.g., Europe, Australia, China, Japan, etc.).

8.8.1 Smart Grid Architecture Developments

Therefore, NIST and the Smart Grid Interoperability Panel (SGIP) are coordinating with these groups to harmonize the architectures that will exist within the Smart Grid architectural framework, evaluating how well they support the architectural goals [NIST SP1108r3]. These architectures are evaluated against the conceptual reference model, the semantic framework, the standards and architecture evaluation criteria, and the conceptual business services. Also, there are International efforts to align Smart Grid architectures.

In USA, NIST and the SGIP committees are coordinating with several Smart Grid stakeholder groups developing different architectures. The coordination helps align or harmonize evolving architectures existing within the Smart Grid architectural framework, evaluating how well they support the architectural goals listed in the road map to Smart Grid.

Architecture alignment efforts are under way with but are not limited to the following groups [NIST SP1108r3]:

- The Institute of Electrical and Electronic Engineers (IEEE) P2030 developed a logical level view of the Smart Grid organized into three major areas: physical, communications, and information. This logical architecture conforms to the NIST Conceptual Reference Model and provides a set of defined interfaces for the Smart Grid. A SGAC/P2030 harmonization activity completed in late 2011 provided the IEEE P2030 team with several recommendations that were incorporated into their work.
- The European Commission's Mandate 490 (EU-M490) for Smart Grid with the European Telecommunications Standards Institute (ETSI), European Committee for Standardization (Comité Européen Normalisation - CEN), and the European Committee for Electrotechnical Standardization (CENELEC), completed their first version of the Smart Grid Reference Architecture in November 2012. The document incorporated comments and suggestions from the Smart Grid Architecture Committee (SGAC). In 2014, this group released the second version with an aligned architecture approach SGAM sharing work developed by both groups. Both groups' goals are to provide the basis for an architectural process leveraging The Open Group Architecture Framework and Service Oriented Ontology. The goal is to provide stakeholders with the tools necessary to quickly identify and define their requirements that include interoperability and application specifications. The work is focused on the requirements of European Union (EU) and NIST/SGIP stakeholders. ETSI/CEN/CENELEC, NIST, and the SGIP are working on a collaborated architecture.

- The Chinese Electrical Power Research Institute (CEPRI) published an initial roadmap that resembles much of the work done in the EU and the United States, with some very specific changes that support differences in the Chinese market.
- The Korea Smart Grid Association (KSGA) has published pieces of the architecture including IT, physical field devices, and interface.
- The European Union SG-CG, the International Electrotechnical Commission (IEC) and NIST teams are working on Common Information Model Reference Architecture and an architectural process consistent with the SGAM including their perspective, recommendations, and artifacts as necessary.

8.8.2 DER Security Architecture Developments

Besides DER security reference architectures described in [NIST SP1800-32], this document [IEC TR 62351-12] provides cybersecurity recommendations and engineering/operational mitigations for improving the resilience of power systems with interconnected DER systems. Improved resilience of the CPS Smart Grid with DER systems is based on tightly intertwined cybersecurity strategies and engineering strategies. System resilience permits a system to operate while under, and recover after, attack, perhaps with degraded performance but with continued delivery of critical functionality. Resilience is an architectural feature that is difficult to provide later in the development life cycle and is very costly after deployment [INCOSE 2015].

Chapter 9

Maintenance and Security Management Best Practices for Smart Grid

9.1 Security Management Practices

Management is fundamentally about deciding and delivering behavior. Researchers want to model and manage the behaviors of hardware, software, and even users with a system. Behavior implies the ability to predict changes in a system, either changes made autonomously or in response to input (events or programming).

Cybersecurity (or security) is a condition that results from the establishment and maintenance of security controls that enable a system to perform its mission or critical functions despite risks posed by threats to its use. Security controls (or protective measures) may involve a combination of deterrence, avoidance, prevention, detection, recovery, and correction that should form part of the enterprise's risk management approach.

Security management is the core of an organization's business and information security structure [Harris 2013]. Although security management has changed over the years because of changes in the computing environment (e.g., from mainframes to personal computers, communication networks, Internet, social networks, etc.) and proliferation of malware, it encompasses all the activities that are needed to keep a security program up and running and evolving.

Security management includes risk management, documentation, security controls implementation and management, processes and procedures, personnel security, auditing, and continual security awareness training [Harris 2013]. Basically, security management is the process of achieving objectives using a given set of resources including a security management infrastructure. As defined in [RFC 4949], security management infrastructure consists of system components and activities that support security policy by monitoring and controlling security services and mechanisms, distributing security information, and reporting security events.

The Energy Independence and Security Act of 2007 [EISA 2007] articulates the critical role of standards for the Smart Grid. As noted in [NIST SP1108r3], the June 2011 and February 2013 National Science and Technology Council (NSTC) reports also advocate the development and adoption of standards to ensure that today's investments in the Smart Grid remain valuable in the future; to catalyze innovations; to support consumer choice; to create economies of scale to reduce costs; to highlight best practices; and to open global markets for Smart Grid devices and systems.

There are many security management models and practices available. Standards can provide guidance to utilities as they face novel cybersecurity, interoperability, and privacy concerns. This international standard [ISO/IEC 27001] provides a model for establishing, implementing, operating, monitoring, reviewing, maintaining, and improving an Information Security Management System. The international standard [ISO/IEC 27002] provides a list of commonly accepted control objectives and best practice controls to be used as implementation guidance when selecting and implementing controls for achieving information security.

Also, NIST published many documents that include models and practices. For example, these standards [NIST SP800-12r1], [NIST SP800-53r5], [NIST SP800-30r1], [NIST SP800-37r2], and [NIST SP800-39] describe best practices and provide information on commonly accepted information security principles that can help security team in the development of a security blueprint. Also, included are specific recommendations for operations/maintenance phase and preparing for contingencies and disasters.

One aspect of information security management for Smart Grid is operations security (OPSEC). Operations security pertains to everything that takes place to keep the information systems (Information Technology (IT) and/or Supervisory Control and Data Acquisition (SCADA) applications) and environments up and running in a secure and protected mode (see Appendix C for the common acronyms). The environment includes all computing resources – hardware, software, media, telecommunications equipment, supporting utilities, personnel, and computer facilities. Operations security is used to identify the controls over resources, and the operators and administrators with access privileges to any of these resources. Auditing and monitoring allow identification of security events. They support subsequent action to identify and report key elements to the appropriate individual, group, or process. The controls needed for secure operations may include these types of controls: directive, preventive, detective, corrective, recovery, deterrent, compensating.

Threats to operations security include disclosure or destruction of the information, interruption or non-availability of resources or information, corruption or modification of the data, theft or removal of equipment. Therefore, organizations should [NIST SP800-12r1]:

- Perform periodic and timely maintenance on systems
- Provide effective controls on the tools, techniques, mechanisms, and personnel used to conduct system maintenance

Maintenance can be performed locally or non locally. Nonlocal maintenance is any maintenance or diagnostics performed by individuals communicating through a network either internally or externally (e.g., the Internet). Examples of maintenance controls include: controlled maintenance, maintenance tools, nonlocal maintenance, maintenance personnel, and timely maintenance.

Controlled maintenance of a system deals with maintenance that is scheduled and performed in accordance with the manufacturer's specifications. Maintenance performed outside of a scheduled cycle, known as corrective maintenance, occurs when a system fails or generates an error condition that must be corrected to return the system to operational conditions.

This document [NIST SP800-18r1] includes management controls for the operations/maintenance phase. In USA, supplemental guidance in security management has been provided by DHS, DOE, ICS-CERT, and many other security organizations and forums.

However, there are unique issues associated with how to manage a system as distributed as the Smart Grid, and still maintain efficiency and reliability factors that imply centralization. Many grid systems are highly distributed, geographically isolated, and require local autonomy as commonly found in modern substations. Also, these systems need to have a measure of centralized security management in terms of event logging/analysis, authentication, etc.

Therefore, standards should be selected to provide the right balance and provide for the appropriate approach (e.g., centralized, distributed, or hybrid approach) to complete the necessary objectives for the Smart Grid. Best practices are highlighted as utilities face new and difficult choices.

9.1.1 Best Security Practices

Security efforts that seek to provide a superior level of performance in the protection of information are referred to as best security practices or simply best practices [Whitman 2016]. Some organizations refer to them as recommended practices. These practices balance the need for information access with the need for adequate protection.

Choosing which recommended practices to implement can pose a challenge for some organizations. First is the issue of which best practices are suitable for an organization, then it is the issue that no two organizations are identical. Most organizations seek lessons that can help them strategically, rather than information about specific technologies they should adopt [Whitman 2016].

Responsibility for privacy and information security management must be assigned, and policies and supporting procedures should be created to apply to the data within each organization. Management activities may be supported by contactors and third party providers including service bureaus and other organizations. They can provide Smart Grid information system operation and maintenance, development, IT services, outsourced applications, and network and security management.

The Interoperability Process Reference Manual (IPRM) [IPRM] developed by Smart Grid Testing and Certification Committee (SGTCC) outlines the conformance, interoperability, and cybersecurity testing and certification requirements for recommended Smart Grid standards. The IPRM has been designed to capture testing and certification processes and best practices needed to verify product interoperability amongst two or more products using the same standards based on communications technology. These processes and best practices are intended for use by an Interoperability Testing and Certification Authority (ITCA) in the design and management of a testing and certification program.

The IPRM manual outlines the roles and requirements of an ITCA and specifies the mandatory testing and certification processes associated with achieving interoperability for a specific standard. The IPRM also includes the recommended best practices for interoperability test constructs. The objective is to address test implementation and execution issues, including qualification criteria for test laboratories and accrediting organizations, and recommend best practices to ensure that test results achieve their desired intent and are used in an appropriate and consistent manner.

The new issue of the IPRM document of 2012 transitioned the first version of the document from an informational focus to an operational focus, providing greater clarity to ITCAs to guide their implementation of the IPRM recommendations. It incorporates internationally recognized quality and performance standards for certification bodies and test laboratories to provide confidence to end purchasers (e.g., utilities) and requirements for testing procedures to assure that testing is comprehensive and rigorous as required to meet deployment expectations.

The requirements for an ITCA as specified in the IPRM are divided into five categories that include governance, lab qualification, technical design for interoperability and conformance program, improvements, and cybersecurity. Cybersecurity covers the requirements that shall be used by the ITCA to validate the security related components of the interoperability testing program.

Another perspective on the certification is proposed in this report [T&D 2016]. The authors argue that a cybersecurity certification framework across European member states is needed to be system oriented and be focused on processes instead on products. This means harmonization of cybersecurity requirements across European member states based on key standards that include series of [ISO/IEC 27000], [IEC 62443], and [IEC 62351]. Certification should be focused on processes and not components, such as [ISO/IEC 27001] standard, which focuses on the management of security, while [IEC 62443-2-4] standard focuses on the process of a system integrator. It specifies requirements for security capabilities for IACS service providers that they can offer to the asset owner during integration and maintenance activities of an Automation Solution.

A certification framework should rely on functional requirements rather than short lived technical requirements. For example, encryption shall be a functional requirement, but the technical implementation of this countermeasure should evolve over time and should not be a part of a regulation or standard. In addition, the cybersecurity certification should support flexible use cases based on context.

The advanced power grid relates to key scientific and technological areas [NIST SP1108r3]. These include power quality, reliability, and resilience; widespread integration of grid tied renewables along with attendant large scale storage; widespread deployment of grid sensors; and secure cyber based communication within the grid.

The resiliency is the attribute that allows a grid to better sustain and more quickly recover from adverse events such as attacks or natural disasters. Grid resiliency includes hardening, advanced capabilities, and recovery/reconstitution. Although most attention is placed on best practices for hardening, resiliency strategies must also consider options to improve grid flexibility and control.

9.1.2 Best Industry Practices - Examples

In addition to meeting the requirements defined in five categories, ITCAs should also leverage industry's best practices in their implementations. The IPRM includes a list of recommended best practices and guidelines for ITCAs in their development and operation of interoperability and conformance testing programs.

The recommendations provided in the IPRM were generated based on input from experienced testing organizations that have evolved interoperability and conformance programs through lessons learned in executing tests for both software and hardware applications.

The recommendations may not apply directly to all testing applications; however, NIST and the SGTCC recommend that ITCAs consider them for interoperability and conformance test programs, as these practices have proven to be valuable in executing a broad cross section of program types. Each ITCA should evaluate how these recommendations, observations, and practices apply to their specific programs and should incorporate the recommendations into their programs where applicable.

The recommended best practices in interoperability test constructs in the IPRM address three main areas:

- General test policies—includes policies related to information that product vendors need to know relative to a specific testing and certification program.
- Test suite specification (TSS) —includes the need to establish a common TSS for use by multiple testing labs; a TSS that is test tool agnostic; and revision control of TSS.
- Attributes of a test profile in lieu of complete test suite specification.

As pointed in [NIST SP1108r3] roadmap, improving critical infrastructure cybersecurity described in [NIST 2014] (updated by [NIST 2018]) document draws heavily on the successful SGIP experience described in [IPRM]. This document enables organizations – regardless of size, degree of cybersecurity risk, or cybersecurity sophistication – to apply the principles and best practices of risk management to improving the security and resilience of critical infrastructure. Ultimately, it provides organization and structure to today's multiple approaches to cybersecurity by assembling standards, guidelines, and practices that are working effectively in industry today. Also, this association [CIS] maintains cybersecurity best practices to safeguard operating systems, software and networks that are most vulnerable to cyber attacks. These guidelines are continuously verified by a volunteer IT community to combat evolving cybersecurity challenges.

However, one challenge for the Smart Grid organizations is scalability, since a Smart Grid involves millions of users. A sound solution in a small scale may not be scalable when applying into such a huge scale. Therefore, California Energy Commission documents best practices in this document [CEC 2014].

Also, this document [INL 2007] includes best practices for control systems that are recommended by the ICS-CERT, Department of Homeland Security, and Department of Energy in U.S. Additional practices are included in [DHL 2016].

In U.S., the ICS-CERT organization provides a section for control systems security recommended practices at [ICS-CERTRP]. Several recommended practices are available for reading and download, including improving industrial control systems (ICS) cybersecurity with Defense in Depth strategies. Currently, the Cybersecurity and Infrastructure Security Agency (CISA) works with partners to defend against today's threats and collaborating with industry to build more secure and resilient infrastructure for the future (see https://www.cisa.gov/uscert/).

Also, ICS-CERT reminds organizations to perform proper impact analysis and risk assessment prior to deploying defensive measures. Additional mitigation guidance and recommended practices are publicly available in this technical report [ICS TIPS]. Organizations observing any suspected malicious activity should follow their established internal procedures and report their findings to ICS-CERT for tracking and correlation against other incidents.

In addition, the ICS-CERT assists critical infrastructure asset owners in protecting their control systems and networks against cyber threats. The program provides three types of voluntary cybersecurity assessments: cyber security evaluation tool (CSET®) review, design architecture review (DAR) and network architecture verification and validation (NAVV) review. ICS-CERT provides facilitated CSET services for asset owners, where the team will guide an asset owner through a step by step process to assess their environment, based on a series of questions derived from industry recognized standards, guidelines and best practices. The CSET exists as a downloadable application (free of charge), which can be installed locally on a standalone workstation or laptop (see downloading and installing CSET, https://ics-cert.us-cert.gov/Downloading-and-Installing-CSET).

Secure key management is essential to the effective use of cryptography in deploying a Smart Grid infrastructure. This document [NIST SP800-57P1r5] recommends best practices for developers and administrators on secure key management. These recommendations are as applicable for the Smart Grid as for any other infrastructure that makes use of cryptography, and they are a starting point for Smart Grid key management.

This document [NISTIR 7628r1] recommends an organization establishes and maintains contact with security groups and associations to stay up to date with the latest recommended security practices, techniques, and technologies and to share current security-related information including threats, vulnerabilities, and incidents. Security groups and associations can include special interest groups, specialized forums, professional associations, news groups, and/or peer groups of security professionals in similar organizations. The groups and associations selected are consistent with the organization's mission/business requirements.

9.1.3 Practicing Ongoing Risk Determination and Acceptance

The ongoing risk determination and acceptance help to determine whether the current risk is acceptable and forwards appropriate direction to the information system owner or common control provider. These activities are facilitated by using:

- Automated support tools to capture, organize, quantify, visually display, and maintain security status information promotes the concept of near real time risk management regarding the overall risk posture of the organization.
- Metrics and dashboards increase an organization's ability to make risk based decisions by consolidating data from automated tools and providing it to decision makers at different levels within the organization in an easy to understand format.
- An effective organization wide continuous monitoring program based on a monitoring strategy to assess risk, awareness of vulnerabilities, up to date threat information, and mission/business impacts.
- A monitoring program of all implemented controls at the organization defined frequency.
- Analysis and reporting problems to appropriate organizational officials.

Another issue is adequately mitigating unacceptable risks arising from the use of external information system services. An organization needs to establish an appropriate chain of trust with external service providers when dealing with the many issues associated with information system security. A chain of trust requires that the organization establish and retain a level of confidence that each participating service provider in the potentially complex consumer provider relationship provides adequate protection for the services rendered to the organization. When a sufficient level of trust cannot be established in the external services and/or service providers, the organization should employ compensating controls or accept a greater degree of risk or does not obtain the service.

9.2 Continual and Daily Maintenance Tasks

The National Institute of Standards and Technology (NIST) outlines System Development Life Cycle (SDLC) model that is breakdown in phases. The fourth phase of the SDLC is the operations and maintenance. Maintenance phase addresses all types of maintenance to all components (hardware, software, firmware, systems, applications). Also, it addresses who is performing the maintenance, maintenance tools, and remote access.

9.2.1 Managing Security Tasks

The traditional SDLC has been adapted to Security Systems Development Life Cycle (SecSDLC), which supports the specialized tasks for security. While this process may differ in several specific activities, the overall methodology is almost the same. The SecSDLC assists in the management of the risk and supports a coherent program for Security Management tasks and operations.

In this phase (operations/maintenance), systems are in place and operating, enhancements and/or modifications to the system are developed and tested, and hardware and/or software is added or replaced. The system is monitored for continued performance in accordance with security requirements and needed system modifications are incorporated. The operational system is periodically assessed to determine how the system can be made more effective, secure, and efficient. Operations continue as long as the system can be effectively adapted to respond to an organization's needs while maintaining an agreed upon risk level. When necessary modifications or changes are identified, the system may reenter a previous phase of the SDLC.

9.2.2 Security Management Models

Management models must be adopted to manage and operate a security program. Security management models also assist in planning for ongoing operations. Also, maintenance models are used to structure the tasks of managing particular set of activities or business functions. The maintenance model can be used to monitor external and internal environment. The recommended maintenance model is based on five areas [Ritchie 2016]:

- External monitoring
- Internal monitoring
- Planning and risk assessment
- Vulnerability assessment and remediation
- Readiness and review.

Therefore, the maintenance of the information security program is needed on an ongoing basis. Key security activities for the maintenance phase include:

- Conduct an operational readiness review.
- Manage and control the configuration of the system.
- Institute processes and procedures for assured operations and continuous monitoring of the information system's security controls.
- Perform reauthorization as required.

Thus, maintenance is also a phase in the risk control cycle. Guidance on developing maintenance plans and tasks is available in standards, manuals, books, etc. For example, these international standards can be useful:

- The standard [ISO/IEC 27002] deals specifically with system acquisition, development, and maintenance.
- The standard [ISO/IEC 27003] provides information for monitoring, reviewing, maintaining and improving an ISMS in accordance with the standard [ISO/IEC 27001].
- This standard [ISO/IEC 27010] provides guidelines in addition to guidance given in the [ISO/IEC 27000] family of standards for implementing information security management within information sharing communities and additionally provides controls and guidance specifically relating to maintaining and improving information security in inter organizational and inter sector communications.

In particular, it may be applicable to information exchanges and sharing relating to the provision, maintenance and protection of an organization's or state's critical infrastructure.

9.3 Smart Grid Maintenance

As recommended in [NISTIR 7628r1], security is most effective when it is designed into the Smart Grid information system and sustained, through effective maintenance, throughout the life cycle of the Smart Grid information system. Maintenance activities encompass appropriate policies and procedures for performing routine and preventive maintenance on the components of a Smart Grid information system. This includes the use of both local and remote maintenance tools and management of maintenance personnel.

An organization develops, implements, reviews, and updates on an organization defined frequency in a documented Smart Grid information system maintenance security policy that addresses [NISTIR 7628r1]:

- The objectives, roles, and responsibilities for the Smart Grid information system maintenance security program as it relates to protecting the organization's personnel and assets.
- The scope of the Smart Grid information system maintenance security program as it applies to all organizational staff, contractors, and third parties.
- Procedures to address the implementation of the Smart Grid information system maintenance security policy and associated Smart Grid information system maintenance protection requirements.

Important is the management commitment to ensure compliance with the organization's security policy and other regulatory requirements. Also, the organization ensures that the Smart Grid information system maintenance security policy and procedures comply with applicable federal, state, local, tribal, and territorial laws and regulations. The Smart Grid information system maintenance policy can be included as part of the general information security policy for the organization. Smart Grid information system maintenance procedures can be developed for the security program in general and for a particular Smart Grid information system when required. Common governance, risk, and compliance (GRC) requirements for Smart Grid information system maintenance are included in [NISTIR 7628r1].

One important aspect of the Smart Grid security is the maintenance of ICS. Most ICS are not stand alone or maintained by a single entity. Often organization rely on third party services to deliver products and services to maintain the ICS operation [Piètre-Cambacédès 2011]. As pointed in this work, these products and services vary from support and maintenance of application software or hardware to services, such as security management (e.g., firewall configuration) or provision of telecommunication services. These products and services can be provided by other entities within the organization (e.g., corporate IT) or by external entities. In each case, there are risks associated with dealing with third parties. Therefore, the risks should be assessed and mitigated (e.g., contractual and formalized relationships, service level agreements, etc.). Not only must the ICS and field devices be secured, their ongoing management and maintenance need to be secured as well, and must be capable of managing changes, such as the disclosure of new vulnerabilities or the emergence of new threats.

Therefore, a secured Smart Grid must have maintenance and efficient system planning (especially for future expansion) [Otuoze 2018]. This document [NISTIR 7628r1] recommends strategic planning that has the purpose of to maintain optimal operations and to prevent or recover from undesirable interruptions to Smart Grid information system operation. Interruptions may take the form of a natural disaster (e.g., hurricane, tornado, earthquake, flood, etc.), an unintentional manmade event (e.g., accidental equipment damage, fire or explosion, operator error, etc.), an intentional manmade event (e.g., attack by bomb, firearm or vandalism, hacker or malware, etc.), or an equipment failure.

The types of planning considered are security planning to prevent undesirable interruptions, continuity of operations planning to maintain Smart Grid information system operation during and after an interruption and planning to identify mitigation strategies. In addition, an organization should have included additional responsibilities for the operations/maintenance team (e.g., security personnel) to protect the security controls (e.g., firewalls, access management systems, intrusion detection and prevention systems, cryptographic systems, etc.). Operational responsibility covers the secure operation and monitoring of security controls. For example, maintaining the effectiveness of a firewall system involves many activities such as [Pohlman 2002]:

- Monitoring
- Access rights management

- Analysis of logged data
- Configuration of new services
- Reporting unusual behavior
- Performance evaluation and strength analysis
- Binding analysis to show no conflicting security mechanisms
- Configuration management.

Because of the higher risk to an organization from the security personnel, extra precautions should be taken. Security personnel require special authorizations and permissions to properly respond to security incidents.

9.3.1 Smart Grid Continuous Monitoring

The ultimate objective of continuous monitoring is to determine if the security controls in the information system continue to be effective over time even when inevitable changes occur in the system as well as the environment in which the system operates. The continuous monitoring process can effectively transform an otherwise static security control assessment and risk determination process into a dynamic process that provides essential, near real-time security status information to appropriate organizational officials. This document [CEC 2014] can be used to take appropriate risk mitigation actions and make credible, risk based authorization decisions regarding the continued operation of the information system and the explicit acceptance of risk that results from that decision [CEC 2014].

The ongoing monitoring of security controls effectiveness can be accomplished in a variety of ways, including security reviews, self assessments, configuration management, antivirus management, patch management, security testing and evaluation, or audits. Automation should be leveraged where possible to reduce level of effort and ensure repeatability [CEC 2014].

A review [Otuoze 2018] of Smart Grid threats discusses threats classification by sources such as technical (infrastructure, operations, system data management) and non technical (environment, government regulatory policies and implementation). Threats against Smart Grid can be classified according to different criteria. Threats can be defined as the range of possible actions that can be taken against a system [Mendel 2017].

9.3.2 Data Maintenance Challenges

One objective of system development cycle is to maintain the security of applications and data including computing environment. Smart Grid, particularly with a full deployment of smart meters and expected market penetrations of advanced distribution automation devices, premise computing devices and distributed energy resources (DERs), adds a massive volume of data that need to be managed effectively. The data includes asset installation location and other attributes, device configuration, equipment performance, inspection and maintenance history and pending work orders as well as measurements and controls of Smart Grid devices.

To prepare for the inrush of Smart Grid data that would be generated by the integration of DER into the Smart Grid, even at the pilot stage, the utility should develop holistic enterprise architecture and integration plans to cover the following, equally important, four areas of need in order of implementation timing [CEC 2014]:

- Deployment of Smart Grid systems and devices, including smart meter and in premise computing devices as well as advanced system automation equipment - to ensure efficient installation, and timely and accurate asset data capture during installation.

- Management of the collected data on Smart Grid asset, system and device configurations – to ensure quality control of the data and that the systems and applications that need the data are updated timely.
- Operation and maintenance of the Smart Grid assets, including hardware, firmware, and software – to ensure that the system, equipment and devices will be properly maintained from day one.

Besides key management challenges, other examples of data challenges may include encryption issues. The following are examples of requirements for the implementation and use of the encryption technology:

- Search methods as traditional search methods may not work on encrypted data. To search a big database by decrypting every entry as needed, it is computationally infeasible. The answer is to use a special algorithm to, for example, create a unique hash value for each entry and include it alongside with the entry key. The hash value can't be converted back to entry key (e.g., a credit card number or customer id) if someone improperly gains access to it, but it can be used for searching the database as easily as any other data element.
- Dependencies are needed to ensure that the sender domain does not pass encrypted data to a receiver domain that expects it to be unencrypted; needed to know all the dependencies where the data is stored and used.
- Performance data processing overhead of encrypted data has to be counted and when processing large amounts of data and the mathematical algorithms used in the encryption and decryption are computationally intensive and may take longer time than the expected delay of the data transfer.

9.3.3 Maintenance Tasks: Examples

Beyond baseline cryptographic services, systems and processes are also needed to manage assets in a secure fashion. For instance, system keys must be protected from disclosure through physical and policy based mechanisms. Role based access controls should authenticate utility operations personnel authorized to manage the system and provide a secure audit trail when system management or maintenance tasks, such as updating of keys is performed. In addition, other tasks should be performed to avoid insecure activities. Examples include:

- Audit log maintenance should ensure that the audit log is maintained in such a way as to prevent unauthorized access, modification, deletion or overflow conditions.
- Online maintenance should be provided with a limited capability to observe the usage of specified services or resources as necessary.
- Establish strong security controls over any medium that is used as a backdoor into the SCADA network.
- Strong authentication methods (e.g., challenge response, hashing algorithms, digital signatures) must be implemented to ensure secure communications where maintenance hooks (called also backdoors) for vendors are allowed.

Another aspect of Smart Grid is the maintenance and repair of industrial control and information system components that are performed consistent with policies and procedures. Mitigating risk and anticipating vulnerabilities on utility grids is not just about installing once at the latest technology. Special attention is necessary to implement regular or routine maintenance of equipment, which includes patch management and necessary security updates in the system.

9.3.4 Patch Management

One important issue for control systems is to ensure that the updates for SCADA systems are done securely since they will likely be done online to prevent expensive physical visits to equipment. It is critical to assure that firmware update mechanisms are not used to install malware. This can be done with a combination of strong authentication/authorization mechanisms for the person performing updates, integrity mechanism to ensure that the firmware is secure, and remote attestation mechanisms to ensure that the correct version is being updated.

Managing the lifecycle of all software based assets (embedded devices, network infrastructures, host based systems) implies to have the full knowledge of their components, applications or operating systems. The impacted products shall be unambiguously identified, the end user shall always have the exact visibility of its installed base and the associated implementation risk shall be properly assessed. Every day, corrective patches, impacting both functionalities and security are likely to be published.

This whitepaper [GE 2016] presents an overview of GE's hardening capabilities that meet patch management [IEC 62443-2-4] standard requirements. It describes a methodology used by GE that hardens customer systems using a combination of technical and procedural measures that have been certified to meet [IEC 62443-2-4] security standard.

Therefore, one issue is implementing an efficient patch management process for control systems [Gauci 2017]. This work describes an improved methodology and a smart patch management system based on [IEC62443-2-4] standard, comprising a process for vulnerability detection, a tool to automatically assess system components, and a corrective patch. The proposed methodology supports an ongoing analysis of many variables as well as an iterative vulnerability management process making an operations/maintenance service more secure and ready to protect the system from possible known threats in a planned and iterative manner. The proposed patch management system establishes processes and tools aiming to efficiently drive the user in implementing the patches. It combines several functions such system assets inventory management, security bulletin, vulnerability assessment, patch installation and testing, etc. It allows an organization to get a structured assessment and corrective method to:

- Progressively move from uncontrolled security state to more controlled security state.
- Optimize the electronic critical assets life cycle management and move to supported and secured versions of components.
- Provide an assessment for the exposure of each asset to cybersecurity risks.

However, there could be legacy devices with limited resources that do not allow upgrades to software when vulnerabilities are identified and expose devices to threat. For example, a recent vulnerability (see [ICSA-16-147-01b]) discovered in a data controller has no available code space to make any additional security patches; so, if a firmware update is not possible, then other mitigation solutions are needed. The vendor has released an advisory that identifies compensating controls to reduce risk of exploitation of the reported vulnerabilities.

9.3.5 Predictive Maintenance

With the benefits of increased connectivity and real time data collection, traditional maintenance models are evolving from repair & replace approach to predict & prevent strategy. Predictive maintenance shortens service interruptions and can avoid downtime altogether, resulting in enormous cost savings and higher customer satisfaction.

Since the EMS is the primary source of the information used by system operators to gather their real time situational awareness, it is important to determine the deficiencies with these systems that contribute to blackouts. Also, it is important to improve these systems to reduce maintenance costs and maximize system uptime.

Blackouts are major catastrophic failures in large interconnected power systems. Predicting the occurrence of such phenomena is difficult at best. When they do occur, however, the socio-economic impact is devastating. The North American (US and Canada power system) blackout of August 14, 2003 started shortly after 4 PM EDT and resulted in the loss of 61,800 MW of electric load that served more than 50 million people including many businesses and manufacturers. A cost estimate of the August 14, 2003 blackout in US prepared by the Anderson Economic Group (AEG) estimates the total cost to have been between $4.5 and $8.2 billion, with a mid point of $6.4 billion [ELCON 2004].

Some of the clear root causes for these catastrophic events are:

- Lack of reliable and synchronized real time data.
- Insufficient time to take decisive and appropriate remedial action against unfolding events on the system.
- No properly automated and coordinated controls to take immediate and decisive action against system events to prevent cascading outages.

The most recent major U.S. blackout, the Arizona Southern California Outages on September 8, 2011, identified that the system failure stemmed primarily from weaknesses in two broad areas - operations planning and real time situational awareness. Similarly, another report found out a principal cause of the August 14 blackout was a lack of situational awareness, which was the result of inadequate reliability tools and backup capabilities.

Therefore, an EMS should offer services for diagnosis, troubleshooting and predictive analytics, as well as maintenance strategy recommendations and coordination of the on site events. Predictive maintenance diagnostic tools could provide insight into the state of equipment and expert advice on how to optimize it, enabling operators to manage proactively.

A perspective on improved interfaces and decision support systems are essential technologies that must be implemented if grid operators and managers are to have the tools and training they will need to operate a modern grid (see more in [NETL 2007].

Also, EMS should be updated to be a reliable platform to integrate large scale renewable energy and support high electric vehicles penetration. As the architecture of future EMS is distributed and autonomous, interactive and coordinated, a new concept, called EMS family, and architecture of future EMS for smart girds is proposed in [Sun 2012]. Some new members of EMS family are introduced for autonomous and fast energy management of Smart Grids, for example, substation EMS, wind farm EMS and electric vehicle (EV) EMS. Interaction among EMS family members is introduced to do coordinated control for accumulating large scale renewable energy as much as possible and to enhance entire gird security, which forms an EMS family network.

Also, utilities are moving towards [ISO 14001] model which allows one set of environmental aspects under one EMS. The [ISO 14001] standard outlines specific requirements for an environmental management system including EMS internal audits, surveillance audits, and recertification every three years. For example, the Baltimore Gas & Electric, Maryland, USA took the initiative towards a Smart Grid project, a Certification Strategy/Model, to pursue one comprehensive certification of BGE versus by facility. This strategy allows to maintain one certification and expand registration scope to other lines of business: Gas Distribution, Electric Transmission & Distribution, Fleet Operations, Shared Services.

In addition, software and hardware manufacturers need to include capabilities based on standards and industry recommendations that support operations. Examples include:

- Energy industry standards and guidelines, such as [IEC 62351] and [IEC 62443], are tools to define cybersecurity capabilities for control systems products including operations, maintenance, and processes.
- This standard [IEC 62443-2-4] defines cybersecurity capabilities that Industrial Automation and Control System (IACS) service providers may implement and offer; the standard can help asset owners consistently procure and manage control systems security; among these capabilities, the standard specifies a comprehensive set of security requirements for IACS installation and maintenance.
- This whitepaper [GE 2016] presents an overview of GE's hardening capabilities that meet [IEC 62443-2-4] standard.

9.4 Physical and Environmental Security Tasks

Among the wide range of security topics covered in the former standard [BS 7799], physical and environmental security is concerned to protect the organization's assets by properly choosing a facility location, erecting and maintaining a security perimeter, implementing access control, and protecting equipment [Harris 2013]. One requirement for Smart Grid is to provide an appropriate level of physical and environmental security. Topics on physical security are discussed elsewhere. We describe a few aspects related to physical and environmental security in this section.

9.4.1 Smart Grid Case

As described in [NISTIR 7628r1], physical and environmental security encompasses protection of physical assets from damage, misuse, or theft. Also, it addresses protection from environmental threats. An organization should have a security program that includes physical and environmental security policy. A documented physical and environmental security policy addresses [NISTIR 7628r1]:

- The objectives, roles, and responsibilities for the physical and environmental security program as it relates to protecting the organization's personnel and assets.
- The scope of the physical and environmental security program as it applies to all of the organizational staff, contractors, and third parties.
- Procedures to address the implementation of the physical and environmental security policy and associated physical and environmental protection requirements.

An organization should consider the following [NISTIR 7628r1]:

- Limit physical access to systems, equipment, and the respective operating environments to authorized individuals only.
- Protect the physical plant and support infrastructure for systems.
- Provide supporting utilities for systems.
- Protect systems against environmental hazards.
- Provide appropriate environmental controls in facilities containing systems.

The organization ensures that the physical and environmental security policy and procedures comply with applicable federal, state, local, tribal, and territorial laws and regulations. Physical access control, physical boundaries, and surveillance are examples of security practices used to ensure that only authorized personnel is allowed to access Smart Grid information systems and components.

This document [NISTIR 7628r1] provides guidance for Smart Grid physical and environmental security requirements. A system should provide system features that prevent, detect, and resist physical tampering of a system component, and use those features to limit security breaches. Common governance, risk, and compliance requirements (GRC) for physical and environmental security are included in [NISTIR 7628r1] document. Supplemental guidance is provided in other documents such as [NIST SP800-53r5], [DHS CAT], [NERC CIP].

An organization should consider the risk associated with physical and environmental hazards when planning new Smart Grid information system facilities or reviewing existing facilities. Physical and environmental hazards include flooding, fire, tornados, earthquakes, hurricanes, acts of terrorism, vandalism, electrical interference, and electromagnetic radiation [NISTIR 7628r1] (see more examples of environmental threats are included in [Otuoze 2018).

However, physical and environmental security that encompasses protection of physical assets from damage is addressed by this document [NISTIR 7628r1] only at a high level. Therefore, additional requirements should be defined and implemented by each organization. Different environments demand different security [Harris 2013]. Also, the use of standards that cover these environmental threats should be considered. Provision of environmental security is very crucial in Smart Grid implementations as it helps control and avoid potential catastrophic effects on the infrastructures [Otuoze 2018].

Improper environmental controls can cause damage to services, hardware, and lives. For example, power, heating, ventilation, air conditioning, and air quality controls can be complex and contain many variables. They all need to be operating properly and to be maintained regularly to ensuring environmental conditions that do not endanger media. Each media type may be susceptible to damage from one or more environmental influences.

In determining the appropriate level of physical protections required for a device, it is important to consider both the operating environment and the value and sensitivity of the data protected by the device [NISTIR 7628r1]. For example, when deploying Smart Grid equipment employing cryptographic modules, the environment, the value of the information, and the functionality protected by the module should be considered when assessing the level of module physical security required. Therefore, the specification of cryptographic module physical protections is a management task in which both environmental hazard and data value are taken into consideration.

The power system's response to failure in its ability to within the shortest possible time, restore service (by using an improvised alternative feeder as may be applicable) or provide sufficient data which will aid system operators in restoring the service is of utmost priority in Smart Grid systems.

Smart response based on environmental considerations is basically achieved by sending appropriate threat alert based on received data and providing alternate feeder for critical infrastructure in such emergencies [Otuoze 2018]. This work proposes a conceptual framework, for identification and clearance of detected threats. This framework is entirely concentrated on using the technique of threat tracing to the source in a comprehensive and a more efficient means of combating the dangers posed by the identified threats. It proposes the use of input data that includes alerts from AMI, sensors, and Geographic Information System (GIS). This data is processed, analyzed and cross checked with the control and support system to identify the source of threats and provide feedback to restore the service. This system involves intelligent data gathering and analysis with time. The GIS based instantaneous data is key in this analysis especially in relation to natural disasters forecast and analysis which are used in determining environmental threat alerts.

9.4.2 NIST Controls

This document [NIST SP800-53r5] provides guidance on how to implement physical and environmental protection control family. Also, it specifies controls for physical and environmental security. The term physical and environmental security refers to measures taken to protect systems, buildings, and related supporting infrastructure against threats associated with their physical environment. Physical and environmental controls cover three broad areas: physical facility, facility's general geographic operating location, and supporting facilities.

The physical facility is typically the building, other structure, or vehicle housing the system and network components. Systems can be characterized, based upon their operating location, as static, mobile, or portable. Static systems are installed in structures at fixed locations. Mobile systems are installed in vehicles that perform the function of a structure, but not at a fixed location. Portable systems may be operated in a wide variety of locations, including buildings, vehicles, or in the open. The physical characteristics of these structures and vehicles determine the level of physical threats such as fire, roof leaks, or unauthorized access.

The facility's general geographic operating location determines the characteristics of natural threats, which include earthquakes and flooding; man made threats such as burglary, civil disorders, or interception of transmissions and emanations; and damaging nearby activities, including toxic chemical spills, explosions, fires, and electromagnetic interference from emitters (e.g., radars).

Supporting facilities are those services (both technical and human) that maintain the operation of the system. The system's operation usually depends on supporting facilities such as electric power, heating and air conditioning, and telecommunications. The failure or substandard performance of these facilities may interrupt operation of the system and cause physical damage to system hardware or stored data. Examples of physical and environmental controls include: physical access authorizations, physical access control, monitoring physical access, emergency shutoff, emergency power, emergency lighting, alternate work site, information leakage, and asset monitoring and tracking.

9.5 Scheduled Maintenance Tasks

As described in [NISTIR 7628r1], maintenance activities encompass appropriate policies and procedures for performing routine and preventive maintenance on the components of a Smart Grid information system. This includes the use of both local and remote maintenance tools and management of maintenance personnel.

An organization should schedule and reviews records of maintenance and repairs on Smart Grid information system components in accordance with manufacturer or vendor specifications and/or organizational requirements [NISTIR 7628r1]. Also, the maintenance team checks all potentially impacted security requirements to verify that the requirements are still functioning properly following maintenance or repair actions; and makes and secures backups of critical Smart Grid information system software, applications, and data for use if the operating system becomes corrupted or destroyed.

All maintenance activities to include routine, scheduled maintenance and repairs, and unplanned maintenance are controlled, whether performed on site or remotely and whether the equipment is serviced on site or removed to another location.

An organization benefits (e.g., more economical, saving resources such as effort and time) and more effective (e.g., faster, better performance) if automated mechanisms are used to schedule and document maintenance and repairs as required, producing up to date, accurate, complete, and available records of all maintenance and repair actions needed, in process, and completed. The use of information system maintenance tools for diagnostic and repair actions is restricted to authorized personnel only and should be checked for malicious code and monitored.

If remote maintenance is allowed, then the planned date and time must be approved by the management and all activities during maintenance should be recorded. The remote maintenance sessions should be protected by using a strong authentication credential.

9.5.1 Regular Backups

Any Smart Grid organization should have a backup strategy that is communicated to maintenance team and incident recovery team. A backup strategy must consider that a failure can occur during the backup or restoration process that could corrupt the data. There should be a safe way of backing out or reconstructing the data from the beginning. Since related information to backups is available elsewhere (e.g., [Harris 2013]) we just discuss briefly a few key points.

In Smart Grid, backups of the data, code, file system, procedures, process documentation are crucial in restoring a failed service or lost data due to an incident (e.g., security event, failure due to an error, natural disaster). Some information may require multiple backups using at least two different methods. The operations/maintenance team is responsible for defining which data is backed up and how often. Critical data (e.g., file systems, device configurations) should be backed up and stored at on site area and an offsite area, and often it is required to have more than one offsite backup.

Different types of media can be used and there should be a media protection strategy. An organization should develop, implement, review, and update on an organization defined frequency (backup early and often) and documented media protection security policy [NISTIR 7628r1]. This policy should address objectives, roles, and responsibilities for the media protection security program and procedures. The organization should ensure that the media protection security policy and procedures comply with applicable federal, state, local, tribal, and territorial laws and regulations.
As defined in [NIST SP800-53r5], an organization should conduct backups at a defined frequency consistent with recovery time and recovery point objectives for:

- User information contained in the information system
- System information
- Information system documentation including security related documentation.

System information includes, for example, system state information, operating system and application software, and licenses. User level information includes any information related to user (e.g., accounts, permissions, authorization, constraints of using information).

Also, an organization should protect the confidentiality, integrity, and availability of backup information at storage locations. Mechanisms employed by organizations to protect the integrity of information system backups include, for example, digital signatures and cryptographic hashes.

Backing up control system software and having industrial network backup hardware devices are also important maintenance tasks for control systems. An organization should be able to restore critical control data in many different scenarios, including if a hard drive fails, a computer resource fails, a physical disaster takes place, or some type of software corruption renders in use data unavailable [INL 2007]. The overarching security policy (one that has used OPSEC components) should be developed to indicate what data gets backed up, how often it is backed up, and how these processes should occur. This practice and the corresponding specifics of the OPSEC plan can be reused from preexisting (and proven) policies and directives. However, as in other cyber OPSEC components, practices need to be tuned and created to accommodate for the nuances and requirements of the control systems domain.

If control system users have important information on their workstations, the operations department needs to develop a method that indicates that backups include certain directories on users' workstations, or that users move their critical control data to a server at the end of each shift to ensure it gets backed up. Backups may occur once or twice a week, every day, or every hour depending on the criticality and the dynamics of the asset under control. It is up to the organization to determine this routine and readers are encouraged to review sector specific guidance. Generally, the more frequent the backups, the more time will be dedicated to the process; therefore, a balance is needed between backup costs and the actual risk of potentially losing critical control data [INL 2007].

9.5.2 Redundancy

The requirement for continuous availability of utility grid operations necessitates a high degree of reliability within and across domains. The legacy infrastructure should gain a layer of redundancy and security [NISTIR 7628r1].

Redundancy in Smart Grid is multiple instances of the same software, firmware, devices, and/or data configured in an active/passive or load sharing mode [NISTIR 7628r1]. Redundancy for data and logs needs to be consistent with the organization's data retention plan and continuity of operations plan.

Backing up software and having backup hardware devices are commonly found in network availability. Generally, contingency plans should be prepared to address how to deal with small incidents that do not qualify as disasters, as in power outages, server failures, a down communication link to the Internet, or the corruption of software. However, business continuity plans should address how to keep organization in business after a disaster occurs.

Therefore, the redundancy strategy is applied to each type of plan according to an organization's risk strategy and security risk strategy. Also, risk analyses, business impact studies, and risk tolerance profiles contain sensitive organization information and need to be protected. This information should be protected and verified that the appropriate versions are retained.
If the Smart Grid architecture does not provide for the sufficient redundancy, then there is a danger of exposing the system to intentional or unintentional denial of service. For example, lack of redundancy for critical networks may affect the availability of services. The concept of high availability refers to a level of fault tolerance and redundancy in computing and communications. Also, architecture designs with one single firewall to protect both the internal network and any systems that are accessible from the Internet may not provide the needed security for protection of industrial control systems. Therefore, the architecture based on dual firewalls adds another layer of security to control systems.

On the other hand, an organization should mitigate or limit the effects of denial of service (DoS) attacks based on an organization defined list of DoS attacks. Thus, the organization should manage the excess capacity, bandwidth, or other redundancy to limit the effects of information flooding types of DoS attacks. For example, network perimeter devices can filter certain types of packets to protect devices on an organization's internal network from being directly affected by DoS attacks.

Redundancy solutions for control systems should be carefully selected. Examples of redundancy include multiple paths between routers in case one router goes down. Dynamic routing protocols should be used so that each router will be informed when a change to the network takes place. Also, an uninterruptible power supply (UPS) and redundant array of inexpensive disks (RAID), or other redundant fault tolerant backup methods can be used. RAID provides fault tolerance for hard drives and can improve system performance (more information in [Lynn 2014]). Breaking up the data and writing it across several disks so that different disk heads can work simultaneously to retrieve the requested information provides redundancy and speed. Control data is also spread across each disk (called parity) so if one disk fails, the other disks can work together and restore the critical control and configuration data. However, these redundant devices should be properly configured.

Another solution recommended in [INL 2007] is fault tolerance and clustering. Clustering is widely used in IT networks as a fault tolerant server technology that is similar to redundant servers, except each server takes part in processing services that are requested. A server cluster is a group of servers that are viewed logically as one server to users and can be managed as a single logical system for use in the enterprise network as well as the control or SCADA LAN. Clustering provides for availability and scalability. It groups physically different systems and combines them logically, which provides immunity to faults and improves performance.

If one of the systems within the cluster fails, processing continues since the other systems pick up the load. This is more appealing than having a secondary control system data historian server that waits in the wings in case a primary server fails. When clustering is used, all systems are processing all time requests and none sits idle in the background waiting for another to fail. This solution improves the performance, avoids the server unavailability during switching from failed server to another server. Another aspect to be considered is that real time interactions may entail short acceptable time latencies and may limit the security technology choices for mitigating on going attacks.

9.5.3 Preventive Maintenance

The goal of predictive maintenance is to perform maintenance at a scheduled point in time when the maintenance activity is most cost effective and before the equipment loses performance within a threshold. This approach uses principles of statistical process control to determine at what point in the future maintenance activities will be appropriate. It is recommended in [NISTIR 7628r1] that an organization performs preventive maintenance on organization defined information system components at defined time intervals. For example, an organization should perform testing of the backup information on a defined frequency to verify media reliability and information integrity.

The organization can use a sample of backup information in the restoration of selected information system functions as part of contingency plan testing. The contingency plan is tested, and personnel receive plan implementation training. Contingency plan testing is a critical element of an effective contingency capability, to identify and address plan deficiencies and to confirm the viability of the plan to support recovery operations. A contingency plan is an alternative Information Systems Security (INFOSEC) plan that is implemented when normal business operations are interrupted by emergency, failover, or disaster. Contingency plans should be exercised and maintained during the Operations/Maintenance phase. The contingency plan is continuously updated for effectiveness.

This document [NIST SP800-100] highlights contingency planning activities involved in each step that should be addressed during all phases of the system development life cycle. The capability to recover and reconstitute data should be integral to the information system design concept during the initiation phase. Recovery strategies should be built into the system and architecture during the Development phase. The contingency processes should be tested and maintained during the Implementation phase; contingency plans should be exercised and maintained during the Operations/Maintenance phase. Maintenance needs are addressed by a combination of corrective, adaptive, perfective, and preventive maintenance [NIST SP800-160v1]. This document includes information about system life cycle processes (acquisition, organizational, technical management, technical processes).

As the organization supports the predictable failure prevention, there should be some controls such as [NIST SP800-53r5]:

- Determines the mean time to failure (MTTF) for defined information system components in specific environments of operation.
- Provides substitute information system components and a means to exchange active and standby components based on defined MTTF substitution criteria.

While MTTF is primarily a reliability issue, these controls address potential failures of specific information system components that provide security capability. Failure rates reflect installation specific issues. An organization may define criteria for substitution of information system components based on MTTF value with consideration for resulting potential harm from component failures (e.g., security service failure, safety hazard). Transfer of responsibilities between active and standby components should not compromise safety, operational readiness, or security capability (e.g., preservation of state variables). Therefore, the standby components should be always available except for maintenance issues or recovery failures in progress. The organization takes information system components out of service by transferring component responsibilities to substitute components no later than defined fraction or percentage of mean time to failure.

Therefore, the primary goal of preventive maintenance is to provide reliability reports and use these in a way to avoid/mitigate the consequences of equipment failures. Preventive maintenance includes proactive care and servicing of organizational information systems components for maintaining equipment and facilities in satisfactory operating condition. Such maintenance provides for the systematic inspection, tests, measurements, adjustments, parts replacement, detection, and correction of incipient failures either before they occur or before they develop into major defects.

9.6 Maintenance Tools

The organization approves and monitors the use of Smart Grid information system maintenance tools. The maintenance plans should address security related issues when the hardware, firmware, and software are brought into the Smart Grid information system for diagnostic and repair actions. The employed security controls require evaluation and maintenance tools too. A lack of firewall maintenance and security testing as major causes for compliance breaches (see [Verizon 2016] report).

9.6.1 Restricting Use of Tools

An organization should employ automated mechanisms to restrict the use of maintenance tools to authorized personnel only. Use of maintenance accounts based on factory set or easily guessed passwords to break into computer systems is a security violation. There are concerns about using many mobile applications with inadequate cybersecurity, in general. This is proving to be a big risk.

In addition, maintenance applications have been hacked in the past. If a hacker gains access to a maintenance application, it could easily comprise the business. In addition, competition from unauthorized service and maintenance providers could pose a substantial threat to the business model. Manufacturers must also consider that their connected maintenance applications may well contain their intellectual property, which if compromised, will allow hackers to make money by selling it on the Dark Web or holding the business to ransom. However, monitoring and analysis tools should be always considered a target for the hackers because they can be used to identify unusual situations and make the right security decisions.

While the spectrum of maintenance tools is vast, the various categories used for security maintenance are documented elsewhere. The choice of tools is unlimited, all depends on the organization's security needs and maintenance team's willingness to employ the appropriate tools. Many tools are available via open source, commercial products, and published books that provide detailed information for security design, architecture, implementation, and maintenance.

Therefore, different layers of security controls have to be applied to protect maintenance tools, service maintenance applications require robust, multi-layered software security. This should include data security at rest and in transit, network/API security and robust software protection for IP addresses. The applications can also be protected by code hardening tools to conceal proprietary algorithms and secrets, including cryptographic keys, private and personal data, and credentials.

9.6.2 Backdoor Tools

Maintenance hook is a hardware or software mechanism that was installed. Maintenance hooks are instructions in software code allowing easy maintenance. They are commonly referred to as trapdoors or backdoors. They are not typically defined during access or design specifications and are generally intended to be removed prior to the completion of the code, which does not always occur. Because hooks can provide entry to the software code at unusual points or without the usual checks, they pose a serious security risk. Essentially, an authorized user who knows that the entry point exits could gain access to information or insert malicious code into the software.

However, maintenance hooks should never be inserted into the program during development. At a minimum, maintenance hooks should be removed prior to deploying a system into production. More, use of maintenance hooks should be constrained and eliminated. A maintenance hook is a hardware or software mechanism that was installed to permit system maintenance and to bypass the system's security protections [Krutz 2004].

Users can apply some preventive measures against backdoors. Examples include [Harris 2013]:

- Use of host intrusion detection systems to monitor for any attempts to access a system using backdoors.
- Use of file system encryption to protect sensitive information.
- Implement auditing to detect the use of backdoors.

9.7 Future Activities

Control centers for power industry have sophisticated monitoring and control systems and are staffed by operators 24 hours per day, 365 days per year. These operators are responsible for several key functions, including balancing power generation and demand, monitoring flows over transmission lines to avoid overloading, planning and configuring systems to operate reliably, maintaining system stability, preparing for emergencies, and placing equipment is and out of service for maintenance and during emergencies. A control center may have SCADA and DCS systems. SCADA systems and DCS monitor the flow of electricity from generators through transmission and distribution lines. These electronic systems enable efficient operation and management of electric systems through use of automated data collection and equipment control. Thus, problems to be resolved in the future Smart Grid include resilience and intelligent threat analysis.

9.7.1 Resilience

With the actual complex networks and networked control systems (e.g., DCS and SCADA systems), the application of control is moving toward a more decentralized state. Therefore, constrained controllers that can react rapidly to maintain resilience for rapidly changing conditions improve the ability to operate these systems to achieve a global optimum for multiple considerations in a Smart Grid. The complex interconnected nature of individual homes, commercial facilities and diverse power generation and storage creates an opportunity and a challenge to ensuring that the resulting system is more resilient to threats. A resilient control system is one that maintains state awareness and an accepted level of operational normalcy in response to disturbances, including threats of an unexpected and malicious nature [Rieger 2009].

Addressing the complex control system interdependencies, including the human systems interaction and cybersecurity are recognized challenges for control systems resilience. To address these issues, an integrated approach for the design and secure operation of the control systems is considered. Also, establishing metrics for resilience requires reflection on these characteristics [Rieger 2014]. However, establishing a metric that can capture the resilience attributes is also complex. An alternative to traditional metrics of reliability and dependability is to define a metric based on a normalizing dynamic attribute, such a performance characteristic that can be impacted by degradation.

To integrate the cognitive and cyber physical aspects that should be considered when defining solutions for resilience, terminology and metrics basis are provided in [Rieger 2014]. Specifically, applications of base metrics to resilience characteristics are given as follows for type of disturbance:

- Physical disturbances
  - Time latency affecting stability
  - Data integrity affecting stability.
- Cyber disturbances
  - Time Latency
  - Availability, integrity, and confidentiality.
- Cognitive disturbances
  - Time latency in response
  - Data digression from desired response.

Such performance characteristics exist with both time and data integrity. Time, both in terms of delay of mission and communications latency, and data, in terms of corruption or modification, are normalizing factors. In general, the idea is to base the metric on what is expected and not necessarily the actual initiator to the degradation.

Disturbances consider both the cyber and physical aspects as impacts to the control system performance, and the human is a necessary attribute in the feedback response that can also impact performance. Disturbances to the control system exhibit impacts in the form of time and data as characterized in the diagram shown in Figure 9.1. Considering time as a metrics basis, resilient and unresilient systems can be observed in this diagram.

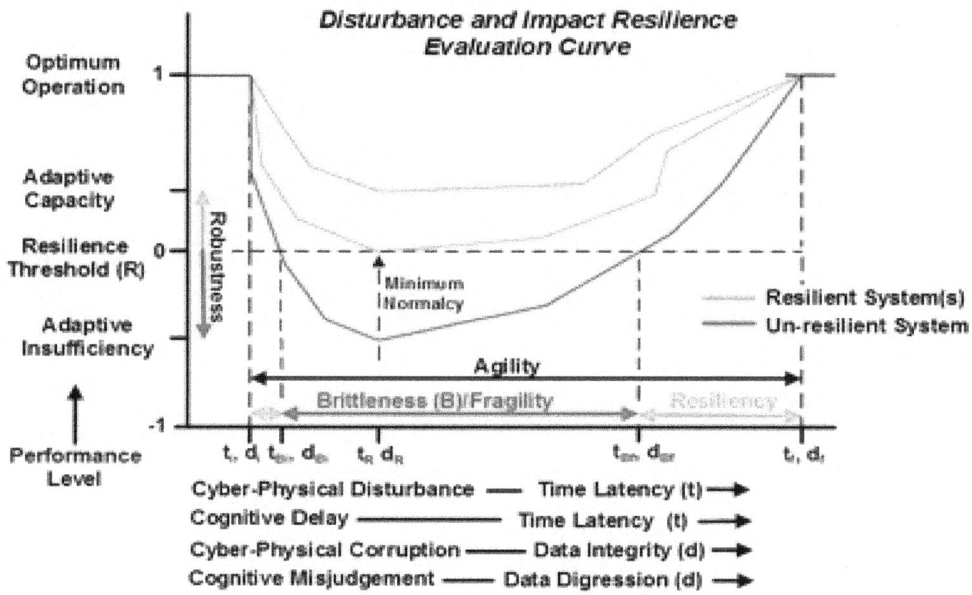

Figure 9.1 Resilience base metrics. *Source*: [Rieger 2014]. © 2014, IEEE.

The diagram reflects a generalization of the resiliency of a system. Several common terms are represented on this diagram, including robustness, agility, adaptive capacity, adaptive insufficiency, resiliency and brittleness. These terms are defined in [Rieger 2014]. A summary includes:

- Adaptive capacity is the ability of the system to adapt or transform from impact and maintain minimum normalcy; considered a value between 0 and 1, where 1 is fully operational and 0 is the resilience threshold.
- Adaptive insufficiency is the inability of the system to adapt or transform from impact, indicating an unacceptable performance loss due to the disturbance; considering a value between 0 and -1, where 0 is the resilience threshold and -1 is total loss of operation.
- Brittleness is the area under the disturbance curve as intersected by the resilience threshold; it indicates the impact from the loss of operational normalcy.
- Resiliency is the converse of brittleness, which for a resilience system is 0 or loss of minimum normalcy.
- Robustness is quantified as a positive or negative number associated with the area between the disturbance curve and the resilience threshold, indicating either the capacity or insufficiency, respectively.

It is recognized that cyber and cognitive influences can affect both the data and the time, which underscores the relative importance of recognizing these forms of degradation in resilient control designs. From a cybersecurity perspective, a single cyber attack can degrade a control system in multiple ways. For example, given the impact to system dynamics or data, the performance of the control loop can be directly ascertained and show approach to instability and operational impact. However, there are some challenges on using thresholds. One technical challenge is determining the thresholds for the desired optimum performance and minimum normalcy. Although some methods are suggested in [Rieger 2014], there should be monitoring systems that support data collection and processing to calculate the thresholds. If some data is omitted or not collected, the calculated threshold could result in misleading characterization of the system as resilient or no resilient. In addition, to be useful to power operators these thresholds should be calculated in real time.

A unified approach to determine the resiliency metrics for a power protection system is described in [Eshghi 2015]. A unifying approach to evaluate such performance characteristics based on physical, cyber-physical, and cognitive disturbances is dependent on both time and data integrity [Rieger 2014]. Time, both in terms of delay of mission and communications latency, and data, in terms of corruption or modification, are normalizing factors. Ultimately, these metrics evolve to evaluate distance from resilience boundaries during normal and disturbing conditions. Time impacts that are reflected through destabilizing latencies may be the result of cyberattacks. Data impacts may be reflected through integrity compromises resulting from cyber attacks or power grid instability. How these are integrated and weighed depend upon the overall impact to power grid performance.

The authors of this work [Eshghi 2015] apply the resilience principles to Smart Grid. They propose the deployment of multi agent systems (MAS), which offer a framework for controlling not only physical systems but also cyber systems and the interconnections between them. The authors present the power system attributes that can be measured in terms of time and/or data to recognize a disturbance for the protection system. These attributes include power system small signal stability (dependent on data), power system transient stability (dependent on data), power system communication latency (dependent on time), and power system physical degradation (dependent on data and time). The attributes can be integrated into distributed agents that collectively evaluate the resiliency status of a power system during abnormal events and measure the system performance against minimum performance expectations. The performance expectation is dependent on the levels of system operating limits and power system contingency status.

During a real time event, a power system operator needs to conservatively reduce operating limits while the changing system conditions are analyzed. The time it takes to develop new operating limits could affect millions of transmission system users, especially if this event is classified by NERC as a Category D type event (extreme events resulting in the loss of two or more bulk electric system elements) [Eshghi 2015]. Therefore, the authors recommend controls must be able to perform real time analysis, identify new reliability risks, and set a new system operating limit for real time operations. Despite disturbances or threats, the system should operate at an acceptable level of normalcy.

Another work [Friedberg 2017] argues that [Rieger 2014] method for cyber-physical systems specifies a resilience metric based on control loop performance and the metric is computed with the use of game theoretic approaches. The authors argue that focusing on control systems there is the problem of flexibility loss, as performance measures that are not control loop specific cannot be considered. The authors compare different approaches including [Rieger 2014] with respect to the criteria based on seven characteristics to include comparable prediction, measurements evaluation, resilience potentials analysis, flexibility, interdependencies, and scalability for an effective resilience metric.

While no discussion of these criteria attributes is clearly described in this work [Friedberg 2017], the authors claim that there are shortcomings of current approaches where significant factors such as time, the dependencies between performance measures or resilience potentials are ignored which leads to an incomplete view of resilience. This work evaluates resilience metrics for microgrids. The resilience metric is applied to a power island operated synchronized to an optimal main grid. Intentional islanding describes the condition in which a microgrid, which consists of a load and a distributed generation (DG) system, is isolated from the remainder of the utility system. In this situation, it is important for the microgrid to continue to provide adequate power to the load (see more information in [Balaguer 2008]). Although microgrids are proposed in different scenarios to increase grid resilience, the authors discuss the islanded operation of microgrids, specifically their reconnection to the main grid.

The reconnection is a problem when the system needs to reliably prevent out of sync reclosure between Smart Grid and microgrid. The out of sync happens if the difference in phase angle between the microgrid and the main grid is too large and can cause severe damage to power equipment. This work assumes that phasor measurement units are deployed in the microgrid and the main grid. While the general performance evaluation of the connection approaches has been previously evaluated by different works, the authors are recognizing the need to introduce a study of the cyber attacks' consequences. They describe a method to evaluate the resilience of synchronous islanding control defined in [Best 2011] to prevent out-of-sync reclosure during denial of service (DoS) attacks.

However, this study [Friedberg 2017] has shortcomings. It assumes that DoS attacks introduce network delay, although the main effect of a DoS attack is no availability of the data, service, or network temporarily or indefinitely disrupting services of a host connected to the Internet. Therefore, the results of the study may need to be reevaluated if the focus of the study are the impacts of DoS attack.

9.7.2 Intelligent Analysis

The need for more effective protection against sophisticated cybersecurity threats is pushing an organization for using more advanced and intelligent analysis functions. Data mining activities help an organization to harness organization wide data for comparative decision making, workflow automation, and/or competitive advantage. Besides information aggregation, an organization management goal is to understand different aspects of the business value and to increase productivity. Data mining is a process of analyzing data using tools that look for trends, correlations, relationships, and anomalies.

Broad spectrum of data mining tools offer functionality such as [Poletto 2014]:

- Looking for causes and effects (e.g., graph traversal)
- Extracting meaning from noisy data (e.g., graph summarization, clustering)
- Triaging malware (classification)

Numerous case studies described in this book [Dua 2011] provide a clear understanding of the design and application of data mining and machine learning techniques in cybersecurity. The authors survey data mining and machine learning solutions to detection problems. From basic concepts in machine learning and data mining to advanced problems in the machine learning domain, data mining and machine learning in cybersecurity provides a unified reference for specific machine learning solutions to cybersecurity problems.

As an organization continues to shift operations and applications into a new Smart Grid infrastructure, it will need a security program that looks at how the network and cybersecurity strategy will keep up with convergence of OT and IT technologies and emerging technologies (e.g., IoT, CPS, DERs). The move to cloud services for cybersecurity and to third party services should be decided by each organization based on their size of business (small, medium, or large) and their stakeholders security needs.

9.8 Standards, Guidelines, and Recommendations

The need for standards, guidelines, and recommendations for Smart Grid and DER systems development is supported by many standard organizations and associations. Several initiatives are taken by professional societies and associations to support different aspects of Smart Grid. Appendix C includes a list of common acronyms.

9.8.1 Correlating Cybersecurity with Standards

Cybersecurity is a cross-cutting issue and should be reflected in requirements at all levels. Cybersecurity policies and procedures mainly cover the GWAC Stack Organizational and Informational levels, while cybersecurity technologies generally address those requirements at the Technical level. OSI 7-Layer Reference Model includes these layers - Application Presentation Session Transport Network Data Link Physical – to be correlated with GWAC stack [GridWise 2008].

Correlating cybersecurity with specific information exchange standards, including functional requirements standards, object modeling standards, and communication standards, is very complex. There is rarely a one-to-one correlation, with more often a one-to-many or many-to-one correspondence.

First, communication standards for the Smart Grid are designed to meet many different requirements at many different layers in the reference model. Two commonly used reference models are the International Organization for Standardization (ISO) / Open Systems Interconnection model (OSI) 7-layer reference model and the GridWise Architecture Council (GWAC) Stack (see Figure 9.2), where the OSI 7-layer model maps to the Technical levels of the GWAC Stack.

Some standards address the lower layers of the reference models, such as wireless media, fiber optic cables, and power line carrier. Others address the transport layer for passing the messages from one location to another. Still others cover the application layer, the semantic structures of the information as it is transmitted between software applications. In addition, there are communication standards that are strictly abstract models of information – the relationships of pieces of information with each other.

Second, regardless of what communications standards are used, cybersecurity must address all layers – end-to-end – from the source of the data to the ultimate destination of the data. In addition, cybersecurity must address those aspects outside of the communications system in the upper GWAC Stack layers that may be functional requirements or may rely on procedures rather than technologies, such as authenticating the users and software applications, and screening personnel. Cybersecurity must also address how to cope during an attack, recover from it afterwards, and create a trail of forensic information to be used in post-attack analysis.

For the elicitation and management of Smart Grid requirements, GWAC structure is used as follows [Danekas 2013]:

- Level 8 of GWAC stack stands the economic and regulatory policy, e.g., the stakeholders of the DER system must adhere to rules and regulations determined by governments and regulators.
- Level 7 is concerned with high level collaboration between businesses concerning strategic and tactical business objectives, e.g., DERs owners interact with utilities, aggregators, markets, service providers, etc.
- Layer 6 refers to business objectives and procedures to help the utility to further elaborate business functions and to identifying which parts of these functions lie outside the organization's boundaries.
- Layer 5 builds on the organizational aspects by transferring them into appropriate information models, e.g., identifying the need for information connected to the business functions embedded in processes and procedures.
- Layer 4 supports a common understanding regarding the semantics within the information models has to be established, e.g., semantic understanding the information contained in the models such as at the plug and automate paradigm.

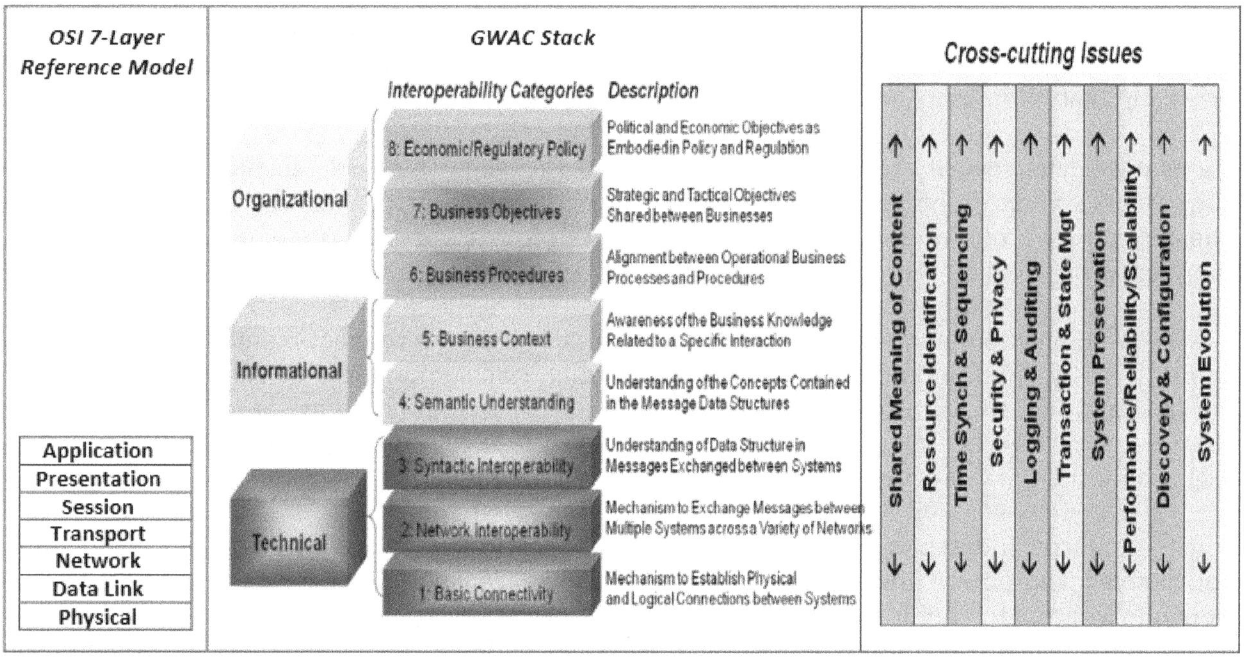

Figure 9.2 ISO/OSI 7-Layer Reference Model and GWAC Stack Reference Model. *Source:* [Cleveland 2012]. © 2012, Xanthus International.

- Layer 3 provides syntactic interoperability concerning the messages which are exchanged between applications and assets in the Smart Grid, e.g., GWAC compares this layer to the seven layer ISO/OSI model [ISO/IEC 7498-1], stating this layer subsumes its application and presentation layers, including character translation, content structure and message exchange patterns; protocols relevant to this layer, among others, include HTML, XML and SOAP.
- Layer 2 deals with network interoperability that enables the exchange of messages within the Smart Grid, e.g., with reference to the ISO/OSI model this layer addresses the network, transport, session and in parts the application layer; it includes functions like resolving logical into physical addresses or management of the exchange of data messages; communication standards relevant to this layer, among others, include TCP, UDP, FTP, and IP(v6) protocols.

- Layer 1 addresses the basic connectivity needed to establish networked communication between distributed assets and applications on this layer, e.g., physical and data link layers including functions like electrical connectivity, character encoding, data flow control or error correction; relevant protocols and standards concerning this layer among others include Ethernet, Wi-Fi, or PPP (Point-to-Point Protocol).

9.8.2 Standardizing Resilience

The IEEE society has established a technical committee for standardizing resilience and resilience of control systems. The purpose of this committee is to establish metrics and standards associated with codifying promising technologies that promote resilience in automation. This effort is distinct from supply chain community focus on resilience and security, such as the efforts of ISO (ISO/TC 29 Security and Resilience, http://www.isotc292online.org/) and NIST (Community Resilience, https://www.nist.gov/topics/community-resilience).

Also, ISO work in progress include these publications:
- ISO 22340 Security and resilience - Protective security - Guidelines for security architecture, framework and controls
- ISO 22342 Security and resilience - Protective security - Guidelines for the development of a security plan for an organization
- ISO 22343 Security and resilience – Vehicle security barriers – Performance requirement, vehicle impact test method and performance rating

This document [NIST SP800-160v2r1] is focused on cyber resiliency engineering applied in conjunction with Systems Security Engineering and Resilience Engineering to develop survivable, trustworthy secure systems. Cyber resiliency engineering provides the foundations to design, develop, implement, maintain, and sustain the trustworthiness of systems with the capability to anticipate, withstand, recover from, and adapt to adverse conditions including cyber attacks. Organizations can use or adapt some or all of the cyber resiliency constructs (e.g., goals, objectives, techniques, approaches, and design principles) described in this publication and apply the constructs to the technical, operational, and threat environments for which systems need to be engineered. However, the resiliency engineering framework is also complex, with relationships among the constructs of goals, objectives, design principles, techniques, and approaches, as discussed in Appendix D of this document.

This document [NIST SP800-160v1] can be used in conjunction with other documents such as [ISO/IEC/IEEE 15288], Systems and software engineering—Systems life cycle processes; [NIST SP800-160v1], Systems Security Engineering—Considerations for a Multidisciplinary Approach in the Engineering of Trustworthy Secure Systems; [NIST SP 800-37r2], Risk Management Framework for Information Systems and Organizations—A System Life Cycle Approach for Security and Privacy; and [NIST SP 800-53r5], Security and Privacy Controls for Information Systems and Organizations.

Also, this technical report [IEC TR 62351-12] provides cybersecurity recommendations and engineering/operational strategies for improving the resilience of power systems with interconnected Distributed Energy Resources (DER) systems. It covers the resilience requirements for the many different stakeholders of these dispersed cyber-physical generation and storage devices, with the goal of enhancing the safety, reliability, power quality, and other operational aspects of power systems, particularly those with high penetrations of DER systems. It addresses the resilience issues for cyber-physical DER systems interconnected with the power grid, building on the CPS concepts and a proposed DER hierarchical architecture.

9.8.3 European Harmonized Cybersecurity Standards

As systems are integrated, there is a need for agreement on both a syntactical and semantical level on the concepts, profiles and serialization of the data to be exchanged between the systems of Smart Grid. Therefore, a common information model is of highest importance. A perspective on the seamless integration architecture based on [IEC TR 62357] standard is presented in [Uslar 2013]. The [IEC TR 62357] technical report provides a clear and comprehensive map of all standards which are contributing to support interactions, in an open and interoperable way, between actors, components and systems in the field of electricity grids from generation to consumers, including transmission and distribution.

A harmonized cybersecurity standards perspective for the European Union electric grid is proposed in this position paper [T&D 2016]. The key basic security standards shown in Figure 9.3 include ISO/IEC 270xx, IEC 62443, and IEC 62351 series. These standards cover security management, systems security and communication security aspects for the T&D industry. Taken into consideration the maturity and depth of expertise invested in these standards, these standards are referred to in European cybersecurity policy and mandates.

For operators of critical infrastructure, harmonization can be described as the need to provide a consistent level of cybersecurity functionality to meet protection goals [[T&D 2016]. One key target is the support of interoperability in a multi vendor environment without jeopardizing protection goals. Suppliers, integrators, and service providers of components of critical infrastructure are at the heart of the effort to support harmonization and implementation of cybersecurity by making contributions toward international standardization and by providing components for critical infrastructures that follow such standards. International standards are forming the common language of cybersecurity used by all actors in the energy value chain. This explains why transmission and distribution vendors are involved (together with utilities) in defining the cybersecurity standards for the electrical grid.

Furthermore, regulatory authorities within the European Union are discussing the possibility to define a minimum set of security requirements to guarantee a minimum level of protection for critical infrastructure. In this context, European Union supports harmonization efforts and is promoting its participation in European regulation activities to ensure that requirements are harmonized with industry initiatives and standards.

The harmonization of security requirements would directly benefit the economy by lowering implementation costs while supporting the European Commission's Digital Single Market strategy with products and systems deployable in all member states. Additionally, a minimum level of protection could be defined equally for all critical infrastructures within the European Union. Furthermore, building on international standards as a common language in cybersecurity with harmonized security requirements is improving interoperability and mitigating the risk of adhoc security requirements seen in the market.

The following is a brief introduction to the standards included in Figure 9.2.

9.8.4 Standards Focused on Power Systems

These are examples of standards focused on power systems.

IEEE C37.240

[IEEE C37.240] standard provides technical requirements for substation cybersecurity. It presents sound engineering practices that can be applied to achieve high levels of cybersecurity of automation, protection, and control systems independent of voltage class or criticality of cyber assets. Cybersecurity includes trust and assurance of data in motion, data at rest, and incident response.

IEC 62056

[IEC 62056] suite of standards provide information on the DLMS/COSEM protocol for meter reading, tariff and load control, smart metering use cases and on architectures supported by the DLMS/COSEM series of standards specifying electricity meter data exchange. The suite includes a framework how standards shall be developed; ways the existing standards shall be extended to support new use cases and to accommodate new communication technologies, while maintaining coherency; aspects of interoperability and information security. It also provides guidance for selecting the suitable standards for a specific interface within the smart metering system. The DLMS/COSEM protocol is not specific to electricity metering, it is also used for gas, water and heat metering.

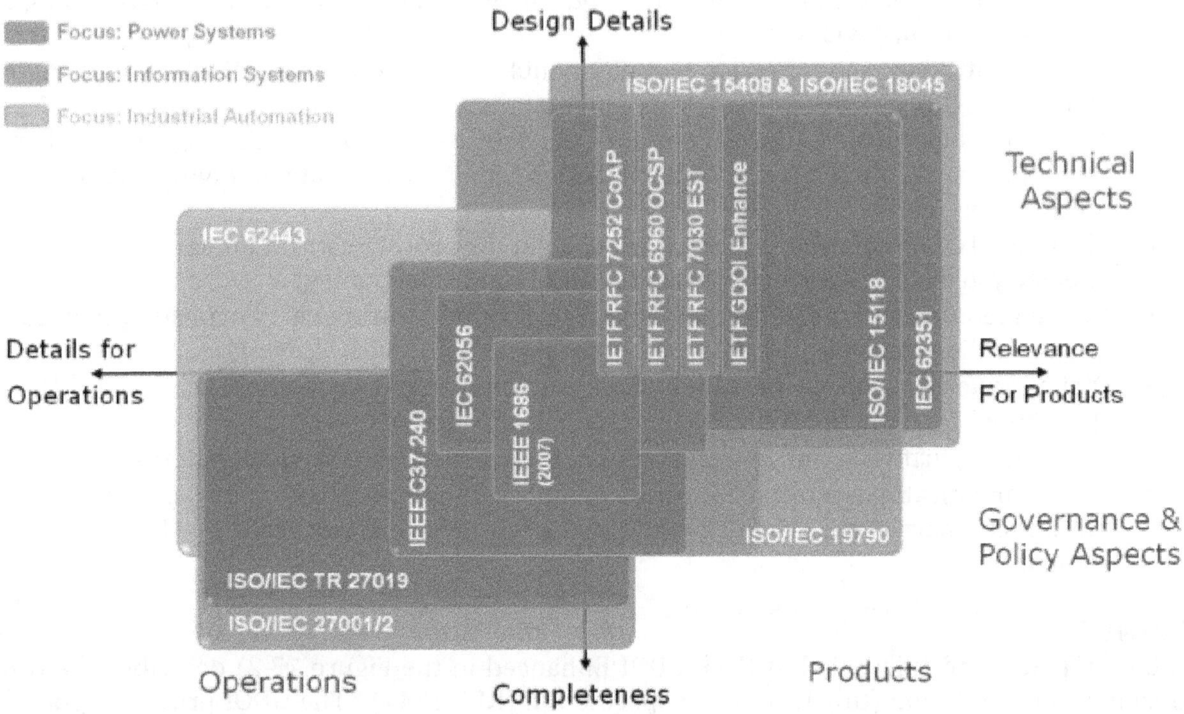

Figure 9.3 Cybersecurity standards. *Source:* [CENELEC 2014]. © 2014, CENELEC.

IEC TR 62051-1

[IEC TR 62051-1] part 1 provides a glossary of terms used in the data exchange with metering equipment.

IEEE 1686

[IEEE 1686] standard defines the functions and features to be provided in intelligent electronic devices (IEDs) to accommodate critical infrastructure protection (CIP) programs. The standard addresses security issues to include access, operation, configuration, firmware revision, and data retrieval from an IED. The standard also allows the user to define a security program around these features and alert the user if an IED does not meet this standard as to the need for other defensive measures (technical and/or procedural) that may need to be taken. The encryption for the secure transmission of data to the IED is also part of this standard. Communications for the purpose of power system protection are not addressed in this standard.

ISO/IEC 27019

[ISO/IEC 27019] provides guiding principles based on [ISO/IEC 27002] standard for information security management applied to process control systems as used in the energy utility industry. The aim is to extend the [ISO/IEC 27000] set of standards to the domain of process control systems and automation technology, thus allowing the energy utility industry to implement a standardized information security management system (ISMS) in accordance with [ISO/IEC 27001] that extends from the business to the process control level. The report covers process control systems used by the energy utility industry for controlling and monitoring the generation, transmission, storage and distribution of electric power, gas and heat in combination with the control of supporting processes. This includes the following systems, applications and components:

- The overall IT supported central and distributed process control, monitoring and automation technology as well as IT systems used for their operation, such as programming and parameterization devices.
- Digital controllers and automation components such as control and field devices or PLCs, including digital sensor and actuator elements.
- All further supporting IT systems used in the process control domain, e.g., for supplementary data visualization tasks and for controlling, monitoring, data archiving and documentation purposes.
- The overall communications technology used in the process control domain, e.g., networks, telemetry, telecontrol applications and remote control technology.
- Digital metering and measurement devices, e.g., for measuring energy consumption, generation or emission values.
- Digital protection and safety systems, e.g., protection relays or safety PLCs.
- Distributed components of future Smart Grid environments.
- All software, firmware and applications installed on above mentioned systems.
- Telecommunication systems and components used in the process control environment are not directly part of the scope of [ISO/IEC 27019]. These are covered by [ISO/IEC 27011] standard.

RFC 6407

[RFC 6407**]** standard (identified as IETF GDOI Enhanced in the Figure 13.2) describes the Group Domain of Interpretation (GDOI) protocol specified in [RFC 3547]. The GDOI provides group key management to support secure group communications according to the architecture specified in [RFC 4046]. The GDOI manages group security associations, which are used by IPsec and potentially other data security protocol.

ISO 15118

[ISO 15118] is an international standard suite defining a vehicle to grid (V2G) communication interface for bidirectional charging/discharging of electric vehicles. The standard establishes the internal communication between battery and charging components and the supply equipment communication controller in electric vehicles (including battery and plug-in electric vehicles). The standards specify the communication between Electric Vehicles (EV), including Battery Electric Vehicles and Plug-In Hybrid Electric Vehicles, and the Electric Vehicle Supply Equipment (EVSE). As the communication parts of this generic equipment are the Electric Vehicle Communication Controller (EVCC) and the Supply Equipment Communication Controller (SECC), the standards describe the communication between these components. Although [ISO 15118] is oriented to the charging of electric road vehicles, it is open for other vehicles as well. The standard does not specify the vehicle internal communication between battery and charging equipment and the communication of the SECC to other actors and equipment (beside some dedicated message elements related to the charging). All connections beyond the SECC, and the method of message exchanging are to be out of the scope as specific use cases.

ISO 15118-1

[ISO 15118-1] specifies terms and definitions, general requirements and use cases as the basis for the other parts of [ISO 15118]. It provides a general overview and a common understanding of aspects influencing the charge process, payment and load levelling.

IEC 62351

[IEC 62351] standard addresses information security for power systems control operations, and the overall objective is to preserve the properties of confidentiality, integrity, availability and non-repudiation in a system, mainly through the introduction of authentication mechanisms. The standard package includes many parts with different security objectives. The scope of the IEC 62351 series is information security for power system control operations. Its primary objective is to undertake the development of standards for security of the communication protocols defined by IEC TC 57, specifically the IEC 60870-5 series, the IEC 60870-6 series, the IEC 61850 series, the IEC 61970 series, and the IEC 61968 series.

IEC TS 62351-1

[IEC TS 62351-1] (part 1) is a technical specification, which provides an introduction to the remaining parts of the IEC 62351 series, primarily to introduce the reader to various aspects of information security as applied to power system operations.

9.8.5 Standards Focused on Information Systems

The following is a list of standards that are focused on information systems.

ISO/IEC 15408

[ISO/IEC 15408] is also known as Common Criteria (CC) for Information Technology Security Evaluation. In brief, the CC is a useful guide for the development of products and systems with IT security functions and a guide for procurement of commercial products and systems with security functions. CC philosophy will provide assurance based upon an evaluation (active investigation) of the IT product or system that is to be trusted. The validity of documentation, and resulting IT product or system, is measured by Expert Evaluators with increasing emphasis on scope, depth, and severity. As of 2012, CC v3.1 version 3.1 consists of the following parts:

- Part 1: Introduction and general model
- Part 2: Security functional components
- Part 3: Security assurance components

ISO/IEC 18045

[ISO/IEC 18045] is a companion document to CC ([ISO/IEC 15408]) standard. It defines the minimum actions to be performed by an evaluator to conduct a CC evaluation, using the criteria and evaluation evidence defined by CC. However, the [ISO/IEC 18045] does not define evaluator actions for certain high assurance CC components, where there is no generally agreed guidance.

RFC 7252

[RFC 7252] describes the Constrained Application Protocol (CoAP), which is a specialized Web transfer protocol for use with constrained nodes and constrained networks (e.g., low power, lossy).

The nodes often have 8-bit microcontrollers with small amounts of ROM and RAM, while constrained networks such as IPv6 over Low Power Wireless Personal Area Networks (6LoWPANs) often have high packet error rates and a low throughput. The protocol is designed for machine-to-machine (M2M) applications such as smart energy and building automation. The protocol provides a request/response interaction model between application endpoints, supports built-in discovery of services and resources, and includes key concepts of the Web such as URIs and Internet media types.

RFC 6960

[RFC 6960] specifies a protocol useful in determining the current status of a digital certificate without requiring Certificate Revocation Lists (CRLs). Additional mechanisms addressing PKIX operational requirements are specified in separate RFC documents.

RFC 7030

[RFC 7030] profiles certificate enrollment for clients using Certificate Management over CMS (CMC) messages over a secure transport. This profile, called Enrollment over Secure Transport (EST), describes a simple, yet functional, certificate management protocol targeting Public Key Infrastructure (PKI) clients that need to acquire client certificates and associated Certification Authority (CA) certificates. It also supports client generated public/private key pairs as well as key pairs generated by the CA.

ISO/IEC 19790

[ISO/IEC 19790] specifies the security requirements for a cryptographic module utilized within a security system protecting sensitive information in computer and telecommunication systems. This standard specifies four security levels for each of 11 requirement areas with each security level increasing security over the preceding level. It defines four security levels for cryptographic modules to provide for a wide spectrum of data sensitivity (e.g., low value administrative data, million dollar funds transfers, life protecting data, personal identity information, and sensitive information used by government) and a diversity of application environments (e.g., a guarded facility, an office, removable media, and a completely unprotected location).

ISO/IEC 27001

[ISO/IEC 27001] specifies the requirements for establishing, implementing, maintaining and continually improving an Information Security Management System (ISMS) within the context of the organization. It also includes requirements for the assessment and treatment of information security risks tailored to the needs of the organization. The requirements are generic and are intended to be applicable to all organizations, regardless of type, size or nature.

ISO/IEC 27002

[ISO/IEC 27002] is a code of practice - a generic, advisory document, not a formal specification such as [ISO/IEC 27001]. It recommends information security controls addressing information security control objectives arising from risks to the confidentiality, integrity and availability of information. Organizations that adopt ISO/IEC 27002 must assess their own information risks, clarify their control objectives and apply suitable controls (or other forms of risk treatment) using the standard for guidance. The standard is structured logically around groups of related security controls.

9.8.6 Standards Focused on Industrial Automation

The following includes examples of standards focused on industrial automation.

IEC 62443

[IEC 62443] standard, formerly known as ISA 99, is a global standard for the security of Industrial Control System (ICS) networks. The framework of this standard guides operators of ICS networks through requirements, controls and best practices necessary for a secure industrial network. It helps organizations to reduce both the risk of failure and exposure of ICS networks to cyber threats. The [IEC 62443] standard consists of several parts, which are organized into four groups: General, Policies and Procedures, System, and Component. The first two groups describe concepts, uses cases, policies and procedures associated with ICS security, whereas the two latter groups, System and Component, define the technical requirements for networks and system components. Figure 9.4 depicts a graphical representation of the categories and scope of [IEC 62443] series of standards (as of December 2019).

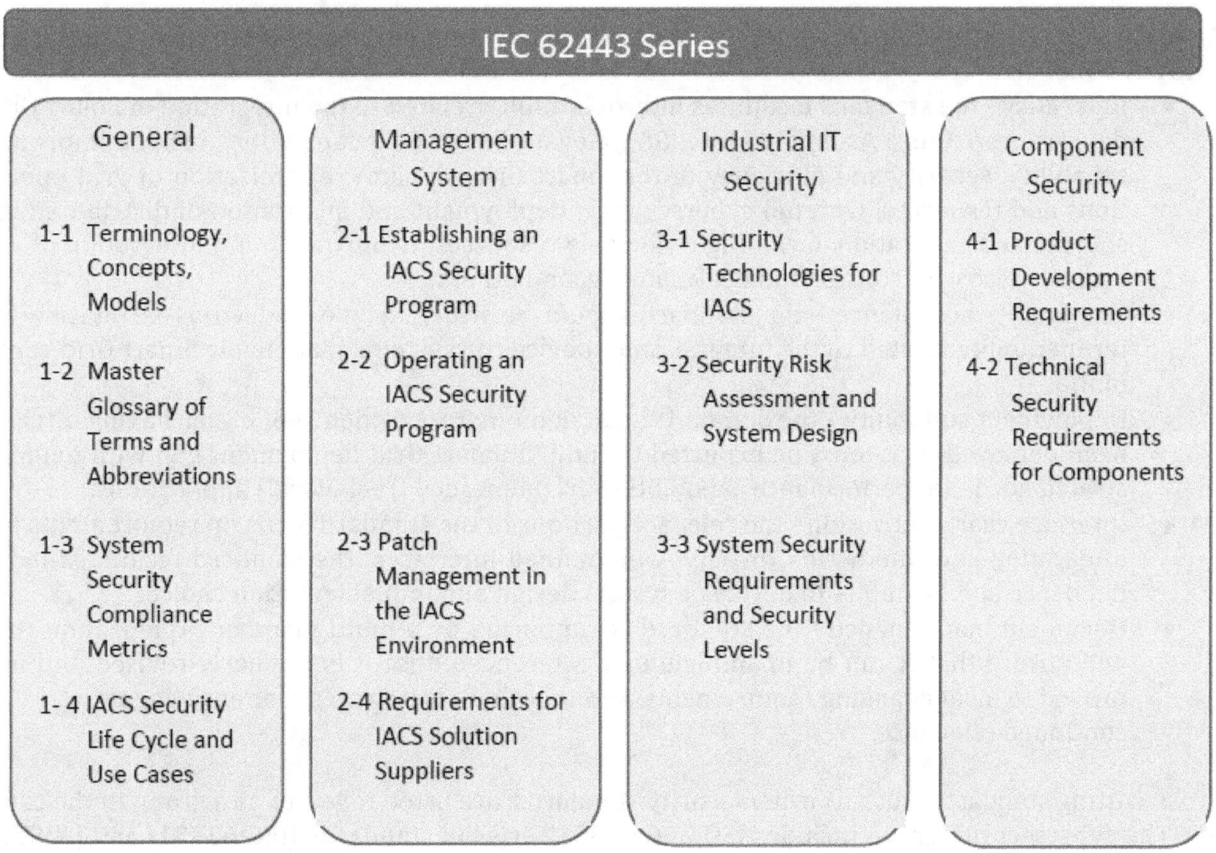

Figure 9.4 IEC 62443 Series.

9.8.7 Standards Focused on Cybersecurity

There is a convergence towards two main series of standards, IEC 62443 and the ISO/IEC 27000 family of standards. The IEC 62443 series focuses on operational technology (OT), which is concerned with keeping cyber-physical systems operating as intended, while the ISO/IEC 27000 family of standards focuses on information technology (IT), which is concerned about the flow and accuracy of data, data privacy, etc. Therefore, for a complete cyber security strategy both are needed, as well as sector specific standards, including for example IEC 62645 for the nuclear industry or the IEC 62351 series of standards for the electrical energy sector.

As part of its charter objectives, the SGIP produced a Catalog of Standards (CoS), which is a compendium of standards and practices considered to be relevant for the development and deployment of a robust and interoperable Smart Grid. Although compliance with a standard does not guarantee interoperability, standards facilitate interoperability. The CoS contains both standards and guidelines from Smart Grid areas including smart metering, substation automation, electric vehicle grid integration, Internet and wireless protocol usage, precision time synchronization, synchrophasors, customer energy usage (e.g., Green Button), cybersecurity, calendaring/scheduling models, and pricing models.

While CoS is not anticipated to be mandatory, the information included for each item of the CoS is a useful resource for utilities, manufacturers, regulators, consumers, and other Smart Grid stakeholders.

Also, the CoS is a trusted source since the inclusion of any item in the CoS is based on five criteria [NISTIR 7628r1]:

- Relevancy - the standard facilitates interoperability related to the integration of Smart Grid devices or systems. As defined by EISA, relevant Smart Grid capabilities include improved reliability, security and efficiency of the Smart Grid; dynamic optimization of grid operations and resources, with full cybersecurity; deployment and integration of distributed resources and generation, including renewable resources; deployment and integration of advanced electricity storage and peak shaving; and others.
- Community acceptance - the standard should be widely acknowledged as facilitating interoperability related to the integration of devices or systems that enable Smart Grid capabilities.
- Deployment suitability - the standard must demonstrate evidence of either having already been deployed or it must be expected to fulfill a Smart Grid deployment gap with demonstrated adequate performance capabilities in commercial (real world) applications.
- Interface characterization - the relevant portions of the standard focus on requirements for integration and interaction through well defined interfaces; the standard facilitates independence and flexibility in device or system design and implementation choices.
- Document maintenance - the standard is supported by a multi member organization that will ensure that it can be unambiguously referenced, that it is regularly revised and improved to meet changing requirements, and that there is a strategy for ensuring its continued relevance.

Cross-cutting standards, such as cybersecurity standards, are listed together as a group in the catalog. The cybersecurity group includes ISO/IEC 62351 series of standards [IEC 62351] and [NISTIR 7628r1] as guidelines.

ISO/IEC 62351

This is an open standard, developed and maintained by a trusted standard development organization. The scope of the IEC 62351 series (many parts) is information security for power system control operations. The standard defines security requirements for power system management and information exchange, including communications network and system security issues, Transmission Control Protocol (TCP) and Manufacturing Messaging Specification (MMS) profiles, and security for Inter-Control Center Protocol (ICCP) and substation automation and protection. The IEC 62351 standard addresses security concerns in power systems, providing in part authentication, integrity and confidentiality of data. The standard proposes both standard technologies (e.g., TLS) and proprietary extensions to industrial protocols.

Currently, the IEC 62351 family of standards depicts the architecture of a secure power system and standardizes its protocols and components. According to [Carrulo 2019], to truly obtain effective end-to-end security, secure protocols must:

- Establish secure connections based on some trusted private key of the actors.
- Have a repository of actors allowed to act inside the system.

The former is standardized in Key Management described in [IEC 62351-9], while the latter is standardized in [IEC 62351-8] as Role-based Access Control (RBAC), and further reviewed and explained in RBAC Guidelines of [IEC/TR 62351-90-1]. Table B.1 of Appendix B includes a list of key parts of the standard that were published by April 2020, when this table was revised. Each part may include several subparts. The complete list of parts is available on IEC Web site.

The family of standards is for use in conjunction with related IEC standards, but has not been widely adopted yet [NISTIR 7628r1]. Although the [IEC 62351] standard contains provisions to ensure the integrity, authenticity and confidentiality for different protocols used in power systems, there are problems with the algorithms or parameters chosen in parts of the standard such as [Schlegel 2017]:

- Some inaccuracies (e.g., cipher suite designations), and unconventional choices (e.g., RSA signatures for IEC 61850.
- No considerations for newer cryptographic algorithms that could provide the same security guarantees at a lower performance cost (e.g., elliptic curve cryptography).

The authors conclude that the standard can significantly improve security in power systems if applied comprehensively. Also, the authors point that the need to preserve (partial) backwards-compatibility has led to some design choices that provide less security than could have been achieved with a more ambitious approach.

As described in [NISTIR 7628r1], homomorphic encryption and signature schemes tailored for efficient in-network aggregation are needed. Cryptographic in-network aggregation schemes have the potential of improving the efficiency of many-to-one communications in the Smart Grid, like those generated from multiple sensors to a single or a small number of designated collection points. To achieve efficient in-network aggregation, intermediate nodes in the routing protocol need to modify data packets in transit; for this reason, standard signature and encryption schemes are not applicable, and it is a challenge to provide resilience to tampering by malicious nodes.

NISTIR 7628

This is a guideline [NISTIR 7628] published by NIST in 2010, a revised version [NISTIR 7628r1] was published in 2014. It was developed through a participatory public process that, starting in March 2009, included several workshops as well as weekly teleconferences, all of which were open to all interested parties. The guidelines are not prescriptive, nor mandatory. Rather they are advisory, intended to facilitate each organization's efforts to develop a cybersecurity strategy effectively focused on prevention, detection, response, and recovery. The [NISTIR 7628r1] document can be used as:

- Tool for organizations that are researching, designing, developing, implementing, and integrating Smart Grid technologies – established and emerging.
- Framework for assessing risks to smart grid components and systems during design, implementation, operation, and maintenance.
- Guide to assist organizations as they craft a Smart Grid cybersecurity strategy that includes requirements to mitigate risks and privacy issues pertaining to Smart Grid customers and uses of their data.

Chapter 10

Measuring Performance of Security Program

10.1 Security Program Performance

Security is a condition that results from the establishment and maintenance of protective measures that enable an enterprise to perform its mission or critical functions despite risks posed by threats to its use of information systems. Protective measures may involve a combination of deterrence, avoidance, prevention, detection, recovery, and correction that should form part of the enterprise's risk management approach [CNSSI 4009].

An organization needs to measure the performance of its security program. The purpose of measuring performance is to monitor the status of measured activities and facilitate improvement in those activities by applying corrective actions based on observed measurements.

According to [NIST SP800-55r1] document, management should use measures to review performance by observing trends, identifying and prioritizing corrective actions, and directing the application of those corrective actions based on risk mitigation factors and available resources. This document describes an approach to identify and evaluate non-productive security controls and prioritize security controls for continuous monitoring. It describes a quantitative approach to measuring and analyzing security controls implementation and effectiveness at the information system and program levels. This is an approach applied across multiple individual efforts, but from an enterprise level perspective.

Information security measures can assist with determining the effectiveness of implemented security processes, procedures, and security controls by relating results of information security activities and events to security controls and information security investments.

As an organization's security program evolves and performance data becomes more available, different types of measures (implementation, effectiveness/efficiency, and impact) can be obtained to be used for performance improvement.

10.1.1 Using Measures

The results of effectiveness/efficiency measures can be used to ascertain whether selected security controls are functioning properly and are helping facilitate corrective action prioritization. Effectiveness/efficiency measures address two aspects of security control implementation results: the robustness of the result itself, referred to as effectiveness, and the timeliness of the result, referred to as efficiency.

Effectiveness/efficiency measures may require fusing information security program activities data with the data obtained from automated monitoring and evaluation tools in a manner that can be directly tied to security controls implementation. For example, the effectiveness/efficiency measure - percentage of information security incidents caused by improperly configured access controls - relies on information regarding implementation and effectiveness of the following security controls: incident monitoring; audit monitoring, analysis, and reporting; and monitoring configuration changes. Additionally, the effectiveness/efficiency measure - the percentage of system components, which undergo maintenance on schedule - relies on information regarding the efficiency of the following security controls: periodic maintenance and life cycle support. This measure, percentage of system components that undergo maintenance in accordance with the formal maintenance schedule, can be determined by using maintenance schedules, maintenance logs, incident reports, monitoring reports, and operational logs. This measure can be reported in a pie chart format. The chart can show a comparison of the percentage of system components receiving maintenance in accordance with the formal maintenance schedule versus the percentage of system components not receiving maintenance in accordance with the formal maintenance schedule over the specified period. The frequency of data collection and analysis (e.g., quarterly or more often) and frequency of reporting (e.g., annually or more often) are often based on the security management strategies.

The measures development process described in this document [NIST SP800-55r1] ensures that measures are developed with the purpose of identifying causes of poor performance and pointing to appropriate corrective actions. The process consists of six phases, which, when fully executed, will ensure use of these measures for security control performance monitoring and improvement (see process diagram shown in Figure 10.1).

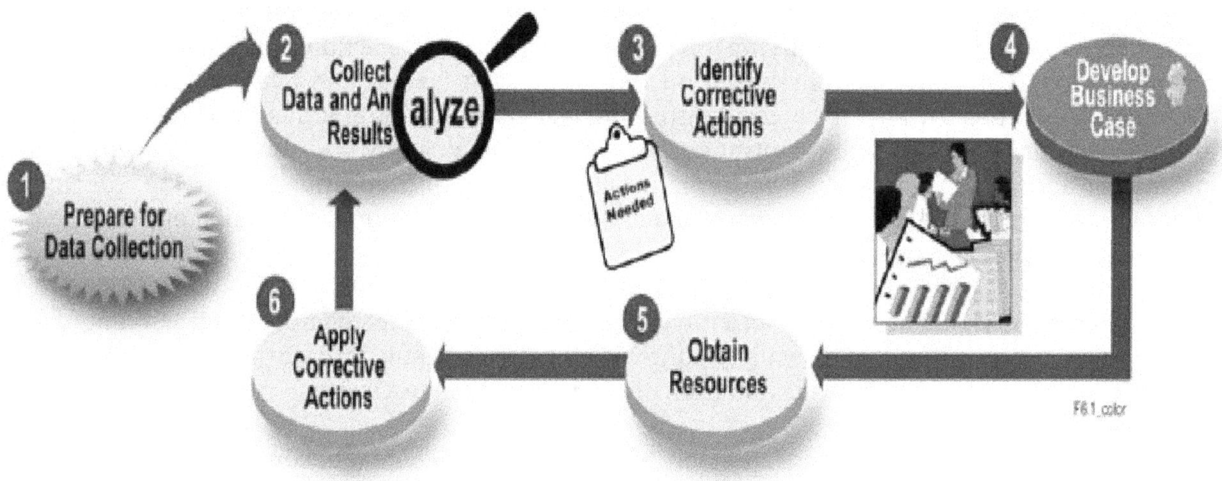

Figure 10.1 Information Security measurement program implementation. *Source:* [NIST SP800-55r1]. Public Domain.

Applying measures for monitoring information security controls performance and using the results to initiate performance improvement actions are goals of the information security measurement implementation process. Information security measures must also be repeatable, provide relevant performance trends over time and be useful for tracking performance and directing resources.

10.1.2 Identify Corrective Actions

The phase to identify corrective actions involves development of a plan to serve as the roadmap for closing the gaps identified in the analysis phase. It includes the following activities [NIST SP800-55r1]:

- Determine range of potential corrective actions using the analysis results and causation factors for each performance issue.
- Prioritize corrective actions based on overall risk mitigation goals for each performance issue; several corrective actions may apply to a single performance issue; however, some

may be inappropriate if they are too costly or are inconsistent with the magnitude of the problem.
- Alternatively, actions priorities may be assigned based on the criticality of implementing specific corrective actions, cost of the actions, and the magnitude of their impact on the organization's information security posture.
- Selecting most appropriate corrective actions from the prioritized list should be finally based on the cost benefit analysis for various business cases.

10.1.3 Directing the Application of Corrective Actions

The results of an effective information security measurement program can provide useful data for directing the allocation of information security resources and should simplify the preparation of performance related reports. To be useful in tracking performance and directing resources, measures need to provide relevant performance trends over time and point to improvement actions that can be applied to problem areas.

As shown in Figure 10.1, phase 4 of the process (develop business case), and phase 5 (obtain resources) address the budgeting cycle for acquiring resources needed to implement remediation actions identified in phase 3. Phase 6 is concerned to apply corrective actions, which involves implementing corrective actions in the security (technical, management, and operational) areas of security controls. For example, this vulnerability measure - percentage of high vulnerabilities mitigated within organization's defined time periods after discovery - helps on improving patch management process and effectiveness of security controls.

Other security measures (see more information in [NIST SP800-55r1]) could provide substantive justification for decisions that directly affect the information security posture of an organization. These decisions may include budget and personnel requests and allocation of available resources. The information security measures can support risk based decision making by contributing quantifiable information to the risk management process.

The measures of successes and failures of past and current information security investments could provide quantifiable data that supports resource allocation for future investments. Using the results of the measures analysis, management and security personnel can isolate problems, use collected data to justify investment requests, and then target investments specifically to the areas in need of improvement. By using measures to target security investments, an organization can obtain the best value from available resources.

10.2 Failure Recovery and Restoration

An organization should provide the capability to recover and reconstitute the Smart Grid information system to a known secure state after a disruption, compromise, or failure [NISTIR 7628r1]. To achieve this capability, an organization should develop and implement a continuity of operations plan dealing with the overall issue of maintaining or reestablishing operations in case of an undesirable interruption for a Smart Grid information system.

Similarly, this standard [NIST SP800-53r5] recommends that an organization develops a contingency plan for the information system that:
- Identifies essential missions and business functions and associated contingency requirements.
- Provides recovery objectives, restoration priorities, and metrics.

- Addresses maintaining essential missions and business functions despite an information system disruption, compromise, or failure.
- Addresses eventual, full information system restoration without deterioration of the security safeguards originally planned and implemented.

10.2.1 Smart Grid System Recovery

Smart Grid information system recovery and reconstitution to a known secure state means that [NISTIR 7628r1]:

- All Smart Grid information system parameters (either default or organization established) are set to secure values.
- Security critical patches are reinstalled.
- Security related configuration settings are reestablished.
- Smart Grid information system documentation and operating procedures are available.
- Application and Smart Grid information system software is reinstalled and configured with secure settings.
- Information from the most recent, known secure backups is loaded.
- The Smart Grid information system is fully tested.
- The organization provides compensating security controls (including procedures or mechanisms) for the organization defined circumstances that inhibit recovery to a known, secure state.
- The organization provides the capability to reimage Smart Grid information system components in accordance with organization defined restoration time periods from configuration controlled and integrity protected media images representing a secure, operational state for the components.

As described in [NISTIR 7628r1] document, a continuity of operations plan addresses both business continuity planning and recovery of Smart Grid information system operations. Development of a continuity of operations plan is a process to identify procedures for safe Smart Grid information system operation while recovering from a Smart Grid information system disruption. The plan requires documentation of critical Smart Grid information system functions that need to be recovered. The organization performs a root cause analysis for the event and submits any findings from the analysis to management.

However, it is assumed that the Smart Grid information system fails to a known state for defined failures. The Smart Grid information system preserves defined system state information in failure. If a failure is not defined, then the organization should have a plan how to handle it. There are different ways that systems can be designed to fail. A fail secure state is where all access is blocked when a system stops working properly. Fail known state means fail secure, trusted recovery at the firmware and system levels, and software quality assurance program [NISTIR 7628r1].

Failure in a known state can be interpreted by organizations in the context of safety or security in accordance with the organization's mission/business/operational needs. Failure in a known secure state helps prevent a loss of confidentiality, integrity, or availability in the event of a failure of the smart grid information system or a component of the smart grid information system. Failure to a known state may include digital, analog, or other modes of operation.

Restoration processes and procedures for control systems should be documented to the extent needed to recover the systems in accordance with requirements defined by an organization to ensure the system is recovered in the same state it was in prior to loss [INL 2007]. Inherent with the need for backup documentation, data storage media, and physical separation of the backup media from the operational system is the process for restoration of part or all control system after loss due to hardware, software or facility malfunction.

10.2.2 Trusted Recovery

Trusted recovery is ability to ensure recovery without compromise after a system failure [CSRC 2015]. A system failure is a serious security risk because the security controls might be overridden when the system is not functioning properly. Trusted recovery is designed to prevent this type of corruption in the event of such a system failure. Concepts of trusted recovery are described in [Gligor 1991]. Portions of the [Rainbow Series] (e.g., the Orange book and the Red Book) have been superseded by the Common Criteria Evaluation and Validation Scheme (CCEVS). Common Criteria Evaluation and Validation Scheme (CCEVS) is a United States program administered by the National Information Assurance Partnership (NIAP) to evaluate security functionality of an IT product conformance to the Common Criteria international standard. The new standard uses Protection Profiles and the Common Criteria Standards [ISO/IEC 15408] to certify the product.

The properties of trusted recovery are defined in terms of two notions: secure states and secure state transitions [Gligor 1991]. A system state is secure whenever consistency invariants derived from valid interpretations of security and accountability models are satisfied. A state transition is secure if both its input state and its output state are secure, and it satisfies the constraints placed on it by valid interpretations of security policy and accountability policy models. Accountability models include models of user authentication, trusted path, and audit.

Recovery mechanisms of a trusted computing base (TCB) become trusted only if they maintain secure states in the normal mode of operation or detect insecure states and reconstruct secure states in the maintenance mode of operation, despite the occurrence of failures and discontinuities. To detect insecure states after system failures or to verify that recovered states are secure, recovery mechanisms must check whether security invariants (security variables of security model) are satisfied. All security invariants are relevant to the recovery mechanisms handling state transition failures. These failures usually leave the system in the normal mode of operation, instead of causing the system to enter maintenance mode. Therefore, all invariants which must hold in the normal mode of operation also must hold after recovery from state transition failures.

There are several different types of trusted recovery that are specified in the Common Criteria standard. A manual recovery is where the system requires human interaction to provide recovery. Automated recovery is where the system has a mechanism that requires no human intervention for at least one type of service being discontinued. An automated recovery without undue loss is where the system can automatically recover itself, and it also has additional mechanisms to protect specific objects on the system. A function recovery is where a system has a recovery mechanism for a specific function either to make sure that it is successfully completed, or to make sure that it rolls back to a previous secure state.

10.3 Reliability Reporting

Power system operations pose many security challenges that are different from most other industries. Specifically, there are strict performance and reliability requirements that are needed by power system operations. For example, this document specifies [NISTIR 7628r1]:

- Operation of the power system must continue 24/7 with high availability for SCADA and higher for protective relaying regardless of any compromise in security or the implementation of security measures that hinder normal or emergency power system operations.

- Power system operations must be able to continue during any security attack or compromise (as much as possible).
- Power system operations must recover quickly after a security attack or the compromise of an information system.
- Testing of security measures cannot be allowed to impact power system operations.

Reliability reporting is an activity that covers aspects of both scheduled and unscheduled maintenance. Reliability reporting includes results of methods of determining what preventive (or other) failure management policies to apply. For example, the results include the original equipment manufacturer (OEM) recommendations, statistical failure records, requirements of codes, legislation, or regulations within a jurisdiction, expert opinion, maintenance that has already been conducted on similar equipment or measured values and performance indications.

10.3.1 Ensuring Reliable Information

This document [NISTIR 7628r1] stresses the importance of cybersecurity in ensuring the reliability of the grid and the confidentiality of specific information. It also discusses the cybersecurity strategy for the Smart Grid and the specific tasks within this strategy to ensure reliable information for any business need.

Cybersecurity needs to be appropriately applied to the combined power system and IT communication system domains to maintain the reliability of the Smart Grid and privacy of consumer information. To avoid disruption, an organization should perform maintenance activities. Maintenance activities encompass appropriate policies and procedures for performing routine and preventive maintenance on the components of a Smart Grid information system.

10.3.2 Automated Reporting Mechanisms

The reliability reporting should be included to maintenance management system. This document [NISTIR 7628r1] recommends that an organization should employ automated mechanisms to transfer predictive maintenance data to a computerized maintenance management system. A computerized maintenance management system maintains a computer database of information about the maintenance operations of organizations and automates processing equipment condition data to trigger maintenance planning, execution, and reporting [NISTIR 7628r1].

10.3.3 Information Sharing

In order to advance its overall cybersecurity strategy and help an organization more effectively manage cyber related risk, an organization should establish an information sharing plan. The plan should address the collection and analysis of threat information from both internal and external sources and the use of this information in the development and deployment of protective measures.

NIST encourages sharing of cyber threat information among organizations, both in acquiring threat information from other organizations and in providing internally generated threat information to other organizations. An organization can use the combined knowledge and experience of its own personnel and others, such as members of cyber threat information sharing organizations, to share threat information while operating per its security, privacy, regulatory, and legal compliance requirements. Sharing cyber threat could be in real time or after a cyber incident occurred (see Figure 10.2).

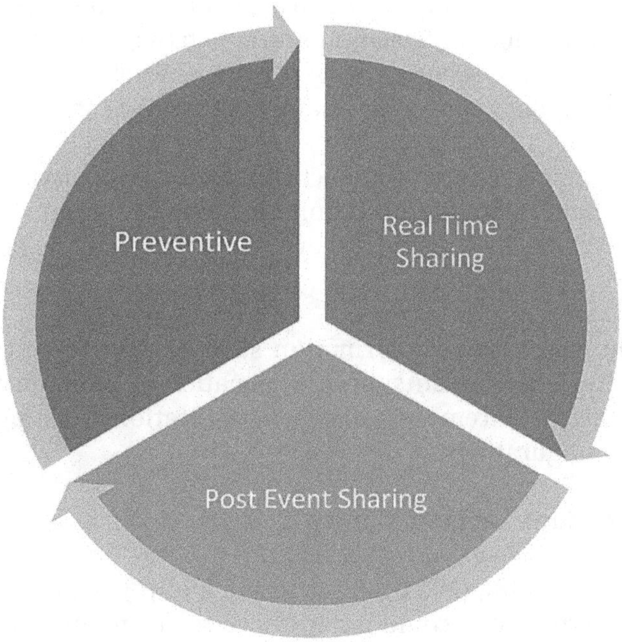

Figure 10.2 Sharing information.

One of the best techniques to address threats, especially the advanced persistent threat (APT), organizations could share threat information, including threat events (e.g., tactics, techniques, and procedures) that organizations have experienced, mitigations that organizations have found are effective against certain types of threats, and threat intelligence (e.g., indications and warnings about threats). By sharing cyber threat information, an organization can proactively build preventive capabilities. The objective is to track and disrupt cyber adversaries as early as possible in the attack sequence and to measurably improve the speed and accuracy of organizational responses. Cyber threat information is any information that can help an organization identify, assess, monitor, and respond to cyber threats. Cyber threat information includes indicators of compromise; tactics, techniques, and procedures used by threat actors; suggested actions to detect, contain, or prevent attacks; and the findings from the analyses of incidents. Organizations that share cyber threat information can improve their own security postures as well as those of other organizations.

10.4 Operations Security Controls

Operations security (OPSEC) process involves five steps: identification of critical information, analysis of threats, analysis of vulnerabilities, assessment of risks, and application of appropriate countermeasures, called operations security controls.

An organization should employ operations security controls to protect key organizational information [NIST SP800-53r5]. The safeguards are applied to both organizational information systems and the environments in which those systems operate. The security controls help to protect the confidentiality of key information including, for example, limiting the sharing of information with suppliers and potential suppliers of information system components, information technology products and services, and with other non organizational elements and individuals. Information critical to mission/business success includes, for example, user identities, element uses, suppliers, supply chain processes, functional and security requirements, system design specifications, testing protocols, and security control implementation details.

Technical operational security covers infrastructure installations and operational procedures, control limitations (manual or automatic) based on system status, reliability and resiliency of operations, system's level of intelligence, security of system's data and analysis, personnel's qualification status and technical skills, regular routine check and maintenance schedule, etc.

An organization should employ operations security controls in accordance with classification guides to protect supply chain related information for the information system, system component, or information system service. This control enhancement expands the scope of OPSEC process to include suppliers and potential suppliers.

Supply chain information may include information such as user identities; uses for information systems, information system components, and information system services; supplier identities; supplier processes; security requirements; design specifications; testing and evaluation results; and system/component configurations.

10.4.1 Examples of Operations Security Controls

Several standards [ISO/IEC 27002], [NIST SP800-53r5], and [ISO/IEC 15408] describe operations security controls. Although the scope could differ, there are common controls that are included in more than one standard (see mapping tables in [NIST SP800-53r5]). Examples of operations security controls recommended in [ISO/IEC 27002] that are described in [NIST SP800-53r5]) include:

- Operational procedures and responsibilities to include documented operating procedures, change management, capacity management, separation of development, testing, and operational environments.
- Protection from malware includes controls against malware.
- Backup information.
- Logging and monitoring include event logging, protection of log information, administrator and operator logs, clock synchronization.
- Control of operational software includes installation of software on operational systems.
- Technical vulnerability management includes management of technical vulnerabilities, restrictions on software vulnerabilities.
- Assurance related controls that support evidence to obtain a degree of confidence that the system satisfies the stated trustworthiness and assurance requirements.
- Information systems audit considerations include information system audit controls.

Although there is a correspondence between operations security controls defined in [ISO/IEC 27001] and security requirements described in [ISO/IEC 15408] or [NIST SP800-53r5], the Security Engineer has to evaluate and validate if the security requirements of an information system or product are satisfied (fully or partially) or not satisfied by the associated operations security controls of the organization. Therefore, it should be clearly understood that satisfying [ISO/IEC 15408] security requirements for a particular product does not imply that such requirements are also satisfied for an information system (which may consist of multiple, integrated individual component products). Also, the implementation of the mapped controls should result in an equivalent information security posture. In some cases, similar topics are addressed in these standards, but the security controls may provide a different context, perspective, or scope. For example, this standard [NIST SP800-53r5] addresses information flow control broadly in terms of approved authorizations for controlling access between source and destination objects, whereas this standard [ISO/IEC 27002] addresses information flow more narrowly as it applies to interconnected network domains.

Also, this standard [ISO/IEC 27002] provides a reference set of generic information security controls including operations controls and implementation guidance. This document is designed to be used by organizations within the context of an information security management system (ISMS) based on ISO/IEC27001; for implementing information security controls based on internationally recognized best practices; for developing organization specific information security management guidelines.

The information security controls are generally regarded as best practice means of achieving those objectives. For each of the controls, implementation guidance is provided in [ISO/IEC 27002] standard. This standard [ISO/IEC 27002] has directly equivalent to National standards in several countries. Security controls are not mandated since each organization is expected to undertake a structured information security risk assessment process to determine its specific requirements before selecting controls that are appropriate to its particular circumstances.

10.4.2 Utilities Case

As utilities are housing their operations in their own or authorized contracted agent facilities, physical and logical security should be in place. If there is equipment that is not under the utility's physical control which contains personal data, physical security is dependent on the customer or an electrical service provider. All personal data collected and created during these activities must be appropriately safeguarded to ensure unauthorized access to the data does not occur, to preserve integrity of the data, and to allow for appropriate availability.

It is important to highlight that industrial control systems (ICS) related technologies are being considered part of the whole set of information communications technology (ICT) supporting the operations of the Smart Grid. The list includes:

- Operational systems such as generators, transformers, SCADA Systems, Energy/Distribution Management Systems (EMS/DMS), programmable logic controllers (PLCs), substations, smart meters, and other intelligent electrical devices (IEDs).
- IT systems such as PCs, servers, mainframes, applications, databases, web sites, web services, etc., among which include the components of corporate infrastructure.
- Communications networks and protocols such as Ethernet, Wi-Fi, PRIME, DLMS/COSEM, Zigbee, 4G, DNP3, etc.
- End points such as smart meters, EVs, smart phones and other mobile devices.

This document [INL 2007] reviews several key operational cybersecurity elements that are important for control systems and industrial networks. It provides guidance and direction for developing operational security strategies such as creating OPSEC plans, embedding cybersecurity into the operations life cycle, creating technical and non-technical security mitigation strategies. Also, this document [NIST SP800-82r2] recommends monitoring and reporting controls for power grid communications. For example, the lack of monitoring and control capabilities could cause a large generating unit to be taken offline, an event that would lead to loss of power at a transmission substation. This report [ICS-CERT 2014] describes cyber attacks to utilities that resulted from weak network configuration, lack of perimeter security, weak remote access, etc. The report also provides guidance to improving cybersecurity protection for ICS including SCADA systems such as applying appropriate architecture and network segmentation, security controls, monitoring, and detection capabilities.

As cybersecurity programs in modern information are beginning to address the issues related to deploying cybersecurity strategies for industrial control architectures, this document [INL 2007] reviews how operational controls for ICS can drive the creation of a cybersecurity sensitive culture.

Personnel responsible for the cybersecurity of control systems may have little knowledge to accurately determine which products and actions are most appropriate for their specific control network or how to reduce their cyber risk in the most efficient way. Mapping an existing IT centric plan into the control domain without proper impact analysis can have negative consequences. As such, key foundations must be prepared in developing the cyber OPSEC plan that can accommodate for these differences as described in [INL 2007].

While creating a separate OPSEC plan for ICS is not recommended, defining an OPSEC plan based on risk management and an integrated strategy to mitigate threats and vulnerability of ICT systems could benefit the organization's business.

Table 10.1 illustrates some of the more common IT security elements that would be expected to be found under a cyber OPSEC plan for ICS and highlights some of the differences in requirements for a control system. Therefore, the differences should be addressed based on the specific security needs for each organization. However, augmenting and migrating a proven cybersecurity program from the business domain into a control domain is not always the most effective solution.

Table 10.1 Comparing operations security requirements.

Security Topic	**Information Technology Systems**	**Industrial Control Systems**
Antivirus & mobile code countermeasures	Common and widely used	Uncommon and difficult to deploy
Support technology lifetime	3-5 years	Up to 20 years
Outsourcing	Common and widely used	Rarely used
Application of patches	Regular/scheduled	Adhoc (vendor specific)
Change management	Regular/scheduled	Legacy based – unsuitable for modern security
Time critical content	Delays are generally accepted	Time critical
Availability	Not always a concern	Required 24x7x365
Security awareness	Good and present in both private and public sector	Generally poor regarding cybersecurity
Security testing/audit	Scheduled and mandated	Occasional testing for outages
Physical security	Commonly employed, but not consistent	Employed on the site but often equipment is remote and unmanned

Adapted from [INL 2007]. Public Domain.

Also, it is important to ensure that the security controls are in place and functioning properly during the day-to-day operations of a SCADA system. An organization's cybersecurity strategy should protect the assets that it deems critical to successful operation. It requires a layered approach known as Defense in Depth. Defense in Depth operations is a defense strategy in multiple places and locations. The approach is not one thing, but a combination of people, technology, operations, and adversarial awareness.

Defense in Depth as a concept originated in military strategy to provide barriers to impede the progress of intruders from attaining their goals while monitoring their progress and developing and implementing responses to the incident in order to repel them. In the cybersecurity paradigm, Defense in Depth correlates to detective and protective measures designed to impede the progress of a cyber intruder while enabling an organization to detect and respond to the intrusion with the goal of reducing and mitigating the consequences of a breach.

Defense in Depth is the approach to protect all assets, while taking into consideration its interconnections and dependencies to provide effective layers of monitoring and protection based on the organization's exposure to cybersecurity risks. In order to apply Defense in Depth to ICS and SCADA environments, an organization must understand the relationship of intruders (threats) and vulnerabilities to the controls (standards and security controls) employed to protect the business, operations, personnel, and technologies that make up an ICS or SCADA system.

Security controls, best practices, and standards, protect ICS critical assets through multiple layers of defense thus improving the protection. Organizations must constantly adjust and refine security countermeasures to ensure protection against known and emerging threats (see Figure 10.3). Because of the complexity of ICS architectures, potential vulnerabilities and/or exploits that introduce new and evolving categories of threats to the ICS environment can have lasting consequences, and without a layered, multitier strategy could result in long term exposure without detection.

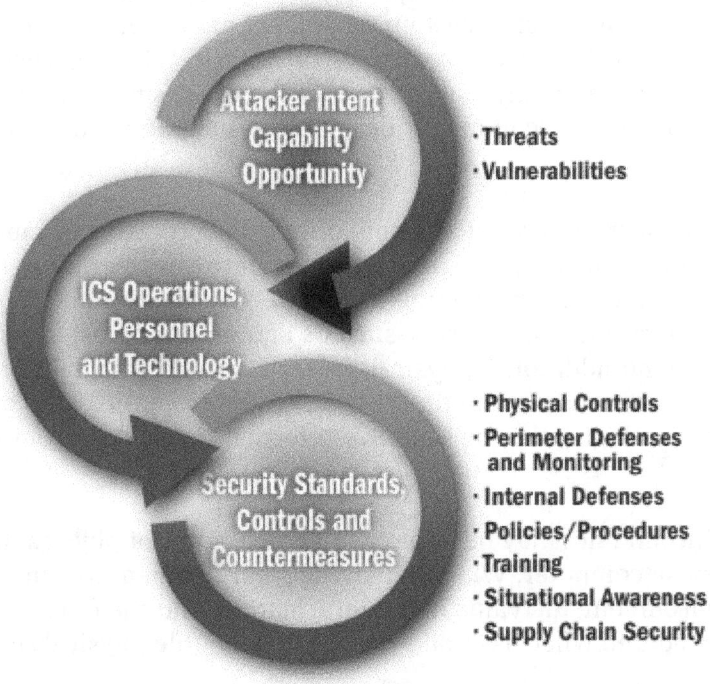

Figure 10.3 Defense in Depth planning. *Source:* [DHS 2016]. Public Domain.

In order to apply Defense in Depth to ICS environments, an organization must understand the relationship of intruders (threats) and vulnerabilities to the controls (standards and countermeasures) put in place to protect the operations, personnel, and technologies that make up an ICS. However, the new ICS technologies replacing the ICS legacy systems are offering better features for updates, change management, security testing, etc.

10.5 Physical Security Environment

In determining the appropriate level of physical protections required for a device, it is important to consider both the operating environment and the value and sensitivity of the data protected by the device.

In unprotected or lightly protected environments, it is common to deploy cryptographic modules with some form of physical security. Even at the consumer level, devices that process and contain valuable or sensitive personal information often include physical protection. Cable television set top boxes, DVD players, gaming consoles, and smart cards are examples of consumer devices. Smart grid equipment, such as smart meters, deployed in similar environments will, in some cases, process information and provide functionality that can be considered sensitive or valuable. In such cases, management responsible for meter functionality and security may determine that meters must include cryptographic modules with a level of physical protection.

The protection of critical security parameters, such as keying material and authentication data, is necessary to maintain the security provided by cryptography. To protect against unauthorized access, modification, or substitution of this data, as well as device tampering, cryptographic modules can include features that provide physical security. There are multiple embodiments of cryptographic modules that may provide physical security, including: multichip standalone, multichip embedded, and single chip devices.

Physical protection is an important aspect of a module's ability to protect itself from unauthorized access to critical security parameters and tampering. A cryptographic module implemented in software and running on an unprotected system such as general purpose computer, commonly does not have the ability to protect itself from physical attack. When discussing cryptographic modules, the term firmware is commonly used to denote the fixed, small, programs that internally control a module. Such modules are commonly designed to include a range of physical security protections and levels.

Therefore, the specification of cryptographic module physical protections is a management task in which both environmental hazard and data value are taken into consideration. For example, management may conclude that a module protecting low value information and deployed in an environment with physical protections and controls, such as equipment cages, locks, cameras, and security guards, etc., requires no additional physical protections and may be implemented in software executing on a general purpose computer system. However, in the same environment, cryptographic modules protecting high value or sensitive information, such as root keys, may require strong physical security.

In summary, cryptographic modules may be implemented in a range of physical forms, as well as in software on general purpose computer. When deploying smart grid equipment employing cryptographic modules, the environment, the value of the information, and the functionality protected by the module should be considered when assessing the level of module physical security required.

10.6 Accountability and Auditing

Information security measures can increase accountability for information security by helping to identify specific security controls that are implemented incorrectly, are not implemented, or are ineffective. Data collection and analysis processes can facilitate identification of the personnel responsible for security controls implementation within specific organizational components or for specific information systems. Audit and accountability are one of the eighteen families of security controls catalog defined in [NIST SP800-53r5].

This document [NISTIR 7628r1] recommends an organization develops, implements, reviews, and updates on an organization defined frequency a documented audit and accountability security policy. This policy should address the objectives, roles, and responsibilities for the audit and accountability security program as it relates to protecting the organization's personnel and assets. This policy should apply also to contractors and third parties service providers. The organization should develop procedures to address the implementation of the audit and accountability security policy and associated audit and accountability protection requirements. The audit and accountability policy can be included as part of the general security policy for the organization. Procedures can be developed for the security program in general and for a specific Smart Grid information system when required. The policy and procedures should comply with applicable federal, state, local, tribal, and territorial laws and regulations.

In supporting accountability, an organization should develop, document, and maintain an inventory of the components of the Smart Grid information system that accurately reflects the current configuration. The inventory should provide the proper level of granularity for tracking and reporting and for effective property accountability. The organization defines a framework of management accountability that establishes roles and responsibilities to approve cybersecurity policy, assign security roles, and coordinate the implementation of cybersecurity across the organization.

The organization identifies the roles responsible for component inventory. Also, it updates the inventory of system components as an integral part of component installations, system updates, and removals. It is recommended the organization employs automated mechanisms to maintain an up to date, complete, accurate, and readily available inventory of information system components. Also, an organization should employ automated mechanisms to detect the addition of unauthorized components or devise into the environment and disables access by components or devices or notifies designated officials.

Periodic audits and logging of the Smart Grid information system should be implemented to validate that the security mechanisms are installed and operating correctly. The security audits review and examine a Smart Grid information system's records and activities to determine the adequacy of the system's security requirements and to ensure compliance with established security policy and procedures. Audits also are used to detect breaches in security services through examination of the Smart Grid system logs. Logging is necessary for anomaly detection as well as forensic analysis. With the convergence of power systems and traditional IT systems, proper analysis of event information is necessary to understand what occurred during the event. Analysis teams need to evaluate power systems logging data and cyber event logs in order to properly ascertain the actual causes of an event.

The information system should generate audit records containing information that establishes what type of event occurred, when the event occurred, where the event occurred, the source of the event, the outcome of the event, and the identity of any individuals or subjects associated with the event. The event outcomes can include indicators of event success or failure and event specific results (e.g., the security state of the information system after the event occurred).

Audits should not rely exclusively on interviews with system administrators; rather, be holistic reviews of processes, procedures, personnel actions, physical and network based resources that can be accomplished using automated mechanisms.

To promote accountability, organizations should identify and address gaps in privacy compliance, management, operational, and technical controls by conducting regular assessments. These assessments result in reports on compliance gaps identified in programs, projects, and information systems. In addition to auditing for effective implementation of all privacy controls (e.g., those controls identified in this guide [NISTIR 7628r1]), the organization should assess whether it has implementing a process to embed privacy considerations into the life cycle of personally identifiable information (PII).

User accountability should be enforced for control systems and users should be accountable for their actions. The specific recommendations include [INL 2007]:

- User access to resources must be limited and properly controlled to ensure that excessive privileges do not provide the opportunity to cause damage to information assets.
- A user's access attempts while using a specific control system resource need to be properly monitored, audited, and logged.
- Individual user ID needs to be included in the audit logs to enforce individual responsibility.
- Users should understand their responsibility when using control system and industrial network resources.

In addition, logs should be monitored and reviewed, through either manual or automatic methods, to uncover suspicious activity and to identify an industrial network environment that is shifting away from its original baselines [INL 2007]. Also, a very important aspect of operations security is administrative management.

10.6.1 Administrative Management

Administrative management is an important, but often overlooked aspect of the OPSEC program in control systems and their associated industrial networks. Another aspect of administrative management is the mitigation of threats that are already inherent in the operational domain, such as the trusted insider [INL 2007].

One aspect of administrative management is dealing with the security administrator's performance. The security administrator is responsible for implementing and maintaining specific network devices and software in the enterprise. A security administrator defines, configures, and maintains the security mechanisms protecting an organization. Also, an administrator configures what actions and events are to be audited and logged.

Besides technical and physical controls, administrative controls, commonly referred to as soft controls, are more management oriented. Examples of administrative controls are policies and procedures, security documentation, risk management, personnel security, and training.
This document [NISTIR 7628r1] includes requirements such as:

- The Smart Grid information system should provide the capability for a privileged administrator to configure, enable, and disable the organization defined security policy filters.
- An organization should provide training on security administration to security administrators.

As utilities may house their operations in their own or authorized contracted facilities, physical, administrative, and technical security should be in place under their existing information security program. For example, an organization should manage authentication credentials for users and devices by [NISTIR 7628r1]:

- Defining initial authentication credential content, such as defining password length and composition, tokens
- Establishing administrative procedures for initial authentication credential distribution; lost, compromised, or damaged authentication credentials; and revoking authentication credentials

If there is equipment that is not under the utility's physical control that contains personal data, then aspects of physical, administrative and technical security will be dependent on the customer or a service provider.

In addition, all personal data collected and created during these activities must be appropriately safeguarded to ensure unauthorized access to the data does not occur, to preserve integrity of the data, and to allow for appropriate availability.

10.6.2 Administrative Controls

Security capabilities must be turned on to be useful. There are many examples of operating systems where protections such as firewalls are configured but not enabled out-of-the-box. If protections are not enabled, the system may be unexpectedly vulnerable to attacks. If the administrator does not realize that protections are disabled, the system may continue in an unprotected state for some time until the omission is noticed.

An organization needs to support application partitioning [NIST SP800-53r5]. The information system separates user functionality (including user interface services) from information system management functionality. Organizations implement separation of system management related functionality from user functionality by using different methods. Separation of system and user functionality may include isolating administrative interfaces on different domains and with additional access controls.

The information system should prevent the presentation of information system management related functionality at an interface for nonprivileged users [NIST SP800-53r5]. This control ensures that administration options (e.g., administrator privileges) are not available to general users.

10.6.3 Separation of Duties

One aspect of administrative management is dealing with personnel issues. This includes separation of duties and job rotation. Separation of duties addresses the potential for abuse of authorized privileges and helps to reduce the risk of malevolent activity without collusion.

An organization should separate defined duties of individuals, document separation of duties of individuals, and define information system access authorizations to support separation of duties. Separation of duties includes, for example [NIST SP800-53r5]:

- Dividing mission functions and information system support functions among different individuals and/or roles.
- Conducting information system support functions with different individuals (e.g., system management, programming, configuration management, quality assurance and testing, and network security).
- Ensuring security personnel administering access control functions do not also administer audit functions.

An organization should employ the principle of least privilege, allowing only authorized accesses for users (or processes acting on behalf of users) which are necessary to accomplish assigned tasks in accordance with organizational missions and business functions [NIST SP800-53r5]. An organization should employ least privilege for specific duties and information systems. The principle of least privilege is also applied to information system processes, ensuring that the processes operate at privilege levels no higher than necessary to accomplish required organizational missions/business functions.

An organization explicitly authorizes access to organization defined security functions (deployed in hardware, software, and firmware) and security relevant information [NIST SP800-53r5]. Security functions include, for example, establishing system accounts, configuring access authorizations (e.g., permissions, privileges), setting events to be audited, and setting intrusion detection parameters. Security relevant information includes, for example, filtering rules for routers/firewalls, cryptographic key management information, configuration parameters for security services, and access control lists. Explicitly authorized personnel include, for example, security administrators, system and network administrators, system security officers, system maintenance personnel, system programmers, and other privileged users.

An organization should restrict privileged accounts on the information system to organization defined personnel or roles [NIST SP800-53r5]. Privileged accounts, including super user accounts, are typically described as system administrator for various types of commercial off the shelf operating systems. Restricting privileged accounts to specific personnel or roles prevents users from having access to privileged information or functions.

An organization should ensure that users of information system accounts, or roles, with access to security functions or security relevant information, use non-privileged accounts or roles, when accessing non-security functions [NIST SP800-53r5]. This control enhancement limits exposure when operating from within privileged accounts or roles.
The need for certain assigned user privileges may change over time reflecting changes in organizational missions/business function, environments of operation, technologies, or threat. Periodic review of assigned user privileges is necessary to determine if the rationale for assigning such privileges remains valid. If the need cannot be revalidated, organizations take appropriate corrective actions.

An organization should prohibit privileged access to the Smart Grid information system by non-organizational users. The organization should [NIST SP800-53r5]:

- Review periodically the privileges assigned to defined roles or classes of users to validate the need for such privileges.
- Reassign or remove privileges, if necessary, to correctly reflect organizational mission/business needs.

An information system should prevent non-privileged users from executing privileged functions including disabling, circumventing, or altering implemented security countermeasures [NIST SP800-53r5]. Privileged functions include, for example, establishing information system accounts, performing system integrity checks, or administering cryptographic key management activities. Non-privileged users are individuals that do not possess appropriate authorizations. Circumventing intrusion detection and prevention mechanisms or malicious code protection mechanisms are examples of privileged functions that require protection from non-privileged users.

10.6.4 Smart Grid Separation of Duties

Roles and responsibilities are assigned during system lifecycle. As recommended in [NISTIR 7628r1], an organization should:

- Establish and document divisions of responsibility and separate functions as needed to eliminate conflicts of interest and to ensure independence in the responsibilities and functions of individuals/roles.
- Enforce separation of information system functions through assigned access authorization.
- Restrict security functions to the least number of users necessary to ensure the security of the information system.
- Assign auditor and information system administration functions to separate personnel.

The information system separates user functionality (including user interface services) from management functionality. The intent of this additional consideration is to ensure that administration options are not available to general users. For example, administration options are not presented until the user has appropriately established a session with administrator privileges.

Smart Grid system management functionality includes, for example, functions necessary to administer databases, network components, workstations, or servers, and typically requires privileged user access. The separation of user functionality from information system management functionality is either physical or logical. The information system should prevent the presentation of information system management related functionality at an interface for general (e.g., nonprivileged) users.

Regarding the least privilege role, the Smart Grid guidelines [NISTIR 7628r1] recommend that an organization should:

- Assign the most restrictive set of rights and privileges or access needed by users for the performance of specified tasks.
- Configure the information system to enforce the most restrictive set of rights and privileges or access needed by users.

An organization should authorize network access to organization defined privileged commands only for compelling operational needs and documents the rationale for such access in the security plan for the information system. Also, an organization should authorize access to organization defined list of security functions (deployed in hardware, software, and firmware) and security relevant information.

10.6.4.1 Control Systems

The concept of separation of duties ensures that one person acting alone cannot compromise the critical infrastructure or the cybersecurity via its control system.

Each control system operator's role needs to have a complete and well defined job description, and cybersecurity personnel should use these job descriptions when assigning control system access rights and permissions for specific control system operations.

In addition, it needs to be recognized that in some cases the concept of separation of duties is simply not feasible, as the operations environment is administered remotely by a managed services entity or a peer site. Yet, in these cases, due diligence should dictate that the operators have access to only the control network resources required to carry out their tasks.

For instance, to help define OPSEC parameters and instill cultural component, a control system configuration specialist/engineer should not be the only person to test their own configuration and any associated programmed code. Another person with a different job and agenda, who has familiarity with the control systems and the assets the code is automating, should perform functionality and integrity testing on the original engineer's configuration. This should be done because the original configuration programmer may have a biased and narrow view of what the control program is to accomplish, and he/she may only test functions and input values for certain situations.

This whitepaper [ISAGCA 2020] defines roles and responsibilities for each phase of the automation solution lifecycle. It is recommended a periodic assessment and review of the organizational security controls and associated security controls during operation phase. Operation phase of the automation solution lifecycle is when the automation solution is placed into service and all of the organizational and technical security measures (security controls) are executed. So, regarding roles and responsibilities, asset owner is accountable and responsible for operations. Examples of key activities include [ISAGCA 2020]:

- Operate the automated solution and the equipment under control.
- Perform organizational security measures for operations, such as incident response and recovery.
- Periodically reassess the organizational and technical security measures.
- Trigger maintenance requests.

10.6.4.2 Use of Personal Devices

As a key component of any security program, policies and procedures should cover all aspects of the organization's operations. It is highly likely that various aspects of the organization's environment were not considered or possibly not in existence when the original policies or procedures were created, so it is critical to create or modify existing policy regarding the use of personal devices.

Also, an organization should define policies for the use of mobile devices performing either personal applications or performing enterprise tasks. If an employee is working from home and if a personal device is breached, both employee's personal data and employer's data is at risk. The same applies when an employee or customer connects with a utility via own personal device, then once that device is breached, hackers can do the same thing to home network and utility's network. Therefore, security on home devices is of such paramount importance. Besides strong policies, adequate security controls have to be considered (e.g., updated software, strong authentication mechanisms, cryptography). One important measure is to separate Wi-Fi network for the smart home devices. This helps to isolate devices such as smart speakers, thermostats and other appliances from personal computers and mobile devices, which are much more likely to access sensitive enterprise data. These policies should be defined and made known to the users.

Today, technologies have merged to enable Wireless Sensor Networks (WSNs) on providing multifunctional miniature devices with interfaces to connect with a multitude of sensors. These devices can gather sensed information from the environment and communicate it over a short distance that can be further routed by means of multi-hop to a central monitoring station. Applications of WSN have now found their way into the organization, home and environment enabling more user control and convenience.

This work [Ali 2015] argues that more recently, a changing trend has been to benefit from the physical and virtual environment synergy provided by Cyber Physical Systems (CPSs) to perform the conventional activities of WSN more reliably and with ease. WSN enabled with (CPS) can provide remote control over the network devices with scattered network elements instead of a standalone system. By definition, CPS is meant to provide a virtual environment that incorporates an interacting network of system elements with physical inputs and outputs at both ends (see Figure 10.4).

As discussed in [Ali 2015], CPS engineering promises to impact system condition monitoring for a diverse range of fields from healthcare, manufacturing, and transportation to aerospace and warfare. A recent trend is to gain assistance from mergers between virtual networking and physical actuation to reliably perform all conventional and complex sensing and communication tasks. For example, utilities use these technologies to monitor the heating and air conditioning units from the buildings. Oil and gas pipeline monitoring is providing a reliable remote monitoring platform to leverage environment, strategic and economic benefits.

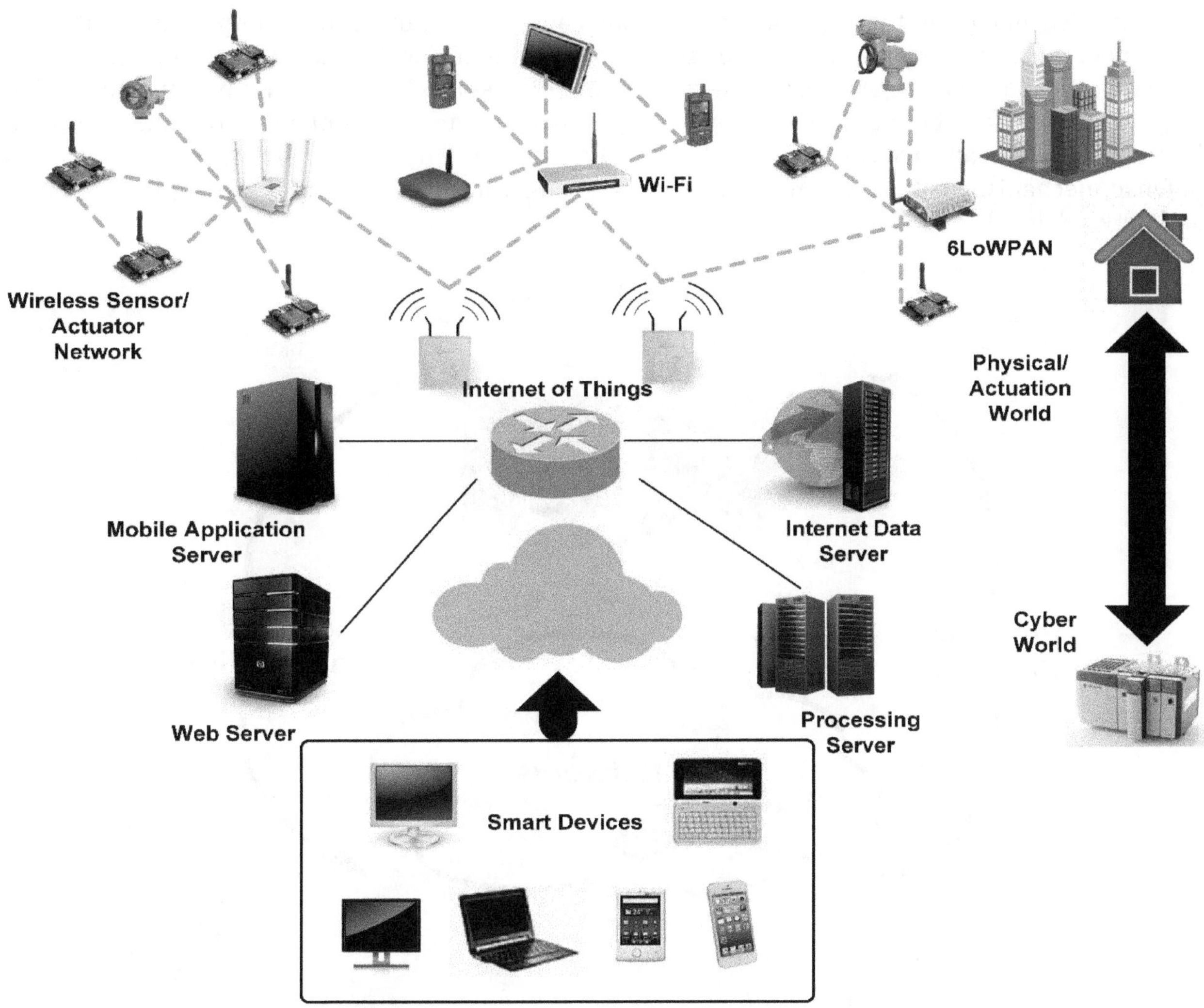

Figure 10.4 Network elements in CPS architecture. *Source*: [Ali 2015]. Licensed under CC BY 4.0.

In addition, the synergy of computational and physical network components leading to the Internet of Things, Data and Services has been made feasible by the use of CPSs. Contrary to traditional embedded systems, CPS provides an interconnected interaction with outputs and inputs that pertain to physical existence and are standalone devices as shown in Figure 10.5.

Also, this Figure 10.5 depicts CPS applications with interconnecting boundary between cyber and object domain. The main layers of CPS are the virtual layer and physical layer. For the physical layer an intelligently deployed network of actuators and sensors collects information and actually controls the physical world. The information is sent to a virtual layer input which serves as the decision making setup. This information is further used to calculate abstract computations that feed into the real world actuation system to drive and control physical world outputs or objects.

The authors [Ali 2015] have reviewed the applications of a Cyber Physical Sensor Network (CPSN) environment from a reliability perspective and demonstrate how the physical information collected from different sensing planes be exploited to abridge the cyber space and real world. Also, they identified the challenges and architecture design issues of CPSN.

The major challenges of these types of applications and interactions (one example - user interacting via personal devices) are security related problems due to the decentralized nature and heterogeneity of protocol and hardware used to interconnect devices. There should be considered that physical interactions between devices over an untrusted environment open up potential vulnerabilities. Therefore, all such concerns need to be addressed with detailed analysis, assessment and a defense mechanism at the personal device level as suggested earlier and at the application, CPS, and network level.

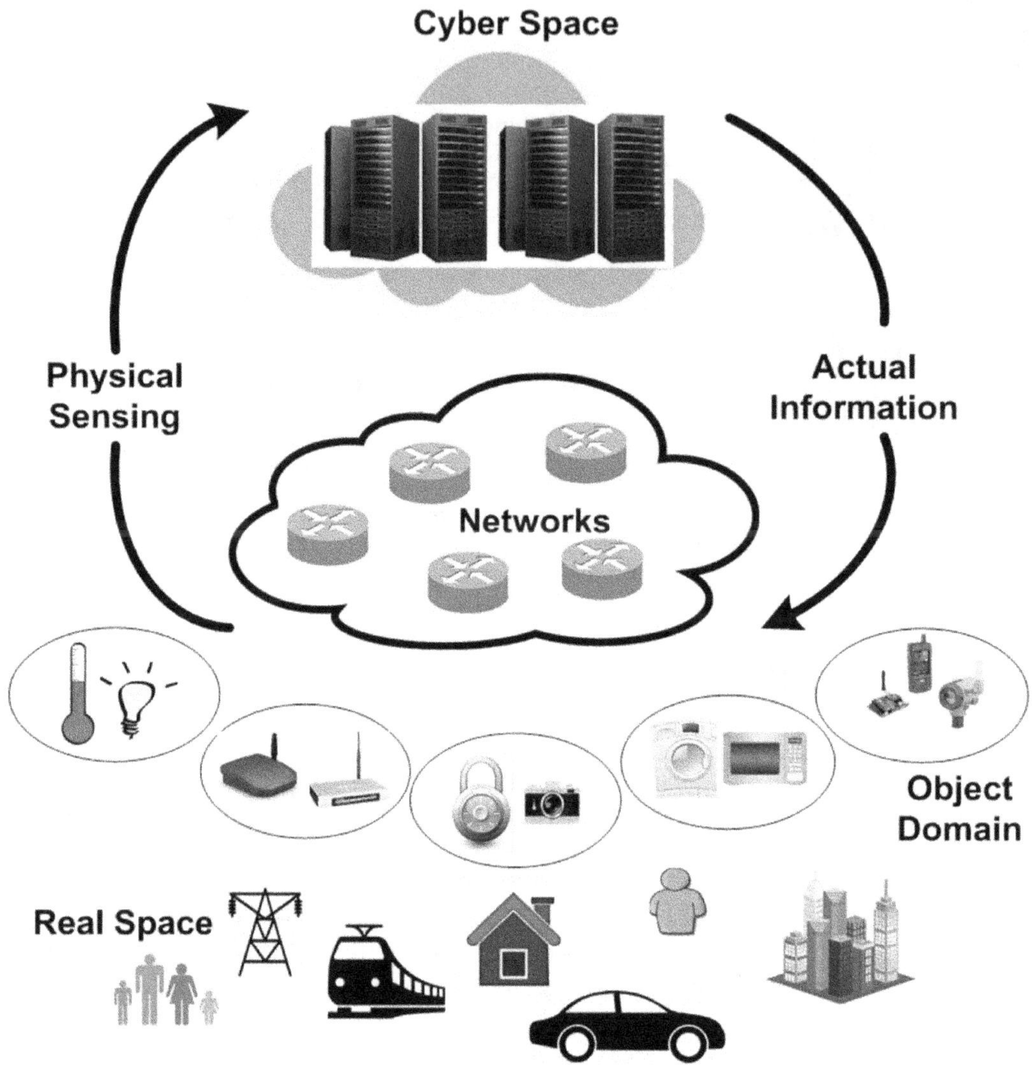

Figure 10.5 Cyber Physical System applications. *Source:* [Ali 2015]. Licensed under CC BY 4.0.

10.6.4.3 Training

A comprehensive program for education, training, practical experience, and awareness, is necessary for skilled staff. Professionalization and certification licensing provide a validated and recognized expert cadre of system administrators.

Therefore, an organization needs to determine the content of security training based on assigned roles and responsibilities and the specific requirements of the organization and the Smart Grid information system to which personnel have authorized access. Also, information system managers, Smart Grid information system and network administrators, and other personnel having access to Smart Grid information system level software should be provided with security related training to perform their assigned duties.

Also, training should include topics recommended for both IT and ICS systems. With more than 40 standards organizations worldwide working on guides that either directly or indirectly impact the security of control systems [INL 2007], many organizations are involved with the security of IT systems or converged systems (IT and ICS). For example, The Information Technology Laboratory (ITL) responsibilities include the development of management, administrative, technical, and physical standards and guidelines for the cost-effective security and privacy of other than national security related information in federal information systems.

An organization should maintain an inventory of personally identifiable information (PII) [NIST SP800-53r5]. This requires an organization to establish, maintain, and update regularly an inventory that contains a listing of all programs and information systems identified as collecting, using, maintaining, or sharing personally identifiable information (PII). The PII inventory enables organizations to implement effective administrative, technical, and physical security policies and procedures to protect PII and to mitigate risks of PII exposure.

10.6.4.4 System and Media Protection Controls

Security measures should be provided during distribution and handling as well as storage, transport, sanitization (removal of information from digital media), destruction, and disposal of the media. Sanitization is the process of removing information from media such that data recovery is not possible. For example, any media on which unencrypted keying material requiring confidentiality protection is stored shall be erased in a manner that removed all traces of the keying material so that it cannot be recovered by either physical or electronic means.

Media assets used in Smart Grid systems include compact discs; digital video discs; erasable, programmable read only memory; tapes; printed reports; and documents [NISTIR 7628r1]. This document [NISTIR 7628r1] includes security requirements for the policy and procedures for limiting access to media to authorized users. An organization is recommended to:

- Develop, implement, review, and update on an organization defined frequency a documented media protection security policy that addresses the objectives, roles, and responsibilities for the media protection security program as it relates to protecting the organization's personnel and assets.
- Physically manage and store Smart Grid information system media within protected areas; the sensitivity of the material determines how the media are stored.
- Check all media containing diagnostic and test programs for malicious code before the media are used in the Smart Grid information system.
- Sanitize Smart Grid information system media before disposal or release for reuse.
- Ensure that the media protection security policy and procedures comply with applicable federal, state, local, tribal, and territorial laws and regulations.
- Establish usage restrictions and implementation guidance for organization controlled mobile devices, including the use of writeable, removable media and personally owned removable media.
- Control the use of writable, removable media in Smart Grid information systems.
- Controls the use of personally owned, removable media in Smart Grid information systems.

- Impose restrictions on authorized individuals with regard the use of organization controlled removable media on external information systems.
- Log access, events, use, and activities related to media.
- Maintain accountability records for media use.

Controlling mobile computing includes remote location activities and media installations. The Smart Grid information system should produce audit records on hardware-enforced write-once media.

10.7 Information System Maintenance

An information system is a discrete set of information resources organized expressly for the collection, processing, maintenance, use, sharing, dissemination, or disposition of information. Information systems also include specialized systems such as industrial/process controls systems, telephone switching/private branch exchange (PBX) systems, and environmental control systems [NIST SP800-53r5]. An information system includes components such as hardware, software, and firmware.

10.7.1 Smart Grid Information System

The Smart Grid information system should isolate security functions from non-security functions [NISTIR 7628r1]. Security functions are the hardware, software, and/or firmware of the information system responsible for enforcing the system security policy and supporting the isolation of code and data on which the protection is based. The information system should isolate security functions (e.g., functions enforcing access and information flow control) from both non-security functions and from other security functions. The information system can employ underlying hardware separation mechanisms to facilitate security function isolation.
When discussing cryptographic modules for Smart Grid systems, the term firmware is commonly used to denote the fixed, small, programs that internally control a module. Such modules are commonly designed to include a range of physical security protections and levels of cyber protection [NISTIR 7628r1], [NIST SP800-53r5].

Software and firmware are the programmable components of a computing environment. Software often contains bugs and vulnerabilities. For example, a lack of prompt security patches from software vendors needs to be considered during acquisition and evaluation process of new products. When a vulnerability is disclosed, there is often a race between adversaries and system administrators to either exploit or close the vulnerability. The security of the system using the software depends on vendors' ability to provide patches in a timely manner, and on administrators' ability to implement those patches. As zero day exploits become more widespread, administrators may be faced with the choice of taking a system offline or leaving it vulnerable.

An organization should establish agreements for the exchange of information, firmware, and software between the organization and external parties such as third parties, vendors and contractors [NISTIR 7628r1]. While this document describes cybersecurity requirements for Smart Grid, there is no mandatory compliance for an organization. Other documents such as [NIST SP800-53r5] and [INL 20007] are providing additional security requirements and guidance for implementing and operating an information system. As we focus more on specific issues for Smart Grid systems, we discuss key maintenance areas that an organization should ensure the protection of the information system as well as the efficiency and effectiveness of the security controls.

An organization should establish terms and conditions for installing any hardware, firmware, or software on Smart Grid information system devices [NISTIR 7628r1]. As recommended in this document, an organization should assign the most restrictive set of rights and privileges or access needed by users for the performance of specified tasks. Security personnel should configure the information system to enforce the most restrictive set of rights and privileges or access needed by users. An organization should authorize network access to organization defined privileged commands only for compelling operational needs and documents the rationale for such access in the security plan for the Smart Grid information system. The organization should authorize access to organization defined list of security functions (deployed in hardware, software, and firmware) and security relevant information.

In addition, maintenance personnel are required to understand several issues and some architecture solutions are specific to control systems. These include handling errors and failures, addressing single points of failure, redundancy, fault tolerance and clustering, access authorization, user actions, etc. The best defenses against being vulnerable to these single points of failure are proper maintenance, regular backups, and redundancy [INL 2007].

Errors or oversights in software and firmware design, development, and deployment may result in unintended functionality that allows adversaries or other conditions to affect, via programmatic means, the confidentiality, integrity, and/or availability of information; these errors and oversights are discovered and reported as vulnerability instances in platform software and firmware; while taxonomies such as Common Vulnerability and Exposures (CVE), Common Weakness Enumeration (CWE), and The Open Web Application Security Project (OWASP) provide descriptions of common errors or oversights that can result in vulnerability instances, the impact of these vulnerabilities may differ in a Smart Grid environment compared to a traditional IT.

The convergence of the control technology with open networking and computing services often results in accelerated business capabilities and unique demands set by the management resulting in single points of failure. These single points of failure often pose high potential risk to a control network. If a device fails, a large segment or even the entire industrial network is negatively affected. Devices that could represent single points of failure are firewalls, routers, network access servers, T1 lines, switches, bridges, hubs, various control system computers, and authentication servers. Therefore, addressing single points of failure in ICS networks is a major objective.

Redundancy is multiple instances of the same software, firmware, devices, and/or data configured in an active/passive or load sharing mode. Redundancy for data and logs needs to be consistent with the organization's data retention plan and continuity of operations plan [NISTIR 7628r1]. However, these solutions should be built in the architecture of the IT and control systems before a system becomes operational. When control system hardware, software, and cybersecurity products are evaluated for specific levels of assurance they provide, many times operational assurance and life cycle assurance are part of the evaluation process.

For example, operational assurance concentrates on a specific ICS network product's architecture, embedded features, and functionality, which enable an industry customer to continually obtain the necessary level of protection when using the product. Examples of operational assurances examined in the evaluation process are access control mechanisms, the separation of privileged and user program code, auditing and monitoring capabilities, covert channel analysis, and trusted recovery when and if the product experiences unexpected circumstances.

Therefore, an organization should require Smart Grid information system developers/integrators to provide an integrity check of delivered software and firmware. Also, the information system should protect audit information and audit tools from unauthorized access, modification, and deletion.

Currently, several initiatives are underway to guide the owner/operator community in the strategic procurement of security sensitive control systems to reduce software supply chain risks [CIS], [NIST 2018]. The supply chain represents a significant risk to ICS systems. ICS manufacturers and software developers create their products in many different locations around the world. Purchasing commercial off-the-shelf (COTS) technologies increases the likelihood of receiving non-genuine equipment. ICS-CERT published several reports_related to equipment with embedded unauthorized code in its firmware or operating system that provides a back door into the equipment or allows the program to "call home" once installed. To help mitigate these threats, asset owners must consider processes to procurement contract arrangements, quality control, and validation of performance to hardware and software specifications. In addition, testing, including vulnerability scanning are necessary to be performed before installing systems in production environments.

10.7.2 Control Systems Updates

One area that deserves special attention within management is related to the maintenance of control systems. Upgrading field technology with current versions of software or firmware is a change management challenge. However, maintaining accurate records of how upgrades are administered is a tremendous task, especially if the number of devices requiring upgrades is significant. One of the key issues in maintaining an effective security posture in control system domains is ensuring that key devices are operating with the most recent and up to date software or firmware.

Missing patches on firmware and software have the potential to present serious risk to the affected system without additional mitigations. Modern control system domains have redundancy built in, with back up networks and key resources mirrored to accommodate for any catastrophic failure. Key to operations and cybersecurity is ensuring that these secondary systems are fully compliant with current configurations and, if needed, can become operational with the exact same configuration and system upgrades as the primary system. This way, if an event requiring a switch over to the redundant system is required, operation can and will be maintained as if the main system was still online.

Organizations need to ensure key secondary systems are upgraded and configured to the same cybersecurity standards. There have been instances where secondary systems have been vulnerable [INL 2007].

10.7.3 Data Removal and Disposal

An organization should consider the disposal, destruction, or loss of data. Some of the factors related to these aspects include the sensitivity of information, the quantity of information, the timing, the medium which the information, and the legal and regulatory requirements for its protection [Hansche 2004]. For example, the secure disposal of all key material after its intended use and lifetime may be done by erasing the information or making key material unavailable as an acceptable alternative for systems where destruction is not possible. One concern is the susceptibility of data to unauthorized access by dumpster diving.

The specific issues revolve around the data whether is government classified or sensitive data in the private or public sector. Another specific issue related to storage media is its reassignment, either within the computer or outside on removable media. Reassignment (object reuse) is the reuse of the media for another group of users. Removal of the information is needed to ensure that any sensitive information is not available to unauthorized users. As recommended in [NISTIR 7628r1], an organization should:

- Implement policy and procedures to address the addition, removal, and disposal of all Smart Grid information system equipment.
- Develop, document, and maintain an inventory of the components of the Smart Grid information system that updates the inventory of system components as an integral part of component installations, system updates, and removals.
- Sanitize information system media before disposal or release for reuse.
- Test sanitization equipment and procedures to verify correct performance on an organization defined frequency.

Sanitization is the process of removing information from media such that data recovery is not possible. An organization should sanitize Smart Grid media before disposal or release for reuse. The organization should test sanitization equipment and procedures to verify correct performance on an organization defined frequency. The organization should track, document, and verify media sanitization and disposal actions.

Also, this document [NIST 7628r1] recommends that third parties should have clear policies and practices on how long data will be retained, as well as when and how information will be disposed of media. Specific care should be detailed in the privacy notice given to the customer information. The policies and procedures should consider the sensitivity of critical security parameters such as passwords, cryptographic keys, and personally identifiable information such as name and social security numbers.

Maintenance procedures that require the physical removal of any Smart Grid information system component needs to be documented, listing the date, time, reason for removal, estimated date of reinstallation, and name personnel removing components. One example is meter removal. Detection of meter removal can occur in different ways including meter sends information where the exception case indicates no contact with the meter or meter sends an alarm where the self-protection capability of the meter notes a tamper event. There is also the case that includes meter not communicating (disassociated from network), where a meter that has been associated or registered on the network is no longer performing necessary activities to maintain registration.

There are potential unintended consequences of seemingly anonymous Smart Grid data being compiled, stored, and cross linked. While current privacy and security anonymization practices tend to focus on the removal of specific personal information data items, the studies show that reidentification and linking to an individual may still occur. This issue of data reidentification becomes potentially more significant as the amount and granularity of the data being gathered during Smart Grid operations increases with the deployment of more Smart Grid components. It then becomes important, from a privacy standpoint, for utilities and Third Parties participating in the Smart Grid to determine which data items will remove the ability to link to specific addresses or individuals whenever they perform their data anonymization activities.

10.7.4 Tracking Changes

A disciplined and structured approach to managing, controlling, and documenting changes to an information system or its environment of operation is an essential element of an effective security control monitoring program [NIST SP800-37r2]. Strict configuration management and control processes should be established by an organization to support such monitoring activities. It is important to record any relevant information about specific changes to hardware, software, or firmware such as version or release numbers, descriptions of new or modified features/capabilities, and security implementation guidance. It is also important to record any changes to the environment of operation for the information system (e.g., modifications to hosting networks and facilities, mission/business use of the system, threats), or changes to the organizational risk management strategy. The information system owner and common control provider can use this information in assessing the potential security impact of the changes.

Documenting proposed or actual changes to an information system or its environment of operation and subsequently assessing the potential impact those changes may have on the security state of the system or the organization is an important aspect of security control monitoring and maintaining the security authorization over time.

Information system changes are generally not undertaken prior to assessing the security impact of such changes. An organization should maximize the use of automation when managing changes to the information system or its environment of operation. Security impact analysis conducted by the organization, determines the extent to which proposed or actual changes to the information system or its environment of operation can affect or have affected the security state of the system. Changes to the information system or its environment of operation may affect the security controls currently in place (including system specific, hybrid, and common controls), produce new vulnerabilities in the system, or generate requirements for new security controls that were not needed previously. If the results of the security impact analysis indicate that the proposed or actual changes can affect or have affected the security state of the system, corrective actions should be initiated. Also, the appropriate documents should be revised and updated (e.g., the security plan, security assessment report, and plan of action and milestones).

10.7.5 Changes to Control Systems - Recommendations

Changes to control systems and industrial networks are critical to successful OPSEC programs, especially in large dynamic environments. Regulated industries such energy have strict guidelines regarding what specifically can be done to their control systems and industrial networks and at exactly what time and under which conditions.

Changes to control software configurations and network devices can take place in automation environments. Keeping these details properly organized is essential. Revision control, or the function that ensures the proper version of the firmware or software is installed, is a core component in ensuring effective, robust, and secure operations.

A well structured change management process should be put into place to aid administrators and change control policy should be implemented for control systems [DHS 2016], [INL 2007]. A change control policy impacts network communications, security, or information systems (see examples of elements to be included in [DHS 2016], [INL 2007]). A full report summarizing the control system change should be submitted to management. This report can be submitted on a periodic basis to keep management up to date and ensure their continual support. The types of changes are usually expensive and can have lasting effects on critical information infrastructures or control systems.

Without strict controls and guidelines on changes, vulnerabilities can inadvertently be introduced into an industrial control systems network. Control system users can make changes that others do not know about and may not have been approved. Although it is uncommon for changes at the industrial level to go undocumented (such as upgrading field devices or swapping in a new human machine interface), last minute changes to the cyber infrastructure may be done out of necessity and can go undocumented and have negative effects. Additionally, without change control, auditing the changes after implementation is a very complicated task.

Organizations operating industrial systems may have a change control policy as it relates to the maintenance phase of the system life cycle, but they may not apply the change management practices to the actual automated processes [NIST SP800-82r2], [DHS 2016], [INL 2007]. This policy is usually applicable to the changing of major system components such as communications systems, operator interfaces, diagnostic tools, and others that relate to overall system reliability. As systems become connected, extensions to traditional change control processes are required to accommodate for the required protection of the information assets that are networked or are critical to safety and process control. Although the types of changes vary, a standard list of procedures can help keep the process manageable and ensure that it is carried out in a predictable and repeatable manner.

However, smaller changes should also go through some type of change control process. It is also critical that the operations department create approved back out plans before implementing changes to systems or the network. Components of the change control that are recommended in several documents include:

- How changes take place within a facility.
- Who can make the changes within a control system network.
- How the changes are approved.
- How the changes are documented and communicated to other control system users and other employees.
- How the changes are backed up to support system restoration.

Wherever possible, organizations should create working groups and outreach teams to ensure effective and timely reporting of changes to the information infrastructure. Requests should be presented to an individual or group that is responsible for approving control system changes and overseeing the activities of changes that take place within a control system environment.
Numerous changes can take place in a control systems environment, examples include [NISTIR 7628r1]:

- New control hardware, end device, server, computer, or any hardware additions to the control or SCADA LANs are installed, including sensors and control devices.
- New applications and operating system platforms are installed.
- Different configurations are implemented.
- Patches and updates are installed.
- New technologies and processes are integrated.
- Policies, procedures, and standards are updated.
- New regulations and requirements are implemented.
- New networking devices, such as temporary or new remote control LAN access devices are integrated into the network wire or wireless system.
- Additional internal and external data connections are added for user access of components, applications and data.

Therefore, planned or unplanned changes to the hardware, software, and/or firmware components of the Smart Grid information system may affect the overall security of the Smart Grid information system. As recommended in [ICS-CERT 2014] and [NISTIR 7628r1], an organization should:

- Define, document, and approve individual access privileges and enforce access restrictions associated with configuration changes to the Smart Grid information system.

- Generate, retain, and review records reflecting all such changes.

Only authorized individuals should be allowed to obtain access to Smart Grid information system components for purposes of initiating changes, including upgrades, and modifications. For example, for the AMI system, some of the security requirements are authentication of the meter to the collector, confidentiality for privacy protection, and integrity for firmware updates. The organization should conduct audits of Smart Grid system changes at an organization defined frequency and if/when suspected unauthorized changes have occurred.

Also, authentication and authorization mechanisms for IoT and IIoT devices should be planned accordingly. An organization should have an IoT authorization and authentication strategy that begins with understanding how IoT devices are used and how devices communicate with their network. It is necessary to ensure that devices only have access and permission to do exactly what they need. Only authorized devices can interact with other devices, applications, cloud accounts and gateways. The mechanisms could be centralized or globally distributed using an open source infrastructure with local authentication and authorization entities deployed on the edge devices.

Access control and physical access for industrial control systems must also include consideration for the installation and management of system centric media (e.g., software/firmware upgrades) in remote facilities [ICS-CERT 2014, [DHS 2016]. In many system architectures, the locations where control equipment require servicing are often geographically disperse and may not have appropriate physical security. However, these locations are often accessed by workers, so upgrades or modifications to the resident technology can be performed. This concern for loading and unloading of media into the systems may justify the cross correlation of cyber activity with physical access, as in many cases the remote facilities are effective connection points into the master control network. As such, logging activities pertaining to physical access may be cross referenced with cyber specific activities to provide a more robust perspective on overall system access.

Maintaining records is important for supporting after the fact actions should the organization become aware of an unauthorized change to a Smart Grid system. Also, all information system components and information should be documented, identified, and tracked so that their location and function are known. An organization may consider these recommendations [NISTIR 7628r1]:

- Establish configuration settings for components within an information system.
- Monitors and control changes to the configuration settings in accordance with organizational policies and procedures.
- Document changed configuration settings.
- Identify, document, and approve exceptions from the configuration settings.
- Enforce the configuration settings in all components of the information system.
- Employ automated mechanisms to centrally manage, apply, and verify configuration settings.
- Employ automated mechanisms to respond to unauthorized changes to configuration settings.
- Incorporate detection of unauthorized, security relevant configuration changes into the organization's incident response capability to ensure that such detected events are tracked, monitored, corrected, and available for historical purposes.

In another scenario, an information system can include components specifically designed to be the target of malicious attacks for detecting, deflecting, analyzing, and tracking such attacks. There are now global positioning system (GPS) devices that track fuel use as it relates to driving behavior.

Law enforcement and investigators have been tracking vehicle activities through the use of GPS for several years to help with cases and solving crimes. Charging plug-in electric vehicles (PEVs) raises the possibility of tracking physical location through new energy consumption data. If these types of monitoring tools are expanded to tracking (PEVs), and then connected to other networks that are part of the Smart Grid, the related privacy issues need to be addressed.

Similarly, if any other types of mobile energy applications using appliances or other devices are connected to a home area network (HAN) or other Smart Grid components, the impact of combining the GPS and related locational data with the energy usage data should be assessed for new privacy risks. The second category includes instances where personal information is available from other sources, and the Smart Grid may present a new source for that same information. For example, an individual's physical location can be tracked through their credit card and cell phone records today. Other specific recommendations for an organization include [NISTIR 7628r1]:

- Track, document, and verify media sanitization and disposal actions.
- Develop, document, and maintain an inventory of the components of the Smart Grid information system that provides the proper level of granularity deemed necessary for tracking and reporting and for effective property accountability.
- Employ automated mechanisms to assist in the tracking of security incidents and in the collection and analysis of incident information.

10.7.6 Update Documentation

Organization owned documentation includes Smart Grid information system administration manuals, field device configurations and operational information, information system network documentation, keys, identification cards, building passes, computers, cell phones, and personal data assistants, documentation of all anomalies for later analysis and risk assessment, etc.

Smart Grid information system recovery and reconstitution to a known secure state means that information system documentation and operating procedures are needed for recovery and reconstitution of systems after failures. Absent or deficient equipment implementation guidelines description and unclear implementation guidelines can lead to unexpected behavior.

A system needs to be configured correctly to provide the desired security properties. This applies to both hardware and software configuration. Different inputs and outputs, both logical and physical, could have different security properties, and an interface that is intended for internal use may be more vulnerable than an interface designed for external use.

Guidelines for installers, operators, and managers should be clear about the security properties expected of the system and how the system is to be implemented and configured to obtain those properties. Regarding information system documentation, an organization should obtain administrator documentation for the information system, system component, or information system service that describes [NIST SP800-53r5]:

- Secure configuration, installation, and operation of the system, component, or service.
- Effective use and maintenance of security functions/mechanisms.
- Known vulnerabilities regarding configuration and use of administrative (e.g., privileged) functions.

This document [NISTIR 7628r1] recommends that an organization should:

- Ensure information system documentation to include how to configure, install, and use the system and its security features.

- Obtain from the contractor/third party information describing the functional properties of the security controls employed within the information system.
- Use software and associated documentation in accordance with contract agreements and copyright laws.
- Control the use of software and associated documentation protected by quantity licenses and copyrighted material.

A continuity of operations plan addresses both business continuity planning and recovery of information system operations. Development of a continuity of operations plan requires documentation of critical Smart Grid information system functions that need to be recovered.

10.7.7 Maintenance Records

All maintenance activities to include routine, scheduled maintenance and repairs, and unplanned maintenance are controlled, whether performed on site or remotely and whether the equipment is serviced on site or removed to another location.

An organization should employ automated mechanisms to schedule and document maintenance and repairs as required, producing up to date, accurate, complete, and available records of all maintenance and repair actions needed, in process, and completed for Smart Grid. It is recommended that an organization should maintain maintenance records for the information system that include [NISTIR 7628r1]:

- Date and time of maintenance.
- Name of the individual performing the maintenance.
- Name of escort, if necessary.
- Description of the maintenance performed.
- List of equipment removed or replaced (including identification numbers, if applicable).

Maintenance procedures that require the physical removal of any Smart Grid information system component should be documented, listing the date, time, reason for removal, estimated date of reinstallation, and name personnel removing components.

Similarly, an organization should include maintenance records for remote maintenance and diagnostic activities. Also, an organization should employ appropriate measures to ensure that long term records and information can be retrieved (e.g., converting the data to a newer format, retaining older equipment that can read the data).

It is also required that remote maintenance or diagnostic services be performed from an information system that implements a level of security at least as high as that implemented on the Smart Grid information system being serviced. Also, an organization should remove the component to be serviced from the Smart Grid information system and prior to remote maintenance or diagnostic services, sanitize the component (regarding organizational information) before removal from organizational facilities and after the service is performed, sanitize the component (regarding potentially malicious software) before returning the component to the Smart Grid information system.

10.8 Standards, Guidelines, and Recommendations

An analysis of various frameworks and Smart Grid roadmaps on standards is provided in [Specht 2013]. The authors argue that many published standards were focused on different parts of Smart Grid realizations and were mainly devised independently from each other.

10.8.1 Common Standardization Issues

Commonly, problems with standards include complexity (e.g., backward compatibility, optimization for various cases, barriers for evaluation, market entry, etc.) and maintenance (e.g., context changes, new technical vulnerabilities, fast changes incompatible with slow consensus procedures) that make security implementation very difficult. An immediate consequence of the diversity of the current standardization ecosystem, and because of the extremely rapid pace of change, is that it is increasingly difficult to authoritatively determine if gaps in standardization or in capability exist.

Standards evolve in time and keeping current with the changes is crucial for maintaining the interoperability and capabilities of the systems. For example, many NIST documents are being revised or withdrawn. The generally accepted principles and practices for securing information technology systems can be found in these documents: [NIST SP800-12r1], [NIST SP800-53r5], [NIST SP800-30r1], [NIST SP800-37r2], [NIST SP160v1], and [NIST SP800-39]. Similarly, the ISO and IEC standards are periodically reviewed.

10.8.2 Supporting Guidance and Best Practice Recommendations

Supporting guidance can be discussed from many perspectives. One example is by country, location, or organization.

10.8.2.1 U.S. Guidance

In U.S., several organizations are dedicated on providing guidance and best practice recommendations on cybersecurity. Examples include:

- US Department of Energy Cybersecurity. https://www.energy.gov/national-security-safety/cybersecurity

- Cybersecurity & Infrastructure Security Agency (CISA), former United States Computer Emergency Readiness Team (US-CERT). https://www.cisa.gov/uscert

- National Cybersecurity and Communications Integration Center (NCCIC). https://www.dhs.gov/national-cybersecurity-and-communications-integration-center

- Cybersecurity & Infrastructure Security Agency (CISA) Industrial Control Systems, former Cyber Emergency Response Team (ICS-CERT). https://www.cisa.gov/uscert/ics

- US Department of Defense Cyber Crime Center (DC3) A Federal Cyber Center. https://www.dc3.mil/

- Science and Technology Directorate Research & Development Advisor to the US Department of Homeland Security Homeland Open Security Technology. https://www.dhs.gov/topics/cybersecurity

- Information Sharing and Analysis Centers (ISACs) National Council of ISACs. https://www.nationalisacs.org/

- SysAdmin Audit Network and Security (SANS) Institute. htpps://www.sans.org/

10.8.2.2 Smart Grid Standardization Organizations

A glimpse of the listed standardization organizations that are listed in [ENISA 2017] identifies European and global bodies involved in global cybersecurity standards for Smart Grid and cybersecurity. The standards are classified in these categories:

- International (e.g,, ISO, IEC, ITU, IETF, IEEE)
- European (e.g., CEN, ETSI, ICTSB)
- US (e,g., NIST, ANSI)
- Industry initiatives (e.g., OASIS, W3C, WS-Security, ISA).

However, the list shows the global cybersecurity standards ecosystem used by the ICT industry as of November 2016. Even the report recognizes that this ecosystem is, however, so complex and rapidly evolving that it is probably incomplete. While there is a worldwide perspective on Smart Grid integration and standardization promoted by IEC TC 57 Technical Committee, there are national perspectives too. These cannot be ignored. One example is the European Smart Grid (see more information in [T&D 2016]). Also, there are other aspects such as a lack of missing or no wide adoption of standards.

10.8.2.3 Global Guidance on Cybersecurity

The following includes examples of important organizations recognized globally that provide guidance on cybersecurity:

- Forum for Incident Response and Security Teams (FIRST). https://www.first.org/
- International Telecommunications Union Cybersecurity. https://www.itu.int/en/action/cybersecurity/Pages/default.aspx
- Organisation of Economic Cooperation and Development Information Security and Privacy https://www.oecd.org/sti/ieconomy/information-security-and-privacy.htm)

10.8.3 Information Sharing Guidance

Cybersecurity Information Sharing Act (CISA) of December 2015 is proposed legislation that allows United States government agencies and non-government entities to share information with each other as they investigate cyberattacks. Sharing is voluntary for participating organizations outside the government. This document [NIST SP800-150] includes guidance on establishing information sharing capabilities. Information Sharing and Analysis Organization (ISAO) and Information Sharing and Analysis Centers (ISAC) are collaboration entities created or employed by public or private sector organizations, for gathering and analyzing critical cyber threat or cyber related information.

Chapter 11

Ethics and Codes of Ethics

11.1 Ethics Definition

In a general acceptance, according to Merriam-Webster dictionary [Ethics], ethics is the discipline dealing with what is good and bad and with moral duty and obligation. It relates thus to the question of what is good or bad in terms of human actions. The ethical actions of a person can be described in general terms as those actions which are performed within the criterium of what is regarded as good.

Ethics or moral philosophy is a branch of philosophy that involves systematizing, defending, and recommending concepts of right and wrong conduct [Philosophy]. According to [Quinn 2006], ethics is the philosophical study of morality, a rational examination into people's moral beliefs and behavior. Morality comprises the rules of conduct describing what people ought and not ought to do in various situations. When ethics are referring to the principles of conduct governing a group are called professional ethics.

Maner [Maner 1980] observed that the adoption of computers generated new ethics problems. He proposed a new field of ethics, called computer ethics, that studies ethical problems aggravated, transformed or created by computer technology. According to [Stanford 2015], Maner is recognized as the first scholar that developed an experimental computer ethics course designed primarily for students in university level computer science programs.

The society has been changing rapidly as it incorporated the latest advances in information technology (IT). People are making greater use of IT in their everyday lives. Some of these uses create new issues and rules of conduct that need to be resolved and the study of ethics is important in the information age [Quinn 2006].

This author [Schultz 2006] argues that ethics seems to be an inclusive term for concerns also referred to by morality, value, and justice. Besides character and action, ethics in this inclusive sense is also concerned with the value or goodness of things and situations and with the justness of institutions (both formal and informal). The basic idea is that ethical problems arise because they involve conflicts between different interests that cannot be resolved on the level of interests alone. Therefore, higher level principles need to be applied.

Information theory shifted the world view that sees reality organized around bits of information rather than atoms and molecules [Floridi 2010].

11.1.1 Computer and Information Ethics Perspectives

The adoption of computers raised specific moral principles called computer ethics that regulate the use of computers. Some common issues of computer ethics include intellectual property rights (such as copyrighted electronic content), privacy concerns, and how computers affect society. Computer ethics is a concept in ethics that addresses the ethical issues and constraints that arise from the use of computers, and how they can be mitigated or prevented. Computer ethics deals with the procedures, values and practices that govern the process of consuming computing technology and its related disciplines without damaging or violating the moral values and beliefs of any individual, organization or entity.

As the world of computers evolves, computer ethics continues to create ethical standards that address new issues raised by new technologies. As technology advances, computers continue to have a greater impact on society. Therefore, computer ethics promotes the discussion of how much influence computers should have in areas such as Artificial Intelligence (AI) and human communication.

According to [Spinello 1995], the purpose of ethics is to help us behave honorably and attain those basic goods that make us more fully human. The collection of case studies and theoretical frameworks included in this book are designed to provoke readers to reflect upon the social and ethical ramifications of managing information.

Computer ethics primarily enforces the ethical implementation and use of computing resources. It includes methods and procedures to avoid infringing copyrights, trademarks and the unauthorized distribution of digital content. Computer ethics also entails the behavior and approach of a human operator, workplace ethics and compliance with the ethical standards that surround computer use.

The core issues surrounding computer ethics are based on the scenarios arising from the use of the Internet, such as Internet privacy, the publication of copyrighted content and user interaction with WEB sites, software and related services.

During the past several decades, the robust and rapidly growing field of computer and information ethics has generated university courses, research professorships, research centers, conferences, workshops, professional organizations, curriculum materials, books and journals.

This work [Gotterbarn 1997] is a brief review of many views of what computer ethics comprises. As described in the review, computer ethicists have been concerned to address areas related to:

- Abuses committed with computers including fraud and theft.
- Computer's impacts on changing society.
- Ethical issues related to the development of software systems.

The authors discuss about the opinion of several works related to computer ethics. They conclude that those who study views of various computer ethicists are often misled by the simple claim that computer ethics is unique. They argue that this view is probably due to a difference in perception of computing characteristics, the adopted focus - broadly split between the outcome/symptom of computing or the undertaking of computing, and the disciplinary background. The authors suggest avoiding the continuation of these difficulties. They argue that it is necessary for those who assert this claim to clearly define their concepts using these categories [Gotterbarn 1997]:

- Social as considering the computing impact on society.
- Professional as considering the manner of professional activity.
- Universal activist as inclusive of communities around the globe.
- Parochial as restricted consideration of current and future issues.
- Ethics only as drawing only from ethics theory.
- Multi-disciplinary as a blending theory from several disciplines.

Therefore, this work suggests that perspectives are not mutually exclusive and can be analyzed using several subcategories. However, the field evolved in time and scholars approached various subjects.

Terms such as computer ethics, information ethics, cyber ethics, Internet ethics have been commonly used in different ways (e.g., [Spinello 2001]). For example, these terms could refer to professional ethics for the computer professionals to apply code of ethics and standards of good practice within their profession.

Besides technological aspects, Information and Communication Technology (ICT) involves epistemology since the main component of ICT is information which represents data, information, and knowledge. Globalization and digital convergence in the emerging knowledge society has raised complex ethical, legal and societal issues. The issue of ICT ethics takes an additional significance on some parts of the world (e.g., Asia Pacific) as presented in [Sembok 2003]. This author views ethics as moral standards that help guide behavior, actions, and choices.

11.1.2 Reviewing Computer Ethics

In another work [Spinello 2012], authors review the intriguing history of information and computer ethics. Information ethics began as a branch of applied ethics concerned with the responsible management of information resources, while computer ethics was originally concerned with the training of computer professionals. Since these two fields, Internet and Web merged together, the access and communications issues became more prominent and ICT has affected - in both good ways and bad ways - community life, family life, human relationships, education, careers, freedom, and democracy. Specifically, this work discusses computer and information ethics that is understood as that branch of applied ethics which studies and analyzes such social and ethical impacts of ICT.

Responding to the ongoing debate about this discipline's structural foundations, philosophers like Floridi have given information ethics a new meaning by focusing on information's ontological status and the intrinsic worth of all information objects [Spinello 2012].

Besides the progress of developing the field of computer and information ethics, it has been observed a tendency to split the whole into distinct subfields such as computer ethics and information ethics (see more in Preface of [Handbook 2008]. As pointed in the Preface of this book, this publication is particularly important because it aims to keep the field whole. This book describes foundational issues, theoretical perspectives, and analyses of a range of applied and practical issues.

This report [Stanford 2015] defines Computer and information ethics as that branch of applied ethics which studies and analyzes such social and ethical impacts of ICT.

A review of new technologies and analysis of impacts to include social costs, moral problems as well as legal disputes that have emerged by the expanding use of the Internet are covered by Spinello in this book [Spinello 2017]. The quick penetration of the Internet affected every society. As reported by US Internet Council, the Internet reached 30% penetration in US under 7 years. This is a very short period of time to adopt the Internet technology, compared to other technologies like TV, phone, and electricity (see Table 11.1).

11.2 Ethical Aspects

Three major areas of study within ethics recognized today are [Philosophy]:

- Metaethics, concerning the theoretical meaning and reference of moral propositions, and how their truth values (if any) can be determined.

- Normative ethics, concerning the practical means of determining a moral course of action.
- Applied ethics, concerning what a person is obligated (or permitted) to do in a specific situation or a specific domain of action.

Table 11.1 Length of time to reach 30% penetration in U.S.

Technology	Number of years to reach 30% technology penetration
Internet	7
TV	17
Phone	38
Electricity	46

Source: US Internet Council, April 1999. Public Domain.

The basic idea is that ethical problems arise because there are conflicts between different interests. The context of IT, ethical issues within organizations are seen to arise from three points of view: IT professionals, IT users, and general managers.

The lines of distinction between metaethics, normative ethics, and applied ethics are often blurry [Philosophy]. Applied ethics involves examining specific controversial issues. By using the conceptual tools of metaethics and normative ethics, discussions in applied ethics try to resolve controversial issues.

11.2.1 Information Ethics - Core Issues

For twenty-first century, the ethics of information technology is only a small part of information ethics. The core issues of information ethics include intellectual freedom, equitable access to information, information privacy, and intellectual property. Advances in information technology have made the general population more aware of these issues [Britz 1996]. The era of information age and knowledge society contribute to globalization and open communication among people. This paradigm shift brings new ethical and juridical problems which are mainly related to issues such as the right of access to information, the right of privacy which is threatened by the emphasis on the free flow of information, and the protection of the economic interest of the owners of intellectual property.

This paper [Britz 1996] is an assessment of technology impact on the private lives of people. The author discusses the privacy issues from a socio-ethical perspective with specific emphasis on the implication for the information profession. The issues include the role of technology on the processing of personal and private information, the relevance of this influence for the information profession, and proposed solutions to these ethical issues for the information profession.

The right to privacy is therefore protected under private law. Most democratic societies protect the legal right to privacy. This constitutional right is expressed in a variety of legislative forms in many countries. Although technology has a major impact on the gathering, storage, retrieval and dissemination of information its main ethical impact relates to accessibility/inaccessibility and the manipulation of information [Britz 1996]. The technology allows a wider and simultaneous access to information. Thus, it becomes easier to access a person's private information by more people. On the other hand, a person can be excluded from necessary information in electronic format by means of a variety of security measures such as passwords. In addition, the author [Britz 1996] discusses the challenges these ethical problems pose to the information professional. Based on ethical norms, the author proposes a set of practical guidelines. The author argues that:

- The ethical dilemmas do not involve new information technology to any large degree.

- The ethical dilemmas that do involve new information technology (e.g., whether to use Internet filters) are clearly special cases of much broader issues in Information Ethics (e.g., intellectual freedom).
- The Information Ethics is essentially concerned with the question of who should have access to what information.

These authors [Lee 2016a] argue that viewing privacy from the perspective of ethics can help enterprises establish and improve their code of conduct. Considering privacy from an ethical point of view and establishing a code of conduct makes all individuals in an organization, not just security personnel, accountable for protecting valuable data.

Pieters [2009] argues that technology could stimulate desirable or undesirable actions. Instead of considering actions as completely determined by either the people or the technology, technology could invite or inhibit people's actions. Technology can amplify or reduce certain aspects of people's experience. Table 11.2 shows examples of actions in the context of desirable and undesirable effects.

Table 11.2 Examples of technology actions.

Actions	Desirable	Undesirable
Amplify experiences that are	+	-
Reduce experiences that are	-	+
Invite actions that are	+	-
Inhibit actions that are	-	+

Source: [Pieters 2009]. © 2009, IEEE.

Apart from accountability in terms of programming errors that cause safety or security issues, the precautionary principle allows defining accountability for impacting people's intentions [Pieters 2009]. As discussed further, Software Engineers must add measures to their applications that invite morally sound use and inhibit undesirable or controversial actions. Also, lack of certainty about the use of software shall not be used to refrain from implementing measures in software design that invite desirable behavior, amplify desirable experiences, and reduce users' undesirable experiences.

Another aspect is the information protection. Traditionally, information protection has been directed at securing an organization at the perimeter of its systems and network, typically in the form of a firewall. The implicit assumption behind this approach is that the inside of the security perimeter is more or less trusted, whereas the outside is not. Due to changes in technologies, business processes, and their legal environments, this assumption is no longer valid and thus security becomes a concern.

Business collaboration at an unprecedented scale, any data accessed from anywhere and at any time via mobile devices, outsourcing of IT services, demand to work from home, microchips installed in smart buildings communicating constantly with each other are those that had accelerated deperimeterization process. Deperimeterization means the disappearing of boundaries between systems and organizations as well as becoming connected and fragmented in the same time. The most obvious problem of deperimeterization is how to reorganize the security [Pieters 2009]. Deperimeterization implies not only that the border of the organization's IT infrastructure becomes blurred, but also that the border of the organization's accountability fades. Deperimeterization has consequences for the moral and legal accountability of people and organizations that develop software.

Pervasive computing and ubiquitous capturing of information on such a massive scale portends a very different world, with important ethical and social implications [Schmidt 2010]. The ubiquitous access to the Internet coupled with innovations such as social networking and WEB 2.0 have given rise to the digital world phenomenon [Rashid 2009]. Also, social networking sites, virtual worlds, and massively multiplayer online games have blurred the boundary between the online world and the physical world.

The vision of Internet of Things (IoT) creates more problems. With IoT, everything real becomes virtual, which means that each person and thing has a locatable, addressable, and readable counterpart on the Internet. These virtual entities can produce and consume services and collaborate toward a common goal. Such extreme interconnection may bring unprecedented convenience and economy, but it also requires novel approaches to ensure its safe and ethical use [Roman 2011]. Therefore, this work considers several significant obstacles that remain to fulfill the IoT vision, among them security. The Internet and its users are already under continual attack, and a growing economy—replete with business models that undermine the Internet's ethical use—is fully focused on exploiting the current version's foundational weaknesses (e.g., the IoT incorporates many constrained devices). Thus, realizing the IoT vision is likely to spark novel and ingenious malicious models. The challenge is to prevent the growth of such models or at least to mitigate and limit their impact. Meeting this challenge requires understanding the characteristics of things and the technologies that empower the IoT.

However, without strong security foundations, attacks and malfunctions in the IoT could outweigh any of its benefits. In addition, privacy is one of the most sensitive subjects in any discussion of IoT protection. The data availability explosion has created entities that profile and track users without their consent. The IoT's anywhere, anything, anytime nature could easily turn such practices into a dystopia. Users would have access to an unprecedented number of personalized services, all of which would generate considerable data, and the environment itself would be able to acquire information about users automatically.
Applying ethical principles to encryption, video games, apps, the Internet of things, and other topics, we can examine many facets of Internet ethics. Technologists, engineers, students, businesspeople, and the general public can use several published resources for identifying the ethical ramifications of information technology.

Tzafestas [Tzafestas 2018] describes ethics is a system of moral principles which deals with what is good or bad for individuals and the society. It is a collection of fundamental concepts and principles on an ideal human character that enable people to make decisions regarding what is right or wrong. Ethics is a code of conduct agreed and adopted by people in a society, which sets the norms of how a person should live and interact with other people. Ethics is a must for all areas of human activity, including those that fall under the umbrella of 'information' which includes the Internet/IoT area. The four ethics theories that may find profound application in the field of Internet/IoT are the utilitarian, the justice as fairness, the social contract, and the case-based theories. The following is a brief introduction of the UNESCO's code of ethics.

11.2.2 UNESCO Code of Ethics for the Information Society

The United Nations Educational, Scientific and Cultural Organization (UNESCO) underlines the need to further universal respect for justice, for the rule of law and for the human rights and fundamental freedoms which are affirmed for the peoples of the world, without distinction of race, sex, language or religion. UNESCO has engaged with major stakeholders and academia, to conduct research and to develop awareness about the ethical dimensions of the information society.

UNESCO supports and promotes the ethical dimensions of the information society, which is one of the priorities of the Organization in its overall effort to implement the decisions of the World Summit on the Information Society (WSIS) [UNESCO Ethics]. Recognizing the needs of information and knowledge society including the importance of universal access to information in the public domain, the UNESCO InfoEthics Program has been established to address the areas of ethical, legal, and societal aspects of the applications of ICT. It addresses the challenges of cyberspace as well as privacy and security concerns in cyberspace.

UNESCO invites member states and all concerned stakeholders to take into consideration and apply the Code of Ethics [UNESCO Code] in the process of elaboration and implementation of policies, plans, programs and strategies for building equitable and inclusive knowledge societies. This document is addressed to all stakeholders of the information and knowledge societies and outlines foundational universal values and guiding principles. The Code of Ethics underlines that the Information Society should be subject to universally held values and promote the common good and to prevent abusive uses of ICTs.

The Code of Ethics emphasizes that ethical principles are relevant to all stakeholders of the information society, collectively or individually, and that existence and implementation, at all levels, of the ethical principles is essential to ensure an all inclusive information society. Also, it underlines that everyone has a right to the protection of personal data and private life on the Internet and other ICTs. Therefore, this document includes recommendations such as:

- Member states and respective stakeholders should take all steps to develop trustworthy Internet and other ICTs ensuring security, reliability and stability of critical and pervasive applications and services.
- Users should be protected against the unlawful storage, abuse or unauthorized disclosure of personal data, and against the intrusion of their privacy.
- The basic technical standards used on the Internet and other ICTs must always be open to allow interoperability and innovation.

Therefore, UNESCO, in collaboration with its partner institutions, and through its Information For All Programs (IFAP) seeks to address these challenges towards a fair and multicultural information society. The UNESCO Code of Ethics [UNESCO Code] was published in October 2011. The IFAP is committed to the full implementation of the human rights and fundamental freedoms proclaimed in the Universal Declaration of Human Rights [UDHR 1948] and other universally recognized legal instruments. It is based on the Preamble of the Constitution of the United Nations Educational, Scientific and Cultural Organization (UNESCO) [UNESCO Preamble] which underlines the need to further universal respect for justice, for the rule of law and for the human rights and fundamental freedoms which are affirmed for the peoples of the world, without distinction of race, sex, language or religion. The UNESCO Code of Ethics is addressed to all stakeholders of the information and knowledge societies and outlines universal values and guiding principles.

11.3 Dual-use Dilemma

Science is primarily used to benefit humanity, but it can be misused, presenting scientists and others with an ethical quandary known as the dual-use dilemma [DualUse]. Dual-use dilemma is the problem of the same technology (or science) having the capacity for both good and ill [Rashid

2009]. The dual-use dilemma occurs when a piece of research or technology can both benefit and harm humanity.

Miller and Selgelid [Miller 2007] state that the dual-use dilemma is an ethical dilemma since it is about promoting good in the context of the potential for also causing harm. These authors argue that the dilemma arises because researchers usually only intend their research to be used for good and it is the actions of others that often cause harm.

The dual-use dilemma is primarily an ethical dilemma for researchers who have to decide whether the ethical benefits are justified given the risks [EthicsAU]. Ethical dilemmas are situations in which moral reasons come into conflict, or in which the applications of moral values are unclear, and it is not immediately obvious what should be done [Martin 2010].

11.3.1 Examples of Cases

The dual-use dilemma arises as a consequence of the fact that the same piece of scientific research sometimes has the potential to be used for harm as well as for good [Rashid 2009]. The same can probably be said of Software Engineers developing software systems for the digital world. They intend their products to be used for good, but others sometimes use them for ill. Examples of cases that illustrate the dual-use dilemma in the context of the digital world phenomenon include [Rashid 2009]:

- Google maps: good - useful for house hunting, planning holidays; bad – invasion of privacy.
- Chat rooms: good- allow geographically distant people to communicate with each other; bad – facilitation of crimes, terrorism, etc.
- Peer-to-peer networks: good- media sharing (e.g., music); bad – illegal use of copyrighted media, illegal content.
- Social networks (e.g., Facebook): good- keeping communication among people; bad – distraction and harmful uses.

Social networking sites may connect people, but they can be vehicle for unethical behavior such as bullying, harassment, terrorism, fraud, etc. [Karahasanovic 2009]. WEB2.0 is the second generation of the Web based on user communities and a range of special services, such as social networking sites, blogs, wikis, video – and photo sharing sites, and folksonomies, that foster – each in its own way – collaboration, participation, and the new options, they also introduce obvious ethical dilemmas and risks.

Dual-use dilemmas arise when the same scientific work can be used to do good or be misused, and it is unclear how to prevent misuse without foregoing beneficial applications [DualUse]. This note examines three scientific areas posing a significant risk of misuse and considers how to tackle dual-use dilemmas in these and other areas. Strategies to prevent misuse could operate at the level of scientific practice, information dissemination or technology applications. It is recommended a risk benefit assessment approach to mitigating risks of misuse.

11.3.2 Responsibilities

Considering software development in the context of the dual-use dilemma, this work [Rashid 2009] raises the question of the Software Engineers' responsibilities in this area (more information is provided in Section 5, Software Engineering Code of Ethics and Professional Practice, of this chapter). Also, social impacts of information system failures include damages on safety, financial, environment, privacy, etc. impact almost any aspect of human life in society [Tamai 2009].

Other several typical and interesting software issues include human interface design, fail safety issues, design anomalies, error injection by fixing code, ambiguous requirements specification, insufficient regression testing, subcontracting, product liability, and corporate governance. There is a general worldwide trend demanding stricter liability for software and thus Software Engineers must be more knowledgeable about software product liability issues. While Software Engineers'actions could be qualified as legal at some extent, some actions and decisions to ignore software liability can be both illegal and unethical.

The Software Engineering Code of Ethics and Professional Practice encourages Software Engineers to undertake positive actions and to resist pressures to act unethically [Gotterbarn 2009]. The Software Engineer is obligated to act in the public's best interest, even if those actions oppose the interests of the company.

Also, people who have access to information about system failures having significant social impact should be encouraged to analyze and report those cases [Tamai 2009]. Therefore, public should be encouraged to report software failures. As computing becomes increasingly intertwined in the details of our lives, the public good depends more and more on Software Engineers [Pieters 2009]. The precautionary principle states that parties should refrain from actions in the face of scientific uncertainties about serious or irreversible harm to public health or the environment. These authors argue that the public good is even more than physical safety.

The emergence of Web 2.0 applications requires reconsideration of ethical norms governing their development and use [Karahasanovic 2009]. Developers can promote social norms or etiquette within the Web 2.0 environment to create a basis for trust and privacy. Case studies and surveys taken in different countries indicate that the most important requirements of WEB2.0 applications are related to trust, privacy, and etiquette. Etiquette is a code that influences expectations and social behavior. A rule of etiquette reflects often unwritten expectations and may reflect an underlying ethical code.

One example is public welfare. Many participants of the surveys report concerns about technology use and exploitation due to the fact that:

- An increasing number of devices are used, each requiring more energy.
- Devices require constant updates and provisioning of novel power cables, chargers, and adapters, which become more difficult to handle and reuse.

These issues are ecological and therefore relate to ethics: users expect future technology to be green – to consume more efficiently and accommodate energy related interactions as well as reuse of old technology.

11.4 Intellectual Property

Intellectual Property is a category of property that includes intangible creations of the human intellect, and primarily encompasses copyrights, patents, and trademarks defined as follows:

- Copyright is a legal right, existing in many countries, that grants the creator of an original work.
- A trademark is a word, phrase, symbol, and/or design that identifies and distinguishes the source of the goods of one party from those of others.
- Patent is the exclusive right granted by a government to an inventor to manufacture, use, or sell an invention for a certain number of years.

Also, Intellectual Property includes other types of rights, such as trade secrets, publicity rights, moral rights, and rights against unfair competition defined as follows:

- A trade secret is a formula, practice, process, design, instrument, pattern, commercial method, or compilation of information not generally known or reasonably ascertainable by others by which a business can obtain an economic advantage over competitors or customers.
- The right of publicity, often called personality rights, is the right of an individual to control the commercial use of his or her name, image, likeness, or other unequivocal aspects of one's identity.
- Moral rights are rights of creators of copyrighted works generally recognized in civil law jurisdictions and, to a lesser extent, in some common law jurisdictions. They include the right of attribution, the right to have a work published anonymously or pseudonymously, and the right to the integrity of the work.
- Unfair competition laws are designed to protect consumers and businesses alike against deceptive business practices. Some common examples of unfair competitive practices in commercial law include the following: trademark infringements, trade defamation, and misappropriation of business trade secrets.

11.4.1 Smart Grid Intellectual Property

Considered as the third industrial revolution, Smart Grid not only revolutionizes the energy industry with new technologies, but it brings new legal issues regarding intellectual property. Smart Grid technologies have experienced significant patent activity over the last decade in areas such as [Davis 2011]:

- Advanced metering
- IT systems and hardware for distribution automation
- Business methods for efficient selection of power sources

Examples of applications with submitted patents include [Davis 2011]:

- Real time performance monitoring and management system.
- Optimized Energy Management System (EMS).
- Resolving energy imbalance requirements in real-time.
- Management of a bulk electric power market.
- Aggregation of Distributed Energy Resources (DERs).
- Detection of islanding in power grids.
- Method of communicating between a utility and its customer locations.

There are issues on how to protect its own novel technologies (e.g., smart meters, solar panels, wind turbines, energy storage systems, electrical vehicles, etc.) and how to avoid exposing to allegations of patent infringement for Information Technologies used in Smart Grid developments (e.g., Wi-Fi, Bluetooth, etc.). Many patent infringement lawsuits implicating businesses in the energy sector were filed recently in US [Park 2017], [Davis 2011]. Therefore, the energy companies are advised to seek legal advice on IP rights.

According to [Reed 2009], an important milestone is arriving at a mutual understanding of Intellectual Property rights. Issues include:

- Different views on the proper role of Intellectual Property rights in new markets, such as the Smart Grid, not only on the global level, but within the United States as well.
- Complexity of Intellectual Property rights issues.

- Concerns that a few patent holders can erect barriers that inhibit incentives for research and development in new technologies.

One view [Reed 2009] is a central and immediate objective for an agreement on those Smart Grid enabling technologies that should be made available on a royalty free and nondiscriminatory basis, versus those that may be bought and sold. Thus, denying one patent and not the other is not a solution. Making the Smart Grid a reality requires an understanding of how to provide incentives for developing enabling technologies. While patents can serve this purpose, there could be an opposite effect when there is a risk/reward imbalance. The author suggests that risk and reward imbalances can be addressed by refining the patent laws such that to distinguish between those innovations that are entitled to a patent right from those that are not patent worthy.

Another aspect is colliding patents in the IoT world. Even a small IoT system could have thousands of active patents covering a myriad of features, Protecting an Intellectual Property in the IoT is likely to be both more complicated and lead to more patent infringement lawsuits than ever before [Bloom 2017].

This analysis report [Davis 2011], finds patent protection is an important element relative to continued development of the Smart Grid. The Federal Energy Regulatory Commission acknowledges that Smart Grid will require new technologies and widely adopted standards. Standards are needed for the Smart Grid to work everywhere. Without widely adopted standards, entities will attempt to establish proprietary versions. Patent owners may agree or disagree to royalty free licensing of rights. Therefore, remaining issues are Open Source type of licensing and patent protection. Innovators need an expectation of a reasonable return to justify investments for developing new technologies.

11.4.2 UNESCO Code of Ethics for Intellectual Property

Regarding information systems, UNESCO code of ethics recommends [UNESCO Code]:

- Intellectual property of the creations in a digital environment should be a subject of and shall be protected under the intellectual property rights legislation.
- Unauthorized copying and distribution of copyrighted materials must not be condoned.
- Legal frameworks facilitating owners of intellectual property to share their knowledge and creations should be supported to promote open access to knowledge and foster creativity.
- Application of international intellectual property conventions should be based on the fair balance between the interests of the rights holders and of the public.

In a society, standards are defined for moral behavior of a specific group. One example is Software Engineering code of ethics.

11.5 Software Engineering Code of Ethics and Professional Practice

The Software Engineering code of ethics are standards for practicing Software Engineering profession. The joint ACM/IEEE-CS Software Engineering Code was published in 1997 [Gotterbarn 1997]. It was developed by a multinational task force with additional input from other professionals from – industry, government posts, military installations, and educational professions. It has been adopted by many professional bodies and companies worldwide and has been translated into Arabic, Croatian, French, German, Hebrew, Italian, Mandarin, Japanese and Spanish.

Software Engineers are those who contribute, by direct participation or by teaching, to the analysis, specification, design, development, certification, maintenance, and testing of software systems.

Software Engineers have a key role in the development of software systems. However, they have significant opportunities to do good or cause harm, to enable others to do good or cause harm, or to influence others to do good or cause harm [Gotterbarn 2001]. To ensure, as much as possible, that their efforts will be used for good, Software Engineers must commit themselves to making Software Engineering a beneficial and respected profession. In accordance with that commitment, Software Engineers shall adhere to the Code of Ethics and Professional Practice.

The code of ethics (available in short and long version) prescribes these ethics as obligations of anyone claiming to be or aspiring to be a Software Engineer. The code of ethics contains eight principles related to the behavior of and decisions made by professional Software Engineers, including practitioners, educators, managers, supervisors, and policy makers, as well as trainees and students of the profession. Each of eight principles includes clauses that detail the application of those principles that are enumerated with the highest priority as the following [Gotterbarn 2001]:

1. Public – Software Engineers shall act consistently with the public interest.
2. Client and employer – Software Engineers shall act in a manner that is in the best interests of their client and employer, consistent with the public interest.
3. Product – Software Engineers shall ensure their products and related modifications meet the highest professional standards possible.
4. Judgment – Software Engineers shall maintain integrity and independence in their professional judgment.
5. Management – Software Engineering managers and leaders shall subscribe to and promote an ethical approach to the management of software development and maintenance.
6. Profession – Software Engineers shall advance the integrity and reputation of the profession consistent with the public interest.
7. Colleagues – Software Engineers shall be fair to and supportive of their colleagues.
8. Self – Software Engineers shall participate in lifelong learning regarding the practice of their profession and shall promote an ethical approach to the practice of the profession.

In ethics-aware Software Engineering, ethical considerations should be included in the software development life cycle and should be an integral part of risk assessment and acceptance criteria. Software Engineers are required to accept responsibility in making decisions consistent with the safety, health, and welfare of the public (first principle) and work to develop software and related documents that respect the privacy of those who will be affected by the software.
The code translates ethical principles into practical guidance. It encourages positive action and resistance to act unethically. In some situations, standards may be in tension with each other or with standards from other sources. These situations require the Software Engineer to use ethical judgment to act in a manner that is most consistent with the spirit of the Code of Ethics and Professional Practice, given the circumstances. Although the code of ethics defines the ethical obligations, there could be challenges on proper implementation and use in practice.

11.5.1 Ethical Challenges

Examples of ethical challenges for Software Engineers include [Rashid 2009]:

- Right to privacy versus the need to protect vulnerable user groups – for example, the need to balance the development of tools for monitoring individuals by law enforcement against the need of protecting children from predatory advances.
- Freedom of choice versus protection from harm – for example, Software Engineers must account for both an individual's freedom to make choices on software (e.g., gaming or gambling) and the notion of common good.
- Context dependent definitions of ethics – ethical considerations driven by geographical, cultural, ethnic, religious, and even historical standpoints are serious challenges for Soft-

ware Engineers and good engineering requires collaboration with ethicists, law makers, and social scientists.

The dynamic and demanding context of Software Engineering requires a code that is adaptable and relevant to new situations as they occur. According to [Miller 2007], ethical dilemmas should be solved not by simply weighing the potential benefits against the potential harms, but rather by finding other alternatives. This also applies to dual-use dilemmas. Alternative options should be found if possible, and this is where Software Engineers have a role to play [Rashid 2009]. The dual-use dilemma suggests that Software Engineers have a duty to take care with respect to harmful uses of their systems. Balancing the ethical considerations when engineering software systems for the digital world is nontrivial. However, Software Engineering researchers and software ethics experts must work together to advance the state of the art to provide ethics aware Software Engineering practices. Software Engineering Code of Ethics must be used in decision making and software development.

In the case of Software Engineering, the code of ethics provided by the IEEE Computer Society and the ACM identifies several different interests at stake, and the public interest is put first in the list. This is analogous to the codes of other professions that give priority to the interests of the users or clients. However, there are issues about how the public interest is to be identified.

A problem with this field, as in the case of other new or rapidly developing technologies, is that the technologies themselves have the potential to impact values. Privacy is a case in point. The debates about what is an ethical conflict and how to set priorities is one issue; incentives and mechanisms for raising ethical awareness constitute another [Rashid 2009]. Also, this work remarks that a difficult challenge is geographical dispersion. For example, in the context of globalization, there are calls for harmonization of ethics and regulation in various contexts, such as international data sharing. However, there are problems in making people accountable in situations of geographical dispersal. This work suggests that some issues can be addressed by better design, some by better testing, some by legislation, and some by education of users.

Another issue in Software Engineering is the product liability. The extent to which software is regarded as a product amenable to product liability laws may depend on legal and cultural boundaries, but there is a general worldwide trend demanding stricter liability for software. More lawsuits are being filed, and thus Software Engineers must be more knowledgeable about software product liability issues [Tamai 2009]. Referring to the use of outsourcing and chain of subcontracting services, the Software Engineering Code of Ethics and Professional Practice recommends Software Engineers to:

- Provide service in their areas of competence, being honest and forth right about any limitations of their experience and education.
- Ensure that they are qualified for any project on which they work or propose to work by an appropriate combination of education and training, and experience.

As discussed in [Gott5erbarn 2009], the quality of Software Engineers' work exists deep inside artifacts, and those artifacts embody the values Software Engineers employ during their development. Although the Software Engineers can not immediately see all the consequences of their work, they have responsibility for the public good.

11.5.2 Unethical Conduct

One case point is the role of Software Engineers in the Volkswagen emissions scandal. This paper [Rogerson 2018] provides an analysis of the unethical conduct of those involved in the creation of the so-called defeat software to hide the gas emission testing results for the Volkswagen diesel motors exported to US. Although the creation of this software was suggested by business managers, the Software Engineers failed to report the breaches of codes of conduct to appropriate authorities. The unethical behavior of all involved with the project affected the company reputation and caused financial loss (see [US DOJ], [Overly 2017]).

Software failures and quality issues including mismanagement are the cause of a new federal government payroll system (called Phoenix) in Canada as described in The Catalogue of Catastrophe [Catalogue]. The catalogue includes a list of failed or troubled projects from around the world. Another example is inadequate testing by Transportation Security Administration before deploying a software system to detect threats in the airports in US in 2015.

Another software glitch in an automated load shedding software system causes black out power for 90,000 people in Southern Australia in February 2017 [Chirgwin 2017]. This report [Jacobs 2013] summarizes the largest power outages in the US. For example, the August 2003 Northeast Blackout resulted from a combination of many factors including key monitoring systems that were offline and lack of system communications and cooperation. No coordination between operations planning and real time situational awareness left operators vulnerable to a technician's mistake switching major equipment that caused the Arizona California outage that affected 2.7 million people in September 8, 2011.

More software failures are reported by Tricentis [Tricentis 2016], [Tricentis 2017]. This study [Olumoye 2013] reviews the effects of ethical and social concerns of information systems on the society. The author reveals key technological trends that led to ethical issues and the moral dimensions of information systems on the society. The potential increase of ethical violations has been attributed to changes in computing power, reduction in data storage costs, better data analysis, and improvement in networks and the Internet. Also, the author recommends means to improve the effectiveness of ethical and social impacts of information systems in the society such as:

- Analysis of ethical concerns (e.g,, identifying and clearly describing the facts, clearly defining the conflict in ethical concerns and identify the ones with higher values, finding the identity of the stakeholders as every ethical, social and political issue that stakeholders might have, identifying the reasonable options to select, identify the potential consequences of selected options).
- Understand the social impact of information system (e.g., positive and negative effects, for example, continuous increase of computer crimes, social networks being used to promote democracy as well as intimidation and rebelling citizens, computer video games may be used in a positive way for training and education, but they can be used to spread and increase violence on people).

As presented in this work [Olumoye 2013], the effect of technology on individuals can be beneficial or detrimental. But in most cases an alteration in technology may help a group of individuals and be harmful to another set of people. The benefits include lower prices, better products and services. Just like other technologies such as steam engines, electricity, the telephone and radios, information system can be used to achieve social progress, but can also be used to commit crimes and threaten treasured social values [Laudon 2018]. The development of information systems can produce benefits for many and costs for others. Typical challenges include loss of privacy, depersonalization and change in incentives [Olumoye 2013].

These authors [Laudon 2018] argue that the new information technology has a ripple effect, raising new ethical, social, and political issues that must be dealt with on the individual, social, and political levels. These issues have five moral dimensions: information rights and obligations, property rights and obligations, system quality, quality of life, and accountability and control (see Figure 11.1).

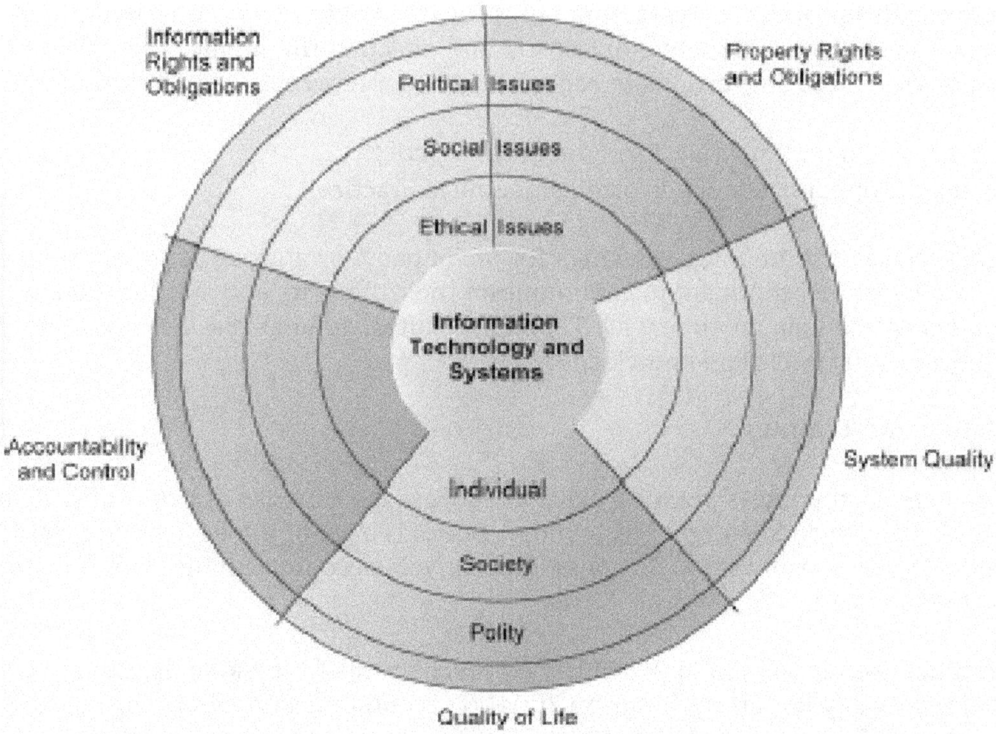

Figure 11.1 Five moral dimensions of the information age. *Source:* [Laudon 2018]. © 2018, Pearson.

Another study [White 2016] discusses the major ethical issues caused by big data analytics, another disruptive technology. As more information regarding individuals' health, financials, location, electricity use and online activity is collected, concerns arise regarding profiling, tracking, discrimination, exclusion, government surveillance and loss of control [Tene 2013]. Other issues include security of the data, loss of privacy and ownership as well as other issues due to evidence-based decision making. The algorithms to make decisions using big data can create discriminatory situations such as perpetuating old prejudices. Data scientist uses predictive analysis to make decisions about individuals or populations, but when sensitive topics are included, it can result in discriminatory situations. Also, removing personal aspects of data compartmentalizes individuals or groups, ignoring personal characteristics might force a different decision. The authors recommend specific ethical standards and regulations regarding the collection, protection, manipulation, use of big data in decision making.

This author [Booch 2014] discusses how every line of code represents a moral decision and how every bit of data collected, analyzed, and visualized has moral implications. This author [Kapucu 2007] recommends that in the rapidly evolving environments of digital technology, code of ethics should be implemented in digital government services to protect privacy and confidentiality of the information collected about citizens. Public administrators should be well equipped with information technology and code of ethical conduct for ethical decision and behavior. Similarly, this author [Cook 2009] discusses the need to apply code of ethics in protecting privacy and confidentiality of information as well as in decision making as a result of data mining based on factual or transactional data collected about people.

In this paper [Rogerson 2015], the author argues that there has been much detailed observation and analysis of the ethical challenges surrounding ICT but the transformation of this into widespread practical positive action remains elusive. It explores why progress has been difficult. The author calls for action as a starting point for entering the next phase of ICT ethics.
The consequences of software disasters are costly and affecting the public. The paper [Rogerson 2018] concludes with these suggestions to reduce the risk of reoccurring unethical practice in the future:

- Resisting the temptation to perform unethical practice.
- Reducing the opportunity of performing unethical practice.

The best insurance against unethical activities for a company, or a group is to foster an environment where employees feel confident to communicate their thoughts and ask questions if they feel that something is not done in an ethic way. Therefore, Software Engineers need a broader education to gain the necessary skills and knowledge.

11.5.3 Education and Training

Software Engineers must possess practical skills to address the complex ethical and societal issues which surround evolving and emerging technology. Ethics is a set of issues involving personal values such as honesty, justice and trust. Business issues involve economic indicators such as income, profit, production units, etc.

For example, Biometrics is an area in which having mechanisms for making decisions that respect human rights is especially important [Johnson 2004]. This author argues about the role of Software Engineer versus the effectiveness of the biometric systems. How developers design, build, protect, and maintain a biometric system will determine its effectiveness and the degree to which it poses a threat to civil liberties. Thus, application designers and developers must understand the tremendous effect of their decisions and actions can have on a society. Therefore, training and education on ethics and social and professional responsibility are crucial to raise the sensitivity of future system designers and developers to the social impact of the systems they create.

11.6 Computer Ethics an Evolving Field

On overview of the field of Computer Ethics is provided in [Bynum 2008]. This article was published in the Summer 2000 issue of the American Philosophical Association's Newsletter on Philosophy and Computing. Following this study, we learn about the beginnings and remarkable vision about this field.

The Foundation of Computer Ethics Computer ethics as a field of study was founded by MIT professor Norbert Wiener during World War Two (early 1940s) while helping to develop an antiaircraft cannon capable of shooting down fast warplanes.

In 1950, Wiener published this book [Wiener 1950]. According to [Bynum 2008], a later edition the book, published in 1954, is better and more complete from a computer ethics point of view.
This book established him as the founder of computer ethics. In this book, Wiener presents a comprehensive computer ethics foundation which remains today - half a century later – a powerful basis for computer ethics research and analysis. While Wiener did not use the term Computer Ethics, his book includes these topics:

- An account of the purpose of a human life.
- Four principles of justice.
- A powerful method for doing applied ethics.

- Discussions of the fundamental questions of computer ethics.
- Examples of key computer ethics topics.

Wiener viewed that the integration of computer technology into society will constitute the remaking of society - the second industrial revolution – destined to affect every major aspect of life. The computer revolution will be a multifaceted, ongoing process that will take decades of effort and will radically change everything. Such a vast undertaking will necessarily include a wide diversity of tasks and challenges. Workers must adjust to radical changes in the workplace; governments must establish new laws and regulations; industry and business must create new policies and practices; professional organizations must develop new codes of conduct for their members; sociologists and psychologists must study and understand new social and psychological phenomena; and philosophers must rethink and redefine old social and ethical concepts.

In 1978, the term Computer Ethics as a separate field of applied ethics was first coined by Walter Maner in his self-published work [Maner 1978], later published as [Maner 1980]. Maner defines Computer Ethics as that branch of applied ethics which studies ethical problems aggravated, transformed or created by computer technology [Bynum 2008]. By the early 1980s, the Computer Ethics had caught on, and other scholars began to develop this new field of applied ethics.

11.6.1 Information Security and Ethics

Computer ethics has received considerable attention not only from ethicists but also from information technology researchers and professionals. According to [Nemeti 2007], most of this interest is the natural consequence of the rapid development and change in computer technology, its uses and its implications. This has a profound implication for the way the information security and ethics should be viewed. The author recommends to fundamentally rethink the way we view and approach any of the following:

- Management techniques
- Technologies
- Organizational policies
- Societal laws and regulations, and
- Professional codes of conduct.

Specifically, the management should insist on an agile, comprehensive, integrated approach to information security. It is the responsibility of information security professionals and management to create an environment where the technology is used in an ethical manner. Therefore, one cannot discuss information security without discussing the ethical issues fundamental in the development and use of the technology. Information technology operates in a dynamic environment. Considerations of dynamic factors such as advances in new technologies, the dynamic nature of the user, the information latency and value, systems' ownerships, the emergence of a new threat and new vulnerabilities, dynamics of external networks, changes in the environment, the changing regulatory landscape should be viewed as important.

11.6.2 Future Directions

The evolution of Computer Ethics is depicted in other studies that describe the stages, milestones, topics, debates and hypotheses, topics, and future directions (e.g., [Stanford 2001], [Vacura 2015]). While the field of Computer Ethics is redefined since the term was coined by Maner in 1980, the best way to understand the nature of the field is through some representative examples of the issues and problems that have attracted research and scholarship [Stanford 2001]. Examples of topics include:

- Computers in the workplace
- Computer crime
- Privacy and anonymity
- Intellectual property
- Professional responsibility
- Globalization
- Metaethics of Computer Ethics.

The rapid development of ICT not only opened tremendous opportunities to humankind but also posed unprecedented ethical challenges. Along with the benefits of a digitally connected world came the threats of misuse and abuse.

Ensuring that information society is based upon principles of mutual respect and the observance of human rights is one of the major ethical challenges of the 21st century. Already countries are building mechanisms to protect their people against these risks, for example to ensure the safety of children on the Internet, but clearly a lot more needs to be done to address the ethical implications of the information society.

11.6.3 Information Security Professional Ethics

Informally, a profession is o vocation that requires a high level of education and practical experience in the field [Quinn 2006]. According to this author, while knowledge and skills are certainly necessary, they are insufficient. Professionals have a special obligation to ensure their actions are for the good of those who depend on them.

As an Information Security Professional is vital to understand the scope of an organization's legal and ethical responsibilities. Laws, ethics, and investigations are very important parts of computer and information security. The Information Security Professionals should be aware of, and be well versed in, computer security laws and regulations that apply in their environment [Harris 2013]. The Information Security Professionals play an important role in an organization's approach to controlling liability for privacy and security risks [Whitman 2014].

Based on the Institute of Information Security Professional (IISP) (https://www.iisp.org/), information security professionals are distinguished by certain characteristics as follows:

- Mastery of a particular security skill, acquired by professional training, education, certification, experience or combination of them.
- Adherence by its members to a common set of values.
- Acceptance of a duty to society as whole.

Information security and ethics is defined as an encompassing term that refers to all activities needed to secure information and systems that support it in order to facilitate its ethical use [Nemeti 2007].

In any profession, rules and regulations enforce professional ethics. Those codes of ethics might come from the certifying agencies. Also, more regulations are requiring organizations to have an ethical statement and potentially an ethical program in place. Ethics should be implemented in the real corporate world through proper business processes.

Ethical concepts in Information Security are based on many different issues and foundations. They can be relative to similar situations and interpreted differently from individual to individual. With no universal framework of ethics for Information Security Professionals, we introduce different codes of ethics as promoted by different associations. A code of ethics helps to ensure professionalism.

11.6.4 Information Security Code of Ethics

Several published codes apply to Information Security. For example, the Ten Commandments of Computer Ethics published by Computer Ethics Institute are useful recommendations to the ethical standards for the industry [Computer Ethics]:

- Thou shalt not use a computer to harm other people.
- Thou shalt not interfere with other people's computer work.
- Thou shalt not snoop around in other people's computer files.
- Thou shalt not use a computer to steal.
- Thou shalt not use a computer to bear false witness.
- Thou shalt not copy or use proprietary software for which you have not paid.
- Thou shalt not use other people's computer resources without authorization or proper compensation.
- Thou shalt not appropriate other people's intellectual output.
- Thou shalt think about the social consequences of the program you are writing or the system you are designing.
- Thou shalt always use a computer in ways that ensure consideration and respect for your fellow humans.

Also, the Internet Architecture Board (IAB) summarizes the policy concerning the proper use of resources of the Internet in [RFC 1087]. The statements of this document include a list of unethical and unacceptable practices of the Internet use by someone who:

- Seeks to gain unauthorized access to the resources of the Internet.
- Disrupts the intended use of the Internet.
- Wastes resources (people, capacity, computer) through such actions.
- Destroys the integrity of computer based information, and/or
- Compromises the privacy of users.

These statements summarize the tone of most security related codes of ethics. Ethics are critical part of any sound cybersecurity defense strategy. Without clear ethical standards and rules, cybersecurity professionals are almost indistinguishable from the black hat criminals against whom they seek to protect systems and data [Knowles 2016]. The study of cybersecurity ethics, which encompasses a wide array of approaches and schools of thought, does not offer a simple solution to the many complex ethical dilemmas IT professionals, chief information security officers (CISOs) and organizations face daily.

When discussing emerging trends in information security and ethics, this author [Nemeti 2007] views information security and ethics to be everyone's business, not just the IT. This vision and approach on information security will be driven primarily due to consumer demand. The consumers will demand more security of the information about them and will insist on more ethical uses of that information.

Therefore, most of the professional organizations have adopted a set of ethical code of conducts. For example, associations such as ISSA [ISSA Code], ISC2 [ISC2 Code], AITP [AITP Code], SANS [SANS Code] are providing their own ethical guidelines in IT and cybersecurity that are requirements for their own members.

Many certification organizations require adherence to a code of ethics. If a member violates the rules, there is the risk of losing the certification or license. On the other hand, those professionals that do not hold a certification, may not be required to follow strictly codes of ethics. Quite often, ethics can be subjective, influenced by an individual's background, culture, education, personality and other factors. There are many issues that cannot be handled completely in any code of ethics. An analysis of the content of the ACM Code of Ethics, the AITP Code of Ethics, and the Software Engineer's Code of Ethics shows on their composite ethical value and differences.

The Institute of Information Security (IISP) publishes a Code of Ethics and disciplinary process for its members [IISP Code] (IISP is now The Chartered Institute of Information Security (CIISec), https://www.ciisec.org/about). This institute promotes the objectives of the information security profession, and its members should work to the highest standards of professionalism and that their work should fully satisfy the needs of all stakeholders and those of society as a whole. These objectives require that three basic needs are satisfied:

- Trust - for employers, clients, regulators, other interested parties and for society as a whole there is a need for trust in information and information systems and in the practitioners working in those fields.
- Quality - there is a need for assurance that all services obtained from an information security professional are carried out to the highest levels of performance.
- Standards - users of the services of information security professionals should be confident that a framework of professional ethics and technical standards exists, which governs the provision of those services.

Many of the international professional bodies such as GIAC [GIAC Code], EC Council [EC Code] and ISC2 [ISC2 Code] use the code of ethics to provide benchmark to their security professional members for self evaluation and also to establish a framework for professional's behavior and responsibilities as certified members. As pointed in this document [Malaysia Code], a scenario in Malaysia indicates that there are many people acquired their information security professionals through mastering the intellectual skill, training, education and experience without receiving certification from the well known international professional bodies. However, these information security professionals do not have any common framework of ethical that they can channel or attach themselves to so the need to develop the Code of Ethics for Information Security Professionals. Therefore, the published [Malaysia Code] is addressing many challenges that are not commonly addressed by other codes.

11.6.5 Malaysia Code of Ethics for Information Security Professionals

The Malaysia Code of Ethics for Information Security Professionals are established with these objectives [Malaysia Code]:

- Guide information security professionals on how to align behavior, action and decision with highest standards of professionalism.
- Provide a benchmark for information security professionals to use for self-evaluation.
- Minimize problems with ethical behaviors and encourages responsible behavior.
- Help professionals identify and resolve the inevitable ethical dilemmas that they will confront during their information security career.

The Code of Ethics is derived from the Unified Framework for Information Security Professionals which emphasizes on three core ethic values: Integrity, Objectivity, and Professional Competence & Due Care which required all members to observe.

Derived from three core ethic values, the code of ethics consist of seven guiding principles where the information security professional shall adhere to [Malaysia Code]:

1. Act all times in accordance with existing laws, association values and exercising the highest moral principles.
2. Protect and maintain appropriate level of confidentiality, integrity and availability of sensitive information in any course of professional activities.
3. Conduct the service with fairness, courtesy and good faith towards clients, colleagues and others, give credit where it is due and accept, as well as give, honest and fair professional comments.
4. Do not engage in any crimes or improper practices.
5. Perform all professional activities and duties in accordance with the highest ethical principles.
6. Avoid professional association with those whose practices or reputation might diminish the profession.
7. Provide service with competence, honesty and forthrightness about limitations, experience and education.

11.6.6 OECD Privacy Guidelines

The OECD guidelines [OECD 1980] on privacy and personal data protection are often the standard that data protection laws of many countries reference. The OECD privacy guidelines are based on eight basic principles:

1. Collection limitation
2. Data quality
3. Purpose specification
4. Use limitation
5. Security safeguards
6. Openness
7. Individual participation
8. Accountability.

This document [OECD 2012a] provides an overview of the history of the OECD Guidelines for the Security of Information Systems and Networks: Towards a Culture of Security (Security Guidelines) since the adoption of their first version in 1992. It explains that the 2002 revision of the Guidelines introduced a fundamental paradigm shift in the way IT security was previously addressed, in order to take into account the emergence of the open Internet and the generalization of interconnectivity. Further, this document [OECD 2015] recommends digital security risk management should be based on ethical conduct which respects and recognizes the legitimate interests of others and of the whole society. Organizations should have a general policy of transparency about their practices and procedures to manage digital security risk.

Information security and ethics are fundamentally related to information privacy. According to [Lee 2016a], different organizations have their own unique policies and a unique code of conduct. However, the authors argue there can be no universal recipe, only a general guideline. They recommend viewing privacy from the perspective of ethics can help enterprises establish and improve their code of conduct. Also, establishing a code of conduct makes all individuals in an organization, not just security personnel, accountable for protecting valuable data.

As stated in [OECD 2015] document, governments, public and private organizations, as well as individuals share responsibility, based on their roles and the context, for managing digital security risk and for protecting the digital environment; and that cooperation is essential at domestic, regional and international levels.

11.6.7 Unethical Case Scenarios

There has been news with stories of security breaches and ethical lapses that have led to criminal behavior [Gupta 2012]. IT security personnel often have access to confidential data and knowledge about individuals' and companies' networks and systems that give them a great deal of power. That power can be abused, either deliberately or inadvertently [Shinder 2005]. As pointed out in this work, there are no standardized training requirements for working as an IT security consultant or inhouse security specialist. Many associations and organizations for IT professionals have started to address the ethical side of the job, but there is no requirement for IT security personnel to belong to those organizations.

Discussions about the information ethics are the focus of many published materials. This work [Shinder 2005] argues that most ethical issues that IT and security professionals confront have not been codified into law, nor is there a standard mandatory oversight body. The author compares IT and security professionals with older and more established professions such as medicine and law national and state medical association or bar association, that have established a detailed code of ethics.

This report [Knowles 2016] argues that without codified cybersecurity ethics guidelines in place at the industry and employer levels, it is largely up to the individual to determine the most ethically sound response to a given incident [Knowles 2016].

11.7 Differences in Ethical Concepts

Ethics are based on many different issues and foundations [Harris 2013]. They are often a topic a debate. However, some ethics are less controversial than others, and these types of ethics are easier to expect of all people.

Ethics is a set of issues involving personal values such as honesty, justice and trust. Business issues involve economic indicators such as income, profit, production units, etc. Referring to ethical concepts in Information Security, these authors [Whitman 2014] identify these perspectives:

- Some consider ethics to be the organized study of how humans ought to act.
- Others refer to ethics to be considerations of rules we should live by.

Recognizing that there are differences in ethical concepts, the authors reveal these facts:

- Studies show that individuals of different nationalities have different perspectives on the ethics of computer use.
- More difficulties arise when one nationality's ethical behavior does not correspond to that of another national group (e.g., clashes between Westerners and Asians views on some issues, for example computer software piracy). Many countries have a similar attitude toward software piracy, while Netherlands is more permissive and United Nations is significantly less tolerant.

However, differences in computer use are not exclusively cultural. Differences are found among individuals within the same country, within the same social class, and within the same organization [Whitman 2014]. These differences can be leveled by educating and training people, as many may not have the formal training to understand that their behavior is unethical or even illegal.

11.7.1 Overcoming Differences

The best insurance against unethical activities for an organization, or a group is to foster an environment where employees feel confident to communicate their thoughts and ask questions if they feel that something is not done in an ethic way.

Also, an organization must develop a work environment that allows an Information Security Professional to carry out his duties in a manner that conforms to accepted standards. This can be accomplished as an organization provides these:

- A written standard of conduct that it expects employees in carrying out their tasks.
- A set of development standards and guidelines on how work must be carried out.
- A set of means by which an employee can express its disapproval on how the company conducted a business transaction.

11.7.2 Ethics Foundations Debate

While researchers study various ethical theories, this author [Quinn 2006] compares workable ethical theories. The workable theories are those that make it possible for a person to present a persuasive, logical argument to a diverse audience of skeptical, yet open minded people. These theories can be used to determine whether an action is right or wrong. By comparing these theories - Kantianism, act utilitarianism, rule utilitarianism, and social contract theory – this study finds out that these theories:

- Share the viewpoint that moral good and moral percepts are objective using reasoning from facts or commonly held values.
- Use different assumptions about decision making when evaluating the consequences of an action.

While no one theory is clearly superior to other, the author concludes that these four theories have a lot of power to propose solutions to many problems arising from the introduction of Information Technology into society.

11.7.3 Engineering Ethics Foundations

Engineering is an important and learned profession. As members of this profession, engineers are expected to exhibit the highest standards of honesty and integrity. Engineers must perform under a standard of professional behavior that requires adherence to the highest principles of ethical conduct. Engineering is a professional career that impacts lives. When ethics is not followed, disaster often occurs; these disasters not only include huge monetary costs and environmental impacts, but also often result in the loss of human life. Engineering ethics applies to every engineer and is very important. Engineering ethics is the field of applied ethics and system of moral principles that apply to the practice of engineering. The field examines and sets the obligations by engineers to society, to their clients, and to the profession.

Engineering ethics is defined in [Schinzinger 2000] as the study of the moral values, issues, and decisions involved in engineering practice. Engineering ethics consists of the responsibilities and rights that ought to be endorsed by those engaged in engineering, and also of desirable ideals and personal commitments in engineering [Martin 2010].

Another researcher [Baura 2006] discusses the engineering ethics foundation to include some of the classic ethical theories such as utilitarianism, duty ethics, rights ethics, and virtue ethics. The researcher discusses how these theories could influence the engineering decision process. Therefore, these theories and others could provide a framework for the ethical decisions in engineering [Baura 2006].

An ethical theory seeks to provide a comprehensive perspective on morality that clarifies, organizes, and guides moral reflection. If successful, it provides a framework for making reasonable moral choices and resolving moral dilemmas—not a simple formula, but rather a unifying way to identify and integrate moral reasons [Martin 2010]. These authors [Martin 2010] discuss five types of ethical theories (and traditions) that have been especially influential to Engineering Ethics. These theories include the rights ethics, duty ethics, utilitarianism, virtue ethics, and self-realization ethics. Briefly they are described as follows [Martin 2010]:

- Rights ethics says we ought to respect human rights, and duty ethics says we ought to respect individuals' rational autonomy.
- Utilitarianism says that we ought to maximize the overall good, taking into equal account all those affected by our actions.
- Virtue ethics says that good character is central to morality.
- Self-realization ethics emphasizes the moral significance of self-fulfillment.

There are many fields of engineering such as:

- Civil
- Mechanical
- Electrical
- Software
- Industrial.

Engineers must perform under a standard of professional behavior that requires adherence to the highest principles of ethical conduct. The Engineering Code of Ethics has three components:

- Fundamental canons articulate the basic components of ethical engineering.
- Rules of practice clarify and specify in detail the fundamental canons of ethics in engineering.
- Professional obligations elaborate the obligations that engineers have.

The code of ethics is not a legally binding document. Ethical behavior is about practice and virtue. It is about going beyond the codes and practicing behavior that leads to an ethical life. Codes of engineering ethics identify a specific precedence with respect to the engineer's consideration for the public, clients, employers, and the profession.

The code of ethics plays at least eight essential roles [Martin 2010]: serving and protecting the public, providing guidance, offering inspiration, establishing shared standards, supporting responsible professionals, contributing to education, deterring wrongdoing, and strengthening a profession's image.

Decisions made by engineers usually have serious consequences to people - often to multitudes of people. Ethics and ethical reasoning guide decision making. Ethical reasoning is a type of practical reasoning that concerns certain societal or life form goals, such as justice, equality, freedom, health and safety. For example, the codes of the National Society of Professional Engineers (NSPE) and other professional codes establish shared minimum standards and provide guidance and support for responsible engineers.

11.7.3.1 NSPE Code of Ethics

NSPE is the only engineering society that represents engineers across all disciplines. Its original code of ethics was approved in 1946. Engineers must perform under a standard of professional behavior that requires adherence to the highest principles of ethical conduct. Engineers, in the fulfillment of their professional duties, shall follow the fundamental cannons [NSPE Code]:

1. Hold paramount the safety, health, and welfare of the public.
2. Perform services only in areas of their competence.
3. Issue public statements only in an objective and truthful manner.
4. Act for each employer or client as faithful agents or trustees.
5. Avoid deceptive acts.
6. Conduct themselves honorably, responsibly, ethically, and lawfully so as to enhance the honor, reputation, and usefulness of the profession.
7. Besides these fundamental canons, the code of ethics includes rules of practice and professional obligations.

11.7.3.2 IEEE Code of Ethics

The IEEE code of ethics provides guidelines to protect both engineers and the public. Its origins date back to the American Institute of Electrical Engineers code adopted in 1912. The code of conduct describes commitment to the highest standards of integrity, responsible behavior, and ethical and professional conduct for IEEE members as follows [IEEE Code]:

We, the members of the IEEE, in recognition of the importance of our technologies in affecting the quality of life throughout the world, and in accepting a personal obligation to our profession, its members, and the communities we serve, do hereby commit ourselves to the highest ethical and professional conduct and agree:

1. to hold paramount the safety, health, and welfare of the public, to strive to comply with ethical design and sustainable development practices, and to disclose promptly factors that might endanger the public or the environment;
2. to avoid real or perceived conflicts of interest whenever possible, and to disclose them to affected parties when they do exist;
3. to be honest and realistic in stating claims or estimates based on available data;
4. to reject bribery in all its forms;
5. to improve the understanding by individuals and society of the capabilities and societal implications of conventional and emerging technologies, including intelligent systems;
6. to maintain and improve our technical competence and to undertake technological tasks for others only if qualified by training or experience, or after full disclosure of pertinent limitations;
7. to seek, accept, and offer honest criticism of technical work, to acknowledge and correct errors, and to credit properly the contributions of others;
8. to treat fairly all persons and to not engage in acts of discrimination based on race, religion, gender, disability, age, national origin, sexual orientation, gender identity, or gender expression;
9. to avoid injuring others, their property, reputation, or employment by false or malicious action;
10. to assist colleagues and coworkers in their professional development and to support them in following this code of ethics.

However, ethical dilemmas arise in engineering because moral values are many and varied and can make competing claims.

11.7.3.3 INCOSE Code of Ethics

The International Council on Systems Engineering (INCOSE) code of ethics refers to Systems Engineers conduct and ethics for the Systems Engineering profession. The INCOSE Code of Ethics includes [INCOSE Ethics]:

- Fundamental principles
  - Being honest and impartial
  - Maintaining the highest levels of integrity and keeping abreast of the knowledge of their disciplines
  - Striving to increase the competence and prestige of the engineering profession
  - Supporting the educational institutions, the professional societies and technical societies of their disciplines.
- Duties to society and public infrastructure
  - Guard the public interest and protect the environment, safety and welfare of those affected by engineering activities and technological artifacts
  - Accept responsibility for your actions and engineering results, including being open to ethical scrutiny and assessment
  - Proactively mitigate unsafe practice
  - Manage risk using knowledge granted by a whole system viewpoint and understanding of systemic interfaces
  - Promote the understanding, implementation, and acceptance of prudent Systems Engineering measures.
- Rules of practice to help the Systems Engineer in practical application of ethics to their work such as:
  - Act legally, honorably, honestly, justly, and responsibly
  - Provide diligent and competent services to the best of your ability
  - Ss Respect the trust and the privileges granted to you
  - Avoid conflicts of interest and the appearance thereof.

As stated in the code of ethics, the practice of Systems Engineering can result in significant social and environmental benefits, but only if unintended and undesired effects are considered and mitigated. The role of Systems Engineer is to stand up for the interests of the customers, community, society, and profession against the temptation and pressure to cut corners to deliver products faster or at lower costs (see also the handbook [INCOSE 2015]).

11.7.4 Ethical Dilemmas in Engineering

As described in [Martin 2010], professionalism in engineering would be threatened at every turn in a corporation devoted primarily to powerful egos. Beginning in 2001, revelation of a wave of scandals shook Americans' confidence in corporations. In that year, Enron became the largest bankruptcy in U.S. history, erasing approximately $60 billion in shareholder value. Illegal financial practices of ENRON revealed after the bankruptcy filing and indictment of top executives and many managers, there were questions about trust and an ethical corporation climate. The employee betrayal, investments in the area where the company lacked the engineering expertise are just examples of unethical climate within ENRON corporation. The types of ethical dilemmas engineers encounter at work include public and safety welfare, data integrity and representation, trade secrets and industrial espionage, gift giving and bribery, principle of informed consent [Baura 2006]. There could be unethical situations that engineers cannot control, so it is up to the engineer to decide what action to take (e.g., leaving the company, leaving the field, fighting for internal change, etc.). Many engineering ethics textbooks advocate internal or external whistleblowing (e.g., ENRON, WORLDCOM, FBI).

Engineers can make a vital contribution to an ethical climate, especially as they move into technical management and then more general management positions. A working environment is conducive to morally responsible conduct. Within corporations it is produced by a combination of formal organization and policies, informal traditions and practices, and personal attitudes and commitments. The defining features of an ethical corporate climate include at least these four aspects in this order:

- Ethical values in their full complexity are widely acknowledged and appreciated by managers and employees alike; responsibilities to all constituencies of the corporation are affirmed—not only to stockholders, but also to customers, employees, and all other stakeholders in the corporation.
- Ethical language is honestly applied and recognized as a legitimate part of corporate dialogue; one way to emphasize this legitimacy is to make prominent a corporate code of ethics; another way is to explicitly include a statement of ethical responsibilities in the job descriptions of all layers of management.
- Top management sets a moral tone in words, in policies, and by personal example; official pronouncements asserting the importance of professional conduct in all areas of the corporation must be backed by support for professionals who work according to the guidelines outlined in professional codes of ethics; workshops on ethics or formal brochures on social responsibility distributed to all employees; fostering confidence that management is serious about ethics.
- There should be procedures for conflict resolution such as confidential discussions about moral concerns and educating managers about conflict resolution.

These authors [Martin 2010] describe reasonable solutions to ethical dilemmas to be clear, informed, and well-reasoned:

- Clear refers to moral clarity about which moral values are at stake and how they pertain to the situation; it also refers to conceptual clarity—precision in using the key concepts (ideas) applicable in the situation.
- Informed means knowing and appreciating the implications of morally relevant facts; it means being aware of alternative courses of action and what they entail.
- *Well-reasoned* means that good judgment is exercised in integrating the relevant moral values and facts to arrive at a morally desirable solution.

In summary, these authors recommend these steps in resolving ethical dilemmas:

- Moral clarity requires to identify the relevant moral values.
- Conceptual clarity means to be clear about key concepts.
- Informed about the facts means to obtain relevant information.
- Informed about the options by considering all (realistic) options.
- Well-reasoned refers to make a reasonable decision.

11.7.5 Differences in Codes of Ethics

However, an engineer might be a member of one or more professional societies and customs of these societies (e.g., a weapons development corporation and a pacifist religious tradition) might be pointing in different directions. Although codes of ethics are designed for guiding professional conduct, codes are not always the complete and final word. Also, Codes can be flawed, both by omission and commission.

This blog [KAYDEE 2009] discusses the Software Engineering profession and Engineering profession by comparing education requirements and code of ethics. It finds differences in the education requirements and code of ethics. Examples include:

- The education and training to do Software Engineering is markedly different than the typical training that other kinds of engineers receive, although Software Engineers are involved in most other kinds of engineering too.

- The code of ethics for Software Engineers is almost similar with the code of ethics for engineers with one difference the social responsibility. Since software is used in any engineering design and process and too many software projects disasters still occur, the author recommends that those who write software for sale or use by others bear the exact same type of responsibility as engineers who design bridges or rockets.

- The text for the Software Engineers Code of Ethics contains over 2400 words compared to the text of IEEE code for engineers that is more comprehensive.

While more important than such superficialities as the length of the codes of ethics, the author raises the question why Software Engineers need a separate set of ethical principles in the first place.

11.7.6 Technical and Ethical Issues in Engineering Smart Grid

As proposed in [ETP 2010], a high level requirement for the Smart Grid is the ability to cost effectively integrate the behavior and actions of all connected users – energy producers, consumers and prosumers – in order to ensure economically efficient, sustainable power system with low losses and high levels of quality and security of supply and safety.

The development of an intelligent grid system and the exploitation of available flexibility in demand from load units are key to confronting the challenges of integrating larger percentages of renewable resources [Neupane 2017]. In this regard, various technologies and solutions have been proposed in different areas, such as demand/response (DR) schemes, energy storage systems, energy management systems, etc. Specifically, DR schemes propose various incentives to motivate energy consumers (prosumers) to actively participate in the DR so that their energy demand (or supply) can be varied to obtain a dynamic energy balance in the market. Information security and privacy are concerns that are not addressed in the majority of this kind of applications.

In addition, the integration of renewable energy sources and energy storage technologies in the grid will result in dynamic electricity demand being an active factor within the electrical system. At the heart of a DR application is the dynamic nature of generation and consumption patterns, as well as their organization and asset boundaries, which is where the traditional closed loop, centralized energy management infrastructure becomes inadequate [Mourshed 2015].

However, Smart Grid deployment must include not only technology, market and commercial considerations, environmental impact, regulatory framework, standardization usage, ICT (Information and Communication Technology) and migration strategy but also societal requirements and governmental edicts [ETP 2010].

While the economic benefit of Smart Grid is apparent and of general importance, the ethical component should be analyzed [Makienko 2017]. The special aspect in adoption of the complex systems such as the Smart Grid is the uncertainty about the full range of the potential hazards when implementing such technology because the social and technological risks are revealed after a long period only. Renewable sources such as hydroelectric power, wind power, solar power, geothermal heat, agricultural biomass and tides do not cause the environmental hazards that fossil fuels do.

However, the renewable resources must be balanced with the amount of energy needed to produce and maintain them. Consequently, when analyzing the environmental hazards, currently, the energy required to manufacture and install solar energy systems comes from fossil fuels. For the sake of clean energy, these opposed uses of energy resources may be addressed in the future.

The ethical issues of the technology implementation such as human security, preserving the personal and cultural identity, privacy right, development of traditional forms for preserving the sociality are problems for our future. Hence, it is necessary to place greater emphasis both on individual values, which are of general importance as well, and on social values which will determine the life of humanity in the future. The components of the ethical evaluation can be used to analyze the requirements considered when adopting the Smart Grid as well as the problems of the ethical competence development in training the engineering undergraduates.

11.7.6.1 Production and Consumption of Electricity Overview

Electricity conservation is a major avenue of limiting electricity related greenhouse gas emissions. Therefore, electricity production and electricity consumption compared with past data are the focus of many reports by OECD, IEA, and others. The consumers, public and other sectors of economy, need to understand the implications of production and consumption of electricity such that policies and information technologies align with the goals of a secure Smart Grid.

As reported in [IEA 2018], generation from combustible fuels accounted for 67.3% of total world gross electricity production in 2016. Combustible fuels include coal and coal products, oil and oil products, natural gas, biofuels including solid biomass and animal products, gas/liquids from biomass, industrial waste and municipal waste.

However, the report [IEA 2018] shows that electricity generation from fossil fuels fell for the fifth consecutive year in 2017, with generation from total combustible fuels accounted for 59.4% of total OECD gross electricity production (compared to 72.8% for non OECD). Electricity generation from renewable sources such as wind (+15.1%) and solar (+21.9%) registered robust growth. Also, the report includes statistics of electricity consumption by sector.

Much of the growth in OECD electricity consumption since 1974 has taken place in the residential, and commercial and public service sectors. In 2016, industry was still the largest end-use sector for electricity consumption, but its share of consumption has been in long term decline.
The remaining consumption sectors – transport, agriculture and fishing – are relatively small consumers of electricity. However, road transport has recently experienced strong growth in electricity consumption as electric vehicles gain market share across OECD countries, particularly in Europe.
As regards the Europe's production of electricity in 2018 (see Figure 11.2), the production of conventional thermal electricity decreased by 8.3 % in the EU-28 in 2018 and accounted for 45.9 % of the total production, while the production from renewables increased [Eurostat]. The electricity production from wind sources increased by 5.0 % and represented 12.2 % of the total production, while the production by hydro which includes pumped hydro increased by 11.4 % and represented 11.8 % of the total production. The electricity production by solar increased by 3.9 % and represented 4.0 % of the total production.

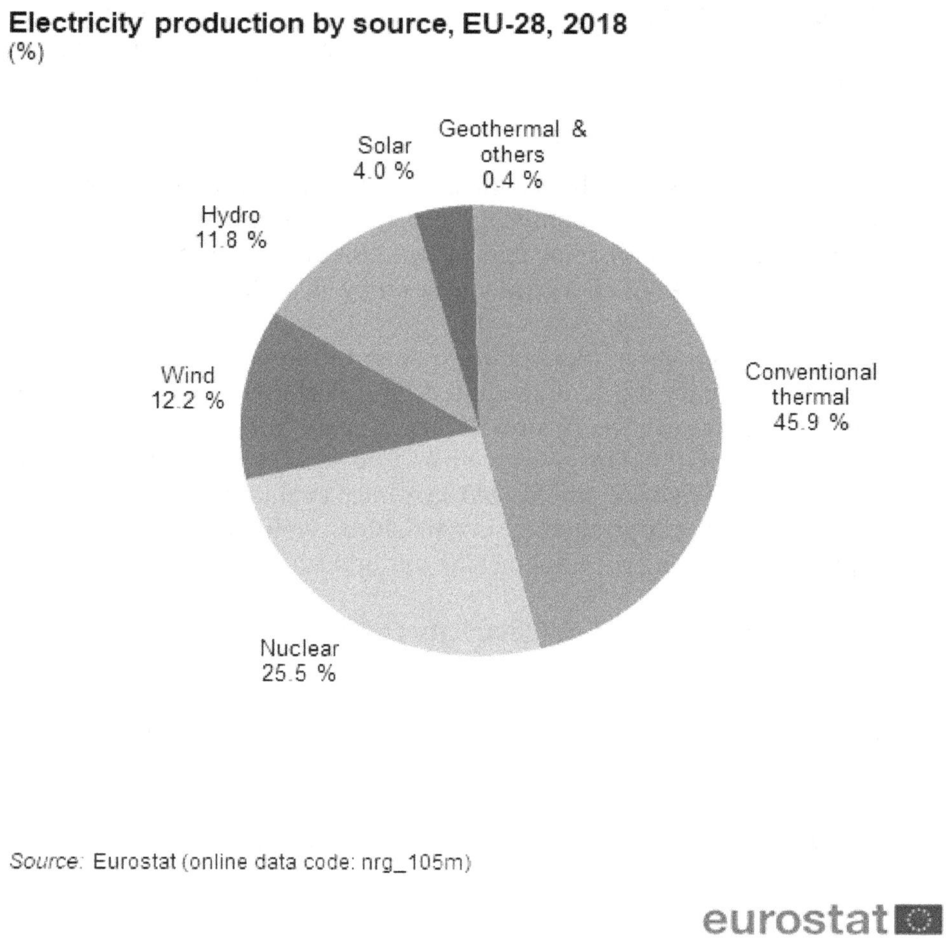

Figure 11.2 Europe Electricity Production in 2018. *Source:* [Eurostat]. © 2018, European Communities.

11.7.6.2 The Demand for Electricity in US

The U.S. electric power system has provided highly reliable electricity for more than a century. Electric utilities now find themselves making transformations such as improvement of infrastructure, addition of the digital layer, and business process transformation. Much of the modernization work that has been going on in the electric grid, especially substation and distribution automation, is now included in the general concept of the Smart Grid, but additional capabilities are evolving as well.

The Smart Grid is evolving from the traditional electrical grid. An electrical grid (also referred to as an electricity grid or electric grid) is an interconnected network for delivering electricity from suppliers to consumers. It consists of generating stations that produce electrical power, high voltage transmission lines that carry power from distant sources to demand centers, and distribution lines that connect individual customers.

Currently, electricity consumed in the United States is divided roughly evenly across three major sectors: residential, commercial, and industrial. A relatively small proportion of electricity is consumed by a fourth sector, transportation, primarily for powering trains and electric vehicles. Figure 11.3 is a representative snapshot of electricity use within a home as reported in [EIA 2018].

Consumers of electricity do not demand a constant, predictable level of electricity. Driving factors of electricity demand include weather conditions, energy efficiency regulations, macroeconomic conditions, and personal preferences [Guo 2015]. For example, factors such as varying high temperatures during summer and cold temperatures during winter increase demand for electricity are leading to daily and seasonal variations in demand. More, services that are used only intermittently throughout the day (e.g., lighting) can lead to variations in demand hour by hour and even second by second, though daily patterns are fairly stable.

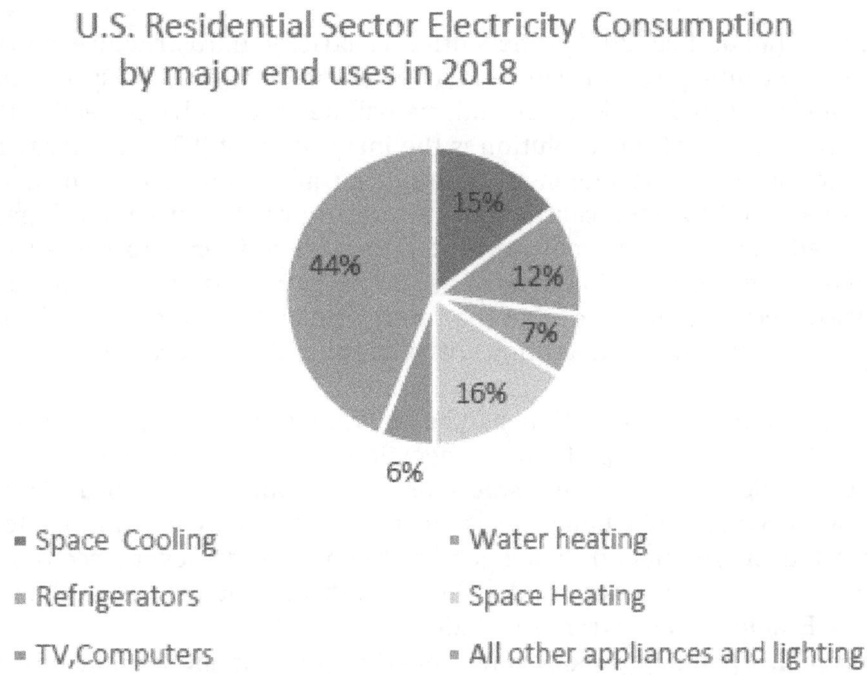

Figure 11.3 Residential electricity consumption based on data reported in [EIA 2018].

At the distribution level, most consumers (residential, commercial, and industrial) purchase electricity from monopoly suppliers known as utilities. In USA, utilities can be investor owned, can be owned by municipalities, or can be operated as cooperatives. By law, utilities must supply electricity to customers within their service regions via universal service obligations and plan future investments.

11.7.6.3 Understanding the Roles of Smart Grid

The definitions for Smart Grid vary. The Smart Grid must be seen as a continuous process of modernizing existing electricity grids and of designing future grids [OECD 2012b]. The Smart Grid is meant to address key challenges – environmental and economic – that the electricity sector is facing. The use of ICTs and Internet applications are at the center of this modernization. Smart Grids can essentially be defined by their functions and their components. Environmental and economic challenges in the electricity sector transcend individual steps in the value chain.

The Smart Grid is therefore expected to address the key challenges stakeholders in the sector are facing: mitigation of climate change, disruptions in supply of conventional energy sources, exploding global demand for electricity, wider diffusion of renewable energy sources, accelerating use of electric vehicles and louder consumer demands for greater transparency. One important task of Smart Grid is to provide a dynamic platform for free and safe interconnection of renewable energy generation to power systems [IEC Roadmap]. According to IEC Roadmap, the key market drivers behind Smart Grid solutions are:

- Need for more energy
- Increased usage of renewable energy resources
- Sustainability
- Competitive energy prices
- Security of supply
- Ageing infrastructure and workforce.

As described in [Mourshed 2015], the entire electricity infrastructure and associated socio-technical system including transmission and distribution networks, the system operator, suppliers, generators, consumers and market mechanisms will need to evolve to realize the full potential of Smart Grids. At the heart of this evolution is the integration of ICT and energy infrastructures for increasingly decentralized development, monitoring and management of a resilient grid. The Smart Grid is expected to significantly lower electricity related greenhouse gas emissions in the future. Direct reductions are typically attributed to more efficient processes in electricity generation, transmission and distribution as well as to energy conservation at customer sites. Indirect reductions are expected in the sense that the Smart Grid will facilitate wider diffusion of renewable energy sources as well as their integration with wider uptake of electric vehicles.

Climate change and its impacts continues to be an issue faced by countries around the world. Concerns over safety, security, air pollution, congestion, infrastructure reliability and transport efficiency provide public pressure to find solutions. These concerns significantly impact public opinion and thus are driving public policy decisions to invest heavily in solutions for improved public infrastructure and environmental challenges. For the past years, several sectors of the global economy have been driving regulatory and policy initiatives that will drive Smart Systems growth [Harbor 2011]. Examples of key trends include:

- Government spending packages and massive government stimulus are promoting the use and adoption of pervasive technologies. Huge amounts of money have been committed by governments around the world to support the development of Smart Grids including the installation of 200-300 million smart meters by 2015 [Harbor 2011].
- An eco friendly push by the governments looking for ways to curb emissions and control pollution in response to increased reports of negative impacts on citizens and resources. Many studies have concluded that buildings around the world account for 50-70% of annual greenhouse gas emissions [Harbor 2011]. With this statistic in mind, one way is to limit emissions from buildings and to consider means to optimize the performance of the buildings. One example of this trend can be found in the EU's Directive on the energy performance of buildings [EU 2018/844].

Smart Systems are a new generation of systems architecture (hardware, software, network technologies and managed services) that provide real time awareness based on inputs from machines, people, video streams, maps, newsfeeds, sensors, and more that integrate people, processes, and knowledge to enable collective awareness and better decision making [Harbor 2011]. The Smart Systems opportunity is not just about people communicating with people or machines communicating with machines: it also includes people communicating with machines, and machines communicating with people. Examples of Smart Systems include Smart Grid, Smart City, Health Care, etc.

The Smart Grid is meant to address key challenges – environmental and economic – that the electricity sector is facing [OECD 2012b]. The Smart Grid has great potential for driving innovation in the way electricity is produced, managed and consumed.

Applications of information and communication technologies (ICTs) and especially the opportunities provided by the Internet can help sustain electricity supply while limiting environmental impacts. Also, the Internet gives rise to a new generation of businesses providing services around electricity, adding further value and innovation to the energy sector value chain. Figure 11.4 shows a simplified view of the energy sector value chain.

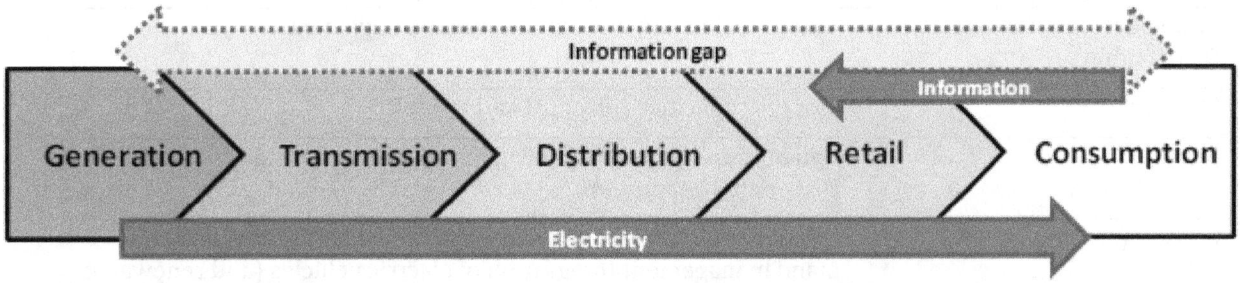

Figure 11.4 Stylized electricity sector value chain. *Source*: [OECD 2012b]. © 2012, OECD.

The Smart Grid is expected to help tackle a number of structural challenges on global energy supply and demand. Table 11.3 provides an overview of areas where electricity sector challenges relate to tackling climate change. It uses the value chain steps highlighted above complemented by energy storage, which is becoming an important issue for coordination between stakeholders.

An important characteristic of the Smart Systems opportunity is the importance of vertically focused solutions, increasingly be delivered as a managed or value added application service [Harbor 2011]. For example, a combination of monitoring, diagnostic, control and tracking functions could be configured to provide basic functionality required to enable an energy management application. Value added application services are solutions that integrate people, business processes and assets and are delivered as managed services. Other examples of value added applications services include:

- Asset management and optimization
- Supply chain integration & business process management
- Customer support
- Security management.

With an increased understanding of consumer behavior there will be new opportunities for the development and marketing of user centric value added products and services [Mourshed 2015]. In addition, consumers will benefit from the increased competition in a liberalized energy market that is integrated with other infrastructure components.

However, implementing Smart Grid infrastructure is investment intensive. Buy-in from all stakeholders is essential to unlock the full potential of the evolving energy value chain, as suggested in this diagram of several entities depicted in Figure 11.5. However, markets and energy supply chains are also based on regulations that are specific to each country and specific geographic areas that can influence the market model.

Connections to Home Area Networks (HAN) are needed so the protection of these networks has to be addressed. HANs are important as means for utility companies to extend their reach beyond meters and incorporate smart thermostat, direct load control appliances, smart appliances and in home energy displays into utility systems, as well as enabling DR and energy efficiency programs. Also, an Advanced Metering Infrastructure (AMI) uses a smart electric meter or other energy gateway to enable continuous two way communications between utilities and HAN based devices.

Table 11.3 Major electricity sector challenges along the value chain.

Sector	Challenges
Generation	Renewable energy generation Distributed, small scale electricity generation
Transport (Transmission & Distribution)	Transmission and distribution grid management
Storage	Storage capacities (physical and logical)
Retail	Dynamic and real-time pricing for electricity consumption and Distributed generation
Consumption	Electricity conservation and energy-efficiency (Automated) demand management Integration of electric vehicles (and renewable energy sources) Facilitate access to electricity in developing countries (Electrification)

Source: [OECD 2012b]. © 2012, OECD.

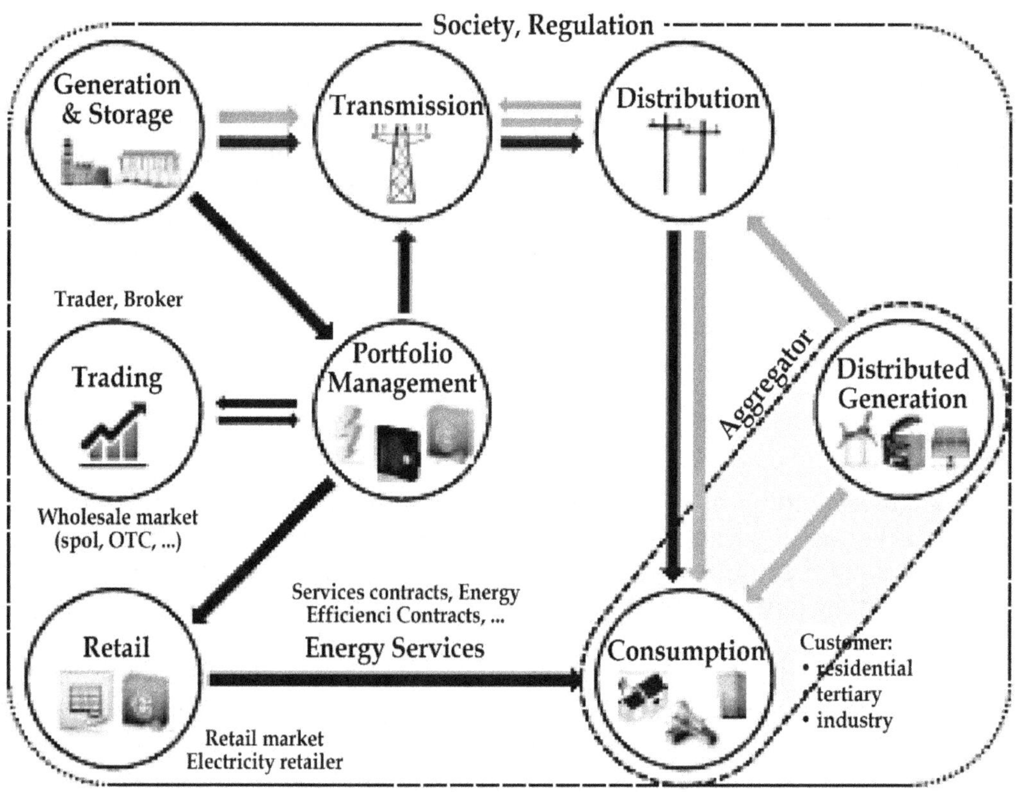

Figure 11.5 Energy value chain in a smart grid. *Source:* [Mourshed 2015]. © 2015, Elsevier.

11.7.6.4 ICT Roles

A look at the traditional energy sector value chain translates global energy sector challenges to tangible areas for action where ICT technologies already provide solutions or might be able to do so in the future. Thus, the use of ICTs and Internet applications are at the center of this modernization. ICTs promote [OECD 2012b]:'

- A wider integration of renewable energy sources, promoting low carbon transport options including electric vehicles and inducting structural shifts in electricity consumption. Innovative applications for final consumers clearly revolve around the smart meter (e.g., over 10% of an individual household's electricity consumption can be cut by simply providing better information).
- Improved monitoring and networked IT systems help limit losses of electricity along the way and thereby improve capacity utilization and avoid pollution; such losses represent on average 8% of production worldwide but over 15% in individual countries.
- Emergence of new value chain entrants and business models; one example are electricity supply aggregators operating virtual power plants. Specialized IT services and infrastructure providers develop targeted solutions for the electricity sector. However, the emergence of Smart Systems solutions for different verticals will present challenges for IT services vendors. For example, smart meter solutions for utilities are not just a collection of different products from many different vendors that a Software Integrator would assemble as a general contractor as they have in the past. These are vertical solutions requiring more industry driven experience [Harbor 2011].

11.8 Standards, Guidelines, and Recommendations

As our societies become increasingly dependent on information technologies, effective practical legal means will have to be employed to help manage the associated risks. Regulatory and compliance requirements will continue to plague the organizations. Regulations and laws will have direct impact on IT implementations and practices. Management teams will be held accountable. Civil and criminal penalties may apply for noncompliance. Security audits will become more widespread as companies are forced to comply with new regulations and laws.

11.8.1 Laws and Regulations

Currently, there are U.S. federal laws that have direct impact on the information security communities. Examples of significant laws include Gramm-Leach-Bliley Act (GLB), Sarbanes-Oxley Act (SOX), and Health Insurance Portability and Accountability Act (HIPAA). These laws address different topics of information security, such as protecting confidentiality and privacy of information or requiring the documentation and audit trails for financial data and transactions. Noncompliance with these laws can bring significant financial and legal liability. However, more needs to be done to ensure the security of information and people using the digital devices. In this regard, the European Union Parliament published a new legislation.

11.8.2 General Data Protection Regulation (GDPR)

The mutually agreed General Data Protection Regulation (GDPR) [EU GDPR] came into force on May 25, 2018 and was designed to modernize laws that protect the personal information of individuals. The Europe's GDPR is the most important change in data privacy regulation in 20 years. The key changes include:

- Increased Territorial Scope (extraterritorial applicability) as it applies to all companies processing the personal data of data subjects residing in the European Union, regardless of the company's location.
- Penalties may be applied if organizations breach GDPR.

- Consent and the conditions for consent have been strengthened and it must be as easy to withdraw consent as it is to give it.
- Right to breach notifications are now mandatory in all member states where a data breach is likely to result in a risk for the rights and freedoms of individuals within 72 hours of first having become aware of the breach.
- Right to access data and to obtain confirmation from the data controller as to whether or not personal data concerning them is being processed, where and for what purpose ensures transparency.
- Right to be forgotten (data erasure) t to be forgotten entitles the data subject to have the data controller erase his/her personal data, cease further dissemination of the data, and potentially have third parties halt processing of the data.
- Right to receive the personal data concerning an individual and transfer data (data portability) to another controller.
- Privacy by design means inclusion of data protection from the onset of the designing of systems; the controller shall implement appropriate technical and organizational measures in an effective way to protect personal information.

The EU GDPR does not impose data transfer restrictions on data flows within the European Union (registration and notification requirements still apply). However, the EU GDPR has a global effect. It does regulate transfers of personal data to and from non-EU countries. Personal information can be transferred only to jurisdictions outside of the European Union that can demonstrate an adequate level of protection for the personal information or have another basis for the transfer. Therefore, the GDPR legislation requires action from all organizations and customers doing business across Europe.

Chapter 12

Ethics and Awareness - Smart Grid Challenges

12.1 Smart Grid Challenges

Smart Grid incorporates numerous advanced technologies to deal with issues prevailing to conventional electric networks. Smart metering could significantly improve knowledge of what is happening in the distribution grid, which nowadays is operated rather blindly [IEC Roadmap].

As described in [Farhangi 2010], Smart Grid is the collection of all technologies, concepts, topologies, and approaches that allow the silo hierarchies of generation, transmission, and distribution to be replaced with an end-to-end, organically intelligent, fully integrated environment where the business processes, objectives, and needs of all stakeholders are supported by the efficient exchange of data, services, and transactions. It empowers consumers to interact with the energy management system to adjust their energy use and reduce their energy costs. Therefore, Smart Grid deployment requires active customers with flexible consumption.

Several reports and research papers discuss challenges and issues in Smart Grid implementation including consumer's acceptance. These challenges can be classified as technical and socio-ethical challenges that are associated with the adoption of emerging technologies, socio-economic issues, lack of policies and awareness. Although considerable acceptance of technical advancements has been achieved in Smart Grid developments, still there are issues that need to be addressed.

12.1.1 Deployment Challenges

While the benefits of the Smart Grid are widely recognized, Smart Grid is still facing challenges in deployment in many countries (e.g., India). Some evidence suggests that either the net benefits have been overestimated or incentives are not aligned for current utilities and customers to fully modernize the grid [Guo 2015]. The authors argue that real world experiences with Smart Grid technologies have not always lived up to expectations, and the benefits cited by Smart Grid proponents have sometimes failed to materialize. The authors argue if these barriers can be overcome, however, existing utilities could have significant opportunity for entrepreneurial activity. The development, deployment, and adoption of different technologies affect several groups in different ways [Guo 2015]:

- Power sector manufacturers are affected by the shift in demand for new technologies.
- Electric utilities need to find ways to economically upgrade their legacy systems.
- Consumers are faced with decisions about whether and the extent to which they adopt emerging technologies.

Due to increasing complexity, the grid requires numerous technological and business process changes [GridWise 2014].

This work [Moursheda 2015] stresses the need of the integration of Information and Communication Technology (ICT) and energy infrastructures as one of the most critical factors for the successful evolution of Smart Grids. The authors propose a decentralized grid architecture based on interoperability standards as the basis for future proofing the process of integration. A decentralized grid architecture can be implemented to account for increasingly dynamic and distributed nature of the assets (generation, consumption, monitoring, control and management) including development of innovative business models to unlock the full potential of the evolving energy value chain.

12.1.2 Policy Challenges

The following challenges are overarching in nature, they provide the context for a closer look at ICT specific policy challenges as discussed in [OECD 2012b]:

- Information and communication asymmetries across the electricity sector value chain remain an important issue to tackle, and with them the need for effective and reliable communications channels. Final electricity consumers, in special residential consumers, have little effective means of obtaining information about the current state of electricity production, its availability, cost and environmental impacts. Utilities have relatively little information about disaggregated electricity consumption patterns below the distribution system level and losses of electricity along transmission and distribution lines are not always accounted for systematically.
- Economic and financial hurdles may impact the needs for high investments in research, development and deployment because they are necessary to modernize national electricity systems. National grid assets in OECD countries are often several decades old and function close to maximum capacity. In many emerging economies, electricity infrastructures are greenfield investments – this means more advanced technological levels from the start, but it requires higher investments too.
- Consumer acceptance, engagement and protection require improved information on energy use and better access to it can bring substantial social, economic and environmental benefits. Mediated or automatic control of electric devices can help manage electricity demand and lower electricity bills. Benefits of the latter sort can be important to low income groups spending relatively large parts of their household income on energy. Although the smart meter could provide valuable information, it also adds a level of complexity to an area of consumption that so far used relatively simple tariff structure. Dynamic pricing of electricity will change that, making electricity prices dependent on levels of electricity supply, demand and their environmental impacts. These changes are necessary, but they will require well designed interfaces between the user and the technology. This requires behavior changes that can come about through guidance and education.
- Consumer concerns around the Smart Grid focus also on electricity provision to poorer and vulnerable parts of the population. Consumer rights groups highlight that smart meters potentially lower the operational barriers for utilities wishing to remotely turn off electricity supply or switch customers to more expensive prepaid tariffs. It therefore seems necessary to meet these new technical possibilities with improved legal and other safeguards for concerned customers. Otherwise, the smart meter risks facilitating wider diffusion of controversial business practices.
- Who bears the initial costs of smart meter deployments is causing substantial debate among utilities, consumer associations and policy makers. The costs per installed meter are estimated to be around USD 500, including the required ICT infrastructure in the back end. The role of the customer in supporting this cost is far from clear.
- Risks of converging Information Technology (IT) and Operational Technology (OT) generate more scenario that impact security, resilience, privacy and exploitation of personal data. Operational systems that exchange data with IP based networks need to be designed for security and resilience. Critical infrastructures converging with information infrastructures require scenario building that includes consideration of highly unlikely types of events. Moreover, IT security considerations need to be integrated within the wider risk management framework of the organization or the operational system.

- Upholding these attributes, availability, integrity, confidentiality and authenticity, is a key concern. The smart meter is likely to become a key node for managing information about the electricity system (e.g., grid loads) and about final customers (e.g., preference profiles for charging of electric vehicles). Questions must be addressed about unauthorized access to electricity data, the prevention of malware in the smart meter and connected devices and other potential security threats. Trends towards greater automation and remote control need to be accompanied by policies that can guarantee integrity and authenticity of information. The smart meter sends and receives control signals that directly impact the functioning of associated devices, e.g., electric vehicles, domestic appliances and small scale energy generators. Availability, integrity and authenticity therefore need to be assured across the entire value chain of the control signal. This includes pricing information sent by utilities, capacity information sent by grid operators, control signals sent by aggregators and virtual power plants operators. But it also includes communications channels that enable exchanges of information between the smart meter and connected devices.
- Personal data access, possession and ownership have both economic and ethical impacts. Since the economic value of personal data is rapidly increasing, both in personally identifiable and aggregated forms, safeguards are necessary to prevent the illicit use of certain data for commercial and criminal purposes. Privacy protection principles need to be adhered to when designing Smart Grid projects, including domestic legislation and international policy instruments. Furthermore, early involvement of consumers and consumer associations is important in order to generate "buy-in". Surveys of consumer opinion show that privacy issues are of major concern and have the potential to steer the public debate on Smart Grids and smart metering in ways that would overshadow many of the socio-economic benefits. Regulation in various countries, for example, requires that personal data be distinguished from non personal data because of the limitations on what utilities and grid operators are allowed to do with personal data without prior consent by consumers. Possession, e.g., storage, of personal data needs to be guided by precautionary principles because a growing number of parties (e.g., aggregators, applications developers, IT service providers) are interested in exploiting the business potential of access to personal data. Finally, ownership (and interoperability) questions need to be addressed to avoid lock-in of customers to specific utilities and service providers. In this context, should there be opt-out clauses whereby customers can choose not to engage in schemes for remote control. Development and implementation of effective policies in these areas require coordination with user associations, consumer protection and law enforcement agencies.
- The environmental and commercial success of the electric vehicle will be a function of individual and systemic variables. On the micro level, consumers will mainly compare affordability, comfort and operating costs with those of traditional cars. On the macro level, electric vehicles will be measured in their contribution to lower transport related greenhouse gas emissions, reduced road congestion and smooth integration with existing installed capacities for electricity production. Other stakeholder interests can be the source of further tradeoffs (see Figure 12.1).

However, overarching policy issues need to be addressed to improve coordination and flows of information between Smart Grid stakeholders, to explore sustainable financing options for Smart Grids and to ensure acceptance by and engagement of consumers and society at large. Also, the concepts differ greatly in the main regions and this is also reflected in the respective roadmaps and studies.

12.1.3 Technical Challenges

Technical challenges are basically difficulties in the integration of several devices, technologies, and business models with the existing grid network such as:

- Inadequacies in grid infrastructure in some countries where the grid infrastructure is still evolving, the integration of renewables is challenging in many aspects.
- Data management is critical due to handling high volume of data collected more frequently from meters, sensors, and controllers.

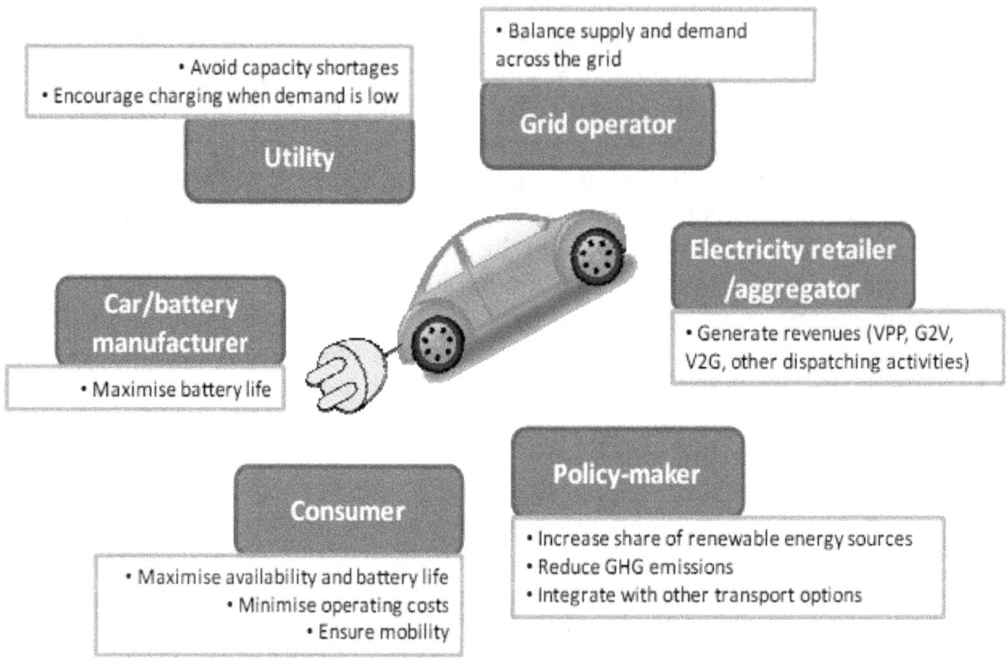

Figure 12.1 Stakeholder interests in operating electric vehicles. *Source:* [OECD 2012b]. © 2012, OECD.

- Storage concerns are related to the various technologies due to the availability, efficiency, portability, costs, safety hazards (e.g., materials for chemical batteries), and life.
- Ease of use installations – e.g., rooftop solar PV system with and without battery backup, which could be considered user friendly, but consumers without technical expertise still need to understand the pros and cons of these configurations.
- Communication issues due to a wide range of communication technologies for home networks with their own limitations (e.g., bandwidth, reliable networks, etc.).
- Stability concerns due to high penetration of renewables and microgrids.
- Energy management and efficient utilization of electric vehicle require well defined controls and standards.

Power management relates to more than just coordinated power state management. It has been observed that the energy required to create an IT gadget can often exceed its energy consumption during its useful life [Kant 2009]. The grand challenge is to develop a science that can address energy management in the entire life cycle context, including manufacturing distribution, operation, and disposal. A formal understanding of energy and computation tradeoffs will lead to significantly more energy efficient hardware and software designs. A formal approach will hopefully lead to energy management solutions that are not only effective but also easy to implement and validate. To improve energy efficiency, there should be considered the end-to-end energy use of a task involving multiple computer systems. Redefining energy efficiency is also discussed in this work [Lu 2011].

Similarly, the solutions to manage and harness the value from distributed energy resources (DER) across stakeholders becomes a more urgent problem [Kane 2017]. Visibility of DER value at different locations and times is an enabler for enhanced DER participation in distribution networks [Li 2014] and the full development of the system flexibility market for DER.

12.1.4 Socio-Economic Challenges

While economy is the primary factor for developing any technology, social acceptance also becomes crucial for its success. Socio-economic scenarios play a vital role in implementation and success of any technology. A technology becomes irrelevant if it fails to attract the investors or users, leading to failure of pilot projects, rejection of new technology, etc. Sometimes such issues may arise as a result of some economic or technological and some sometimes due to lack of appropriate awareness among stakeholders. Following are some examples of socio-economic challenges and ways to address may include:

- High capital investment could be unaffordable for some countries, so awareness programs and incentives are essential to encourage utilities, organizations and individuals to understand the Smart Grid benefits and cost burden.
- Flexible approach is needed from place to place for Smart Grid deployment; operational aspects such as billing, tariff structure and operational strategies allow utilities to influence the Smart Grid deployment.
- Utilities should focus on consumer awareness programs to teach about power delivery system and the role of Smart Grid in building economy and efficiency.
- Consumers need to be informed about economic and environmental benefits of the technology.
- Policy makers and regulators also need appropriate awareness training; they must understand the potentials of the present and future uses of the technology.
- Privacy concerns should be addressed to increase the confidence and trust; therefore, sound security measures and regulations should be mandatory for Smart Grid acceptance.
- Technology obsolescence and eventually higher costs associated with the new technologies are concerns to developers of Smart Grid, so these should be addressed appropriately.
- Consumers need to understand the price schemes for electricity use; consumers need to understand about new tariff approaches pursued by the utilities and government.
- While low tariff at off peak loads is attracting some consumers, some consumers do not like constraints to their electricity usage, so they are opposing tariff hike during peak periods.
- Lack of regulations and policies slow the process of Smart Grid development.
- Coordination among stakeholders and various industries is crucial. Smart Grid is a collective enterprise of power, information and communication industries, policy makers, system operators, regulators, government, utilities, manufacturers, business market, economists as well as consumers.
- Another concern is lack of Smart Grid expertise that is - expert of one sector is oblivious to other domains; proper coordination of training and education among different groups is vital including joint agendas of ICT organizations and utilities, ICT and energy policy makers, as well as bridging dispersed academic and civil society communities around the Smart Grid benefits.

12.2 Ethical Smart Grid

As more technologies including Internet of Things (IoT) and Artificial Intelligence (AI) are used for Smart Grid applications can result in benefits to the environment and stakeholders, but also to undesired effects to the customers and society.

12.2.1 A Smart Grid Path Review

As Smart Grid programs have been initiated, utilities believed that investing in distribution automation could provide them with increasing capabilities over time. Within the context of these new capabilities, communication and data management play an important role. These basic ingredients enable the utilities to place a layer of intelligence over their current and future infrastructure, thereby allowing the introduction of new applications and processes in their businesses. Figure 12.2 depicts a graphical representation of the increasing capabilities expected by utilities.

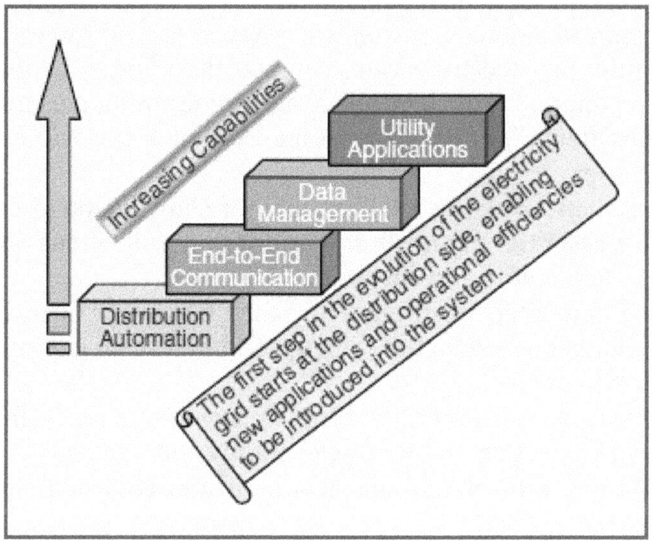

Figure 12.2 Utility desired capabilities. *Source:* [Farhangi 2010]. © 2010, IEEE.

One approach towards Smart Grid implementation was the distribution automation and introduction of advanced meters. As Figure 12.3 shows, the metering side of the distribution system has been the focus of most recent infrastructure investments. The earlier projects in this sector saw the introduction of automated meter reading (AMR) systems in the distribution network. AMR lets utilities read the consumption records, alarms, and status from customers' premises remotely.

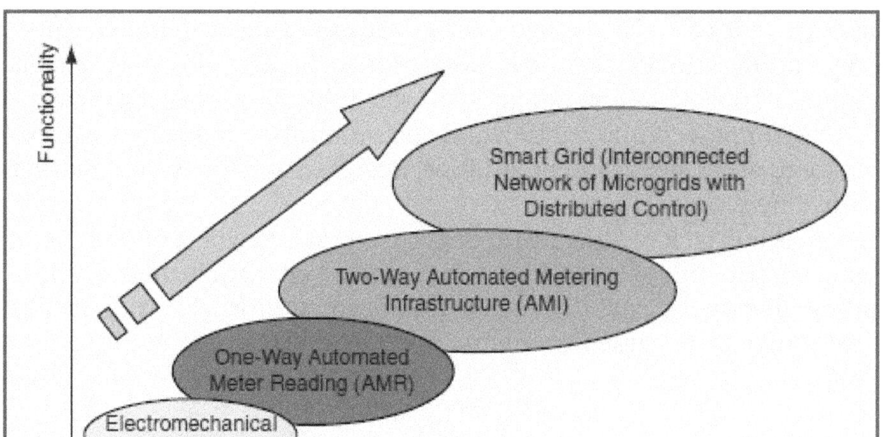

Figure 12.3 The evolution of the smart grid. *Source:* [Farhangi 2010]. © 2012, IEEE.

Since AMR is using one way communication system, the AMR capability is restricted to reading meter data. It does not let utilities take corrective action based on the information received from the meters. Therefore, utility organizations have realized that AMR does not address the major issue - demand side management, which they need to solve.

Instead of investing in AMR, utilities across the world moved towards advanced metering infrastructure (AMI). AMI provides utilities with a two way communication system to the meter, as well as the ability to modify a customer's service level parameters. Through AMI, utilities can meet their basic targets for load management and revenue protection.

As transitioning to the Smart Grid, the next logical step is the need to leverage the AMI infrastructure and implement its distributed command and control strategies over the AMI backbone. The pervasive control and intelligence that elevates the Smart Grid has to reside across all geographies, components, and functions of the system. Distinguishing these elements is significant, as it determines the topology of the Smart Grid and its comprising components.

Furthermore, the organic growth and evolution of the Smart Grid is expected to come through the plug-and-play integration of certain basic structures called intelligent (or smart) microgrids [Farhangi 2010]. Microgrids are defined as interconnected networks of distributed energy systems (loads and resources) that can function whether they are connected to or separate from the electricity grid. Figure 12.4 depicts the topology of a smart microgrid. A smart microgrid can be characterized as follows:

- Network that can operate in both grid tied as well as islanded modes.
- Typically integrates power plants and renewable sources of energy, such as wind, solar, and biomass.
- Capable of meeting local demand as well as feeding the unused energy back to the electricity grid.
- Services a variety of loads, including residential, office and industrial loads. It makes use of local and distributed power storage capability to smooth out the intermittent performance of renewable energy sources.
- Incorporates smart meters and sensors capable of measuring a multitude of consumption parameters (e.g., active power, reactive power, voltage, current, demand, and so on) with acceptable precision and accuracy.
- Smart meters should be tamper resistant and capable of soft connect and disconnect for load and service control.
- Incorporates a communication infrastructure that enables system components to exchange information and commands securely and reliably.
- Incorporates smart end points, loads, and appliances capable of communicating their status and accepting commands to adjust and control their performance and service level based on user and/or utility requirements.

Farhangi [Farhangi 2010] describes the emergence of these smart microgrids and the degree of their interplay and integration as a function of rapidly escalating Smart Grid capabilities and requirements. However, not all microgrids are being created equal. It is expected that microgrids could be built with different capabilities, assets, and structures, depending on some factors. Examples of factors include diversity of load, mix of primary energy sources, geography and economics at work in particular areas, among other factors.

12.2.2 Smart Grid Implementations and Ethics

Currently, worldwide Smart Grid programs are focused in objectives such as energy efficiency, energy independence and climate change. One of the Smart Grid goals is the integration of the increasing share of renewable generation while efficiently accommodating their variability and limited predictability through the actuation of consumer flexibility.

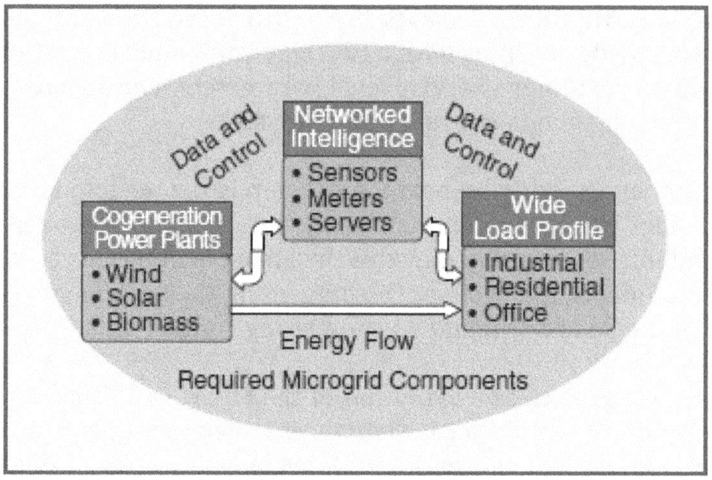

Figure 12.4 The topology of a Smart Microgrid. *Source:* Farhangi 2010]. © 2010, IEEE.

The World Energy Council's definition of energy sustainability is based on three core dimensions: Energy Security, Energy Equity, and Environmental Sustainability of Energy Systems. Balancing these three goals constitutes a 'Trilemma' and balanced systems enable prosperity and competitiveness of individual countries. Figure 12.5 is a graphical representation of the dimensions for the balanced Energy Trilemma Index. The three dimensions are defined as follows:

- Energy Security is defined as effective management of primary energy supply from domestic and external sources, reliability of energy infrastructure, and ability of energy provider to meet current and future demand.
- Energy equity is the accessibility and affordability of energy supply across the population.
- Environmental sustainability encompasses achievement of supply and demand side energy efficiencies and development of energy supply from renewable and other low carbon sources.

The World Energy Council considers that achieving high performance on all three dimensions entails ethics and complex interwoven links between public and private actors, governments and regulators, economic and social factors, national resources, environmental concerns, and individual consumer behaviors. Countries take different pathways in the context of energy transition, driven by changes in markets, politics, and society.

12.2.3 Addressing Smart Grid Ethics

According to this study [Hatzakis 2019], the use of Smart Grid technologies can help solve the Energy Trilemma defined based on this goal: how to secure (energy security) affordable energy for all (energy equity) in a sustainable manner (environmental sustainability).

One of the Smart Grids' goals is the integration of a growing share of renewable energy sources while efficiently accommodating their variability and limited predictability through actuation of consumer flexibility. Smart grid management systems require the analysis of real time energy consumption data and energy production data.

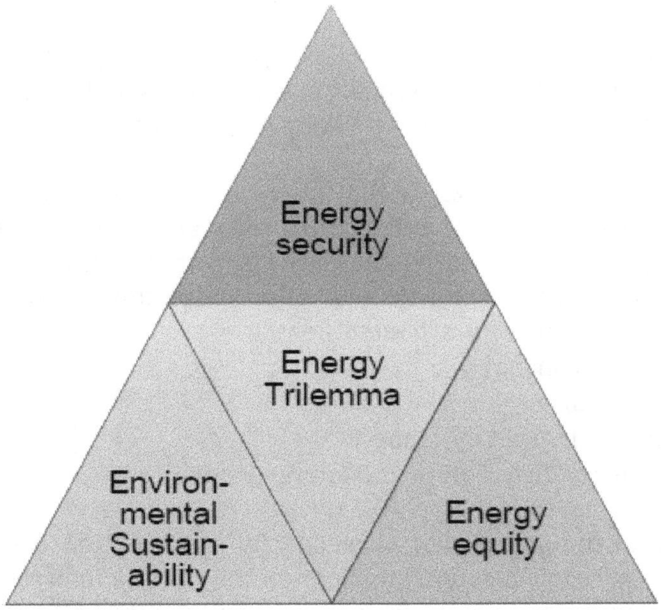

Figure 12.5 The three dimensions of the energy trilemma. *Source:* World Energy Council. © World Energy Council.

To reach the 20/20/20 targets, the European Union aims at modernizing power grids to make them smart by collecting close to real time data and subsequently operate grids more optimally [Le Ray 2020]. To do so, 80% of the customers throughout Europe were planned to be equipped with smart meters by 2020. Achieving this target is dependent on customers' acceptance to install smart meters on their site. The European Union regulation defines a legal framework in which the utilities can implement the deployment and exploitation of smart meters with more or less focus on customers. Bad practices during the roll out (e.g., installing smart meters without informing customers) are not gaining the customer's trust and confidence in a new technology. In addition, bad practices give a negative reaction to a customer, at the end not accepting the technology. Contrary, ethics define good practice for a positive relationship between customers and utilities.

The success of energy programs relies on customer involvement in altering their energy consumption through the use of analytics and incentives on demand side management. The fact that it is required to improve the use of renewables or that it is imposed by legislation are not arguments supporting the implementation of the Smart Grid goals. Customers do not adopt or at least accept the technology in a first time if they do not find any benefits for themselves.

Fact is that utilities are arguably showing limited ethics in the way smart meters are rolled out and on the use of collected data [Le Ray 2020]. The authors argue that beyond legal binds and technical obstacles, the deployment of smart meters and the use of collected data is unclear in terms of how to deal with customers. Therefore, ethics should be an import driver for decision making to guarantee the sustainability of Smart Grid programs which relies on the active participation of customers. Also, the perspective of the long term development of Smart Grid can change the status of consumers to prosumers, so they should be treated as collaborators providing services to the grid. Hence, the authors recommend a balance between the necessity of data to build the Smart Grid and the respect of the customers as partners has to be determined. A balance between risks and the benefits for customers and the utilities has to be considered. The authors highlight that only the practices that exhibit ethical behavior of the utilities towards customers and consider them as stakeholders in smart grids will lead to a fruitful and long lasting relationship between customers and utilities.

In addition, utilities need to evaluate the diverse energy needs for specific communities. This work [Podder 2017] shows the importance of simulations and analysis of energy needs before deployment of solar system with a battery backup for a community in Bangladesh.

The literature review in this work [Hatzakis 2019] reveals mainly ethical issues around health and safety, privacy and informed consent, cybersecurity risks and energy security, affordability and equity, and sustainability. The authors also explore the principal ethical issues that occur in the use of smart information systems in smart grids and offer suggestions as to how they might be addressed. The following ethical concerns are identified:

- Privacy and informed consent.
- Security of the Smart Grid.
- Data stewardship and market dynamics.
- Prioritization of energy distribution and energy justice.

Privacy concerns relate to the granularity of electricity consumption data that can be collected about a household and a person's lifestyle information that can be inferred from it. As reading intervals of data collected by smart meters is reduced (e.g., at intervals of 1 minute), more information about a household's activities or appliances use can be inferred. For instance, smart meters can tell when someone is at home, whether they are cooking, or watching TV based on the monitoring the energy consumptions patterns (loads) of such appliances. Therefore, the ability to detect specific activities depend on the time resolution of the consumption of the data and the frequency of data collected by smart meters. In addition, the use of tools to process the data make a distinct outcome. For example, quarterly statistics may reveal the mean energy usage per household, while intelligent systems based on Artificial Intelligence techniques can infer more private information about a customer.

The authors identify the presence of an underlying mistrust towards energy players and governments about their positions and commitments towards practices that benefit the society as a whole once the implementation takes place. The analysis of granular information on household energy usage via smart meters is another issue. The smart meters are perceived that can be used to reveal detailed information about people's private lives within the home, raising serious questions at a technical and policy level about in home surveillance and how to address consumers' privacy interests. Therefore, mechanisms for addressing ethical issues need to be included in the strategy of the organizations when engaging with the customers to support the Smart Grid developments. Also, the authors recommend further exploration of ethical issues in the use of smart information systems in the energy sector. Piloting research needs to take place to understand how Smart Grid developments change social life, interactions between people in a community, their values and local culture.

Another work [Tzafestas 2018] argues that regulations in the European Union and United States for the Internet of Things (IoT) and their characteristics have the potential to create ethical problems such as: IoT ethical questions and principles; IoT security, privacy, and trust aspects; and the ethical culture of IoT related companies. The range of IoT applications is continuously increasing in several domains and each domain involves several distinct applications. Examples of applications include smart home, smart grids, connected cars, industrial IoT, smart supply chains, smart retail, smart banking, smart investment, smart insurance, and smart farming. While IoT technologies have potential to solve many real life problems, they cause serious ethical concerns and legal challenges related to protection of privacy, data security, data usability, data user experience, trust, safety, etc.

The analysis of recent surveys [Tzafestas 2018] across IoT related companies and IoT users in several developed and developing countries revealed that the large majority of IoT consumers lack trust in the IoT devices and favor more government regulations to protect data access/use, privacy, and safety. These issues are becoming increasingly crucial because of the continuous expansion and penetration of the IoT in public affairs, private enterprises, and personal human life/activity. This research combined with the possible impact of Artificial Intelligence in modern digital society, calls for urgent reactions on the part of the scientific community, governments, and professional societies, to quickly take proper measures or enhance existing ones to slow down and minimize negative implications.

12.3 Smart Grid Work Force

Smart Grid has many obstacles to overcome including the scarce work force. There is this view that finds IT and Industrial Control Systems (ICS) personnel as opposing forces due to different cultures. Apart from their differences in functionality and technology, ICS and IT historically have been managed by separate organizational departments. IT people tend not to talk about physical processes, while control engineers are focused on keeping the processes running. The number of people understanding and bridging both domains is limited. The number of people understanding cybersecurity for both IT and ICS, and how to adopt an integrated security approach, is even more limited. However, as more IT technologies are being used in ICS, closer cooperation between the ICS and IT domains is desirable.

Meeting the needs of Smart Grid stakeholders requires consideration not only of the end state when the Smart Grid vision is realized, but the evolutionary period to that state during which the legacy infrastructure will be used side by side with new technologies. To transition legacy networks into intelligent and more secure electrical infrastructures, utilities will need to define the Smart Grid objectives best suited to their own customers and community and develop the best path to achieving standards based Smart Grids based on interoperable solutions and flexible business processes [NISTIR 7628r1]. Also, this document stresses it is important that all parties take the time to understand the traditional approach the other has taken and learn how those principles can be applied in the new system.

This publication [NIST SP800-181r1] from the National Initiative for Cybersecurity Education (NICE) describes the Workforce Framework for Cybersecurity (NICE Framework). Figure 12.6 depicts de conceptual building blocks of the framework which could help organizations overcome the barrier of describing their workforce to multiple stakeholders by presenting a building block approach. It presents a common language for organizations to use internally and with others.

The framework is based on concepts such as:

- Work describes an organization needs to achieve cybersecurity risk management objectives.
- Tasks are activities to describe the work performed by organization; there could be common tasks as well as some context-unique tasks; for example, every organization has some form of management tasks, but only some organizations have tasks to deploy bulk energy systems securely.
- Knowledge and skills is a way to describe an organization their work.
- Learner is the person who has knowledge and skills. A learner can be a student, job seeker, employee, or other people within the workforce.

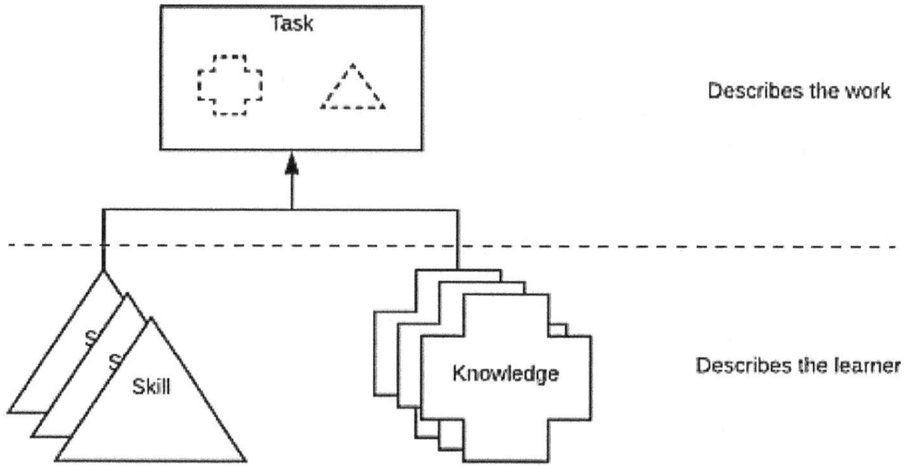

Figure 12.6 NICE framework building blocks approach. *Source:* [NIST SP800-181r1]. Public Domain.

The NICE Framework considers the cybersecurity workforce to include not only those whose primary focus is on cybersecurity, but also those who need specific cybersecurity related knowledge and skills to properly manage cybersecurity related risks to the enterprise.

12.3.1 Expected Skills

The implementation of the Smart Grid blends skills from the generation, transmission and distribution groups with the skills of information and communication technology groups. Other than technical and socio-economic challenges, there are few more issues that demand concerns from some general perspectives.

Addressing these issues requires to meet expectations of every sector involved including consumers who help shape the environment in which the electric market operates. These issues cannot be resolved only by electricity industry professionals, consumers of electricity, researchers, and policymakers. Therefore, Smart Grid implementation requires skilled staff.

Therefore, a comprehensive program of education, training, practical experience, and awareness is necessary [NISTIR 7628r1]. Also, professionalization and certification licensing provide a validated and recognized expert cadre of system administrators. There is a need for legitimate focus on Smart Grid work force creation and awareness initiatives [Kappagantu 2018].

These authors [Guo 2015] point a lack of customer awareness of basic rate structures and price changes and a need for better customer education about smart meter technology. If consumers are more engaged in demand/response (DR) program, they could be increasingly important to enhance the overall system efficiency and effectiveness. Having detailed information on electricity consumption, consumers could make more informed decisions on how and when they can save energy, either by changing their behavior or by engaging with an energy efficiency service provider.

Also, the workforce demand in the electrical power industry is also evolving. For example, policy makers, and legal regulators need to understand the technical as well as social aspects of the technology (see also [Arnold 2011], [OECD 2012a]). This document [GridWise 2014] recommends developing a workforce with the right combination of information technology and operational technology skills to support the grid's new functions. Grid operators, particularly at the distribution level, will need further education to manage the increasing complexity of operations.

In this work [Hou 2008], the authors recommend that protection engineers need to understand and have more computer and networks knowledge in order to gain full benefit of the new ingredients of the [IEC 61850] series that are targeted to protection and real time controls. The authors argue that communications standards and protocols are no longer only for substation integration engineers. To effectively use the available features for protection, protection engineers must understand the communications and information systems. For example, engineers need to understand that the [IEC 61850] communications standard is based on popular Ethernet technology (the IEEE 802.3) and other protocols are involved in the communication of the messages between various devices. This standard has gained worldwide applications in power substation integration, automation, and control. It brings the benefits of cost savings in Engineering design, installation, commissioning tests, operation, and maintenance. Therefore, power protection engineers will have to learn the new technology, terms, and substation paradigms.

The participants at the workshop organized by National Science Foundation in April 2009 identified the need for more scientific investigation of power/thermal issues and development of good metrics and models. In addition, it was stressed the importance of multidisciplinary education so that future computer scientists are better prepared to deal with energy issues and numerous sustainability challenges inherent in IT equipment and infrastructure design.

The increased connectivity to Internet and mobile device technology has a major impact on control systems software design and system architecture. These systems/products must evolve very rapidly to take advantage of new market opportunities and emerging technologies. Control systems are becoming very complex software applications, and the software engineers responsible for designing and maintaining these control systems must be trained innovatively [Hentea 2007]. This paper discusses the complexity of new generation of control systems and the need for greater compliance to safety, quality of service, and security of systems and data. All these issues impact the education and training of software developers and control engineers. Software developers are required to ensure that software is stable and reliable to avoid hazard conditions. Besides mastering control methods, the control engineers must be able to deal with the challenges created by emerging technologies, and the capabilities of the users of these complex control systems. These challenges demand adoption of new emerging technologies and new competencies and skills. These competencies allow open possibilities for applying a wider range of control principles and design methods as well as productivity control and risk management involving decision making under uncertainty with increased levels of decision support.

Also, the software intensive system design skills for the construction of control systems are often misunderstood. In control industry, two separate groups of engineers are typically involved in the development of any nontrivial controller: control engineers and programmers. These two groups tend to have very different perspectives and working practices, and both lack the global picture needed for the task [Rinaldi 2001].

The industry is demanding power engineering education to broaden in a cross function of areas such as networking, control systems, power electronics, data analytics along with information security and business, that will contribute to match the technical engineering ingenuity with the dynamics of the Smart Grid [Shireen 2013]. These authors recommend changes to the current power engineering education to be more relevant to future industry needs. In the 21st century, a graduate must have multiple skills in cross disciplinary areas and capable to adapt to the future work environment.

Promising areas for new ICT jobs and competencies include cloud computing, green ICTs and smart applications. The last two have been promoted in government green growth stimulus packages. ICTs are needed in areas such as Research and Development, production and deployment of green ICTs, manufacturing semiconductors for energy efficiency and clean technologies such as photovoltaics and wind power and in ICT recycling services, as well as in the development and use of virtualization software. In the long term, demand for ICT related employment will continue to rise as economies become smarter and ICTs ever more pervasive.

The implementation of Smart Grids will require field specific knowledge of legal frameworks, environmental impacts, etc. [OECD 2012b]. According to this report, the emerging skills requirements are affected by a decline of attracting students to so called STEM subjects (science, technology, engineering, mathematics) in some OECD countries while immediate skills needs are accelerating at the same time. Moreover, the growing need for an operational understanding of electricity, transport and buildings management might require adapting curricula for engineering courses and other IT oriented education programs (see more information in Chapter 3 of [OECD 2010]).

There are now mainstream policies underpinning growth and jobs, increasing productivity, enhancing the delivery of public and private services, and achieving broad socio-economic objectives in the areas of health care and education, climate change, energy efficiency, employment and social development.

12.3.2 Information Communications Technology Impacts

As ICT applications and services have become ubiquitous, they are essential for ensuring sustainability throughout the economy [OECD 2010]. As described in this report, ICTs are key enablers of green growth in all sectors of the economy and offer means of tackling environmental challenges and climate change. ICTs affect the environment at three levels: direct impacts, enabling impacts and systemic impacts, etc. (see Figure 12.7).

ICTs have considerable direct environmental impacts in terms of energy use, materials throughput and end of life treatment. Users and consumers can spearhead more sustainable growth through informed consumption decisions based on easy access to reliable environment related information. They also require information about how to use ICTs to improve the environment.

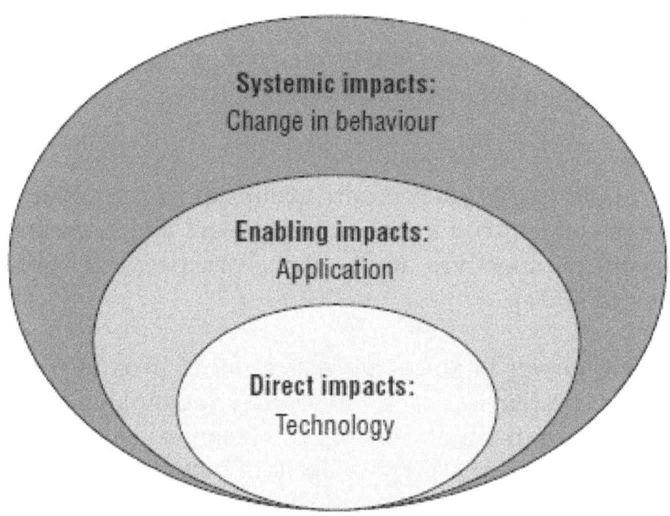

Figure 12.7 Three levels of environmental impacts of ICTs. *Source*: [OECD 2010]. © 2012, OECD.

ICT systems enable more sustainable production and consumption across the economy, ranging from product specific improvements (embedded ICTs for energy efficient vehicles) to entire systems (ICTs for smarter transport management). ICTs can lead to significant environmental benefits in buildings, transport and energy.

Sensor technology can help improve environmental performance, reduce greenhouse gas emissions and underpin green growth. Sensor technology and sensor network applications show particular promise for tackling environmental challenges in energy, transport, industrial applications, and smart buildings. In smart buildings, minimum standards of energy efficiency coupled with sensor technology can be a major factor in reducing electricity use and greenhouse gas emissions.

According to this report [OECD 2010], three categories of ICT competencies are distinguished as follows:

- ICT specialists have the ability to develop, operate and maintain ICT systems; ICTs constitute the main part of their job.
- Advanced users are competent users of advanced, and often sector specific, software tools; ICT is not the main job but a tool.
- Basic users are competent users of generic tools (e.g., Word, Excel, Outlook, PowerPoint) needed for the information society, e-government and working life (ICT is not the main job but a tool).

The first category covers those who supply ICT tools (hardware and software), and the second and third categories those who use them. ICT specialists are increasingly expected to have additional skills including business skills. Similarly, non-ICT related professions increasingly require at least basic ICT user skills.

Statistics (based on data around 2010) included in this report [OECD 2010] show a divergence between ICT specialists and ICT sector employment. This suggests that there is ongoing occupational specialization as higher level ICT skills are required. These skills are needed in the ICT sector as it restructures around more advanced products and activities, but they are also used to a greater extent across the wider non-ICT economy. This is because ICT specialist skills are required to produce both ICT products such as software in non-ICT sectors (e.g., financial services) and non-ICT products such as automobile systems with ICTs embedded in them.

Also, the OECD report includes information about women employment in the ICT sector and ICT specialist occupations that is significantly lower than men, but their share in employment is increasing in most countries. In 2009, the share of women employed in the ICT sector was around 30% in selected countries. This is almost double the share of women employed as ICT specialists (around 18%).

12.3.3 Education on Sustainable Development

As defined in [ETP 2010], a Smart Grid is an electricity network that can intelligently integrate the actions of all users connected to it - generators, consumers and those that do both – in order to efficiently deliver sustainable, economic and secure electricity supplies.

This paper [Makienko 2017] discusses the need of educating a work force in Russia to gain skills and knowledge not only for developing and implementing the Smart Grid technologies, but for assisting consumers to understand the benefits and limitations of the Smart Grid paradigm. While consumers can benefit from using the Smart Grid technologies, the studies show that in some countries, they are not aware of possible positive and negative impacts of the technology adoption.

A consumer is entitled to the information provided as well as to the information which they are interested in, such as resources, the economic feasibility, the impact on environment, business models, policies and regulations, the production management, etc.

Therefore, there is a need for educating consumers. As discussed in this paper, engineers can interact with the society and clarify possible impacts of the technology, so supporting the informed consent. That is the aspect which strengthens the social constituent of the engineer's activity in the modern world, and this means changing the goals of the engineering education as declared in the learning for the future concept [Learning 2012]. As knowledge and understanding of problems and solutions are formed through the education, this document states that the education is a condition for the sustainable development as the appropriate world view. Smart Grid technology is no exception, it is subject to the analysis against the sustainable development background.

The informed consent should be a condition for the adoption of the global technology such as the Smart Grid and stimulate its expansion among populations. Implementing this principle is possible on condition of the changing the goals of the engineering education according to the sustainable development. This need requires to change the structure of the Engineering education that is proposed in Russia - to include other goals such as social responsibility, stress resistance, ability to forecast the future. Thus, besides technical skills, an energy engineer should gain communication and social skills. When developing the necessary competencies of the future engineers, the authors raise the issue whether the combination of the sustainable requirements is possible in the training process. The authors argue that there is a conflict of the values in the modern society with the simultaneous commitment to consumption and saving.

According to the new concept of education, the training of social and humanity competencies is to be reflected in the qualifications of a graduate who should be able to make the professional decisions with the principles of social responsibility, meet social challenges and demands, be an efficient worker both in national and international teams.

12.4 Smart Grid Awareness

This document [GridWise 2014] recommends DOE should launch a national education campaign that lays the foundation for state and regional grid modernization efforts and builds a greater awareness of the need and value of grid modernization, resilience, security, and other pressing energy issues, ultimately resulting in an energy conscious culture. Using various methods and approaches, federal agencies, in partnership with states and regions, should facilitate collaborative exchanges to leverage lessons learned in order to expedite and lower the cost of modernizing the electric system.

12.4.1 Broader Education

As solutions should be driven through stakeholder engagement and education, the document recommends [GridWise 2014]:

- Defining a process for educating industry, policy, and regulatory stakeholders throughout the transition. Education will be critical for these stakeholders to stay abreast of the rapidly changing landscape and to understand the nuances of operations and how policies and regulations might affect utility operations. It will be critical for stakeholders to learn from each other and from existing and past efforts.
- Addressing common challenges and sharing more optimal and cost-effective solutions across different jurisdictions help on breaking down the stakeholders silos and perceptions that can occur.

- Convening stakeholders in a collaborative environment for open discussions will be instrumental in helping to develop innovative solutions. Examples of important topics to address include:
  - Single /multi customer microgrid policies, rules, and societal implications, including interconnection rules and obligation to serve
  - Ancillary services at the distribution level
  - Regulated versus unregulated services at the distribution level
  - Rate reform: volumetric versus a fee based structure, cross subsidies, etc.
- Educating consumers to understand the changes that could take place during the transition to the future grid as well as the value of energy and how to use it wisely. Consumers need to be informed about their options for energy use in order to understand, accept, and manage the changes in how they will receive and pay for electricity. It will be important that consistent messages are provided from multiple sources to avoid confusion and to garner trust.
- National level education provides the umbrella for empowering an energy conscious culture. Regulators and utilities should play important roles in consumer education: regulators can help consumers understand why investments are necessary and beneficial, and utilities need to educate consumers on the specific offerings available to them.
- Utilities have the role to keep consumers informed (in their language, not in technical terms) at the forefront as changes are implemented and consumers become more involved. Utilities need to incorporate marketing and educations efforts to: better understand consumer preferences so products and services can be created that meet their needs/desires; explain to consumers the options available to them and the accompanying benefits and value.

12.4.2 Deterring Unethical and Illegal Behavior

Information Security Professionals have the responsibility to deter unethical and illegal acts, using policy, education and training, and technology as controls to protect the information systems. Although most of security professionals understand technological means of protection, still there are many that underestimate the policy. In general, organizations should seek to eliminate these unethical categories of behavior [Whitman 2014]:

- Ignorance of the code of ethics, regulations, and laws.
- Accident of unintentional use of the authorization and permission credentials to access or misuse information or negligence on using controls to protect the information.
- Intent is the use of tools with the purpose to damage the security of information (e.g., computer abuse).

For preventing the unethical or illegal activities, organizations can use different measures to protect a system. The methods should include deterrence, prevention, and detection. This document [RFC 4949] specifies that providing a condition of system security may involve six basic functions (some overlapping may occur) as described in [Parker 1981]:

- Deterrence is reducing an intelligent threat by discouraging action, such as by fear or doubt.
- Avoidance is reducing a risk by either reducing the value of the potential loss or reducing the probability that the loss will occur.
- Prevention is impeding or thwarting a potential security violation by deploying a countermeasure.
- Detection is determining that a security violation is impending, is in progress, or has recently occurred, and thus make it possible to reduce the potential loss.

- Recovery is restoring a normal state of system operation by compensating for a security violation, possibly by eliminating or repairing its effects.
- Correction is changing a security architecture to eliminate or reduce the risk of reoccurrence of a security violation or threat consequence, such as by eliminating a vulnerability.

As evidence showed that information systems are vulnerable to dangerously high and persistent abuse, managers perceived this threat to be high. As managers increasingly decided to treat computer abuse as a behavioral and people problem rather than just a technical one, the function of security administration has been gradually incorporated into the life stream of American business. The organizational response to abusive potential has been to implement a computer security administrative unit with the charge of deterring and preventing computer abuse. This author [Whitman 2014] argues that abusive damage is significantly inhibited by the certainty and severity of punishment for abuse, preventive software, environmental motivational factors.

Despite of several standards, regulations, and laws including content labeling services, unacceptable content is still widely available. Recognizing the need for accepting and deterring WEB content from WEB sites has been approached by World Wide Web Consortium (W3C) more than two decades ago. The Platform for Internet Content Selection (PICS) was an attempt to provide a mechanism that supports a diversity of attitudes towards content and censorship [Chapman 2004]. However, it restricts the reception, not distribution. PICS aims to allow the users themselves to prevent unacceptable material reaching them.

PICS is a specification created by W3C that used metadata to label webpages to help parents and teachers control what children and students could access on the Internet. However, this specification was claimed to be too complex, so it was superseded by the W3C Protocol for Web Description Resources (POWDER) project, which integrates PICS concepts with Resource Description Framework (RDF). It specifies a protocol for publishing metadata about Web resources using these W3C specifications: Resource Description Framework (RDF), Web Ontology Language (OWL), and Hypertext Transfer Protocol (HTTP). The Protocol for Web Description Resources (POWDER) is the (W3C) recommended method for describing Web resources. According to [Chapman 2004], there are no regulation for labeling services, so there is room for abuse.

12.4.3 Organizational Liability

Security is developed and implemented to protect the organization's valuable resources; thus, appropriate safeguards need to in place to protect its mission and ensure the protection of its tangible and intangible assets, reputation, employees, customers, business partners, shareholders, and legal position.

Organizations need to develop not only their preventive, detective, and corrective approaches to computer crimes, but also their liability and responsibility approaches. When organizations work together in an integrated business, special care must be taken to ensure each party promises the necessary level of protection, liability, and responsibility. In US, service organizations such as hosted data centers, application service providers, managed security providers are required to have an auditing and testing, called SAS 70 (Statement on Auditing Standards No. 70: Service Organizations) [SAS 70]), performed to ensure that each party is indeed holding up its side of the agreement. The SAS 70 auditing is carried by a third party.

Liability is the legal obligation of an organization and can be applied to conduct even when no law or contract has been breached. Even if there is no criminal conduct that can be prosecuted under criminal code, there can be liability. Liability for a wrongful act includes the obligation to make a compensation for wrongdoing. If an employee, acting with authorization or without authorization, performs an illegal or unethical act, causing some degree of harm, the organization can be held financially liable for that action. An organization increases its liability if refuses to take measures - due care- such that every employee knows about acceptable behavior including the consequences of illegal or unethical actions [Whitman 2014]. Due care ensures that a minimal effort is allocated to ensure basic protection in accordance with the best practices in the industry. Also, an organization should take measures - due diligence – requiring an organization makes a valid and ongoing effort to protect others and properly investigated all its possible weaknesses and vulnerabilities [Harris 2013].

If an organization does not have security policies, the necessary countermeasures implemented, and security awareness training in place, it is not practicing due care and can be found negligent. The affected organization can sue the negligent organization. In the context of information security, due care means that a company did all it could have reasonably done, under the circumstances, to prevent security breaches, and also it took reasonable steps to ensure that if a security breach was realized, proper controls or countermeasures were in place to mitigate the damages [Harris 2013].

Organizations are acknowledged to be distinct legal persons that bear liability both for organizational failures and misconduct of their agents. Following the common law jurisdictions, in which notions of corporate criminal liability were introduced already in the early twentieth century, many civil law countries also recognize the possibility of holding companies criminally liable [Heine 2016]. However, distinctions in the use and theoretical underpinning of organizational liability remain across countries. Also, studies of ethical sensitivity to computer use have determined that individuals of different nationalities have various perspectives on the ethics of computer use [Whitman 2014].

12.5 Information Security Specialist Education

In today's technology based world, information security specialists prevent computers and digital data from being stolen, infiltrated, or corrupted. Information security specialists assist businesses and organizations in maintaining computer security by preventing hacks, viruses and unauthorized user access. To achieve this, they design, implement and maintain various computer programs and applications according to their roles within an organization.

Education, training, and awareness are countermeasures that effectively reduce exposure to a variety of known risks. The minimum education requirement for this occupation is typically an undergraduate degree or work experience combined with certification. However, employers generally prefer education requirements that include a bachelor's degree related to the computer sciences, information systems or information technology, and technology certification is often expected.

Beyond education, professionals may need to be certified in the use of specific technology, programming, and/or professional information security practices. Also, professionals in this industry must have problem solving skills and be able to pay attention to details. Demand for these professionals will remain high, as governments, private sectors, and societies become digital and huge amounts of data are collected and stored requiring more information security specialists to keep it safe.

12.5.1 Security Education, Training, And Awareness

Keeping employees up to date and educated on security matters is an important part of an organization security strategy. A security education, training, and awareness (SETA) program is one function of the security program within an organization. SETA is one entity that is managed in an organization to provide critical information to employees to either improve their knowledge or maintain it. SETA targets all users in an organization with specific programs for their jobs and level of technical expertise. SETA is designed to reduce the incidence of accidental security breaches by members of an organization including employees, contractors, consultants, vendors, and business partners who have access to information; to help hold employees accountable for their actions by communicating policy to all users.

A SETA program consists of three elements: security education, security training, and security awareness. SETA's main objectives include:

- Build an in-depth knowledge base to design, implement, or operate security programs for organizations and systems.
- Develop skills and knowledge so that users can perform their jobs using IT systems more securely.
- Improve awareness of the need to protect system resources.
- Continually provide specific and effective training programs for all organization's employees.

However, a dedicated training professional staff may be required (depending on the training deliver method chosen). Also, the professional staff must assess organizational needs, plan effective programs, implement these programs, and evaluate their effectiveness.

12.5.1.1 Implementing a Successful SETA

In order to achieve the desired results of its security program, it must be communicated to employees what is a security program, how it is implemented, and why security is important to employees. These objectives require the establishment of a training programs for security awareness, knowledge management, and system development. In addition, security education and training are important issues that must be included in a security program to ensure the appropriate and needed skills within an organization.

12.5.1.2 Why Security Education?

Security education is an issue that requires more attention. Governments recognized that education is an important part of protecting their critical infrastructures. Therefore. there is need for graduates with education on security. Many education institutions teach networking, programming, and engineering, but security is not often offered or it is offered as an elective. On the other hand, security courses are offered by many departments of computer science, information technology, or information systems, but the teaching of security in many education institutions does not follow a standardized curriculum. A standardized curriculum enables graduates with the appropriate knowledge and skills needed for a security specialist. A security curriculum should be balanced and include courses that cover areas of systems engineering, software engineering, management, and social sciences. Many courses are offered as specializations that cover limited areas of security [Hentea 2007]. Although there is progress in educating more graduates in information security, courses on Supervisory Control And Data Ackquisition (SCADA) and ICS security are not the norms in these programs.

This work [Hentea 2007] discusses the complexity of new generation of control systems and the need for greater compliance to safety, quality of service, and security of systems and data. All these issues impact the education and training of software developers and control engineers. Software developers are required to ensure that software is stable and reliable to avoid hazard conditions. Besides mastering control methods, the control engineers must be able to deal with the challenges created by emerging technologies, and the capabilities of the users of these complex control systems. These challenges demand adoption of new knowledge in the emerging technologies, new competencies and skills. These competencies allow open possibilities for applying a wider range of control principles and design methods as well as productivity control and risk management involving decision making under uncertainty with increased levels of decision support.

12.5.1.3 Continuous Security Training

Since there is a need for continuous training, many security forums and associations developed courses to train information security professionals on the most current topics. For example, The Committee on National Security Systems (CNSS) has defined the minimum course content for the training of information systems security (INFOSEC) professionals in the disciplines of telecommunications security and automated information systems security. Several education institutions have been certified compliant with the National Training Standard for Information Systems Security Professionals (e.g, NSTISSI 4011 Certification). The Committee on National Security Systems (CNSS) is a United States intergovernmental organization that sets policy for the security of the US security systems.

Also, the US Department of Energy (DOE) offers the Cybersecurity Awareness and Training (CSAT) program [DOE CSAT] provides Departmental personnel with the knowledge and skills necessary to protect DOE information and information systems. DOE organizes CyberForce Competition including Webinars [DOE CyberForce] to help participants and volunteers increase their knowledge and understanding of cyber-physical threats, vulnerabilities, and consequences.

12.5.2 Expected Knowledge and Skills

Additional knowledge and skills are needed for the current and future applications in digital age. The following is a perspective on how to address many education gaps.

12.5.2.1 Security Knowledge Management

Organizations need to build a strategy that supports security knowledge management and training competencies. Security knowledge is likely to include policies, standards, design and attack patterns, threat models, code samples, reference architecture and secure development framework [Steven 2006] or risk management practices.

12.5.2.2 Risk Management Knowledge

Managers and practitioners with experience in risk management recognize the need to monitor the information security risk management process itself and acknowledge that, for companies to rely on their risk management processes, refinement through constant review and updating is critical to create effective information security management system. Also, people engaged in risk management tasks in different functional areas of a domain should be able to exchange knowledge. One of the characteristics of a successful risk management is the capability to diffuse knowledge in the right directions among domains of the power grid.

People within a domain working in different functional areas but sharing the knowledge of practicing risk management would considerably benefit by participating and collaborating in a network of practice.

One solution could be a network of practice for risk management. By being part of this network, people from the information security function can gain access to the knowledge necessary for improving their risk management approach.

A collaboration and knowledge sharing platform based on WEB 2.0 technologies could support the network of practice for risk management. In order to be beneficial for the improvement of information security risk management methodologies, the participation to this network should not have an ad hoc character. It is crucial to identify a process that will ensure the effective sharing of existing explicit and tacit knowledge including the efficient organization and dissemination of created knowledge. The process of developing or improving these methods generates new knowledge about information security risk management, which constitutes valuable intelligence. Therefore, it is very important to have a systematic process, which ensures that the acquired knowledge will be elicited, shared and managed appropriately.

12.5.2.3 Smart Grid Knowledge Diffusion

Examples of features needed in order to allow codification, sharing, and utilization of risk management knowledge, and to enable collaboration between stakeholders include:

- Forum (a virtual community) is needed to allow participating members to easily communicate and collaborate. As a result, other stakeholders can quickly acquire information about the current status of the discussion. One important motivation for people to participate in forums is the ability to create a community feeling. Knowledge is essential to effective security and to foster communication between stakeholders, and to motivate them to share risk management.
- Reusable knowledge refers to potentially relevant risk management knowledge, which is much more valuable if it is codified as a reusable asset. Techniques can be employed to enrich unstructured risk management knowledge present in each domain.
- Repositories allow to store reusable knowledge such as results from practicing risk management methods including open issues or the problems identified in risk management methods.
- E-mails are the format of communication to push relevant risk management knowledge to certain stakeholders, or to notify subscribed stakeholders if new knowledge emerges. E-mails are complementary to the more traditional pull mechanism of using the repositories. Users can subscribe to certain topics of interest and get updated without having to search the methods form themselves, which is a lightweight approach to share risk management knowledge among relevant stakeholders.
- Experts support allows users to easily find colleagues based on experience, interests or projects on which they work. By connecting people in the risk management process, there are benefits within the domain and intra-domains across several entities that can result in higher quality solutions.

Actors participating in the creation of new risk management knowledge, for example by sharing information or by publishing content, are improving the risk methodologies and are extending more beyond the limits of a specific domain to other domains in the power grid. It would be mutually beneficial to create an intra-domain network across which information security risk management knowledge could flow. For example, the National Electric Sector Cybersecurity Organization (NESCO) was established in 2010 to serve as a gathering point for utilities, federal agencies, regulators, researchers, and academics. Partially funded by the US DOE, this organization could help facilitate better information sharing. NESCO could help focus cybersecurity research and development priorities, identify effective common practices, and organize the collection, analysis, and dissemination of infrastructure vulnerabilities and threats.

Also, the Cybersecurity Risk Information Sharing Program (CRISP) enables and manages the near real time sharing of IT network information between private sector electricity utilities and DOE. Recently, NERC's Electricity Information Sharing and Analysis Center (E-ISAC) partnered with DOE on an expansion of the CRISP program to include operational technology. The vision of E-ISAC is to a world class, trusted source of quality and rapid sharing of security information for the electricity industry. While the mission to provide insights, leadership, and collaboration is a genuine goal, the information posted by this Web site is an adversary's goal to easily learn about vulnerabilities and exploit the vulnerabilities to launch cyber attacks. source of quality analysis and rapid sharing of security information for the electricity industry.
The CRISP expansion, to include two operational technology pilots, marks a major milestone to improve E-ISAC capabilities that strengthen grid security in North America. CRISP leverages DOE's unique intelligence capabilities, expertise, and advanced threat detection technologies to collect, analyze, and distribute actionable threat information to the energy sector.

In US, the National Electric Sector Cybersecurity Organization Resource (NESCOR) is intended to strengthen the cybersecurity posture of the electric sector by establishing a broad based public and private partnership with the DOE for collaboration and cooperation. NESCOR works collaboratively with NESCO, DOE, and other federal agencies to identify and disseminate best practices.

12.5.2.4 Reinforcement of System Development Skills

There are specific issues that need to be addressed by SCADA systems developers. First, the evaluation of the software is crucial. Software plays an increasingly important role in all types of controllers in SCADA systems. Developers of control systems require programming skills in addition of engineering skills to be able to understand the broad picture and detailed control problems [Hentea 2007]. A common perspective and security solution to the control systems is essential in the design of these systems. New approaches and skills for software development of distributed control systems architectures are required to ensure that software is stable and reliable to avoid hazard conditions [Heck 2003]. Control methods are essential in distributed software development and E-commerce applications [Hellerstein 2005].

12.5.3 Smart Grid - Emerging Skills Requirements

Immediate skills needs are accelerating when Smart Grid developments intensify. The growing need for an operational understanding of electricity, transport and buildings management might require adapting curricula for engineering courses and other IT oriented education programs (see Chapter 3 of [OECD 2010]). Also, the implementation of Smart Grids requires field specific knowledge of legal frameworks, environmental impacts, etc.

To help reduce risk and potential cyber attack surface, distributed system operators must implement a security organization and adopt active security processes [Gauci 2017]. Cybersecurity is about people, processes, and technology, in exactly that order. Distribution system operators need to deploy a complete program consisting of proper organization, processes, and procedures in order to fully take advantage of cybersecurity protection technologies. Countering flexible and persistent human adversaries requires properly trained and equipped human defenders [Lee 2016b].

The convergence of IT and operational technologies for the Smart Grid implementation and operations requires new approaches. Developing competencies and cross functional teams for OT and IT integration becomes a key factor for distributed systems operators (DSO) to address the unique challenge of securing their modern control systems [Gauci 2017]. The operational technology is related to the real time operation of the electric grid, it needs to ensure high reliability and availability. The information technology is related to computers and servers for corporate control rooms. The cybersecurity integration for the OT and IT across the DSO system is discussed in [Gauci 2017]. Figure 12.8 depicts the integrated view of cybersecurity for both OT and IT. Therefore, DSOs need to develop competencies and cross functional teams to address the unique challenge of securing their modern control systems.

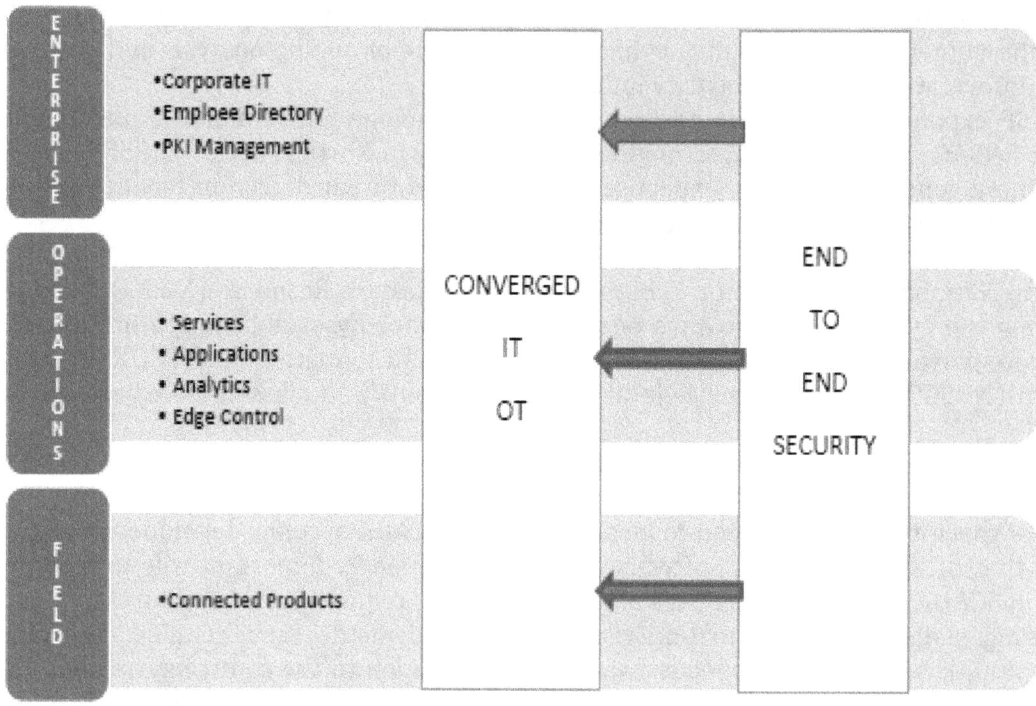

Figure 12.8 An integrated cybersecurity view across the DSO system. Adapted from [Gauci 2017]. Licensed under CC BY 3.0.

The outmost important aspect is to be aware that cybersecurity relies strongly on processes and humans, therefore generating the need for adequate services such as an efficient patch management program and automated tools.

The operational technology and information technology collaboration is largely education and awareness of each other's perspective, so together the resulting solution best meets the business needs and provides adequate security [Hawk 2014]. This alignment of skills drives thorough evaluation of the requirements, product capabilities and integration needed to provide the right capability for the business and security capabilities.

The OT, IT, and supply chain collaboration ensures that the right foundational capabilities (e.g., network security, authentication, monitoring, configuration management, etc.) are in the procured component or solution, as described in [ESCSWG 2014]. This document complements other cybersecurity efforts by providing organizations that acquire, integrate, and supply energy delivery systems with guidance on how to communicate cybersecurity expectations in a clear and repeatable manner. Energy delivery systems are used to monitor and control the production, transfer, and distribution of energy. These systems include SCADA systems, Energy Management Systems, Distributed Management Systems, and Distributed Control Systems. Energy delivery systems comprise the following components:

- Sensors and actuators used for monitoring and controlling energy delivery processes.

- Computer systems that analyze and store data.
- Communication pathways and networks that interconnect the various computer systems.

Addressing cybersecurity aspects for distribution grids is challenging but unavoidable. Key technical features and precautions at device level enable to reduce significantly the risks, but main aspects rely in grid global information system, relevant process deployment and people awareness and skills, through whole grid lifecycle [Batard 2017].

As tomorrow's systems will combine hardware and software to provide security and trustworthiness, engineers must understand the design principles and techniques that relate to both [Tehranipoor 2011].

Also, engineers should have hardware assurance expertise and security skills to design and maintain these systems. Utilities are guided to plan and implement an Integrated Security Operations Center (ISOC) that includes corporate systems, control systems, and physical security [EPRI 2013]. However, recruiting and retaining ISOC personnel are difficult given the required breadth and depth of knowledge and experience that are required. Moreover, ISOC staff must receive regular training to refresh their skills and learn new skills needed to operate new tools, or identify and respond to new threats. Management must allocate a significant training budget as the ISOC is being set up and once it has moved into an operational state. The ISOC director should also routinely run internal tabletop and training exercise. ISOC staff requires expertise and continuous training in the following areas:

- Operating systems
- Multiple hardware platforms
- Networking systems (routers, switches, firewalls) and protocols
- Industrial Control Systems (SCADA, DCS)
- Control theories
- Database technologies
- IT systems and tools
- Applications for corporate and OT domains
- Power system protocols
- Malware analysis
- Intrusion detection and prevention systems
- Programming or scripting
- Investigative/forensics processes
- Chain of custody issues
- Ethics
- Corporate policy.

ENISA certifications require a high level of knowledge areas that need to be developed for ICS/SCADA cybersecurity professionals.

12.6 Addressing Security Awareness Program

Societal and cultural factors influence how individuals understand and use systems which consequently impact the information security of the system and organization. Individuals perceive, reason, and make risk based decisions in different ways. To address this, organizations make information security functions transparent, easy to use, and understandable. Additionally, providing regularly scheduled security awareness training mitigates individual differences of risk perception [NIST SP800-12r1].

Security awareness is a comprehensive training program, tailored for specific groups, and organization wide. The goal is to communicate to each employee the importance of security, expected responsibilities and behavior including non-compliance repercussions. Various methods can be employed to reinforce the concepts of security awareness. Examples include videos, lectures and conferences, posters and banners, newsletters, brochures and flyers, bulletin boards, etc. As described in [Monitor May2014], ICS-CERT offers an introductory course in ICS Cybersecurity for all organizations that need ICS security awareness for their employees.

The training program is based on a blended learning approach to teach ICS Cybersecurity course that blends WEB based (e.g., multimedia approaches such as videos, audio, PowerPoint, Flash) and instructor led methods. Also, this ICS-CERT program ensures course availability on mobile device (iPads, iPhones or Android devices) or desktop. Trainees can access the courses via this WEB site [ICS Learning] from several US states and countries.

The NERC CIP compliance program for power grid requires staff who interact with BES Cyber Systems to have both Security Awareness Training and Cyber Security Training appropriate to their roles.

An awareness program should be formal, planned and documented [Harris 2013]. An organization should ensure that all people understand the security risks associated with their activities and the applicable laws, executive orders, directives, policies, standards, instructions, regulations, or procedures related to the security of organizational systems.

Effective training and awareness programs make employees accountable for their actions. If employees understand how to properly access resources, why access controls are in place, and the consequences for not using the access controls, an organization can reduce many types of security incidents.

An important aspect of security awareness and training program is the preparation of the awareness material. Awareness must be designed with the organization mission in mind supporting the business needs of the organization and be relevant to the security architecture. The most successful programs are those that users feel are relevant to the subject matter and issues presented. An awareness program may be designed, developed, and implemented in many different ways (see more in [NIST SP800-50]). Security awareness efforts are designed to change behavior or reinforce good security practices. Awareness is defined in [NIST SP800-16] as follows:

Awareness is not training. The purpose of awareness presentations is simply to focus attention on security. Awareness presentations are intended to allow individuals to recognize IT security concerns and respond accordingly.

A significant number of topics can be mentioned and briefly discussed in any awareness session or campaign. Examples of topics may include: protection from viruses, worms, Trojan horses, and other malicious code; social engineering; shoulder surfing, handheld device security issues - address both physical and wireless security issues; use of encryption and the transmission of sensitive/confidential information over the Internet, etc.

Key steps leading to the implementation of the awareness program are depicted in Figure 12.9. Once the awareness program has been implemented, processes must be put in place to monitor compliance and effectiveness. An automated tracking system should be designed to capture key information regarding program records (e.g., presented materials, dates, audience, costs, sources, success measurements, failure measurements, attendance, etc.). Reports can be generated and used to identify gaps or problems. Corrective action and necessary follow-up are needed. This may take the form of formal reminders to management; additional awareness, training, or education offerings; and/or the establishment of a corrective plan with scheduled completion dates. A feedback strategy needs to incorporate elements that will address quality, scope, deployment method (e.g., web-based, onsite, offsite), level of difficulty, ease of use, duration of session, relevancy, currency, and suggestions for modification.

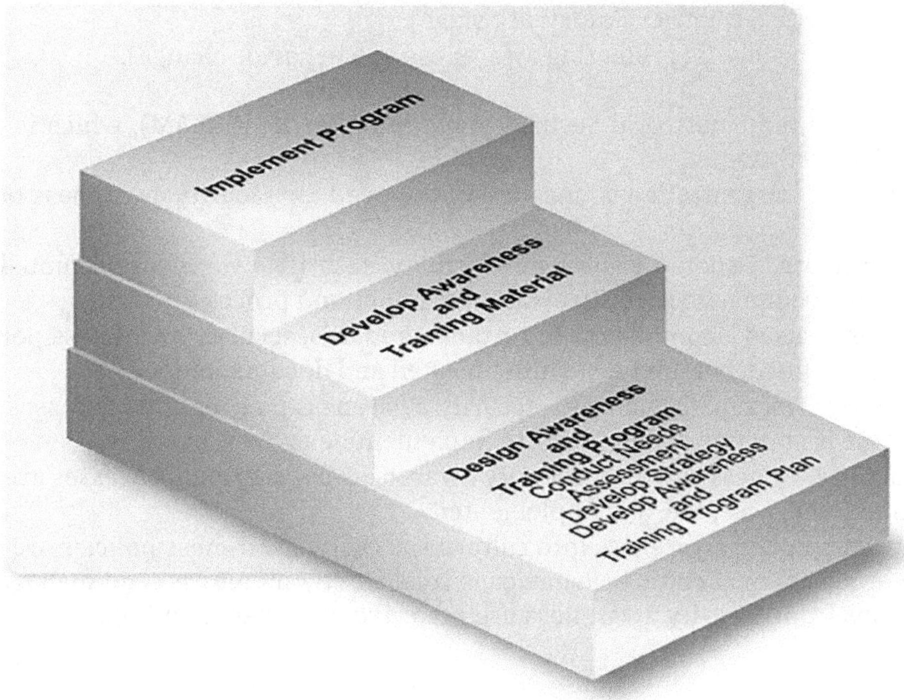

Figure 12.9 Key steps leading to program implementation. *Source:* [NIST SP800-50]. Public Domain.

12.6.1 Evaluating Security Awareness Impacts

By using the SysAdmin, Audit, Network, and Security (SANS) Security Awareness Maturity Model© [Spitzner 2016], an organization can identify and benchmark the current maturity level of their security awareness program and determine a path to improvement. Established in 2011, the model is the key measure of program impact and success as established by the level of measurable human risk that can be mitigated by changing end-user behavior. The most successful, most mature, security awareness programs not only change behavior and culture but can also measure and demonstrate their worth via a metrics framework. This model is based on 5 stages:

- Nonexistent.
- Compliance focused.
- Promoting awareness and behavior change.
- Long term sustainment and culture change.
- Metrics framework program is robust enough to provide metrics, such as progress and behavioral impact.

This SANS 2018 Security Awareness Report [SANS 2018] analyzes the data submitted by 1,718 security awareness professionals from around the world to identify and benchmark how organizations are managing their human cybersecurity risk. The analysis includes how various factors including security awareness program maturity, funding, and staffing combine to make successful programs. According to this report, participants to an organized survey agree that security awareness program has a positive impact on the security of their organization.

Organizations identify themselves in the stage 3 (promoting awareness and behavior change) or greater in the Security Awareness Maturity Model. Utilities responses can be classified in these categories:

- Few are in stage 1 (nonexistent security awareness program).
- Few are in stage 5 (metrics framework stage).
- Majority are in stage 3 (promoting awareness and behavior change).

Figure 12.10 depicts the Smart Grid Security Awareness Model (SGSAM), which is a refined CMI model with 6 identified stages:

- Non-existent - Organization is unaware of the need for security awareness policies or processes.
- Adhoc promoted - sporadic awareness activity, security awareness promoted occasionally, there is no documented security awareness policy and process.
- Standards focused – compliance to certain standards, security awareness policies and processes are defined, but not fully communicated and documented.
- Defined behavioral path changes – security awareness policies and processes are standardized; there is an active commitment to implement behavioral changes.
- Promoted awareness culture – security awareness policies and processes are communicated, understood, accepted, and implemented.
- Integrated security awareness into culture – security awareness policies are integrated into the organizational culture; behavior is consistently monitored and managed; behavioral deviations from security awareness are recorded, reported, and improvements are implemented.

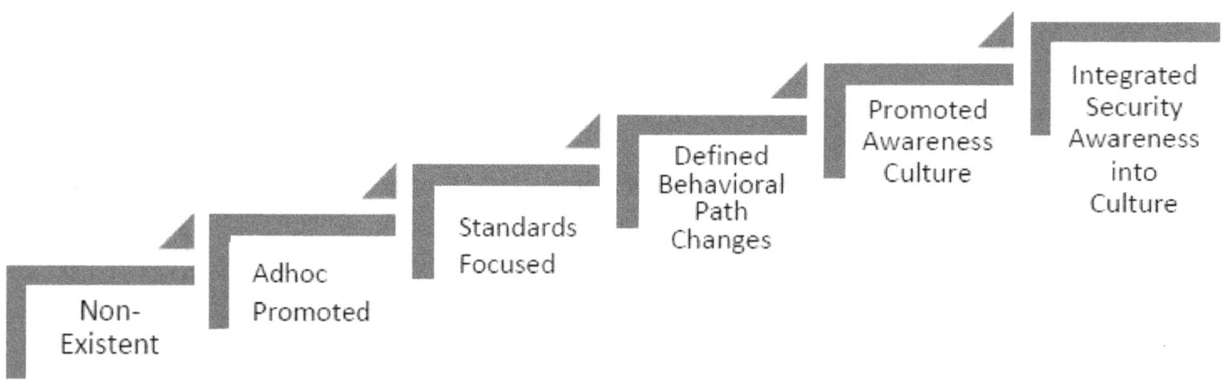

Figure 12.10 Smart Grid security awareness model.

This model is supporting a continuous cycle process to manage security awareness for a diverse community of Smart Grid stakeholders. For example, utilities have the task of promoting security awareness for employees, customers, partners, vendors, and service providers.

12.6.2 Information Security Culture

Research works stress the importance of a strong information security culture to address the threat that employee behavior poses to the protection of information assets [Da Veiga 2009]. Various proposed solutions discuss how an acceptable level of information security culture should be cultivated, and how to assess this culture to determine whether it is on an acceptable level. Information security culture may be used in organizations to construct the appropriate security beliefs and values and to guide the employees' security behavior in order to achieve a secure environment for organizational information assets.

An information security culture is defined as the attitudes, assumptions, beliefs, values and knowledge that employees/stakeholders use to interact with the organization's systems and procedures at any point in time. The interaction results in acceptable or unacceptable behavior (e.g., incidents) evident in artifacts and creations that become part of the way things are done in an organization to protect its information assets. This information security culture changes over time. This work [AlHogail 2015] proposes an adapted framework of the Information Security Culture Framework (ISCF) to create a questionnaire to assess the level of an information security culture in organizations.

The ISCF is a comprehensive framework that consists of five dimensions and incorporates change management and human factor in information security [Da Veiga 2009]. The researcher developed a Comprehensive Information Security Framework (CISF) that equips organizations with a holistic approach to the implementation of information security. The ISCF can be used by organizations to cultivate an information security culture conducive to the protection of information assets.

In addition, this work [Da Veiga 2017] recommends considering the possibility of several information security subcultures that could be present in the organization when considering an information security culture in an organization. This means that different geographical, ethnic or age groups of employees could have different assumptions, values and beliefs about the protection of information, resulting in unique information security subcultures. This work discusses how dominant information security cultures and subcultures develop and how they can be influenced positively over time through targeted interventions.

An information security culture is a critical component of an organization's Information Security program. It has to be embedded in the organization, changed and influenced to direct employee, contractor and third party behavior, thereby reducing risk to the organization's information assets. Therefore, an organization needs to identify the dominant Information Security culture, and then to identify the factors (e.g., management, awareness, training, education, policies, compliance, national culture, etc.) that influence the Information Security culture.

12.6.3 Continuous Refinement and Improvement

Organizations need to support a continuous refinement for the Smart Grid security education and awareness program. As described in [NISTIR 7628r1] document, sufficiently trained personnel is critical to ensure that everyone in organization has a clear understanding of the importance of cybersecurity, understands their role in cybersecurity, and the importance of each role in supporting cybersecurity within the organization.

Lack of an adequate security training and security awareness program can result in insufficiently trained personnel that do not know or understand an organization's policy framework to guard against vulnerabilities, leading to the risk of mishandled or inappropriately used information, unauthorized access to information and systems, and potentially damage to profit and organizational reputation. This document [NISTIR 7628r1] recommends:

- Throughout the entire organization, all personnel should have a level of security awareness training based on the individual organizational and/or the critical asset responsibilities.

- Providing privacy awareness and training for all entities within the Smart Grid. Providing effective information security and privacy training and awareness not only supports privacy principles but also helps to ensure that workers, throughout all entities within the Smart Grid, have the knowledge necessary to keep personal information and energy usage data assets appropriately secured during their daily work activities.
- Providing consumer education on privacy awareness, in addition to employee education. This training supports informed decisions related to participating in the deployment of Smart Grid technologies and granting access to the information such technologies enable. Concerns related to privacy can result in consumers opting out of smart meter deployment or in limiting access to customer energy usage data collected using Smart Grid technologies.

As recommended in this document, all stakeholders have an important role in educating consumers on their rights to know about their data collected and to promote confidence in the way that such information is used and safeguarded from unauthorized use. To promote these awareness objectives, information on privacy protections should be incorporated conspicuously into communications with consumers.

Raising awareness of privacy concerns for customer and energy usage data, and showing how those concerns are being addressed, could help on managing relationships between various stakeholders. The audience for this training could include consumer advocates, legislators, state regulatory commissions, and utility companies.

Also, the [NISTIR 7628r1] recommends security training and security awareness programs should be an ongoing effort. Also, these programs should include a continuous retraining effort over an organization defined period of time to reflect new procedures, new technologies, and reinforcement of the importance of the cybersecurity program.

While some utilities are trying to set up their security programs, others may have better plan. As reported in [Nash 2011], one utility has two cybersecurity teams within the IT organization: One focuses on strategy and the other on implementation. The two departments work together to ensure that the company has the appropriate policies in place and that the necessary training and communication are provided to employees on how to securely use IT tools.

Also, privacy is an essential issue of trust. Without privacy there is no trust. When implementing information security privacy, both employees and customers must be considered, and controls must be implemented to protect the personal identifiable information of an individual. An organization must ensure that adequate controls are in place to protect personal information of employees, contractors, customers and third parties. See more information about privacy regulation in the recent European Privacy document [EU GDPR]. Also, ISACA supports information, guidance, and privacy certification; (see ISACA Privacy initiatives at [ISACA Privacy]). Also, this guide [Herold 2016] helps to understand and apply privacy principles as defined in this document [ISACA Principles].

Recently, the National Initiative for Cybersecurity Education (NICE) [NICE] was created to establish an operational, sustainable and continually improving cybersecurity education program for the US. The National Institute of Standards and Technology (NIST) provides the leadership for NICE in US. The overall aim of the NICE program is to ensure, coordination, cooperation, focus, public engagement, technology transfer and sustainability. For example, NICE can be used to identify resources for the certification and education.

NICE provides a definition of roles for secure software professionals. Each role is mapped to an area that contains a range of software engineering functions and associated knowledge, skills, and abilities (KSAs). These KSAs represent a solid first attempt to create a single point of reference for understanding the range of professional certifications, since those KSAs can be mapped directly to the knowledge requirements of each certification. In that respect, it is possible to identify which certificate would satisfy a job role in the field.

While education for Smart Grid cybersecurity is approached by different entities (research organizations, academia, government agencies, or private companies), these authors [Yardley 2014] argue that the training courses do not cover the full needs of the electricity sector. They propose a modular learning platform to provide the essential base knowledge and hands on teaching by means of exercises for understanding Smart Grid cybersecurity.

12.7 Certifications and Professional Security Organizations

Many organizations seek job applicants who have professional certifications so they can more easily identify their knowledge. Some employers accept certification along with work experience in lieu of a bachelor's degree. Certification is typically offered through industry groups and associations, training institutions and product vendors as validation of an individual's demonstrated understanding of an area of expertise.
Most existing certifications are not fully understood by the hiring organizations. There are many certifying bodies, so employers need to understand the match between certifications and the job position requirements. Research has shown that certifications in information security have an effect on the economic prospects of certificate holders. In the long term, the value of an information security certification adds leverage to the certification granting authority to exert influence over its members, including influence in matters of ethical responsibility.

The American National Standards Institute (ANSI) (www.ansi.org) accredits standards developers, certification bodies and technical advisory to industry. ANSI administers accreditation programs for personnel certification agencies that is based on the international standard [ISO/IEC 17024] (see also [ANSI/ISO 17024]).

The international standard [ISO/IEC 17024] specifies criteria for the operation of a Personnel Certification Body (also known as a certification body for persons). The standard includes requirements for the development and maintenance of the certification scheme for persons upon which the certification is based. The standard provides a global benchmark for personnel certification programs to ensure that they operate in a consistent, comparable and reliable manner worldwide, thereby allowing individuals to have skills that translate across national lines. It helps organizations that certify individuals in a variety of occupations and professions protect the integrity and ensure the validity of individual certification programs. It also promotes consumer and public confidence in the capabilities and competence of the people who provide specialized services or who create the products that support our daily lives and livelihoods.

12.7.1 Information Security Certifications

Examples of most known security certifications currently (as of December 2018) listed in the ANSI accreditation directory (see [ANSI Directory] include:

- CompTIA Security+™ Certification is offered by the Computing Technology Industry Association (CompTIA) (see certification.comptia.org/security/). This certification is a popular, entry that validates knowledge of systems security, network infrastructure, access control, assessments and audits, cryptography and organizational security. CompTIA is a non profit trade association, issuing vendor neutral professional certifications for the information technology (IT) industry in over 120 countries.

- Certified Information Systems Security Professional (CISSP) and Systems Security Certified Professional (SSCP) are considered as the most prestigious certification for security managers and CISOs. These certifications are offered by the International Information Systems Security Certification Consortium (ISC2) (www.isc2.org), a leading accreditation organization for information security professionals. CISSP is aimed at providing information security professionals with an objective measure of competence.
- Certified Ethical Hacker certification (CEH) is a core certification (see others advanced or expert certifications) offered by the International Council of E-Commerce Consultants, also known as EC-Council (https://www.eccouncil.org/), the world's largest cybersecurity technical certification body that currently (as of December 2018) operates in 145 countries globally. The Certified Ethical Hacker is a skilled professional who understands and knows how to look for the weaknesses and vulnerabilities in target systems and uses the same knowledge and tools as a malicious hacker. The CEH Program certifies individuals in the specific network security discipline of ethical hacking from a vendor neutral perspective.
- Global Information Assurance Certification (GIAC) is offered by the SANS Institute (www.sans.org), a private US company. GIAC offers more than 30 information security certifications that correspond to specific job duties at GIAC (https://www.giac.org/). Each certification focuses on specific job skills and requires unmatched and distinct knowledge. Besides [ANSI/ISO 17024] standards accreditation, GIAC Certifications align with the National Initiative for Cybersecurity Education (NICE) framework. The NICE framework, described in [NIST SP800-181r1], is a reference resource for information about cybersecurity work and the knowledge, skills, and abilities needed to complete tasks that can strengthen the cybersecurity posture of an organization.

12.7.2 Secure Software Certifications

Current events related to cybersecurity encourage a fundamental shift in the way we think about educating and training a workforce prepared to address security issues in all phases of a software system. Secure software is a critical component of secure systems and secure software certifications are thus an important validation of competencies in Software Engineering.

- Certified Application Security Engineer (CASE), offered by EC-Council, aims to provide the essential and fundamental skills to programmers and application developers in secure programming. The credential tests the critical security skills and knowledge required throughout a typical software development life cycle (SDLC), focusing on the importance of the implementation of secure methodologies and practices in today's insecure operating environment. It is developed in partnership with large application and software development experts globally. Beyond just the guidelines on secure coding practices, it includes secure requirement gathering, CASE training robust application design, and handling security issues in post development phases of application development.
- GIAC Secure Software Programmers (GSSP) certification (e,g., GSSP-JAVA) allows candidates to demonstrate mastery of the security knowledge and skills needed to write secure code and recognize security shortcomings in existing code (e.g., common programming errors) that lead to most security problems. It is developed in a joint effort involving the SANS Institute, CERT/CC, several US government agencies, and leading companies in the US, Japan, India, and Germany.
- ISSECO®Certified Professional for Secure Software Engineering is offered by the International Secure Software Engineering Council (ISSECO) and certification exam thru iSQI, Inc., a division of the iSQI GmbH in Potsdam, Germany. The certification focuses on providing and validating the skills necessary to produce secure software during the entire software

development lifecycle. The certificate is designed for all involved in Software Engineering, such as designers, architects, testers, requirements engineers and project managers. The International Secure Software Engineering Council is a German non-profit organization that focuses on software security.

- Certified Secure Software Lifecycle Professional (CSSLP) aims to provide application security skills and knowledge necessary for authentication, authorization and auditing throughout the software development lifecycle using best practices, policies and procedures established by the cybersecurity experts at ISC2.

12.7.3 Software Assurance Certifications

Software assurance provides a focus for secure software components, software security in services, and secure software supply chain risk management. Certifications help identify resources to supplement the efforts as well as identify strategies to inject software assurance related topics to a system during the SDLC.

Certified Expert Software Assurance Specialist (CEAS) demonstrates ability to exhibit key practices for mitigating the most egregious exploitable software weaknesses. Also, it validates the extensive knowledge on software security testing; secure coding, architecture and design as well as knowledge on software supply chain risk management and due diligence. This certification is offered in U.S. by the National Initiative for Cybersecurity Careers and Studies (NICCS) [NICCS CEAS]. Also, [NICCS] provides training materials and a course supporting the requirements for this certification.

12.7.4 Information Security Management Certifications

Certified Information Security Manager (CISM) is the leading credential for information security managers and those who have information security management responsibilities. The CISM certification aims to enhance and validate the skills of the individual who manages, designs, oversees and/or assesses an enterprise's information security. It is offered by ISACA, previously known as the Information Systems Audit and Control Association (ISACA), see [ISACACert]. ISACA is a nonprofit, global association with members currently (as of December 2018) in 180 countries. ISACA engages in the development, adoption and use of globally accepted, industry leading knowledge and practices for information systems.

12.7.5 Industrial Control Systems: Cybersecurity Certifications

The industrial world is constantly evolving, including new technologies adapting to market requirements. One of the most transcendental adaptations the industrial world is currently experiencing is the convergence between Operations Technology (OT), the operations needed to carry out the industrial processes, and Information Technology (IT), the use of computers to manage data needed by the organization's enterprise processes.

This convergence has many advantages (optimization of operations, better use of resources, cost savings, etc.), but it also raises additional issues, such as the need for cybersecurity of industrial control systems (ICS) and Supervisory Control and Data Acquisition (SCADA) systems.

This convergence, which affects hundreds of thousands of systems worldwide, implies that professionals with knowledge of cybersecurity for ICS/SCADA is needed. However, currently, there are very few professionals with the proven skills available to do this work.

This document [ANSSI 2014] defines categories of ICS personnel in Europe. Personnel are certified if they have received specific training regarding their role in the ICS concerned and in awareness of IT security. This training must be officially recorded by the ICS's responsible entity. This document classifies users into these categories:

- Class 1 All users should be certified.
- Class 2 All users shall be certified.
- Class 3 All users shall be certified and controlled. Cybersecurity training for this certification shall be carried out by certified providers.

Cybersecurity levels are defined in terms of consequences for the Nation, rather than consequences for responsible entities. Personnel are considered to be controlled if they have been explicitly authorized to intervene and if their actions can be tracked. The following includes examples of current ICS/SCADA certifications.

12.7.5.1 Examples of ICS/SCADA Cybersecurity Certifications

There are different certifications for ICS/SCADA offered by different organizations, each applying its own topic and requirements.

- Most common ICS cybersecurity include: ISA/IEC 62443 Cyber Security Certificate Program [ISA 62443Certify] includes certificates that validate knowledge from basic level 1 to level 4. Individuals that achieve all certifications are designated as Cybersecurity Experts (ISA/IEC 62443 Cybersecurity Expert):
    - Certificate 1: ISA/IEC 62443 Cybersecurity Fundamentals Specialist)
    - Certificate 2: ISA/IEC 62443 Cybersecurity Risk Assessment Specialist
    - Certificate 3: ISA/IEC 62443 Cybersecurity Design Specialist
    - Certificate 4: ISA/IEC 62443 Cybersecurity Maintenance Specialist.

  These certifications are offered by the offered by the International Association of Automation (ISA). This is a nonprofit professional association that sets the standard for those who apply engineering and technology to improve the management, safety, and cybersecurity of modern automation and control systems used across industry and critical infrastructure. However, the certification scope is limited to the ISA standards; as such it lacks comprehensive coverage of the ICS/SCADA security landscape that includes other standards, that are currently used and needed in the future.
- Certified ICS/SCADA Security Architect (CSSA) validates adequate knowledge to properly secure SCADA systems. It is designed to be relevant for power transmission, oil and gas and water treatment industries. CSSA is targeted towards network administrators and their managers, as well as IT professionals and their managers. This certification is offered by the Information Assurance Certification Review Board (http://www.iacertification.org/), a non profit organization formed by information security professionals.
- Global Industrial Cyber Security Professional (GICSP) certification aims to bring together IT, engineering and cybersecurity capabilities to achieve security for industrial control systems from design through to retirement. GICSP is offered by SANS Institute and it is targeted at professionals, worldwide and across industries, who engineer or support control systems and share responsibility for the security of these environments.
- Certified Automation Professional (CAP) validates the skills and knowledge for professionals of the automation industry worldwide. CAP provides very comprehensive and detailed contents about automation including security and safety topics; cybersecurity topics are addressed just from a very general point of view without detailing the specifics of cybersecurity in Industrial Control Systems. CAP is targeted at people working in automation at the

professional level, technical supervisors, technical sales personnel as well as automation educators. The certification is offered by ISA. CAP Certification is used by companies to evaluate hiring candidates and making contracting decisions.

12.7.5.2 European ICS Cybersecurity Certification

The problem of understanding the variety of certificates represents one issue in using as a proof of capability. Because there are so many certificates and because they mean different things, it is difficult for the general population to tell what they are getting when they hire a CISA versus a CISSP, or a CompTIASecurity Plus versus a GIAC.

This document [ICS Certification]_explores how current initiatives on certification of professional skills are related to the topic of ICS/SCADA cyber security. It also identifies the challenges and proposes a series of recommendations towards the development of certification schemes for ICS/SCADA cybersecurity professionals. Challenges identified in relation to operational issues of certifications include the need to address:

- The confluence of contents, objectives and needs of two very different topics such as cybersecurity and industrial automation within a number of different industrial sectors (energy, oil & gas, automated manufacturing, water treatment, chemical, pharmaceutical, etc.).
- The complexity of including topics and content relevant to different roles and profiles. In the field of cybersecurity for ICS/SCADA systems there are implications ranging from the lower operative levels to the top management.
- The important differences of these sectors (automation, energy, chemical, pharmaceutics, energy, etc.) in their processes and operational procedures as well as health, safety and environmental consequences due to a failure of a system or component.
- The challenges related to the societal aspects such as:
  - Avoiding commercial interests that could impair the credibility of certifications
  - Obtaining stakeholders' support to underscore the relevance, credibility and strength of future certifications
  - Ensuring that the future certifications will improve compared with existing ones in regards the level of knowledge related to cybersecurity for ICS/SCADA
  - Exploring the professional roles and specific knowledge needed by professionals for ICS/SCADA.

The knowledge areas for ICS/SCADA cybersecurity professionals have been identified by taking into account the existing ICS/SCADA Cyber Security Certification schemes and other relevant studies. One of its subgroups has focused on defining the competences, qualifications and experience, needed by ICS Cyber Security Professionals. The result is a high level view of the knowledge areas that need to be developed (see Figure 12.11).

These knowledge areas are the foundation for all the technical ICS/SCADA cybersecurity roles. All roles with accountability or responsibility of ICS/SCADA cybersecurity would need the basic competence across these knowledge areas. Where more specialist roles are defined, for example ICS/SCADA forensic analyst, ICS/SCADA Incident Response professional, more detailed knowledge, skills and abilities on that specific topic will be needed.

12.7.6 Safety Certifications

Traditionally, in the industrial world, most of the effort and resources invested in security were related to safety. Today, due to convergence between operations technology and information technology, cybersecurity is achieving an importance equal to that of safety among industrial companies including failure of systems via cyber attacks may have a safety impact. The industrial security market has supported a number of initiatives related to the training and certification of safety professionals. The main features of safety certifications offered by the Board of Certified Safety Professionals (BCSP) [BCSP] are:

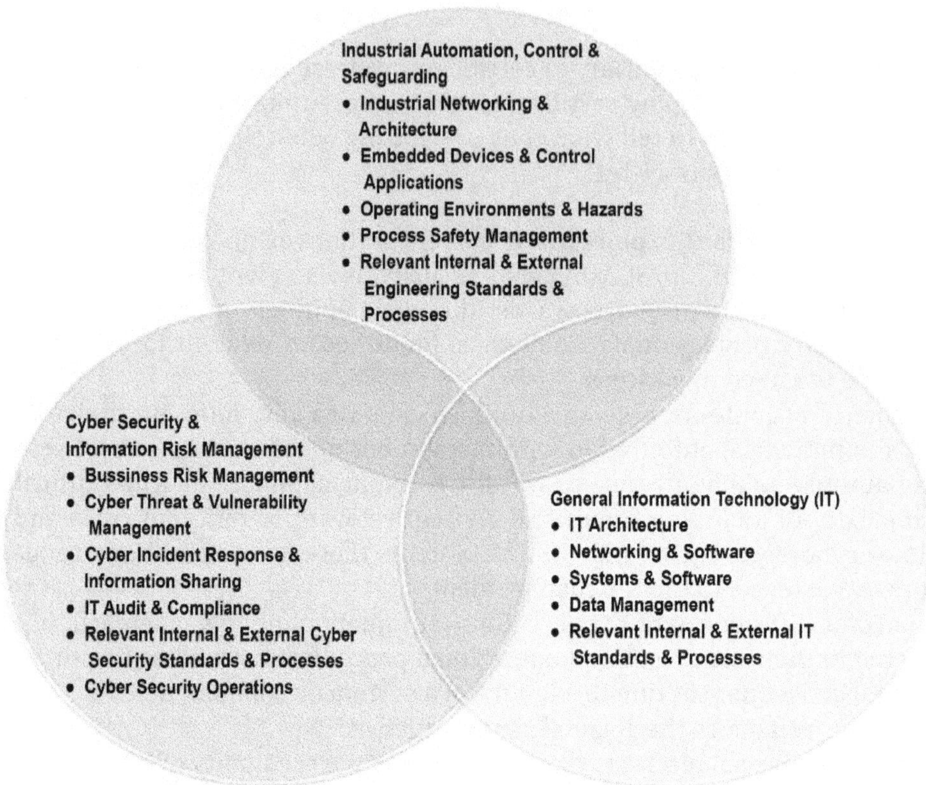

Figure 12.11 Knowledge areas. *Source:* [ICS Certification]. © 2014, ENISA.

- Complexity due to factors such as different disciplines such as safety, health, environment protection being involved; different environments include emergency management, engineering, ergonomics, fire protection, risk management, occupational safety, etc.

- Multiple levels of certification allowing safety professionals to achieve professional recognition regardless their education and giving priority to experience.

The Board of Certified Safety Professionals (BCSP) began in 1969 as a peer certification board whose objectives are to set standards for professional, technician, technologist, and supervisory level safety practices.

12.7.7 Cybersecurity Training and Resources

Many National and global organizations support cybersecurity training. Examples include:

- **National Cyber Security Workforce Framework**
  The National Cyber Security Workforce Framework, as published by the United States National Institute of Standards and Technology (NIST), is seen in the U.S., as a national resource that categorizes, organizes and describes cybersecurity work. It has been developed by the National Initiative for Cyber Security Education (NICE) in order to provide educa-

tors, students, employers, employees, training providers and policy makers with a systematic way to approach cybersecurity work, and describe what is required of the cybersecurity workforce. This framework is a thorough approach to describe cybersecurity workforce related initiatives. It is backed by the United States Government and it is being followed by different federal agencies as such it is becoming a de facto standard. It is not a certification scheme, but a comprehensive definition of cybersecurity workforce related contents that should be considered in any certification scheme development in order to guarantee a proper definition and scope of contents.

- **National Cyber Security Alliance (NCSA)**
  The National Cyber Security Alliance (NCSA) is a unique partnership among the Federal government, leading private sector companies, trade associations and educational organizations. NCSA aims to increase awareness about the risks associated with using Internet technologies and how to help protect against them, while also providing free tips, checklists and best practices for remaining safe while online. Through collaboration with the government, corporate, nonprofit and academic sectors, the mission of NCSA is to create a culture of cybersecurity and safety awareness by providing the knowledge and tools necessary to prevent cybercrime and attacks.

- **International Information System Security Certification Consortium (**ISC2)
  ISC2 is a non-profit organization which specializes in training and granting certifications for cybersecurity professionals. It has been described as the world's largest IT security organization. Besides training for certification exams, the organization offer customized security training to enterprises, organizes conferences and events to advance security leaders by arming them with the knowledge, tools, and expertise to protect their organization.

- **SANS Institute**
  The System Administration, Networking and Security Organization, known as SANS Institute, is a private U.S. company that specializes in cybersecurity training and granting cybersecurity certifications among others. It provides training materials, courses, newsletters, conferences, and research. The SANS Institute also sponsors the Internet Storm Center, an Internet monitoring system staffed by a global community of security practitioners, and the SANS Reading Room, a research archive of information security policy and research documents, delivers over one million downloads per year to professionals globally.

- **Cloud Security Alliance (CSA)**
  The Cloud Security Alliance is a not for profit organization with a mission to promote the use of best practices for providing security assurance within cloud computing, and to provide education on the uses of cloud computing services and controls to secure all other forms of computing.

- **Internet Security Alliance (ISA)**
  The Internet Security Alliance was founded in 2001 as a non-profit collaboration between the Electronic Industries Alliance, a federation of trade associations, and Carnegie Mellon University's CyLab (Pittsburg, Pennsylvania, USA), focusing on cyber security. ISA is a dedicated participant in government developments, laws, policies and programs for cyber security. ISA provides expert testimony and thought leadership in government and serve as an expert witness to the press.

- **Industrial Control Systems - Cyber Emergency Response Team (ICS-CERT)**

The former ICS-CERT, now CISA, provides advisories and timely information about current security issues, vulnerabilities, and exploits to industrial control system. It provides assessments and alerts with timely notification to critical infrastructure owners and operators concerning threats or activity with the potential to impact critical infrastructure computing networks. Also, it publishes the Monitor Newsletter when an adequate amount of pertinent information has been collected. In addition, ICS-CERT supports training materials, research and white papers, standards information, quarterly and annual reports (Year in Review) that are of interest to persons engaged in protecting industrial control systems.

12.8 Standards, Guidelines, and Recommendations

While many Smart Gid and cybersecurity initiatives are identified around the globe, specific initiatives are supported by European Union and world decisions.

12.8.1 European Legislation

Recently, the European Union Parliament and Council have made a bold decision in drafting the cybersecurity legislation and in considering the issues related to the security evaluation of ICT products and services.

The European Union (EU) legislation [EU COM450] calls for adoption of high standards for building strong cybersecurity and increased resilience to cyber attacks as ICT products and services would be formally evaluated against a defined set of cybersecurity standards. This regulation proposes to set up a cybersecurity certification framework aiming towards a single cybersecurity market. The framework provides the procedure for the creation of EU wide cybersecurity certification schemes, covering products, services and/or systems, which adapt the level of assurance to the use involved: be it critical infrastructures or consumer devices.

The Commission highlights the specific issues concerning liability raised by new digital technologies. At the beginning, the certification schemes would be voluntary and would not create any immediate regulatory obligations on vendors or service providers. The schemes would not contradict any applicable legal requirements, such as the EU legislation on data protection.
After, the framework is established, the EU Commission proposes the relevant stakeholders to focus on three priority areas:

- Provide security in critical or high risk applications. These are systems that people depend on their daily activities, from cars to the machinery in factories, from the largest of systems such as airplanes or power plants to the smallest such as medical devices, that are becoming increasingly digital and interconnected. Therefore, core ICT components in such products and systems require rigorous security assessments.
- Apply robust cybersecurity and regulatory obligations to defend against attacks. Cybersecurity protections must be widely deployed in digital products, networks, systems and services used by private and public sector alike; examples of controls include Email encryption, firewalls and Virtual Private Networks; it is critical that the spreading use of such tools does not lead to new sources of risk or new vulnerabilities.
- Use of Security by Design methods in the production of low cost, digital, interconnected mass consumer devices which make up the IoT. The manufacturers should ensure that adequate security testing has been performed to their products. Also, they should commit to update their software in the event of newly discovered vulnerabilities or threats.

The regulation recommends that these priorities should consider the evolving cybersecurity threat landscape, as well as the importance of essential sectors such as transport, energy, health care, banking, financial market infrastructures, drinking water or digital infrastructure. Different sectors may have specific needs that have to be considered. Therefore, general cybersecurity strategies should be complemented by sector specific cybersecurity strategies in areas like financial services, energy, transport and health.

The framework recommends third party security researchers discovering vulnerabilities in existing products and services needs to be acknowledged. Building on best practices and relevant standards, proper coordination of the vulnerability disclosure should be created across Europe's member states.

It is envisioned that this cybersecurity certification framework will bring benefits to businesses by avoiding the need to go through several certification processes when trading across borders, thereby limiting administrative and financial costs. Also, the use of schemes developed under this framework would help build consumers' confidence, with a certificate of conformity to inform and reassure purchasers and users about the security properties of the products and services they buy and use [Lipner 2018].

Therefore, the EU cybersecurity certification framework is intended to further strengthen Europe's international position that could be complemented by continued efforts towards the development of high security global standards and mutual recognition agreements.

One recommendation is the adoption of international standards for software development. Specifically, the EU security certification schemes should be aligned with appropriate international standards for secure development, and in particular with the [ISO/IEC 27034] series of standards and allow for mutual recognition.

The standard [ISO/IEC 27034] offers guidance on information security to those specifying, designing and programming or procuring, implementing and using application systems, in other words business and IT managers, developers and auditors, and ultimately the end users of ICT. The aim is to ensure that computer applications deliver the desired or necessary level of security in support of the organization's Information Security Management System, adequately addressing many ICT security risks. This multi part standard provides guidance on specifying, designing/selecting and implementing information security controls through a set of processes integrated throughout an organization's Systems Development Life Cycle/s (SDLC). It is process oriented. It covers software applications developed internally, by external acquisition, outsourcing/offshoring or through hybrid approaches. The standard is applicable to inhouse developed applications, applications acquired from third parties, and where the development or the operation of the application is outsourced. It addresses all aspects from determining information security requirements, to protecting information accessed by an application as well as preventing unauthorized use and/or actions of an application. The standard is SDLC method agnostic. It does not mandate one or more specific development methods, approaches or stages but is written in a general manner to be applicable to them all. In this way, it complements other systems development standards and methods without conflicting with them.

In addition, privacy raises greater concerns as users realize how much information can be collected without their knowledge. The rights of the individuals are at risk, so the recent European Union legislation General Data Protection Regulation (GDPR) is designed to enable individuals to better control their personal data.

12.8.2 The World Energy Council Strategy

The World Energy Council (www.worldenergy.org) is the principal impartial network of energy leaders and practitioners promoting an affordable, stable and environmentally sensitive energy system for the greatest benefit of all. Formed in 1923, the Council is the United Nations' accredited global energy body, representing the entire energy spectrum, with over 3,000 member organizations in over 90 countries, drawn from governments, private and state corporations, academia, NGOs and energy stakeholders.

The World Energy Trilemma Index has been prepared annually since 2010 by the World Energy Council in partnership with others (global consultancy Oliver Wyman, along with Marsh & McLennan). It presents a comparative ranking of 128 countries' energy systems. It provides an assessment of a country's energy system performance, reflecting balance and robustness in the three Trilemma dimensions.

Accessing the complete Index results, national Trilemma profiles and the interactive Trilemma Index tool allows to find out more about countries' Trilemma performance and what it takes to build a sustainable energy system (see https://trilemma.worldenergy.org).

References Part 1

[Anderson 2008] Anderson, R.J. (2008). *Security Engineering: A Guide to Building Dependable Distributed Systems* (2nd Edition). Wiley.

[ANSI 1471] *ANSI/IEEE 1471:2000 Recommended Practice for Architectural Description for Software-Intensive Systems.* *superseded by ISO/IEC/IEEE 42010

[Barman 2002] Barman, S. (2002). *Writing Information Security Policies.* New Riders.

[Bayuk 2011] Bayuk, J.L. (2011). Systems security engineering. *IEEE Security & Privacy, 9*(2), 72–74. https://doi.org/10.1109/MSP.2011.41[Berger 2012] Berger, L.T., Iniewski, K. (2012). *Smart Grid Applications, Communications, and Security.* Wiley.

[Bishop 1996] Bishop, M., Bailey, D. (1996). *A Critical Analysis of Vulnerability, Taxonomies* [Research Report CSE-96-11]. http://nob.cs.ucdavis.edu/bishop/notes/1996-cse-11/1996-cse-11.pdf

[Cardenas 2008] Cárdenas, A.A., Amin, S., Sastry, S. (2008). Research Challenges for the Security of Control Systems

[Cavusoglu 2007] Cavusoglu, H., Cavusoglu, H., Raghunathan, S. (2007). Efficiency of Vulnerability Disclosure Mechanisms to Disseminate Vulnerability Knowledge. *IEEE Transactions On Software Engineering, 33*(3), 171-185.

[CC] Commoncriteriaportal. (n.d.). *The Common Criteria for Information Technology Security Evaluation V3.1 R4* (September 2012). http://www.commoncriteriaportal.org/cc/

[Cencini 2005] Cencini, A., Yu, K., Chan, T. (2005). *Software Vulnerabilities: Full, Responsible, and Non-Disclosure.* http://courses.cs.washington.edu/courses/csep590/05au/whitepaper_turnin/software_vulnerabilities_by_cencini_yu_chan.pdf

[CENELEC 2014] CENELEC. (2014). *SG-CG/M490/H_ Smart Grid Information Security* (December 2014). [M490-SGCG-SGIS-Intermediate-Report-V1.pdf]. CEN-CENELEC-ETSI Smart Grid Coordination Group Document for the M/490 Mandate Smart Grid Information Security. ftp://ftp.cencenelec.eu/EN/EuropeanStandardization/HotTopics/SmartGrids/SGCG_SGIS_Report.pdf

[CENELEC 2017] CENELEC. (2017). *SEGCG/M490/G_Smart Grid Set of Standards Version 4.1* (Jan 6th 2017).

[CERT-ICS Disclosure] CERT-IST. (n.d.). *Vulnerability disclosure policy.* http://www.cert-ist.com/public/en/politiquededivulgationdesfailles

[Chesla 2015] Chesla, A. (2015, March 4). *Taking Back the Reigns: Proactive Security Design vs. Reactive Administration* [News]. SecurityWeek Internet and Enterprise Security News, Insights & Analysis. http://www.securityweek.com/taking-back-reigns-proactive-security-design-vs-reactive-administration

[Cho 2014] Cho, D-E., Yeo, S-S., Kim, S-J. (2014). Authentication Method for Privacy Protection in Smart Grid Environment. *Journal of Applied Mathematics*, Article ID280978, 10 pages. https://doi.org/10.1155/2014/280978

[Cisco SAFE] CISCO. (2000). Cisco SAFE: A Security Blueprint for Enterprise Networks [White Paper]. http://www.cisco.com/c/dam/en/us/products/collateral/wireless/mobile-office-net-software/product_implementation_design_guide09186a00800a3016.pdf

[Clark 2016] Clark, B. (2016). *Incident Response Planning in Industrial Control Systems ICTN 6823 Information Security Management* (July 2016) [White Paper]. http://www.infosecwriters.com/Papers/WClark_IRinICS.pdf

[Cleveland 2012] Cleveland, F. (2012). *IEC TC57: IEC 62351 Security Standards the Power System Information Infrastructure* (June 2012). http://iectc57.ucaiug.org/wg15public/Public%20Documents/White%20Paper%20on%20Security%20Standards%20in%20IEC%20TC57.pdf

[Cleveland 2013] Cleveland, F. (2013). *List of Standards that Include Cybersecurity* (May 2013) [Report to IEC TC 57 WG15]. http://xanthus-consulting.com/Publications/documents/List_of_Smart_Grid_Standards_with_Cybersecurity.pdf

[COBIT v5] ISACA. (n.d.). *COBIT 5 Framework.* https://www.isaca.org/resources/cobit

[Conte de Leon 2002] Conte de Leon, D., Alves-Foss, J., Krings, A., Oman, P. (2002). Modeling Complex Control Systems to Identify Remotely Accessible Devices Vulnerable to Cyber Attack. *Proceedings of First Workshop on Scientific Aspects of Cyber Terrorism*, Washington, D.C., USA, (November 2002).

[Creel 2013] Creel, R., Ellison, R.J. (2013). *System-of-Systems Influences on Acquisition Strategy Development.* https://www.us-cert.gov/bsi/articles/best-practices/acquisition/system-of-systems-influences-on-acquisition-strategy-development#several-approaches-risk-management

[Deiuliis 2012] Deiuliis, K. (2012). An Integrated Approach to Emerging Technologies. *Electric Perspectives, January/February 2012*, 57-60. Edison Institute.

[Denning 1999] Denning, D.E. (1999). *The Limits of Formal Security Models, National Computer Systems Security Award Acceptance Speech* (October 18, 1999). http://faculty.nps.edu/dedennin/publications/National%20Computer%20Systems%20Security%20Award%20Speech.htm

[Dhondge 2021] Dhondge, K. (2021). *Lifecycle IoT Security for Engineers.* Artech House.

[DHS 2011] DHS. (2011). *Blueprint for a Secure Cyber Future: The Cybersecurity Strategy for Homeland Security Enterprise* (November 2011). http://www.dhs.gov/xlibrary/assets/nppd/blueprint-for-a-secure-cyber-future.pdf

[DHS 2016] DHS. (2016, September). *Recommended Practice: Improving Industrial Control System Cybersecurity with Defense-in-Depth Strategies.* Industrial Control Systems Cyber Emergency Response Team. https://www.cisa.gov/uscert/sites/default/files/recommended_practices/NCCIC_ICS-CERT_Defense_in_Depth_2016_S508C.pdf

[DoD 2001] DoD. (2001). *Systems Engineering Fundamentals* (January 2001). Department of Defense Systems Management College. Supplementary Text Prepared By The Defense Acquisition University Press, Fort Belvoir, Virginia, USA. https://ocw.mit.edu/courses/aeronautics-and-astronautics/16-885j-aircraft-systems-engineering-fall-2005/readings/sefguide_01_01.pdf

[DoD 2007] DoD. (2007). *Systems Engineering Plan Preparation Guide Version 2.0* (October 2007). Systems Engineering Plan Preparation Guide, Version 2.0, October 18, 2007 (navy.mil)

[DoD 2008] DoD. (2008). *Systems Engineering Guide for Systems of Systems Version 1.0* (August 2008). US Department of Defense (DoD), Washington, DC, USA. http://www.acq.osd.mil/se/docs/SE-Guide-for-SoS.pdf

[DOE 2010] DOE. (2010). *High-Impact, Low-Frequency Event Risk to the North American Bulk Power System A Jointly-Commissioned Summary Report of the North American Electric Reliability Corporation and the U.S. Department of Energy's November 2009 Workshop.* https://www.energy.gov/sites/prod/files/High-Impact%20Low-Frequency%20Event%20Risk%20to%20the%20North%20American%20Bulk%20Power%20System%20-%202010.pdf

[DOE 2015a] DOE. (2015a). *United States Electricity Industry Primer* (July 2015) [DOE/OE-0017]. Office of Electricity Delivery and Energy Reliability U.S. Department of Energy. http://www.energy.gov/sites/prod/files/2015/12/f28/united-states-electricity-industry-primer.pdf

[DOE 2015b] DOE. (2015b). *Energy Sector Cybersecurity Framework Implementation Guidance* (January 2015). Office of Cybersecurity, Energy Security, and Emergency Response.

https://energy.gov/oe/downloads/energy-sector-cybersecurity-framework-implementation-guidance

[DOE C2M2] DOE. (n.d.). *Cybersecurity Capability Maturity Model (C2M2)*. Office of Cybersecurity, Energy Security, and Emergency Response. https://www.energy.gov/ceser/energy-security/cybersecurity-capability-maturity-model-c2m2-program

[DOE ES-C2M2] DOE. (2014). *Electricity Subsector Cybersecurity Capability Maturity MODEL (ES-C2M2) Version 1.1* (February 2014). U.S. Department of Energy Office of Electricity. https://www.energy.gov/oe/downloads/electricity-subsector-cybersecurity-capability-maturity-model-v-11-february-2014

[DOE Policy] DOE. (2011). Cyberspace Policy Review: Assuring a Trusted and Resilient Information and Communications Infrastructure (June 16, 2011). U.S. Department of Energy Office of the Chief Information Officer. http://energy.gov/cio/downloads/cyberspace-policy-review-assuring-trusted-and-resilient-information-and-communications

[Dong 2011] Dong, Y., Kezunovic, M. (2011). Communication Infrastructure for Emerging Transmission-Level Smart Grid Applications. *2011 IEEE Power and Energy Society General Meeting* (pp. 1-7), Detroit, Michigan, USA, (July 2011). https://doi.org/10.1109/PES.2011.6039640.

[Dundas 2012] Dundas, M., Hurley, J. (2012). RSA Security Anatomy of an Attack – Lessons learned [Presentaton]. *2012 RSA Conference Europe*, London, United Kingdom, (October 2012). https://docplayer.net/10528257-Rsa-security-anatomy-of-an-attack-lessons-learned.html

[EAC 2014] Energy. (2014). *Implementing Effective Enterprise Security Governance Outline for Energy Sector Executives and Boards* (March 2014). U.S. Department of Energy Electricity Advisory Committee. http://energy.gov/sites/prod/files/Mar2014EAC_Recs-CyberGovernance.pdf

[Easter 2015] Easter, R.J. (2015). Accreditation, Validation and Recognition based on ISO Standards. *International Cryptographic Module Conference (ICMC) ICMC 2015Certifications Programs Track*, Washington, D.C., USA, (November 2015). https://icmconference.org/wp-content/uploads/C12EasterR.ISO-Standards-JIC.pdf

[ENISA 2011a] ENISA. (2011a). Protecting Industrial Control Systems Annex V: Key Findings (December 14, 2011). European Union Agency for Network and Information Security. https://www.enisa.europa.eu/publications/annex-v

[ENISA 2011b] ENISA. (2011b). Protecting Industrial Control Systems Annex III: ICS Security Related Standards, Guidelines and Policy Documents (December 9, 2011). European Union Agency for Network and Information Security. https://www.enisa.europa.eu/activities/Resilience-and-CIIP/critical-infrastructure-and-services/scada-industrial-control-systems/annex-iii

[ENISA 2012] ENISA. (2012). *Smart Grid Security: Recommendations for Europe and other Member States*. European Union Agency for Network and Information Security. https://www.enisa.europa.eu/publications/ENISA-smart-grid-security-recommendations

[ENISA 2014a] ENISA. (2014a). Proposal for a list of security measures for smart grids (December 2013). SMART GRID TASK FORCE 5 EG2 DELIVERABLE. European Union Agency for Network and Information Security. http://www.tpeb.cz/wp-content/uploads/2014/04/20140409_enisa.pdf

[ENISA 2014b] ENISA. (2014b). *Security Guide for ICT Procurement ICT Procurement Security Guide for Electronic Communications Service Providers* (December 2014). European Union Agency for Network and Information Security. https://www.enisa.europa.eu/publications/security-guide-for-ict-procurement

[ENISA 2015] ENISA. (2015). *Analysis of ICS-SCADA Cyber Security Maturity Levels in Critical Sectors* (December 2015). European Union Agency for Network and Information Security. http://doi.org/10.2824/835661

[ENISA 2016] ENISA. (2016). *Communication network interdependencies in smart grids - Annexes, Methodology for the identification of Critical Communication Networks Links and Components* (January 2016). European Union Agency for Network and Information Security. https://www.enisa.europa.eu/topics/critical-information-infrastructures-and-services/communication-network-interdependencies-in-smart-grids-annexes

[ENISA ISMS] ENISA. (n.d.). *Threat and Risk Management ISMS Framework.* European Union Agency for Network and Information Security. https://www.enisa.europa.eu/topics/threat-risk-management/risk-management/current-risk/risk-management-inventory/files/isms_framework.jpg/view

[EO 13636] WhiteHouse. (2013, February 12). *Executive Order (EO) 13636 Improving Critical*

[EPRI 2015] EPRI. (2015). *Distribution Modeling Guidelines: Recommendations for System and Asset Modeling for Distributed Energy Resource Assessments* (3002006115). https://www.epri.com/research/products/3002006115

[ETSI ISG] ETSI. (2015). Information Security Indicators - measuring the impact of incidents affecting electronic communications (December 20915). European Union Agency for Network and Information Security. http://doi.org/10.2824/887699. http://www.etsi.org/technologies-clusters/technologies/information-security-indicators

[Fang 2012] Fang, X., Misra, S., Xue, G., Yang, D. (2012). Smart grid – the new and improved power grid: a survey. *IEEE Communications Surveys and Tutorials, 14*(4), 944–980. https://doi.org/10.1109/SURV.2011.101911.00087

[Feng 2008] Feng, C., Jin-Shu, S. (2008). A Flexible Approach to Measuring Network Security Using Attack Graphs. 2008 International Symposium on Electronic Commerce and Security (ISECS) (pp. 426-431), Guangzhou City, China, (August 2008). http://www.jinshusu.net.cn/static/fengflexible.pdf

[FIPS 199] *FIPS PUB 199 Standards for Security Categorization of Federal Information and Information Systems, February 2004.* Computer Security Division Information Technology Laboratory. U.S. Department of Commerce. National Institute of Standards and Technology. http://csrc.nist.gov/publications/fips/fips199/FIPS-PUB-199-final.pdf

[FIPS 31] *FIPS PUB 31 Guidelines for Automatic Data Processing Physical Security and Risk Management, June 1974.* National Bureau of Standards. U.S. Department of Commerce.

[FISMA 2002] *Federal Information Security Management Act of 2002, Title III of the E-Government Act of 2002, Pub. L. No. 107-347 – Dec 17, 2002 116 Stat. 2899.*

[Foreman 2010] Foreman, P. (2010). *Vulnerability Management* (1st Edition). CRC Press Auerbach Publications.

[Fragkos 2005] Fragkos, G., Blyth, A. (2005). Security Threat Assessment across Large Network Infrastructures. *First Workshop on Safeguarding National Infrastructures* (pp. 22-29), Editor: Chris Johnson.

[GAO 2005] GAO. (2005). *Department of Homeland Security's (DHS's) Role in Critical Infrastructure Protection (CIP) Cybersecurity* (May 2005), [GAO-05-434]. Government Accountability Office (GAO).

[Garnaut 2011] Garnaut, P., Thompson, J. (2011). *Review of Data Integrity Models in Multi-Level Security Environments (February 2011)* [DSTO–TN–0971]. Command, Control, Communications and Intelligence Division Defense Science and Technology Organization. Australian Government. Department of Defense.

[Gartner 2015] Gartner. (2015). Innovation Tech Insight for Security Operations, Analytics and Reporting (November 11, 2015), [Report, Editors: Rochford, O., Proctor, P.E.].

[Gartner Glossary] Gartner. (n.d.). Gartner IT Glossary. https://www.gartner.com/en/information-technology/glossary/it-information-technology

[Ghernouti 2010] Ghernouti-Hélie, S. (2010). A National Strategy for an Effective Cybersecurity Approach and Culture. *ARES 2010 Fifth International Conference on Availability, Reliability, and Security* (pp. 370-373), Krakow, Poland, (February 2010). IEEE. http://doi.org/10.1109/ARES.2010.119

[Greene 2014] Greene, S.S. (2014). Governance and Risk Management. Chapter in *Security Program and Policies: Principles and Practices* (Second Edition) (pp. 92-123). Pearson IT Certification.

[Hansche 2004] Hansche, S., Berti, J., Hare, C. (2004). *Official (ISC)2 Guide to The CISSP Exam.*

[Harris 2013] Harris, S. (2013). *All in One CISSP Exam Guide* (6th Edition). McGraw Hill Education.

[Hasselbring 2006] Hasselbring, W., Reussner, R. (2006). Toward Trustworthy Software Systems. *Computer, 39*(4), 91-92. http://doi.org/10.1109/MC.2006.142

[Hentea 2005] Hentea, M. (2005). Information security management. In M. Pagani (Ed.), *Encyclopedia of Multimedia Technology and Networking* (2-volume set, pp. 390–395). IGI Global. https://doi.org/10.4018/978-1-59140-561-0.ch056

[Hentea 2008] Hentea, M. (2008). Improving security for SCADA control systems. *Interdisciplinary Journal of Information, Knowledge, and Management, 3*, 73–86. https://doi.org/10.28945/3185

[Hentea 2021] Hentea, M. (2021). Planning Security Protection. Chapter in *Building an Effective Security Program for Distributed Energy Resources and Systems Understanding Security for Smart Grid and Distributed Energy Resources and Systems*, pp. 165-196. Wiley. https://doi.org/10.1002/9781119070740.ch5

[Hilliard 2012] Hilliard, R., Malavolta, I., Muccini, H., Pellicione, P. (2012). On the Composition and Reuse of Viewpoints across Architecture Frameworks. *Proceedings the Joint 10th Working IEEE/IFIP Conference on Software Architecture & 6th European Conference on Software Architecture (WICSA/ECSA)* (pp. 131-140), Helsinki, Finland, (August 2012). http://doi.org/10.1109/WICSA-ECSA.212.21

[Hirst 1996] Hirst, E., Kirby, B. (1996). Ancillary Services (February 1996) [Technical Report ORNL/CON 310]. Oak Ridge National Laboratory. U.S. Department of Energy.

[Huxham 2006] Huxham, H. (2006). One View of Enterprise Information Security Architecture (EISA) Framework (September 18, 2006). Public Domain. https://en.wikipedia.org/wiki/File:Huxham_Security_Framework.jpg

[ICS-CERT 2016] ICS-CERT. (2016). *ICS-CERT Vulnerability Coordination.* Monitor (ICS-MM201610), September-October 2016. https://us-cert.cisa.gov/sites/default/files/Monitors/ICS-CERT_Monitor_Sep-Oct2016_S508C.pdf

[ICS-CERT Threat] ICS-CERT. (n.d.). *Cyber Threat Source Descriptions.* https://us-cert.cisa.gov/ics/content/cyber-threat-source-descriptions

[IDC 2007] IDC. (2007, September). *Unified Threat Management Appliances and Identity-based Security: The Next Level in Network Security.* IDC Go-to Market Services.

[IEC 60870-6] *IEC 60870-6-503:2014 Telecontrol equipment and systems - Part 6-503: Telecontrol protocols compatible with ISO standards and ITU-T recommendations - TASE.2 Services and protocol.*

[IEC 60870] *IEC 60870 Telecontrol equipment and systems, series of several parts.*

[IEC 61508] *IEC 61508 Functional Safety of Electrical/Electronic/Programmable Electronic Safety-related Systems (E/E/PE, or E/E/PES).*

[IEC 61850] *IEC 61850 Communication networks and systems in substations, series of several parts.*

[IEC 61968] *IEC 61968 Application integration at electric utilities - System interfaces for distribution management, series of several parts.*

[IEC 61970] *IEC 61970 Energy management system application program interface (EMS-API), series of several parts.*

[IEC 62351] *IEC 62351 Power systems management and associated information exchange - Data and communications security, series of several parts.*

[IEC 62351] *IEC 62351:2017 Series Power systems management and associated information exchange – Data and communications security - ALL PARTS.*
[IEC 62443] *IEC 62443 Industrial communication networks - Network and system security, series of several parts.*
[IEC TS 62443-1-1] *IEC TS 62443-1-1:2009 Industrial communication networks - Network and system security - Part 1-1: Terminology, concepts and models, Edition 1.02009-07.*
[IETF 2002] IETF.(2002). *Draft Request for Comments, Responsible Vulnerability Disclosure Process February 2002*. https://tools.ietf.org/id/draft-christey-wysopal-vuln-disclosure-00.txt
[ISA 62443-1-1] *ISA 62443-1-1:2007 Industrial Automation and Control Systems Security Terminology Concepts and Models*. International Society of Automation. https://www.isa.org/isa99/
[ISA 62443-1-1] ISA. (n.d.). *ISA 62443-1-1 Industrial Automation and Control Systems Security Concepts and Models.* International Society of Automation.
[ISA 62443SER] ISA. (2016). *The 62443 series of standards Industrial Automation and Control Systems Security* (December 2016). International Society of Automation. https://cdn2.hubspot.net/hubfs/3415072/Resources/The%2062443%20Series%20of%20Standards.pdf
[ISA TR62443-1-2] ISA. (n.d.). *ISA TR62443-1-2 Industrial Automation and Control Systems Security Master glossary of terms and abbreviations.* International Society of Automation.
[ISA TR62443-2-2] ISA. (n.d.). *ISA TR62443-2-2 Establishing and Operating an IACS Security Program.* International Society of Automation.
[ISA-62443-x-y*]* ISA. (n.d.). *ISA-62443-x-y series of standards Industrial Automation and Control Systems Security.* International Society of Automation.
[ISA/IEC 62443] *ISA/IEC 62443* Industrial Automation and Control Systems Security, *series of several parts.* International Society of Automation.
[ISA99 Glossary] *ISA TR62443-1-2 Industrial Automation and Control Systems Security Master glossary of terms and abbreviations.*
[ISA99] ISA. (n.d.). *ISA99 Industrial Automation and Control Systems Security.* ISA99 Standards Development Committee. International Society of Automation. https://www.isa.org/isa99/
[ISACA Glossary] ISACA. (n.d.). *Glossary, Model.* http://www.isaca.org/Pages/Glossary.aspx?tid=2027&char=M
[ISAGCA 2020] ISASECURE. (2020, October). *Security Lifecycles in the ISA/IEC 62443 Series Security of Industrial Automation and Control System* [White Paper]. International Society of Automation Global Cybersecurity Alliance. https://www.isasecure.org/en-US/Documents/Articles-and-Technical-Papers/ISAGCA-Security-Lifecycles-whitepaper
[ISC2 2006] ISC2. (2006). *The (ISC2) CISSP CBK Review SeminarVersion 5.0 Student Manual.* Pearson Custom Publishing.
[ISI Portal] ETSI. (2015). *Information Security Indicators (ISI); Indicators (INC); Part 1: A full set of operational indicators for organizations to use to benchmark their security posture* [ETSI GS ISI 001-1 V1.1.1 (2013-04) DGS/ISI-001-1]. ETSI Industry Specification Group.
[ISM3 2009] Seclists. (2009, April 2). *STANDARD FOR INFORMATION SECURITY MANAGEMENT UPDATED Information Security Management Maturity Model (ISM3) update, April 2009* [News]. InfoSec News. https://seclists.org/isn/2009/Apr/7[ISO 12207] *ISO/IEC 12207:2008 Systems and software engineering -- Software life cycle processes.*
[ISO 14001] *ISO 14001:2015 Environmental management systems - Requirements with guidance for use.*
[ISO 15288] *ISO/IEC/IEEE 15288:2015 Systems and software engineering—System life cycle processes Version 4.0.*
[ISO 15289] *ISO/IEC/IEEE 15289:2011 Systems and software engineering -- Content of life-cycle information items.*

[ISO 15408] *ISO 15408:2012 Information technology – Security techniques – Evaluation criteria for IT security - P1-3 CC v3.1 Release 4.*

[ISO 157004] *ISO 157004:2000 Industrial Automation Systems – Requirements for enterprise-reference architectures and methodologies.*

[ISO 27k] ISO. (n.d.*). About the ISO27k Standards.* International Standards Organization. https://www.iso27001security.com/ISO27k_Standards_listing.pdf

[ISO 29148] *ISO/IEC/IEEE 29148:2011 - ISO/IEC/IEEE International Standard - Systems and software engineering -- Life cycle processes --Requirements engineering*

[ISO 42010] *ISO/IEC/IEEE 42010:2011 Systems and software engineering – Architecture description, A Conceptual Model of Architecture Description.*

[ISO 9000] *ISO 9000:2015 Quality management systems — Fundamentals and vocabulary.*

[ISO/IEC 15026-2] *ISO/IEC 15026-2:2011 Systems and software engineering -- Systems and software assurance -- Part 2: Assurance case.*

[ISO/IEC 15026-3] *ISO/IEC 15026-3:2015 Systems and software engineering -- Systems and software assurance -- Part 3: System integrity levels.*

[ISO/IEC 15504] *ISO/IEC 15504 Information technology – Process assessment, also termed Software Process Improvement and Capability Determination (SPICE).*

[ISO/IEC 18367] *ISO/IEC 18367:2016 Information technology -- Security techniques -- Cryptographic algorithms and security mechanisms conformance testing.*

[ISO/IEC 19790] *ISO/IEC 19790:2012 Information technology -- Security techniques -- Security requirements for cryptographic modules.*

[ISO/IEC 19896-1] *ISO/IEC 19896-1:2018 IT security techniques -- Competence requirements for information security testers and evaluators --Part 1: Introduction, concepts and general requirements.*

[ISO/IEC 19896-2] *ISO/IEC 19896-2:2018 IT security techniques -- Competence requirements for information security testers and evaluators -- Part 2: Knowledge, skills and effectiveness requirements for ISO/IEC 19790 testers.*

[ISO/IEC 19896] ***ISO/IEC 19896****, IT security techniques -- Competence requirements for information security testers and evaluators, series of several parts.*

[ISO/IEC 21827] *ISO/IEC 21827:2008 Information technology -- Security techniques -- Systems Security Engineering -- Capability Maturity Model® (SSE-CMM®).*

[ISO/IEC 24759] *ISO/IEC 24759:2017 Information technology -- Security techniques -- Test requirements for cryptographic modules.*

[ISO/IEC 27001] *ISO/IEC 27001:2013 Information technology – Security techniques – Information security management systems – Requirements.*

[ISO/IEC 27002] *ISO/IEC 27002:2013, Information technology – Security techniques – Code of practice for information security controls.*

[ISO/IEC 27005] *ISO/IEC 27005:2011 Information technology — Security techniques — Information security risk management.*

[ISO/IEC 29147] *ISO/IEC 29147:2018 Information Technology – Security techniques - Vulnerability Disclosure.*

[ISO/IEC 31000] *ISO 31000:2009* Risk management – Principles and guidelines.

[ISO/IEC 38500] *ISO/IEC 38500:2015 Information technology -- Governance of IT for the organization.*

[ISO/IEC TR 19791] *ISO/IEC TR 19791:2010 Information technology -- Security techniques -- Security assessment of operational systems.*

[ISO/IEC TR 27019] *ISO/IEC TR 27019: 2013 Information technology - Security techniques - Information security management guidelines based on ISO/IEC 27002 for process control systems specific to the energy utility industry.* *revised by ISO/IEC 27019:2017.

[ITU T] ITU. (2011). *Deliverable on Smart Grid Terminology* (December 2011, Geneva, Switzerland) [Smart-O-30Rev.6]. Focus Group on Smart Grid. International Telecommunication Union Telecommunication Standardization Sector. https://studylib.net/doc/7891519/terminology-deliverable-updating

[Jaakkola 2011] Jaakkola, H., Thalheim, B. (2011). Architecture-Driven Modelling Methodologies. In Heimbürger, A., Heimbürger, A., Kiyoki, Y., Tokuda, T., Jaakkola, H., Yoshida, N. (Eds.), *Proceedings of the 2011 Conference on Information Modelling and Knowledge Bases XXII* (pp. 97-116). IOS Press.

[Kadrich 2007] Kadrich, M.S. (2007). *Endpoint Security*. Addison-Wesley.

[Kezunovic 2011] Kezunovic, M., Dutta, P. (2011). The Role of Data Exchange Standards in Developing Automated Fault Disturbance Monitoring. *6th International Workshop on Deregulated Electricity Market Issues in South-Eastern Europe (DEMSEE)*, Bled, Slovenia, (September 2011).

[Knowles 2015] Knowles, W., Prince, D.D.C., Hutchison, D., Disso, J.F.P., Jones, K. (2015). A survey of cyber security management in industrial control systems. *International Journal of Critical Infrastructure Protection, 9*(C), 52-80. https://doi.org/10.1016/j.ijcip.2015.02.002

[Komninos 2014] Komninos, N., Philippou, E., Pitsillides, A. (2014). Survey in smart grid and smart home security: issues, challenges and countermeasures. *IEEE Communications Surveys & Tutorials, 16*(4), 1933–1954. https://doi.org/10.1109/COMST.2014.2320093

[Krutz 2004] Krutz, R.L., Vines, R.D. (2004). *The CISSP Prep Guide: Mastering the CISSP and ISSEP Exams* (Second Edition). Wiley.

[Krutz 2006] Krutz, R.L. (2006). *Securing SCADA Systems*. Wiley.

[Lal 2013] Lal, B., Chavan, C.C., Bajaj, J. (2013). An Optimization Approach to Analysis of Requirement Pre-Processing in Software Engineering. *International Journal of Advanced Research in Computer Science and Software Engineering, 3*(2), 311-316.

[Laplante 2007] Laplante, P.A. (2007). *What Every Engineer Should Know about Software Engineering*. CRC Press Taylor & Francis Group.

[Laplante 2014] Laplante, P.A. (2014). *Requirements Engineering for Software and Systems* (Second Edition). CRC Press, Auerbach Publishing, Taylor & Francis Group.

[Marley 2003] Marley, S. (2003). *Architectural Framework*. NASA/SCI.

[Meyers 2013] Meyers, J. (2013). *How the convergence of IT and OT enables smart grid development* (998-2095-08-13-13R0). Schneider Electric. http://cdn.iotwf.com/resources/10/How-the-Convergence-of-IT-and-OT-Enables-Smart-Grid-Development_2013.pdf

[MITRE Guide] MITRE. (n.d.). Architectural Frameworks, Models, and Views. *Systems Engineering Guide*. https://www.mitre.org/publications/systems-engineering-guide/se-lifecycle-building-blocks/system-architecture/architectural-frameworks-models-and-views

[Mo 2012] Mo, Y., Kim, T. H-J.,Brancik, K., Dickinson, D., Lee, H., Perrig, A., Sinopoli, B. (2012). Cyber–Physical Security of a Smart Grid Infrastructure. *Proceedings of the IEEE, 100*(1), 195-209. https://doi.org/10.1109/JPROC.2011.2161428

[NERC 2010] Energy. (2010). *High-Impact, Low-Frequency Event Risk to the North American Bulk Power System* (June 2010). A Jointly-Commissioned Summary Report of the North American Electric Reliability Corporation and the U.S. Department of Energy's November 2009 Workshop. http://energy.gov/sites/prod/files/High-Impact%20Low-Frequency%20Event%20Risk%20to%20the%20North%20American%20Bulk%20Power%20System%20-%202010.pdf

[NERC CIP] NERC. (n.d.). *CIP Standards*. North American Electric Reliability Corporation. http://www.nerc.com/pa/Stand/Pages/CIPStandards.aspx

[NERC Glossary] NERC. (n.d.) *Glossary of Terms Used in NERC Reliability Standards*. http://www.nerc.com/files/Glossary_of_Terms.pdf

[NIAC 2004] NIAC. (2004). Vulnerability Disclosure Framework: Final Report and Recommendations (January 2004). National Infrastructure Advisory Council. https://www.cisa.gov/publication/niac-vulnerability-framework-final-report

[NIST 2014a] *NIST Framework for Improving Critical Infrastructure Cybersecurity Version 1.0* (February 2014).

https://www.nist.gov/system/files/documents/cyberframework/cybersecurity-framework-021214.pdf

[NIST 2014b] NIST. (2014b). *NISTIR 7628 User's Guide A White Paper developed by the Smart Grid Interoperability Panel Smart Grid Cybersecurity Committee* (February 2014). https://tsapps.nist.gov/publication/get_pdf.cfm?pub_id=915613

[NIST SP1108r1] *NIST Special Publication (SP) 1108r1 NIST Framework and Roadmap for Smart Grid Interoperability Standards* (Release 1.0, January 2010). https://doi.org/10.6028/NIST.sp.1108

[NIST SP1108r3] *NIST Special Publication (SP) 1108r3 NIST Framework and Roadmap for Smart Grid Interoperability Standards* (Release 3.0, September 2014). http://dx.doi.org/10.6028/NIST.SP.1108r3

[NIST SP1108r4] *NIST Special Publication (SP) 1108r4 NIST Framework and Roadmap for SmartGrid Interoperability Standards Release 4.0* (February 2021). https://dx.doi.org/10.6028/NIST.SP.1108r4

[NIST SP800-12] *NIST Special Publication (SP) 800-12 An Introduction to Computer Security: the NIST Handbook* October 1995).

[NIST SP800-161] *NIST Special Publication (SP) 800-161* Supply Chain Risk Management Practices for Federal Information Systems and Organizations (April 2015). http://dx.doi.org/10.6028/NIST.SP.800-161

[NIST SP800-16r1] *NIST Special Publication (SP) 800-16 A Role-Based Model for Federal Information Technology/Cybersecurity Training Revision 1* (March 2014).

[NIST SP800-18r1] *NIST Special Publication (SP) 800-18 Guide for Developing Security Plans for Federal Information Systems Revision 1* (February 2006).

[NIST SP800-33] *NIST Special Publication (SP) 800-33 Underlying Technical Models for Information Technology Security* (December 2001).

[NIST SP800-37r1] *NIST Special Publication (SP) 800-37 Guide for Applying the Risk Management Framework to Federal Information Systems: a Security Life Cycle Approach Revision 1* (Updated June 2014).

[NIST SP800-39] *NIST Special Publication (SP) 800-39 Managing Information Security Risk: Organization, Mission, and Information System View* (March 2011)

[NIST SP800-53Ar4] *NIST Special Publication (SP) 800-53A Assessing Security and Privacy Controls in Federal Information Systems and Organizations: Building Effective Assessment Plans Revision 4* (December 2014)

[NIST SP800-53r5] *NIST Special Publication (SP) 800-53 Revision 5 Security and Privacy Controls for Information Systems and Organizations* (December 2020). https://doi.org/10.6028/NIST.SP.800-53r5

[NIST SP800-82r2] *NIST Special Publication (SP) 800-82 Guide to Industrial Control Systems Revision 2 (May 2015).*

[NISTIR 7628r1] *NIST Interagency Report (NISTIR) 7628r1 Guidelines for Smart Grid Cybersecurity* (Revision 1, September 2014). http://dx.doi.org/10.6028/NIST.IR.7628r1

[NSF 2011] Energy. (2011). *A Policy Framework For The 21st Century Grid: Enabling Our Secure Energy Future* (June 13, 2011). NATIONAL SCIENCE AND TECHNOLOGY COUNCIL. https://www.energy.gov/sites/default/files/oeprod/DocumentsandMedia/nstc-smart-grid-june2011.pdf

[Nuseibeh 2000] Nuseibeh, B., Easterbrook, S. (2000). Requirements Engineering: A Roadmap. *ICSE'00 Proceedings of the International Conference* on The Future of Software Engineering (pp. 35-46), Limerick, Ireland, (June 2000). https://doi.org/10.1145/336512.336523

[O-ISM3] Opengroup. (2011, April 11). *Open Information Security Management Maturity Model (O-ISM3)* [News]. The Open Group. https://www.opengroup.org/open-group-releases-maturity-model-information-security-management

[OECD 2013] OECD. (2013). *The OECD Privacy Framework*. Organisation for Economic Co-operation and Development. https://www.oecd.org/sti/ieconomy/oecd_privacy_framework.pdf

[OIS 2004] Oisafety. (2004). *Guidelines for Vulnerability Reporting and Response by the Organization for Internet Safety (OIS) Version 2.0* (2004, September 1). Organization for Internet Safety. https://docs.broadcom.com/doc/ois-guidelines-for-responsible-disclosure-en

[OMB 130] OMB. (2016, July 28). *Revision of OMB Circular No A-130 Transmittal 3 Appendix III Security of Federal Automated Information Resources Managing Information as a Strategic Resource Management of Federal Information Resources.* Office of Management and Budget. https://obamawhitehouse.archives.gov/omb/circulars_a130_a130appendix_iii

[OWASP] OWASP. (n.d.). *Security Positive Model.* Open Web Application Security Project (OWASP). https://www.owasp.org/index.php/Positive_security_model

[P2030 2011] *IEEE Std 2030-2011 - IEEE Guide for Smart Grid Interoperability of Energy Technology and Information Technology Operation with the Electric Power System (EPS), End-Use Applications, and Loads.*

[Pacyna 2009] Pacyna, P., Rutkowski, A., Sarma, A., Takahashi, K. (2009). Trusted Identity for All: Toward Interoperable Trusted Identity Management System Guest Editors' Introduction, Special Issue on Identity Management in Computer. *Computer, 42*(5), 30-32. https://doi.org/10.1109/MC.2009.168

[Prince 2013] Prince, B. (2013, July 1). *Cyber Attacks Against Energy Sector Jump in 2013: ICS-CERT* [News]. SecurityWeek, Internet and Enterprise Security News, Insights & Analysis.

[Privacy 1974] *US Privacy Act of 1974 5 U.S.C. § 552a (2012).* U.S. Department of Justice.

[R2GS] Cloudsecurityalliance. (2015). Towards a standardised pan-European approach to SIEM. ClubR2GS Workshop for the European chapters of Club R2GS (20 January 2015 – 7th International Forum on Cybersecurity – Lille). https://cloudsecurityalliance.it/wp-content/uploads/2015/01/FIC-2015_Workshop-for-the-European-chapters-of-Club-R2GS_Underlying-ideas.pdf

[R2GS] R2GS, Club R2GS Europe - European SIEM User Community and international cyber security expert group, http://www.clubr2gs.eu/, https://www.gsdays.fr/Club-R2GS.html

[Ray 2010] Ray, P.D., Harnoor, R., Hentea, M. (2010). Smart power grid security: a unified risk management approach. In D.A. Pritchard, L.D. Sanson (Eds.), *44th Annual 2010 IEEE International Carnahan Conference on Security Technology Proceedings* (pp. 276–285), San Jose, California, USA, (October 2010). https://doi.org.10.1109/CCST.2010.5678681

[Report]. Forrester. http://mobility-sp.com/images/gallery/FORRESTER-Understand-The-State-Of-Data-Security-And-Privacy-2013-To-2014.pdf

[Research Report]. *HOTSEC'08: Proceedings of the 3rd conference on Hot topics in security* (pp. 1-6). https://www.usenix.org/legacy/events/hotsec08/tech/full_papers/cardenas/cardenas_html/hotsecHTML.html

[RFC 2196] *IETF Request for Comments (RFC) 2196 Site Security Handbook (September 1997).*

[RFC 4949] *IETF Request for Comments (RFC) 4949 Internet Security Glossary* (Version 2, August 2007).

[RFC 6272] *IETF Request for Comments (RFC) 6272 Internet Protocols for the Smart Grid (June 2011).*

[Robles 2008] Robles, R.J, Choi, M-K., Cho, E-S., Kim, S-S., Park, G-C., Lee, J-H. (2008). Common threats and vulnerabilities of critical infrastructures. *International Journal of Control and Automation, 1*(1), 17–22.

[Ross 2012] Ross, B. (2012). *Security Architecture Frameworks.* INFOSECFORCE. *revised *Cybersecurity Architecture Development.* http://www.academia.edu/9715942/Cybersecurity_Architecture_Development

[Rowe 2006] Rowe, B.R., Gallaher, M.P. (2006). *Private Sector Cyber Security Investment Strategies: An Empirical Analysis* (March 2006) [Report Study funded by U.S. Department of Homeland Security]. https://econinfosec.org/archive/weis2006/docs/18.pdf

[Schneier 2013] Schneier, B. (2013, January 30). People, Process, and Technology [Blog]. *Schneier on Security.* https://www.schneier.com/blog/archives/2013/01/people_process.html

[SDD 2010] EDSOFORSMARTGRIDS. (2010, April). Strategic Deployment Document for Europe's Electricity Networks of the Future. European Technology Platform SmartGrids. https://www.edsoforsmartgrids.eu/policy/eu-steering-initiatives/smart-grids-european-technology-platform/

[SEGCG/M490/G – version 4.1]. CEN-CENELEC-ETSI Coordination Group on Smart Energy Grids (CG-SEG). ftp://ftp.cencenelec.eu/EN/EuropeanStandardization/Fields/EnergySustainability/SmartGrid/CGSEG_Sec_0042.pdf

[SGIP CoS] SGIP. (n.d.). Catalog of Standards. https://www.nist.gov/programs-projects/smart-grid-national-coordination/catalog-standards

[Shei 2013] Shei, H. (2013, October 1). *Understand the state of data security and privacy: 2013 to 2014*

[Shukla 2015] Shukla, V., Pandey, D., Shree, R. (2015). Requirements Engineering: A Survey. *Communications on Applied Electronics (CAE), 3*(5), 28-31. Foundation of Computer Science (FCS). https://doi.org/10.5120/cae2015651947

[SonicWall 2022] SonicWall. (n.d.). 2022 SonicWall Cyber Threat Report Threat Intelligence. https://www.sonicwall.com/2022-cyber-threat-report/

[Sridhar 2012] Sridhar, S, Hahn, A., Govindarasu, M. (2012). Cyber–Physical System Security for the Electric Power Grid. *Proceedings of the IEEE, 100*(1), 210-224. https://doi.org/10.1109/JPROC.2011.2165269

[Stamp 2012] Stamp, P. (2012). Implementing an Intelligent SOC. *2012 RSA Conference Europe,* London, United Kingdom, (October 2012). https://www.rsaconference.com/writable/presentations/file_upload/spo-208.pdf

[Steinberger 2015] Steinberger, R. (2015). *Proactive vs. Reactive Security.* http://www.crime-research.org/library/Richard.html

[Stevanović 2011] Stevanović, B. (2011). Maturity Models in Information Security. *International Journal of Information and Communication Technology Research, 1*(2), 44-47.

[Stoneburner 2006] Stoneburner, G. (2006). Toward a Unified Security-Safety Model. *Computer, 39*(8), 96–97.

[Summerville 2005] Summerville, J. (2005). *Software Engineering.* Pearson Publishing.

[Symantec 2016] Symantec. (2016). *2016 Internet Security Threat Report (ISTR)* (April 2016). https://www.symantec.com/security-center/threat-report

[Tabor 2015] Tabor, M.L. (2015). The Information Security Management System in Healthcare Computer Networks [ICTN 6870]. InfosecWriters. http://www.infosecwriters.com/Papers/MTabor_Healthcare2.pdf

[TechTarget] TechTarget. (n.d.). Security Models and Architecture. Chapter 5 in Harris, S., *CISSP Certification All-in-One Exam Guide* (pp. 184-252). http://media.techtarget.com/searchSecurity/downloads/29667C05.pdf

[TISN 2007] TISN. (2007). *Secure Your Information: Information Security Principles for Enterprise Architecture* [Report]. Trusted Information Sharing Network. http://www.tisn.gov.au/Documents/Secure_Your+Information+-+Information+Security+Principles+for+Enterprise+Architecture+-+Report.pdf

[Torres 2015] Torres, A. (2015). *Building a World-Class Security Operations Center: A Roadmap* (May 2015) [White Paper]. InfoSec Reading Room. SANS Institute. https://www.sans.org/reading-room/whitepapers/analyst/building-world-class-security-operations-center-roadmap-35907

[Urbaczewski 2006] Urbaczewski, L., Mrdalj, S. (2006). A Comparison of Enterprise Architecture Frameworks. *Issues in Information Systems, 7*(2), 18–23.

[Uslar 2014] Uslar, M., Rosinger, C., Schlegel, S. (2014). Security by Design for the Smart Grid: Combining the SGAM and NISTIR 7628. *Proceedings of the IEEE 38th Annual International Computers, Software and Applications Conference Workshops (COMPSAC)* (pp. 110-115), Vasteras, Sweden, (July 2014). https://doi.org/10.1109/COMPSACW.2014.23

[Vavra 2021] Vavra, C. (2021, February 9). *SolarWinds attack changing nature of cybersecurity for ICSs.* IndustrialCybersecurity Pulse. https://www.industrialcybersecuritypulse.com/changing-winds-of-cybersecurity-for-icss/

[Viega 2008] Viega, A. (2008). Cultivating and Assessing Information Security Culture [Thesis, Doctor of Philosophy in the subject of Information Technology]. University of Pretoria. South Africa.

[VMware] VMware. (n.d.). About Us. https://www.vmware.com/company.html

[Wang 2010] Wang, J. (2010). Anatomy of a Security Operations Center. *Conference Government Forum of Incident Response and Security Teams (GFirst)*, Hill Country, Texas, USA, (August 2010).

[Wang 2013] Wang, Z., Yang, K., Wang, X. (2013). Privacy-Preserving Energy Scheduling in Microgrid Systems. *IEEE Transactions on Smart Grid, 4*(4), 1810 – 1820. https://doi.org/10.1109/TSG.2013.2274466

[WEF 2016] Weforum. (2016). *Recommendations for Public-Private Partnership against Cybercrime*. World Economic Forum. http://www3.weforum.org/docs/WEF_Cybercrime_Principles.pdf

[Weiss 2010] Weiss, J. (2010). *Protecting Industrial Control Systems from Electronic Threats.* Momentum Press.

[Whitman 2003] Whitman, M. (2003). *Enemy at the Gates: Threats to Information Security.* Communications of the ACM, 46(8), 91-95.

[Whitman 2004] Whitman, M.E., Mattord, H.J. (2004). *Management of Information Security.* Thomson Course Technology.

[Whitman 2011] Whitman, M.E., Mattord, H.J. (2011). *Management of Information Security* (Third Edition). Thomson Course Technology.

[Whitman 2014] Whitman, M.E., Mattord, H.J. (2014). *Management of Information Security* (Fourth Edition). Cengage Learning.

[Wiki Disclosure] Wikipedia. (n.d). Full Disclosure. https://en.wikipedia.org/wiki/Full_disclosure (computer_security)

[Wiki Threat] Wikipedia. (n.d.). Threat. http://en.wikipedia.org/wiki/Threat_(computer)

[Wiki Unified] Wikipedia. (n.d.). Unified Smart Grid. http://en.wikipedia.org/wiki/Unified_Smart_Grid

[Xen] Citrix. (n.d.). Xen Server. https://www.citrix.com/products/xenserver/

[Yu 2006] Yu, X., Tolbert, L.M. (2006). Ancillary Services Provided from DER with Power Electronics Interface. *IEEE Power Engineering Society General Meeting*, Montreal, Canada, (June 2006). https://doi.org/10.1109/PES.2006.1709460

[Zafar 2013] Zafar, N., Arnautovic, E., Diabat, A., Svetinovic, D. (2013). System Security Requirements Analysis: A Smart Grid Case Study, July 2013, Wiley Online Library (wileyonlinelibrary.com)

[Zave 1997] Zave, P. (1997). Classification of Research Efforts in Requirements Engineering. *ACM Computing Surveys, 2*(4), 315-321.

[Zhang 2008] Zhang, J., Yun, L., Zhou, Z. (2008). Research of BLP and Biba Dynamic Union Model Based on Check Domain. *Proceedings of 2008 International Conference on Machine Learning and Cybernetics, 7*, (pp. 3679-3683), Kunming, China, (July 2008).

[Zetter 2016] Zetter, K. (March 3, 2016). *Inside the Cunning, Unprecedented Hack of Ukraine's Power Grid*. WIRED. https://www.wired.com/2016/03/inside-cunning-unprecedented-hack-ukraines-power-grid/

References Part 2

[Adnan 2006] Adnan, R., Bassem, M. (2006). A New Software Quality Model for Evaluating COTS Components. *Journal of Computer Science, 2*(4), 373-381.

[Akkaya 2015] Akkaya, I., Liu, I., Lee, E.A. (2015). Modeling and Simulation of Network Aspects for Distributed Cyber-Physical Energy Systems. In S.K. Khaitan, J.D. McCalley, C.C. Liu (Eds.), *Cyber Physical Systems Approach to Smart Electric Power Grid. Power Systems*. Springer-Verlag. https://doi.org/10.1007/978-3-662-45928-7_1

[Allen 2002] Allen, J.H., Sledge, C.A. (2002). Information Survivability: Required Shifts in Perspective. *Cross Talk The Journal of Defense Software Engineering, 15*(4). 7-9.

[Al-Qutaish 2010] Al-Qutaish, R.E. (2010). Quality Models in Software Engineering Literature: An Analytical and Comparative Study. *Journal of American Science 2010, 6*(3), 166-175.

[American] Ahdictionary. (2018). *The American Heritage Dictionary of the English Language* (Fifth Edition). Houghton Mifflin Harcourt. www.ahdictionary.com

[Anderson 2008] Anderson, R.J. (2008). *Security Engineering: A Guide to Building Dependable Distributed Systems* (Second Edition). Wiley.

[Anderson 2012] Anderson, D., Zhao, C., Hauser, C., Venkatasubramanian, V., Bakken, D., Bose, A. (2012). Intelligent Design Real-Time Simulation for Smart Grid Control and Communications Design. *IEEE Power & Energy Magazine, 10*(1), 49-57.

[Andradottir 1998] Andradottir, S. (1998). Simulation Optimization. In J. Banks (Ed.), *Handbook of Simulation Principles, Methodology, Advances, Applications, and Practice*. Wiley.

[Arghandeh] Arghandeh, R., von Meier, A., Mehrmanesh, L., Mili, L. (n.d.). On the definition of cyber-physical resilience in power systems. https://arxiv.org/ftp/arxiv/papers/1504/1504.05916.pdf

[Arghandeh 2014] Arghandeh, R., Brown, M., Del Rosso, A., Ghatikar, G., Stewart, E., Vojdani, A., von Meier, A. (2014). The local team: leveraging distributed resources to improve resilience. *IEEE Power Energy Magazine 12*(5), 76–83. https://doi.org/10.1109/MPE.2014.2331902

[Arghandeh 2016] Arghandeh, R., von Meier, A., Mehrmanesh, L., Mili, L. (2016). On the definition of cyber-physical resilience in power systems. *Renewable and Sustainable Energy Reviews, 10*(58), 1060–1069. https://doi.org/10.1016/J.RSER.2015.12.193

[AS IEC 62628] *AS IEC 62628:2014 Guidance on software aspects of dependability*. Standards Australia.

[Avizienis 2001] Avizienis, A, Laprie, J-C., Randell, B. (2001). Fundamental Concepts of Dependability [Research Report 1145 LAAS-CNRS] (April 2001).

[Avizienis 2004] Avizienis, A., Laprie, J-C., Randell, B., Landwehr, C. (2004). Basic Concepts and Taxonomy of Dependable and Secure Computing. *IEEE Transactions On Dependable and Secure Computing, 1*(1), 11-33. https://doi.org/10.1109/TDSC.2004.2

[Bai 2007] Bai, L., Biswas, S. (2007). Survivability – An Information Fusion Process Metric from An Operational. *2007 10th International Conference on Information Fusion* (pp. 1-8), Quebec, QC, Canada, (July 2007).

[Bakken 2000] Bakken, D., Bose, A., Bhowmik, S. (2000). Survivability and Status Dissemination in Combined Electric Power and Computer Communications Networks. *Proceedings of the Third Information Survivability Workshop (ISW-2000)*, Boston, Massachusetts, USA, (October 2000)

[Bakken 2002] Bakken, D., Bose, A., Dyreson, C., Bhowmik, S., Dionysiou, I., Gjermundrod, H., Xu, L. (2002). Impediments to Survivability of the Electric Power Grid and Some Collaborative EE-CS Research Issues to Solve Them. *Proceedings of the Fourth Information Survivability Workshop.*

[Banks 1998] Banks, J. (1998). Principles of Simulation. In J. Banks (Ed.), *Handbook of Simulation Principles, Methodology, Advances, Applications, and Practice*. Wiley.

[Barbacci 1995] Barbacci, M., Klein, M.H., Longstaff, T.A., Weinstock, C.B. (1995). *Quality Attributes* [CMU/SEI-95-TR-021 ADA307888]. Software Engineering Institute. Carnegie Mellon University, Pittsburgh, Pennsylvania, USA.

[Bayuk 2011a] Bayuk, J.L. (2011a). Systems Security Engineering. *IEEE Security & Privacy*, 9(2), 72-74. https://doi.org/10.1109/MSP.2011.41

[Bayuk 2011b] Bayuk J.L., Horowitz, B. (2011b). An Architectural Systems Engineering Methodology for Addressing Cyber Security. *Systems Engineering, 14*(3), 294-304. Wiley Online.

[Benson 2000] Benson, C. (2000). *Security Strategies.* Microsoft TecNet.

[Bieliková 2001] Bieliková, M., Krajcovic, T. (2001). Ambient Intelligence within a Home Environment. *ERCIM News Online edition* 47. http://www.ercim.eu/publication/Ercim_News/enw47/bielikova.html

[Billinton 1992] Billinton, R., Allan, R.N. (1992*). Reliability Evaluation of Engineering Systems Concepts and Techniques* (2nd edition). Plenum Press.

[Billinton 2003] Billinton, R., Cui, Y. (2003). Reliability evaluation of small stand-alone wind energy conversion systems using a time series simulation model. *IEE Proceedings-Generation, Transmission and Distribution, 150*(1), 96-100. IET Digital Library.

[Bobba 2012] Bobba, R.B., Dagle, J., Heine, E., Khurana, H., Sanders, W. H., Sauer, P., Yardley, T. (2012). Enhancing Grid Measurements: Wide Area Measurement Systems, NASPInet, and Security. *IEEE Power & Energy Magazine, 10*(1), 67-73. https://doi.org/10.1109/MPE.2011.943133

[Boehm 2004] Boehm, B., Huang, L., Jain, A., Madachy, R. (2004). The Nature of Information System Dependability: A Stakeholder/Value Approach [USC-CSE Report], University of Southern California Center for Software Engineering, Los Angeles, USA, (October 2004).

[Botella 2004] Botella, P., Burgués, X., Carvallo, J.P., Franch, X., Grau, G., Marco, J., Quer, C. (2004). ISO/IEC 9126 in practice: what do we need to know? Proceedings of the 1 st Software Measurement European Forum (SMEF), Rome, Italy, (January 2004). http://www.essi.upc.edu/~webgessi/publicacions/SMEF%2704-ISO-QualityModels.pdf

[Bouhafs 2012] Bouhafs, F., Mackay, M., Merabti, M. (2012). Links to the Future: Communication Requirements and Challenges in the Smart Grid. *IEEE Power & Energy Magazine, 10*(1), 24-32.

[Brown 2008] Brown, R.E. (2008). Impact of Smart Grid on Distribution System Design. *Proceedings 2008 IEEE Power & Energy Society General Meeting* (pp. 986-989), Pittsburgh, Pennsylvania, USA, (July 2008).

[Brown 2009] Brown, H.E., Suryanarayanan, S. (2009). A Survey Seeking a Definition of a Smart Distribution System. *Proceedings 41st North American Power Symposium* (pp. 1-7), Starkville, Mississippi, USA, (October 2009). https://doi.org/10.1109/NAPS.2009.5484078

[Brown 2010] Brown, H.E., Suryanarayanan, S., Heydt, G.T. (2010). Some Characteristics of Emerging Distribution Systems Considering the Smart Grid Initiative. *The Electricity Journal, 23*(5), 64-75. Elsevier Inc.

[Bugeja 2016] Bugeja, J., Jacobsson, A., Davidsson, P. (2016). On Privacy and Security Challenges in Smart Connected Homes. 2016 European Intelligence and Security Informatics Conference (EISIC) (pp. 172-175), Uppsala, Sweden, (August 2016).

[BS 25999] *BS 25999 Business Continuity Management (BCM)* (series of two parts). British Standards Institution (BSI).

[BS 25999-1] *BS 25999-1:2006 Business Continuity Management. Code of Practice.* *Replaced by ISO 22313:2013 Societal Security — Business continuity management systems — Guidance.

[BS 2599-2] *BS 25999-2:2007 Specification for Business Continuity Management.* *Replaced by ISO 22301:2012 Societal Security — Business continuity management systems — Requirements.

[BS EN ISO 22313] *BS EN ISO 22313:2020 Security and resilience. Business continuity management systems. Guidance on the use of ISO 22301.*

[BS EN ISO 22301] *BS EN ISO 22301:2014 Societal security. Business continuity management systems. Requirements.*

[Cantor 2006] Cantor, M., Roose, G. (2006). The Six Principles of Systems Engineering IBM's Rational Rules Developed over 10-Year Period. *WEBSPHERE* (March 18 2006).

[Cavoukian 2011] Cavoukian, A. (2011). *Privacy by Design in Law, Policy and Practice.* Office of the Privacy Commissioner, Ontario, Canada.

[CENELEC 2012] EC. (2012). *Smart Grid Reference Architecture Version 3.0* [Technical Report], (November 2012). CEN-CENELEC-ETSI Smart Grid Coordination Group. http://ec.europa.eu/energy/sites/ener/files/documents/xpert_group1_reference_architecture.pdf

[CENELEC 2017] EC. (2017). *SEGCG/M490/G_Smart Grid Set of Standards Version 4.1* [SEGCG/M490/G – version 4.1], (Jan 6th 2017). CEN-CENELEC-ETSI Coordination Group on Smart Energy Grids (CG-SEG).

[CERT Cyber] [CERT. (n.d.). *Cybersecurity Engineering.* http://www.cert.org/cybersecurity-engineering/index.cfm

[CERT IRP] ICS-CERT. (n.d.). *Abstract: ICS Cyber Incident Response Plan RP.* https://ics-cert.us-cert.gov/Abstract-ICS-Cyber-Incident-Response-Plan-RP

[CERT SAF] CERT. (n.d.). *Survivability Analysis Framework.* http://www.cert.org/cybersecurity-engineering/research/survivability-analysis-framework.cfm

[Cho 2014] Cho, D-E., Yeo, S-S., Kim, S-J. (2014). Authentication Method for Privacy Protection in Smart Grid Environment. *Journal of Applied Mathematics, Volume 2014,* Article ID 280978, CC BY 3.0. https://doi.org/10.1155/2014/280978

[Cleveland 2012] Cleveland, F. (2012). *IEC TC57: IEC 62351 Security Standards the Power System Information Infrastructure.* IEC TC57 WG15 Security Standards ver 14 (June 2012). International Electrotechnical Commission. http://iectc57.ucaiug.org/wg15public/Public%20Documents/White%20Paper%20on%20Security%20Standards%20in%20IEC%20TC57.pdf

[CNSSI 4009] CNSSI. (2015). *Committee on National Security Systems (CNSS) Glossary* [CNSSI No. 4009].

[Coulouris 2011] Coulouris, G., Dollimore, J., Kindberg, T., Blair, G. (2011). *Distributed Systems: Concepts and Design* (5th Edition). Addison-Wesley.

[CPS 2012] Asare, P., Bakirtzis, G., Bernard, R., Broman, D., Lee, E.A., Prinsloo, G., Torngren, M., Sunder, S.S. (2012). *Cyber-Physical Systems.* https://ptolemy.berkeley.edu/projects/cps/

[CSIIRW06] ORNL. (2006). *CSIIRW06: Cyber Security and Information Infrastructure Research Workshop,* Oak Ridge National Laboratory, Tennessee, USA, (May 2006). http://www.csm.ornl.gov/~sheldon/CSIIRW-06.htm

[Cuadra 2015] Cuadra, L., Salcedo-Sanz, S., Del Ser, J., Jiménez-Fernández, S., Geem, Z.W. (2015). A Critical Review of Robustness in Power Grids Using Complex Networks Concepts, *Energies 2015, 8*(9), 9211-9265. https://doi.org/10.3390/en8099211

[Curry 2012] Curry, E. (2012). System of Systems Information Interoperability using a Linked Dataspace. *IEEE 7th International Conference on System of Systems Engineering (SoSE 2012)* (pp.101–106), Genova, Italy, (July 2012). https://doi.org/10.1109/SYSoSE.2012.6384200

[Cyber Tool] CISA. (n.d.). *Cyber Security Evaluation Tool (CSET).* https://us-cert.cisa.gov/ics/Downloading-and-Installing-CSET

[DHS 2009] US-CERT. (2009). *Recommended Practices: Developing an Industrial Control Systems Cybersecurity Incident Response Plan* (October 2009). Homeland Security Control Systems Security Program. National Cyber Security Division. https://ics-cert.us-cert.gov/sites/default/files/recommended_practices/final-RP_ics_cybersecurity_incident_response_100609.pdf

[Dictionary] Cambridge. (n.d.). *Cambridge Dictionary American Language.* http://dictionary.cambridge.org/us/dictionary/american-english/

[DOE 2007] Energy. (2007). *Critical Infrastructure and Key Resources, Sector-Specific Plan as input to the National Infrastructure Protection Plan* (May 2007). U.S. Department of Energy. https://www.energy.gov/oe/downloads/energy-critical-infrastructure-and-key-resources-sector-specific-plan-input-national

[DOE 2014] Energy. (2014). *Smart Grid Investments Improve Grid Reliability, Resilience, and Storm Responses* (November 2014). Smart Grid Inves tment Program. https://www.energy.gov/sites/prod/files/2016/10/f33/SGI_Improve_Storm_Response_Nov_2014_0.pdf

[Dondossola 2013] Dondossola, G., Garrone, F., Szanto, J. (2013). Cyber risks in energy grid ICT infrastructures. In P. Theron, S. Bologna (Eds.), *Critical Information Infrastructure Protection and Resilience in the ICT Sector* (pp. 198–219). IGI Global. https://doi.org/10.4018/978-1-4666-2964-6.ch010

[DoT 2005] Osborne, L., Brummond, J., Hart, R., Zarean, M., Conger, S. (2005). *Systems Engineering Process.* https://commons.wikimedia.org/wiki/File:Systems_Engineering_Process_II.svg

[Dromey1995] Dromey, R.G. (1995). A Model for Software Product Quality. *IEEE Transactions on Software Engineering, 21*(2), 146-162.

[EAC 2008] Energy. (2008). *Smart Grid: Enabler of the New Energy Economy A Report by the Electricity Advisory Committee* (December 2008). https://www.energy.gov/sites/prod/files/oeprod/DocumentsandMedia/final-smart-grid-report.pdf

[ECD/114] Europe. (2008). *COUNCIL DIRECTIVE 2008/114/EC of 8 December 2008 on the identification and designation of European critical infrastructures and the assessment of the need to improve their protection.* European Council Directive 2008/114/EC. https://eur-lex.europa.eu/LexUriServ/LexUriServ.do?uri=OJ:L:2008:345:0075:0082:EN:PDF

[EISA 2007] EISA. (2007). *U.S. Energy Independence and Security Act of 2007 Public Law No: 110-140 Title XIII, Sec. 1301.* http://www.gpo.gov/fdsys/pkg/BILLS-110hr6enr/pdf/BILLS-110hr6enr.pdf

[Electropedia] Electropedia. (n.d.). *Dependability / Basic concepts IEV ref 192-01-23.* International Electrotechnical Commission. https://www.electropedia.org/iev/iev.nsf/display?openform&ievref=192-01-23

[Ellison 1999] Ellison, R.J., Fisher, D.A., Linger, R.C., Lipson, H.F., Longstaff, T., Mead, N.R. (1999). *Survivable Network Systems: An Emerging Discipline* [Revised Report CMU/SEI-97-TR-013], (May 1999). Software Engineering Institute. Carnegie Mellon University, Pittsburgh, USA. http://resources.sei.cmu.edu/asset_files/TechnicalReport/1998_005_001_16598.pdf

[Ellison 2008] Ellison, R.J., Goodenough, J., Weinstock, C., Woody, C. (2008). Survivability Assurance for System of Systems. TECHNICAL REPORT CMU/SEI-2008-TR-008], (May 2008). Software Engineering Institute. Carnegie Mellon University, Pittsburgh, USA. http://resources.sei.cmu.edu/asset_files/TechnicalReport/2008_005_001_14978.pdf

[Ellison 2010] Ellison, R.J., Woody, C. (2010). Survivability Analysis Framework. TECHNICAL REPORT CMU/SEI-2010-TN-013, (June 2010). Software Engineering Institute. Carnegie Mellon University, Pittsburgh, USA. https://doi.org/ 10.1184/R1/6584474.v1

[Englander 2003] Englander, I. (2003). *The Architecture of Computer Hardware and Systems Software An Information Technology Approach.* Wiley.

[ENISA 2012] ENISA. (2012). Cloud Computing Benefits, risks and recommendations for information security Version 2.0 (December 2012). European Network and Information Security Agency. https://resilience.enisa.europa.eu/cloud-security-and-resilience/publications/cloud-computing-benefits-risks-and-recommendations-for-information-security

[ENISA 2015] ENISA. (2015). Communication network interdependencies in smart grids - Annexes, Methodology for the identification of Critical Communication Networks Links and Components (October 2015). European Network and Information Security Agency. https://www.enisa.europa.eu/topics/critical-information-infrastructures-and-services/communication-network-interdependencies-in-smart-grids-annexes

[EO 13636] WhiteHouse, 2013, Executive Order (EO) 13636 Improving Critical Infrastructure Cybersecurity, February 12 2013, https://www.whitehouse.gov/the-press-office/2013/02/12/executive-order-improving-critical-infrastructure-cybersecurity

[EO 13636] WhiteHouse. (2013, February 12). *Executive Order (EO) 13636 Improving Critical Infrastructure Cybersecurity.* https://www.whitehouse.gov/the-press-office/2013/02/12/executive-order-improving-critical-infrastructure-cybersecurity

[EPRI 2011] EPRI. (2011). *Estimating the Costs and Benefits of the Smart Grid A Preliminary Estimate of the Investment Requirements and the Resultant Benefits of a Fully Functioning Smart Grid* [Technical Report 1022519], (March 2011). Electric Power Research Institute.

[EPRI 2013] EPRI. (2013). *Cybersecurity for DER Systems Version 1.0* (July 2013). Electric Power Research Institute. http://smartgrid.epri.com/doc/der-rpt-07-25-13.pdf

[EU Glossary] Eqavet. (n.d.). *Quality Assurance Glossary*. EQAVET - the European Quality Assurance in Vocational Education and Training. https://www.eqavet.eu/EU-Quality-Assurance/Glossary

[Fabian 2010] Fabian, B., Gurses, S.F., Heisel, M., Santen, T., Schmidt, H. (2010). A comparison of security requirements engineering methods. *Requirements Engineering Journal, 15*(1), 7–40. https://doi.org/10.1007/s00766-009-0092-x

[FailRecovery] Encyclopedia. (n.d.). *Failure recovery*. Computing Dictionaries thesauruses pictures and press releases. Oxford University Press. https://www.encyclopedia.com/computing/dictionaries-thesauruses-pictures-and-press-releases/failure-recovery

[Farhangi 2010] Farhangi, H. (2010). The Path of the Smart Grid. *IEEE Power & Energy Magazine, 8*(1), 18-28. https://doi.org/10.1109/MPE.2009.934876

[FaultRes] PCMag. (n.d.). *Fault resilient*. https://www.pcmag.com/index.php/encyclopedia/term/fault-resilient

[Firesmith 2003] Firesmith, D.G. (2003). Common Concepts Underlying Safety, Security, and Survivability Engineering [Technical Report CMU/SEI-2003-TN-033], (December 2003). http://resources.sei.cmu.edu/asset_files/TechnicalNote/2003_004_001_14198.pdf

[FreeDictionary] Thefreedictionary. (n.d.). *Fault Tolerance*. http://encyclopedia.thefreedictionary.com/Fault+Tolerance

[GAO 2011] GAO. (2011). Electricity Grid Modernization: Progress Being Made on Cybersecurity Guidelines, but Key Challenges Remain to Be Addressed [GAO Report 11-117] (January 2011). http://www.gao.gov/products/GAO-11-117

[Ghansah 2009] Ghansah, I. (2009). *Smart Grid Cyber Security Potential Threats, Vulnerabilities and Risks* [Interim Project Report CEC-500-2012-047], (May 2012). Public Interest Energy Research (PIER) Program. California Energy Commission. University State of California Sacramento.

[Graydon 2007] Graydon, P.J., Knight, J.C., Strunk, E.A. (2007). Achieving Dependable Systems by Synergistic Development of Architectures and Assurance Cases. In R. de Lemos, C. Gacek, A. Romanovsky, *Lecture Notes in Computer Science Architecting Dependable Systems 4615*(IV), 362-382. https://doi.org/10.1007/978-3-540-74035-3_16. Springer.

[Graydon 2010] Graydon, P.J., Knight, J.C., Yin, X. (2010). Practical Limits on Software Dependability: A Case Study. In J. Real, T. Vardanega (Eds.), *Reliable Software Technology –*

Ada-Europe 2010. Lecture Notes in Computer Science, 6106, 83-96. Springer. https://doi.org/10.1007/978-3-642-13550-7_6

[Gunther 2012] Gunther, E.W. (2012). *NIST Conceptual Model Overview and Evolution.* http://www.ietf.org/proceedings/10mar/slides/intarea7.pdf

[Hale 1998] Hale, J., Bose, A. (1998). Information Survivability in the Electric Utility Industry. *1998 Information Survivability Workshop*, Orlando, Florida, USA, (October 1998).

[Hansche 2004] Hansche, S., Berti, J., Hare, C. (2004). *Official (ISC2)2 Guide to the CISSP Exam.* Auerbach Publications CRC Press.

[Hanseth 2002] Hanseth, O. (2002). From systems and tools to networks and infrastructures – from design to cultivation, towards a theory of ICT solutions and its design methodology implications. Publications University of Oslo, Oslo, Norway.

[Hanseth 2010] Hanseth, O. (2010). From Systems and Tools to Networks and Infrastructures-from Design to Cultivation: Towards a Design Theory of Information Infrastructures. In J. Holmström, M. Wiberg, M., A. Lund (Eds.), *Industrial Informatics Design, Use and Innovation: Perspectives and Services* (pp. 122-156). IGI Global. http://doi:10.4018/978-1-61520-692-6.ch011

[Hardeveld 2012] Hardeveld, T.V., Kiang, D. (2012). Achieving Dependability Value for Pipelines and Facilities. *Proceedings of the 9th International Pipeline Conference IPC2012* (pp. 531-540), Calgary, Alberta, Canada, (September 2012). https://doi.org/10.1115/IPC2012-90234

[Harris 2005] Harris, S. (2005). *All in One CISSP Exam Guide* (Third Edition). McGraw-Hill/Osborne.

[Harris 2013] Harris, S. (2013). *All in One CISSP Exam Guide* (Sixth Edition). McGrawHill Education.

[Huxham] Huxham, H. (2006). Enterprise Architecture as a Service orientated architecture. http://en.wikipedia.org/wiki/File:BITS.jpg

[Heegaard 2008] Heegaard, P.E., Trivedi, K.S. (2008). Survivability Quantification of Communication Services. *2008 IEEE International Conference on Dependable Systems and Networks with FTCS and DCC (DSN)* (pp. 462-471), Anchorage, Alaska, USA, (June 2008). IEEE Proceedings DSN 2008. https://doi.org/10.1109/DSN.2008.4630117

[Heegaard 2009] Heegaard, P.E., Trivedi, K.S. (2009). Network Survivability Modeling. *Computer Networks, 53*(8),1215-1234.

[Hentea 2005] Hentea, M. (2005). A Perspective on Achieving Information Security Awareness. *Journal of Issues in Informing Science and Information Technology, 2,* 169-178. https://doi.org/10.28945/819

[Hentea 2006] Hentea, M. (2006). Enhancing information security risk management with data mining and fuzzy logic techniques. *Proceedings of 19th International Conference on Computer Applications in Industry and Engineering* (pp. 132-139), Las Vegas, Nevada, USA, (November 2006).

[Hentea 2008a] Hentea, M. (2008a). Improving Security for SCADA Control Systems. *Interdisciplinary Journal of Information, Knowledge, and Management, 3*, 73-86. https://doi.org/10.28945/3185

[Hentea 2008b] Hentea, M. (2008b). A Perspective on Security Risk Management for SCADA Systems. *Proceedings of the ISCA 23rd International Conference on Computers and Their Applications (CATA 2008)*, (pp. 222-227), Cancun, Mexico, (April 2008).

[Heyerman 2009] Heyerman, D. (2009). *The Smart Grid Frontier: Wide Open* (May 3, 2009). http://tinycomb.com/29/05/03

[Hilliard 2007] Hilliard, R. (2007). All About IEEE Std 1471 (June 2007). http://www.iso-architecture.org/ieee-1471/docs/all-about-ieee-1471.pdf

[Hilliard 2012] Hilliard, R. (2012). On the Composition and Reuse of Viewpoints across Architecture Frameworks. *Proceedings of the Joint 10th Working IEEE/IFIP Conference on*

Software Architecture & 6th European Conference on Software Architecture (WICSA/ECSA), Helsinki, Finland, (August 2012). https://doi.org/10.1109/WICSA-ECSA.212.21

[Hirst 1999] Hirst, E.; Kirby, B.; Hadley, S. (1999). *Generation and Transmission Adequacy in a Restructuring U.S. Electricity Industry*. Edison Electric Institute.

[Hong 2005] Hong, J.I, Landay, J.A. (2005). *An architecture for privacy-sensitive ubiquitous computing* [Research-PhD Thesis] (Spring 2005). University of California, Berkeley, California, USA. https://www.cs.cmu.edu/~jasonh/publications/jihdiss.pdf

[IEC 60050-192] *IEC 60050-192:2015/AMD1:2016 International electrotechnical vocabulary (IEV) - Part 192: Dependability.*

[IEC 60834-1] *IEC 60834-1:1999 Teleprotection Equipment of Power Systems - Performance and Testing - Part 1: Command Systems.*

[IEC 62508] *IEC 62508:2010 Guidance on human aspects of dependability Edition 1.0.*

[IEC 62628] *IEC 62628:2012 Guidance on software aspects of dependability Edition 1.0.*

[IEC 62673] *IEC 62673:2013 Methodology for communication network dependability assessment and assurance Edition 1.0.* International Electrotechnical Commission.

[IEC TC56] IEC. (n.d.). *Technical Committee (TC) 56 Dependability*. International Electrotechnical Commission. https://tc56.iec.ch/

[IEEE 1366] *IEEE 1366:2012 Guide for Electric Power Distribution Indices.*

[INCOSE 2015] D.D. Walden, G.J. Roedler, K.J. Forsberg, R. D. Hamelin, T.M. Shortell (Eds.). (2015). *INCOSE System Engineering Handbook—A Guide for System Engineering Life Cycle Processes and Activities* Fourth Edition. International Council on Systems Engineering. Wiley.

[IRENA 2013] IRENA. (2013). *Smart Grids and Renewables: A Guide for Effective Deployment.* International Renewable Energy Agency (IRENA). https://www.irena.org/publications/2013/Nov/Smart-Grids-and-Renewables-A-Guide-for-Effective-Deployment

[Irvine 2000] Irvine, C. E., Levin, T. E. (2000). Quality of Security Service. *NSPPW '00 Proceedings of the 2000 workshop on New Security Paradigms* (pp. 91-99), Ballycotton, Ireland, (September 2000). ACM. https://doi.org/10.1145/366173.366195

[Irvine 2001] Irvine, C.E., Levin, T. (2001). *Teaching Security Engineering Principles.* https://www.hsdl.org/?view&did=19229

[ISACA 2010] ISACA. (2010). *Principles for Information Security Practitioners.* Information Systems Audit and Control Association. https://www.isaca.org/resources/

[ISACA 2019] ISACA. (2019). *ISACA Glossary of Terms.* https://www.isaca.org/pages/glossary.aspx

[ISC2 Blog] ISC2. (2011, January 7). ISF, (ISC)2 and ISACA Release Information Security Principles [Blog]. *ISC2 BLOG.* The International Information System Security Certification Consortium. https://blog.isc2.org/isc2_blog/2011/01/isf-isc2-and-isaca-release-information-security-principles.html

[ISO 9000] *ISO 9000:2015 Quality Management systems – Fundamentals and vocabulary.*

[ISO 9001] *ISO 9001:2015 Quality management systems – Requirements.*

[ISO 9004] *ISO 9004:2009 Managing for the sustained success of an organization -- A quality management approach.*

[ISO 22301] *ISO 22301:2019 Security and resilience — Business continuity management systems — Requirements.* [ISO 22313] ISO 22313:2020 Security and resilience — Business continuity management systems — Guidance on the use of ISO 22301.

[ISO/IEC 9126] *ISO/IEC 9126:2001, Software engineering — Product quality.* *Replaced by ISO/IEC 25010:2011.

[ISO/IEC 9126-1] *ISO/IEC 9126-1:2001 Software engineering -- Product quality -- Part 1: Quality model.* *Replaced by ISO/IEC 25010:2011.

[ISO/IEC 20000] *ISO/IEC 20000:2018 Information technology -- Service management, series of several parts.*

[ISO/IEC 25010] *ISO/IEC 25010:2011 Systems and software engineering -- Systems and software Quality Requirements and Evaluation (SQuaRE) -- System and software quality models.*

[ISO/IEC 25012] *ISO/IEC 25012:2008 Software engineering -- Software product Quality Requirements and Evaluation (SQuaRE) -- Data quality model.*

[ISO/IEC 27001] *ISO/IEC 27001:2013 Information technology – Security techniques – Information security management systems – Requirements.*

[ISO/IEC 27002] *ISO/IEC 27002:2013 Information technology — Security techniques — Code of practice for information security controls.*

[ISO/IEC 27031] *ISO/IEC 27031:2011 Information Technology – Security techniques – Guidelines for information and communication technology readiness for business continuity.*

[ISO/IEC 27035] *ISO/IEC 27035:2016 Information technology — Security techniques — Information security incident*

[ISO/IEC 21827] *ISO/IEC 21827:2008 Information technology - Security techniques - Systems Security Engineering - Capability Maturity Model® (SSE-CMM).*

[ISO/IEC/IEEE 15288] *ISO/IEC/IEEE 15288:2015 Systems and software engineering — System life cycle processes.*

[ISO/IEC/IEEE 24765] *ISO/IEC/IEEE 24765:2017 Systems and software engineering – Vocabulary.*

[Jackson 2013] Jackson, S. (2013). Resilience principles for the ICT sector. In P. Theron, S. Bologna (Eds.), *Critical Information Infrastructure Protection and Resilience in the ICT Sector* (pp. 36–49). IGI Global. https://doi.org/10.4018/978-1-4666-2964-6.ch002

[Johnson 2013] Johnson, C.W. (2013). The telecoms inclusion principle: the missing link between critical infrastructure protection and critical information infrastructure protection. In P. Theron, S. Bologna (Eds.), *Critical Information Infrastructure Protection and Resilience in the ICT Sector* (pp. 277–303). IGI Global. https://doi.or/10.4018/978-1-4666-2964-6.ch014

[Ju 2013] Ju, X., et al., 2013). On Fault Resilience of OpenStack. *SOCC '13: Proceedings of the 4th annual Symposium on Cloud Computing* (Article no. 2, pages 1-16). Santa Clara, California, USA, (October 2013). ACM. https://doi.org/10.1145/2523616.2523622

[Karnouskos 2011] Karnouskos, S. (2011). Cyber-Physical Systems in the SmartGrid. *2011 9th IEEE International Conference on Industrial Informatics* (pp. 20-23), Caparica, Lisbon, Portugal, (July 2011). https://doi.org/10.1109/INDIN.2011.6034829

[Kelly 2010] Kelly, J.F., Meiners, M., Rouse, G. I. (2010). *IIT Perfect Power Prototype* [Report], (February 2010). http://www.ippconnect.com

[Kim 2007] Kim, Y-G., Jeong, D., Park, S-H., Lim, J., Baik, D-K. (2007). Modeling and Simulation for Security Risk Propagation in Critical Information Systems. In Y. Wang, Y. Cheung, H. Liu (Eds.), *Computational Intelligence and Security. CIS 2006. Lecture Notes in Computer Science Computational Intelligence and Security, 4456*, 858-868. Springer. https://doi.org/10.1007/978-3-540-74377-4_90

[Knight 2000] Knight, J.C., Sullivan, K.J. (2000). *On the Definition of Survivability* [Technical Report]. University of Virginia, Charlottesville, Virginia, USA.

[Knight 2003] Knight, J.C., Strunk, E.A., Sullivan, K.J. (2003). Towards a Rigorous Definition of Information System Survivability. *Proceedings of the DARPA Information Survivability Conference and Exposition (DISCEX'03)* (pp. 78-89), Washington DC, USA, (April 2003).

[Kotonya 1998] Kotonya, G., Sommerville, I. (1998). *Requirements Engineering: Processes and Techniques* (1st edn). Wiley.

[Kossiakoff 2011] Kossiakoff, A., Sweet, W.N., Seymour, S.J., Biemer, S.M. (2011). *Systems Engineering Principles and Practice* (2nd Edition). Wiley.

[Kreizman 2006] Kreizman, G., Robertson, B. (2006). *Incorporating Security into the Enterprise Architecture Process* [Research Report G00137028]. http://www.gartner.com

[Krutz 2004] Krutz, R.L., Vines, R.D. (2004). *The CISSP Prep Guide* (Second Edition). Wiley.

[Kumar 2015] Kumar, R., Khan, S.A., Khan, R.A. (2015). Revisiting Software Security: Durability Perspective. *International Journal of Hybrid Information Technology (SERSC), 8*(2), 311-322.

[KSGI] Hwng, W. (2020). *Microgrids for Electricity Generation in the Republic of Korea Gapa Island Microgrid Model* [NAPSNet Special Reports]. *Nautilus Institute for Security and Sustainability.* https://nautilus.org/napsnet/napsnet-special-reports/microgrids-for-electricity-generation-in-the-republic-of-korea/

[Laprie 1992] Laprie J.C. (1992). Dependability: Basic Concepts and Terminology. In J.C. Laprie (Ed.), *Dependability: Basic Concepts and Terminology. Dependable Computing and Fault-Tolerant Systems, 5.* Springer. https://doi.org/10.1007/978-3-7091-9170-5_1

[Laprie 1995] Laprie, J-C. (1995). Dependable Computing: Concepts, Limits, Challenges [Invited paper]. *FTCS-25 the 25th IEEE International Symposium on Fault-Tolerant Computing. 1995 Special Issue* (pp. 42-54), Pasadena, California, USA, (June 1995).

[Laprie 2008] Laprie, J-C. (2008). From dependability to resilience. *Proceedings of IEEE International Conference on Dependable Systems and Networks DSN 2008*, G8–G9.

[Lathi 2009] Lathi, B.P., Ding, Z. (2009). *Modern Digital and Analog Communication Systems* (Fourth Edition). Oxford University Press.

[Law 2007] Law, A.M. (2007). *Simulation Modeling & Analysis* (4th Edition). McGraw Hill.

[Levin 2006] Levin, T.E., Irvine, C.E., Spyropoulu, E. (2006). Quality of Security Service: Adaptive Security. In H. Bidgoli (Ed.), *Handbook of Information Security Threats, Vulnerabilities, Prevention, Detection, and Management Volume 3* (pp. 1016-1025). Wiley.

[Madachy 2005] Madachy, R., Boehm, B. (2005). Software Dependability Applications in Process Modeling. In S.T. Acuna, N. Juristo, *Software Process Modeling* (pp. 65-86). International Series in Software Engineering. Springer. https://doi.org/10.1007/0-387-24262-7_3

[Marksteiner 2019] Marksteiner, S., Vallant, H., Nahrgang, K. (2019). Cyber security requirements engineering for low-voltage distribution smart grid architectures using threat modeling. *Journal of Information Security and Applications, 49*, Article 102389. https://doi.org/10.1016/j.jisa.2019.102389

[Martin 1985] Martin, J., McClure, C. (1985). *Structured Techniques for Computing*. Prentice-Hall.

[Marris 2008] Marris, E. (2008). Energy: Upgrading the grid. *Nature, 454* (7204), 570-573. https://doi.org/10.1038/454570a

[McGraw-Hill 2003] McGraw-Hill. (2003). *McGraw-Hill Dictionary of Scientific and Technical Terms.* McGraw-Hill.

[McGraw-Hill 2005] McGraw-Hill. (2005). *McGraw-Hill Concise Encyclopedia of Engineering.* McGraw-Hill.

[Meier 2009] Meier, J.D., Homer, A., Hill, D., Taylor, J., Bansode, P, Wall, L., Boucher, R.Jr., Bogawat, A. (2009). *Microsoft Application Architecture Guide Patterns & Practices* (2nd Edition). Microsoft Press.

[Mekkanen 2012] Mekkanen, M., Virrankoski, R., Elmusrati, M., Antila, E. (2012). Reliability and Availability Investigation for Next-Generation Substation Function Based on IEC 61850. *A4 Conference Proceedings, 3rd Workshop on Wireless Communications and Applications (WoWCA2012)*, Vaasa, Finland, (April 2012).

[Meyer 2009] Meyer, J.F. (2009). Defining and Evaluating Resilience: A Performability Perspective. *Proceedings of the International Workshop on Performability Modeling of Computer and Communication Systems (PMCCS)*, Eger, Hungary, (September 2009).

[Mohammad 2006] Mohammad, A.J., Hutchison, D., Sterbenz, J.P.G. (2006). Poster: Towards Quantifying Metrics for Resilient and Survivable Networks. *14th IEEE International Conference on Network Protocols (IEEE ICNP 2006)* (pp. 17-18), Santa Barbara, California, USA, (January 2006).

[Mohsenian-Rad 2010] Mohsenian-Rad, A., Wong, V.W.S., Jatskevich, J., Schober, R., Leon-Garcia, A. (2010). Autonomous Demand-Side Management Based on Game-Theoretic Energy

Consumption Scheduling for the Future Smart Grid. *IEEE Transactions on Smart Grid, 1*(3), 320-331. https://doi.org/ 10.1109/TSG.2010.2089069

[Moore 1996] Moore, T. (1996). Tighter Security for Electronic Information. *EPRI Journal, 21*(6).

[Moslehi 2010] Moslehi, K., Kumar, R. (2010). A Reliability Perspective of the Smart Grid. *IEEE Transactions on Smart Grid, 1*(1), 57-63.

[NDIA 2008] NDIA. (2008). *Engineering for System Assurance Version 1.0* (October 2008). National Defense Industrial Association (NDIA) System Assurance Committee. http://www.acq.osd.mil/sse/docs/ SA-Guidebook-v1-Oct2008.pdf

[NERC CIP] NERC. (n.d.). *Critical Infrastructure Protection (CIP) v5 series of standards.* North American Electric Reliability Corporation (NERC). https://www.nerc.com/pa/Stand/Pages/CIPStandards.aspx

[NERC Glossary-16] NERC. (2016). *Glossary of Terms Used in NERC Reliability Standards Updated August 17, 2016.* North American Electric Reliability Corporation (NERC). *See Update of April 2, 2021. http://www.nerc.com/files/glossary_of_terms.pdf

[NESCOR 2013a] NESCOR. (2013a). *NESCOR Guide to Penetration Testing for Electric Utilities Version 3.* National Electric Sector Cybersecurity Organization Resource (NESCOR). https://smartgrid.epri.com/doc/NESCORGuidetoPenetrationTestingforElectricUtilities-v3-Final.pdf

[NESCOR 2013b] NESCOR. (2013b). *Electric Sector Failure Scenarios and Impact Analyses Version 1.0* (September 2013). National Electric Sector Cybersecurity Organization Resource (NESCOR) Technical Working Group 1. https://smartgrid.epri.com/doc/NESCOR%20failure%20scenarios09-13%20finalc.pdf

[NESCOR 2015] NESCOR. (2015). *Electric Sector Failure Scenarios and Impact Analyses Version 3.0* (December 2015). National Electric Sector Cybersecurity Organization Resource (NESCOR). https://smartgrid.epri.com/doc/NESCOR%20Failure%20Scenarios%20v3%2012-11-15.pdf

[NETL 2007a] NETL. (2007a). *A Systems View of the Modern Grid V2.0* (January 2007). The NETL Modern Grid Initiative Conducted by the National Energy Technology Laboratory for the U.S. Department of Energy Office of Electricity Delivery and Energy Reliability. https://www.smartgrid.gov/files/Systems_View_Modern_Grid_200712.pdf

[NETL 2007b] NETL. (2007b). *A Vision for the Modern Grid* v*1.0* (March 2007). The NETL Modern Grid Initiative Conducted by the National Energy Technology Laboratory for the U.S. Department of Energy Office of Electricity Delivery and Energy Reliability.

[NETL 2009] NETL. (2009). A vision for the Modern Grid v2.0 (June 209). The Modern Grid Strategy Developed for the U.S. Department of Energy Office of Electricity Delivery and Energy Reliability by the National Energy Technology Laboratory. https://netl.doe.gov/sites/default/files/Smartgrid/Whitepaper_The-Modern-Grid-Vision_APPROVED_2009_06_18.pdf

[NIST 2001] NIST. (2001). *Engineering Principles for IT Security (A baseline for achieving security) Recommendations of the National Institute of Standards and Technology /* Gary Stoneburner, Clark Hayden, and Alexis Feringa. U.S. Dept. of Commerce Technology Administration. National Institute of Standards and Technology.

[NIST SP800-30r1] *NIST Special Publication (SP) 800-30 Guide for Conducting Risk Assessments Revision 1* (September 2012).

[NIST SP800-34r1] *NIST Special Publication (SP) 800-34 Contingency Planning Guide for Federal Information Systems Revision 1* (May 2010).

[NIST SP800-39] *NIST Special Publication (SP) 800-39 Managing Information Security Risk Organization, Mission, and Information System View* (March 2011).

[NIST SP800-53r5] *NIST Special Publication (SP) 800-53 Security and Privacy Controls for Federal Information Systems and Organizations Revision 5* (December 2020, includes updates as December 10, 2020). https://doi.org/10.6028/NIST.SP.800-53r5

[NIST SP800-61r2] *NIST Special Publications (SP) SP800-61 Computer Security Incident Handling Guide Revision 2* (August 2012).

[NIST SP800-82r2] *NIST Special Publication (SP) 800-82 Guide to Industrial Control Systems (ICS) Security Supervisory Control and Data Acquisition (SCADA) systems, Distributed Control Systems (DCS), and other control system configurations such as Programmable Logic Controllers (PLC) Revision 2* (May 2015).

[NIST SP800-83] *NIST Special Publication (SP) 800-83 Guide to Malware Incident Prevention and Handling for Desktops and Laptops* (July 2013). http://dx.doi.org/10.6028/NIST.SP.800-83r1

[NIST SP800-86) *NIST Special Publication (SP) 800-83 Guide to Integrating Forensic Techniques into Incident Response* (August 2006).

[NIST SP800-92] *NIST Special Publication* (SP) 800-92 Guide to Computer Security Log Management (September 2006).

[NIST SP800-100] *NIST Special Publication (SP) 800-100 Information Security Handbook* (October 2006, updated March 2007).

[NIST SP800-137] *NIST Special Publication (SP) 800-137 Information Security Continuous Monitoring (ISCM) for Federal Information Systems and Organizations* (September 2011).

[NIST SP800-160v1] *NIST Special Publication (SP) 800-160 Systems Security Engineering Considerations for a Multidisciplinary Approach in the Engineering of Trustworthy Secure Systems Volume 1* (March 2018). https://doi.org/10.6028/NIST.SP.800-160v1

[NIST SP800-160v2] *NIST Special Publication (SP) 800-160 Developing Cyber-Resilient Systems: A Systems Security Engineering Approach Volume 2 Revision 1* (September 2021). https://doi.org/10.6028/NIST.SP.800-160v2r1

[NIST SP1500-201] *NIST Special Publication 1500-201 Framework for Cyber-Physical Systems: Volume 1, Overview Version 1.0* (June 2017). https://doi.org/10.6028/NIST.SP.1500-201

[NIST SP1500-202] *NIST Special Publication 1500-202 Framework for Cyber-Physical Systems: Volume 2, Version 1.0, (June 2017).* https://doi.org/10.6028/NIST.SP.1500-202

[NISTIR 7298r3] *NIST Interagency Report (NISTIR) 7298 Glossary of Key Information Security Terms Revision 3* (July 2019). https://doi.org/10.6028/NIST.IR.7298r3

[NISTIR 7628] *NIST Interagency Report (NISTIR) 7628* Guidelines *for Smart Grid Cyber Security* (August 2010).

[NISTIR 7628r1] *NIST Interagency Report (NISTIR) 7628r1 Guidelines for Smart Grid Cybersecurity* (Revision 1, September 2014). http://dx.doi.org/10.6028/NIST.IR.7628r1

[NISTIR 7761r1] *NIST Interim Report (NISTIR) 7761 Guidelines for Assessing Wireless Standards for Smart Grid Applications Revision 1* (June 2014). NIST Smart Grid Interoperability Panel Priority Action Plan 2.

[OECD 1992] OECD. (1992). *Recommendation of the Council Concerning Guidelines for the Security of Information Systems.* Organisation for Economic Co-operation and Development. https://www.oecd.org/internet/ieconomy/oecdguidelinesforthesecurityofinformationsystems1992.htm

[OECD 2002] OECD. (2002, July 25). *Guidelines for the Security of Information Systems and Networks: Towards a Culture of Security.* http://www.oecd.org/sti/ieconomy/oecdguidelinesforthesecurityofinformationsystemsandnetworkstowardsacultureofsecurity.htm

[OECD 2015] OECD. (2015). *Digital Security Risk Management for Economic and Social Prosperity: OECD Recommendation and Companion Document.* OECD Publishing. http://dx.doi.org/10.1787/9789264245471-en

[OpenGroup 2011] Opengroup. (2011). *Security Principles for Cloud and SOA* [White Paper W119] (December 2011). www.opengroup.org/bookstore/catalog/w119.htm

[OpenGroup 2015] OpenGroup. (2015). The Open Group IT4IT™ Reference Architecture Version 2.0 an Open Group Standard (October 2015). https://pubs.opengroup.org/it4it/refarch20/

[OSA Framework] Opensecurityarchitecture. (n.d.). *Open Security Architecture (OSA) Landscape Security Architecture Landscape.* CC BY-SA 3.0. http://www.opensecurityarchitecture.org/cms/foundations/osa-landscape

[OSA Taxonomy] Opensecurityarchitecture. (n.d.). *Open Security Architecture (OSA) IT Security Requirements.* CC BY-SA 3.0. http://www.opensecurityarchitecture.org/cms/definitions/it_security_requirements

[Osborn 2001] Osborn, J.G., Kawann, C. (2001). *Reliability of the U.S. Electricity System: Recent Trends and Current Issues* [Report LBNL-47043] (August 2001). Prepared for the Assistant Secretary for Energy Efficiency and Renewable Energy U.S. Department of Energy. Berkeley Lab. https://emp.lbl.gov/publications/reliability-us-electricity-system

[P2030 2011] *IEEE Std 2030-2011- IEEE Guide for Smart Grid Interoperability of Energy Technology and Information Technology Operation with the Electric Power System (EPS), End-Use Applications, and Loads.*

[Panke 2002] Panke, R.A. (2002). *Energy Management Systems and Direct Digital Control.* The Fairmont Press.

[Pasik-Duncan 2006] [Pasik-Duncan, B., Patton, R., Schilling, K., Camacho, E.F. (2006). Four focused forums. Math, science, and technology in control engineering education. *IEEE Control Systems, 26*(4), 93-98.

[Port 2003] Port, D., Huang, L.. (2003). Software Dependability Risks and the Insurance Process. The Fifth International Workshop on Economics-Driven Software Engineering Research (EDSER-5), 2003, pp. 66–70.

[Poslad 2009] Poslad, S. (2009). *Ubiquitous Computing Smart Devices, Smart Environments and Smart Interaction*. Wiley.

[PPD-21] WhiteHouse. (2013, February 12). *The White House Presidential Policy Directive/PPD-21 – Critical Infrastructure Security and Resilience.* http://www.whitehouse.gov/the-pressoffice/2013/02/12/presidential-policy-directive-critical-infrastructure-security-and-resil

[PPD-8] WhiteHouse. (2011, March 30). *The White House Presidential Policy Directive/PPD-8 – National Preparedness.* http://www.dhs.gov/xlibrary/assets/presidentialpolicy-directive-8-national-preparedness.pdf

[Praxiom] Praxiom. (n.d.). *ISO 9000 2015 Plain English Definitions.* Praxiom Research Limited. http://www.praxiom.com/iso-definition.htm

[Pressman 2005] Pressman, R.S. (2005). *Software Engineering A Practitioner's Approach* (Sixth Edition). McGraw Hill Higher Education.

[Reed-Mohn 2007] Reed-Mohn, A. (2007). *Incident Reporting Systems* [Master's Thesis]. Master of Science in Information Security. Gjøvik University College, Norway. https://ntnuopen.ntnu.no/ntnu-xmlui/bitstream/handle/11250/143837/Reed-Mohn%20-%20Incident%20Reporting%20Systems1.pdf?sequence=1

[Reitman 1984] Reitman, J. (1984). Simulation. In A.H. Seidman, I. Flores (Eds.), *The Handbook of Computers and Computing.* Van Nostrand Reinhold Company.

[RFC 4949] *IETF Request for Comments (RFC) 4949 Internet Security Glossary* (Version 2, August 2007).

[Rieger 2008] Rieger,C.G., Grosshans, R.R. (2008). *Smart is Not Enough: Resilience and Securing the Power Grid* [Report]. Idaho National Laboratory, Idaho Falls, Idaho, USA.

[Rieger 2009] Rieger, C.G., Gertman, D.I., McQueen, M.I. (2009). Resilient Control Systems: Next Generation Design Research. *Proceedings HSI '09: 2nd International Conference on Human System Interaction* (pp. 632-636), Catania, Italy, (May 2009). IEEE. https://doi.org/doi: 10.1109/HSI.2009.5091051

[Rieger 2012] Rieger, C.G. (2012). *Resilient Control System.* File:RCS Architecture 09-50368.jpg https://en.wikipedia.org/wiki/File:RCS_Architecture_09-50368.jpg

[Rieger 2013] Rieger, C.G., Moore, K.L., Baldwin, T.L. (2013). Resilient Control Systems: A Multi-Agent Dynamic Systems Perspective. *2013 IEEE International Conference on Electro/Information Technology (EIT)* (pp. 1-16), Rapid City, South Dakota, USA, (May 2013). https://doi.org/10.1109/EIT.2013.6632721

[Roberts 2008] Roberts, G.R. (2008). *Evaluating Security and Quality Of Service Considerations In Critical Infrastructure Communication Networks* [Master Thesis]. Theses and Dissertations. 2740. Air Force Institute of Technology. https://scholar.afit.edu/etd/2740

[Ross 2008] Ross, A.M., Rhodes, D.H., Hastings, D.E. (2008). Defining Changeability: Reconciling Flexibility, Adaptability, Scalability, Modifiability, and Robustness for Maintaining System Lifecycle Value. *Systems Engineering, 11*(3), 246-262.

[Roudier 2014] Roudier, Y., Idrees, M.S., Apvrille, L. (2014). Improved Security Requirements Engineering using Knowledge Representation. *SAR-SSI 2014, 9ème Conférence sur la Sécurité des Architectures Réseaux et des Systèmes d'Information*, Saint-Germain-Au-Mont-D'Or, France, (May 2014).

[Saadeh 2014] Saadeh, O. (2014). *Distributed Energy Resource Management Systems 2014: Technologies, Deployments and Opportunities* [White Paper] (September 2014). GTM Research.

[SAF] Ellisson, R.J., Woody, C. (2010). *Survivability Analysis Framework* [Report CMU/SEI-2010-TN-013]. Software Engineering Institute. Carnegie Mellon University, Pittsburgh, Pennsylvania, USA. https://doi.org.10.1184/R1/6584474.v1

[Salles 2012] Salles, R.M., Marino Jr., D.A. (2012). Strategies and Metric for Resilience in Computer Networks. *The Computer Journal, 55(6)*, 728-739. https://doi.org/10.1093/comjnl/bxr110

[Sanders 2012] Sanders, B. (2012). *The Path to a Secure and Resilient Power Grid Infrastructure* [Presentation]. http://www.acs.org/content/dam/acsorg/policy/acsonthehill/briefings/cybersecurity/sanders-presentation-apr2012.pdf

[Sayama 2010] Sayama, H.D.Sc. (2010). *Complex systems organizational map.* CC BY-SA 3.0. https://en.wikipedia.org/wiki/Complex_system#/media/File:Complex_systems_organizational_map.jpg

[SEBoK 2019] BKCASE Editorial Board (2019) The Guide to the Systems Engineering Body of Knowledge (SEBoK), v. 2.0, ed Cloutier RJ (The Trustees of the Stevens Institute of Technology, Hoboken, NJ). BKCASE is managed and maintained by the Stevens Institute of Technology Systems Engineering Research Center, the International Council on Systems Engineering, and the Institute of Electrical and Electronics Engineers Computer Society. https://www.sebokwiki.org/wiki/Guide_to_the_Systems_Engineering_Body_of_Knowledge_(SEBoK)

[Simpson 2011a] Simpson, S. (2011a). Software Assurance Forum for Excellence in Code (SAFECode) [Presentation] (March 1, 2011). https://www.yumpu.com/en/document/read/48437706/software-assurance-forum-for-excellence-in-code-build-security-in

[Simpson 2011b] [Simpson, S. (2011b*). Fundamental Practices for Secure Software Development 2ND EDITION A Guide to the Most Effective Secure Development Practices in Use Today* (February 8, 2011). https://safecode.org/publication/SAFECode_Dev_Practices0211.pdf

[Sollecito 2009] Sollecito, L. (2009). Smart grid: the road ahead. *Protection & Control Journal, 8th Edition*, 15–19.

[Sommerville 2004] Sommerville, J. (2004). *Software Engineering* (7th Edition). Pearson Addison-Wesley.

[Sommerville 2007] Sommerville, I. (2007). *Software Engineering* (8th Edition). Addison-Wesley.

[Sommerville 2016] Sommerville, I. (2016). *Software Engineering* (Tenth Edition). Pearson Education Limited.

[Speer 2015] Speer, B., Miller, M., Schaffer, W., Gueran, L., Reuter, A., Jang, B., Widegren, K., (2015). *The Role of Smart Grids in Integrating Renewable Energy ISGAN Synthesis Report Annex 4 Task 3.2* [Technical Report NREL/TP-6A20-63919] (May 2015). National Renewable Energy Laboratory. https://www.nrel.gov/docs/fy15osti/63919.pdf

[Sproge 2012] Sproge, S., Cevere, R. (2012). Quality Models in Software Product Development Life Cycle. *6th International Conference on Application of Information and Communication Technologies (AICT 2012)*, Tbilisi, Georgia, (October 2012).

[Stamper 2003] Stamper, D.A., Case, T.L. (2003). *Business Data Communications* (Sixth Edition). Prentice Hall.

[Steeley 2011] Steeley, W. (2011). *Functional Requirements for Electric Energy Storage Applications on the Power System Grid* [Report 1022544] (May 2011). EPRI.

[Stevens 1998] Stevens, W.R. (1998). *UNIX Network Programming Networking APIs: Sockets and XTI Vol.1* (Second Edition). Prentice Hall PTR.

[Strigini 2009] Strigini, L. (2009). Fault tolerance and resilience: meanings, measures and assessment Report (June 2009). European Framework Programme 7 (FP7) (June 2009). https://citeseerx.ist.psu.edu/viewdoc/download?doi=10.1.1.1002.235&rep=rep1&type=pdf

[Surakhi 2017] Surakhi, O.M., Hudaib, A., AlShraideh, M., Khanafseh, M. (2017). A Survey on Design Methods for Secure Software Development. INTERNATIONAL JOURNAL OF COMPUTERS & TECHNOLOGY, 16(7), 7047-7064. https://doi.org/10.24297/ijct.v16i7.6467

[SYSTEM] EventHelix. (n.d.). *System Reliability and Availability*. http://www.eventhelix.com/RealtimeMantra/FaultHandling/system_reliability_availability.htm#.U-YhkuNdWSp

[Tanenbaum 2003] Tanenbaum, A.S. (2003). *Computer Networks* (Fourth Edition). Prentice Hall PTR.

[Tanenbaum 2007] Tanenbaum, A,S., Steen, M.V. (2007). *Distributed Systems Principles and Paradigms* (2nd Edition). Pearson Prentice Hall.

[Tanenbaum 2015] Tanembaum, A.S., Bos, H. (2015). *Modern Operating Systems* (4th Edition). Pearson Education Limited.

[Technopedia] [Techopedia. (n.d.). *Technology Dictionary*. https://www.techopedia.com/dictionary

[TechTarget] TechTarget. (n.d.). *Enterprise Architecture*. https://searchcio.techtarget.com/definition/enterprise-architecture

[Telecom 1996] ITS. (1996). *Telecommunications: Glossary of Telecommunication Terms*. Federal Standard 1037C. http://www.its.bldrdoc.gov/fs-1037/fs-1037c.htm

[Theron 2013] Théron, P. (2013). ICT Resilience as Dynamic Process and Cumulative Aptitude. In P. Théron, S. Bologna (Eds.), *Critical Information Infrastructure Protection and Resilience in the ICT Sector* (pp. 1-35). IGI Global. https://doi.org/10.4018/978-1-4666-2964-6.ch001

[Theron 2016] Théron, P. (2016,). Towards a general theory of resilience. Lessons from a multi-perspective research [cel-01342846v2]. *ERNCIP training for professionals in CIP: From risk management to resilience* (pp. 43), Bruxelles, Belgium, (July 2016). https://hal.archives-ouvertes.fr/cel-01342846/

[TISN 2007] TISN. (2007). *Secure Your Information: Information Security Principles for Enterprise Architecture* [Report]. Trusted Information Sharing for Critical Infrastructure Protection.

Networkhttps://www.tisn.gov.au/Documents/Secure_Your+Information+-+Information+Security+Principles+for+Enterprise+Architecture+-+Report.pdf

[Trivedi 2009] Trivedi, K.S., Kim, D.S., Ghosh, R. (2009). Resilience in Computer Systems and Networks. *International Conference on Computer Aided Design - Digest of Technical Papers (ICCAD 2009)* (pp. 74-79), San Jose, California, (November 2009).

[Tsiakis 2009] Tsiakis, T., Katsaros, P. (2009). Hands on Dependability Economics. D '09: Proceedings of the 2009 Second International Conference on Dependability (pp. 117–121), Athens, Greece, (June 2009). https://doi.org/10.1109/DEPEND.2009.24

[Vacca 2012] Vacca, J.R. (2012). *Computer and Information Security Handbook* (2nd Edition). Morgan Kaufmann.

[Vellaithurai 2015] Vellaithurai C.B., Biswas S.S., Liu R., Srivastava A. (2015). Real Time Modeling and Simulation of Cyber-Power System. In S.K. Khaitan, J.D. McCalley, C.C. Liu (Eds.), *Cyber Physical Systems Approach to Smart Electric Power Grid. Power Systems.* Springer. https://doi.org/10.1007/978-3-662-45928-7_3

[Vos 2009] Vos, A. (2009). Demand Response Management Systems: The Next Wave of Smart Grid Software. *Electric Energy T&D Magazine, 13*(5), 35-37.

[Westmark 2004] Westmark, V.R. (2004). A Definition for Information System Survivability. *Proceedings of the 37th Annual Hawaii International Conference on System Sciences (HICSS'04)* (pp. 1-10), Big Island, Hawaii, USA, (January 2004). https://doi.org/10.1109/HICSS.2004.1265710

[Whitman 2004] Whitman, M.E., Mattord, H.J. (2004). *Management of Information Security.* Thomson Course Technology.

[Whitman 2007] Whitman, M.E., Mattord, H.J. (2007). *Principles of Incident Response and Disaster Recovery*. Thomson Course Technology.

[Whitman 2011] Whitman, M.E., Mattord, H.J. (2011). *Management of Information Security* (Third Edition). Course Technology Cengage Learning.

[Whitman 2012] Whitman, M.E., Mattord, H.J. (2012). *Principles of Information Security* (4th Edition). Course Technology Cengage Learning.

[Whitman 2014] Whitman, M.E., Mattord, H.J. (2014). *Management of Information Security* (Fourth Edition). Course Technology Cengage Learning.

[Wiki HighAvail] Wikipedia. (n.d.). *High availability.* http://en.wikipedia.org/wiki/High_availability

[Wiki List] Wikipedia. (n.d.). List of system quality attributes. http://en.wikipedia.org/wiki/List_of_system_quality_attributes

[Wiki OS] Wikipedia. (n.d.). *Comparison of operating systems.* https://en.wikipedia.org/wiki/Comparison_of_operating_systems#Operating_system_comparisons

[Wiki Res] Wikipedia. (n.d.). *Resilience*. Wikipedia. http://simple.wikipedia.org/wiki/Resilience

[Wiki Surv] Wikipedia. (n.d.). *Survivability.* https://wiki.ittc.ku.edu/resilinets_wiki/index.php/Survivability

[Wiki Unavailability] Wikipedia. (n.d.). *Unavailability*. http://en.wikipedia.org/wiki/Unavailability.

[Winkelvos 2011] Winkelvos, T., Rudolph, C., Repp, J. (2011). A Property Based Security Analysis Through Weighted Simulation. *Proceedings of Conference Information Security South Africa (ISSA)* (pp. 1-8), Johannesburg, South Africa, (August 2011).

[Wolf Davis 2015] Wolf Davis, K. (2015, June 1). *Defending SCADA Systems Against the Growing Cyber Threat* [News]. Energy Central. https://energycentral.com/c/iu/defending-scada-systems-against-growing-cyber-threat

[Youn 2011] Youn, H., Park, C., Lee, E. (2011). Security based survivability risk analysis with extended HQPN. *Proceedings of the 5th International Conference on Ubiquitous Information Management and Communication (ICUIMC 2011)* (Article No: 25, 10 pages), Seoul, Republic of Korea, (February 2011). https://doi.org/10.1145/1968613.1968644

[Young 1970] Young, G. I. M. (1970). Feasibility studies. *Appraisal Journal, 38*(3), 376-383.

[Zafar 2013] Zafar, S., Mehboob, M., Naveed, A., Malik, B. (2013). Security quality model: an extension of Dromey's model. *Software Quality Journal*, 23, 29-54. Springer. https://doi.org/10.1007/s11219-013-9223-1

[Zhu 2011] Zhu, K., Chenine, M., Nordstrom, L. (2011). ICT Architecture Impact on Wide Area Monitoring and Control Systems' Reliability. *IEEE Transactions on Power Delivery, 26*(4), 2801-2808. https://doi.org/10.1109/TPWRD.2011.2160879

References Part 3

[Abi-Samra 2012] Abi-Samra, N. (2012, December 13). *Distribution Systems Automation & Optimization - Part 3* [Newsletter]. EEWeb. https://www.eeweb.com/distribution-systems-automation-optimization-part-3/

[Ai Ling 2011a] Ai Ling, A.P., Mukaidono, M. (2011a). Smart Grid Information Security (IS) Functional Requirement. *International Journal of Emerging Sciences, 1*(3), 371-386.

[Ai Ling 2011b] Ai Ling, A.P., Mukaidono, M. (2011b). Selection of Model in Developing Information Security Criteria on Smart Grid Security System, Smart Grid Security and Communications. *The Ninth International Symposium on Parallel and Distributed Processing with Applications (ISPA)* (Article No.108, pp. 91-98), Busan, South Korea (May 2011). *Journal of Convergence, 2*(1), 39-46. https://arxiv.org/abs/1109.2697

[Anderson] [Anderson, R.J. (n.d.). *Economics and Security Resource Page.* https://www.cl.cam.ac.uk/~rja14/econsec.html

[Anderson 2001] Anderson, R.J. (2001). Why Information Security is Hard-An Economic Perspective. *Proceedings of the 17th Annual Computer Security Applications Conference (ACSAC01)* (pp. 358-365), Washington, DC, USA, (December 2001). IEEE Computer Society.

[Anderson 2005] Anderson, N., Nolte, W. (2005). Systems Engineering Principles Applied to Basic Research and Development [Presentation]. *Space Systems Engineering Conference (1st Atlanta, November 2005) Session Title: Systems Analysis and Systems Engineering* (Paper No. GT-SSEC.D.5), Atlanta, Georgia, USA, (November 2005). Publisher Georgia Institute of Technology. https://smartech.gatech.edu/handle/1853/8034

[Anderson 2007] Anderson, C.W., Santos, J.R., Haimes, Y.Y. (2007). A risk-based input-output methodology for measuring the effects of the August 2003 Northeast blackout. *Economic Systems Research, 19*(2), 183-204.

[Anderson 2008a] Anderson, R.J. (2008a). *Security Engineering, A Guide to Building Dependable Distributed Systems* (Second Edition). Wiley.

[Anderson 2008b] Anderson, R.J., Bohme, R., Clayton, R., Moore, T. (2008b). *Security economics and the internal Market* [Report]. European Network and Information Security Agency. https://www.enisa.europa.eu/publications/archive/economics-sec/

[Anderson 2012] Anderson, R.J., Barton, C., Bohme, R., Clayton, R., van Eeten, M.J.G., Levi, M., Moore, T., Savage, S. (2012). Measuring the cost of cybercrime. *Proceedings of the 11th Workshop on the Economics of Information Security (WEIS2012)*, Berlin, Germany, (June 2012). In R. Bohme (Ed.), *The Economics of Information Security and Privacy* (pp. 265-300). Springer. https://doi.org/10.1007/978-3-642-39498-0_12

[ANSI/EIA 632] *ANSI/EIA-2003 Processes for Engineering a System.* American National Standards Institute

[Apvrille 2013] [Apvrille, L., Roudier, Y. (2013). SysML-Sec: A SysML Environment for the Design and Development of Secure Embedded Systems. *Proceedings of Conference APCOSEC 2013, Asia-Pacific Council on Systems Engineering*, Yokohama, Japan, (September 2013).

[ARP4761] *Guidelines and Methods for Conducting the Safety Assessment Process on Civil Airborne Systems and Equipment ARP4761 (*December 1996). SAE International.

[Ashrafi 2009] Ashrafi, N., Ashrafi, H. (2009). *Object-Oriented Systems Analysis and Design.* Pearson Education. [Assurance] US-CERT. (n.d.). *Software Assurance Resources.* https://buildsecurityin.us-cert.gov/swa/resources

[Avolio 2000] Avolio, F.M. (2000, March 20). Best practices in network security. *Network Computing*, 60-66.

[Axelsson 2019] Axelsson, J. (2019). Game theory applications in systems-of-systems engineering: A literature review and synthesis. *Procedia Computer Science 153*, 154-165. Elsevier. https://doi.org/10.1016/j.procs.2019.05.066

[Baldwin 2009] Baldwin, K. (2009). Systems Security Engineering: A Critical Discipline of Systems Engineering. *INCOSE INSIGHT, 12*(2), 11-13.

[Barbacci 1995] Barbacci, M., Klein, M.H., Longstaff, T.A. Weinstock, C.B. (1995). *Quality Attributes* [Technical Report CMU/SEI-95-TR-021. ESC-TR-95-021]. Software Engineering Institute.

[Barbacci 1999] Barbacci, M. (1999, March). *Analyzing quality attributes* [News]. Column in SEI newsletter. The Architect. Software Engineering Institute.

[Barman 2002] Barman, S. (2002). *Writing Information Security Policies*. New Riders.

[Bass 2012] Bass, L., Clements, P., Kazman, R. (2012). Why is Software Architecture Important? *Software Architecture in Practice The SEI Series in Software Engineering* (Third Edition) (pp. 25-37). Addison-Wesley. Software Engineering Institute.

[Bayuk 2011a] Bayuk, J.L. (2011). Systems Security Engineering. *IEEE Security & Privacy, 9*(2), 72-74. https://doi.org/10.1109/MSP.2011.41

[Bayuk 2011b] Bayuk J.L., Horowitz, B. (2011). An Architectural Systems Engineering Methodology for Addressing Cyber Security. *Systems Engineering, 14*(3), 294-304. Wiley Online.

[BD 2004] Uni-kl-de. (2004). *V-Modell ® XT Part 1: Fundamentals of the V-Modell* (Version 1.1.0). Bundesrepublic Deutschland. http://ftp.uni-kl.de/pub/v-modell-xt/Release-1.1-eng/Dokumentation/pdf/V-Modell-XT-eng-Teil1.pdf

[Behrens 2018] Behrens, A. (2018). What are Security Patterns? A Formal Model for Security and Design of Software. *ARES 2018: Proceedings of the 13th International Conference on Availability, Reliability and Security* (pp. 1-6), Hamburg, Germany, (August 2018). https://doi.org/10.1145/3230833.3230834

[Bentley 2007] Bentley, L.D., Whitten, J.L. (2007). *Systems Analysis and Design for the Global Enterprise* (7th Edition). Irwin/McGraw-Hill.

[Berthier 2010] Berthier, R., Sanders, W.H., Khurana, H. (2010). Intrusion Detection for Advanced Metering Infrastructures: Requirements and Architectural Directions. *Proceedings of 2010 IEEE First International Conference on Smart Grid Communications (SmartGridComm)* (pp. 350-355), Gaithersburg, Maryland, USA, (October 2010). https://doi.org/10.1109/SMARTGRID.2010.5622068

[Bishop 2005] Bishop, M. (2005). *Introduction to Computer Security*. Addison-Wesley Professional.

[Blackout 2003] Wikipedia. (n.d.). Northeast Blackout of 2003. https://en.wikipedia.org/wiki/Northeast_blackout_of_2003

[Blackout 2004] Energy. (2004*). Final Report on the August 14, 2003 Blackout in the United States and Canada: Causes and Recommendations* (April 2, 2004). https://www.energy.gov/oe/downloads/blackout-2003-final-report-august-14-2003-blackout-united-states-and-canada-causes-and

[Bodeau 2012] Bodeau, D., Graubart, R., Picciotto, J., McQuaid, R. (2012). *Cyber Resiliency Engineering Framework (January 2012).* The MITRE Corporation.

[Bouwens 2013] Bouwens, C.L. (2013). *Systems Geometry: A Methodology For Analyzing Emergent System Of Systems Behaviors* [Doctoral Dissertation]. Electronic Theses and Dissertations, 2004-2019. 2800. University of Central Florida Orlando, Florida, USA. https://stars.library.ucf.edu/etd/2800.

[Bruinenberg 2012] Bruinenberg, J., Colton, L., Darmois, E., Dorn, J., Doyle, J., Elloumi, O., Englert, H., Forbes, R., Heiles, J., Hermans, P., Uslar, M. (2012). *CEN -CENELEC - ETSI: Smart Grid Coordination Group - Smart Grid Reference Architecture Report 2.0* [Technical Report] (January 2012). Smart Grids Coordination Group. CEN -CENELEC – ETSI.

[Bugeja 2016] Bugeja, J., Jacobsson, A., Davidsson, P. (2016). On Privacy and Security Challenges in Smart Connected Homes. *2016 European Intelligence and Security Informatics Conference (EISIC)* (pp. 172-175), Uppsala, Sweden, (August 2016).

[BuildSecurity] US-CERT. (n.d.). Build Security In. https://buildsecurityin.us-cert.gov

[Camp 2000] Camp, L.J., Wolfram, C. (2000). Pricing Security. *Proceedings of the CERT Information Survivability Workshop* (pp. 31-39), Boston, Massachusetts, USA, (October 2000).

[Cantor 2006] Cantor, M., Roose, G. (2006, March 18). The Six Principles of Systems Engineering, IBM's Rational Rules Developed over 10-Year Period. *WEBSPHERE.*

[Carpenter 2009] Carpenter, M, Goodspeed, T., Singletary, B., Skodius, E., Wright, J., (2009). *Advanced Metering Infrastructure Attack Methodology Version 1.0* (January 5, 2009). http://inguardians.com/pubs/AMI_Attack_Methodology.pdf

[Carter 2016] Carter, B. (2016). *Systems Theory Based Architecture Framework for Complex System Governance* [PhD Thesis]. Engineering Management Old Dominion University. https://doi.org/10.25777/eaj5-sj38

[Cavoukian 2011] Cavoukian, A. (2011). *Privacy by Design in Law, Policy and Practice A White Paper for Regulators, Decision-makers and Policy-makers* [White Paper] (August 2011). Office of the Privacy Commissioner Ontario, Canada. https://studylib.net/doc/14438149/privacy-by-design-in-law--policy-and-practice-decision-ma...

[Cavoukian 2013] Cavoukian, A., Chanliau, M. (2013). Privacy and Security by Design: A Convergence of Paradigms. http://www.ipc.on.ca/site_documents/PbDBook-From-Rhetoric-to-Reality-ch8.pdf

[CC Arrangement] Commoncriteriaportal. (2014, July 2). *Arrangement on the Recognition of Common Criteria Certificates in the field of Information Technology Security.* https://www.commoncriteriaportal.org/files/CCRA%20-%20July%202,%202014%20-%20Ratified%20September%208%202014.pdf

[CC News] Commoncriteriaportal. (2014, September 8). *Twenty-six countries agree on reform to improve cyber security certification through international public-private collaboration* [News]. http://www.commoncriteriaportal.org/news/

[CDSE 2012] CDSE. (2012). *Glossary of Security Terms, Definitions, and Acronyms.* Center for Development of Security Excellence.

[CENELEC 2012] EC. (2012). *CEN-CENELEC-ETSI Smart Grid Coordination Group Smart Grid Reference Architecture* (November 2012). https://ec.europa.eu/energy/sites/ener/files/documents/xpert_group1_reference_architecture.pdf

[CENELEC 2014] CENCENELEC. (2014). *SG-CG/M490/H_ Smart Grid Information Security* [Technical Report M490-SGCG-SGIS-Intermediate-Report-V1] (December 2014). CEN-CENELEC-ETSI Smart Grid Coordination Group. Smart Grid Coordination Group Document for the M/490 Mandate Smart Grid Information Security. https://ftp.cencenelec.eu/EN/EuropeanStandardization/HotTopics/SmartGrids/SGCG_SGIS_Report.pdf

[CERT Response] US-CERT. (n.d.). *Industrial Control Systems (ICS-CERT).* http://www.us-cert.gov/control_systems/

[CERT] CERT. (n.d.). *The CERT Coordination Center.* http://www.cert.org

[CERT-Response] US-CERT. (n.d.). *Industrial Control Systems. Cyber Emergency Response Team.* https://ics-cert.us-cert.gov/

[Chan 2003] Chan, H., Perrig, A. (2003). Security and Privacy in Sensor Networks. *IEEE Computer, 36*(12), 103-105. https://doi.org/10.1109/MC.2003.1236475

[Chen 2006] Chen, D., Daclin, N. (2006). Framework for Enterprise Interoperability. *IFAC TC5.3 Workshop EI2N (2006)*, Bordeaux, France, (March 2006). https://doi.org/10.1002/9780470612200.ch6

[Chen 2008] Chen, D., Doumeingts, G., Vernadat, F. (2008). Architectures for enterprise integration and interoperability: Past, present and future. *Computers in Industry, 59*(7), 647-659. https://doi.org/10.1016/j.compind.2007.12.016

[Chen] Chen, D. (n.d.). *Enterprise Interoperability Framework.* http://ceur-ws.org/Vol-200/19.pdf

[Chinnow 2013] Chinnow, J., Tonn, J., Bsufka, K., Konnerth, T., Albayrak, S. (2013). A Tool Set for the Evaluation of Security and Reliability in Smart Grids. In J. Cuellar J. (Ed.) *Smart Grid Security. SmartGridSec 2012 Lecture Notes in Computer Science, 7823,* 45-57. Springer. https://doi.org/10.1007/978-3-642-38030-3_3

[Clements 2010] Clements, P., Bachmann, F., Bass, L., Garlan, D., Ivers, J., Little, R., Merson, P., Nord, R., Stafford, J. (2010). *Documenting Software Architectures: Views and Beyond* (Second Edition). Addison-Wesley.

[Cleveland 2012] Cleveland, F. (2012). *IEC TC57 WG15: IEC 62351 Security Standards for the Power System Infrastructure* [White Paper]. http://iectc57.ucaiug.org/wg15public/Public%20Documents/White%20Paper%20on%20Security%20Standards%20in%20IEC%20TC57.pdf

[Cleveland 2013] Cleveland, F., Lee, A. (2013). *Cybersecurity for DER Systems Version 1.0* (July 2013). Electric Power Research Institute. http://smartgrid.epri.com/doc/der-rpt-07-25-13.pdf

[CMMI 2013] CMMI. (2013). *Security by Design with CMMI for Development Version 1.3 An Application Guide for Improving Processes for Secure Products* (May 2013). Siemens AG Corporate Technology. CMMI Institute powered by Carnegie Mellon.

[CNSSI 4009] CNSSI. (2015). *Committee on National Security Systems (CNSS) Glossary* [CNSSI No. 4009].

[CodingStandards] US-CERT. (n.d.). SEI CERT Coding Standards. https://wiki.sei.cmu.edu/confluence/display/seccode/SEI+CERT+Coding+Standards

[Conference 2006] Conference Reports. (2006). Four focused forums. *IEEE Control Systems Magazine, 26*(4). 93-98.

[Connolly 2010] Connolly, D., Leahy, M. (2010). *A Review of Energy Storage Technologies for the Integration of Fluctuating Renewable Energy Version 4* (October 2010). University of Limerick, Limerick, Ireland.

[Consulting 2003] Firesmith, D. (2003). *Global Personal Marketplace System Requirements Specification (SRS) Version 1.0* [GPM-SRS]. Firesmith Consulting.

[Corbin 2015] Corbin, K. (2015, January 8). *Internet of Things Demands Security by Design* [News]. CSO Online. https://www.csoonline.com/article/2866446/internet-of-things-demands-security-by-design.html

[Common Criteria] Commoncriteriaportal. (n.d.). *The Common Criteria for Information Technology Security Evaluation V3.1 R4* (September 2012). http://www.commoncriteriaportal.org/cc/

[CORDIS 2016] Wolthuis, R. (2016). Security for Smart Electricity GRIDs Technical periodic report Year 2 [FP7-SEC-2013-1] (November 30, 2016). http://cordis.europa.eu/docs/results/607/607109/periodic2-segrid-technical-periodic-report-year-2-v1-0-20161207.pdf

[Craft 1998] Craft, R., Vandewart, R., Wyss, G., Funkhouser, D. (1998). An open framework for risk management. *21st National Information Systems Security Conference Proceedings*, Crystal City, Virginia, USA, (October 1998). http://csrc.nist.gov/nissc/1998/proceedings/paperE6.pdf

[Crawley 2004] Crawley, E., De Weck, O., Magee, C. et al. (2004). *The Influence of Architecture in Engineering. Engineering Systems Monograph.* http://esd.mit.edu/symposium/pdfs/monograph/architecture-b.pdf

[Crossley 2010] Crossley, P., Beviz, A. (2010). Smart energy systems: Transitioning renewables onto the grid. *Renewable Energy Focus, 11*(5), 54-56, 58-59. https://doi.org/10.1016/S1755-0084(10)70118-6

[CSIIRW06] ORNL. (2006). *CSIIRW06: Cyber Security and Information Infrastructure Research Workshop May 10-11 2006 Oak Ridge National Laboratory.* https://www.csm.ornl.gov/~sheldon/CSIIRW-06.htm

[Cummings 2000] Cummings, R. (2002). The Evolution of Information Assurance. *Computer, 35*(12), 65-72. https://doi.org/10.1109/MC.2002.1106181

[Curry 2012] Curry, E. (2012). System of Systems Information Interoperability using a Linked Dataspace. *IEEE 7th International Conference on System of Systems Engineering (SOSE 2012)* (pp. 101–106), Genova, Italy, (July 2012). https://doi.org/10.1109/SYSoSE.2012.6384200

[CVE] Mitre. (n.d.). *CVE*. cve.mitre.org/index.html

[CWE] Mitre. (n.d.). *Common Weakness Enumeration (CWE)*. http://cwe.mitre.org/

[Dandashi 2006] Dandashi, F., Siegers, R., Jones, J., Blevins, T. (2006). *The Open Group Architecture Framework (TOGAF) and the US Department of Defense Architecture Framework (DoDAF)* [White Paper] (November 2006). http://pubs.opengroup.org/onlinepubs/7699939899/toc.pdf

[Danekas 2013] Dänekas, C., Gonzalez, J.M. (2013). Requirements Engineering for Smart Grids. In M. Uslar, M. Specht, C. Dänekas, J. Trefke, S. Rohjans, J.M. Gonzalez, C. Rosinger, R. Bleiker (Eds.), *Standardization in Smart Grids Introduction to IT-Related Methodologies, Architectures and Standards* (pp. 15–38). Springer. https://doi.org/10.1007/978-3-642-34916-4_2

[Defense 2003] *Department of Defense Instruction 8500.2: Information assurance (IA) implementation* (February 6, 2003). U.S. Department of Defense, Washington, DC., USA. https://fas.org/irp/doddir/dod/d8500_2.pdf

[Delange 2013] Delange, J., Feiler, P. (2013). *Supporting the ARP4761 Safety Assessment Process with AADL*. https://juli1pb.files.wordpress.com/2014/06/article-erts2014-safety.pdf

[DeLaurentis 2006] DeLaurentis, D., Sindiy, O., Stein, W. (2006). Developing Sustainable Space Exploration via a System-of-Systems Approach. *AIAA Space 2006 Conference*, San Jose, California, USA, (September 2006). https://doi.org/10.2514/6.2006-7248

[Dennis 2012] Dennis, A., Wixom, B.H., Roth, R.M. (2012). *Systems Analysis Design* (Fifth Edition). Wiley.

[Deshmukh 2009] Deshmukh, M. (2009). *Seminar Report Security Requirements Engineering Process* (July 29, 2009)., Universitat Duisburg Essen, Germany. https://citeseerx.ist.psu.edu/viewdoc/download?doi=10.1.1.477.3708&rep=rep1&type=pdf

[Devanbu 2000] Devanbu, P.T., Stubblebine, S. (2000). Software Engineering for Security: A Road Map. *Proceedings of the Conference on the Future of Software Engineering* (pp. 227-239), Limerick, Ireland, (June 2000).

[DHHS 2005] DHHS. (2005). *DHHS Policies and Procedures 2005 Section VIII: Privacy and Security Application Security Policy Security Manual* (June 15, 2005). https://schs.dph.ncdhhs.gov/units/ccr/documents/AcceptableUseforDHHSInformationSystems.pdf

[DHS 2011] DHS. (2011). *Risk Management Fundamentals Homeland Security Risk Management Doctrine* (April 2011). https://www.dhs.gov/xlibrary/assets/rma-risk-management-fundamentals.pdf

[DHS 2013a] DHS. (2013a). *NIPP 2013 Partnering for Critical Infrastructure Security and Resilience*. U.S. Department of Homeland Security. https://www.dhs.gov/sites/default/files/publications/National-Infrastructure-Protection-Plan-2013-508.pdf

[DHS 2013b] [DHS. (2013b). *National Infrastructure Protection Plan Risk Management Framework*. U.S. Department of Homeland Security. https://www.dhs.gov/xlibrary/assets/NIPP_RiskMgmt.pdf

[DHS 2013c] DHS. (2013c). *Supplemental Tool: Executing A Critical Infrastructure Risk Management Approach*. U.S. Department of Homeland Security.

https://www.cisa.gov/sites/default/files/publications/NIPP-2013-Supplement-Executing-a-CI-Risk-Mgmt-Approach-508.pdf

[DHS 2016] DHS. (2016, September). *Recommended Practice: Improving Industrial Control System Cybersecurity with Defense-in-Depth Strategies.* Industrial Control Systems Cyber Emergency Response Team. https://www.cisa.gov/uscert/sites/default/files/recommended_practices/NCCIC_ICS-CERT_Defense_in_Depth_2016_S508C.pdf

[DoD 1995] DoD. (1995). *Handbook for Systems Security Engineering Program Management Requirements.* U.S. Department of Defense. Headquarters Air Force Systems Command Office of the Chief of Security Police.

[DoD 2001] DoD. (2001). *Systems Engineering Fundamentals Supplementary Text Prepared By the Defense Acquisition University Press.* U.S. Government Printing Office.

[DOE 2010] DOE. (2010). *DOE energy sector specific plan: an annex to the national infrastructure protection plan.* U.S. Department of Energy. http://energy.gov/oe/downloads/energy-sector-specific-plan-annex-national-infrastructure-protection-plan

[DOE 2012] DOE. (2012, May). *Electricity subsector cybersecurity risk management process* [DOE/OE-0003]. U.S. Department of Energy. http://energy.gov/sites/prod/files/Cybersecurity%20Risk%20Management%20Process%20Guideline%20-%20Final%20-%20May%202012.pdf

[DoT 2005] DoT. (2005). *Clarus Concept of Operations a Nationwide Surface Transportation Weather Observing and Forecasting System* [Publication No. FHWA-JPO-05-072] (October 2005). U.S. Department of Transportation Federal Highway Administration (FHWA). https://rosap.ntl.bts.gov/view/dot/3710

[EABOK] EABOK. (n.d.). *EABOK Glossary.* https://eabok.org/glossary/EISA 2007] EISA. (2007). *U.S. Energy Independence and Security Act of 2007 Public Law No: 110-140 Title XIII, Sec. 1301.* http://www.gpo.gov/fdsys/pkg/BILLS-110hr6enr/pdf/BILLS-110hr6enr.pdf

[ELCON 2004] ELCON. (2004). *The Economic Impacts of the August 2003 Blackout* (February 9, 2004]. Electricity Consumers Resource Council (ELCON). https://elcon.org/wp-content/uploads/Economic20Impacts20of20August20200320Blackout1.pdf

[Ellison 2008] Ellison, R.J., Goodenough, J.B., Weinstock, C.B., Woody, C. (2008). *Survivability Assurance for System of Systems* [Technical Report CMU/SEI-2008-TR-008] (May 2008). Software Engineering Institute.

[ENISA 2006] ENISA. (2006). *Risk Management: Implementation principles and Inventories for Risk Management / Risk Assessment methods and tools* (June 2006). European Network and information Security Agency (ENISA). http://www.enisa.europa.eu/activities/risk-management/current-risk/risk-management-inventory/files/deliverables/risk-management-principles-and-inventories-for-risk-management-risk-assessment-methods-and-tools

[ENISA 2012] ENISA. (2012). *Introduction to Return on Security Investment Helping CERTs Assessing the Cost of (Lack of) Security* (December 2012). European Network and information Security Agency (ENISA). https://www.enisa.europa.eu/publications/introduction-to-return-on-security-investment

[ENISA 2015] ENISA. (2015). *Communication network interdependencies in smart grids – Annexes Methodology for the identification of Critical Communication Networks Links and Components ANNEX C Distributed Energy Resources (DER).* European Union Agency for Network and Information Security (ENISA). https://www.enisa.europa.eu/topics/critical-information-infrastructures-and-services/communication-network-interdependencies-in-smart-grids-annexes

[ENISA 2016] ENISA. (2016). *Inventory of Risk Management / Risk Assessment Methods.* European Union Agency for Network and Information Security (ENISA).

https://www.enisa.europa.eu/topics/threat-risk-management/risk-management/current-risk/risk-management-inventory/rm-ra-methods

[EO 14028] TheWhiteHouse. (2021, May 12). *Executive Order 14028 Improving the Nation's Cybersecurity*. https://www.federalregister.gov/d/2021-10460

[EPRI 2002] EPRI. (2002). IntelliGrid Architecture Report Volume 1 IntelliGrid User Guidelines and Recommendations [Report No. 1012160]. Electric Power Research Institute (EPRI).

[EPRI 2013] EPRI. (2013). *Cybersecurity for DER Systems Version 1.0* (July 2013). Electric Power Research Institute. http://smartgrid.epri.com/doc/der-rpt-07-25-13.pdf

[EPRI-DOE 2003] EPRI. (2003). *EPRI-DOE Handbook of Energy Storage for Transmission & Distribution Applications Final Report* [Report No. 1001834] (December 2003). Electric Power Research Institute (EPRI). https://www.sandia.gov/ess-ssl/publications/ESHB 1001834 reduced size.pdf

[ESCSWG 2014] Energy. (2014, April). *Cybersecurity Procurement Language for Energy Delivery Systems*. Energy Sector Control Systems Working Group (ESCSWG). https://www.energy.gov/sites/prod/files/2014/04/f15/CybersecProcurementLanguage-EnergyDeliverySystems_040714_fin.pdf

[Fabian 2010] Fabian, B., Gürses, S., Heisel, M., Santen, T., Schmidt, H. (2010). A comparison of security requirements engineering methods. *Requirements Eng* 15, 7–40. https://doi.org/10.1007/s00766-009-0092-x

[Failure] Thefreedictionary. (n.d.). *Failure*. http://encyclopedia.thefreedictionary.com/Failure

[Fang 2012] Fang, X., Misra, S., Xue, G., Yang, D. (2012). Smart grid – the new and improved power grid: a survey. *IEEE Communications Surveys and Tutorials, 14*(4), 944–980. https://doi.org/10.1109/SURV.2011.101911.00087

[Farah 2004] Farah, G. (2004). *Information Systems Security Architecture A Novel Approach to Layered Protection A Case Study GSEC Practical Version 1.4b* (January 2005). SANS Institute, USA. https://www.sans.org/reading-room/whitepapers/auditing/paper/1532

[Fassois 2007] [Fassois, S.D., Rivera, D.E. (2007). Applications of System Identification. *IEEE Control Systems Magazine, 27*(5), 24-26. https://doi.org/10.1109/MCS.2007.904658

[Fault] Thefreedictionary. (n.d.). *Fault*. http://encyclopedia.thefreedictionary.com/fault

[FaultPower] Thefreedictionary. (n.d.). Fault Power. http://encyclopedia.thefreedictionary.com/Fault+(power+engineering)

[FaultTech] Thefreedictionary. (n.d.). Fault Technology. http://encyclopedia.thefreedictionary.com/Fault+(technology)

[FEAPO 2018] FEAPO. (2018). *Common Perspective on Enterprise Architecture Developed and Endorsed by the Federation of Enterprise Architecture Professional Organizations*. https://feapo.org/wp-content/uploads/2018/10/Common-Perspectives-on-Enterprise-Architecture-Final-1-copy.pdf

[Feng 2012] Feng, X., Stoupis, J., Mohagheghi, S., Larsson, M. (2012). Introduction to smart grid applications. In L.T. Berger, I. Krzysztof (Eds.), *Smart Grid Applications, Communications, and Security* (pp. 3–48). Wiley.

[FERC 2006] FERC. (2006). *Assessment of Demand Response and Advanced Metering Staff Report* [Docket AD-06-2-000] (August 2006, Revised 2008). Federal Energy Regulatory Commission. https://www.ferc.gov/sites/default/files/2020-04/demand-response_4.pdf

[FERC 2012] FERC. (2012). *Assessment of Demand Response and Advanced Metering* [Report]. Federal Energy Regulatory Commission. https://www.ferc.gov/sites/default/files/2020-04/12-20-12-demand-response.pdf

[FIPS 199] *FIPS PUB 199 Standards for Security Categorization of Federal Information and Information Systems, February 2004*. Computer Security Division Information Technology Laboratory. U.S. Department of Commerce. National Institute of Standards and Technology. http://csrc.nist.gov/publications/fips/fips199/FIPS-PUB-199-final.pdf

[Firesmith 2003] Firesmith, D.G. (2003). *Common Concepts Underlying Safety, Security, and Survivability Engineering* [Technical Report CMU/SEI-2003-TN-033], (December 2003). http://resources.sei.cmu.edu/asset_files/TechnicalNote/2003_004_001_14198.pdf

[Firesmith 2004] Firesmith, D. (2004). Specifying Reusable Security Requirements. *Journal of Object Technology 3*(1), 61-65. https://doi.org/10.5381/jot.2004.3.1.c6

[Firesmith 2008] [Firesmith, D.G., Capell, P., Hammons, C.B., Latimer, D., Merendino, T., Falkenthal, D. (2008). *The Method Framework for Engineering System Architectures* (1st Edition). Auerbach Publication.

[Fisher 2000] Fisher, M. (2000). Protecting Binary Executables. *Embedded Systems Programming, 13*(2), 24-30.

[Forsberg 2005] Forsberg, K., Mooz, H., Cotterman, H. (2005). *Visualizing Project Management* (3rd edition). Wiley.

[Frameworks] ISO-Architecture. (n.d.). *Survey of Architecture Frameworks.* http://www.iso-architecture.org/ieee-1471/afs/frameworks-table.html

[Galloway 2013] Galloway, B., Hancke, G.P. (2013). Introduction to Industrial Control Networks. *IEEE Communications Surveys & Tutorials, 15*(2), 860–880. https://doi.org/10.1109/SURV.2012.071812.00124

[Gao 2012] Gao, J., Xiaoa, Y., Liua, J., Liang, W., Chen, C.L.P. (2012). A survey of communication/networking in smart grids. *Future Generation Computer Systems, 28*(2), 391–404. https://doi.org/10.1016/j.future.2011.04.014

[Gartner 2009] Gartner. (2009). *Gartner IT Glossary Enterprise Architecture.* Gartner Group. http://www.gartner.com/it-glossary/enterprise-architecture-ea/

[GDPR 2016] EU. (April, 2016). General Data Protection Regulation. European Union. https://gdpr-info.eu/

[Geer 2001] Geer, D.E. (2001). Return on security investment: calculating the security investment equation. *Secure Business Quarterly, 1*(2).

[Gegick 2005] Gegick, M., Williams, L. (2005). Matching attack patterns to security vulnerabilities in software intensive system designs. *Proceedings of the 2005 Workshop on Software Engineering for Secure Systems International Conference on Software Engineering* (pp. 1-7), St. Louis, Missouri, USA, (July 2005). https://doi.org/10.1145/1082983.1083211

[Ghansah 2009] Ghansah, I. (2009). *Smart Grid Cyber Security Potential Threats, Vulnerabilities and Risks* [Interim Project Report CEC-500-2012-047], (May 2012). Public Interest Energy Research (PIER) Program. California Energy Commission. University State of California Sacramento.

[Glavin 2003] Glavin, H.E. (2003). A Risk Modeling Methodology. *Computer Security Journal, XIX*(3), 1-29.

[Goertzel 2008] Goertzel, M.K., Winograd, T. (2008*). Enhancing the Development Life Cycle to Produce Secure Software: A Reference Guidebook on Software Assurance Version 2.0* (October 2008). Publisher: Data and Analysis Center for Software for Department of Homeland Security. https://www.academia.edu/1383839/Enhancing_the_Development_Life_Cycle_to_Produce_Secure_Software_A_Reference_Guidebook_on_Software_Assurance

[Golden] Golden, B. (n.d.). *What is Systems Architecture*? https://borisgolden.com/#Publications

[González 2013] González, J.M., Trefke, J. (2013). Management of Information Models in the Energy Sector. In M. Uslar, M. Specht, C. Dänekas, J. Trefke, S. Rohjans, J.M. González, C. Rosinger, R. Bleiker (Eds.) *Standardization in Smart Grids Introduction to IT-Related Methodologies, Architectures and Standards. Power Systems* (pp. 79-96). Springer. https://doi.org/10.1007/978-3-642-34916-4_5

[Gordon 2002] Gordon, L.A., Loeb, M.P. (2002). The Economics of Information Security Investment. *ACM Transactions on Information and System Security, 5*(4), 438- 457.

[Grady 2008] Grady, J.O. (2008, September 11). Universal Architecture Description Framework. *Wiley Online.* Wiley InterScience. https://doi.org/10.1002/sys.20112

[Grady 2009] Grady, J.O, 2009, Universal Architecture Description Framework (UADF). *Systems Engineering, 12*(2), 91-116. Wiley Periodicals.

[GridPrinciples] GridWise. (2011). *GridWise Alliance Principles for Grid Interoperability* (September 2011). The GridWise Alliance. https://www.smartgrid.gov/files/documents/GridWise_Allice_Principles_for_Grid_Interoperability.pdf

[GridWise] GridWise. (n.d.). *GridWise Architecture Council Mission & Structure.* https://www.gridwiseac.org/about/mission.aspx

[GridWise 2005] GridWise. (2005). *Interoperability Path Forward v1.0 [*Whitepaper], (November 30, 2005). GridWise Architecture Council.

[GWAC 2008] GridWise. (2008*). GridWise® Interoperability Context- Setting Framework v1.1* (March 2008). The GridWise Architecture Council. https://www.gridwiseac.org/pdfs/interopframework_v1_1.pdf

[Hansche 2004] Hansche, S., Berti, J., Hare, C. (2004). *Official (ISC)2 Guide to The CISSP Exam.* Auerbach Publications.

[Harris 2013] Harris, S. (2013). *All in One CISSP Exam Guide* (6th Edition). McGraw Hill Education.

[Heartbleed] Wikipedia. (n.d.). *Heartbleed.* http://en.wikipedia.org/wiki/Heartbleed

[Heck 2003] Heck, B. S., Wills, L. M., Vatchtsevanos, G. J. (2003). Software Technology for Implementing Reusable, Distributed Control Systems. *IEEE Control Systems Magazine, 23*(2), 21-35. https://doi.org/10.1109/MCS.2003.1172827

[Hellerstein 2005] Hellerstein, J. L., Diao, Y., Parekh, S., Tilbury, D.M. (2005). Control engineering for computing systems - industry experience and research challenges. *IEEE Control Systems Magazine, 25*(6), 56-68. https://doi.org/10.1109/MCS.2005.1550153

[Hentea 2005b] Hentea, M. (2005b). Information security management. In M. Pagani (Ed.), *Encyclopedia of Multimedia Technology and Networking* (2-volume set, pp. 390–395). IGI Global. https://doi.org/10.4018/978-1-59140-561-0.ch056

[Hentea 2006] Hentea, M. (2006). Enhancing information security risk management with data mining and fuzzy logic techniques. *Proceedings of 19th International Conference on Computer Applications in Industry and Engineering (CAINE 2006)* (pp. 132-139), Las Vegas, Nevada, USA, (November 2006).

[Hentea 2008a] Hentea, M. (2008a). Improving security for SCADA control systems. *Interdisciplinary Journal of Information, Knowledge, and Management, 3,* 73–86. https://doi.org/10.28945/3185

[Hentea 2008b] Hentea, M. (2008b). A Perspective on Security Risk Management for SCADA Systems. *Proceedings of the ISCA 23rd International Conference on Computers and Their Applications (CATA 2008),* Cancun, Mexico, (April 2008).

[Hernan 2006] Hernan, S., Lambert, S., Ostwald, T., Shostack, A. (2006). Uncover Security Design Flaws Using the STRIDE Approach. *MSDN Magazine* (November 2006). http://msdn.microsoft.com/en-us/magazine/cc163519.aspx

[Higginbotham 1998] Higginbotham, M.D., Milheizler, A.J., Maley, J.G., Suskie, B.J. (1998). Integrating Information Security Engineering with System Engineering with Systems Engineering Tools. *Proceedings of WETICE '98 Conference* (pp. 320-326), Palo Alto, California, USA, (July 1998).

[Higgins 2012] Higgins, K.J. (2012, July 19). *Smart Grid Researcher Releases Open Source Meter-Hacking Tool* [News]. Dark Reading. https://www.darkreading.com/vulnerabilities---threats/smart-grid-researcher-releases-open-source-meter-hacking-tool/d/d-id/1138052

[Hilliard 2007] Hilliard, R. (2007). *All About IEEE Std 1471* (June 2007). http://www.iso-architecture.org/ieee-1471/docs/all-about-ieee-1471.pdf

[Hilliard 2012] Hilliard, R., Malavolta, I., Muccini, H., Pellicione, P. (2012). On the Composition and Reuse of Viewpoints across Architecture Frameworks. *Proceedings the Joint 10th Working IEEE/IFIP Conference on Software Architecture & 6th European Conference on Software Architecture (WICSA/ECSA)* (pp. 131-140), Helsinki, Finland, (August 2012). http://doi.org/10.1109/WICSA-ECSA.212.21

[Hoglund 2004] Hoglund, G., McGraw, G. (2004). Attack patterns. Computer Security Journal, XX(2), 15-32.

[Hohlbaum 2009] Hohlbaum, F., Braendle, M., Alvarez, F. (2009). *Cyber Security Practical considerations for implementing IEC 62351*. ABB Switzerland. http://library.e.abb.com

[Holzmann 2006] Holzmann, G.J. (2006). The power of 10: rules for developing safety-critical code. *Computer, 39*(6), 95-97.

[Hou 2010] Hou, D., Dolezilek, D.D. (2010). IEC 61850 – What It Can and Cannot Offer to Traditional Protection Schemes. *SEL Journal of Reliable Power, 1*(2), 1-11.

[Howard 2006] Howard, M., Lipner, S. (2006*). Security Development Lifecycle SDL: A Process for Developing Demonstrably More Secure Software*. Microsoft Press.

[Howard 2014] Howard, M., LeBlanc, D. (2014). *Writing Secure Code* (Second Edition). Microsoft Press.

[Huxham 2006a] Huxham, H. (2006a). *One dimensional view of Enterprise Architecture as a Service orientated architecture BITS* (September 4, 2006). http://en.wikipedia.org/wiki/File:BITS.jpg

[Huxham 2006b] Huxham, H. (2006b). Own View of Enterprise Information Security Architecture (EISA) Framework (September 18, 2006). https://en.wikipedia.org/wiki/File:Huxham_Security_Framework.jpg

[IA] Iainstitute. (n.d.). *What is Information Architecture?* Information Architecture Institute. https://www.iainstitute.org/what-is-ia

[ICS-CERT 2012a] ICS-CERT. (2012a). *Malware Infections in the Control Environment.* Monitor (ICS-MM201212), October-December. https://us-cert.cisa.gov/sites/default/files/Monitors/ICS-CERT_Monitor_Oct-Dec2012.pdf

[ICS-CERT 2012b] ICS-CERT. (2012b). *RISK MANAGEMENT PLAN FOR THE ELECTRICITY SECTOR—A SUMMARY.* Monitor (ICS-MM201204), April 2012. https://us-cert.cisa.gov/sites/default/files/Monitors/ICS-CERT_Monitor_Apr2012.pdf

[ICS-CERT] ICS-CERT. (n.d.). *Industrial Control Systems Cyber Emergency Response Team.* https://homeport.uscg.mil/Lists/Content/Attachments/1094/ICS-CERT_140520.pdf

[iEAi] Internationaleainstitute. *iEAi Glossary EA Terms & Definitions.* International Enterprise Architecture Institute. http://internationaleainstitute.org/ea-terms-definitions/

[IEC 61850] *IEC 61850 Communication networks and systems in substations, series of several parts.*

[IEC 62264] *IEC 62264 Enterprise-control system integration, series of several parts.*

[IEC 62351] *IEC 62351, Power systems management and associated information exchange - Data and communications security - ALL PARTS.*

[IEC TS 62351-1] *IEC TS 62351-1:2007 Power systems management and associated information exchange - Data and communications security - Part 1: Communication network and system security - Introduction to security issues.*

[IEC 62351-9] *IEC 62351-9:2017 Power systems management and associated information exchange – Data and communications security – Part 9: Cyber security key management for power system equipment.*

[IEC TR 62351-12] *IEC TR 62351-12:2016 Power systems management and associated information exchange – Data and communications security – Part 12: Resilience and security recommendations for power systems with distributed energy resources (DER) cyber-physical systems, Edition 1.0, 2016-04.*

[IEC 62443] *IEC 62443 Industrial communication networks – Network and system security, series of several parts.*

[IEC 62443-1-1] *IEC/TS 62443-1-1:2009 Industrial communication networks – Network and system security –Part 1-1: Terminology, concepts and models.*

[IEC 62443-3-3] *IEC 62443-3-3:2013 Industrial communication networks – Network and system security – Part 3-3: System security requirements and security levels.*

[IEC TR 62357] *IEC TR 62357 Power systems management and associated information exchange, series of several parts.*

[IEEE 1220] *IEEE Std 1220-1998 Standard for Application and Management of the Systems Engineering Process -Description.*

[IEEE 1471] *IEEE Std 1471-2000 Recommended Practice for Architectural Description of Software-Intensive Systems.*

[IEEE 1490] *IEEE 1490-2011 - IEEE Guide--Adoption of the Project Management Institute (PMI(R)) Standard A Guide to the Project Management Body of Knowledge (PMBOK(R) Guide)--Fourth Edition. * Withdrawn.*

[IEEE 1547] *IEEE 1547-2018 – IEEE Standard for Interconnection and Interoperability of Distributed Energy Resources with Associated Electric Power Systems Interfaces.*

[IEEE Design] IEEE. (2014). *Avoiding the Top 10 Software Security Design Flaws.* IEEE Cybersecurity. Computer Society Center for Secure Design. https://cybersecurity.ieee.org/blog/2015/11/13/avoiding-the-top-10-security-flaws/

[IEEE Std 610.12] *IEEE Std 610.12-1990 IEEE Standard Glossary of Software Engineering Terminology.*

[IETF SGIP] IETF. (n.d.). *SGIP NIST Conceptual Model Overview and Evolution, Copiously cribbed from presentation by Eric W. Gunther.* https://www.ietf.org/proceedings/77/slides/intarea-7.pdf

[INCOSE 2015] D.D. Walden, G.J. Roedler, K.J. Forsberg, R. D. Hamelin, T.M. Shortell (Eds.). (2015). *INCOSE System Engineering Handbook—A Guide for System Engineering Life Cycle Processes and Activities* Fourth Edition. International Council on Systems Engineering. Wiley.

[ISE] Infosecon. (n.d.). *ISE Information Security Economics. Bibliography.* http://infosecon.net/workshop/bibliography.php

[ISO 73 Guide] *ISO 73 Guide:2009 Risk Management – Vocabulary.*

[ISO 13335-2] ISO 13335-2:1997 Information technology -- Guidelines for the management of IT Security -- Part 2: Managing and planning IT Security.

[ISO 31000] *ISO 31000:2009, Risk management – Principles and guidelines.*

[ISO ARCH] ISO. (n.d.). *A Conceptual Model of Architecture Description.* http://www.iso-architecture.org/ieee-1471/cm/

[ISO/IEC 7498-1] *ISO/IEC 7498-1:1994 Information Technology - Open Systems Interconnection - Basic Reference Model: The Basic Model (1996).*

[ISO/IEC 10746-3] *ISO/IEC 10746-3:2009 Information technology -- Open distributed processing -- Reference model: Architecture.*

[ISO/IEC 12207] *ISO/IEC 12207:2008 Systems and software engineering -- Software life cycle processes.*

[ISO/IEC 15026-2] *ISO/IEC 15026-2:2011 Systems and software engineering -- Systems and software assurance –Part 2: Assurance case.*

[ISO/IEC 15026-3] *ISO/IEC 15026-3:2015 Systems and software engineering -- Systems and software assurance – Part 3: System integrity levels.*

[ISO/IEC 15408] *ISO/IEC 15408 Information technology - Security techniques - Evaluation criteria for IT security, Parts 1-3.*

[ISO/IEC 17000] *ISO/IEC 17000:2004 Conformity assessment — Vocabulary and general principles.*

[ISO/IEC 17025] *ISO/IEC 17025:2005 General requirements for the competence of testing and calibration laboratories.*

[ISO/IEC 17065] *ISO/IEC 17065:2012 Conformity assessment -- Requirements for bodies certifying products, processes and services.*

[ISO/IEC 21827] *ISO/IEC 21827:2008, Information Technology – Systems Security Engineering – Capability Maturity Model (SSE-CMM®).*

[ISO/IEC 27000] *ISO/IEC 27000:2016 Information technology — Security techniques — Information security management systems — Overview and vocabulary, ISO/IEC 27000 Fourth Edition, 2016.*

[ISO/IEC 27001] *ISO/IEC 27001:2013 Information technology — Security techniques — Information security management systems — Requirements.*

[ISO/IEC 27002] *ISO/IEC 27002:2013 Information technology — Security techniques — Code of practice for information security controls.*

[ISO/IEC 27005] *ISO/IEC 27005:2011 Information technology - Security techniques - Information security risk management.*

[ISO/IEC 27013] *ISO/IEC 27013:2021 Information technology — Security techniques — Guidance on the integrated implementation of ISO/IEC 27001 and ISO/IEC 20000-1.*

[ISO/IEC 27014] *ISO/IEC 27014:2013 Information technology - Security techniques - Governance of information security.*

[ISO/IEC 27019] *ISO/IEC 27019:2017 Information technology — Security techniques — Information security controls for the energy utility industry.*

[ISO/IEC 27031] *ISO/IEC 27031:2011, Information Technology – Security techniques – Guidelines for information and communication technology readiness for business continuity.*

[ISO/IEC 27036] *ISO/IEC 27036:2013 Information technology — Security techniques — Information security for supplier relationships (four parts).*

[ISO/IEC TS 17961] *ISO/IEC TS 17961:2013 Information technology -- Programming languages, their environments and system software interfaces -- C secure coding rules standard.*

[ISO/IEC/IEEE 15288] *ISO/IEC/IEEE 15288:2015 Systems and software engineering — System life cycle processes First Edition* (May 2015).

[ISO/IEC/IEEE 24748-1] *ISO/IEC/IEEE 24748-1:2018 Systems and software engineering — Life cycle management — Part 1: Guidelines for life cycle management* (November 2018).

[ISO/IEC/IEEE 42010] *ISO/IEC/IEEE 42010:2011 Systems and software engineering -- Architecture description.*

[ISO Survey] ISO. (2019). *Survey of Architecture Frameworks* (last updated November 9, 2019). http://www.iso-architecture.org/ieee-1471/afs/frameworks-table.html

[ISORTO 2010] ISORTO. (n.d.). *IRC Wholesale Interaction Demand Response & Energy Storage Technical Requirements.* http://www.isorto.org

[ISS] ISS. (2005). *Assessment and Remediation of Vulnerabilities in the SCADA and Process Control Systems of Utilities.* Internet Security Systems. https://docplayer.net/13025091-Assessment-and-remediation-of-vulnerabilities.html

[ITU-T X.901-X.904] *ITU-T Rec. X.901-X.904 Information technology - Open Distributed Processing - Reference Model, series of standards.*

[ITU-T X.1520] *ITU-T X.1520 Recommendation Common vulnerabilities and exposures.*

[Jacobson 1999] Jacobson, I., Booch, G., Rumbaugh, J. (1999). *The Unified Software Development Process.* Addison-Wesley.

[Jajodia 1993] Jajodia, S., Millen, J. (1993). Editors' preface. *Journal of Computer Security, 2*(2/3), 85.

[Järventausta 2010] Järventausta, P., Repo, S., Rautiainen, A., Partanen, J. (2010). Smart grid power system control in distributed generation environment. *Annual Reviews in Control, 34*(2), 277-286.

[Jiwnami 2002] Jiwnami, K., Zelkowitz, M. (2002). Maintaining software with a security perspective. *Proceedings of the International Conference on Software Maintenance* (pp. 194- 203), Montreal, Quebec, Canada, (October 2002).

[Juárez 2011] Juárez, N.E., Ortega, E.R., López, S., Juárez, J., Torres, L., Santiago, M., Alcázar, E., Altuve, H.J. (2011). Ethernet-Based Protection and Control System for the Tabasco Substation: Design Concepts, Testing, and Field Performance. *38th Annual Western Protective Relay Conference*, Spokane, Washington, USA, (October 2011).

[Kadrich 2007] Kadrich, M.S. (2007). *Endpoint Security*. Addison-Wesley.

[Kaur 2010] Kaur, R., Singh, T. (2010). Analysis and Need of Requirements Engineering. *International Journal of Computer Applications, 7*(14), 27-32.

[Kazman 2013] Kazman, R., Nielsen, C., Schmid, K. (2013). *Understanding Patterns for System-of-Systems Integration* [Report CMU/SEI-2013-TR-017] (December 2013). Software Engineering Institute.

[Keromythis 2004] Keromythis, A.D. (2004). "Patch on Demand" saves even more time? *Computer, 37*(8), 94-96. https://doi.org/10.1109/MC.2004.71

[Khan 2012] Khan, K.K. (2012). *Developing and Evaluating Security-Aware Software Systems*. IGI Global. https://doi.org/10.4018/978-1-4666-2482-5

[Kiely 2008] Kiely, L., Benzel, T. (2008). Systemic Security Management: A new conceptual framework for understanding the issues, inviting dialogue and debate, and identifying future research needs. *IEEE Security and Privacy Magazine, 4*(6), 74-77. https://doi.org/1.1109/MSP.2006.167

[Knowles 2015] Knowles, W., Prince, D.D.C., Hutchison, D., Disso, J.F.P., Jones, K. (2015). A survey of cyber security management in industrial control systems. *International Journal of Critical Infrastructure Protection, 9*(C), 52-80. https://doi.org/10.1016/j.ijcip.2015.02.002

[Kossiakoff 2011] Kossiakoff, A., Sweet, W.N., Seymour, S., Biemer, S.M. (2011). *Systems Engineering Principles and Practice*. Wiley.

[Kotonya 1998] Kotonya, G., Sommerville, I. (1998). *Requirements Engineering: Processes and Techniques* (1st edn). Wiley.

[Kotov 1997] Kotov, V. (1997). *Systems of Systems as Communicating Structures* (October 1997). Computer Systems Laboratory Hewlett Packard. http://www.hpl.hp.com/techreports/97/HPL-97-124.pdf

[Koziolek 2012] Koziolek, A., Happe, L., Avritzer, A., Suresh, S. (2012). A common analysis framework for smart distribution networks applied to survivability analysis of distribution automation. *Proceedings of the First International Workshop on Software Engineering Challenges for the Smart Grid (SE4SG'12)* (pp. 23-29), Piscataway, New Jersey, USA, (June 2012). IEEE Press.

[Kreizman 2006] Kreizman, G., Robertson, B. (2006). *Incorporating Security into the Enterprise Architecture Process.* http://www.gartner.com

[Krutz 2004] Krutz, R.L., Vines, R.D. (2004). *The CISSP Prep Guide* (Second Edition). Wiley.

[Kunreuther 2003] Kunreuther, H., Heal, G. (2003). Interdependent Security. *Journal of Risk and Uncertainty, 26*(2—3), 231-249.

[Laplante 2014] Laplante, P.A. (2014). *Requirements Engineering for Software and Systems* (Second Edition). Applied Software Engineering Series. Taylor & Francis Group.

[Lawrence 1987] Lawrence, P.D., Mauch, K. (1987). *Real-time Microcomputer System Design: An Introduction.* McGraw-Hill.

[Li 2010] Li, F., Qiao, W., Sun, H., Wan, H., Wang, J., Xia, Y., Xu, Z., Zhang, P. (2010). Smart Transmission Grid: Vision and Framework. *IEEE Transactions on Smart Grid, 1*(2), 168-177. https://doi.org/doi: 10.1109/TSG.2010.2053726

[Long 2013] Long, F., Mohindra, D., Seacord, R.C., Sutherland, D.F., Svoboda, D. (2013). *Java Coding Guidelines: 75 Recommendations for Reliable and Secure Programs.* Addison-Wesley Professional.

[Maier 2009] Maier, M.W., Rechtin, E. (2009). *The Art of Systems Architecting (Systems Engineering)* (Third Edition). CRC Press.

[Mannion 1995] Mannion, M., Keepence, B. (1995). SMART Requirements. *ACM SIGSOFT Software Engineering Notes, 20*(2), 42-47.

[Manthorpe 1996] Manthorpe Jr., W.H. (1996). The Emerging Joint System-of-Systems: A Systems Engineering Challenge and Opportunity for APL. *Johns Hopkins APL Technical Digest, 17*(3), 305–310.

[Martin 1985] Martin, J., McClure, C. (1985). *Structured Techniques for Computing*. Prentice-Hall.

[Martinez 2014] Martinez, C. (2014). *Enterprise Architectures A Just-in-Time Approach for Decision-Making* (March 24, 2014). The MITRE Corporation. https://eapad.dk/article/enterprise-architectures-a-just-in-time-approach-for-decision-making/

[McAllister 2002] McAllister, R.K. (2002). *INFORMATION SYSTEMS SECURITY ENGINEERING The Need for Education* [Position Paper]. http://www.asac.org

[McGraw 2013] McGraw, G. (2013, January 17). *Thirteen principles to ensure enterprise system security* [Opinion]. TechTarget. http://searchsecurity.techtarget.com/opinion/Thirteen-principles-to-ensure-enterprise-system-security

[Mead 2007] Mead, N.R. (2007). *How to Compare the Security Quality Requirements Engineering (SQUARE) Method with Other Methods* [TECHNICAL NOTE CMU/SEI-2007-TN-021] (August 2007).

[Mellado 2006] Mellado, D., Fernández-Medina, E., Piattini, M. (2006). Applying a Security Requirements Engineering Process. In D. Gollman, J. Meier, A. Sabefeld (Eds.) *Computer Security – ESORICS 2006. ESORICS 2006. Lecture Notes in Computer Science, 4189*, 192-206. Springer. https://doi.org/10.1007/11863908_13

[Menzies 2006] Menzies, T., Richardson, J. (2006). Making Sense of Requirements, Sooner. *IEEE Computer, 39*(10), 112-114.

[Merriam] M-W. (n.d.). *Merriam-Webster Dictionary*. http://www.m-w.com

[Meier 2009] Meier, J.D., Homer, A., Hill, D., Taylor, J., Bansode, P, Wall, L., Boucher, R.Jr., Bogawat, A. (2009). *Microsoft Application Architecture Guide Patterns & Practices* (2nd Edition). Microsoft Press.

[Microsoft] Microsoft. (n.d.). *How Microsoft's intelligence is protecting small businesses.* https://www.microsoft.com/en-us/security/business/security-fundamentals

[Minkel 2008] Minkel, J.R. (2008, August 13). *The 2003 Northeast Blackout--Five Years Later* [News]. Scientific American. http://www.scientificamerican.com/article/2003-blackout-five-years-later/

[Mitigating] US-CERT. (2019). *2019 CWE Top 25 Most Dangerous Software Errors*. Cybersecurity & Infrastructure Security Agency (September 17, 2019). https://us-cert.cisa.gov/ncas/current-activity/2019/09/17/2019-cwe-top-25-most-dangerous-software-errors

[MITRE 2011] [MITRE. (2011). *CWE/SANS Top 25 Most Dangerous Software Errors* (September 13, 2011). https://cwe.mitre.org/top25/archive/2011/2011_cwe_sans_top25.pdf

[Momoh 2009] Momoh, J.A. (2009). Smart Grid Design for Efficient and Flexible Power Networks Operation and Control. *IEEE/PES Power Systems Conference and Exposition (PSCE "09)* (pp. 1-8), Seattle, Washington, USA, (March 2009).

[Moore 2011] Moore, T., Anderson, R.J. (2011). *Economics and Internet Security: a Survey of Recent Analytical, Empirical and Behavioral Research* [Technical Report TR-03-11]. Harvard Library. https://dash.harvard.edu/handle/1/23574266

[Muller 2011] Muller, G. (2011). *Systems Architecting: A Business Perspective*. CRC Press.

[NCCIC] DHS. (n.d.). *National Cybersecurity and Communications Integration Center (NCCIC).* https://www.dhs.gov/taxonomy/term/946/all/feed

[NDIA 2008] NDIA. (2008). *Guidebook on Engineering for System Assurance Version 1.0*. National Defense Industrial Association System Assurance Committee. https://www.acqnotes.com/Attachments/NDIA%20Engineering%20for%20System%20Assurance%20Guidebook.pdf

[NESCOR 2013a] NESCOR. (2013a). *NESCOR Guide to Penetration Testing for Electric Utilities Version 3.* National Electric Sector Cybersecurity Organization Resource (NESCOR). https://smartgrid.epri.com/doc/NESCORGuidetoPenetrationTestingforElectricUtilities-v3-Final.pdf

[NESCOR 2013b] NESCOR. (2013b). *Electric Sector Failure Scenarios and Impact Analyses Version 1.0* (September 2013). National Electric Sector Cybersecurity Organization Resource (NESCOR) Technical Working Group 1. https://smartgrid.epri.com/doc/NESCOR%20failure%20scenarios09-13%20finalc.pdf

[NESCOR 2015] NESCOR. (2013b). *Electric Sector Failure Scenarios and Impact Analyses - Version 3.0* (December 2015). National Electric Sector Cybersecurity Organization Resource (NESCOR). https://smartgrid.epri.com/doc/NESCOR%20Failure%20Scenarios%20v3%2012-11-15.pdf

[Neumann 2004] Neumann, P. (2004). *Principled Assuredly Trustworthy Composable Architectures.* CDRL A001 Final Report, SRI International, Menlo Park, CA, December 28, 2004

[Neumann 2006] Neumann, S., Sioshansi, F., Vojdani, A., Yee, G. (2006). How to get more response from demand response. *The Electricity Journal, 19*(8), 25–31.

[Newmann 2007] Newmann, S., Sioshansi, F., Vojdani, A., Yee, G. (2007). The Missing Link. *Public Utilities Fortnightly* (March 2007, pp. 52—66).

[News 2011] Federalnewsradio. (2011, October 10). *The Cybersecurity Dilemma: Incentives to Drive Change Rather than Reactive Response* (October 10, 2011, 10:48 PM) [News]. Federal News Network. Federal News Staff. http://federalnewsradio.com/technology/2011/10/the-cybersecurity-dilemma-incentives-to-drive-change-rather-than-reactive-response/

[News 2014] Tmcnet. (2014, August 14). *Markets and Markets : Demand Response Management System (DRMS) Market worth $24.64 Billion by 2019* [News]. Contact Center Solutions Enews. Industry News. http://callcenterinfo.tmcnet.com/news/2014/08/14/7971273.htm

[Nhlabatsi 2010] Nhlabatsi, A., Nuseibeh, B., Yu, Y. (2010). Security Requirements Engineering for Evolving Software Systems: A Survey. *International Journal of Secure Software Engineering, 1*(1), 54-73.

[NIST 2009] NIST. (2009). *NIST Framework and Roadmap for Smart Grid Interoperability, Interoperability Standards Release 1.0.* Office of the National Coordinator for Smart Grid Interoperability.

[NIST 2010] NIST. (2010). *Testing and Conformance Framework Development Guide* (August 2010).

[NIST 2014] *NIST Framework for Improving Critical Infrastructure Cybersecurity Version 1.0* (February 2014). https://www.nist.gov/system/files/documents/cyberframework/cybersecurityframework-021214.pdf

[NIST 7628 Guide] NIST. (2014). *NISTIR 7628 User's Guide A White Paper developed by the Smart Grid Interoperability Panel (SGIP)* (February 2014). https://tsapps.nist.gov/publication/get_pdf.cfm?pub_id=915613

[NIST 2016] NIST. (2016). *Smart Grid Testing and Certification.* https://www.nist.gov/programs-projects/smart-grid-testing-and-certification

[NIST 2018] *NIST Framework for Improving Critical Infrastructure Cybersecurity Version 1.1* (April 2018). https://doi.org/10.6028/NIST.CSWP.04162018

[NIST 2042TN] NIST Technical Note 2042 Review of Smart Grid Standards for Testing and Certification Landscape Analysis (April 2019). https://doi.org/10.6028/NIST.TN.2042

[NIST 2051TN] *NIST Technical Note 2051 Cybersecurity Framework Smart Grid Profile* (July 2019). https://doi.org/10.6028/NIST.TN.2051

[NIST 2182TN] *NIST Technical Note 2051 Distributed Energy Resource Security: Potential Guidelines and Research Topics* (October 2021). https://doi.org/10.6028/NIST.TN.2182

[NISTIR 7628] *NIST Interagency Report (NISTIR) 7628* Guidelines *for Smart Grid Cyber Security* (August 2010).

[NISTIR 7628r1] *NIST Interagency Report (NISTIR) 7628r1 Guidelines for Smart Grid Cybersecurity* (Revision 1, September 2014). http://dx.doi.org/10.6028/NIST.IR.7628r1

[NIST Guide] NIST. (2012). *Guide for Assessing the High-Level Security Requirements in NISTIR 7628 Version 1.0* (August 24, 2012).

[NIST IKB] NIST. (n.d.). NIST Smart Grid Collaboration Site IKB Use Cases. http://collaborate.nist.gov/twikisggrid/ bin/view/SmartGrid/IKBUseCases

[NIST ManagControl] NIST. (n.d.). *Glossary. Management controls.* Information Technology Laboratory Computer Security Resource Center (CSRC). https://csrc.nist.gov/glossary/term/management_controls

[NIST RiskManag] NIST. (n.d.). *Glossary. Risk management.* Information Technology Laboratory Computer Security Resource Center (CSRC). https://csrc.nist.gov/glossary/term/risk_management

[NIST Roadmap] NIST. (2014). *Roadmap for Measurement of Security Technology Impacts for Industrial Control Systems* (September 2014). https://www.nist.gov/sites/default/files/documents/el/isd/cs/NIST_ICS-Workshop-FinalReport.pdf

[NIST SP800-12] *NIST Special Publication (SP) 800-12r1 An Introduction to Computer Security* (Revision 1, June 2017). https://doi.org/10.6028/NIST.SP.800-12r1

[NIST SP800-18r1] *NIST Special Publication (SP) 800-18r1 Guide for Developing Security Plans for Federal Information Systems (Revision 1.0, February 2006).*

[NIST SP800-27rA] *NIST Special Publication (SP) 800-27 Rev A Engineering Principles for Information Technology Security (A Baseline for Achieving Security) Revision A* (June 2004). * superseded by NIST SP800-160 (November 2016).

[NIST SP800-30r1] *NIST Special Publication (SP) 800-37r1 Guide for Conducting Risk Assessments,* (Revision 1, September 2012).

[NIST SP800-33] *NIST Special Publication (SP) 800-33 Underlying Technical Models for Information Technology Security. *Withdrawn, August 2018.*

[NIST SP800-37r2] *NIST Special Publication (SP) 800-37r2* Risk Management Framework for Information Systems and Organizations A System Life Cycle Approach for Security and Privacy Revision 2 (December 2018). https://doi.org/10.6028/NIST.SP.800-37r2

[NIST SP800-39] *NIST Special Publication (SP) 800-39 Managing Information Security Risk: Organization, Mission, and Information System View* (March 2011).

[NIST SP800-53Ar4] *NIST Special Publication (SP) 800-53Ar4 Assessing Security and Privacy Controls in Federal Information Systems and Organizations: Building Effective Assessment Plans* Revision 4 (December 2014).

[NIST SP800-53r5] *NIST Special Publication (SP) 800-53 Revision 5 Security and Privacy Controls for Information Systems and Organizations* (December 2020). https://doi.org/10.6028/NIST.SP.800-53r5

[NIST SP800-55r1] *NIST Special Publication (SP) 800-55r1 Performance Measurement Guide for Information Security* (Revision 1, July 2008).

[NIST SP800-82r2] *NIST Special Publication 800-82r2 Guide to Industrial Control Systems (ICS) Security Supervisory Control and Data Acquisition (SCADA) systems, Distributed Control Systems (DCS), and other control system configurations such as Programmable Logic Controllers (PLC)* (Revision 2, May 2015).

[NIST SP800-115] *NIST Special Publication (SP) 800-115 Technical Guide to Information Security Testing and Assessment* (September 2008).

[NIST SP800-160v1] *NIST Special Publication (SP) 800-160 Systems Security Engineering Considerations for a Multidisciplinary Approach in the Engineering of Trustworthy Secure Systems* Volume 1 (March 2018). https://doi.org/10.6028/NIST.SP.800-160v1

[NIST SP800-160v2] *NIST Special Publication (SP) 800-160 Developing Cyber-Resilient Systems: A Systems Security Engineering Approach* Volume 2 (December 2021). https://doi.org/10.6028/NIST.SP.800-160v2r1

[NIST SP800-207] *NIST Special Publication (SP) 800-160 Zero Trust Architecture* (August 2020). https://doi.org/10.6028/NIST.SP.800-207

[NIST SP1108r1] *NIST Special Publication (SP) 1108r1 NIST Framework and Roadmap for Smart Grid Interoperability Standards, Release 1.0,* (January 2010). https://doi.org/10.6028/NIST.sp.1108

[NIST SP1108r3] *NIST Special Publication (SP) 1108r3 NIST Framework and Roadmap for Smart Grid Interoperability Standards, Release 3.0,* (September 2014). https://dx.doi.org/10.6028/NIST.SP.1108r3

[NIST SP1108r4] *NIST Special Publication (SP) 1108r4 NIST Framework and Roadmap for Smart Grid Interoperability Standards Release 4.0* (February 2021). https://dx.doi.org/10.6028/NIST.SP.1108r4

[NIST SP1500-201] *NIST Special Publication 1500-201 Framework for Cyber-Physical Systems: Volume 1, Overview Version 1.0* (June 2017). https://doi.org/10.6028/NIST.SP.1500-201

[NIST SP1800-32] *NIST SPECIAL PUBLICATION 1800-32 Securing Distributed Energy Resources: An Example of Industrial Internet of Things Cybersecurity Includes Executive Summary (A); Approach, Architecture, and Security Characteristics (B); and How-To Guides (C)* (February 2022). https://doi.org/10.6028/NIST.SP.1800-32

[NISTIR 7628r1] *NIST Interagency Report (NISTIR) 7628r1 Guidelines for Smart Grid Cybersecurity* Revision 1 (September 2014). http://dx.doi.org/10.6028/NIST.IR.7628r1

[NSF 2011] NSF. (2011, June). *A Policy framework for the 21st century grid: enabling our secure energy future.* U.S. National Science Foundation. https://www.ourenergypolicy.org/wp-content/uploads/2011/12/2011_06_WhiteHouse_PolicyFramework21stCenturyGrid.pdf

[NVD] NIST. (n.d.). *National Vulnerability Database (NVD).* https://nvd.nist.gov/

[OASIS 2006] OASIS. (2006). *OASIS reference model for service oriented architecture* 1.0 (soa-rm-cs). Organization for the Advancement of Structured Information Standards (OASIS). https://www.oasis-open.org/committees/download.php/19679/soa-rm-cs.pdf

[OCTAVE] CMU. (2001). *OCTAVE Method Implementation Guide Version 2.0, Volume 2: Preliminary Activities* (June 2001). Software Engineering Institute. Carnegie Mellon University. https://resources.sei.cmu.edu/asset_files/UsersGuide/2001_012_001_51572.pdf

[OECD 1980] OECD. (1980). *Guidelines on the Protection of Privacy and Transborder Flows of Personal Data.* Organisation for Economic Co-operation and Development. https://www.oecd.org/internet/ieconomy/oecdguidelinesontheprotectionofprivacyandtransborderflowsofpersonaldata.htm

[O-ESA 2011] OpenGroup. (2011). *Open Enterprise Security Architecture (O-ESA) A Framework and Template for Policy-Driven Security.* Van Haren Publishing.

[Op't Land 2009] Op't Land, M., Proper, E., Waage, M., Cloo, J., Steghuis, C. (2009). *Enterprise Architecture Creating Value by Informed Governance.* Springer.

[OpenHAN 2010] UCAIUG. (2010). *UCAIug Home Area Network System Requirements Specification Version 2.0* (August 30, 2010). A Work Product of the OpenHAN Task Force formed by the SG Systems Working Group under the Open Smart Grid (OpenSG) Technical Committee of the UCA®International Users Group. http://osgug.ucaiug.org/sgsystems/openhan/Shared%20Documents/OpenHAN%202.0/UCAIug%20HAN%20SRS%20-%20v2.0.pdf

[OpenSG] UCAIUG. (n.d.). *OpenSG users group.* http://osgug.ucaiug.org/default.aspx

[OSGP] OSGP. (n.d.). *Open Smart Grid Protocol Alliance.* www.osgp.org/
[Oracle] Oracle. (n.d.). *Importance of Software Security Assurance.* http://www.oracle.com/us/support/assurance/overview/index.html
[OSA Framework] Opensecurityarchitecture. (n.d.). *Security Architecture Landscape.* http://www.opensecurityarchitecture.org/cms/foundations/osa-landscape
[OSA IT] Opensecurityarchitecture. (n.d.).*IT Security Requirements.* https://www.opensecurityarchitecture.org/cms/definitions/it_security_requirements
[OSA Taxonomy] Opensecurityarchitecture. (n.d*.). IT Security Requirements, Taxonomy, Open Security Architecture (OSA).* http://www.opensecurityarchitecture.org/cms/definitions/it_security_requirements
[OWASP 2014] OWASP. (2014). *Security-Relevant Requirements The Open Web Application Security Project (OWASP) Overview Document.* https://www.owasp.org/index.php/Document_security-relevant_requirements
[OWASP Top10] OWASP. (n.d.). *OWASP Top Ten.* https://owasp.org/www-project-top-ten/
[OWASP Threat] OWASP. (n.d.). *Threat Modeling.* https://owasp.org/www-community/Threat_Modeling
[Ozier 2002] Ozier, W. (2002). A Framework for Automated Risk Assessment Tools. *The EDP Audit, Control, and Security Newsletter EDPACS 30*(4), 11-19. https:doi.org/10.1201/1079/43287.30.4.20021001/38847.2
[P2030 2011] *IEEE Std 2030-2011- IEEE Guide for Smart Grid Interoperability of Energy Technology and Information Technology Operation with the Electric Power System (EPS), End-Use Applications, and Loads.*
[P2030.2 2015] *IEEE Std 2030.2-2015 - IEEE Guide for the Interoperability of Energy Storage Systems Integrated with the Electric Power Infrastructure.*
[Patel 2006] Patel, S.C., Graham, J.H., Ralston, P.A.S. (2006). Security enhancement for SCADA communication protocols using augmented vulnerability trees. *Proceedings of 19th International Conference on Computer Applications in Industry and Engineering (CAINE 2006)* (pp. 244-251), Las Vegas, Nevada, USA, (November 2006).
[Permann 2005] Permann, M.R., Rohde, K. (2005). Cyber Assessment Methods for SCADA Security. *15th Annual Joint ISA POWID/EPRI Controls and Instrumentation Conference* [Report No NL/CON-05-00093], Nashville, Tennessee, USA, (June 2005).
[PMBOK] [see IEEE 1490]
[PNNL 2003] Gridwiseac. (2003). *Gridwise_™ Architecture Tenets and Illustrations* [PNNL-SA-39480], (October 2003). Pacific Northwest National Laboratory U.S. Department of Energy. https://www.gridwiseac.org/pdfs/tenet_illustrations.pdf
[Pourbeik 2006] Pourbeik, P., Kundur, P.S., Taylor, C.W. (2006). The anatomy of a power grid blackout - Root causes and dynamics of recent major blackouts. *IEEE Power and Energy Magazine, 4*(5), 22 – 29. https://doi.org/10.1109/MPAE.2006.1687814
[Practices] MITRE. (2012). *Key Practices for Mitigating the Most Egregious Exploitable Software Weaknesses Software Assurance Pocket Guide Series: Development, Volume II Version 2.3* (November 1, 2012). https://cwe.mitre.org/documents/KeyPracticesMWV22_20121101.pdf
[Pressman 2005] Pressman, R.S. (2005). *Software Engineering A Practitioner's Approach* (Sixth Edition). McGraw Hill Education.
[Principles] Cavoukian, A. (2011). *Privacy by Design The 7 Foundational Principles Implementation and Mapping of Fair Information Practices.*https://iapp.org/media/pdf/resource_center/pbd_implement_7found_principles.pdf
[Qiu 2003] Qiu, Q. (2003). *Risk Assessment of Power System Catastrophic Failures and Hidden Failure Monitoring & Control System* [PhD dissertation] (December 9, 2003). University of

Virginia, Blacksbourg, Virginia, USA. http://scholar.lib.vt.edu/theses/available/etd-12112003-002453/unrestricted/Dissertation-Qiu.pdf

[Rae 2006] Rae, A., Fidge, C., & Wildman, L. (2006). Fault Evaluation for Security-Critical Communications Devices. *IEEE Computer, 39*(3), 61-68. http://doi.org/10.1109/MC.2006.161

[Rastler 2010] Rastler, D. (2010). *Electricity Energy Storage Technology Options A White Paper Primer on Applications, Costs, and Benefits* [1020676] (Technical Update December 2010). Electric Power Research Institute (EPRI). http://large.stanford.edu/courses/2012/ph240/doshay1/docs/EPRI.pdf

[Rastler 2012] Rastler, D. (2012). Electricity Energy Storage Technology Options: System Cost Benchmarking. *IPHE Workshop Hydrogen - A competitive Energy Storage Medium for large scale integration of renewable electricity*, Seville, Spain, (November 2012). https://www.iphe.net/iphe-workshop

[Ray 2010] Ray, P.D., Harnoor, R., Hentea, M. (2010). Smart power grid security: a unified risk management approach. In D.A. Pritchard, L.D. Sanson (Eds.), *44th Annual 2010 IEEE International Carnahan Conference on Security Technology Proceedings (*pp. 276–285), San Jose, California, USA, (October 2010). https://doi.org/10.1109/CCST.2010.5678681

[Requirements] Wikipedia. (n.d.). *Requirements Analysis.* https://en.wikipedia.org/wiki/Requirements_analysis

[RFC 3198] *IETF Request for Comments (RFC) 3198 Terminology for Policy-Based Management* (November 2001).

[RFC 3444] *IETF Request for Comments (RFC) 3444 On the Difference between Information Models and Data Models* (January 2003).

[RFC 4949] *IETF Request for Comments (RFC) 4949 Internet Security Glossary* Version 2 (August 2007).

[Riahi 2013] [Riahi, A., Challal, Y., Natalizio, E., Chtourou, Z., Bouabdallah, A. (2013). A Systemic Approach for IoT Security. *2013 IEEE International Conference on Distributed Computing in Sensor Systems* (DCoSS) (pp. 351-355), Cambridge, Massachusetts, USA, (May 2013). https://doi.org/10.1109/DCOSS.2013.78

[Rihan 2013] Rihan, M., Ahmad, M., Beg, M.S. (2013). Vulnerability Analysis of Wide Area Measurement System in the Smart Grid. *Smart Grid and Renewable Energy, 4*(6A), 1-7. https://doi.org/10.4236/sgre.2013.46A001

[Roberts 2012] Roberts, B.P., Sandberg, C. (2012). The Role of Energy Storage in Development of Smart Grids. *Proceedings of the IEEE, 99*(6), (pp. 1139-1144). https://doi.org/10.1109/JPROC.2011.2116752

[Rohjans 2012] Rohjans, S., Dänekas, C., Uslar, M. (2012). Requirements for Smart Grid ICT Architectures. *3rd IEEE PES Innovative Smart Grid* http://www.ibm.com/developerworks/library*Technologies (ISGT 2012) Europe Conference* (pp. 1-8), Berlin, Germany, (October 2012).

[Rosenberg 2004] Rosenberg, J., Remy, D. (2004). *Securing Web Services with WS-Security*. SAMS Publishing.

[Ross 2014] Ross, B. (2014). *The Invisible Person .... The Security Architect INFOSECFORCE.* https://www.slideshare.net/infosecforce/the-invisible-person-08-21-2013

[Ross 2018] Ross, R. (2018). Foreward. *NIST Special Publication (SP) 800-160 Systems Security Engineering Considerations for a Multidisciplinary Approach in the Engineering of Trustworthy Secure Systems* Volume 1 (March 2018). https://doi.org/10.6028/NIST.SP.800-160v1

[Roudier 2014] Roudier, Y., Idrees, M.S., Apvrille, L. (2014). Improved Security Requirements Engineering using Knowledge Representation. *SAR-SSI 2014 9ème Conférence sur la Sécurité des Architectures Réseaux et des Systèmes d'Information*, Saint-Germain-Au-Mont-D'Or, France, (May 2014).

[Roy 2010] Roy, S., Ellis, C., Shiva, S., Dasgupta, D., Shandilya, V., Wu, Q. (2010). A Survey of Game Theory as Applied to Network Security. *Proceedings of the 43rd Hawaii International Conference on System Sciences* (pp. 1-10), Honolulu, Hawaii, USA, (January 2010). https://doi.org/10.1109/HICSS.2010.35

[Rozanski 2011] Rozanski, N., Woods, E. (2011). *Software Systems Architecture: Working with Stakeholders Using Viewpoints and Perspectives* (Edition 2). Addison-Wesley.

[Russell 2013] Russell, S. (2013). Distribution Management Systems: Transforming the electric utility industry. *IntelligentUtility 5*(1), 8-11.

[SAF] Ellison, R.J., Woody, C. (2010). *Survivability Analysis Framework* [CMU/SEI-2010-TN-013], (June 2010). Software Engineering Institute. https://resources.sei.cmu.edu/asset_files/TechnicalNote/2010_004_001_15170.pdf

[SAFECode 2008] SAFECode. (2008). *Software Assurance: An Overview of Current Industry Best Practices* (February 2008). Software Assurance Forum for Excellence in Code, Driving Security and Integrity. http://www.safecode.org/publication/SAFECode_BestPractices0208.pdf

[SANDIA 2013] Energy. (2013). *DOE/EPRI 2013 Electricity Storage Handbook in Collaboration with NRECA* (July 2013). Office of Electricity. https://www.energy.gov/oe/downloads/doeepri-2013-electricity-storage-handbook-collaboration-nreca-july-2013

[SANS 2016] Spitzner, L. (2016, March 8). Defining the Security Awareness Maturity Model [Blog]. *SANS.* https://www.sans.org/security-awareness-training/blog/defining-security-awareness-maturity-model

[Sanz 2003] Sanz, R., Arzen, K.E. (2003). Trends in Software and Control. *IEEE Control Systems Magazine, 23*(3), 12-15. https://doi.org/10.1109/MCS.2003.1200238

[Sayama 2010] Sayama, H.D.Sc. (2010). *Collective Dynamics of Complex Systems (CoCo)* (November 2010). Research Group at Binghamton University, New York, New York, USA. http://en.wikipedia.org/wiki/File:Complex_systems_organizational_map.jpg

[Saydjari 2002] Saydjari, O.S. (2002). Defending Cyberspace. *IEEE Computer, 35*(12), 125-127. https://doi.org/10.1109/MC.2002.1106187

[SCADA Steps] Energy. (2007). *21 Steps to Improve Cyber Security of SCADA Networks.* U.S. Department of Energy. https://energy.gov/sites/prod/files/oeprod/DocumentsandMedia/21_Steps_-_SCADA.pdf

[SCALe] Seacord, R.C., Dormann, W., McCurley, J., Miller, P., Stoddard, R., Svoboda, D., Welch, J. [CMU/SEI-2012-TN-013], (April 2012). Source Code Analysis Laboratory (SCALe). Software Engineering Institute. https://resources.sei.cmu.edu/asset_files/TechnicalNote/2012_004_001_15440.pdf

[Scarbo 2013] Scarbo, R. (2013). *Systems Assurance within the Systems Engineering Lifecycle Key Lessons and Benefits* (12th February 2013). ATKINS. https://www.engineersaustralia.org.au.

[Schneier 2002] Schneier, B. (2002). We don't spend enough (on security). *Workshop on the Economics of Information Security (WEIS 2002)*, Berkeley, California, USA, (May 2002).

[Schneier 2006] Schneier, B. (2006). *Beyond Fear*. Springer.

[Schneier 2007] Schneier, B. (2007). The Psychology of Security... a work in progress. *Workshop on Information Security Economics (WEIS 2007),* DIMACS Center, Rutgers, New Jersey, USA, (January 2007).

[Schultz 2007] Schultz, M. (2007). *Architecture principles: Creating the foundation for robust architecture.* http://www.ibm.com/developerworks/library

[Schütz 2021] Schütz, J, Uslar, M., Meister, J. (2021, April). *A case study research on interoperability improvement in Smart Grids: state-of-the-art and further opportunities.* Open Research Europe. https://doi.org/10.12688/openreseurope.13313.1

[Seacord 2010] Seacord, R.C., Dormann, W., McCurley, J., Miller, P., Stoddard, R., Svoboda, D., Welch, J. (2010). *Source Code Analysis Laboratory (SCALe) for Energy Delivery Systems* [CMU/SEI-2010-TR-021], (December 2010). Software Engineering Institute. https://resources.sei.cmu.edu/library/asset-view.cfm?assetid=9561

[Seacord 2014] [Seacord, R.C. (2014*). CERT® C Coding Standard The: 98 Rules for Developing Safe, Reliable, and Secure Systems* (Second Edition). Addison-Wesley Professional.

[SEbook 2010] INCOSE. (2010). *Systems Engineering Handbook Version 3.2* (January 2010). http://www.incose.org

[Secfocus] Securityfocus. (n.d.). *Vulnerabilities.* http://www.securityfocus.com/

[SecReq] Mead, N. (2010). *Security Requirements Engineering.* https://www.us-cert.gov/bsi/articles/best-practices/requirements-engineering/security-requirements-engineering

[Security] Techtarget. (n.d.). *Security by design.* WhatIs.com.

[SEcycle] MITRE. (n.d.). *Systems Engineering Guide SE life cycle building blocks.* http://www.mitre.org/work/systems_engineering/guide/se_lifecycle_building_blocks/

[SGIP] SGIP. (n.d.). Smart Grid Interoperability. http://www.sgip.org/Interoperability-and-the-GWAC-Stack

[SGIP CoS] NIST. (n.d.). *SGIP Catalog of Standards.* https://www.nist.gov/programs-projects/smart-grid-national-coordination/catalog-standards

[Shannon 2015] Shannon, G. (2015). *Five Myths (Debunked) About Security and Privacy for Internet of Things* [Opinion] (January 26, 2015). CSO. https://www.csoonline.com/article/2872360/five-myths-debunked-about-security-and-privacy-for-internet-of-things.html

[Shaw 2001] Shaw, A.C. (2001). *Real-Time Systems and Software.* Wiley.

[Sheldon 2006] Sheldon, F., Krings, A., Yoo, S-M., Mili, A., Trien, J. (2006). Beyond the Maginot Line. *Proceedings Second Annual Cyber Security and Information Infrastructure Research Workshop*, Oak Ridge, Tennessee, USA, (May 2006).

[Sherwood 2009] Sherwood, J., Clark, A., Lynas, D. (2009). Enterprise Security Architecture [White Paper]. *SABSA White Paper*, 6, 43-54.

[Shin 2010] Shin, D-H. (2010). The Effects of Trust, Security and Privacy in Social Networking: A Security-Based Approach to Understand the Pattern of Adoption. *Interacting with Computers, 22*(5), 428-438. Oxford University Press. https://doi.org/10.1016/j.intcom.2010.05.001

[Shostack 2014] Shostack, A. (2014). *Threat Modeling: Designing for Security.* Wiley.

[Simpson 2011a] Simpson, S. (2011a). *Software Assurance Forum for Excellence in Code* (March 1, 2011). https://buildsecurityin.us-cert.gov

[Simpson 2011b] Simpson, S. (Ed.), (2011b). *Fundamental Practices for Secure Software Development 2ND EDITION A Guide to the Most Effective Secure Development Practices in Use Today* (February 8, 2011). https://safecode.org/fundamental-practices-for-secure-software-development-2nd-edition/

[Simpson 2014] Simpson, S. (2014). SAFEcode Whitepaper: Fundamental Practices for Secure Software Development 2nd Edition. In H. Reimer, N. Pohlman, W. Schneider (Eds.), *ISSE 2014 Securing Electronic Business Processes.* Springer. https://doi.org/10.1007/978-3-658-06708-3_1

[Slankas 2013] Slankas, J., Riaz, M., King, J.T., Williams, L.A. (2013). Discovering Security Requirements from Natural Language Project Artifacts [TR-2013-9].

[Sledge 2010] Sledge, C.A. (2010). Reports from the Field on System of Systems Interoperability Challenges and Promising Approaches [CMU/SEI-2010-TR-013 ESC-TR-2010-013] (March 2010). Software Engineering Institute.

[Sollecito 2009] Sollecito, L. (2009). Smart grid: the road ahead. *Protection & Control Journal, 8th Edition*, 15–19.

[Sommer 2011] Sommer, P., Brown, I. (2011). *Reducing systemic cybersecurity risk.* Organisation for Economic Co-operation and Development. http://www.oecd.org/dataoecd/57/44/46889922.pdf

[Sommerville 2007] Sommerville, I. (2007). *Software Engineering* (8th Edition). Addison-Wesley.

[Specht 2013] Specht, M., Rohjans, S. (2013). ICT and Energy Supply: IEC 61970/61968 Common Information Model. In M. Uslar, M. Specht, C. Dänekas, J. Trefke, S. Rohjans, J.M. Gonzalez, C. Rosinger, R. Bleiker (Eds.), *Standardization in Smart Grids Introduction to IT-Related Methodologies, Architectures and Standards* (pp. 99-114). Springer. https://doi.org/10.1007/978-3-642-34916-4_6

[Steeley 2011] Steeley, W. (2011). *Functional Requirements for Electric Energy Storage Applications on the Power System Grid: What Storage Has to Do to Make Sense* [Report No. 1021936] (Technical Update, December 2011). Electric Power Research Institute **(**EPRI**).**

[Steven 2006] Steven, J. (2006). Adopting an Enterprise Software Security Framework. *IEEE Security & Privacy, 4* (2), 84-87. https://doi.org/10.1109/MSP.2006.33

[Stoneburner 2006] Stoneburner, G. (2006). Toward A Unified Security/Safety Model. *IEEE Computer, 39*(8), 96-97. https://doi.org/10.1109/MC.2006.283

[Surakhi 2017] Surakhi, M., Hudaib, O., AlShraideh, A., Khanafseh, M. (2017). A Survey on Design Methods for Secure Software Development. *INTERNATIONAL JOURNAL OF COMPUTERS &Amp TECHNOLOGY, 16*(7), 7047–7064. https://doi.org/10.24297/ijct.v16i7.6467

[Swire 2007] Swire, S. (2007). Security Through Obscurity: When It Works, When It Doesn't. *DIMACS Workshop on Information Security Economics*, Rutgers, New Jersey, USA. (January 2007).

[Tanenbaum 2006] Tanenbaum, A.S., Herder, J.N., Bos, H. (2006). Can We Make Operating Systems Reliable and Secure? *IEEE Computer, 39*(5), 44-51. https://doi.org/10.1109/MC.2006.156

[TechTarget] TechTarget. (n.d.). *Enterprise Architecture.* http://searchcio.techtarget.com/definition/enterprise-architecture

[Telang 2005] Telang, R., Wattal, S. (2005). Impact of Software Vulnerability Announcements on the Market Value of Software Vendors – an Empirical Investigation [Research Paper], (February 2005). *SSRN.* Carnegie Mellon University, Pittsburg, Pennsylvania, USA. http://dx.doi.org/10.2139/ssrn.677427

[Telang 2007] Telang, R., Wattal, S. (2007). An Empirical Analysis of the Impact of Software Vulnerability Announcements on Firm Stock Price. *IEEE Transactions on Software Engineering, 33*(8), 544-557. https://doi.org/10.1109/TSE.2007.70712

[Thomas 2013] Thomas, R.C., Antkiewicz, M., Florer, P., Widup, S., Woodyrd, M. (2013). How Bad Is It? – A Branching Activity Model to Estimate the Impact of Information Security Breaches. *Proceedings of 12th Workshop on the Economics of Information Security (WEIS 2013),* Washington, D.C., USA, (June 2013).

[Thompson 2005] Thompson, H.H., Chase, S.G. (2005). *The Software Vulnerability Guide.* Charles River Media.

[TOGAF] Opengroup. (2011). *The TOGAF® Standard, Version 9.2.* https://www.opengroup.org/togaf/

[TOGAF3] Mitre. (2004). *EABOK Guide to the (Evolving) Enterprise Architecture Body of Knowledge* [A Project of MITRE Corporation], (6 February 2004). http://www2.mitre.org/public/eabok/pdf/glossary.pdf

[Tollefson 2013a] Tollefson, J. (2013a). *US electrical grid on the edge of failure* [News] (August 25, 2013). Nature International weekly journal of science. http://www.nature.com/news/us-electrical-grid-on-the-edge-of-failure-1.13598

[Tollefson 2013b] Tollefson, J. (2013b). US electrical grid on the edge of failure (August 26, 2013). *Scientific American Sustainability.* http://www.scientificamerican.com/article/us-electrical-grid-on-failure/

[Tondel 2008] Tondel, I.A., Jaatun, M.G., Meland, P.H. (2008). Security Requirements for the Rest of Us: A Survey. *IEEE Software, 25*(1), 20-27. https://doi.org/10.1109/MS.2008.19

[Trefke 2012] Trefke, J., Dänekas, C., Rohjans, S., González Vázquez, J.M., (2012). Adaptive Architecture Development for Smart Grids Based on Integrated Building Blocks. *2012 3rd IEEE PES Innovative Smart Grid Technologies (ISGT) Europe Conference* (pp. 1-8), Berlin, Germany, (October 2012). https://doi.org/ 10.1109/ISGTEurope.2012.6465697

[Trefke 2013] Trefke, J., Dänekas, C. (2013). Development of Smart Grid Architectures. In M. Uslar, M. Specht, C. Dänekas, J. Trefke, S. Rohjans, J.M. Gonzalez, C. Rosinger, R. Bleiker (Eds.), *Standardization in Smart Grids Introduction to IT-Related Methodologies, Architectures and Standards* (pp. 59-77). Springer. https://doi.org/10.1007/978-3-642-34916-4_4

[Tzanev 2013] Tzanev, A. (2013). Modeling and Simulation of Systems of Systems – a Survey. *Cybernetics and Information Technologies, 13*(2), 3–36.

[UCAIUG] UCAIUG. (n.d.). *Shared Documents.* OpenSG Users Group. http://osgug.ucaiug.org/Shared%20Documents/Forms/AllItems.aspx

[UCAIUG 2008] UCAIUG. (2008). *AMI System Requirements V1.01* [UCAIUG: AMI-SEC-ASAP], (December 2008). http://energy.gov/sites/prod/files/oeprod/DocumentsandMedia/14-AMI_System_Security_Requirements_updated.pdf

[UCAIUG 2012] UCAIUG. (2012). *Security Profile for Advanced Metering Infrastructure Version 2.1* (October 23, 2012). AMI-SEC Task Force (UCAIug).

[UTC 2015] Bartol, N. (2015). *Cyber Supply Chain Risk Management for Utilities – Roadmap for Implementation.* Utilities Telecom Council. https://utc.org/wp-content/uploads/2018/02/SupplyChain2015-2.pdf

[Uslar 2012] Uslar, M., Specht, M., Rohjans, S., Trefke, J., González, J.M. (2012). *The Common Information Model CIM IEC 61968/61970 and 62325 - A Practical Introduction to the CIM.* Springer.

[Uslar 2013] Uslar, M., Bleiker, R. (2013). Automation for the Smart Grid: IEC 61850 – Substation Automation and DER Communication. In M. Uslar, M. Specht, C. Dänekas, J. Trefke, S. Rohjans, J.M. Gonzalez, C. Rosinger, R. Bleiker (Eds.), *Standardization in Smart Grids Introduction to IT-Related Methodologies, Architectures and Standards* (pp. 115-128). Springer. https://doi.org/10.1007/978-3-642-34916-4_7

[Uslar 2014] Uslar, M., Rosinger, C., Schlegel, S. (2014). Security by Design for the Smart Grid: Combining the SGAM and NISTIR 7628. *2014 IEEE 38th International Computer Software and Applications Conference Workshops*, 110-115. https://doi.org/10.1109/COMPSACW.2014.23

[Uzunov 2012a] Uzunov, A.V., Fernandez, E.B., Falkner, K. (2012a). Engineering Security into Distributed Systems: A Survey of Methodologies. *Journal of Universal Computer Science, 18(*20). 2920-3006. https://doi.org/10.3217/jucs-018-20-2920.

[Uzunov 2012b] Uzunov, A.V., Fernandez, E.B., Falkner, K. (2012b). Securing Distributed Systems using Patterns: A Survey. *Computers & Security, 31*(5), 681-703. https://doi.org/10.1016/j.cose.2012.04.005

[Uzunov 2014a] Uzunov, A.V., Fernandez, E.B. (2014a). An extensible pattern-based library and taxonomy of security threats for distributed systems. Computer Standards & Interfaces, 36(4), 734-747. https://doi.org/10.1016/j.csi.2013.12.008

[Uzunov 2014b] Uzunov, A.V., Fernandez, E.B., Falkner, K. (2014b). A Comprehensive Pattern-Driven Security Methodology for Distributed Systems. *Proceedings of 2014 23rd Australian Software Engineering Conference (ASWEC)* (pp. 142-151), Milsons Point, NSW, Australia, (April 2014). https://doi.org/10.1109/ASWEC.2014.14

[Vacca 2012] Vacca, J.R. (2012). *Computer and Information Security Handbook* (2nd Edition). Morgan Kaufmann.

[Van Haren 2013] Vanharen. (2013, August 22). SABSA – in 3 Minutes [Blog]. *Van Haren Publishing.* https://www.vanharen.net/blog/sabsa-in-3-minutes/

[V-Diagrams] Theprojectmanagementblueprint. (2019, October 28). Using V-Diagrams in Engineering Projects [Blog]. *The Project Management Blueprint.* https://www.theprojectmanagementblueprint.com/blog/procurements-management/the-v-diagram

[Vojdani 2006] Vojdani, A., Neumann, S., Yee, G. (2006). How to Get More Response from Demand Response. *the Electricity Journal, 19*(8), 24-31.

[Vos 2009] Vos, A. (2009). Demand Response Management Systems: The Next Wave of Smart Grid Software [News]. *EE ONLINE.* http://www.electricenergyonline.com/show_article.php?article=444

[Warner 2019] Warner, M.H. (2019, October 28). Using V-Diagrams in Engineering Projects [Blog]. *The Project Management Blueprint.* https://www.theprojectmanagementblueprint.com/blog/procurements-management/the-v-diagram

[Wasson 2015] Wasson, C.S. (2015). *System Analysis, Design, and Development: Concepts, Principles, and Practices* (2nd Edition). Wiley.

[WH 2011] Whitehouse. (2011). *A POLICY FRAMEWORK FOR THE 21st CENTURY GRID: Enabling Our Secure Energy Future* (June 13 2011). https://obamawhitehouse.archives.gov/sites/default/files/microsites/ostp/nstc-

[Whitman 2014] Whitman, M.E., Mattord, H.J. (2014). *Management of Information Security* (Fourth edition). Cengage Learning.

[Wiki CompSEC] Wikipedia. (n.d.). *Computer Security.* https://en.wikipedia.org/wiki/Computer_security

[Wiki EA] Wikipedia. (n.d.). *Enterprise Architecture Framework.* https://en.wikipedia.org/wiki/Enterprise_architecture_framework

[Wiki IA] Wikipedia. (n.d.). *Information Architecture.* https://en.wikipedia.org/wiki/Information_architecture

[Wiki Infmodel] Wikipedia. (n.d). *Information Model.* https://en.wikipedia.org/wiki/Information_model

[Wiki SoS] Wikipedia. (n.d.). System of Systems. https://en.wikipedia.org/wiki/System_of_systems

[Wildberger 1999] Wildberger, A.M. (1999). AI & Simulation. *SIMULATION, 73*(1), 55-56.

[Wu 2007] Wu, D. (2007). *Security Functional Requirements Analysis for Developing Secure Software* [PhD Dissertation UMI Number: 3262759]. University of Southern California, USA.

[Yahya 2019] Yahya, S., Kamalrudin, M., Sidek, S., Jaimin, M., Yusuf, J., Hua, A.K., Gani, P. (2019). A Review Paper: Security Requirement Patterns for a Secure Software Development. *2019 1st International Conference on Artificial Intelligence and Data Sciences (AiDAS)*, (pp. 146-151). Ipoh, Perak, Malaysia, (September 2019). https://doi.org/10.1109/AiDAS47888.2019.8971014

[Yang 2012] Yang, H-S, Kim, S-S, Jang, H-S. (2012). Optimized Security Algorithm for IEC 61850 based Power Utility System. *Journal of Electrical Engineering and Technology 7*(3), 443-450. https://doi.org/10.5370/jeet.2012.7.3.443

[Yoshioka 2008] Yoshioka, N., Washizaki, H., Maruyama, K. (2008). A survey of Security Patterns. *Progress in Informatics, 5*(5), 35-47. https://doi.org/10.2201/NiiPi.2008.5.5

[Zafar 2013] Zafar, N., Arnautovic, E., Diabat, A., Svetinovic, D. (2013). *System Security Requirements Analysis: A Smart Grid Case Study* (July 2013). Wiley Online Library. wileyonlinelibrary.com

References Part 4

[Ali 2015] Ali, S., Qaisar, S.B., Saeed, H., Khan, M.F., Naeem, M., Anpalagan, A. (2015). Network Challenges for Cyber Physical Systems with Tiny Wireless Devices: A Case Study on Reliable Pipeline Condition Monitoring. *Sensors 15*(4), 7172–7205. CC BY-SA 4.0. https://doi.org/10.3390/s150407172

[Balaguer 2008] Balaguer, I.J., Supatti, U., Lei, Q., Choi, N., Peng, F.Z. (2008). Intelligent control for intentional islanding operation of microgrids. *2008 IEEE International Conference on Sustainable Energy Technologies* (pp. 898-903), Singapore, (November 2008). https://doi.org/10.1109/ICSET.2008.4747135

[Best 2011] Best, R.J., Morrow, D.J., Laverty, D.M., Crossley, P.A. (2011). Techniques for Multiple-Set Synchronous Islanding Control. IEEE Transactions on Smart Grid, 2(4), 60–67. https://doi.org/10.1109/TSG.2010.2100833

[BS 7799] *British Standard 7799 Information Security Management* (1995). British Standards Institute (BSI), United Kingdom.

[Carullo 2019] Carullo, M. (2019, January 29). IEC 62351 Standards for Securing Power System Communications [Blog]. *NOZOMI Networks.* https://www.nozominetworks.com/blog/iec-62351-standards-for-securing-power-system-communications/

[CEC 2014] Energy. (2014). *Best Practices for Handling Smart Grid Cyber Security* [CEC-500-2014-049], (May 2014). California Energy Commission.

[CENELEC 2014] CENCENELEC. (2014*). SG-CG/M490/H_ Smart Grid Information Security* [M490-SGCG-SGIS-Intermediate-Report-V1], (December 2014). CEN/CENELEC/ETSI Smart Grid Coordination Group. https://www.cencenelec.eu/standards/Sectorsold/SustainableEnergy/SmartGrids/Pages/default.aspx

[CIS] Cisecurity. (n.d.). Cybersecurity Best Practices CIS Controls™ and CIS Benchmarks™ are global industry best practices endorsed by leading IT security vendors and governing bodies. Center for Internet Security. https://www.cisecurity.org/cybersecurity-best-practices/

[Cleveland 2012] Cleveland, F. (2012). *IEC TC57 WG15: IEC 62351 Security Standards for the Power System Infrastructure* [White Paper]. http://iectc57.ucaiug.org/wg15public/Public%20Documents/White%20Paper%20on%20Security%20Standards%20in%20IEC%20TC57.pdf

[CNSSI 4009] CNSSI. (2015). *Committee on National Security Systems (CNSS) Glossary* [CNSSI No. 4009].

[CSRC 2015] NIST. (2015). *Glossary. Trusted Recovery*. Computer Security Resource Center. https://csrc.nist.gov/Glossary/?term=2279

[DHS 2016] DHS. (2016). *Recommended Practice: Improving Industrial Control Systems Cybersecurity with Defense-In-Depth Strategies Industrial Control Systems Cyber Emergency Response Team* (September 2016). US Department of Homeland Security. Control Systems Security Program. National Cyber Security Division. https://us-cert.cisa.gov/sites/default/files/recommended_practices/NCCIC_ICS-CERT_Defense_in_Depth_2016_S508C.pdf

[DHS CAT] HSDL. (n.d.). *Catalog of Control Systems Security: Recommendation for Standards Developers* (June 2010). Control Systems Security Program. National Cyber Security Division. Homeland Security Digital Library. https://www.hsdl.org/?abstract&did=8002

[Dua 2011] [Dua, S., Du, X. (2011). *Data Mining and Machine Learning Techniques.* CRC Press. Taylor & Francis Group.

[EISA 2007] EISA. (2007). *U.S. Energy Independence and Security Act of 2007 Public Law No: 110-140 Title XIII, Sec. 1301.* http://www.gpo.gov/fdsys/pkg/BILLS-110hr6enr/pdf/BILLS-110hr6enr.pdf

[ELCON 2004] ELCON. (2004). *The Economic Impacts of the August 2003 Blackout* (February 9, 2004]. Electricity Consumers Resource Council (ELCON). https://elcon.org/wp-content/uploads/Economic20Impacts20of20August20200320Blackout1.pdf

[ENISA 2017] [ENISA. (2017). *Gaps in NIS Standardization - Recommendations for improving NIS in EU standardization policy V. 1.0* (March 15, 2017). European Union Agency for Security. https://www.enisa.europa.eu/publications/gaps-eu-standardisation

[Eshghi 2015] Eshghi, K., Johnson, B.K., Rieger, C.G. (2015). Power System Protection and Resilient Metrics [INL/CON-15-35475]. *2015 Resilience Week (RWS)*, (pp. 1-8), Philadelphia, Pennsylvania, USA, (August 2015). https: https://doi.org/10.1109/RWEEK.2015.7287448 https://www.osti.gov/servlets/purl/1357912

[Friedberg 2017] Friedberg, I., McLaughlin, K., & Smith, P. (2017). A Cyber-Physical Resilience Metric for Smart Grids. *Proceedings IEEE Power & Energy Society Innovative Smart Grid Technologies Conference (ISGT) 2017*, Washington D.C., USA, (April 2017). https://doi.org/10.1109/ISGT.2017.8086065

[Gauci 2017] Gauci, A., Michelin, S., Salles, M. (2017). Addressing the Challenge of Cyber Security Maintenance Through Patch Management. *24th International Conference & Exhibition on Electricity Distribution (CIRED)*, Glasgow, Scotland, (June 2017). *IET Journals, Open Access Proc. J. 2017* (1), 2599–2601. https://doi.org/10.1049/oap-cired.2017.025

[GE 2016] GE. (2016). *IEC 62443-2-4 Cyber Security Capabilities* [Whitepaper], [GEA32435A], (March 2016). GE OIL & Gas. https://fdocuments.in/document/iec-62443-2-4-cyber-security-capabilities-nbsppdf-filegea32435a-iec.html

[Gligor 1991] Gligor, V.D. (1991). *Guide to Understanding Trusted Recovery in Trusted Systems* (June 1, 1991). Boston Public Library.

[GridWise 2008] GridWise. (2008). *Interoperability context- setting framework v1.1* (March 2008). GridWise Architecture Council. https://www.gridwiseac.org/pdfs/interopframework_v1_1.pdf

[Hansche 2004] Hansche, S., Berti, J., Hare, C. (2004). *Official (ISC)2 Guide to The CISSP Exam*. Auerbach Publications.

[Harris 2013] Harris, S. (2013). *All in One CISSP Exam Guide* (6th Edition). McGraw Hill Education.

[ICSA-16-147-01b] ICS-CERT. (2016). *Environmental Systems Corporation Data Controllers Vulnerabilities (Update B)* [Advisory (ICSA-16-147-01B)] (December 2016). https://ics-cert.us-cert.gov/advisories/ICSA-16-147-01

[ICS TIPS] ICS-CERT. (2013). *Targeted Cyber Intrusion Detection and Mitigation Strategies (Update B)* (February 2013). https://ics-cert.us-cert.gov/tips/ICS-TIP-12-146-01B

[ICS-CERTRP] ICS-CERT. (n.d.). *Recommended Practices*. http://ics-cert.us-cert.gov/content/recommended-practices

[ICS-CERT 2014] CISA. (2014, April). ICS-CERT Monitor January-April. https://www.cisa.gov/uscert/sites/default/files/Monitors/ICS-CERT_Monitor_Jan-April2014.pdf

[IEC 62056] *IEC 62056 Electricity metering data exchange - The DLMS/COSEM suite.*

[IEC 62351] *IEC 62351:2020 SER Series Power systems management and associated information exchange – Data and communications security –- ALL PARTS.*

[IEC 62351-8] *IEC 62351-8:2020 Power systems management and associated information exchange - Data and communications security - Part 8: Role-based access control for power system management.*

[IEC 62351-9] *IEC 62351-9:2017 Power systems management and associated information exchange - Data and communications security - Part 9: Cyber security key management for power system equipment.*

[IEC 62351-11] *IEC 62351-11:2016 Power systems management and associated information exchange - Data and communications security - Part 11: Security for XML documents.*

[IEC 62443] *IEC 62443 Industrial communication networks - Network and system security, series of several parts.*

[IEC 62443-2-4] *IEC 62443-2-4:2015 Security for industrial automation and control systems - Part 2-4: Security program requirements for IACS service providers.*

[IEC Glossary] IEC. (n.d.). *International Electrotechnical Commission Glossary.* http://std.iec.ch/terms/terms.nsf/ByPub?OpenView&Count=-1&RestrictToCategory=IEC%2062351-2

[IEC TR 62051-1] *IEC TR 62051-1:2004 Electricity metering - Data exchange for meter reading, tariff and load control - Glossary of terms - Part 1: Terms related to data exchange with metering equipment using DLMS/COSEM.*

[IEC TR 62351-12] *IEC TR 62351-12:2016 Power systems management and associated information exchange - Data and communications security - Part 12: Resilience and security recommendations for power systems with distributed energy resources (DER) cyber-physical systems.*

[IEC TR 62351-90-1] *IEC TR 62351-90-1:2018 Power systems management and associated information exchange - Data and communications security - Part 90-1: Guidelines for handling role-based access control in power systems.*

[IEC TR 62357] *IEC TR 62357-1: 2016 Edition 2.0, Power systems management and associated information exchange - Part 1: Reference architecture.*

[IEC TS 62351-1] *IEC TS 62351-1:2007 Power systems management and associated information exchange – Data and communications security Part 1: Communication network and system security – Introduction to security issues Edition 1.0 May 2007.*

[IEC TS 62351-6] *IEC TS 62351-6:2007 Power systems management and associated information exchange - Data and communications security - Part 6: Security for IEC 61850.*

[IEEE 1686] *IEEE Std 1686-2007 IEEE Standard for Intelligent Electronic Devices Cyber Security Capabilities.*

[IEEE C37.240] *IEEE Std C37.240-2014 - IEEE Standard Cybersecurity Requirements for Substation Automation, Protection, and Control Systems.*

[INL 2007] ICS-CERT. (2007). *Using Operational Security (OPSEC) to Support a Cyber Security Culture in Control Systems Environments Version 1.0 Draft Recommended Practice* (February 2007). Idaho National Laboratory. https://ics-cert.us-cert.gov/sites/default/files/recommended_practices/RP_Using%20OpSec_v1_Draft.pdf

[ISAGCA 2020] Isasecure. (2020, October). *Security Lifecycles in the ISA/IEC 62443 Series Security of Industrial Automation and Control Systems.* https://www.isasecure.org/en-US/Documents/Articles-and-Technical-Papers/ISAGCA-Security-Lifecycles-whitepaper

[IPRM] NIST. (2012). SGTCC Interoperability Process Reference Manual (IPRM) Version 2.0 (January 2012). Smart Grid Interoperability Panel. Smart Grid Testing & Certification Committee (SGTCC).

[ISO 14001] *ISO 14001:2015 Environmental Management Systems.*

[ISO 15118] *ISO 15118 Road vehicles -- Vehicle to grid communication interface.*

[ISO 15118-1] *ISO 15118-1:2013 Road vehicles -- Vehicle to grid communication interface -- Part 1: General information and use-case definition.*

[ISO/IEC 15408] *ISO/IEC 15408 Information technology -- Security techniques -- Evaluation criteria for IT security.*

[ISO/IEC 18045] *ISO/IEC CD 18045:2008 Information technology -- Security techniques -- Methodology for IT security evaluation.*

[ISO/IEC 19790] *ISO/IEC 19790:2012 Information technology -- Security techniques -- Security requirements for cryptographic modules, corrected 2015.*

[ISO/IEC 27000] *ISO/IEC 27000:2018 Information technology — Security techniques — Information security management systems - Overview and vocabulary* (Fifth edition).

[ISO/IEC 27001] *ISO/IEC 27001:2013 Information technology — Security techniques — Information security management systems — Requirements.*

[ISO/IEC 27002] *ISO/IEC 27002:2022 Information Technology — Security Techniques — Code of Practice for Information Security Controls.*

[ISO/IEC 27003] *ISO/IEC 27003:2017 Information technology — Security techniques — Information security management systems — Guidance.*

[ISO/IEC 27010] *ISO/IEC 27010:2015 Information technology — Security techniques — Information security management for inter-sector and inter-organizational communications.*

[ISO/IEC 27011] *ISO/IEC 27011:2016 Information technology — Security techniques — Code of practice for Information security controls based on ISO/IEC 27002 for telecommunications organizations.*

[ISO/IEC 27019] *ISO/IEC 27019:2017 Information technology - Security techniques- Information security controls for the energy utility industry.*

[Krutz 2004] Krutz, R.L., Vines, R.D. (2004). *The CISSP Prep Guide* (Second Edition). Wiley.

[Lynn 2014] Lynn, S. (2014, March 27). *RAID Levels Explained* [News]. PC Magazine.

[Mendel 2017] Mendel, J. (2017). Smart Grid Cyber Security Challenges: Overview and Classification. *e-mentor 1*(68), 55-66. http://www.e-mentor.edu.pl/artykul/index/numer/68/id/1282

[NERC CIP] NERC. (n.d.). *CIP Standards.* North American Electric Reliability Corporation. http://www.nerc.com/pa/Stand/Pages/CIPStandards.aspx

[NETL 2007] NETL. (2007). *IMPROVED INTERFACES AND DECISION SUPPORT Version 2.0 Appendix B5: A Systems View of the Modern Grid* (March 2007). Conducted by the National Energy Technology Laboratory for the U.S. Department of Energy Office of Electricity Delivery and Energy Reliability. https://netl.doe.gov/sites/default/files/Smartgrid/Improved-Interfaces-and-Decision-Support_FINAL_v2-0.

[NIST 2014] *NIST Framework for Improving Critical Infrastructure Cybersecurity Version 1.0* (February 2014). https://www.nist.gov/system/files/documents/cyberframework/cybersecurity-framework-021214.pdf

[NIST 2018] *NIST Framework for Improving Critical Infrastructure Cybersecurity Version 1.1* (April 2018). https://nvlpubs.nist.gov/nistpubs/CSWP/NIST.CSWP.04162018.pdf

[NISTIR 7628] *NIST Interagency Report (NISTIR) 7628* Guidelines *for Smart Grid Cyber Security* (August 2010).

[NISTIR 7628r1] *NIST Interagency Report (NISTIR) 7628r1 Guidelines for Smart Grid Cybersecurity Revision 1* (September 2014). http://dx.doi.org/10.6028/NIST.IR.7628r1

[NIST SP1108r3] *NIST Special Publication (SP) 1108r3 NIST Framework and Roadmap for Smart Grid Interoperability Standards* (Release 3.0, September 2014). http://dx.doi.org/10.6028/NIST.SP.1108r3

[NIST SP1108r4] *NIST Special Publication (SP) 1108r4 NIST Framework and Roadmap for SmartGrid Interoperability Standards Release 4.0* (February 2021). https://dx.doi.org/10.6028/NIST.SP.1108r4

[NIST SP800-12r1] *NIST Special Publication (SP) 800-12 Revision 1 An Introduction to Information Security* (June 2017). *https://doi.org/10.6028/NIST.SP.800-12r1*

[NIST SP800-18r1] *NIST Special Publication (SP) 800-18 Revision 1 Guide for Developing Security Plans for Federal Information Systems* (February 2001).

[NIST SP800-30r1] *NIST Special Publication (SP) 800-30 Revision 1 Guide for Conducting Risk Assessments* (September 2012).

[NIST SP800-37r2] *NIST Special Publication (SP) 800-37 Revision 2 Risk Management Framework for Information Systems and Organizations A System Life Cycle Approach for Security and Privacy* (December 2018). https://doi.org/10.6028/NIST.SP.800-37r2

[NIST SP800-39] *NIST Special Publication (SP) 800-39 Managing Information Security Risk Organization, Mission, and Information System View* (March 2011).

[NIST SP800-53r5] *NIST Special Publication (SP) 800-53 Revision 5 Security and Privacy Controls for Information Systems and Organizations* (December 2020). https://doi.org/10.6028/NIST.SP.800-53r5

[NIST SP800-55r1] *NIST Special Publication (SP) 800-55 Revision 1 Performance Measurement Guide for Information Security* (July 2008).

[NIST SP800-57P1r5] *NIST Special Publication (SP) 800-57 Part 1 Revision 5, Recommendation for Key Management Part 1: General* (May 2020). https://doi.org/10.6028/NIST.SP.800-57pt1r5

[NIST SP800-57P3r1] *NIST Special Publication (SP) 800-57 Part 3 Revision 1, Recommendation for Key Management Part 3: Application-Specific Key Management Guidance* (January 2015). http://dx.doi.org/10.6028/NIST.SP.800-57pt3r1

[NIST SP800-150] *NIST Special Publication (SP) 800-150 Guide to Cyber Threat Information Sharing* (December 2016). https://doi.org/10.6028/NIST.SP.800-150

[NIST SP800-160v1] *NIST Special Publication (SP) 800-160 Systems Security Engineering Considerations for a Multidisciplinary Approach in the Engineering of Trustworthy Secure Systems* Volume 1 (March 2018). https://doi.org/10.6028/NIST.SP.800-160v1

[Otuoze 2018] Otuoze, A.O., Mustafa, M.W., Larik, R.M. (2018). Smart grids security challenges: Classification by sources of threats. *Journal of Electrical Systems and Information Technology, 5*(3), 468-483. https://doi.org/10.1016/j.jesit.2018.01.001

[Piètre-Cambacédès] Piètre-Cambacédès, L., Tritschler, M., Ericsson, G.N. (2011). Cybersecurity Myths on Power Control Systems: 21 Misconceptions and False Beliefs. *IEEE Transactions on Power Delivery, 26*(1), 161-172.

[Pohlman 2002] Pohlman, N., Crothers, T. (2002). *Firewall Architecture for the Enterprise*. Wiley.

[Poletto 2014] Poletto, M. (2014, 28 October). *Data mining for security at Google*. Google Security Team. Stanford CS259D. https://web.stanford.edu/class/cs259d/lectures/Session11.pdf

[Rainbow Series] FAS. (1991). *A Guide to Understanding Trusted Recovery in Trusted Systems*. The Rainbow Books. https://fas.org/irp/nsa/rainbow/tg022.htm

[RFC 3547] *IETF Request for Comments (RFC) 3547 The Group Domain of Interpretation* (July 2003).

[RFC 4096] *IETF Request for Comments (RFC) 4096 Policy-Mandated Labels Such as "Adv:" in Email Subject Headers Considered Ineffective At Best* (May 2005).

[RFC 4949] *IETF Request for Comments (RFC) 4949 Internet Security Glossary Version 2 (*August 2007).

[RFC 6407] *IETF Request for Comments (RFC) 6407 The Group Domain of Interpretation (GDOI)* (October 2011).

[RFC 6960] *IETF Request for Comments (RFC) 6960 X.509 Internet Public Key Infrastructure Online Certificate Status Protocol – OCSP* (June 2013).

[RFC 7030] *IETF Request for Comments (RFC) 7030 Enrollment over Secure Transport* (October 2013).

[RFC 7252] *IETF Request for Comments (RFC) 7252 The Constrained Application Protocol (CoAP)* (June 2014).

[Rieger 2009] Rieger, C.G., Gertman, D.I., McQueen, M.A. (2009). Resilient Control Systems: Next Generation Design Research. *2nd IEEE Conference on Human System Interaction* (HSI) (pp. 632-636), Catania, Italy, (May 2009).

[Rieger 2014] Rieger, C.G. (2014). Resilient control systems practical metrics basis for defining mission impact. *7th International Symposium on Resilient Control Systems* (pp. 1–10). Denver, Colorado, USA, (August 2014). https://doi.org/10.1109/ISRCS.2014.6900108

[Ritchie 2016] Ritchie, C. (2016, June 9). *NIST Cybersecurity Framework Protect / Maintenance and Protective Technology* [Presentation]. ISACA Chapter Meeting, Sacramento, California, USA.

[Schlegel 2017] Schlegel, R., Obermeier, S., Schneider, J. (2016). A security evaluation of IEC 62351. *Journal of Information Security and Applications, 34*(Part 2), 197-204. https://doi.org/10.1016/j.jisa.2016.05.007

[Specht 2013] Specht, M., Rohjans, S., Trefke, J., Uslar, M., González, J.M. (2013). International Smart Grid Roadmaps and their Assessment. *EAI Endorsed Transactions on Energy Web, 13*(1-6, e2). http://dx.doi.org/10.4108/trans.ew.2013.01-06.e2

[Sun 2012] Sun, H., Zhang, B., Wu, W., Guo, Q. (2012). Family Of Energy Management System for Smart Grid. *2012 3rd IEEE PES Innovative Smart Grid Technologies Europe (ISGT Europe)*, (pp. 1-5), Berlin, Germany, (October 2012). https://www.doi.org/10.1109/ISGTEurope.2012.6465777

[T&D 2016] TDEUROPE. (2016). *Harmonized Cyber Security standards for the European Union Electric Grid Position Paper* (29 November 2016). T&D Europe. The European Association of the Electricity Distribution and Distribution Equipment and Services Industry. https://www.tdeurope.eu/component/attachments/attachments.html?id=225

[Uslar 2013] Uslar, M. (2013). Introduction and smart grid basics. In M. Uslar, M. Specht, C. Danekas, J. Trefke, S. Rohjans, J.M. Gonzalez, C. Rosinger, R. Bleiker (Eds.), *Standardization in Smart Grids Introduction to IT-Related Methodologies, Architectures and Standards* (pp. 3–12). Springer. https://doi.org/10.1007/978-3-642-34916-4_1

[Verizon 2016]Verizon. (2016, April 29). *Verizon's 2016 Data Breach Investigations Report finds cybercriminals are exploiting human nature* [News]. Verizon News Archive. https://www.verizon.com/about/news/verizons-2016-data-breach-investigations-report-finds-cybercriminals-are-exploiting-human

[Whitman 2016] Whitman, M.E., Mattord, H.J. (2016). *Management of Information Security* (Fifth Edition). Cengage Learning.

References Part 5

[AITP Code] AITP. (n.d.). *Code of Ethics & Standards of Conduct.* Association of Information Technology Professionals (AITP). https://sites.google.com/a/gmatc.matc.edu/aitp/code-of-ethics

[AlHogail 2015] AlHogail, A., Mirza, A. (2015). Organizational information security culture assessment. *Proceedings of the 2015 World Congress in Computer Science, Computer Engineering and Applied Computing (SAM'15)*, (pp 286–292), Las Vegas, Nevada, USA, (July 2015).

[Anderson 2005] Anderson, M., Leigh Anderson, S., Armen, C. (2005). Towards Machine Ethics: Implementing Two Action-Based Ethical Theories [AAAI Technical Report FS-05-06]. *AAAI Fall Symposium on Machine Ethics*, Pittsburgh, Pennsylvania, USA, (July 2005). The AAAI Press.

[Anderson 2011] Anderson, M., Leigh Anderson, S. (2011). *Machine Ethics.* Cambridge University Press.

[ANSI Directory] ANSI. (n.d.). *ANSI Accreditation ANSI/ISO/IEC 17024 (Accredited).* https://www.ansi.org/Accreditation/credentialing/personnel-certification/ALLdirectoryListing?menuID=2&prgID=201&statusID=4

[ANSI/ISO 17024] ANSI. (n.d.). *ANSI Accreditation Program for Personnel Certification Bodies under ANSI/ISO/IEC 17024.* https://www.ansi.org/Accreditation/credentialing/personnel-certification/Default

[ANSI] ANSI. (n.d.). *American National Institute of Standards (ANSI).* www.ansi.org

[ANSSI 2014] SSI. (2014). *Cybersecurity for Industrial Control Systems - Classification Method and Key Measures Version 1.0* (January 2014). Agence nationale de la sécurité des systèmes d'information. http://www.ssi.gouv.fr/IMG/pdf/industrial_security_WG_Classification_Method.pdf

[Arnold 2011] Arnold, G.W. (2011). Challenges and opportunities in smart grid: a position article. *IEEE Proceedings, 99* (6), 922-927. https://doi.org/10.1109/JPROC.2011.2125930

[Batard 2017] Batard, J-L., Chollot, Y., Pipet, P., Lamberti, L., Gauci, A. (2017). Cybersecurity for modern distribution automation grids. *24th International Conference & Exhibition on Electricity Distribution (CIRED),* Glasgow, Scotland, (June 2017). *IET Journals Open Access Proceedings Journal, 2017*(1), 1002-1005. https://doi.org/10.1049/OAP-CIRED.2017.1328

[Baura 2006] Baura, G.D. (2006). *Engineering Ethics An Industrial Perspective.* Academic Press.

[BCSP] BCSP. (n.d.). *Board of Certified Safety Professionals (BCSP).* https://www.bcsp.org/

[Bloom 2017] Bloom, R. (2017, October 5). *Protecting Your Intellectual Property in the Internet of Things.* IP Watchdog. www.ipwatchdog.com/2017/10/05/protecting-intellectual-property-internet-of-things/id=88653/

[Booch 2014] Booch, G. (2014). The Human and Ethical Aspects of Big Data. *IEEE Software, 31*(1), 20-22. https://doi.org/10.1109/MS.2014.16

[Britz 1996] Britz, J.J. (1996). *TECHNOLOGY AS A THREAT TO PRIVACY: Ethical Challenges to the Information Profession.* web.simmons.edu/~chen/nit/NIT%2796/96-025-Britz.html

[Bynum 2008] Bynum, T.W. (2008). *A Very Short History of Computer Ethics.* www.cs.utexas.edu/~ear/cs349/Bynum_Short_History.html

[Catalogue] Calleam. (n.d.). *Catalogue of Catastrophe.* International Project Leadership Academy. calleam.com/WTPF/?page_id=3

[CENELEC 2011] ETSI. (2011). *Final report of the CEN/CENELEC/ETSI Joint Working Group on Standards for Smart Grids* (May 2011). https://www.etsi.org/images/files/Report_CENCLCETSI_Standards_Smart_Grids.pdf

[Chapman 2004] Chapman, N., Chapman, J. (2004). *Digital multimedia* (Second Edition). Wiley.

[Chirgwin 2017] Chirgwin, R. (2017, 20 Feb). *Software glitch, not wind farms, blacked out 60,000 in South Australia Senate committee told SA Power 'sat on' a problem for a while while SA sweltered* (20 Feb 2017 at 23:00) [News]. The Register. https://www.theregister.co.uk/2017/02/20/software_glitch_worsened_south_australian_blackouts/

[CNSSI 4009] CNSSI. (2015). *Committee on National Security Systems (CNSS) Glossary* [CNSSI No. 4009].

[Computer Ethics] CPSR. (2011, September 1). *The Ten Commandments of Computer Ethics Written by Computer Ethics Institute.* http://cpsr.org/issues/ethics/cei/

[Cook 2009] Cook, J. (2009). Ethics of Data Mining. In *Encyclopedia of Data Warehousing and Mining* (Second Edition) (pp. 783-788). IGI Global. https://doi.org/10.4018/978-1-60566-010-3.ch121

[Da Veiga 2009] Da Veiga, A. (2009, April 24). *Cultivating and assessing information security culture* [PhD Thesis]. University of Pretoria, South Africa.

[Da Veiga 2017] Da Veiga, A., Martins, N. (2017). Defining and Identifying Dominant Information Security Cultures and Subcultures. *Computers & Security, 70*, 72-94. https://doi.org/10.1016/j.cose.2017.05.002

[Davis 2011] Davis, P. (2011, July 4). *United States: Smart Grid and the Importance of Intellectual Property* [News]. Mondaq Connecting Knowledge & people. http://www.mondaq.com/unitedstates/x/137536/Patent/Smart+Grid+And+The+Importance+Of+Intellectual+Property

[DOE CSAT] Energy. *DOE Cybersecurity Training and Education.* https://www.energy.gov/cio/training/cybersecurity-awareness-training-warehouse/doe-cybersecurity-training-and-education

[DOE CyberForce] Cyberforcecompetition. (n.d.). *Department of Energy's CyberForce Competition.* https://cyberforcecompetition.com/cyberforce/

[DualUse] Parliament. (2009, July). *Dual-Use Dilemma* [Postnote Number 340]. Parliamentary Office of Science and Technology, London, United Kingdom. https://www.parliament.uk/documents/post/postpn340.pdf

[EC Code] ECcouncil. (n.d.). *Code of Ethics.* EC-Council Certification. https://cert.eccouncil.org/code-of-ethics.html

[EIA 2018] EIA. (n.d.). *How is electricity used in U.S. homes?* U.S. Energy Information Administration. https://www.eia.gov/tools/faqs/faq.php?id=96&t=3

[EPRI 2013] EPRI. (2013). *Guidelines for Planning an Integrated Security Operations Center* [3002000374], (Technical Update December 2013). https://www.metering.com/wp-content/uploads/2014/02/EPRI-Planning-ISOC-report.pdf

[ESCSWG 2014] Energy. (2014). *Cybersecurity Procurement Language for Energy Delivery Systems* (April 2014). Energy Sector Control Systems Working Group (ESCSWG). https://www.energy.gov/sites/prod/files/2014/04/f15/CybersecProcurementLanguage-EnergyDeliverySystems_040714_fin.pdf

[Ethics] Merriam-Webster. (n.d.). *Ethics.* Merriam-Webster Dictionary. https://www.merriam-webster.com/dictionary/ethic

[EthicsAU] Ethics. (2016, September). *Ethics Explainer: The Dual-Use Dilemma.* THE ETHICS CENTRE, Sydney, Australia. https://www.ethics.org.au/on-ethics/blog/september-2016/ethics-explainer-dual-use-dilemma

[ETP 2010] Edsoforsmartgrids. (n.d.). *European Technology Platform (ETP) SmartGrids Strategic Deployment Document for Europe's Electricity Networks of the Future (2010).* https://www.edsoforsmartgrids.eu/policy/eu-steering-initiatives/smart-grids-european-technology-platform/

[EU 2018/844] *Directive (EU) 2018/844 of the European Parliament and of the Council of 30 May 2018 amending Directive 2010/31/EU on the energy performance of buildings and Directive*

2012/27/EU on energy efficiency. https://eur-lex.europa.eu/legal-content/EN/TXT/PDF/?uri=CELEX:32018L0844&from=IT

[EU COM450] *JOINT COMMUNICATION TO THE EUROPEAN PARLIAMENT AND THE COUNCIL Resilience, Deterrence and Defence: Building strong cybersecurity for the EU JOIN/2017/0450 final.* https://eur-lex.europa.eu/legal-content/en/TXT/?uri=CELEX%3A52017JC0450

[EU GDPR] EUGDPR. (n.d.). *General Data Protection Regulation*. https://www.eugdpr.org/

[Eurostat] Europa. (n.d.). *Electricity generation statistics – first results*. Eurostat Statistics Explained. https://ec.europa.eu/eurostat/statistics-explained/index.php/Electricity_generation_statistics_%E2%80%93_first_results#Production_of_electricity

[Farhangi 2010] Farhangi, H. (2010). The path of the smart grid. *IEEE Power and Energy Magazine, 8*(1), 18-28. https://doi.org/10.1109/MPE.2009.934876

[Floridi 2010] Floridi, L. (2010). *The Cambridge Handbook of Information and Computer Ethics*. Cambridge University Press. https://doi.org/10.1017/CBO9780511845239

[Gauci 2017] Gauci, A., Michelin, S., Salles, M. (2017). Addressing the Challenge of Cyber Security Maintenance Through Patch Management. *24th International Conference & Exhibition on Electricity Distribution (CIRED)*, Glasgow, Scotland. *IET Journals Open Access Proceedings Journal, 2017*(1), 2599–2601.

[GIAC Code] GIAC. (n.d.). Ethics. GIAC Certifications. https://www.giac.org/about/ethics

[Gotterbarn 1997] Gotterbarn, D., Rogerson, S. (1997). Computer Ethics: The Evolution of the Uniqueness Revolution. In M.J. van den Hoven (Ed.), *Proceedings of the Conference on Computer Ethics - Philosophical Enquiry: CEPE97*, (pp. 151–159), Rotterdam, The Netherlands, (June 1997). University of Erasmus Press 1998.

[Gotterbarn 2001] Gotterbarn, D., Miller, K.W., Rogerson, S. (2001). Software Engineering Code of Ethics and Professional Practice IEEE-CS/ACM Joint Task Force on Software Engineering Ethics and Professional Practices. *Science and Engineering Ethics, 7*(2), 231-238. https://doi.org/10.1007/s11948-001-0044-4

[Gotterbarn 2009] Gotterbarn, D., Miller, K.W. (2009). The Public is the Priority: Making Decisions Using the Software Engineering Code of Ethics. *IEEE Computer, 42*(6), 66-73. https://doi.org/10.1109/MC.2009.204

[GridWise 2014] GridWise. (2014). *The future of the grid: evolving to meet America's needs Final report an industry-driven vision of the 2030 grid and recommendations for a path forward* (December 2014). GridWise Alliance. http://energy.gov/sites/prod/files/2014/12/f19/Future%20of%20the%20Grid%20December%202014.pdf

[Guo 2015] Guo, C., Bond, C.A., Narayanan, A. (2015). *The Adoption of New Smart-Grid Technologies*. Published by the RAND Corporation.

[Gupta 2012] Gupta, U. (2012, February 4). *Role of Ethics in IT Security Organizations Put New Emphasis on Staff's Ethical Behavior*. Information Security Media Group. InfoRisk today. https://www.inforisktoday.com/role-ethics-in-security-a-4469

[Handbook 2008] Himma, K.E., Tavani, H.T (Eds.). (2008). *THE HANDBOOK OF INFORMATION AND COMPUTER ETHICS*. Wiley.

[Harbor 2011] HarborResearch. (2011). *M2M Smart Systems Machine-To-Machine (M2M) & Smart Systems Market Opportunity 2010-2014*. Market Overview. Harbor Research Inc., San Francisco, California, USA. https://www.researchgate.net/profile/Niraj-Pandey-2/publication/291337455_Harbor_Research-M2M_and_Smart_Sys_Report/links/56a0a0c508aee4d26ad6a23c/Harbor-Research-M2M-and-Smart-Sys-Report.pdf?origin=publication_list

[Harris 2013] Harris 2013] Harris, S. (2013). *All in One CISSP Exam Guide* (Sixth Edition). McGraw-Hill Education.

[Hatzakis 2019] Hatzakis, T., Rodrigues, R., David, W. (2019). Smart Grids and Ethics: A case Study. *ORBIT Journal, 2*(2), 1-28. https://doi.org/10.29297/orbit.v2i2.108

[Hawk 2014] Hawk, C, Kaushiva, A.. (2014). Cybersecurity and the Smarter Grid. *The Electricity Journal, 27*(8), 84-95. https://doi.org/10.1016/j.tej.2014.08.008

[Heck 2003] Heck, B.S., Wills, L.M., & Vatchtsevanos, G.J. (2003). Software technology for implementing reusable, distributed control systems. *IEEE Control Systems Magazine, 23*(2), 21-35. https://doi.org/10.1109/MCS.2003.1172827

[Heine 2016] Heine, K., Grabovets, K. (2016). Organizational Liability. In A. Marciano, G. Ramello (Eds.), *Encyclopedia of Law and Economics*. Springer. https://doi.org/10.1007/978-1-4614-7883-6_605-1

[Hellerstein 2005] Hellerstein, J.L., Diao, Y., Parekh, S., Tilbury, D.M. (2005). Control engineering for computing systems - industry experience and research challenges. *IEEE Control Systems Magazine, 25*(6), 56-68. https://doi.org/10.1109/MCS.2005.1550153

[Hentea 2007] Hentea, M., Dhillon, H. (2007). New competencies for control engineers to meet the market demands in control systems. *Proceedings of ICEE (International Conference in Engineering Education) 2007 Conference*, Coimbra, Portugal, (September 2007).

[Herold 2016] Herold, R. (2016). *Using ISACA's Privacy Principles to Create an Effective Privacy Program*. Data Privacy Asia Singapore 2016.

[Hou 2008] Hou, D., Dolezilek, D.D. (2008). IEC 61850 – What It Can and Cannot Offer to Traditional Protection Schemes. *SEL Journal of Reliable Power, 1*(2), 1-11.

[ICS Certification] ENISA. (2014). *Certification of Cyber Security skills of ICS/SCADA professionals Good practices and recommendations for developing harmonised certification schemes* (December 2014). https://www.enisa.europa.eu/publications/certification-of-cyber-security-skills-of-ics-scada-professionals

[ICS Learning] US-CERT. (n.d.). *Web - Based Training available on the CISA Virtual Learning Portal*. Computer Infrastructure & Security Agency. https://us-cert.cisa.gov/ics/Training-Available-Through-ICS-CERT

[IEA 2018] IEA. (2018). *Statistics: Electricity Information 2018 Overview comprehensive annual data on electricity and heat*. International Energy Agency. https://www.iea.org/events/statistics-electricity-information-2018-overview

[IEC Roadmap] IEC. (2010). *Smart Grid Standardization Road Map Edition 1.0. Prepared by SMB Smart Grid Strategic Group (SG3)* (June 2010). http://www.itrco.jp/libraries/IEC-SmartgridStandardizationRoadmap.pdf

[IEC 61850] *IEC 61850 Communication networks and systems in substations, series of several parts.*

[IEC 62443] *IEC 62443 Industrial communication networks - Network and system security, series of several parts.*

[IEEE Code] IEEE. (n.d.). *IEEE Code of Ethics.* https://www.ieee.org/about/corporate/governance/p7-8.html

[IISP Skills] IISP. (n.d.). *IISP Skills Framework Version 2.3*. The Institute of Information Security Professionals. * Currently is Chartered Institute of Information Security (CIISec). https://www.ciisec.org

[INCOSE 2015] D.D. Walden, G.J. Roedler, K.J. Forsberg, R. D. Hamelin, T.M. Shortell (Eds.). (2015). *INCOSE System Engineering Handbook—A Guide for System Engineering Life Cycle Processes and Activities* Fourth Edition. International Council on Systems Engineering. Wiley.

[INCOSE Ethics] INCOSE. (n.d.). *Code of Ethics*. https://www.incose.org/about-incose/Leadership-Organization/code-of-ethics

[ISA 62443Certify] ISA. (n.d.). *ISA/IEC 62443 Cybersecurity Certificate Programs.* https://www.isa.org/training-and-certification/isa-certification/isa99iec-62443/isa99iec-62443-cybersecurity-certificate-programs

[ISACA Principles] ISACA. (n.d.). *ISACA Privacy Principles and Program Management Guide ©2016.* https://www.isaca.org/bookstore/cobit-5/ipp

[ISACA Privacy] ISACA. (2021). *Privacy Policy*. https://www.isaca.org/privacy-policy

[ISACACert] ISACA. (n.d.). *Credentialing.* https://www.isaca.org/credentialing

[ISC2 Code] ISC2. (n.d.). (ISC)^2 Code of Ethics. https://www.isc2.org/Ethics

[ISO/IEC 17024] ISO/IEC 17024:2012 *Conformity assessment – General requirements for bodies operating certification of persons.*

[ISO/IEC 27034] *ISO/IEC 27034 Information technology -- Security techniques -- Application security, several parts.*

[ISSA Code] ISSA. (n.d.). *ISSA Code of Ethics.* https://www.issa.org/page/codeofethics

[Jacobs 2013] Jacobs, M. (2013, August 8). 13 of the Largest Power Outages in History — and What They Tell Us About the 2003 Northeast Blackout [Blog]. *Union of Concerned Scientists.* https://blog.ucsusa.org/mike-jacobs/2003-northeast-blackout-and-13-of-the-largest-power-outages-in-history-199

[Johnson 2004] Johnson, M.L. (2004). Biometrics and the Threat to Civil Liberties. *IEEE Computer, 37*(4), 90-92. https://doi.org/10.1109/MC.2004.1297317

[Kane 2017] Kane, L., West, R., Taljaard, R., Macdonald, R., Ault, G., Davidson, E., Gooding, A. (2017). Supporting DER customer participation in active distribution networks and local markets. *24th International Conference & Exhibition on Electricity Distribution (CIRED)*, Glasgow, Scotland, (June 2017). *IET Journals Open Access Proceedings Journal, 2017*(1), 2019–2022.

[Kant 2009] Kant, K. (2009). Toward a Science of Power Management. *IEEE Computer, 42*(9), 99-101. https://doi.org/10.1109/MC.2009.303

[Kappagantu 2018] Kappagantu, R., Daniel, S.A. (2018). Challenges and issues of smart grid implementation: A case of Indian scenario. *Journal of Electrical Systems and Information Technology, 5*(3), 453-467. https://doi.org/10.1016/j.jesit.2018.01.002

[Kapucu 2007] Kapucu, N. (2007). Ethics of Digital Government. In A-V. Anttiroiko (Ed.), *Electronic Government: Concepts, Methodologies, Tools, and Applications*, (pp. 130-135). IGI Global.

[Karahasanovic 2009] Karahasanovic, A., Brandtzaeg, P. B., Vanattenhoven, J, Lievens, B., Nielsen, K.T., Pierson, J. (2009). Ensuring Trust, Privacy, and Etiquette in Web 2.0 Applications. *IEEE Computer, 42*(6), 42-49. https://doi.org/10.1109/MC.2009.186

[KAYDEE 2009] KAYDEE. (2009, February 2). Engineering Ethics and "Software Engineering" [Blog]. *Engineering Ethics Blog Comments on current events with an engineering ethics angle.* http://engineeringethicsblog.blogspot.com/2009/02/engineering-ethics-and-software.html

[Knowles 2016] Knowles, A. (2016, October 12). *Tough Challenges in Cybersecurity Ethics* [News]. Security Intelligence. https://securityintelligence.com/tough-challenges-cybersecurity-ethics/

[Laudon 2018] Laudon, K. C., Laudon, J.P. (2018). Ethical and Social Issues in Information Systems. In *Management Information Systems: Managing the Digital Firm* (15th Edition). Pearson.

[Learning 2012] UNECE. (2012*). Learning for the future Competences in Education for Sustainable Development.* United Nations Economic Commission for Europe Strategy for education for Sustainable Development. https://unece.org/fileadmin/DAM/env/esd/ESD_Publications/Competences_Publication.pdf

[Lee 2016a] Lee, W.W., Zankl, W., Chang, H. (2016a). An Ethical Approach to Data Privacy Protection. *ISACA Journal, 6*, 1-9. https://www.isaca.org/Journal/archives/2016/volume-6/Pages/an-ethical-approach-to-data-privacy-protection.aspx

[Lee 2016b] Lee, R. M., Assante, M.J., Conway, T. (2016b). *Analysis of the cyber attack on the Ukrainian power grid.* E-ISAC Electricity Information Sharing and Analysis Center. https://media.kasperskycontenthub.com/wp-content/uploads/sites/43/2016/05/20081514/E-ISAC_SANS_Ukraine_DUC_5.pdf

[Le Ray 2020] Le Ray, G., Pinson, P. (2020). The ethical smart grid: Enabling a fruitful and long-lasting relationship between utilities and customers. *Energy Policy, 140*(2020), Article: 111258. https://doi.org/10.1016/j.enpol.2020.111258

[Li 2014] Li, R., Wu, Q., Oren, S.S. (2014). Distribution locational marginal pricing for optimal electric vehicle charging management. *IEEE Transactions on Power Systems, 29*(1), 203–211. https://doi.org/ 10.1109/TPWRS.2013.2278952

[Lipner 2018] Lipner, S. (2018). *SAFECode Comments on EU Cybersecurity Legislation* (October 2018). https://safecode.org/wp-content/uploads/2018/11/SAFECode_Comments_on_EU_Cybersecurity_Legislation_Oct_2018_v2.pdf

[Lu 2011] Lu, Y-H., Qiu, Q., Cameron, K.W. (2011). End-to-End Energy Management (November 2011). *Green IT* (pp. 75-77). IEEE Computer Society.

[Luiijf 2015] Luiijf, H.A.M., te Paske, B.J. (2015). *Cyber Security of Industrial Control Systems* (March 2015). TNO Publishing. https://doi.org/10.13140/RG.2.1.3797.4566

[MacWilson 2017] MacWilson, A. (2017, May 31). *IISP Launches New Skills Framework for Information Security Professionals* [News]. realwire. https://www.realwire.com/releases/IISP-Launches-New-Skills-Framework-for-Information-Security-Professionals

[Makienko 2017] Makienko, M.M., Kurkan, N.V., Emelyanenko, E.E. (2017). Ethical Implications to The Assessment of The Smart Grid Technology In Russia. *MATEC Web of Conferences, 91*(2017), Article:01020, 5 p. https://doi.org/10.1051/matecconf/20179101020

[Malaysia Code] Cybersecurity. (2010). *Code of Ethics for Information Security Professionals.* Cybersecurity Agency under Ministry of Science, Technology and Innovation (MOSTI). CyberSecurity Malaysia An Agency under MOSTI. https://www.cybersafe.my/pdf/guidelines/Code_of_Ethics_for_Information_Security_Professionals.pdf

[Maner 1978] Maner, W. (1978). *Starter Kit on Teaching Computer Ethics.* Self published in 1978.

[Maner 1980] Maner, W. (1980). *Starter Kit on Teaching Computer Ethics.* Helvetia Press in cooperation with the National Information and Resource Center for Teaching Philosophy.

[Marnay 2007] Marnay, C., Zhou, N. (2007). *Microgrid Research Activities in the U.S* [GE Research Paper], (November 2007). General Electric Inc., USA.

[Martin 2010] Martin, M.W., Schinzinger, R. (2010). *Introduction to Engineering Ethics Basic Engineering Series and Tools* (Second Edition). McGraw-Hill Higher Education.

[Miller 2007] Miller, S. Selgelid, M. (2007). Ethical and Philosophical Consideration of the Dual-Use Dilemma in the Biological Sciences. *Science and Engineering Ethics, 13(*4), 523-580. https://doi.org/10.1007/s11948-007-9043-4

[Monitor May2014] US-CERT. (2014). *ICS-CERT Monitor May-August 2014.* https://us-cert.cisa.gov/sites/default/files/Monitors/ICS-CERT_Monitor_May-Aug2014.pdf

[Mourshed 2015] Mourshed, M., Robert, S., Ranalli, A., Messervey, T., Reforgiato, D., Contreau, R., Becue,A., Quinn, K., Rezgui, Y., Lennard, Z. (2015). Smart grid futures: Perspectives on the integration of energy and ICT services. *The 7th International Conference on Applied Energy – ICAE2015*, Abu Dhabi, United Arab Emirates, (March 2015). *Energy Procedia, 75*, 1132 – 1137. https://doi.org/10.1016/j.egypro.2015.07.531

[Nash 2011] Nash, J. (2011). Protect the Grid. *Electric Perspectives Magazine*, (May/June 2011), 34-43. Edison Electric Institute.

[NCSC Guidance] NCSC. (2018). Guidance to NCSC Certification for Cyber Security/IA Professionals [Issue No: 2.3] (November 2018). https://www.ncsc.gov.uk/files/CCP-Guidance_to_certification_for_IA_professionals_2-3.pdf

[Nemeti 2007] Nemeti, H. (2007). Introductory Chapter: Information Security and Ethics. In (H. Nemeti (Ed.), *Information Security and Ethics: Concepts, Methodologies, Tools, and Applications.* IGI Global. https://doi.org/10.4018/978-1-59904-937-3

[Neupane 2017] Neupane, B. (2017). *Predictive Data Analytics for Energy Demand Flexibility* [PhD Thesis], Aalborg University, Denmark. https://doi.org/10.5278/VBN.PHD.TECH.00023

[NICCS CEAS] NICCS. (n.d.). *NICCS Education and Training Catalog*. National Initiative for Cybersecurity Careers and Studies. https://niccs.cisa.gov/training/search

[NICCS] NICCS. (n.d.). Training. National Initiative for Cybersecurity Careers and Studies. https://niccs.cisa.gov/training

[NICE] NIST. (n.d.). *National Initiative for Cybersecurity Education (NICE)*. Information Technology Laboratory / Applied Cybersecurity Division. https://www.nist.gov/itl/applied-cybersecurity/nice

[NIST SP800-12r1] *NIST Special Publication (SP) 800-12 Revision 1 An Introduction to Information Security* (June 2017). *https://doi.org/10.6028/NIST.SP.800-12r1*

[NIST SP800-16] *NIST Special Publication (SP) 800-16 Information Technology Security Training Requirements: a Role- and Performance-Based Model* (April 1998).

[NIST SP800-50] *NIST Special Publication (SP) 800-50 Building an Information Technology Security Awareness and Training Program* (October 2003).

[NIST SP800-181r1] *NIST Special Publication (SP)800-181 Revision 1 Workforce Framework for Cybersecurity (NICE Framework)* (November 2020). National Initiative for Cybersecurity Education (NICE). https://doi.org/10.6028/NIST.SP.800-181r1

[NISTIR 7628r1] *NIST Interagency Report (NISTIR) 7628r1 Guidelines for Smart Grid Cybersecurity Revision 1* (September 2014). http://dx.doi.org/10.6028/NIST.IR.7628r1

[NSPE Code] NSSPE. (n.d.). *Code of Ethics for Engineers.* https://www.nspe.org/resources/ethics/code-ethics

[OECD 1980] OECD. (1980). *Guidelines on the Protection of Privacy and Transborder Flows of Personal Data.* Organisation for Economic Co-operation and Development. https://www.oecd.org/internet/ieconomy/oecdguidelinesontheprotectionofprivacyandtransborderflowsofpersonaldata.htm

[OECD 2010] OECD. (2010). *OECD Information Technology Outlook* (November 2010).

[OECD 2012a] OECD. (2012a, November 16). *The Role of the 2002 Security Guidelines: Towards Cybersecurity for an Open and Interconnected Economy* (DSTI/ICCP/REG (2012)8/FINAL). OECD Publishing. http://dx.doi.org/10.1787/5k8zq930xr5j-en

[OECD 2012b] OECD. (2012b, January 10). *ICT Applications for the Smart Grid: Opportunities and Policy Implications. OECD Digital Economy Papers* (No. 190) (DSTI/ICCP/REG(2011)12/FINAL). OECD Publishing. http://dx.doi.org/10.1787/5k9h2q8v9bln-en

[OECD 2015] OECD. (2015). *Digital Security Risk Management for Economic and Social Prosperity: OECD Recommendation and Companion Document.* OECD Publishing. http://dx.doi.org/10.1787/9789264245471-en

[Olumoye 2013] Olumoye, M. Y. (2013). Ethics and Social Impact of Information Systems in Ethics and Social Impact of Information Systems. *International Journal of Science and Research (IJSR), 2*(11), 154-158.

[Overly 2017] Overly, S. (2017, January 11). *Six Volkswagen executives indicted in emissions-cheating scandal* [News]. The Washington Post. https://www.washingtonpost.com/news/innovations/wp/2017/01/11/justice-department-volkswagen-to-plead-guilty-to-three-criminal-counts-in-emissions-scandal/?noredirect=on&utm_term=.653078bd764a

[Park 2017] Park, J.Y., Abraham, S., Tamburo, S. (2017, January 26). *Intellectual property in the energy sector* [News]. Power Engineering International. https://www.powerengineeringint.com/digitalization/internet-of-things/intellectual-property-in-the-energy-sector/

[Parker 1981] Parker, D.B. (1981). *Computer Security Management.* Prentice Hall.

[Philosophy] IEP. (n.d). *Internet Encyclopedia of Philosophy A Peer Reviewed Academic Resource.* https://www.iep.utm.edu/ethics/

[Pieters 2009] Pieters, W., van Cleeff, A. (2009). The Precautionary Principle in a World of Digital Dependencies. *IEEE Computer, 42*(6), 50-56. https://doi.org/10.1109/MC.2009.203

[Podder 2017] Podder, A,K. Ahmed, K., Roy, N.K., Biswas, P.C. (2017). Design and simulation of an independent solar home system with battery backup. *2*017 4th International Conference on Advances in Electrical Engineering (ICAEE), (pp. 427-431), Dhaka, Bangladesh, (September 2017). https://doi/org/10.1109/ICAEE.2017.8255394

[Quinn 2006] Quinn, M.J. (2006). *Ethics for the Information Age* (Second Edition). Pearson Addison Wesley.

[Rashid 2009] Rashid, A., Weckert, J., Lucas, R. (2009). Software Engineering Ethics in a Digital World. *IEEE Computer, 42*(6), 34-41.

[Reed 2009] Reed, J.L. (2009). *Building the Smart Grid: Role of Intellectual Property Rights* [Presentation], (November 2009). Squire, Sanders & Dempsey L.L.P. https://www.squirepattonboggs.com/~/media/files/insights/publications/2009/11/building-the-smart-grid-role-of-intellectual-pro__/files/smart-grid-powerpoint-presentation/fileattachment/presentation_on_ipr__smart_grid_by_j_reed_nov_2009.pdf

[RFC 1087] *IETF Request for Comments (RFC) 1087 Ethics and the Internet* (January 1989).

[RFC 4949] *IETF Request for Comments (RFC) 4949 Internet Security Glossary Version 2* (August 2007).

[Rinaldi 2001] Rinaldi, S.M., Peerenboom, J.M., Kelly, T.K. (2001). Identifying, understanding and analyzing critical infrastructure interdependencies. *IEEE Control Systems Magazine, 21*(6), 11–25. https://doi.org/10.1109/37.969131

[Rogerson 2015] Rogerson, S. (2015). Future Vision. *Journal of Information, Communication and Ethics in Society, 13*(3/4), 346–360. Special Issue -20 years of ETHICOMP. https://doi.org/10.1108/JICES-05-2015-0011

[Rogerson 2018] Rogerson, S. (2018). Ethics Omission Increases Gases Emission: A Look in The Rearview Mirror at Volkswagen Software Engineering. *Communications of the ACM, 61*(3), 30-32. https://doi.org/10.1145/3180490

[Roman 2011] Roman, R., Najera, P., Lopez, J. (2011). Securing the internet of things. *Computer, 44*(9), 51–58. https://doi.org/10.1109/MC.2011.291

[SANS 2018] SANS. (2018). *2018 SANS Security Awareness Report Building Successful Security Awareness Program.* https://www.sans.org/sites/default/files/2018-05/2018%20SANS%20Security%20Awareness%20Report.pdf

[SANS Code] SANS. (2004). IT Code of Ethics Version 1.0 (April 24, 2004). https://www.sans.org/security-resources/ethics

[SAS 70] Sciencedirect. (n.d.). *Statements on Auditing Standards Statement on Auditing Standards No. 70: Service Organizations.* ScienceDirect. https://www.sciencedirect.com/

[Schinzinger 2000] Schinzinger, R., Martin, M.W. (2000). *Introduction to Engineering Ethics.* McGraw-Hill.

[Schmidt 2010] Schmidt, A. (2010). Ubiquitous Computing: Are We There Yet? *IEEE Computer, 43*(2), 95-97. https://doi.org/ 10.1109/MC.2010.54

[Schultz 2006] Schultz, R.A. (2006). *Contemporary Issues in Ethics and Information Technology.* IGI Global. https://doi.org/10.4018/978-1-59140-779-9

[Sembok 2003] Sembok, T.M.T. (2003). *Ethics of Information Communication Technology (ICT).* Regional Meeting on Ethics of Science and Technology, Bangkok, Malaysia, (November 2003). https://www.academia.edu/4234405/ETHICS_OF_INFORMATION_COMMUNICATION_TECHNOLOGY_ICT

[Shinder 2005] Shinder, D. (2005, August 2). *Ethical issues for IT security professionals* [News]. Computerworld. https://www.computerworld.com/article/2557944/security0/ethical-issues-for-it-security-professionals.html

[Shireen 2013] Shireen, W., Kotti, R. (2013). Smart Grid, Industry Trends and Power Engineering Education. *120th ASEE Annual Conference* and *Exposition American Society for Engineering Education* (Paper ID: 7132), Atlanta, Georgia, USA, (June 2013). https://doi.org/10.18260/1-2--22454

[Spinello 1995] Spinello, R.A. (1995). *Ethical Aspects of Information Technology*. Prentice-Hall.

[Spinello 2001] Spinello, R.A., Tavani, H.T. (2001). The Internet, Ethical Values, and Conceptual Frameworks: An Introduction to Cyber Ethics. *ACM SIGAS Computers and Society, 31*(2), 5-7. https://doi.org/10.1145/503345.503346

[Spinello 2012] Spinello, Richard A. (2012). Information and Computer Ethics. *Journal of Information Ethics, 21*(2), 17-32. https://doi.org/10.3172/JIE.21.2.17

[Spitzner 2016] Spitzner, L. (2016, March 8). Defining the Security Awareness Maturity Model [Blog]. *SANS*. https://www.sans.org/blog/defining-security-awareness-maturity-model/

[Stanford 2001] Bynum, T.W. (2001). *Stanford Encyclopedia of Philosophy Computer Ethics: Basic Concepts and Historical Overview*. http://plato.stanford.edu/archives/win2001/entries/ethics-computer/

[Stanford 2015] Bynum, T.W. (2015). *Stanford Encyclopedia of Philosophy Computer and Information Ethics* (substantive revision October 26, 2015). https://plato.stanford.edu/entries/ethics-computer/

[Steven 2006] Steven, J. (2006). Adopting an enterprise software security framework. *IEEE Security & Privacy, 4*(2), 84-87. https://doi.org/10.1109/MSP.2006.33

[Straub 1987] Straub, D.W. Jr. (1987). Controlling Computer Abuse: An Empirical Study of Effective Security Countermeasures. *ICIS 1987 Proceedings of International Conference on Information Systems* (Paper 32), Pittsburgh, Pennsylvania, USA. Association for Information Systems. http://aisel.aisnet.org/icis1987/32

[Tamai 2009] Tamai, T. (2009). Social Impact of Information System Failures. *IEEE Computer, 42*(6), 58-65. https://doi.org/ 10.1109/MC.2009.199

[Tehranipoor 2011] Tehranipoor, M., Salmani, H., Zhang, X., Wang, M., Karri, R., Rajendran, J., Rosen, K. (2011). Trustworthy Hardware: Trojan Detection and Design-for-Trust Challenges. *IEEE Computer, 44*(11), 66-74. https://doi.org/ 10.1109/MC.2010.369

[Tene 2013] Tene, O., Polonetsky, J. (2013). Big data for all: Privacy and user control in the age of analytics. *Northwestern Journal of Technology and Intellectual Property, 11*(5), Article 1. https://ssrn.com/abstract=2149364

[Tricentis 2016] Yahoo. (2017, January 17). *Software Fail Watch Says $1.1 Trillion in Assets Affected by Software Bugs in 2016 - Tricentis Software Fail Watch 2016 Quarter One* [News]. Marketwired. https://www.tricentis.com/blog/2016/04/07/software-fail-watch-2016-quarter-one/

[Treicentis 2017] Tricentis. (2017). Software Fail Watch, Q1: 2017 Defects Impact Divorces, the Deceased & the Dark WEB [Blog]. Tricentis. https://www.tricentis.com/blog/real-life-examples-of-software-development-failures/

[Tzafestas 2018] Tzafestas, S.G. (2018). Ethics and Law in the Internet of Things World. *Smart Cities, 1*(1), 98-120. https://doi.org/10.3390/smartcities1010006

[UDHR 1948] UN. (n.d.). *Universal Declaration of Human Rights*. United Nations. http://www.un.org/en/universal-declaration-human-rights/

[UNESCO Code] UNESCO. (2011). *Code of Ethics for The Information Society Proposed by the Intergovernmental Council of the Information for all Programme (IFAP)* (October 2011). General Conference 36th Session, Paris, France, (October 10, 2011). United Nations Educational, Scientific and Cultural Organization. UNESDOC Digital Library. https://unesdoc.unesco.org/ark:/48223/pf0000212696

[UNESCO Ethics] UNESCO. (n.d.). *Ethics of Information.* United Nations Educational, Scientific and Cultural Organization. https://en.unesco.org/themes/ethics-information

[UNESCO Preamble] UNESCO. (n.d.). *UNESCO Constitution.* portal.unesco.org/en/ev.php-URL_ID=15244&URL_DO=DO_TOPIC&URL_SECTION=201.html

[US DOJ] Justice. (2017, January 11). *Volkswagen AG Agrees to Plead Guilty and Pay $4.3 Billion in Criminal and Civil Penalties and Six Volkswagen Executives and Employees Are Indicted in Connection with Conspiracy to Cheat U.S. Emissions Tests* [News]. Justice News. The United States Department of Justice. https://www.justice.gov/opa/video/volkswagen-ag-agrees-plead-guilty-pay-43-billion-criminal-civil-penalties-six-volkswagen

[Vacura 2015] Vacura, M. (2015). The History Of Computer Ethics and Its Future Challenges. In P. Doucek, G. Chroust, V. Oskradal (Eds.) *Proceedings of IDIMT-2015 Information Technology and Society Interaction and Interdependence 23rd Interdisciplinary Information Management Talks (IDIMT 2015)*, Poděbrady, Czech Republic, (September 2015).

[White 2016] White, G., Ariyachandra, T. (2016). Big Data and Ethics: Examining the Grey Areas of Big Data Analytics. *Issues in Information Systems, 17*(IV), 1-7.

[Whitman 2014] Whitman, M.E., Mattord, H.J. (2014). *Management of Information Security* (Fourth Edition). Cengage Learning.

[Wiener 1950] Wiener, N. (1950). *The Human Use of Human Beings: Cybernetics and Society.* Houghton Mifflin Harcourt. According to [Bynum 2008]

[Yardley 2014] Yardley, T., M., Uludag, S., Nahrstedt, K., Sauer, P. (2014). Developing A Smart Grid Cybersecurity Education Platform and A Preliminary Assessment of Its First Application. *2014 IEEE Frontiers in Education Conference (FIE) Proceedings* (pp. 1-9), Madrid, Spain, (October 2014). IEEE Computer Society Digital Library. https://doi.org/10.1109/FIE.2014.7044273

Appendix A

Cybersecurity Terms, Concepts, and Standards

The Appendix A includes definitions for cybersecurity terms and concepts that are recommended including information about various glossaries that include cybersecurity definitions. The wide spread of the cybersecurity glossaries with not unique definitions is creating often confusion and room for the misinterpretation at the expense of a sound guide to provide for adequate security controls. Also, the appendix includes recommendations on applying NIST cybersecurity Framework to power grid (see Table A2), guidance on providing security for industrial control systems (Table A3), and standards that are focused on security or Smart Grid (Table A4).

A1. Cybersecurity Terms

Table A1 includes a list of most common security terms as defined in the standards and glossaries.

Table A1 Recommended definitions of most common security terms.

Term	Definition
Acceptable risk	A risk that is understood and tolerated by a system's user, operator, owner, or accreditor, usually because the cost or difficulty of implementing an effective countermeasure for the associated vulnerability exceeds the expectation of loss. (See: adequate security, risk, "second law" under "Courtney's laws") [RFC 4949]
Access	The ability and means to communicate with or otherwise interact with a system to use system resources either to handle information or to gain knowledge of the information the system Contains [RFC 4949]
Accountability	The property of a system or system resource that ensures that the actions of a system entity may be traced uniquely to that entity, which can then be held responsible for its actions [RFC 4949]
Accountability service	A security service that records information needed to establish accountability for system events and for the actions of system entities that cause them [RFC 4949]
Acquisition	Process of obtaining a system, product, or service [ISO/IEC/15288]
Advanced persistent threat	An adversary with sophisticated levels of expertise and significant resources, allowing it through the use of multiple different attack vectors (e.g., cyber, physical, and deception) to generate opportunities to achieve its objectives, which are typically to establish and extend footholds within the information technology infrastructure of organizations for purposes of continually exfiltrating information and/or to undermine or impede critical aspects of a mission, program, or organization, or place itself in a position to do so in the future; moreover, the advanced persistent threat pursues its objectives repeatedly over an extended period of time, adapting to a defender's efforts to resist it, and with determination to maintain the level of interaction needed to execute its objectives [NIST SP800-39]
Adversary	1. An entity that attacks a system. (Compare: cracker, intruder, hacker) 2. An entity that is a threat to a system [RFC 4949]

Table A1 (Continued)

Term	Definition
Air gap	An interface between two systems at which (a) they are not connected physically and (b) any logical connection is not automated (i.e., data is transferred through the interface only manually, under human control) [RFC 4949]
Anomaly detection	An intrusion detection method that searches for activity that is different from the normal behavior of system entities and system resources. (See: IDS. Compare: misuse detection) [RFC 4949]
Anonymity	The condition of an identity being unknown or concealed. (See: alias, anonymizer, anonymous credential, anonymous login, identity, onion routing, persona certificate. Compare: privacy) [RFC 4949]
Architecture	A set of related physical and logical representations (i.e., views) of a system or a solution. The architecture conveys information about system/solution elements, interconnections, relationships, and behavior at different levels of abstractions and with different scopes [NIST SP800-160v1]
Asset	An item of value to achievement of organizational mission/business objectives. *Note 1:* Assets have interrelated characteristics that include value, criticality, and the degree to which they are relied upon to achieve organizational mission/business objectives. From these characteristics, appropriate protections are to be engineered into solutions employed by the organization. *Note 2:* An asset may be tangible (e.g., physical item such as hardware, software, firmware, computing platform, network device, or other technology components) or intangible (e.g., information, data, trademark, copyright, patent, intellectual property, image, or reputation) [NIST SP800-160v1]
Asset	A system resource that is (a) required to be protected by an information system's security policy, (b) intended to be protected by a countermeasure, or (c) required for a system's mission [RFC 4949]
Assurance	Grounds for justified confidence that a claim has been or will be achieved. *Note 1:* Assurance is typically obtained relative to a set of specific claims. The scope and focus of such claims may vary (e.g., security claims, safety claims) and the claims themselves may be interrelated. *Note 2:* Assurance is obtained through techniques and methods that generate credible evidence to substantiate claims [ISO/IEC 15026-1]
Attack	1. An intentional act by which an entity attempts to evade security services and violate the security policy of a system. That is, an actual assault on system security that derives from an intelligent threat. (See: penetration, violation, vulnerability.) 2. A method or technique used in an assault (e.g., masquerade) [RFC 4949]
Attribute	An attribute is any distinctive feature, characteristic, or property of an object that can be identified or isolated quantitatively or qualitatively by either human or automated means [ISO/IEC 27000]
Authentication	The process of verifying a claim that a system entity or system resource has a certain attribute value. (See: attribute, authenticate, authentication exchange, authentication information, credential, data origin authentication, peer entity authentication, "relationship between data integrity service and authentication services" under "data integrity service", simple authentication, strong authentication, verification, X.509) [RFC 4949]

Table A1 (Continued)

Term	Definition
Authentication information	Information used to verify an identity claimed by or for an entity. (See: authentication, credential, user. Compare: identification information) [RFC 4949]
Authentication service	A security service that verifies an identity claimed by or for an entity. (See: authentication.) [RFC 4949]
Authenticity	The property of being genuine and able to be verified and be trusted. (See: authenticate, authentication, validate vs. verify) [RFC 4949]
Automated security monitoring	Use of automated procedures to ensure security controls are not circumvented or the use of these tools to track actions taken by subjects suspected of misusing the information system [CNSSI 4009]
Authorization	An approval that is granted to a system entity to access a system resource. (Compare: permission, privilege) [RFC 4949]
Availability	The property of a system or a system resource being accessible, or usable or operational upon demand, by an authorized system entity, according to performance specifications for the system; i.e., a system is available if it provides services according to the system design whenever users request them. (See: critical, denial of service. Compare: precedence, reliability, survivability) [RFC 4949]
Availability service	A security service that protects a system to ensure its availability [RFC 4949]
Back door	Un undocumented way of gaining access to computer system. A backdoor is a potential security risk [NIST SP 800-82r1]
Birthday attack	A class of attacks against cryptographic functions, including both encryption functions and hash functions. The attacks take advantage of a statistical property: Given a cryptographic function having an N-bit output, the probability is greater than 1/2 that for 2**(N/2) randomly chosen inputs, the function will produce at least two outputs that are identical. (See: Tutorial under "hash function") [RFC 4949]
Brute force	A cryptanalysis technique or other kind of attack method involving an exhaustive procedure that tries a large number of possible solutions to the problem. (See: impossible, strength, work factor) [RFC 4949]
Buffer overflow	A condition at an interface under which more input can be placed into a buffer or data holding area than the capacity allocated, overwriting other information. Attackers exploit such a condition to crash a system or to insert specially crafted code that allows them to gain control of the system [CNSSI]
Business continuity plan (BCP)	The documentation of a predetermined set of instructions or procedures that describe how an organization's mission/business processes will be sustained during and after a significant disruption [NIST SP800-34r1]
Business impact analysis (BIA)	An analysis of an information system's requirements, functions, and interdependencies used to characterize system contingency requirements and priorities in the event of a significant disruption [NIST SP 800-34r1]
Certificate	1. /general English/ A document that attests to the truth of something or the ownership of something 2. /general security/ See: capability token, digital certificate 3. /PKI/ See: attribute certificate, public-key certificate [RFC 4949]
Cipher	A cryptographic algorithm for encryption and decryption [RFC 4949]

Table A1 (Continued)

Term	Definition
Cipher text	/noun/ Data that has been transformed by encryption so that its semantic information content (i.e., its meaning) is no longer intelligible or directly available. (See: ciphertext. Compare: clear text, plain text) [RFC 4949]
Computer system	Synonym for "information system", or a component thereof (Compare: computer platform) [RFC 4949]
Confidentiality	See: data confidentiality [RFC 4949]
Contingency plan	A plan for emergency response, backup operations, and post-disaster recovery in a system as part of a security program to ensure availability of critical system resources and facilitate continuity of operations in a crisis (See: availability) [RFC 4949]
Countermeasure	An action, device, procedure, or technique that meets or opposes (i.e., counters) a threat, a vulnerability, or an attack by eliminating or preventing it, by minimizing the harm it can cause, or by discovering and reporting it so that corrective action can be taken [RFC 4949]
Covert channel	An unintended or unauthorized intra-system channel that enables two cooperating entities to transfer information in a way that violates the system's security policy but does not exceed the entities' access authorizations [RFC 4949]
Critical	/system resource/ A condition of a system resource such that denial of access to, or lack of availability of, that resource would jeopardize a system user's ability to perform a primary function or would result in other serious consequences, such as human injury or loss of life. (See: availability, precedence. Compare: sensitive) [RFC 4949]
Critical infrastructure	The systems and assets, whether physical or virtual, so vital to society that the incapacity or destruction of such may have a debilitating impact on the security, economy, public health or safety, environment, or any combination of these matters [NICCS]
Cryptanalysis	1. Operations performed in defeating cryptographic protection without an initial knowledge of the key employed in providing the protection. 2. The study of mathematical techniques for attempting to defeat cryptographic techniques and/or information systems security. This includes the process of looking for errors or weaknesses in the implementation of an algorithm or of the algorithm itself [NIST SP 800-57P1r3]
Cryptographic algorithm	An algorithm that uses the science of cryptography, including (a) encryption algorithms, (b) cryptographic hash algorithms, (c) digital signature algorithms, and (d) key-agreement algorithms [RFC 4949]
Cyber attack	Actions taken in cyberspace that create noticeable denial effects (i.e., degradation, disruption, or destruction) in cyberspace or manipulation that leads to denial that appears in a physical domain [CNSSI 4009]
Cyber ecosystem	The interconnected information infrastructure of interactions among persons, processes, data, and information and communications technologies, along with the environment and conditions that influence those interaction [NICCS]
Cyber infrastructure	An electronic information and communications systems and services and the information contained therein Extended Definition: An electronic information and communications systems and services and the information contained therein. Extended Definition: The information and communications systems and services composed of all hardware and software that process, store, and communicate information

Table A1 (Continued)

Term	Definition
Cybersecurity plan	Formal document that provides an overview of the cybersecurity requirements for an IT and ICS and describes the cybersecurity controls in place or planned for meeting those requirements [DOE RMP]
Cybersecurity policy	A set of criteria for the provision of security services [DOE RMP]
Cybersecurity program strategy	A plan of action designed to achieve the performance targets that the organization sets to accomplish its mission, vision, values, and purpose [CERT RMM]
Cybersecurity program management	Provides governance, strategic planning, and sponsorship for the organization's cybersecurity activities in a manner that aligns cybersecurity objectives with the organization's strategic objectives and the risk to critical infrastructure [ES-C2M2]
Cyberspace	The interdependent network of information technology infrastructures, and includes the Internet, telecommunications networks, computer systems, and embedded processors and controllers in critical industries. Common usage of the term also refers to the virtual environment of information and interactions between people [CPR] CPR definition is based on National Security Presidential Directive 54/Homeland Security Presidential Directive 23 (NSPD-54/HSPD23)
Cyber-physical system	A system that includes engineered, interacting networks of physical and computational components [NIST SP800-160v1]
Cybersecurity	Prevention of damage to, protection of, and restoration of computers, electronic communications systems, electronic communications services, wire communication, and electronic communication, including information contained therein, to ensure its availability, integrity, authentication, confidentiality, and nonrepudiation [NIST SP800-160v1]
Data availability	The property that data is available when it is needed [RFC 4949]
Data confidentiality	The property that data is not disclosed unless they have been authorized to know the data. (See: Bell-LaPadula model, classification, data confidentiality service, secret. Compare: privacy) [RFC 4949]
Data confidentiality Service	A security service that protects data against unauthorized disclosure (See: access control, data confidentiality, datagram confidentiality service, flow control, inference control) [RFC 4949]
Data integrity	The property that data has not been changed, destroyed, or lost in an unauthorized or accidental manner. (See: data integrity service. Compare: correctness integrity, source integrity) [RFC 4949]
Data integrity service	A security service that protects against unauthorized changes to data, including both intentional change or destruction and accidental change or loss, by ensuring that changes to data are detectable. (See: data integrity, checksum, datagram integrity service) [RFC 4949]
Data origin authentication	The corroboration that the source of data received is as claimed (See: authentication) [RFC 4949]
Data origin authentication service	A security service that verifies the identity of a system entity claimed to be the original source. (See: authentication) [RFC 4949]
Data security	The protection of data from disclosure, alteration, destruction, or loss that either is accidental or is intentional but unauthorized [RFC 4949]
Defense in breadth	A planned, systematic set of multi-disciplinary activities that seek to identify, manage, and reduce risk of exploitable vulnerabilities at every stage of the system, network, or sub-component lifecycle (system, network, or product design and development; manufacturing; packaging; assembly; system integration; distribution; operations; maintenance; and retirement) [CNSSI 4009]

Table A1 (Continued)

Term	Definition
Defense in depth	An Information Security strategy integrating people, technology, and operations capabilities to establish variable barriers across multiple layers and missions of the organization [NIST SP800-53Rev5]
Denial of service (DoS)	The prevention of authorized access to a system resource or the delaying of system operations and functions. (See: availability, critical, flooding) [RFC 4949]
distributed attack	1a. An attack that is implemented with distributed computing (See: zombie) 1b. An attack that deploys multiple threat agents [RFC 4949]
Distributed denial of service (DDoS)	A denial of service technique that uses numerous hosts to perform the attack [CNSSI 4009]
Disruption	A circumstance or event that interrupts or prevents the correct operation of system services and functions. (See: availability, critical, system integrity, threat consequence) [RFC 4949]
Domain	An environment or context that (a) includes a set of system resources and a set of system entities that have the right to access the resources and (b) usually is defined by a security policy, security model, or security architecture. (See: CA domain, domain of interpretation, security perimeter. Compare: COI, enclave) [RFC 4949]
Eavesdropping	Passive wiretapping done secretly, i.e., without the knowledge of the originator or the intended recipients of the communication [RFC 4949]
Effectiveness	/ITSEC/ A property of a TOE representing how well it provides security in the context of its actual or proposed operational use [RFC 4949]
Encryption	1. Cryptographic transformation of data (called "plain text") into a different form (called "cipher text") that conceals the data's original meaning and prevents the original form from being used. The corresponding reverse process is "decryption", a transformation that restores encrypted data to its original form (See: cryptography) [RFC 4949]
Event	An observable occurrence in an information system/network [NICCS]
Exposure	A type of threat action whereby sensitive data is directly released to an unauthorized entity (See: unauthorized disclosure.) [RFC 4949]
Fail Safe	A mode of termination of system functions that prevents damage to specified system resources and system entities (i.e., specified data, property, and life) when a failure occurs or is detected in the system (but the failure still might cause a security compromise). (See: failure control) [RFC 4949]
Fail Secure	A mode of termination of system functions that prevents loss of secure state when a failure occurs or is detected in the system (but the failure still might cause damage to some system resource or system entity). (See: failure control. Compare: fail-safe) [RFC 4949]
Firewall	An internetwork gateway that restricts data communication traffic to and from one of the connected networks (the one said to be "inside" the firewall) and thus protects that network's system resources against threats from the other network (the one that is said to be "outside" the firewall). (See: guard, security gateway) [RFC 4949]
Firmware	Computer programs and data stored in hardware -- typically in read-only memory (ROM) or programmable read-only memory (PROM) -- such that the programs and data cannot be dynamically written or modified during execution of the programs. (See: hardware, software) [RFC 4949]

Table A1 (Continued)

Term	Definition
Flaw	An error in the design, implementation, or operation of an information system. A flaw may result in a vulnerability (Compare: vulnerability) [RFC 4949]
Flooding	An attack that attempts to cause a failure in a system by providing more input than the system can process properly (See: denial of service, fairness. Compare: jamming) [RFC 4949]
Governance	An organizational process of providing strategic direction for the organization while ensuring that it meets its obligations, appropriately manages risk, and efficiently uses financial and human resources. Governance also typically includes the concepts of sponsorship (setting the managerial tone), compliance (ensuring that the organization is meeting its compliance obligations), and alignment (ensuring that processes such as those for cybersecurity program management align with strategic objectives) [CERT RMM]
Hacker	Someone with a strong interest in computers, who enjoys programming them, and experimenting and otherwise working with them. (See: hack. Compare: adversary, cracker, intruder) [RFC 4949]
Harden	To protect a system by configuring it to operate in a way that eliminates known vulnerabilities (See: default account) [RFC 4949]
Identity	The collective aspect of a set of attribute values (i.e., a set of characteristics) by which a system user or other system entity is recognizable or known. (See: authenticate, registration. Compare: identifier) [RFC 4949]
Incident	See Security incident [RFC 4949]
Information assurance	/U.S. Government/ Measures that protect and defend information and information systems by ensuring their availability, integrity, authentication, confidentiality, and non-repudiation. These measures include providing for restoration of information systems by incorporating protection, detection, and reaction capabilities [RFC 4949]
Information security (INFOSEC)	Measures that implement and assure security services in information systems, including in computer systems (see: COMPUSEC) and in communication systems (see: COMSEC) [RFC 4949]
Information security governance	The set of responsibilities and practices exercised by the board and executive management with the goal of providing strategic direction, ensuring that objectives are achieved, ascertaining that risk is managed appropriately and verifying that the enterprise's resources are used responsibly [ISACA Glossary]
Information security policy	Aggregate of directives, regulations, and rules that prescribe how an organization manages, protects, and distributes information [CNSSI 4009]
Information security program	The overall combination of technical, operational and procedural measures and management structures implemented to provide for the confidentiality, integrity and availability of information based on business requirements and risk analysis [OECD 2002]
Information system	An organized assembly of computing and communication resources and procedures -- i.e., equipment and services, together with their supporting infrastructure, facilities, and personnel – that create, collect, record, process, store, transport, retrieve, display, disseminate, control, or dispose of information to accomplish a specified set of functions. (See: system entity, system resource. Compare: computer platform) [RFC 4949]

Table A1 (Continued)

Term	Definition
Information system resilience	The ability of an information system to continue to: (i) operate under adverse conditions or stress, even if in a degraded or debilitated state, while maintaining essential operational capabilities; and (ii) recover to an effective operational posture in a time frame consistent with mission needs [NIST SP800-39]
Integrity	See: data integrity, datagram integrity service, correctness integrity, source integrity, stream integrity service, system integrity [RFC 4949]
Integrity service	A security service that protects against unauthorized changes to data, including both intentional change or destruction and accidental change or loss, by ensuring that changes to data are detectable [RFC 4949]
Intellectual property (IP)	Creations of the mind such as musical, literary, and artistic works; inventions; and symbols, names, images, and designs used in commerce, including copyrights, trademarks, patents, and related rights. Under intellectual property law, the holder of one of these abstract "properties" has certain exclusive rights to the creative work, commercial symbol, or invention by which it is covered [CNSSI 4009]
Interception	A type of threat action whereby an unauthorized entity directly accesses sensitive data while the data is traveling between authorized sources and destinations (See: unauthorized disclosure) [RFC 4949]
Intruder	An entity that gains or attempts to gain access to a system or system resource without having authorization to do so (See: intrusion. Compare: adversary, cracker, hacker) [RFC 4949]
Intrusion	1.A security event, or a combination of multiple security events, that constitutes a security incident in which an intruder gains, or attempts to gain, access to a system or system resource without having authorization to do so. (See: IDS.) 2. A type of threat action whereby an unauthorized entity gains access to sensitive data by circumventing a system's security protections. (See: unauthorized disclosure) [RFC 4949]
Intrusion detection	Sensing and analyzing system events for the purpose of noticing (i.e., becoming aware of) attempts to access system resources in an unauthorized manner. (See: anomaly detection, IDS, misuse detection. Compare: extrusion detection) [RFC 4949]
Intrusion detection system	1. A process/subsystem, implemented in software/hardware, that automates tasks of (a) monitoring events in a computer network and (b) analyzing them for security problems. (See: intrusion detection.) 2. A security alarm system to detect unauthorized entry [RFC 4949]
Intrusion prevention system	System that can detect an intrusive activity and can also attempt to stop the activity, ideally before it reaches its targets [CNSSI 4009]
Key	1a. /cryptography/ An input parameter used to vary a transformation function performed by a cryptographic algorithm (See: private key, public key, storage key, symmetric key, traffic key. Compare: initialization value) 1b. /cryptography/ Used in singular form as a collective noun referring to keys or keying material. Example: A fill device can be used transfer key between two cryptographic devices. 2. /anti-jam/ An input parameter used to vary a process that determines patterns for an anti-jam measure (See: frequency hopping, spread spectrum) [RFC 4949]
Key management	1a. The process of handling keying material during its life cycle in a cryptographic system; and the supervision and control of that process (See: key distribution, key escrow, keying material, public-key infrastructure) [RFC 4949]

Table A1 (Continued)

Term	Definition
Legacy system	A system that is in operation but will not be improved or expanded while a new system is being developed to supersede it [RFC 4949]
Malicious code	Software or firmware intended to perform an unauthorized process that will have adverse impact on the confidentiality, integrity, or availability of an information system. A virus, worm, Trojan horse, or other code-based entity that infects a host. Spyware and some forms of adware are also examples of malicious code [NIST SP800-53r5]
Management controls	Actions taken to manage the development, maintenance, and use of the system, including system-specific policies, procedures and rules of behavior, individual roles and responsibilities, individual accountability, and personnel security decisions [CNSSI 4009]
Man-in-the-middle attack	A form of active wiretapping attack in which the attacker intercepts and selectively modifies communicated data to masquerade as one or more of the entities involved in a communication association. (See: hijack attack, piggyback attack) [RFC 4949]
Masquerade	A type of threat action whereby an unauthorized entity gains access to a system or performs a malicious act by illegitimately posing as an authorized entity (See: deception) [RFC 4949]
Measure (noun)	Variable to which a value is assigned as the result of measurement NOTE: "measures" is used to refer collectively to base measures, derived measures, and indicators [ISO/IEC/IEE 15939:2017]
Measurement	Set of operations having the object of determining a value of a measure [ISO/IEC/IEE 15939:2017]
Mission critical	A condition of a system service or other system resource such that denial of access to, or lack of availability of, the resource would jeopardize a system user's ability to perform a primary mission function or would result in other serious consequences. (See: Critical. Compare: mission essential) [RFC 4949]
Non-repudiation service	A security service that provides protection against false denial of involvement in an association (especially a communication association that transfers data) [RFC 4949]
Non-repudiation with proof of origin	A security service that provides the recipient of data with evidence that proves the origin of the data, and thus protects the recipient against an attempt by the originator to falsely deny sending the data. (See: non-repudiation service) [RFC 4949]
Non-repudiation with proof of receipt	A security service that provides the originator of data with evidence that proves the data was received as addressed, and thus protects the originator against an attempt by the recipient to falsely deny receiving the data. (See: non-repudiation service) [RFC 4949]
Operational security	System capabilities, or performance of system functions, that are needed either (a) to manage a system or (b) to manage security features of a system (Compare: operations security (OPSEC)) [RFC 4949]
Operations security (OPSEC)	A process to identify, control, and protect evidence of the planning and execution of sensitive operations, and thereby prevent potential adversaries from gaining knowledge of capabilities and intentions. (See: communications cover. Compare: operational security) [RFC 4949]
Personal Identifiable information (PII)	Any information about an individual maintained by an agency, including (1) any information that can be used to distinguish or trace an individual's identity, such as name, social security number, date and place of birth, mother's maiden name, or biometric records; and (2) any other information that is linked or linkable to an individual, such as

Table A1 (Continued)

Term	Definition
	medical, educational, financial, and employment information [NIST SP800-122]
Physical security	Tangible means of preventing unauthorized physical access to a system. Examples: Fences, walls, and other barriers; locks, safes, and vaults; dogs and armed guards; sensors and alarm bells (See: security architecture) [RFC 4949]
Quality of service (QoS)	The measurable end-to-end performance properties of a network service, which can be guaranteed in advance by a Service Level Agreement (SLA) between a user and a service provider, so as to satisfy specific customer application requirements. Note: These properties may include throughput (bandwidth), transit delay (latency), error rates, priority, security, packet loss, packet jitter, etc. [CNSSI 4009]
Reliability	The ability of a system to perform a required function under stated conditions for a specified period of time. (Compare: availability, survivability) [RFC 4949]
Repudiation	1. Denial by a system entity that was involved in an association (especially a communication association that transfers data) of having participated in the relationship. (See: accountability, non-repudiation service) 2. A type of threat action whereby an entity deceives another by falsely denying responsibility for an act. (See: deception) [RFC 4949]
Residual risk	The portion of an original risk or set of risks that remains after countermeasures have been applied. (Compare: acceptable risk, risk analysis) [RFC 4949]
Resilience	The ability to prepare for and adapt to changing conditions and withstand and recover rapidly from disruptions. Resilience includes the ability to withstand and recover from deliberate attacks, accidents, or naturally occurring threats or incidents [PPD 21]
Risk	1. An expectation of loss expressed as the probability that a particular threat will exploit a particular vulnerability with a particular harmful result. (See: residual risk.) 2. /SET/ The possibility of loss because of one or more threats to information (not to be confused with financial or business risk) [RFC 4949]
Risk mitigation	Prioritizing, evaluating, and implementing the appropriate risk-reducingcontrols/countermeasures recommended from the risk management process [CNSSI 4009]
Safeguards	Protective measures prescribed to meet the security requirements (i.e., confidentiality, integrity, and availability) specified for an information system. Safeguards may include security features, management constraints, personnel security, and security of physical structures, areas, and devices. Synonymous with security controls and countermeasures [CNSSI 4009]
Safety	The property of a system being free from risk of causing harm (especially physical harm) to its system entities (Compare: security) [RFC 4949]
Secure state	1a. A system condition in which the system is in conformance with the applicable security policy. (Compare: clean system, transaction.) 1b. /formal model/ A system condition in which no subject can access any object in an unauthorized manner. (See: secondary definition under "Bell-LaPadula model") [RFC 4949]

Table A1 (Continued)

Term	Definition
Security	1a. A system condition that results from the establishment and maintenance of measures to protect the system. 1b. A system condition in which system resources are free from unauthorized access and from unauthorized or accidental change, destruction, or loss. (Compare: safety) 2. Measures taken to protect a system [RFC 4949]
Security architecture	A plan and set of principles that describe (a) the security services that a system is required to provide to meet the needs of its users, (b) the system components required to implement the services, and (c) the performance levels required in the components to deal with the threat environment (e.g., [R2179]). (See: defense in depth, IATF, OSIRM Security Architecture, security controls, Tutorial under "security policy") [RFC 4949]
Security assurance	1. An attribute of an information system that provides grounds for having confidence that the system operates such that the system's security policy is enforced. (Compare: trust.) 2. A procedure that ensures a system is developed and operated as intended by the system's security policy [RFC 4949]
Security by obscurity	Attempting to maintain or increase security of a system by keeping secret the design or construction of a security mechanism [RFC 4949]
Security compromise	A security violation in which a system resource is exposed, or is potentially exposed, to unauthorized access (Compare: data compromise, exposure, violation) [RFC 4949]
Security controls	The management, operational, and technical controls (safeguards or countermeasures) prescribed for an information system which, taken together, satisfy the specified security requirements and adequately protect the confidentiality, integrity, and availability of the system and its information [FP199] (See: security architecture) [RFC 4949]
Security Controls assessment	The testing and/or evaluation of the management, operational, and technical security controls in an information system to determine the extent to which the controls are implemented correctly, operating as intended, and producing the desired outcome with respect to meeting the security requirements for the system [CNSSI 4009]
Security domain	See Domain [RFC 4949]
Security Engineering	An interdisciplinary approach and means to enable the realization of secure systems. It focuses on defining customer needs, security protection requirements, and required functionality early in the systems development lifecycle, documenting requirements, and then proceeding with design, synthesis, and system validation while considering the complete problem [CNSSI 4009]
Security environment	The set of external entities, procedures, and conditions that affect secure development, operation, and maintenance of a system (See: "first law" under "Courtney's laws") [RFC 4949]
Security event	An occurrence in a system that is relevant to the security of the system (See: security incident) [RFC 4949]
Security incident	A security event that involves a security violation. (See: CERT, security event, security intrusion, security violation) [RFC 4949]
Security intrusion	A security event, or a combination of multiple security events, that constitutes a security incident in which an intruder gains, or attempts to gain, access to a system or system resource without having authorization to do so [RFC 4949]

Table A1 (Continued)

Term	Definition
Security management infrastructure	System components and activities that support security policy by monitoring and controlling security services and mechanisms, distributing security information, and reporting security events [RFC 4949]
Security mechanism	A method or process (or a device incorporating it) that can be used in a system to implement a security service that is provided by or within the system. (See: Tutorial under "security policy". Compare: security doctrine) [RFC 4949]
Security model	A schematic description of a set of entities and relationships by which a specified set of security services are provided by or within a system. Example: Bell-LaPadula model, OSIRM. (See: Tutorial under "security policy") [RFC 4949]
Security policy	A set of criteria for the provision of security services. [NIST SP 800-53r5]
Security posture	The security status of an enterprise's networks, information, and systems based on information assurance (IA) resources (e.g., people, hardware, software, policies) and capabilities in place to manage the defense of the enterprise and to react as the situation changes [CNSSI 4009]
Security program plan	Formal document that provides an overview of the security requirements for an organization-wide information security program and describes the program management security controls and common security controls in place or planned for meeting those requirements [CNSSI 4009]
Security requirements	Requirements levied on an information system that are derived from applicable laws, Executive Orders, directives, policies, standards, instructions, regulations, or procedures, or organizational mission/business case needs to ensure the confidentiality, integrity, and availability of the information being stored, or transmitted [FIPS PUB 200]
Security service	1. A processing or communication service that is provided by a system to give a specific kind of protection to system resources. (See: access control service, audit service, availability service, data confidentiality service, data integrity service, data origin authentication service, non-repudiation service, peer entity authentication service, system integrity service). Tutorial: Security services implement security policies, and implemented by security mechanisms. 2. A service, provided by a layer of communicating open systems, [that] ensures adequate security of the systems or the data transfers [RFC 4949]
System	A combination of interacting elements organized to achieve one or more stated purposes [ISO/IEC 15288:2015]
System integrity	An attribute or quality that a system has when it can perform its intended function in a unimpaired manner, free from deliberate or inadvertent unauthorized manipulation (See: recovery, system integrity service) [RFC 4949]
System integrity service	A security service that protects system resources in a verifiable manner against unauthorized or accidental change, loss, or destruction. (See: system integrity) [RFC 4949]
Systems Security Engineering	Systems security engineering is a specialty engineering field strongly related to systems engineering. It applies scientific, engineering, and information assurance principles to deliver trustworthy systems that satisfy stakeholder requirements within their established risk tolerance [CNSSI 4009]

Table A1 (Continued)

Term	Definition
Technical security	Security mechanisms and procedures that are implemented in and executed by computer hardware, firmware, or software to provide automated protection for a system. (See: security architecture. Compare: administrative security) [RFC 4949]
Threat	1a. A potential for violation of security, which exists when there is an entity, circumstance, capability, action, or event that could cause harm. (See: dangling threat, INFOCON level, threat action, threat agent, threat consequence. Compare: attack, vulnerability.) 1b. Any circumstance or event with the potential to adversely affect a system through unauthorized access, destruction, disclosure, or modification of data, or denial of service [C4009] (See: sensitive information) [RFC 4949]
Threat action	A realization of a threat, i.e., an occurrence in which system security is assaulted as the result of either an accidental event or an intentional act. (See: attack, threat, threat consequence) [RFC 4949]
Threat agent	A system entity that performs a threat action, or an event that results in a threat action [RFC 4949]
Threat consequence	A security violation that results from a threat action [RFC 4949]
Trust	1. /information system/ A feeling of certainty (sometimes based on inconclusive evidence) either (a) that the system will not fail or (b) that the system meets its specifications (i.e., the system does what it claims to do and does not perform unwanted functions). (See: trust level, trusted system, trustworthy system. Compare: assurance) 2. /PKI/ A relationship between a certificate user and a CA in which the user acts according to the assumption that the CA creates only valid digital certificates [RFC 4949]
Trusted system	1. /information system/ A system that operates as expected, according to design and policy, doing what is required – despite environmental disruption, human user and operator errors, and attacks by hostile parties -- and not doing other things (See: trust level, trusted process. Compare: trustworthy) 2. /multilevel secure/ A [trusted system is a] system that employs sufficient hardware and software assurance measures to allow its use for simultaneous processing of a range of sensitive or classified information (See: multilevel security mode)) [RFC 4949]
Trustworthiness (system)	The degree to which an information system (including the information technology components that are used to build the system) can be expected to preserve the confidentiality, integrity, and availability of the information being processed, stored, or transmitted by the system across the full range of threats. A trustworthy information system is believed to operate within defined levels of risk despite the environmental disruptions, human errors, structural failures, and purposeful attacks that are expected to occur in its environment of operation [NIST SP800-160v1]
Trustworthy system	A system that not only is trusted, but also warrants that trust because the system's behavior can be validated in some convincing way, such as through formal analysis or code review (See: trust. Compare: trusted) [RFC 4949]
Vulnerability	A flaw or weakness in a system's design, implementation, or operation and management that could be exploited to violate the system's security policy (See: harden) [RFC 4949]

Table A1 (Continued)

Term	Definition
Vulnerability assessment	Systematic examination of an information system or product to determine the adequacy of security measures, identify security deficiencies, provide data from which to predict the effectiveness of proposed security measures, and confirm the adequacy of such measures after implementation [CNSSI 4009]
Zero day attack	An attack that exploits a previously unknown hardware, firmware, or software vulnerability [CNSSI 4009]

A2. Applying NIST Cybersecurity Framework to Power Grid

The goals include: Maintain Safety, Reliability, Resilience, Support Grid Modernization. This document [NIST SP1108r4] includes evaluation of the NIST Cybersecurity Framework (NCF) functions and guidance on how to apply these functions and activities to secure the power grid. A summary of these functions includes:

- Identify function is critical in the development of the foundation for cybersecurity management, and in the understanding of cyber risk to systems, assets, data, and capabilities. Activities in the Asset Management, Business Environment, Risk Assessment, Risk Management Strategy, and Supply Chain Risk Management are the primary security areas that address protections for the four business objectives.
- Protect function is critical to limit the impact of a potential cybersecurity event. Identity Management and Access Control, Awareness and Training, Information Protection Processes, Maintenance, and Protective Technology are the priority security focus areas. Identity Management and Access Control identifies and regulates personnel ingress and egress. Awareness and Training and the Protection Processes prepare the workforce to achieve cybersecurity. Protective technology implements security decisions.
- Detect function enables timely discovery of cybersecurity events. Real time awareness and continuous monitoring of the systems is critical to detect cybersecurity events.
- Respond function supports the ability to contain the impact of a potential cybersecurity event. Rapid and effective response and communication to cybersecurity incidents is critical in protecting personnel and environmental safety. Situational awareness to the event unfolding is needed to properly address it.
- Recover function supports timely recovery to normal operations to reduce the impact of a cybersecurity event. Recovery objectives are needed when recovering from disruptions.

Figure A1 shows the core functions and related concepts of NIST cybersecurity framework for improving critical infrastructure [NIST 2018]. This document is designed for identifying and developing cybersecurity risk frameworks for voluntary use by critical infrastructure owners and operators. Obviously, the power grid is part of energy sector, which is considered a component of critical infrastructure. This document focuses on using business drivers to guide cybersecurity activities and considers cybersecurity risks as part of the organization's risk management processes. It outlines a maturity model of tiers against which organizations can benchmark the efficiency of their security program. The framework complements existing business and cybersecurity operations. It can serve as the foundation for a new cybersecurity program or a mechanism for improving an existing program. It provides means of expressing cybersecurity requirements in an organization and can help on identifying gaps in an organization's cybersecurity practices. It also provides a general set of considerations and processes for considering privacy and civil liberties implications in the context of a cybersecurity program. The framework can be applied throughout the life cycle phases of plan, design, build/buy, deploy, operate, and decommission. The framework organizes the five cybersecurity functions into categories, that can be used to group similar cybersecurity activities that support a particular function.

Function Unique Identifier	Function	Category Unique Identifier	Category
ID	Identify	ID.AM	Asset Management
		ID.BE	Business Environment
		ID.GV	Governance
		ID.RA	Risk Assessment
		ID.RM	Risk Management Strategy
		ID.SC	Supply Chain Risk Management
PR	Protect	PR.AC	Access Control
		PR.AT	Awareness and Training
		PR.DS	Data Security
		PR.IP	Information Protection Processes and Procedures
		PR.MA	Maintenance
		PR.PT	Protective Technology
DE	Detect	DE.AE	Anomalies and Events
		DE.CM	Security Continuous Monitoring
		DE.DP	Detection Processes
RS	Respond	RS.RP	Response Planning
		RS.CO	Communications
		RS.AN	Analysis
		RS.MI	Mitigation
		RS.IM	Improvements
RC	Recover	RC.RP	Recovery Planning
		RC.IM	Improvements
		RC.CO	Communications

Figure A1 Cybersecurity Framework functions and categories. *Source*: [NIST SP1108r4]. Public Domain.

Figure A1 shows the five functions divided into the 23 categories of cybersecurity activities. The framework provides an additional level of granularity by further dividing the 23 Categories of cybersecurity activities into 108 Subcategories, each defining a desired outcome. These Subcategories provide a list of outcomes that — if achieved — would increase the likelihood that an organization could successfully optimize its cybersecurity. Table A2 includes a summary of the security requirements and recommendations based on NIST Cybersecurity Framework [NCF 2018].

Distributed energy resource (DER) system manufacturers are required to support these functions:

- Monitoring function
  - DER provides nameplate, configuration, status, measurements, and other requested data.
  - DER provides status, measurements, alarms, logs, and other data as authorized and requested by users. Examples include connect status, updated capacities, real and reactive

power output/consumption, state of charge, voltage, and other measurements. Also of interest are forecast statuses and expected measurements.

- Collect and provide historical information
  - Collect and provide detailed measurement and performance data which may be valuable to record in an operational historian.
  - Collect and provide detailed measurement and performance data, which may be used to assess the real-time responses to power system events, control commands, and autonomous functions. This data could also be used to determine actual capabilities, impacts, compliance, and other characteristics of DER systems.

Data collected as result of monitoring and collection in historical logs should be secure (protect integrity, availability, and confidentiality) during processing or use for analysis, in transit, at rest in logs or historian; or when data is disposed.

Table A2 NCF core functions and subcategories recommendations for power grid cybersecurity.

NCF Category	NCF Subcategory	Recommendations – Examples
Identity Function (ID)	Asset Management	Inventory of hardware, software, networks, internal and external information systems, devices, personnel, users, stakeholders' roles and responsibilities
	Business Environment	Supply chain for IT and OT products and services business partners, and other relevant third parties, understand magnitude of potential cascading effects, understand variety of state and local regulatory requirements, implement specific requirements to ensure resilient operation of the power system, information security governance, information security policies to cover both IT and OT systems, establish roles and responsibilities
	Risk Assessment	Identifying and documenting asset vulnerabilities, potential threats, security risks, consequences of cyber-attacks, impacts to business owners and partners, electricity delivery and operation of power systems
	Risk Assessment Strategy	Comprehensive strategy to manage risk including integrating the modernized components of the grid into the determination and description of risk tolerance, determining organizational risk tolerance and considering the cascading effects in case of disruption events
	Supply Chain Risk Management	Power system owners/operators should implement robust processes for managing cybersecurity risks occurring from supply chains that include members of a diverse ecosystem (Industrial Control System (ICS) vendors, commercial off-the-shelf (COTS) providers, etc.); power system owners/operators to mutually agree on a set of appropriate security requirements in order to manage security risks for business partners and their third-party partners

Table A2 (Continued)

NCF Category	NCF Subcategory	Recommendations – Examples
Protect Function (PR)	Access Control	Identity management for all users, devices, and processes in both traditional and modernized environments, limited remote access, least privilege for limiting permissions and authorizations to manage connected devices; authentication for all devices connecting to the network grid to ensure that only authorized devices are allowed , network segmentation for containing potential incidents (safety, reliability) and limiting the damage (resilience); identify risks introduced during some situations (e.g., emergency maintenance or need to restore power) in which the binding and proofing of credentials may interfere with safety, reliability, and resilience; establish criteria and work around if those credentials are not required for certain user actions
	Awareness and Training	Security awareness training to all users, including manufacturing system users and managers; Training and responsibilities for physical and information security personnel adjusted to the unique threats and risks of the grid modernization environment and distributed and multi-owner nature of the grid .
	Data Security	Protect the integrity of control information and device configuration at rest and transit; protect integrity of hardware/software/firmware running on system (critical to safety, reliability, resilience); separation of development and testing environments; protect data from data leaks; disposal of configuration data should be under strict control; find alternative solutions to protect confidentiality of data when resource-intensive cryptographic mechanisms interfere with the functional performance of control systems and use additional methods to protect data in transit using less resource intensive cryptographic mechanisms (see more on DER Data Security)
	Information Protection Processes and Procedures	Create and maintain baseline configurations for all devices owned by a power system owner/operator; systems development lifecycle to ensure quality and predictable performance of systems and networks; configuration change control processes; regular backups for configuration files; configuration change control processes; data and device disposal processes, include devices in cybersecurity response and recovery plans and their testing; support collaboration between IT and OT personnel; continuous assessment and improvement of processes, information and physical security policies
	Maintenance	Maintenance of geographical dispersion, and the mission need for remote maintenance of both legacy and modernized devices
	Protective Technology	Create logs to capture information and protect logs from manipulation; protect and restrict the use of removable media on modernized devices; information from logs can be used to trace adversary's activities and find anomalies; power system owners/operators should consider all possible ways to achieve resiliency requirements;

Table A2 (Continued)

NCF Category	NCF Subcategory	Recommendations – Examples
Detect Function (DE)	Anomalies and Events	Create a baseline of network operations and expected data flows in the OT context because information flows are predictable, and control systems generally have few users; use control information flows to monitor and detect unusual network behavior and allow for timely response; analysis of detected cybersecurity events and determine impacts; establish incident alert thresholds
	Security Continuous Monitoring	monitor the grid for cybersecurity events in physical environment and personnel activity; power system owners/operators should use malicious code detection methodologies in both traditional and modernized infrastructure; detect unauthorized mobile code; perform vulnerability scans to identify vulnerabilities; include an agreement to scan 3rd party-owned devices that are connected to their grid
	Detection Processes	Establish roles and responsibilities with respect to detection restoration across power system ownership lines and within a single power system owner/operator with traditional and modernized components and networks; ensure that detection activities comply with jurisdiction-specific safety requirements; identify any potential negative impact to the power system due to testing of detection processes, continuous improvement of detection processes
Response Function (RS)	Response Planning	Create Response plan and execute it in both traditional and modernized environments
	Communications	Create roles and responsibilities with respect to response and grid restoration across power system ownership lines and within a single power system owner/operator; establish criteria for reporting incidents; share the event response information to restoration crews and execute it across traditional and modernized systems and components; share incident information across power system boundaries and external stakeholders to include neighboring power system owners/operators
	Analysis	Investigate notifications from detection events; understand any similarities and differences in impacts between the traditional and modernized environments; preserve evidence for forensics analysis of incidents; categorize incidents; use processes for receiving and analyzing vulnerability information
	Mitigation	Perform containing and mitigation of incidents for both traditional and modernized infrastructures; establish processes to receive vulnerability information from vendors, share vulnerability information with device owners/operators across power systems that may have different ownership
	Improvements	Use of lessons learned for improvement of recovery processes and activities; continuous Updates to recovery strategies to improve future activities and response

Table A2 (Continued)

NCF Category	NCF Subcategory	Recommendations – Examples
Recover Function (RC)	Recovery Planning	Identify consequences to the safety of power system owner/operator workers (e.g., linemen) when the cyber security recovery plan is executed; the plan should include both traditional and modernized parts of the grid
	Improvements	Incorporate lessons-learned into plans; update recovery strategies
	Communications	Coordinate cyber security event recovery activities to ensure safety of power system owner operator workers (e.g., linemen) working on power recovery; coordination of recovery efforts across power systems, some of which may be modernized and some not
Because the grid is a large cyber-physical system, governance and risk management processes should address all risks, not just cybersecurity		
Power system owners/operators should consider all possible ways to achieve resiliency requirements for protective technologies		

Adapted from [NIST SP1108r4]. Public Domain.
*DER Data Security

A3. Catalog of Standards

The SGIP Catalog of Standards (CoS) is a compendium of standards and practices considered to be relevant for the development and deployment of a robust and interoperable Smart Grid, and the Catalog contains more information in [NIST SP1108r4] document. Table A3 include a list of standard organizations with focus on Smart Grid.

Table A3 Organizations and Standards considered in the Catalog of Standards for Smart Grid.

Organization	Name	Recognized -International, Regional, National (USA)	Examples of Protocols, Standards
ANSI	American National Standards Institute	USA	ANSI C12 Suite
ASHRAE	American Society of Heating, Refrigeration, and Air Conditioning Engineers	USA	BacNet
CEA	Consumer Electronics Association	USA	Lon Protocol Suite
DHS	Department of Homeland Security (DHS)	USA	DHS Cyber Security Procurement Language for Control Systems
GWAC	GridWise Architecture Council	USA	GWAC Interoperability Model
IEC	International Electrotechnical Commission	International	IEC 60870, IEC 61850, IEC 62351
IEEE	Institute of Electrical and Electronics Engineers	International	IEEE 1815 (DNP3), IEEE C37.118.1-2011, IEEE 1547

Table A3 (Continued)

Organization	Name	Recognized -International, Regional, National (USA)	Examples of Protocols, Standards
IETF	Internet Engineering Task Force	International	RFC 6272, IPv4/IPv6
ISO	International Organization for Standardization	International	ISO 9000
ITU	International Telecommunication Union	International	ITU-T G.9972
NAESB	North American Energy Standards Board	USA	REQ-21 Energy Services Provider Interface (ESPI)
NEMA	National Electrical Manufacturers Association	USA	SG-AMI 1-2009
NERC	North American Electric Reliability Corporation	Regional	NERC CIP 002-014
NIST	National Institute of Standards and Technology	USA	NIST Interagency Report NISTIR 7628
NRECA	National Rural Electric Cooperative Association	USA	MultiSpeak
OASIS	Organization for the Advancement of Structured Information Standard	International	EMIX
OGC	Open Geospatial Consortium	International	GML
OPC-UA	Open Platform Communications United Architecture	USA	SOA Architecture
OpenADR	OpenADR Alliance	International	OpenADR 2.0
SAE	Society of Automotive Engineers	International	SAEJ2847/1
UCAIug	Utility Communications Architecture International Users Group	International	OpenHAN
W3C	World Wide Web Consortium	International	HTML, XML
ZigBee	ZigBee Alliance	International	Smart Energy Profile 2.0

Note: Some organizations are not included in CoS
Other standard organizations focused on Smart Grid and technologies include:
AFNOR – Association Française de Normalisation (French Standardization Association)
BSI – British Standards Institution
CEN- European Committee for Standardization
CENELEC - European Committee for Electrotechnical Standardization
DIN – Deutsches Institut für Normung (German Standards Institute)
ENISA – European Network for Information Security Agency
ETSI – European Telecommunications Standards Institute
UL – Underwriter's Laboratories.

A4. Supplemental Standards Guidance

A4.1 NIST SP800-82r2

This standard includes 18 security related areas:

- Access control - the process of granting or denying specific requests for obtaining and using information and related information processing services for physical access to areas within the information system environment.
- Access Control Technologies - filter and blocking technologies designed to direct and regulate the flow of information between devices or systems once authorization has been determined.
- Awareness and training - policies and procedures to ensure that all information system users are given appropriate security training relative to their usage of the system and that accurate training records are maintained.
- Audit and accountability - independent review and examination of records and activities to assess the adequacy of system controls, to ensure compliance with established policies and operational procedures, and to recommend necessary changes in controls, policies, or procedures.
- Security assessment and authorization - assurance that the specified controls are implemented correctly, operating as intended, and producing the desired outcome.
- Contingency planning - policies and procedures designed to maintain or restore business operations, including computer operations, possibly at an alternate location, in the event of emergencies, system failures, or disaster.
- Configuration management - policies and procedures for controlling modifications to hardware, firmware, software, and documentation to ensure the information system is protected against improper modifications prior to, during, and after system implementation.
- Identification and authentication - the process of verifying the identity of a user, process, or device, through the use of specific credentials (e.g., passwords, tokens, biometrics), as a prerequisite for granting access to resources in an IT system.
- Incident response - policies and procedures pertaining to incident response training, testing, handling, monitoring, reporting, and support services.
- Maintenance - policies and procedures to manage all maintenance aspects of an information system.
- Media protection - policies and procedures to ensure secure handling of media. Controls cover access, labeling, storage, transport, sanitization, destruction, and disposal.
- Physical and environmental protection - policies and procedures addressing physical, transmission, and display access control as well as environmental controls for conditioning (e.g., temperature, humidity) and emergency provisions (e.g., shutdown, power, lighting, fire protection).
- Planning - development and maintenance of a plan to address information system security by performing assessments, specifying and implementing security controls, assigning security levels, and responding to incidents.
- Personnel security - policies and procedures for personnel position categorization, screening, transfer, penalty, and termination; also addresses third-party personnel security.
- Risk assessment - the process of identifying risks to operations, assets, or individuals by determining the probability of occurrence, the resulting impact, and additional security controls that would mitigate this impact.
- System and services acquisition - allocation of resources for information system security to be maintained throughout the systems life cycle and the development of acquisition policies based on risk assessment results including requirements, design criteria, test procedures, and associated documentation.
- System and communications protection - mechanisms for protecting both system and data transmission components.
- System and information integrity - policies and procedures to protect information systems and their data from design flaws and data modification using functionality verification, data

integrity checking, intrusion detection, malicious code detection, and security alert and advisory controls.

- Program management - provides security controls at the organizational rather than the information system level.

A4.2 NIST SP800-53r5

Appendix J of [NIST SP 800-53r5] document includes a catalog of Privacy Controls. Privacy controls are the administrative, technical, and physical safeguards employed within organizations to protect and ensure the proper handling of personally identifiable information (PII). The privacy control families are each aligned with the Fair Information Practice Principles (FIPPs), which are designed to build public trust in an organization's privacy practices and to help organizations avoid tangible costs and intangible damages stemming from privacy incidents.

The FIPPs are widely accepted in the United States and internationally as a general framework for privacy and are reflected in other federal and international laws and policies. In a number of organizations, FIPPs serve as the basis for analyzing privacy risks and determining appropriate mitigation strategies. The Federal Enterprise Architecture Security and Privacy Profile (FEA-SPP) also provided information and materials in development of the privacy controls

A4.3 Industrial Control Systems Guidance

Table A4 includes a list of standards that provide guidance on security for industrial control systems (ICS).

Table A4 Examples of standards and use recommendations.

Standard	Use Recommendation
NIST SP800-53r5	Describes security controls for each of control categories (families) including privacy controls, provides background information on the control family, access control technologies and their use with ICS as well as implementation considerations for ICS owners
IEC 62443 series	Provides ICS-specific security recommendations and guidance
ISO/IEC 62351 series	Provides information security as applied to power systems
NIST SP800-48	Provides guidance on wireless network security with particular emphasis on the IEEE 802.11b and Bluetooth standards
NIST SP800-97	Provides guidance on IEEE 802.11i wireless network security
NIST SP800-92	Provides guidance on log management (including audit logs)
NIST SP800-100	Provides guidance on information security governance and planning
NIST SP800-53A	Provides guidance on security control assessments
NIST SP800-37	Provides guidance defining the information system boundary and security certification and accreditation of the information system
NIST SP800-53	Describes Contingency Planning (CP) family that includes policies and procedures to implement a contingency plan
NIST SP800-34	Provides guidance on contingency planning

A4.4 Supply Chain Risk Management

NIST SP800-161 describes Cybersecurity-Supply Chain Risk Management (C-SCRM) which is the organized and purposeful management of cybersecurity risk in the supply chain. C-SCRM requires enterprise recognition and awareness and lies at the intersection of security, integrity quality, and resilience as depicted in Figure A2. These dimensions are layers of consideration for enterprises as they approach C-SCRM and should be considered the outputs of effective C-SCRM.

Figure A2 C-SCRM Dimensions. Source: [NIST SP800-161]. Public Domain.

References

[CNSSI 4009] CNSSI. (2015). *Committee on National Security Systems (CNSS) Glossary.* Committee on National Security Systems CNSSI No. 4009.

[ES-C2M2] DOE. (2014). *Electricity Subsector Cybersecurity Capability Maturity MODEL (ES-C2M2) Version 1.1* (February 2014). U.S. Department of Energy. https://www.energy.gov/sites/default/files/2014/02/f7/ES-C2M2-v1-1-Feb2014.pdf

[ISO/IEC/IEEE 15939] ISO/IEC/IEEE 15939:2017 Systems and software engineering — Measurement process, first edition.

[ISO/IEC/IEEE 15288] ISO/IEC/IEEE 15288:2015 Systems and software engineering — System life cycle processes First edition.

[NICCS 2016] NICCS. (2016). *Explore Terms: A Glossary of common cybersecurity terminology.*

[NIST 2018] *NIST Framework for Improving Critical Infrastructure Cybersecurity Version 1.1* (April 2018). https://doi.org/10.6028/NIST.CSWP.04162018

[NIST SP800-53r5] *NIST Special Publication (SP) 800-53 Revision 5 Security and Privacy Controls for Information Systems and Organizations* (December 2020). https://doi.org/10.6028/NIST.SP.800-53r5

[NIST SP800-82r2] *NIST Special Publication (SP) 800-82 Guide to Industrial Control Systems (ICS) Security Supervisory Control and Data Acquisition (SCADA) systems, Distributed Control Systems (DCS), and other control system configurations such as Programmable Logic Controllers (PLC) Revision 2* (May 2015).

[NIST SP800-160v1] *NIST Special Publication (SP) 800-160 Systems Security Engineering Considerations for a Multidisciplinary Approach in the Engineering of Trustworthy Secure Systems* Volume 1 (March 2018). https://doi.org/10.6028/NIST.SP.800-160v1

[NIST SP800-161] *NIST Special Publication (SP) 800-161* Supply Chain Risk Management Practices for Federal Information Systems and Organizations (April 2015). http://dx.doi.org/10.6028/NIST.SP.800-161

[NIST SP1108r4] *NIST Special Publication (SP) 1108r4 NIST Framework and Roadmap for Smart Grid Interoperability Standards Release 4.0* (February 2021). https://dx.doi.org/10.6028/NIST.SP.1108r4

[PPD 21] *Presidential Policy Directive (PPD)-21Critical Infrastructure Security and Resilience* (February 2013).

[RFC 4949] *IETF Request for Comments (RFC) 4949 Internet Security Glossary* (Version 2, August 2007).

Appendix B

Smart Grid and Distributed Energy Resources

B1. Smart Grid - Different definitions, basically the same concept

Smart Grid is an electricity supply network that uses digital communications technology to detect and react to local changes in usage. The establishment of an EuropeanTechnology Plartform (ETP) in this field was for the first time suggested by the industrial stakeholders and the research community at the first International Conference on the Integration of Renewable Energy Sources and Distributed Energy Resources in Europe in 2004. The SmartGrids Platform was started by the European Commission Directorate General for Research of the European Commission in 2005 [SmartGrids 2006]. The initiative aims at boosting the competitive situation of the European Union in the field of electricity networks, especially smart power grids. Based on the initial term of 2006, this European document of 2010 [SDD 2010] defines:

A SmartGrid is an intelligent network that can intelligently integrate the actions of all users connected to it – generation, consumers and those that do both – in order to efficiently deliver sustainable, economic, and secure electricity supplies.

Examples of other definitions include:

Smart Grid is the electric delivery network, from electrical generation to end-use customer, integrated with the latest advances in digital and information technology to improve electric-system reliability, security, and efficiency [DHS 2010].

Smart Grid is the integration of power, communications, and information technologies for an improved electric power infrastructure serving loads while providing for an ongoing evolution of end-use applications [P2030 2011].

Definitions and terminology vary somewhat, but all notions of an advanced power grid for the 21st century include the addition and integration of many varieties of digital computing and communication technologies and services with the power delivery infrastructure. Bidirectional flows of energy and two-way communication and control capabilities will enable an array of new functionalities and applications that go well beyond smart meters for homes and businesses.

While recognizing that the different names used for the future grid have meaningful distinctions to some stakeholders, the term Smart Grid is used to describe a modernized grid that enables bidirectional flows of energy and uses two-way communication and control capabilities that will lead to an array of new functionalities.

Smart Grid is the modernization of the traditional electric grid. As presented by the U.S. Department of Energy, Office of Electricity [DOE 2008], a smarter grid applies technologies, tools and techniques to bring knowledge to power – knowledge capable of making the grid work far more efficiently:

- Ensuring its reliability to degrees never before possible.
- Maintaining its affordability.
- Reinforcing our global competitiveness.
- Fully accommodating renewable and traditional energy sources.
- Potentially reducing our carbon footprint.

- Introducing advancements and efficiencies yet to be envisioned.

Figure B1 shows the characteristics of traditional power grid versus the Smart Grid.

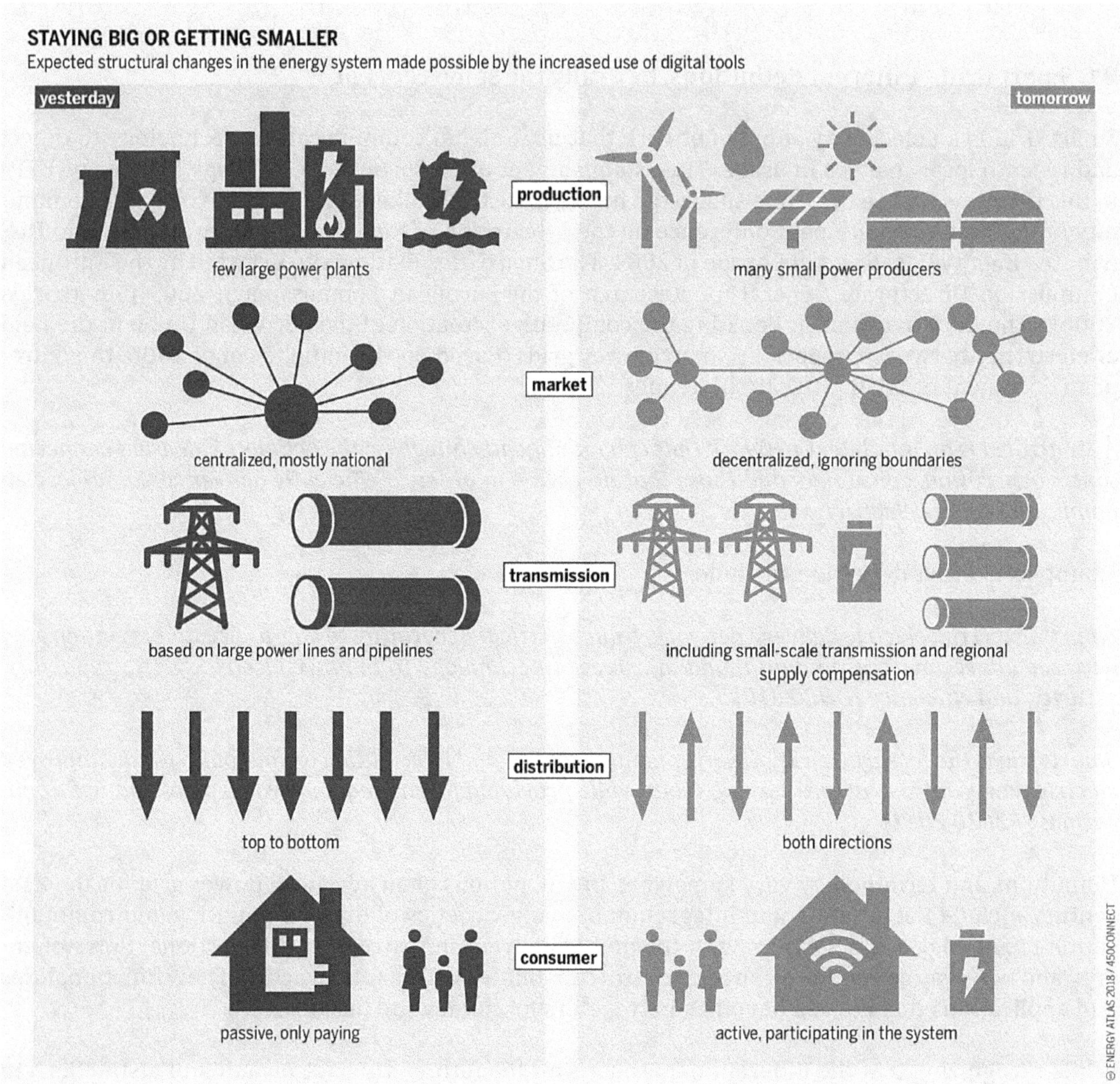

Figure B1 Characteristics of a traditional system (left) versus the smart grid (right). *Source*: [Bartz/ Stockmar 2018]. Licensed under CC BY 4.0.

A Smart Grid system may consist of Information Technology (IT) (e.g., business applications) and/or operational technologies (OT) (e.g., industrial control systems (ICS)). More information about Smart Grid goals and benefits can be found in [Smart Grid 2017], [Smart Grid]).

The current trend to a utility infrastructure containing distributed energy resources (DERs) (also called distributed generation), energy storage systems and no central control may offer future relief from grid transmitted and facility caused disturbances. The financial impacts of less than adequate power quality and reliability and the consequent demand for solutions, coupled with the unique characteristics of distributed energy resources, offer an opportunity for more rapid deployments of DERs as primary and/or backup power supplies. DERs have the potential to overcome utility and site specific supply problems and to help compensate for other developing limitations of the existing distribution grid.

B2. Distributed Energy Resources

Distributed energy resources are dispatchable energy generation and storage technologies, that are interconnected to the distribution grid to provide electric capacity and/or energy to a customer or a group of customers and potentially export the excess to the grid for economical purposes.

DERs are small, modular, energy generation and storage technologies that provide electric capacity or energy where it is needed. A unified and generally accepted definition of DER does not exist, so many definitions have evolved from different perspectives over time. Distributed generation (also known as embedded or local generation) is the term used when electricity is generated from sources, often renewable energy sources, near the point of use instead of centralized generation sources from power plants.

DERs include distributed generation (DG), storage, electric vehicles, and microgrids. DER units are operated either independently of the electrical grid or connected to the low or medium voltage distribution level of the main network. Typically, they are located close to the point of consumption, irrespective of the technology. DERs may be subdivided into:

- Distributed generation (DG) which can be classified as:
  - Renewable generation, e.g., small wind power systems, solar photovoltaic (PV), fuel cells, bio-fuel generators and digesters, nuclear, hydro and other generation resources using renewable fuel.
  - Non-renewable generation, e.g., micro-turbines, small combustion turbines, diesel, natural gas and dual-fueled engines, etc.
- Distributed storage which can be classified as electric storage, mechanical, electrochemical, thermal, etc.:
  - Electric storage, e.g., battery systems and uninterruptible power supplies (UPS), flywheel, superconducting magnetic energy storage (SMES), etc.
  - Thermal storage – these convert electric power to thermal (cooling/heating) energy for later use, e.g., ice based air cooling system, and high capacity brick based air heating system.
- Plug-in electric vehicles (PEVs) and hybrid vehicles (PEV/PHEV) may also be considered as DER; the electric vehicles may be used to supply stored electric energy back to the grid.
- Microgrid is a localized grouping of electricity generation sources (e.g., fuel cells, wind, solar), energy storage, and loads that normally operates to a traditional centralized grid (macrogrid). A connected microgrid can be controlled as if it were one entity. It can function independently as a controlled electric power grid in which a distribution architecture integrates, loads and DERs.

B3. NIST Conceptual Model

The high level Conceptual Model in Figure B2 shows the electrical and communications flows between seven grid domains: generation including DER, transmission, distribution, customer, markets, operations, service provider. The model is useful only for exploring the electrical and communications flows between grid domains. Innovation is occurring within the grid domains, so the exploration of each domain helps to understand the relationships between technology, communications, and interoperability. Underlying the Conceptual Model is a legal and regulatory framework that governs many aspects of the electrical grid [NIST SP1108r4]. These regulations apply to actors and applications, and to their interactions, throughout the system and enable the implementation and management of policies and requirements that keep the power system safe, reliable, and cost effective while maximizing the public good.

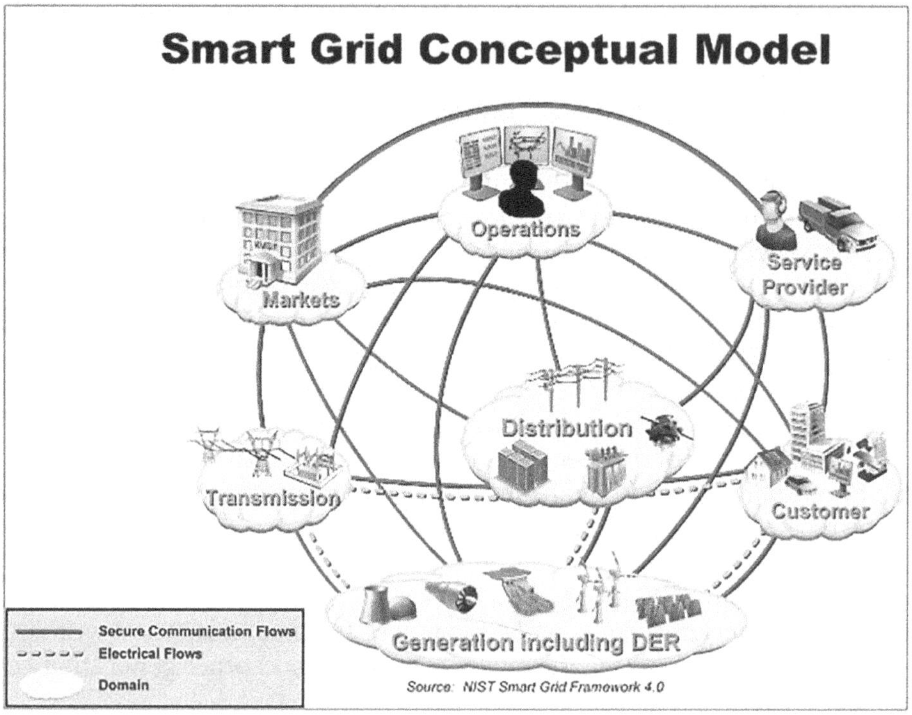

Figure B2 Conceptual Smart Grid Model. *Source*: [NIST SP1108r4]. Public Domain.

The challenge is to develop key interfaces and standards that will enable a dynamic market driven ecosystem while protecting the critical power infrastructure. These interfaces must be able to operate over a variety of networking technologies while maintaining consistent messaging semantics. The services and applications developed must not compromise the cybersecurity, reliability, stability, integrity, or safety of the electrical power network when delivering existing or emerging services. Besides generation domain, it should be observed that DERs are assets in other domains - Transmission, Distribution, Customer - so the model reflects this reality from the electrical and communications standpoints.

B3.1 Generation including DER

Figure B.3 shows the generation domain. The domain name, generation including DER, explicitly acknowledges the growing diversity in scale and utilization of grid resources including:

- Technology diversity with the number and types of generation technologies that has been expanded, to reflect the growing diversity of U.S. generation assets
- Physical siting represents the geographic and topological diversity of included technologies. Icons representing large scale generation technologies are physically closer to the Transmission domain, and smaller scale or more modular technologies are physically closer to the Distribution and Customer domains
- Customer participation refers to resources provided by the customer, whether generation or demand management.

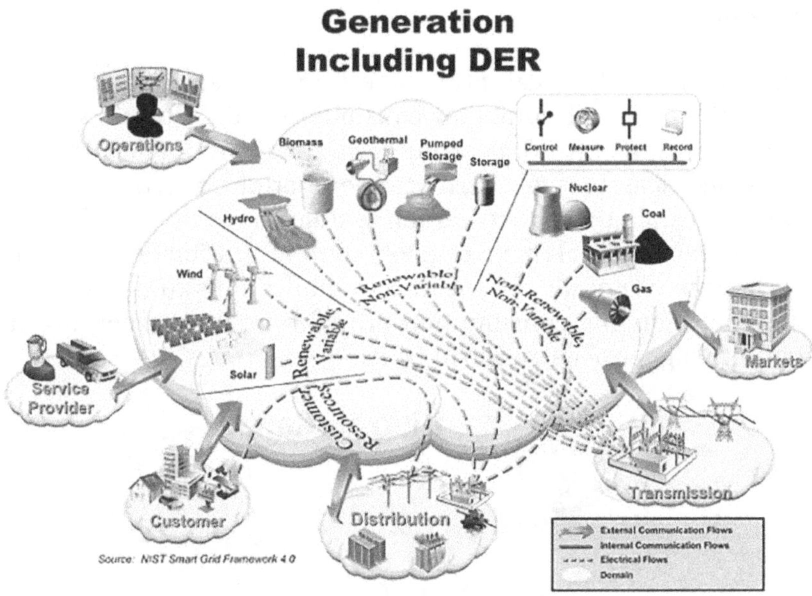

Figure B3 Overview of the Generation including DER domain. *Source*: [NIST SP1108r4]. Public Domain.

B3.2 Transmission Domain

Transmission is the bulk transfer of electrical power from generation sources to distribution through multiple substations. This Figure B.4 depicts both step-up and step-down substations connecting generation and storage with distribution. Substations may also connect two or more transmission lines. Examples of physical actors in this domain include remote terminal units, substation meters, protection relays, power quality monitors, phasor measurement units, sag monitors, fault recorders, and substation user interfaces. Substations may also connect two or more transmission lines. The transmission network is typically monitored and controlled through a SCADA system that uses a communication network, field monitoring devices, and control devices. Typical applications include substation, storage, measurements and control.

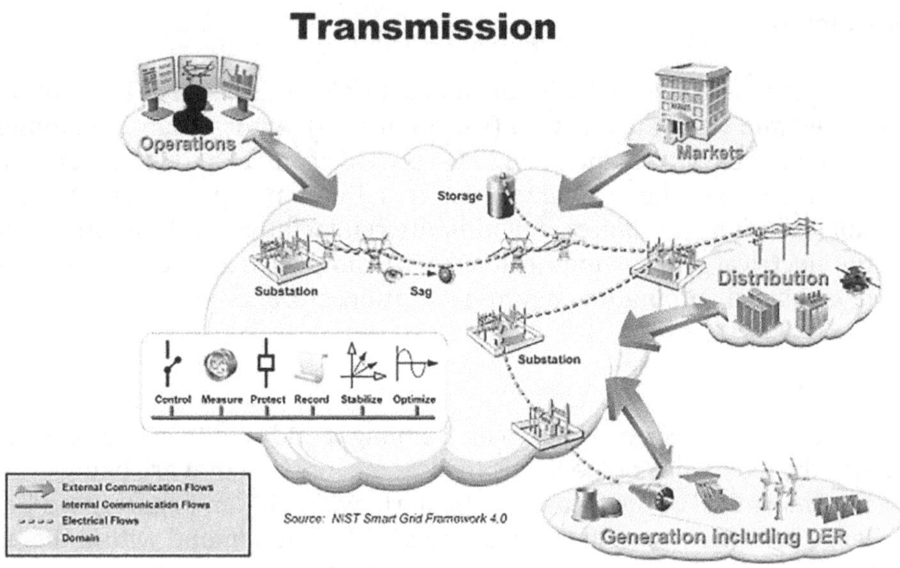

Figure B4 Overview of the Transmission domain. *Source*: [NIST SP1108r4]. Public Domain.

B3.3 Distribution Domain

The Distribution domain is the electrical interconnection between the Transmission domain, the Customer domain, and the metering points for consumption, distributed storage, and distributed generation (see Figure B5). Similar to Generation including DER domain, the Distribution domain contains DERs, such as electrical storage, distributed generation units, or other medium-scale assets such as community solar installations. Distribution systems have the role for optimizing grid function. Distribution actors may use local (peer-to-peer) communication or a more centralized communication path. However, the use of higher speed communications to manage and optimize power flow and electricity generation and consumption in real time is an emerging challenge to be addressed, particularly with higher penetration of DERs.

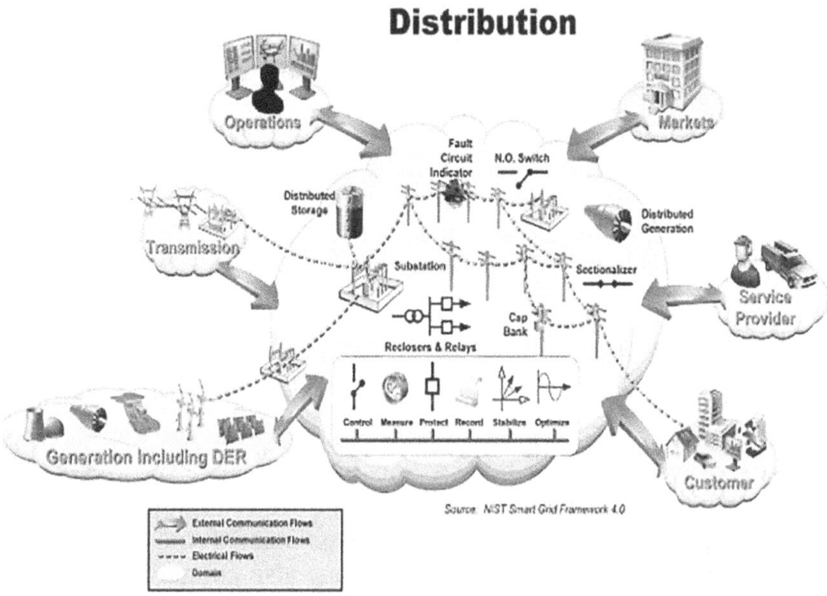

Figure B.5 Overview of the Distribution domain. *Source*: [NIST SP1108r4]. Public Domain.

B3.4 Customer Domain

The customer domain is where electricity is consumed, but is increasingly a domain where electricity is actively managed and generated as well (see Figure B6). Actors enable customers to manage their energy usage and generation. Some actors also provide control and information flow between the Customer domain and the other domains. Customer Domain depicts many types of customers (homes, multi-family dwellings, commercial and industrial facilities, and campuses) served by and interacting with the electrical grid and energy markets. Examples of applications include building or home automation, industrial automation, micro-generation, storage.

B.3.5 Markets

Markets domain is where grid assets and services are bought and sold (see Figure B7). The Markets domain interfaces with all domains of the Smart Grid. Some markets that are being established. Also, they may define the Smart Grid of the future, particularly with DER and aggregated DER. Entities of this domain exchange price information and balance supply and demand within the power system. DERs have an active and growing role in several wholesale markets through aggregation and participation at a greater extent as the Smart Grid becomes more interactive.

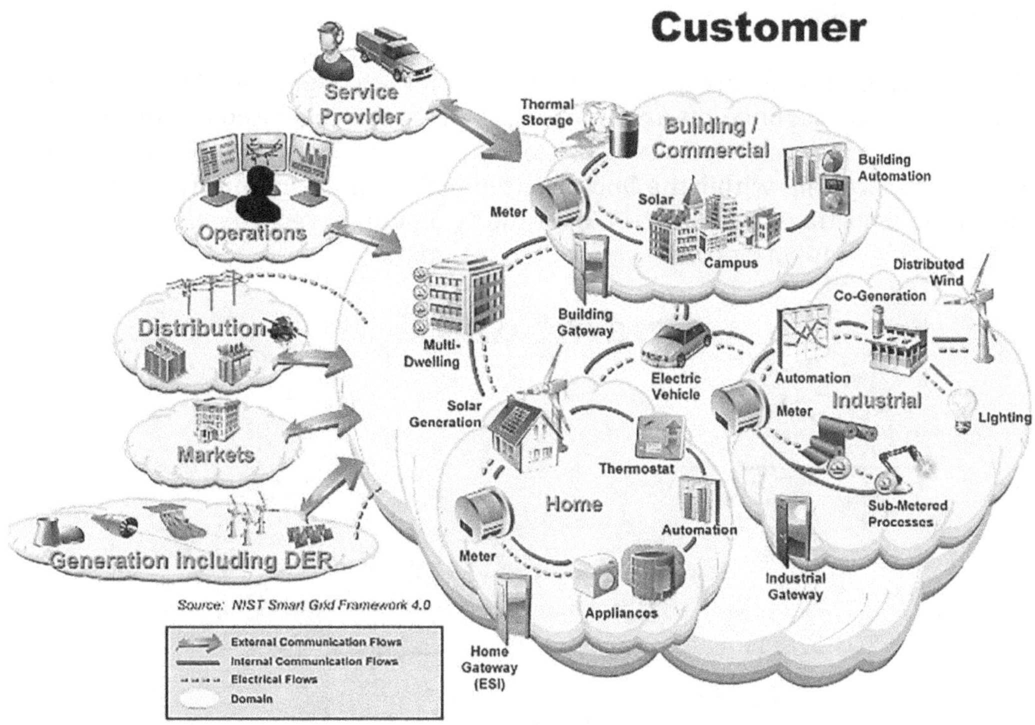

Figure B6 Overview of the Customer domain. *Source*: [NIST SP1108r4]. Public Domain.

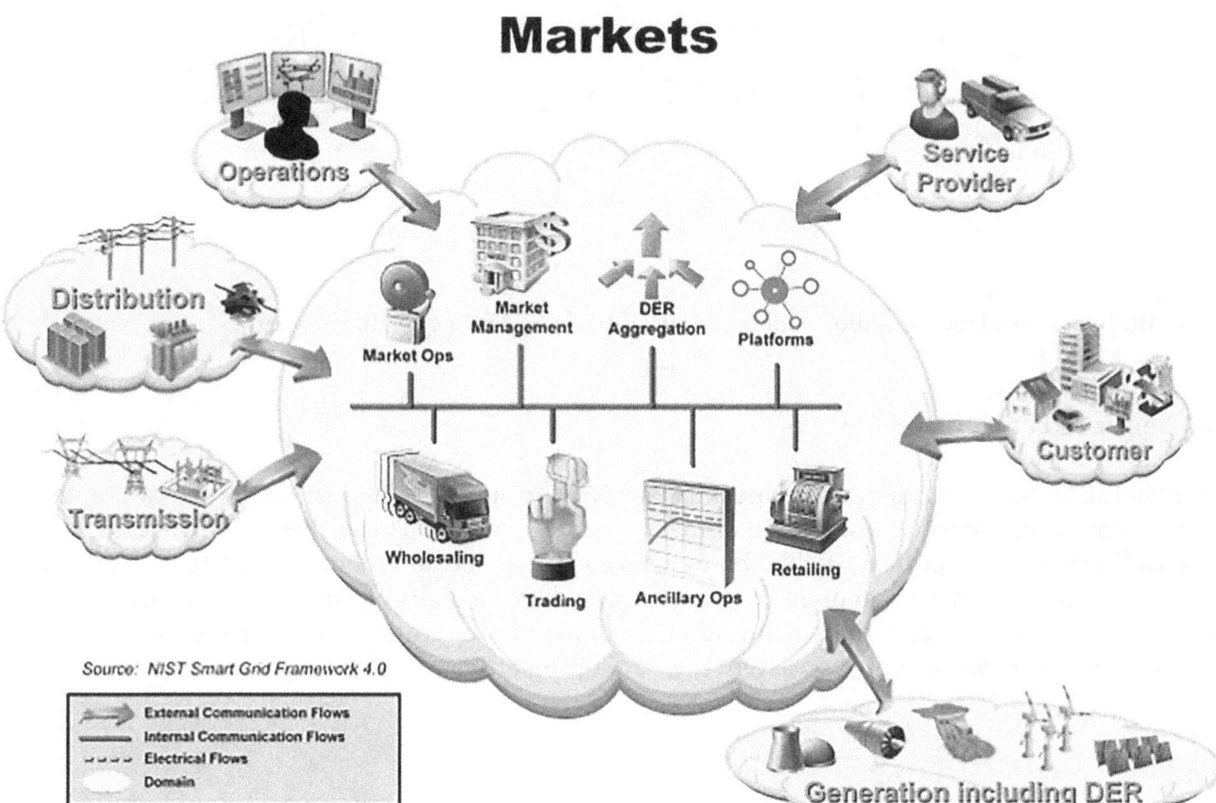

Figure B7 Overview of the Markets domain. *Source*: [NIST SP1108r4]. Public Domain.

B.3.6 Operations Domain

Actors in the Operations domain are responsible for the smooth operation of the power system. Although, the majority of these functions are the responsibility of a regulated utility (see Figure B8), it is visioned that in the future many of these functions to be provided by service providers. Examples of representative applications within the Operations domain include: monitoring, control, fault management, analysis, reporting and statistics, operations planning, records and asset management, maintenance, customer support, power system state estimation.

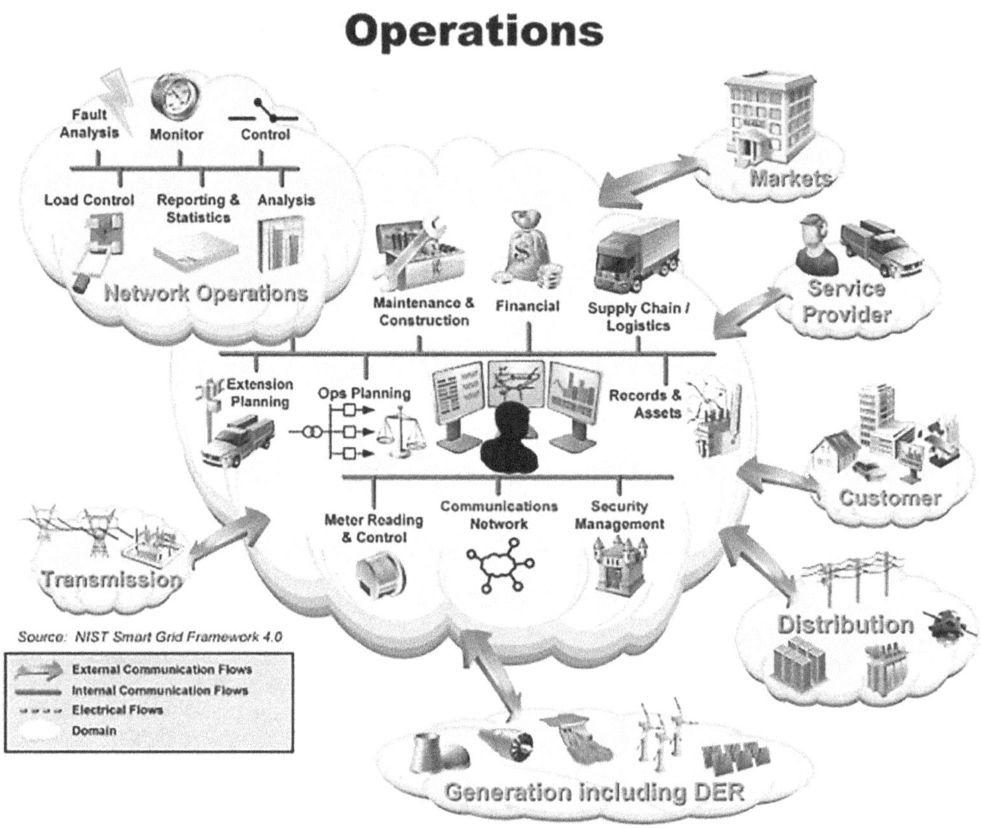

Figure B8 Overview of the Operations domain. *Source*: [NIST SP1108r4]. Public Domain.

B3.7 Service Provider Domain

Actors in the Service Provider domain perform services to support the business processes of power system producers, distributors, and customers (see Figure B9). These business processes range from traditional utility services, such as billing and customer account management, to enhanced customer services, such as management of energy use and home energy generation. The actors of this domain interact with applications from other domains and operate over a variety of networking technologies while maintaining consistent messaging semantics.

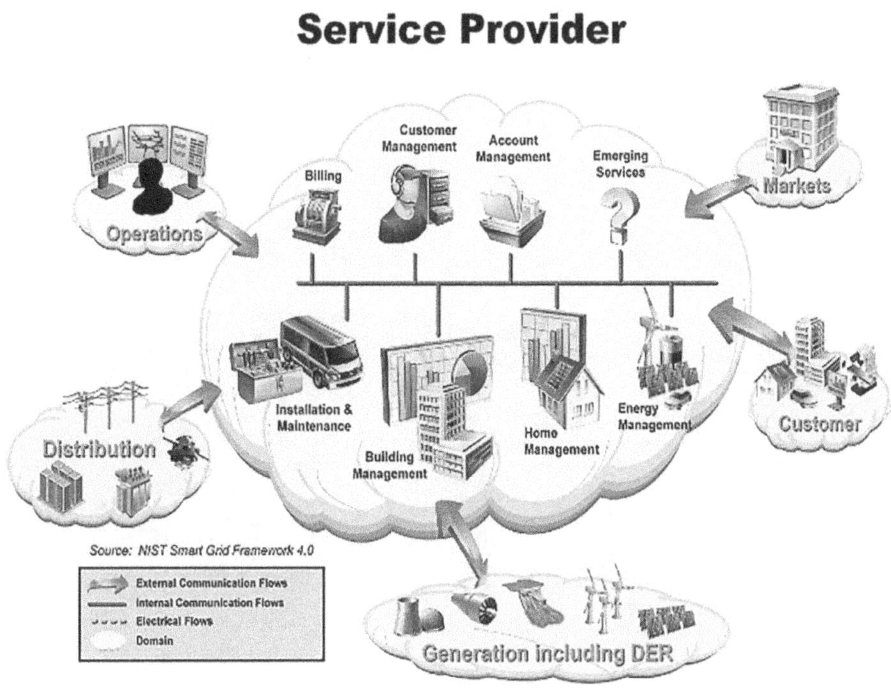

Figure B9 Overview of the Service Provider domain. *Source*: [NIST SP1108r4]. Public Domain.

References

[Bartz/Stockmar] Bartz/Stockmar. (2018, May 24). *Characteristics of a Smart Grid (Right) Versus the Traditional System (Left)*. CC BY-SA 4.0. https://en.wikipedia.org/wiki/Smart_grid#/media/File:Staying_big_or_getting_smaller.jpg

[DOE 2008] Energy. (2008, September 10). *The Smart Grid: An Introduction*. https://www.energy.gov/oe/downloads/smart-grid-introduction-0

[DHS 2010] DHS. (2010). *Appendix 1 Dictionary of Key Terms. SPP 2010 Energy Sector Specific Plan An Annex to the National Infrastructure Protection Plan*. http://www.dhs.gov/xlibrary/assets/nipp-ssp-energy-2010.pdf

[NIST SP1108r4] *NIST Special Publication (SP) 1108r4 NIST Framework and Roadmap for Smart Grid Interoperability Standards Release 4.0*, (February 2021). https://dx.doi.org/10.6028/NIST.SP.1108r4

[P2030 2011] *IEEE Std 2030-2011- IEEE Guide for Smart Grid Interoperability of Energy Technology and Information Technology Operation with the Electric Power System (EPS), End-Use Applications, and Loads.*

[SDD 2010] Edsoforsmartgrids. (2010, April). *European Technology Platform SmartGrids Strategic Deployment Document for Europe's Electricity Networks of the Future*. https://www.edsoforsmartgrids.eu/policy/eu-steering-initiatives/smart-grids-european-technology-platform/

[SmartGrids 2006] EuropeanCommission. (2006). *European SmartGrids Technology Platform Vision and Strategy for Europe's Electricity Networks of the Future* (EUR 22040). http://ec.europa.eu/research/energy/pdf/smartgrids_en.pdf

[Smart Grid 2017] Techopedia. (2017). *Smart Grid*. https://www.techopedia.com/definition/692/smart-grid

[Smart Grid] Smartgrid. (n.d.). *The Smart Grid*. https://www.smartgrid.gov/the_smart_grid/smart_grid.html

Appendix C

Acronyms

1G, 2G, 3G, 4G, 5G First, Second, Third, Fourth, Fifth Generation Wireless
3DES Triple DES
3GPP 3rd Generation Partnership Project
6LoWPAN IPv6 over Low-Power Wireless Personal Area Network
AAA Authentication Authorization and Accounting
AC Access Control
AC Alternating Current
ACID Atomicity, Consistency, Isolation, Durability
ACK Acknowledgment
ACL Access Control List
ACM Association for Computing Machinery
AD Active Directory
ADA Advanced Distribution Automation
ADMS Advanced Distribution Management System
ADR Automated Demand Response
ADSL Asymmetric Digital Subscriber Line
AES Advanced Encryption Standard
AGC Automatic Generation Control
AH Authentication Header
AI Artificial Intelligence
ALE Annualized Loss Expectancy
AMI Advanced Metering Infrastructure
AMR Automatic Meter Reading
AN Access Network
ANN Artificial Neural Network
ANSI American National Standards Institute
AP Access Point
API Application Programming Interface
APT Advanced Persistent Threat
ARO Annualized Rate of Occurrence
ARP Address Resolution Protocol
AS Autonomous System
ASCII American Standard Code for Information Interchange
ASHRAE American Society of Heating, Refrigerating and Air Conditioning
ASIC Application Specific Integrated Circuit
AT Awareness and Training
ATM Asynchronous Transfer Mode
AU Audit and Accountability
AV Autonomous Vehicle
AWL Application White Listing
AWWA American Water Works Association
B2B Business to Business
B2C Business to Consumer
BACNEet Building Automation and Control Networks
BAN Building Area Network

BAS Building Automation System
BC Business Continuity
BCM Business Capability Model
BCM Business Continuity Management
BCP Business Continuity Plan
BES Bulk Electric System
BGP Border Gateway Protocol
BEMS Building Energy Management System
BIOS Basic Input Output System
BLE Bluetooth Low Energy
BMIS Business Model for Information Security
BOOTP Bootstrap Protocol
BPL Broadband Over Powerline
BT Bluetooth
BYOD Bring Your Own Device
C3 Command, Control, and Communications
C2M2 Cybersecurity Capability Maturity Model
C2S Client to Site
CA Certificate Authority
CA Security Assessment and Authorization
CAES Compressed Air Energy Storage
CAG Consensus Audit Guidelines
CAN Controller Area Network
CAPEX Capital Expenditures
CASE Computer Aided Software Engineering tools
CBC Cipher Block Chaining
CBK Common Body of Knowledge
CC Critical Control
CC Common Criteria
CCHP Combined Cooling, Heat, and Power
CCTV Closed Circuit Television
CBM Condition Based Maintenance
CEA Consumer Electronics Association
CD Compact Disc
CDF Cumulative Distribution Function
CDMA Code Division Multiple Access
CEN Comité Européen de Normalisation
CEN European Committee for Standardization
CENELEC Comité Européen de Normalisation Electrotechnique
CENELEC European Committee for Electrotechnical Standardization
CEO Chief Executive Officer
CFB Cipher Feed Back mode
CHAP Challenge Handshake Authentication Protocol
CHP Combined Heat and Power
CIA Confidentiality, Integrity and Availability
CIM Common Information Model
CIO Chief Information Officer
CISO Chief Information Security Officer
CDI Constrained Data Item
CIKR Critical Infrastructures and Key Resources
CIM Common Information Model

CIP Critical Infrastructure Protection
CIR Committed Information Rate
CIS Customer Information Systems
CISSP Certified Information Systems Security Professional
CLI Command Line Interface
CM Configuration Management
CMAC Cipher Based Message Authentication Code
CMIA Cyber Mission Impact Analysis
CMM Capability Maturity Model
CMMI Capability Maturity Model Integration
CNSS Committee on National Security Systems
CNSSI Committee on National Security Systems Instruction
COBIT Control Objectives for Information and Related Technology
COM Component Object Model
CORBA Common Object Request Broker Architecture
COS Catalog of Standards
COSEM Companion Specification for Energy Metering
COTS Commercial-Off-The-Shelf
CP Contingency Planning
CPO Chief Privacy Officer
CPP Critical Peak Pricing
CPS Cyber Physical System
CPTED Crime Prevention Through Environmental Design
CPU Central Processing Unit
CRR Cyber Resilience Review
CRUD Create, Read, Update, Delete
CSA Cloud Security Alliance
CSET cybersecurity evaluation Tool
CSI Computer Security Institute
CSL Coordinated Sampled Listening
CSMA/CA Carrier Sense Multiple Access with Collision Avoidance
CSP Concentrated Solar Power
CSRC Computer Security Resource Center
CSRF Cross Site Request Forgery
CTA Consumer Technology Association
CT-IAP Communications Technology Interoperability Architectural Perspective
CTS Clear to Send
CVE Common Vulnerability Enumeration
CVR Conservation Voltage Reduction
CVSS Common Vulnerability Scoring System
CWE Common Weakness Enumeration
DA Distribution Automation
DAC Discretionary Access Control
DAC Dual Attached Concentrator
DAG Directed Acyclic Graph
DAP Data Aggregation Point
DAR Design Architecture Review
DASD Direct Access Storage Device
DC Direct Current
DCE Data Circuit terminating Equipment
DCL Data Control Language

DCOM Distributed Component Object Model
DCS Distributed Control System
DCU Data Concentrator Unit
DDC Distributed Control System
DDoS Distributed Denial of Service
DEA Data Encryption Algorithm
DER Distributed Energy Resource
DERMS Distributed Energy Resource Management System
DES Data Encryption Standard
DFS Distributed File Service
DG Distributed Generation
DGM Distribution Grid Monitoring
DHCP Dynamic Host Configuration Protocol
DHS U.S. Department of Homeland Security
DIFS Distributed Coordination Function Interframe Spacing
DLC Direct Load Control
DLMS Device Language Message Specification
DLR Dynamic Line Rating
DMS Distribution Management System
DMZ Demilitarized Zone
DNP3 Distributed Network Protocol Version 3
DNS Domain Name Service
DOCIS Data Over Cable Interface Specifications
DODAG Destination-Oriented Directed Acyclic Graph
DOE U.S. Department of Energy
DoD U.S. Department of Defense
DoS Denial of Service
DR Demand Response
DRM Digital Rights Management
DRMS Demand Response Management System
DS Distributed Storage
DSCADA Distribution Supervisory Control And Data Acquisition
DSL Digital Subscriber Line
DSM Demand Side Management
DSN Data Source Name
DSO Distribution System Operator
DSS Digital Signature Standard
DSSS Direct Sequence Spread Spectrum
DSU Data Service Unit
DTLS Datagram Transport Layer Security
DVD Digital Video Disc
EABOK Enterprise Architecture Body of Knowledge
EAL Evaluation Assurance Level
EAM Enterprise Asset Management
EAP Extensible Authentication Protocol
eBIX (European forum for) energy Business Information Exchange
ECB Electronic Code Book mode
ECC Elliptic Curve Cryptosystem algorithm
ECU Electronic Control Unit
EDGE Enhanced Data rates for GSM Evolution
EDR Enhanced Data Rate

EER Equal Error Rate
EF Exposure Factor
EGP Exterior Gateway Protocol
EGx EU Smart Grid Task Force Expert Group x (1 to 3)
EIA Electronic Industries Association
EIFS Extended Interframe Spacing
EIGRP Enhanced Interior Gateway Routing Protocol
EISA Energy Independence and Security Act
EMI Electromagnetic Interference
EMIX Energy Market Information Exchange
EMS Energy Management System
ENISA European Union Agency for Cybersecurity
ENTSO-E European Network of Transmission System Operators for Electricity
EPM Energy Portfolio Management
EPS Electric Power System
EPRI Electric Power Research Institute
ERP Enterprise Resource Planning
ES-C2M2 Electricity Subsector Cybersecurity Capability Maturity Model
ESCO Energy Service Company
ESI Energy Services Interface
ESP Energy Service Provider
EU European Union
eTOM extended Telecom Operations Map
ETSI European Telecommunications Standard Institute
ETX Expected Transmission Count
EU DPD European Union Data Protection Directive
EUMD Energy Usage Metering Device
EV Electrical Vehicle
EVO Electrical Vehicle Operator
EVSE Electric Vehicle Supply Equipment
FACTS Flexible Alternating Current Transmission Systems
FAN Field Area Network
FBI Federal Bureau of Investigation
FDEMS Facilities DER Energy Management System
FDMA Frequency Division Multiple Access
FERC Federal Energy Regulatory Commission
FHSS Frequency Hopping Spread Spectrum
FIPS Federal Information Processing Standard
FISMA Federal Information Security Management Act
FLIR Fault Location, Isolation, and Restoration
FLISR Fault Location Isolation and Service Recovery
FMEA Failure Modes and Effect Analysis
FSGIM Facility Smart Grid Information Model
FTP File Transfer Protocol
G2V Grid to Vehicle
G/W Gateway
GDPR General Data Protection Regulation
GEO Geographic Routing
GF Greedy Forwarding
GIAC Global Information Assurance Certification
GIF Graphic Interchange Format

GIS Geographical Information System
GLBA Gramm Leach Bliley Act
GNSS Global Navigation Satellite System
GOOSE Generic Object Oriented Substation Events
GOTS Government Off-The-Shelf
GPS Global Positioning System
GPSR Greedy Perimeter Stateless Routing
GRE Generic Routing Encapsulation
GSM Global System for Mobile
GSM Global System for Mobile Communications
GW Gigawatt
GWAC GridWise Architecture Council
HAN Home Area Network
HART Highway Addressable Remote Transducer
HDSL High level Data Link Control
HDSL High-bit-rate Digital Subscriber Line
HEMS Home Energy Management System
HEMS Home Energy Management System
HFRT High-Frequency Ride-Through
HIDS Host-based intrusion detection system
HIP Host Identity Protocol
HIPAA Health Insurance Portability and Accountability Act
HMI Human-Machine Interface
HSDPA High-Speed Downlink Packet Access
HSPA High Speed Packet Access
HTTP Hyper Text Transfer Protocol
HTTPS HTTP Secure
HVAC Heat Ventilation Air Conditioning
HVDC High Voltage Direct Current
HVRT High-Voltage Ride-Through
IA Identification and Authenticity
IA Information Assurance
IaaS Infrastructure as a Service
IAB Internet Architecture Board
IACS Industrial Automation and Control System
IAM Identity and Access Management
IAN Industrial Area Network
IAP Interoperability Architectural Perspective
IC Internal Combustion
ICMP Internet Control Message Protocol
ICMPv6 Internet Control Message Protocol version 6
ICS Industrial Control System
ICS-CERT Industrial Control Systems Cyber Emergency Response Team
ICT Information and Communications Technology
ICV Integrity Check Value
ID Identification
ID Identity
IdM Identity Management
IDEA International Data Encryption Standard
IDS Intrusion Detection Systems
IDS/IPS Intrusion Detection System/Intrusion Prevention System

IEA International Energy Agency
IEC International Electrotechnical Commission
IED Intelligent Electronic Device
IEEE Institute of Electrical and Electronics Engineers
IETF Internet Engineering Task Force
IGMP Internet Group Management Protocol
IGP Interior Gateway Protocol
IGRP Interior Gateway Routing Protocol
IIoT Industrial Internet of Things
IKE Internet Key Exchange
INCOSE International Council on Systems Engineering
IOCE International Organization on Computer Evidence
IoT Internet of Things
IOU Investor Owned Utility
IP Internet Protocol
IPS Intrusion Prevention System
IPsec Internet Protocol Security
IPv4 Internet Protocol version 4
IPv6 Internet Protocol version 6
IPRM Interoperability Process Reference Manual
IR Incident Response
IRENA International Renewable Energy Agency
IRM Interface Reference Model
IS Information System
ISA International Society of Automation
ISAC Information Sharing and Analysis Center
ISACA Information Systems Audit and Control Association
ISAE International Standard on Assurance Engagements
ISAO Information Sharing and Analysis Organization
ISC2 International Information Systems Security Certification Consortium
ISM Industrial, Scientific, and Medical
ISMS Information Security Management System
ISO Independent System Operator
ISO International Organization for Standardization
ISP Internet Service Provider
ISSA Information Systems Security Association
ISSAP Information Systems Security Architecture Professional
IT Information Technology
ITCA Independent Testing and Certification Authority
IT-IAP Information Technology Interoperability Architectural Perspective
ITIL Information Technology Infrastructure Library
ITSEC Information Technology Security Evaluation Criteria
ITU International Telecommunication Union
ITU-T International Telecommunications Union for the Telecommunication Standardization Sector
IV Initialization Vector
IVVC Integrated Volt-VAR Control
JWG Joint Working Report for Standards for the Smart Grids
KDC Key Distribution Center
Key Cryptographic Key
kW Kilowatt
kWh Kilowatt-hour

kWp Kilowatt peak
L2TP Layer 2 Tunneling Protocol
LAN Local Area Network
LCP Link Control Protocol
LDAP Lightweight Directory Access ***Protocol***
LEAP Lightweight Extensible Authentication Protocol
LED Light Emitting Diode
LFRT Low Frequency Ride-Through
LIC Logical Interface Category
LIDAR Light Detection and Ranging
Li-ion Lithium –ion
LLC Logical Link Control
LM Load Management
LMS Load Management System
LR-WPAN Low-Rate Wireless Personal Area Network
LTE Long Term Evolution
LTE-A LTE Advanced
LVRT Low Voltage Ride-Through
M2M Machine-to-Machine
MA System Development and Maintenance
MAC Medium Access Control
MAC Message Authentication Code
MAN Metropolitan Area Network
MD5 Message Digest algorithm series 5
MFA Multi-Factor Authentication
MFR Most Forwarding Progress within Radius
MIB Management Information Base
MIME Multipurpose Internet Mail Extensions
MIMO Multiple-Input Multiple-Output
MISO Midcontinent Independent System Operator
MITM Man In The Middle
MitM Man-in-the-Middle
MP Media Protection
MP2P Multi-Point-to-Point
MPDU MAC Protocol Data Unit
MPLS Multiprotocol Label Switching
MPLS-TP MPLS Transport Profile
MQTT Message Queue Telemetry Transport
MTBF Mean Time Between Failures
MTC Machine Type Communications
MTTR Mean Time To Repair
MTX Mean Transmission Time
MTU Maximum Transmission Unit
MWh Megawatt hour
NAC Network Applications Consortium
NAESB North American Energy Standards Board
NAN Neighbor Area Network
NAT Network Address Translation
NAVV Network Architecture Validation and Verification
NCCIC National Cybersecurity & Communication Integration Center
NCP Network Control Protocol

NEC National Electrical Code
NEMA National Electrical Manufacturers Association
NERC North American Electric Reliability Corporation
NERC CIP North American Electric Reliability Corporation Critical Infrastructure Protection
NESCOR National Electric Sector Cybersecurity Organization Resource
NFC Near-Field Communication
NFP Nearest with Forwarding Progress
NIC Network Interface Card
NIST National Institute of Standards and Technology
NISTIR NIST Interagency Report
NRA National Regulatory Authority
NRC Nuclear Regulatory Commission
NSPE National Society of Professional Engineers
NSTC National Science and Technology Council
NYMEX New York Mercantile Exchange
NVD National Vulnerability Database
NWP Numerical Weather Prediction
O&M Operations and Maintenance
OASIS Organization for the Advancemebnt of Structured Information Standards
OCTAVE Operationally Critical Threat, Asset, and Vulnerability Evaluation
OECD Organisation for Economic Co-operation and Development
OEM Original Equipment Manufacturer
O-ESA Open-Enterprise Security Architecture
OF Objective Function
OFDM Orthogonal Frequency Division Multiplexing
OFDMA Orthogonal Frequency Division Multiple Access
OLE Object Linking and Embedding
OMS Outage Management System
OPC OLE for Process Control
OpenADR Open Automated Demand Response
OPEX Operational Expenditure
O-QPSK Offset Quadrature Phase-Shift Keying
OPSEC Operations Security
OS Operating System
OSA Open Security Architecture
OSHA Occupational Safety and Health Administration
OSI Open Systems Interconnection
OSPF Open Shortest Path First
OT Operational Technology
OTN Optical Transport Network
OTP One Time Password
OWASP Open Web Application Security Project
OWL Web Ontology Language
P2MP Point-to-Multi-Point
P2P Point-to-Point
PAP Priority Action Plan
PAP Password Authentication Protocol
PBX Private Branch eXchange
PCC Point of Common Coupling
PCI DSS Payment Card Industry Data Security Standard
PCS Process Control System

PDC Phasor Data Concentrator
PDCA Plan Do Check Act cycle
PDF Portable Document File
PDR Packet Delivery Ratio
PE Physical and Environmental Protection
PEV Plug-in Electric Vehicle
PF Power Factor
PGP Pretty Good Privacy
PHEV Plug-in Hybrid Electric Vehicle
PHY Physical Layer
PII Personally Identifiable Information
PIFS Point Coordination Function Interframe Spacing
PIN Personal Identification Number
PIPEDA Personal Information Protection and Electronic Documents Act
PIV Personal Identity Verification
PKI Public Key Infrastructure
PL Planning
PLC Power Line Carrier
PLC Power Line Communications
PLC Programmable Logic Controller
PM Program Management
PMU Phasor Measurement Unit
PON Passive Optical Network
PPP Point to Point Protocol
PPS Proactive Parent Switching
PPTP Point to Point Tunneling Protocol
PS Personnel Security
PSH Pumped Stored Hydropower
PS-IAP Power Systems Interoperability Architectural Perspective
PSIM Physical Security Information Management
PSTN Public Switched Telephone Network
PUC Public Utility Commission
PV Photovoltaic
QoS Quality of Service
QoSS Quality of Security Service
RA Risk Assessment
RA Request Authority
RAD Rapid Application Development
RADIUS Remote Authentication Dial-In User Service
RAID Redundant Array of Independent Disks
RAM Random Access Memory
RARP Reverse Address Resolution Protocol
RBAC Role Based Access Control
RE Renewable Energy
RE Requirements Engineering
REP Retail Energy Providers
RF Radio Frequency
RFC Request For Comments
RFID Radio Frequency Identification
RG Regulatory Guide
RM Risk Management

RMP Risk Management Process
RNC Radio Network Controller
RoE Rules of Engagement
ROI Return on Investment
ROLL Routing over Low-Power and Lossy Networks
ROM Read Only Memory
RPC Remote Procedure Call
RPL Routing Protocol for Low-Power and Lossy Networks
RTO Regional Transmission Organization
RTP Real-Time Pricing
RTP Real Time Protocol
RTS Request to Send
RTU Remote Terminal Unit
RSVP Resource Reservation Protocol
S/MIME Secure MIME
SA System and Services Acquisition
SaaS Software as a Service
SABSA Sherwood Applied Business Security Architecture
SAE System Architecture Evolution
SAFECode Software Assurance Forum for Excellence in Code
SAIDI System Average Interruption Duration Index
SAML Security Assertion Markup Language
SAN Storage Area Network
SANS SysAdmin, Audit, Networking, and Security Institute
SAS Substation Automation System
SASL Simple Authentication and Security Layer
SC System and communications protection
SCADA Supervisory Control And Data Acquisition
SCAP Security Content Automation Protocol
SC-FDMA Single-Carrier Frequency-Division Multiple Access
SD Secure Digital memory card
SDH Synchronous Optical Networking
SDLC Systems Development Life Cycle
SDN Software-Defined Networking
SDO Standards Developing Organization
SDO Standards Developing Organization
SE Systems Engineering
SecSDLC Security Systems Development Life Cycle
SEP Smart Energy Profile
SEPA Smart Electric Power Alliance
SG Smart Grid
SGAM European Smart Grid Architecture (reference) Model
SGCG Smart Grids Coordination Group
SGCN Smart Grid Communications Network
SGIP Smart Grid Interoperability Panel
SGIRM Smart Grid Interoperability Reference Model
SET Secure Electronic Transaction
SI System and Information Integrity
SEM Security Event Management
SIEM Security Information and Event Management
SIFS Short Interframe Spacing

SIG Special Interest Group
SIM Security Information Management
SIP Session Initiation Protocol
SIS Safety Instrumented Systems
SKIP Simple Key Management Protocol
SLA Service Level Agreement
SM Smart Meter
SMTP Simple Mail Transfer Protocol
SNMPv3 Simple Network Management Protocol
SOA Service Oriented Architecture
SOC Service Organization Control
SOC State of Charge
SOCKS Circuit Level Proxy Firewall
SOAP Simple Object Access Protocol
SODAR Sonic Detection and Ranging
SoS System of Systems
SoSE System of Systems Engineering
SOX Sarbanes Oxley Act
SP Special Publication
SQL Structured Query Language
SQL/SQLi Structured Query Language/Structured Query Language injection
SRE Security Requirements Engineering
SSDLC Secure Software Development Life Cycle
SSE Systems Security Engineering
SSE-CMM Systems Security Engineering-capability Maturity Model
SSH Secure Shell
SSL Secure Socket Layer
ST Slot Time
STATCOM Static Synchronous Compensator
STRIDE Spoofing, Tampering, Repudiation, Information disclosure, Denial of service, and Elevation of privilege
SUN Smart Utility Network
SVC Static VAR Compensator
SysML Systems Modeling Language
T&D Transmission and Distribution
TCB Trusted Computing Base
TCP Transmission Control Protocol
TCP/IP Transmission Control Protocol / Internet Protocol
TDM Time Division Multiplexing
TDMA Time Division Multiple Access
TI Time Interval
TKIP Temporal Key Integrity Protocol
TLP Traffic Light Protocol
TLS Transport Layer Security
TMF Tele Management Forum
TOGAF The Open Group Architecture Framework
TOU Time-of-Use
TPM Trusted Module Platform
TR Technical Report
TSA Transportation Safety Administration
TSO Transmission System Operator

TSS Test Suite Specification
TTL Time To Live
U.S. United States
UBR Unspecified Bit Rate
UCAIug Utility Communication Architecture International Users Group
UDDI Universal Description Discovery and Integration
UDP User Datagram Protocol
UHF Ultra High Frequency
UL Underwriters Laboratories
UML Unified Modeling Language
UMTS Universal Mobile Telecommunications System
UNESCO United Nations Educational, Scientific and Cultural Organization
UPS Uninterruptible Power Supply
URL Uniform Resource Locator
USB Universal Serial Bus
USD United States Dollar
US-CERT United States Computer Emergency Readiness Team
USIM Universal Subscriber Identity Module
V2G Vehicle to Grid
V2G Vehicle-to-Grid
VAR Volt-Ampere Reactive
VHF very high frequency
VLAN Virtual LAN
VM Virtual Machine
VoIP Voice over Internet Protocol
VPN Virtual Private Network
VPP Virtual Power Plant
WAMPAC Wide Area Monitoring Protection And Control
WAMS Wide Area Management Systems
WAN Wide Area Network
WASA Wide Area Situational Awareness
WEB World Wide Web
WEP Wired Equivalent Privacy
Wi-Fi Wireless Fidelity
WiMAX Worldwide Interoperability for Microwave Access
WMN Wireless Mesh Network
WPA Wi-Fi Protected Access
WPA2 Wi-Fi Protected Access version2
WPAN Wireless Personal Area Network
WSAN Wireless Sensor and Actuator Network
WSDL Web Services Description Language
WSIS World Summit on The Information Society
WSN Wireless Sensor Network
XACML Extensible Access Control Markup Language
xDSL Digital Subscriber Line
XML Extended Markup Language
XOR Exclusive OR
XSS Cross Site Scripting
ZT Zero Trust
ZTA Zero Trust Architecture

www.ingramcontent.com/pod-product-compliance
Lightning Source LLC
LaVergne TN
LVHW082007200926
848481LV00010B/2818